# CITIES AND CLIMATE CHANGE

# CITIES AND CLIMATE CHANGE
## GLOBAL REPORT ON HUMAN SETTLEMENTS 2011

### United Nations Human Settlements Programme

**UN⬤HABITAT**

publishing for a sustainable future

London • Washington, DC

First published in 2011 by Earthscan

Earthscan Ltd, Dunstan House, 14a St Cross Street, London EC1N 8XA, UK
Earthscan LLC, 1616 P Street, NW, Washington, DC 20036, USA

Earthscan publishes in association with the International Institute for Environment and Development

For more information on Earthscan publications, see www.earthscan.co.uk or write to earthinfo@earthscan.co.uk

United Nations Human Settlements Programme (UN-Habitat)
PO Box 30030, GPO Nairobi 00100, Kenya
Tel: +254 20 762 3120
Fax: +254 20 762 3477 / 4266 / 4267
Web: www.unhabitat.org

DISCLAIMER
The designations employed and the presentation of the material in this publication do not imply the expression of any opinion whatsoever on the part of the Secretariat of the United Nations concerning the legal status of any country, territory, city or area, or of its authorities, or concerning delimitation of its frontiers or boundaries, or regarding its economic system or degree of development. The analysis, conclusions and recommendations of the report do not necessarily reflect the views of the United Nations Human Settlements Programme, the Governing Council of the United Nations Human Settlements Programme or its Member States.

HS Number:    HS/1/11E (paperback)
               HS/2/11E (hardback)

ISBN:        978-1-84971-371-9 (paperback)
             978-1-84971-370-2 (hardback)
             978-92-1-131929-3 (UN-Habitat series)
             978-92-1-132296-5 (UN-Habitat paperback)
             978-92-1-132297-2 (UN-Habitat hardback)

Typeset by MapSet Ltd, Gateshead
Cover design by Peter Cheseret

A catalogue record for this book is available from the British Library

Library of Congress Cataloging-in-Publication Data

Cities and climate change : global report on human settlements, 2011 / United Nations Human Settlements Programme.
    p. cm.
 Includes bibliographical references and index.
 ISBN 978-1-84971-370-2 (hardback) — ISBN 978-1-84971-371-9 (pbk.)  1.  Human beings—Effect of climate on. 2.  Human settlements. 3.  Urban climatology. 4.  Climatic changes.  I. United Nations Human Settlements Programme.
 GF71.C45 2011
 304.2'5—dc22
                                                     2010050454

At Earthscan we strive to minimize our environmental impacts and carbon footprint through reducing waste, recycling and offsetting our $CO_2$ emissions, including those created through publication of this book. For more details of our environmental policy, see www.earthscan.co.uk.

This book was printed in Malta by Gutenberg Press.
The paper used is FSC certified
and the inks are vegetable based.

# FOREWORD

In the decades to come, climate change may make hundreds of millions of urban residents – and in particular the poorest and most marginalized – increasingly vulnerable to floods, landslides, extreme weather events and other natural disasters. City dwellers may also face reduced access to fresh water as a result of drought or the encroachment of saltwater on drinking water supplies. These are the forecasts, based on the best available science. Yet none of these scenarios needs to occur, provided we act now with determination and solidarity.

This year's edition of UN-Habitat's *Global Report on Human Settlements* elucidates the relationship between urban settlements and climate change, and suggests how cities and towns that have not yet adopted climate change policies can begin to do so. The report details the possible impacts of climate change on cities and towns. It also reviews mitigation and adaptation steps being taken by national and local authorities, and assesses their potential to shape future climate change policy.

Urban development has traditionally been seen as a national concern. This report shows its international relevance. Cities and towns contribute significantly to climate change – from the fossil fuels used for electricity generation, transport and industrial production, to waste disposal and changes in land use.

I commend this report to all concerned with improving the ability of towns and cities to mitigate climate change and adapt to its impacts. How cities and towns are planned affects not just the health and well-being of their inhabitants, but the global environment and our prospects for sustainable development.

**Ban Ki-moon**
*Secretary-General*
*United Nations*

# INTRODUCTION

FOREWORD

The effects of urbanization and climate change are converging in dangerous ways that seriously threaten the world's environmental, economic and social stability. *Cities and Climate Change: Global Report on Human Settlements 2011* seeks to improve knowledge, among governments and all those interested in urban development and in climate change, on the contribution of cities to climate change, the impacts of climate change on cities, and how cities are mitigating and adapting to climate change. More importantly, the Report identifies promising mitigation and adaptation measures that are supportive of more sustainable and resilient urban development paths.

The Report argues that local action is indispensable for the realization of national climate change commitments agreed through international negotiations. Yet most of the mechanisms within the international climate change framework are addressed primarily to national governments and do not indicate a clear process by which local governments, stakeholders and actors may participate. Despite these challenges, the current multilevel climate change framework does offer opportunities for local action at the city level. The crux of the challenge is that actors at all levels need to move within short time frames to guarantee long-term and wide-ranging global interests, which can seem remote and unpredictable at best.

An important finding of the Report is that the proportion of human-induced (or anthropogenic) greenhouse gas (GHG) emissions resulting from cities could be between 40 and 70 per cent, using production-based figures (i.e. figures calculated by adding up GHG emissions from entities located within cities). This is in comparison with as high as 60 to 70 per cent if a consumption-based method is used (i.e. figures calculated by adding up GHG emissions resulting from the production of all goods consumed by urban residents, irrespective of the geographic location of the production). The main sources of GHG emissions from urban areas are related to the consumption of fossil fuels. They include energy supply for electricity generation (mainly from coal, gas and oil); transportation; energy use in commercial and residential buildings for lighting, cooking, space heating, and cooling; industrial production; and waste.

However, the Report concludes that it is impossible to make accurate statements about the scale of urban emissions, as there is no globally accepted method for determining their magnitude. In addition, the vast majority of the world's urban centres have not attempted to conduct GHG emission inventories.

The Report argues that, with increasing urbanization, understanding the impacts of climate change on the urban environment will become even more important. Evidence is mounting that climate change presents unique challenges for urban areas and their growing populations. These impacts are a result of the following climatic changes:

- Warmer and more frequent hot days and nights over most land areas;
- Fewer cold days and nights in many parts of the world;
- Frequency increases in warm spells/heat waves over most land areas;
- Increased frequency of heavy precipitation events over most areas;
- Increase in areas affected by drought;
- Increases in intense tropical cyclone activity in some parts of the world; and
- Increased incidence of extreme high sea levels in some parts of the world.

Beyond the physical risks posed by the climatic changes above, some cities will face difficulties in providing basic services to their inhabitants. These changes will affect water supply, physical infrastructure, transport, ecosystem goods and services, energy provision and industrial production. Local economies will be disrupted and populations will be stripped of their assets and livelihoods.

The impacts of climate change will be particularly severe in low-elevation coastal zones, where many of the world's largest cities are located. Although they account for only 2 per cent of the world's total land area, approximately 13 per cent of the world's urban population lives in these zones – with Asia having a higher concentration.

While local climate change risks, vulnerabilities and adaptive capacity vary across cities, evidence suggests some key common themes. First, climate change impacts may have ripple effects across many sectors of city life. Second, climate change does not impact everyone within a city in the same way: gender, age, race and wealth have implications for the vulnerability of individuals and groups. Third, in terms of urban planning, failure to adjust zoning and building codes and standards with an eye to the future may limit the prospects of infrastructure adaptation and place lives and assets at risk. Fourth, climate change impacts can be long-lasting and can spread worldwide.

In proposing the way forward, following a global review of climate change mitigation and adaptation measures taken by cities all over the world, the Report emphasizes that several principles are fundamental to an integrated, multipartner approach towards climate change action at the urban level:

- No single mitigation or adaptation policy is equally well-suited to all cities;
- It would be beneficial to take an opportunity/risk management approach in a sustainable development perspective, considering not only emissions, but also risks that are present in a range of possible climate and socioeconomic futures;
- Policies should emphasize, encourage, and reward 'synergies' and 'co-benefits' (i.e. what policies can do to achieve both developmental and climate change response goals);
- Climate change policies should address both near-term and longer-term issues and needs; and
- Policies should include new approaches that support multiscale and multisector action, rooted in the different expectations of a wide range of partners.

The Report suggests three main areas in which the international community can support and enable more effective urban mitigation and adaptation responses:

- Financial resources need to be made more directly available to local players – for example, for climate change adaptation in vulnerable cities, for investment in a portfolio of alternative energy options, and in mitigation partnerships between local governments and local private sector organizations;
- Bureaucratic burdens on local access to international support should be eased, with the international community helping to create direct communication and accountability channels between local actors and international donors; and
- Information on climate change science and options for mitigation and adaptation responses should be made more widely available by the Intergovernmental Panel on Climate Change (IPCC), the United Nations and other international organizations, including available knowledge on observed and future climate impacts on urban centres, on urban-based mitigation and adaptation alternatives, and on the costs, benefits, potentials and limits of these options.

With respect to the national level, the Report suggests that national governments should use the following mechanisms to enable mitigation and adaptation actions at the local level:

- Engage in the design and implementation of national mitigation strategies and adaptation planning;
- Offer tax rebates, tax exceptions and other incentives for investments in alternative energy sources, energy-efficient appliances, and climate-proof infrastructure, houses and appliances, among other climate change mitigation and adaptation actions;
- Encourage appropriate climate responses (for example, redesign policies enacted with other issues in mind or in periods prior to climate change, such as flood protection policies that can result in maladaptations);
- Enhance coordination and streamlining between sectoral and administrative entities (for instance, make sure that decisions by one city to protect coastal areas with barriers do not have impacts on basins that are suppliers of fresh water, or wetland ecologies that are important to the economic base of that city or other cities inland);
- Develop partnerships with non-governmental actors to share risks (for example, national governments can work with private insurance providers to offer protection to each city without requiring each to make a sizeable investment in order to reduce risks from a particular kind of low-probability threat); and
- Anticipate and plan for the possibility of much more substantial climate change impacts and adaptation needs in the longer term than those that are currently anticipated in the next decades.

For the local level, the Report suggests, broadly, that urban policy-makers should begin from an awareness of local development aspirations and preferences, local knowledge of needs and options, local realities that shape choices, and local potential for innovation. In this context, urban local authorities should:

- Develop a vision of where they want their future development to go and find ways to relate climate change responses to urban development aspirations;
- Expand the scope of community participation and action by representatives of the private sector, neighbourhoods (especially the poor) and grassroots groups, as well as opinion leaders of all kinds, in order to ensure a broad-based collection of perspectives; and
- Using an inclusive, participatory process, cities should conduct vulnerability assessments to identify common and differentiated risks to their urban development plans and their different demographic sectors, and decide on objectives and ways to reduce those risks.

To achieve more effective policies, local governments need to expand the scope, accountability and effectiveness of participation and engagement with non-governmental organizations (NGOs), such as community and grassroots groups, the academic sector, the private sector and opinion leaders. Effective engagement with NGOs will serve multiple purposes:

- It will become a source of innovative options, as well as both scientific and locally relevant knowledge;
- It will allow participants to understand and mediate the diverse perspectives and interests at play; and
- It will provide broad-based support for decisions and promote knowledge on the causes of emissions and vulnerabilities, as well as mitigation and adaptation options thus achieved.

Partnerships with the private sector and NGOs are of special relevance in this context. For example:

- Resources from international, national and local private organizations can be mobilized to invest in the development of new technologies, housing projects and climate-proof infrastructures, and to assist in the development of climate change risk assessments; and
- The widespread involvement of NGOs in climate arenas as diverse as climate awareness and education and disaster relief should be welcomed – the inputs and perspectives of these organizations can be harnessed to help develop a more integrated urban development planning.

Finally, the Report suggests that broad-based oversight organizations, such as advisory boards, representing the interests of all actors, should be created to help avoid the danger that private or sectarian interests may distort local action (for instance, by investing in technologies, infrastructures and housing that only benefit a minority, or by hijacking the benefits of grassroots funding). This is especially of concern in urban areas within countries that have experienced strong centralized control in the hands of local elites and state agents, but the principle of broad-based oversight can and should be practised everywhere.

Many towns and cities, especially in developing countries, are still grappling with the challenges of how to put in place climate change strategies, how to access international climate change funding and how to learn from pioneering cities. I believe this Global Report will provide a starting point for such towns and cities. More generally, I believe this Report will contribute to raising global awareness of the important role that cities can and should play in the mitigation of greenhouse gas emissions and in adapting to climate change.

**Dr Joan Clos**
*Under-Secretary-General and Executive Director*
United Nations Human Settlements Programme (UN-Habitat)

# ACKNOWLEDGEMENTS

## MANAGEMENT TEAM

Director: Oyebanji O. Oyeyinka.
Chief editor: Naison D. Mutizwa-Mangiza.

## AUTHORS: UN-HABITAT CORE TEAM

Naison D. Mutizwa-Mangiza; Ben C. Arimah; Inge Jensen; Edlam Abera Yemeru; Michael K. Kinyanjui.

## AUTHORS: EXTERNAL CONSULTANTS

Patricia Romero Lankao and Daniel M. Gnatz, Institute for the Study of Society and Environment, National Center for Atmospheric Research, Colorado, US (Chapters 1, 2 and 7); Sebastian Carney, Tyndall Centre for Climate Change Research, Manchester, UK (Chapter 2); Tom Wilbanks, Oak Ridge National Laboratory, Tennessee, US (Chapter 7); David Dodman, David Satterthwaite and Saleem Huq, International Institute for Environment and Development, UK (Chapters 3 and 6); Matthias Ruth, Rebecca Gasper and Andrew Blohm, University of Maryland, US (Chapter 4); Harriet Bulkeley and Vanesa Castán Broto, Durham University, UK, with the assistance of Andrea Armstrong, Anne Maassen and Tori Milledge (Chapter 5).

## PREPARATION OF STATISTICAL ANNEX (UN-HABITAT)

Gora Mboup; Inge Jensen; Ann Kibet; Michael K. Kinyanjui; Julius Majale; Philip Mukungu; Pius Muriithi; Wandia Riunga.

## TECHNICAL SUPPORT TEAM (UN-HABITAT)

Nelly Kan'gethe; Naomi Mutiso-Kyalo.

## INTERNATIONAL ADVISERS (HS-NET ADVISORY BOARD MEMBERS)[1]

Samuel Babatunde Agbola, Department of Urban and Regional Planning, University of Ibadan, Nigeria; Louis Albrechts, Institute for Urban and Regional Planning, Catholic University of Leuven, Belgium; Suocheng Dong, Institute of Geographic Sciences and Natural Resources Research, Chinese Academy of Sciences, China; Ingemar Elander, Centre for Urban and Regional Research, Örebro University, Sweden; József Hegedüs, Metropolitan Research Institute (Városkutatás Kft), Hungary; Alfonso Iracheta, Programme of Urban and Environmental Studies, El Colegio Mexiquense, Mexico; A. K. Jain, Delhi Development Authority, India; Paula Jiron, Housing Institute, University of Chile, Chile; Winnie Mitullah, Institute of Development Studies (IDS), University of Nairobi, Kenya; Aloysius Mosha, Department of Architecture and Planning, University of Botswana, Botswana; Mee Kam Ng, Centre for Urban Planning and Environmental Management, University of Hong Kong; Deike Peters, Centre for Metropolitan Studies, Berlin University of Technology, Germany; Debra Roberts, eThekwini Municipality, Durban, South Africa; Pamela Robinson, School of Urban and Regional Planning, Ryerson University, Toronto, Canada; Elliott Sclar, Centre for Sustainable Urban Development, Columbia University, US; Dina K. Shehayeb, Housing and Building National Research Centre, Egypt; Graham Tipple, School of Architecture, Planning and Landscape, Newcastle University, UK; Iván Tosics, Metropolitan Research Institute (Városkutatás Kft), Budapest, Hungary; Belinda Yuen, School of Design and Environment, National University of Singapore, Singapore.

## OTHER INTERNATIONAL ADVISERS

Titilope Ngozi Akosa, Centre for 21st Century Issues, Lagos, Nigeria; Gotelind Alber, Sustainable Energy and Climate Policy Consultant, Berlin, Germany; Margaret Alston, Department of Social Work, Monash University, Australia; Jenny Crawford, Royal Town Planning Institute, UK; Simin Davoudi, Institute for Research on Environment and Sustainability, Newcastle University, UK; Harry Dimitriou, Bartlett School of Planning, University College London, UK; Will French, Royal Town Planning Institute, UK; Rose Gilroy, Institute for Policy and Planning, Newcastle University, UK; Zan Gunn, School of Architecture, Planning and Landscape, Newcastle University, UK; Cliff Hague, Commonwealth Association of Planners, UK; Collin Haylock, Ryder HKS, UK; Patsy Healey, School of Architecture, Planning and Landscape, Newcastle University, UK; Jean Hillier, Global Urban Research Centre, Newcastle University, UK; Aira Marjatta Kalela, Ministry for Foreign Affairs, Finland; Prabha Kholsa, VisAble Data, Burnaby, Canada; Nina Laurie, School of Geography, Politics and Sociology, Newcastle University, UK; Ali Madanjpour, Global Urban Research Centre, Newcastle University, UK; Michael Majale, School of Architecture, Planning and Landscape, Newcastle University, UK; Peter Newman, Sustainability Policy Institute, Curtin University, Australia; Ambe Njoh, College of Arts and Sciences, University of South Florida, US; John Pendlebury, Global Urban Research Centre, Newcastle University, UK; Christine Platt, Commonwealth Association of Planners, South Africa; Carole Rakodi, Religions and Development Research Programme, University of Birmingham, UK; Diana Reckien, Potsdam Institute for Climate Impact Research, Germany; Maggie Roe, Institute for Policy Practice, Newcastle University, UK; Christopher Rodgers, Newcastle Law School, Newcastle University, UK; Mark Seasons, School of Planning, University of Waterloo, Ontario, Canada; Bruce Stiftel, Georgia Institute of Technology, US; Pablo Suarez, Red Cross/Red Crescent Climate Centre, Boston University, US; Alison Todes, School of Architecture and Planning, University of the Witwatersrand, WITS University, South Africa; Robert Upton, Royal Town Planning Institute, UK; Geoff Vigar, School of Architecture, Planning and Landscape, Newcastle University, UK; Vanessa Watson, School of Architecture, Planning and Geomatics, University of Cape Town, South Africa.

## ADVISERS (UN-HABITAT)

Sharif Ahmed; Karin Buhren, Maharufa Hossain; Robert Kehew, Cecilia Kinuthia-Njenga; Lucia Kiwala, Rachael M'Rabu; Raf Tuts; Xing Quan Zhang.

## AUTHORS OF BACKGROUND PAPERS

Stephen A. Hammer and Lily Parshall, Center for Energy, Marine Transportation and Public Policy (CEMTPP), Columbia University, US; Cynthia Rosenzweig and Masahiko Haraguchi, Climate Impacts Group, NASA Goddard Institute for Space Studies, US ('The contribution of urban areas to climate change: New York City case study'). Carolina Burle Schmidt Dubeux and Emilio Lèbre La Rovere, Center for Integrated Studies on Climate Change and the Environment, Centro Clima, Brazil ('The contribution of urban areas to climate change: The case study of São Paulo, Brazil'). David Dodman, International Institute for Environment and Development, UK; Euster Kibona and Linda Kiluma, Environmental Protection and Management Services, Dar es Salaam, Tanzania ('Tomorrow is too late: Responding to social and climate vulnerability in Dar es Salaam, Tanzania'). Jimin Zhao, Environmental Change Institute, University of Oxford, UK ('Climate change mitigation in Beijing, China'). Heike Schroeder, Tyndall Centre for Climate Change Research and the James Martin 21st Century School, Environmental Change Institute, University of Oxford, UK ('Climate change mitigation in Los Angeles, US'). David Rain, Environmental Studies, George Washington University, US; Ryan Engstrom, Spatial Analysis Lab and Center for Urban Environmental Research, George Washington University, US; Christianna Ludlow, independent researcher, US; and Sarah Antos, World Health Organization ('Accra, Ghana: A city vulnerable to flooding and drought-induced migration'). María Eugenia Ibarrarán, Department of Economics and Business, Universidad Iberoamericana Puebla, Mexico ('Climate's long-term impacts on Mexico City's urban infrastructure'). Rebecca Gasper and Andrew Blohm, Center for Integrative Environmental Research, University of Maryland, US ('Climate change in Hamilton City, New Zealand: Sectoral impacts and governmental response'). Alex Aylett, International Center for Sustainable Cities, Canada ('Changing perceptions of climate mitigation among competing priorities: The case of Durban, South Africa'). Alex Nickson, Greater London Authority, UK ('Cities and climate change: Adaptation in London, UK'). Gotelind Alber, independent consultant, Germany ('Gender, cities and climate change').

## PUBLISHING TEAM (EARTHSCAN LTD)

Jonathan Sinclair Wilson; Hamish Ironside; Alison Kuznets; Andrea Service.

## NOTE

1   The HS-Net Advisory Board consists of experienced researchers in the human settlements field, selected to represent the various geographical regions of the world. The primary role of the advisory board is to advise UN-Habitat on the substantive content and organization of the Global Report on Human Settlements.

# CONTENTS

# STATISTICAL ANNEX

# LIST OF FIGURES, BOXES AND TABLES

## FIGURES

## BOXES

## TABLES

# LIST OF ACRONYMS AND ABBREVIATIONS

| | |
|---|---|
| °C | degrees Celsius |
| BRT | bus rapid transit |
| C40 | Cities Climate Leadership Group |
| CCCI | Cities and Climate Change Initiative (UN-Habitat) |
| CCP | Cities for Climate Protection Campaign (ICLEI) |
| CDM | Clean Development Mechanism |
| cm | centimetre |
| $CO_2$ | carbon dioxide |
| $CO_2$eq | carbon dioxide equivalent |
| Convention, the | United Nations Framework Convention on Climate Change |
| COP | Conference of the Parties (to the UNFCCC) |
| EU | European Union |
| GDP | gross domestic product |
| GEF | Global Environment Facility |
| GHG | greenhouse gas |
| ha | hectare |
| ICLEI | Local Governments for Sustainability |
| IPCC | Intergovernmental Panel on Climate Change |
| kW | kilowatt |
| kWh | kilowatt hour |
| km | kilometre |
| m | metre |
| mm | millimetre |
| MW | megawatt (1MW = 1000kW) |
| MWh | megawatt hour (1MWh = 1000kWh) |
| NAPA | National Adaptation Programme of Action |
| NGO | non-governmental organization |
| OECD | Organisation for Economic Co-operation and Development |
| RMB | Chinese yuan |
| TWh | terawatt hour (1TWh = 1 million MWh) |
| UCLG | United Cities and Local Governments |
| UK | United Kingdom of Great Britain and Northern Ireland |
| UN | United Nations |
| UNDP | United Nations Development Programme |
| UNEP | United Nations Environment Programme |
| UNFCCC | United Nations Framework Convention on Climate Change |
| UN-Habitat | United Nations Human Settlements Programme |
| US | United States |
| WMO | World Meteorological Organization |

# 1

# URBANIZATION AND THE CHALLENGE OF CLIMATE CHANGE

As the world enters the second decade in the new millennium, humanity faces a very dangerous threat. Fuelled by two powerful human-induced forces that have been unleashed by development and manipulation of the environment in the industrial age, the effects of urbanization and climate change are converging in dangerous ways which threaten to have unprecedented negative impacts upon quality of life, and economic and social stability.

Alongside the threats posed by the convergence of the effects of urbanization and climate change, however, is an equally compelling set of opportunities. Urban areas, with their high concentration of population, industries and infrastructure, are likely to face the most severe impacts of climate change. The same concentration of people, industrial and cultural activities, however, will make them crucibles of innovation, where strategies can be catalysed to promote reductions in greenhouse gas (GHG) emissions (mitigation) and to improve coping mechanisms, disaster warning systems, and social and economic equity, to reduce vulnerability to climate change impacts (adaptation).

While some cities are shrinking, many urban centres are seeing rapid and largely uncontrolled population growth, creating a pattern of rapid urbanization. Most of this growth is now taking place in developing countries[1] and is concentrated in informal settlements and slum areas. Therefore, the very urban areas that are growing fastest are also those that are least equipped to deal with the threat of climate change, as well as other environmental and socio-economic challenges. These areas often have profound deficits in governance, infrastructure, and economic and social equity.

People arriving in already overstressed urban centres are forced to live in dangerous areas that are unsuitable for real estate or industrial development, many constructing their own homes in informal settlements on floodplains, in swamp areas and on unstable hillsides, often with inadequate or completely lacking infrastructure and basic services to support human life, safety and development. Many of these slum residents are often blamed by their governments for their own poor living conditions. Even without additional weather-related stresses, such as higher-intensity or more frequent storms, these are dangerous living environments.

Climate change, the second major force unleashed by human industrial development, is quickly building momentum. Climate change is increasing the magnitude of many of the threats to urban areas that are already being experienced as a result of rapid urbanization. Yet, climate change can also be a source of opportunities to redirect the patterns of production and consumption of cities and individuals, at the same time enhancing their capacity to cope with hazards.

Climate change is an outcome of human-induced driving forces such as the combustion of fossil fuels and land-use changes, but with wide-ranging consequences for the planet and for human settlements all over the world. The range of effects include a warming of sea water, and its consequent expansion, that has provided some warning signs, including the collapse of the ice shelves such as Larsen A (1995) and Larsen B (2002) in Antarctica. This melting polar ice threatens to add more water to the already expanding warmer seas, accelerating a dangerous sea-level rise that threatens many coastal urban centres. At the same time, the increasingly warm (and acidic[2]) seas threaten, along with pollution and other anthropogenic or human-related drivers, the very existence of coral reef ecosystems around the world, giving rise to new risks in urban coastal areas that gain protection from the ecosystem services of coral reefs and other aquatic ecosystems. These changes to the natural world gravely threaten the health and quality of life of many urban dwellers.

With sea-level rise, urban areas along the coasts, particularly those in low-elevation coastal zones,[3] will be threatened with inundation and flooding, saltwater intrusion affecting drinking water supplies, increased coastal erosion and reductions in liveable land space. All of these effects, however, will be compounded by other climate impacts, including increase in the duration and intensity of storms such as hurricanes and cyclones, creating extreme hazards for both rich and poor populations occupying low-elevation coastal zones.

Even in non-coastal areas, the convergence of rapid urbanization with climate change can be very dangerous. Poor people living on unstable hillsides could face continuous threats of being swept away or buried by rain-induced mud- and rock-slides. Uncontrolled growth of urban centres into natural forest or brush areas that will dry out with increases in temperatures and in the intensity and duration of droughts will see increases in the frequency of life- and

*The effects of urbanization and climate change ... threaten to have unprecedented negative impacts upon quality of life, and economic and social stability*

*Climate change can also be a source of opportunities to redirect the patterns of production and consumption of cities and individuals*

property-threatening wildfires. Droughts in both coastal and non-coastal cities could disrupt urban water supplies and supplies of forest and agricultural products. These impacts will fall disproportionately upon the urban poor in developing as well as developed countries.

In developed countries, an uneven distribution of political and economic power is the reason why the poor, ethnic and other minorities, and women will bear the brunt of climate change. This uneven distribution of vulnerability can have a destabilizing effect within these countries. This can be seen, for instance, in the racial and social tension that came to the fore in the US when it became evident that African-Americans, the poor and the elderly were disproportionately affected by Hurricane Katrina in 2005.

It is true that destruction of property and loss of life in the coastal areas and elsewhere will certainly not be limited to the poor; but it is also true that affluent segments of the population will be much better protected by insurance, political and economic advantages. It is, however, highly probable that the need for responses to an increased frequency of disasters will stress national economies even in developed countries, also creating much higher stress on the global economy.

The challenges associated with the rapid pace of urbanization will complicate responses to climate change. The other side of the coin, however, is that urbanization will also offer many opportunities to develop cohesive responses in both mitigation and adaptation strategies to deal with climate change. The populations, enterprises and authorities of urban centres will be fundamental players in developing these strategies. In this way, climate change itself will offer opportunities, or it will force cities and humanity, in general, to improve global, national and urban governance to foster the realization of human dignity, economic and social justice, as well as sustainable development.

The purpose of this chapter is to identify the main issues of concern as they relate to urban areas and climate change. It describes, in the section below, key urbanization trends as they relate to climate change, and presents the reasons why it is important to explore the factors shaping urban development and changes in the Earth's climate system. The section after that presents, in summary form, the most important and recent evidence of the causes of climate change, and briefly looks at climate change implications for urban centres. This is followed by a presentation of the framework for exploring linkages between urban areas and climate change used in this Global Report, covering two main issues: drivers of urban contributions to climate change; and urban vulnerability and resilience. The final section contains some concluding remarks and a short description of the main contents of the rest of the report.

## URBANIZATION AND CLIMATE CHANGE

Development and its many environmental impacts are inextricably bound. As such, urbanization and climate change are co-evolving in such a way that populations, often

in densely packed urban areas, will be placed at much higher risk from climate change as well as from other profound societal and environmental changes. The pace of these changes is rapid, and for this reason, many aspects of urban change during recent decades are of importance for this Global Report. There are six primary reasons why it is important to understand the forces shaping the world's growing urban areas in order to be able to mitigate climate change and to cope with its inevitable consequences. *First* among these is the rapid pace of urban population growth. By the end of the last decade the world reached a milestone when, for the first time in human history, half of the world's population lived in urban areas. The pace of urbanization in the world today is unprecedented, with a near quintupling of the urban population between 1950 and 2011.[4]

The *second* important issue bearing on urbanization and climate change is that, unlike urbanization during the early 20th century, which was mostly confined to developed countries, the fastest rates of urbanization are currently taking place in the least developed countries, followed by the rest of the developing countries (see Table 1.1), which now host nearly three-quarters of the world's urban population. In fact, more than 90 per cent of the world's urban population growth is currently taking place in developing countries.[5] This rapid urbanization of developing countries, coupled with the increased intensity and frequency of adverse weather events, will have devastating effects on these countries, which also have lower capacities to deal with the consequences of climate change.[6]

*Third*, while the populations of some cities are shrinking, the number of large cities and the size of the world's largest cities are increasing. The number of cities in the world with populations greater than 1 million increased from 75 in 1950 to 447 in 2011; while during the same period, the average size of the world's 100 largest cities increased from 2.0 to 7.6 million. By 2020, it is projected that there will be 527 cities with a population of more than 1 million, while the average size of the world's 100 largest cities will have reached 8.5 million.[7] However, it is significant that the bulk of new urban growth is taking place in smaller urban areas. For instance, urban centres with fewer than 500,000 people are currently home to just over 50 per cent of the total urban population.[8] The primary disadvantage of this development pattern is that these smaller urban areas are often institutionally weak and unable to promote effective mitigation and adaptation actions. However, there is a possible advantage to be gained here also, as the burgeoning development of these centres may be redirected in ways that reduce their emission levels to a desired minimum (e.g. through the promotion of mono-centric urban structures based on the use of public transportation), and their resilience and ability to cope with climate hazards and other stresses enhanced (e.g. through the development of climate-proof urban infrastructure and effective response systems).

*Fourth*, since urban enterprises, vehicles and populations are key sources of GHGs, gaining an understanding of the dynamics of the forces and systems that drive the urban generation of GHGs is fundamental in helping urban policy-makers, enterprises and consumers target the readily

*An uneven distribution of political and economic power is the reason why the poor, ethnic and other minorities, and women will bear the brunt of climate change*

*Urbanization will also offer many opportunities to develop cohesive responses in both mitigation and adaptation strategies to deal with climate change*

*Urbanization and climate change are co-evolving in such a way that populations, often in densely packed urban areas, will be placed at much higher risk from climate change*

| Region | Urban population (millions) | | | Proportion of total population living in urban areas (%) | | | Urban population rate of change (% change per year) | |
|---|---|---|---|---|---|---|---|---|
| | 2010 | 2020 | 2030 | 2010 | 2020 | 2030 | 2010–2020 | 2020–2030 |
| World total | 3486 | 4176 | 4900 | 50.5 | 54.4 | 59.0 | 1.81 | 1.60 |
| Developed countries | 930 | 988 | 1037 | 75.2 | 77.9 | 80.9 | 0.61 | 0.48 |
| North America | 289 | 324 | 355 | 82.1 | 84.6 | 86.7 | 1.16 | 0.92 |
| Europe | 533 | 552 | 567 | 72.8 | 75.4 | 78.4 | 0.35 | 0.27 |
| Other developed countries | 108 | 111 | 114 | 70.5 | 73.3 | 76.8 | 0.33 | 0.20 |
| Developing countries | 2556 | 3188 | 3863 | 45.1 | 49.8 | 55.0 | 2.21 | 1.92 |
| Africa | 413 | 569 | 761 | 40.0 | 44.6 | 49.9 | 3.21 | 2.91 |
| Sub-Saharan Africa | 321 | 457 | 627 | 37.2 | 42.2 | 47.9 | 3.51 | 3.17 |
| Rest of Africa | 92 | 113 | 135 | 54.0 | 57.6 | 62.2 | 2.06 | 1.79 |
| Asia/Pacific | 1675 | 2086 | 2517 | 41.4 | 46.5 | 52.3 | 2.20 | 1.88 |
| China | 636 | 787 | 905 | 47.0 | 55.0 | 61.9 | 2.13 | 1.41 |
| India | 364 | 463 | 590 | 30.0 | 33.9 | 39.7 | 2.40 | 2.42 |
| Rest of Asia/Pacific | 674 | 836 | 1021 | 45.5 | 49.6 | 54.7 | 2.14 | 2.00 |
| Latin America and the Caribbean | 469 | 533 | 585 | 79.6 | 82.6 | 84.9 | 1.29 | 0.94 |
| Least developed countries | 249 | 366 | 520 | 29.2 | 34.5 | 40.8 | 3.84 | 3.50 |
| Other developing countries | 2307 | 2822 | 3344 | 47.9 | 52.8 | 58.1 | 2.01 | 1.70 |

*Source:* UN, 2010; see also Statistical Annex, Tables A.1, A.2, A.3, B.1, B.2, B.3

**Table 1.1**

**Urban population projections, by region (2010–2030)**

available options to reduce those emissions at the same time that urban resilience to the impacts of climate change is enhanced. For instance, many cities exceed the recommended annual average figure of 2.2 tonnes of $CO_2$ equivalent value ($CO_2$eq) per capita.[9]

*Fifth*, cities are also centres of diverse kinds of innovations that may contribute to reducing or mitigating emissions, adapting to climate change, and enhancing sustainability and resilience. Mechanisms for that purpose include changes in transportation, land-use patterns, and the production and consumption patterns of urban residents. The economies of scale, as well as proximity and concentration of enterprises in cities, make it cheaper and easier to take the actions and provide the services necessary to minimize both emissions and climate hazards.[10]

*Last*, but certainly not least in importance, the dynamics of urban centres are intimately linked to geography. Latitude determines a city's need for more or less energy to run air-conditioning and heating systems within its buildings, industries and houses. However, cities also depend on biodiversity, clean water and other ecosystem services that they have developed over existing ecosystems or 'ecozones', such as coastal areas, wetlands and drylands.[11] Indeed, settling along large bodies of water such as seas, lakes and rivers has historically been a vital factor in the economic and demographic growth of cities, and this trend continues today. For instance, ecozones near water bodies (inland and coastal) have greater shares of population residing in urban areas than other ecozones (see Table 1.2). In developing countries especially, these urban centres are already faced with flooding resulting from a combination of factors (such as impermeable surfaces in the built environment, scarcity of green spaces to absorb water flows and inadequate drainage systems). There are also health-related risks that affect ecozones near water bodies. These include flood-related increases in diarrhoeal diseases, typhoid and cholera.

Many weather-related risks – which, as can be seen in Figure 1.1, already have an urban face – will be exacerbated as climate change progresses and hazards such as sea-level rise, saltwater intrusion and more intense storms become day-to-day realities for the poor and vulnerable populations that inhabit many of the most hazardous areas in urban centres. Drylands are also home to a considerable share of urban populations and, as will be illustrated later, these areas too will see an increase in climate-related impacts, especially in the western parts of the US, the northeast of Brazil and around the Mediterranean (see Table 1.2).

As illustrated in Figure 1.1, many urban dwellers and their livelihoods, property, quality of life and future prosperity are threatened by the risks from cyclones, flooding, landslides and drought: adverse events which climate change is expected to aggravate. Yet, urbanization is not only a source of risks. Certain patterns of urban development can increase resilience. For instance, while large population densities in urban areas create increased vulnerability, they also create the potential for city-scale changes in behaviour that can mitigate human contributions to climate change and encourage adaptation to the inevitable changes that climate change will bring. Furthermore, infrastructure developments can provide physical protection. As illustrated by Cuba's experience, well-designed communications and early warning systems can help to evacuate people swiftly when tropical storms approach.[12] Appropriate urban planning can help to restrict growth of population and activities in risk-prone areas.

Given the above, it is necessary to pay attention to the worsening global problem of climate change in relation to urban centres – the most local of the human systems on Earth – which concentrate more than half of the world's population and have significant potential to perform key roles in the climate change arena.

Many weather-related risks ... will be exacerbated as climate change progresses and hazards such as sea-level rise, saltwater intrusion and more intense storms become day-to-day realities

Appropriate urban planning can help to restrict growth of population and activities in risk-prone areas

| Ecozone | Year | Share of urban population (%) | | | | | | |
|---------|------|--------|------|--------|---------------|---------|---------------|-------|
|         |      | Africa | Asia | Europe | North America | Oceania | South America | World |
| Coastal | 2000 | 62 | 59 | 83 | 85 | 87 | 86 | 65 |
|         | 2025 | 73 | 70 | 87 | 89 | 90 | 92 | 74 |
| Low-elevation coastal zone | 2000 | 60 | 56 | 80 | 82 | 79 | 82 | 61 |
|         | 2025 | 71 | 68 | 85 | 86 | 83 | 90 | 71 |
| Cultivated | 2000 | 38 | 42 | 70 | 75 | 67 | 67 | 48 |
|         | 2025 | 48 | 55 | 75 | 81 | 72 | 80 | 59 |
| Dryland | 2000 | 40 | 40 | 66 | 78 | 49 | 61 | 45 |
|         | 2025 | 51 | 51 | 70 | 84 | 60 | 75 | 55 |
| Forested | 2000 | 21 | 28 | 53 | 64 | 36 | 53 | 37 |
|         | 2025 | 31 | 41 | 59 | 72 | 40 | 68 | 47 |
| Inland water | 2000 | 51 | 47 | 78 | 84 | 77 | 71 | 55 |
|         | 2025 | 62 | 58 | 82 | 88 | 80 | 83 | 64 |
| Mountain | 2000 | 21 | 27 | 46 | 50 | 11 | 54 | 32 |
|         | 2025 | 30 | 40 | 53 | 60 | 13 | 67 | 43 |
| Continent average | 2000 | 36 | 42 | 69 | 74 | 66 | 66 | 49 |
|         | 2025 | 47 | 55 | 75 | 80 | 70 | 78 | 59 |

*Source:* Balk et al, 2009

# EVIDENCE OF CLIMATE CHANGE: IMPLICATIONS FOR URBAN CENTRES

This section presents a brief overview of how the global climate system functions, and what is changing as a result of climate change. It also presents a brief summary of the characteristics of the main causes of climate change (i.e. the GHGs). The last part of this section takes a closer look at the main human activities that cause increasing GHG emissions.

## How the climate system functions and what is changing

Several factors influence the climate of the Earth: the incoming energy from the Sun, the outgoing or radiated energy leaving the Earth, and the exchanges of energy among oceans, land, atmosphere, ice and living organisms (see Figure 1.2). Structure and dynamics within both the carbon cycle (see Box 1.1) and the atmosphere can be equally responsible for alterations in climate. Within the atmosphere, incoming solar radiation and outgoing infrared radiation are affected by some gases and aerosols (see Box 1.1). While most aerosols have some cooling effect, the amount of GHGs present in the Earth's atmosphere before human beings began the large-scale emission of these gases keeps the planet about 33°C warmer than it would be otherwise.[13] This natural greenhouse effect, by providing protection from the loss of heat, has made most life on Earth possible. The functioning of the carbon cycle has provided a good part of this protection; but human activities such as the combustion of fossil fuels, large-scale industrial pollution, deforestation and land-use changes, among others, have led to a build-up of GHGs in the atmosphere together with a reduction of the capacity of oceans and vegetation to absorb GHGs. This attack on the natural carbon cycle on two fronts has reduced the Earth's natural ability to restore balance to the carbon cycle and is now resulting directly in the current global changes in average temperatures.

Looking back to the Earth's history, it is not surprising that its climate system has always changed.[14] Yet, a remarkable stability is also evident, with variations in temperature within a narrow range over thousands of years before the

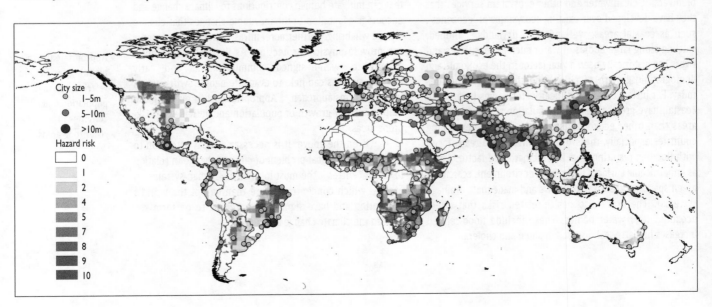

**Box 1.1 Climate change-related terminology**

*Adaptation:* initiatives and measures to reduce the vulnerability of natural and human systems against actual or expected climate change effects.

*Adaptive capacity:* the whole of capabilities, resources and institutions of a country or region to implement effective adaptation measures. Adaptive capacity is the opposite of vulnerability (see below).

*Adaptation deficit:* the lack of adaptive capacity to deal with the problems associated with climate variability. Many cities, and at least some of their populations, already show adaptive deficits within the current range of climate variability without regard to any future climate change impacts. In many such cities and smaller urban centres, the main problem is the lack of provision for infrastructure (all-weather roads, piped water supplies, sewers, drains, electricity, etc.) and the lack of capacity to address this. This is one of the central issues with regard to adaptation because most discussions on this issue focus on adjustments to infrastructure – but infrastructure that is not there cannot be climate-proofed. Funding for 'adaptation' has little value if there is no local capacity to design, implement and maintain the needed adaptation.

*Aerosols:* airborne solid or liquid particles, with a typical size of between 0.01 and 10 micrometres (1 millionth of 1 metre) that reside in the atmosphere for at least several hours. Aerosols may be of either natural or anthropogenic origin. Aerosols may influence climate in several ways: directly through scattering and absorbing radiation, and indirectly through acting as cloud condensation nuclei or modifying the optical properties and lifetime of clouds.

*Anthropogenic:* resulting from or produced by human beings.

*Carbon intensity:* the amount of emission of $CO_2$ per unit of gross domestic product (GDP).

*Climate change:* a change in the state of the climate that can be identified (e.g. by using statistical tests) by changes in the mean and/or the variability of its properties, and that persists for an extended period, typically decades or longer. Climate change may be due to natural processes, or to persistent anthropogenic changes in the composition of the atmosphere or in land use.

*Carbon cycle:* the flow of carbon (in various forms – e.g. as $CO_2$) through the atmosphere, ocean, terrestrial biosphere and lithosphere.

*Carbon footprint:* the total amount of emissions of GHGs caused by a product, an event and an organization. The concept of carbon footprint is a subset of the ecological footprint.

*Carbon sequestration:* the process of increasing the uptake of carbon-containing substances, in particular $CO_2$, by reservoirs other than the atmosphere, such as forests, soils and other ecosystems.

*Climate variability:* variations in the mean state and other statistics (such as standard deviations, the occurrence of extremes, etc.) of the climate on all spatial and temporal scales beyond that of individual weather events. Variability may be due to natural internal processes within the climate system, or to variations in natural or anthropogenic external forcing.

*Ecological footprint:* a measure of human demand on the Earth's ecosystems that compares human demand with planet Earth's ecological capacity to regenerate. It represents the amount of biologically productive land and sea area needed to regenerate the resources that a human population consumes and to absorb and render harmless the corresponding waste.

*Energy intensity:* the ratio of energy use to economic or physical output. At the national level, energy

intensity is the ratio of total primary energy use or final energy use to GDP. At the activity level, one can also use physical quantities in the denominator (e.g. litres of fuel per vehicle kilometre).

*Global warming:* the documented increase in the average temperature of the Earth's near-surface air and sea surface temperatures based on records since the 1880s and the projected continuation of these increasing temperatures.

*Greenhouse gases (GHGs):* those gaseous constituents of the atmosphere, both natural and anthropogenic, that absorb and emit radiation at specific wavelengths within the spectrum of thermal infrared radiation emitted by the Earth's surface, the atmosphere itself and by clouds. This property causes the greenhouse effect.

*Greenhouse effect:* the process by which GHGs trap heat within the surface–troposphere system.

*Mitigation:* technological change and substitution that reduce resource inputs and emissions per unit of output. Mitigation means implementing policies to reduce GHG emissions and enhance sinks.

*Resilience:* the ability of a social or ecological system to absorb disturbances while retaining the same basic structure and ways of functioning, the capacity for self-organization, and the capacity to adapt to stress and change.

*Vulnerability:* the degree to which a system is susceptible to, and unable to cope with, adverse effects of climate change, including climate variability and extremes. Vulnerability is a function of the character, magnitude and rate of climate change and variation to which a system is exposed, its sensitivity and its adaptive capacity.

*Sources:* based on IPCC, 2007b; European Commission, 2007; Nodvin and Vranes, 2010

industrial era.[15] Particularly striking about the current changes are the speed and intensity at which transformations in the greenhouse effect have been fostered by the exponential growth in concentrations of $CO_2$ and other GHGs during the industrial era: the increase of about 100 parts per million since the dawn of industrialization has led to a dramatic alteration of both the carbon cycle and the climate system.[16] An analysis of this period reveals that human actions are pushing the Earth's climate beyond a tipping point where changes in human behaviour and systems will no longer be able to mitigate the effects of climate change.

It is undeniable that the Earth's climate is warming. This is evident from models and observations at global and continental levels (see Figure 1.3), and from the work leading up to and including the Fourth Assessment Report of the Intergovernmental Panel on Climate Change (IPCC), according to which there was an increase of 0.74°C during 1906 to 2005. It has been further validated and strengthened by research published afterwards, according to which the observed increase in global mean surface temperature since 1990 is 0.33°C.[17] Since the onset of the industrial era, concentrations of $CO_2$ and methane ($CH_4$) have increased, with an increase of 70 per cent during the 1970 to 2004 period, and urban centres have played a key – though not yet fully understood – role in this process (see Chapter 3). Most important to this discussion, current research validates that there have been changes in the frequency and severity of

**It is undeniable that the Earth's climate is warming**

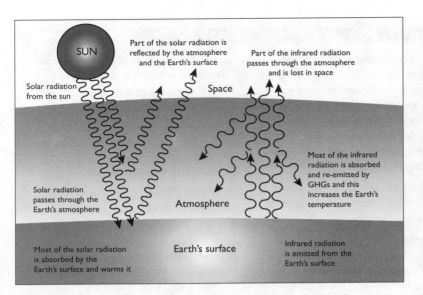

**Figure 1.2**

**Schematic diagram of the greenhouse effect**

*Source:* adapted from http://web.chjhs.tp.edu.tw/~j-bio/warmhouse/images/v1.gif

storms, precipitation, droughts and other weather extremes of relevance, all of which have impacts on urban centres (see Box 1.2).

## The types of greenhouse gases[18]

Various human activities result in the production of GHGs. Water vapour is the most abundant GHG in the atmosphere; but its abundance means that human activities have only a small influence on its concentration. However, human action may generate feedback mechanisms that inadvertently have much larger effects on the concentration of this gas. The four most important types of GHGs produced by human activities are $CO_2$, methane, nitrous oxide ($N_2O$), and the halocarbons (hydrofluorocarbons and perfluorocarbons) and other fluorinated gases.[19] These GHGs are produced from various sources, but can also be removed from the atmosphere by various processes or activities, referred to as 'sinks'.

These gases do not all have the same impacts upon climatic change, so are often described using their $CO_2$ equivalent value ($CO_2$eq). This is a useful tool for comparing emissions, although it does not imply a direct equivalence because of the different time-scales over which these effects take place. Because of this, the gases may be allocated a global warming potential value that takes into account both the time for which they remain in the atmos-

---

**Box 1.2 Recent changes in climate of relevance to urban areas**

**Rising temperatures**

- 11 of the last 12 years rank among the 12 hottest years on record since 1850, when sufficient worldwide temperature measurements began. The eight warmest years have all occurred since 1998.[a]

- Over the last 50 years, 'cold days, cold nights, and frost have become less frequent, while hot days, hot nights, and heat waves have become more frequent'.[c]

**Increasingly severe weather**

- The intensity of tropical cyclones (hurricanes) in the North Atlantic has increased over the past 30 years, which correlates with increases in tropical sea surface temperatures.[a] According to several recent studies, the frequency of strong tropical cyclones has increased during recent decades in all world regions. Other studies suggest that the intensity of strong cyclones will further increase in the future.[b]

- Storms with heavy precipitation have increased in frequency over most land areas. Between 1900 and 2005, long-term trends show significantly increased precipitation in eastern parts of North and South America, Northern Europe, and Northern and Central Asia.[a]

- Between 1900 and 2005, the African Sahel, the Mediterranean, Southern Africa and parts of Southern Asia have become drier, adding stress to water resources in these regions.[a]

- Droughts have become longer and more intense, and have affected larger areas since the 1970s, especially in the tropics and subtropics.[a]

- More recent climate models point to the fact that the difference between humid and arid regions in terms of extreme

events is projected to become even greater under a changing climate.[b]

**Rising sea levels**

- Since 1961, the world's oceans have been absorbing more than 80 per cent of the heat added to the climate, causing ocean water to expand and contributing to rising sea levels. Between 1993 and 2003, ocean expansion was the largest contributor to sea-level rise.[a] More recent figures on sea-level rise are substantially higher than the model-based estimates in the IPCC's Fourth Assessment Report, which did not include ice-sheet dynamics.[b]

- Melting glaciers and losses from the Greenland and Antarctic ice sheets have also contributed to recent sea-level rise (see below).[a]

**Melting and thawing**

- Since 1900, during winters in the Northern Hemisphere, there has been a 7 per cent loss in the seasonally average area covered by frozen ground. According to the United Nations Environment Programme (UNEP) and the World Glacier Monitoring System,[a] the average annual melting rate of mountain glaciers has doubled since 2000, in comparison with the already accelerated melting rates observed in the two decades before. Mountain glaciers and snow cover have declined worldwide.[b]

- Although the current and future contribution to sea-level rise from Antarctica is subject to large uncertainties, recent studies using extensive satellite observations found that loss of Antarctic sea ice increased by 75 per cent during the ten years between 1996 and 2006.[b]

*Sources:* a IPCC, 2007d; b Füssel, 2009; c IPCC, 2007d, p8

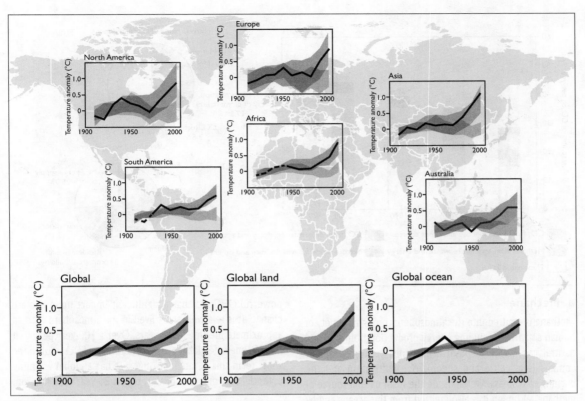

**Figure 1.3**

**Global and continental temperature change**

*Note:* The *black line* in the figures represents observed surface temperature changes. The *light grey band* represents how the climate would have evolved over the past century in response to natural factors only. The dark grey band represents how the climate would have changed in response to both human and natural factors. The overlap of the dark grey band and black line suggests that human activity very likely caused most of the observed increase since the mid 20th century. Lines are dashed where spatial coverage is less than 50 per cent.

*Source:* IPCC, 2007d, p11

phere and their relative effectiveness in causing the greenhouse effect. The global warming potential is a measure of the contribution that different GHGs make to global warming. It takes into account the extent to which these gases absorb warming radiation and the length of time that they remain in the atmosphere. The warming potential of $CO_2$ is used as the baseline against which this is measured (see also Table 1.3).

### ■ Carbon dioxide

Carbon dioxide ($CO_2$) is the most important anthropogenic GHG. Indeed, $CO_2$ emissions are often used synonymously with contributions to climate change. The main sources of atmospheric $CO_2$ are from the burning of fossil fuels, which is responsible for more than 75 per cent of the increase in atmospheric $CO_2$ since pre-industrial times. This energy from fossil fuels is used in transportation, heating and cooling of buildings, and manufacture of cement and other goods – all of which are substantial activities in urban areas. Land-use changes – deforestation and changing agricultural practices – account for the remaining 25 per cent of $CO_2$ emissions. Deforestation also reduces an important sink for the gas, as plants absorb $CO_2$ in the process of photosynthesis. The average annual $CO_2$ emissions from fossil fuels, cement production and gas flaring were 12.5 per cent greater during the period of 2000 to 2005 than during 1990 to 2000. The global atmospheric concentration of $CO_2$ in 2005 was approximately 379 parts per million – an increase from a pre-industrial value of about 280 parts per million. The approximate lifetime of $CO_2$ in the atmosphere is 50 to 200 years.

Carbon dioxide ($CO_2$) is the most important anthropogenic GHG

The main sources of atmospheric $CO_2$ are from the burning of fossil fuels ... used in transportation, heating and cooling of buildings, and manufacture of cement and other goods

| | Carbon dioxide (CO₂) | Methane (CH₄) | Nitrous oxide (N₂O) | Halocarbons[a] | | |
|---|---|---|---|---|---|---|
| | | | | CFC-11 | CFC-12 | HFC-23 |
| **Atmospheric concentration: parts per million (ppm)/billion (ppb)/trillion (ppt):** | | | | | | |
| Pre-industrial times | 280 ppm | 715 ppb | 270 ppb | – | – | – |
| 1998 | 366 ppm | 1763 ppb | 314 ppb | 264 ppt | 534 ppt | 14 ppt |
| 2005 | 379 ppm | 1774 ppb | 319 ppb | 251 ppt | 538 ppt | 18 ppt |
| **Change in atmospheric concentration (%):** | | | | | | |
| Pre-industrial times–2005 | +31 | +147 | +16 | ∞ | ∞ | ∞ |
| 1998–2005 | +4 | +1 | +2 | –5 | +1 | +29 |
| **Approximate lifetime in the atmosphere (years)** | 50–200 | 12 | 114 | 45 | 100 | 270 |
| **Global warming potential relative to CO₂ in 100 years** | 1 | 25 | 298 | 4750 | 10,900 | 14,800 |
| **Radiative forcing 2005 (watts per square metre)** | 1.66 | 0.48 | 0.160 | 0.063 | 0.170 | 0.0033 |
| **Change in radiative forcing 1998–2005 (%)** | +13 | – | +11 | –5 | +1 | – |

*Notes:* a For details on other halocarbons, see IPCC (2007d). ∞ = infinity.

*Sources:* Forster et al, 2007; IPCC, 2007d

**Table 1.3**

**Major characteristics of the most important GHGs**

**Figure 1.4**

**Global anthropogenic GHG emissions**

*Notes:* (a) Global annual emissions of anthropogenic GHGs from 1970 to 2004; (b) share of different anthropogenic GHGs in total emissions in 2004 in terms of $CO_2$ equivalents ($CO_2$eq); (c) share of different sectors in total anthropogenic GHG emissions in 2004 in terms of $CO_2$eq (forestry includes deforestation).

*Source:* IPCC, 2007a

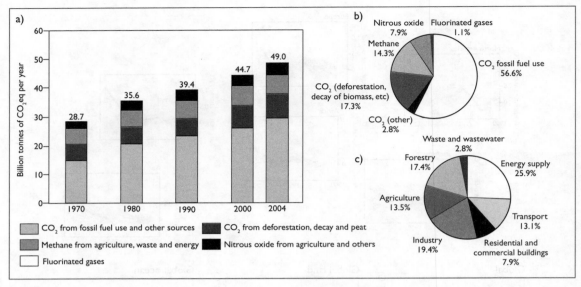

## Methane

Methane is emitted into the atmosphere through a variety of human activities, including energy production from coal and natural gas, waste disposal in landfills, raising ruminant animals (e.g. cattle and sheep), rice cultivation and the burning of biomass. Wetlands are the main natural source of methane, although it is also emitted from the oceans and by the activities of termites. In 2005, methane accounted for about 1774 parts per billion of the atmosphere, more than twice its pre-industrial value – and these current levels are due to the continued human-induced emissions of the gas. Despite this apparently low concentration, methane is a

powerful GHG that has a significant impact upon climate change. It is relatively short lived in the atmosphere with an approximate lifetime of 12 years. Over a 100-year period, it has 25 times the global warming potential of $CO_2$; but in the short term this is much stronger: it has a global warming potential 72 times that of $CO_2$ over a 20-year time horizon.

## Nitrous oxide

Nitrous oxide is emitted from fertilizers and the burning of fossil fuels, and is also released by natural processes in soils and oceans. About 40 per cent of total nitrous oxide emissions result from human activities. In 2005, atmos-

**Table 1.4**

**Total and per capita GHG emissions ('top 20 countries')**

| Country | GHG emissions (2005)[a] | | | $CO_2$ emissions (2007)[b] | | | |
|---|---|---|---|---|---|---|---|
| | Thousand metric tonnes of $CO_2$eq | Percentage of total $CO_2$eq | Metric tonnes of $CO_2$eq per capita | Thousand metric tonnes of $CO_2$ | Percentage of total $CO_2$ | Metric tonnes of $CO_2$ per capita | Percentage change in $CO_2$ (2005–2007) |
| China | 7,303,630 | 18.89 | 5.60 | 6,538,367 | 22.30 | 4.96 | 16.5 |
| US | 7,211,977 | 18.66 | 24.40 | 5,838,381 | 19.91 | 19.38 | –0.1 |
| India | 2,445,328 | 6.33 | 2.23 | 1,612,362 | 5.50 | 1.43 | 14.3 |
| Russian Federation | 2,115,042 | 5.47 | 14.78 | 1,537,357 | 5.24 | 10.82 | 1.4 |
| Japan | 1,446,883 | 3.74 | 11.32 | 1,254,543 | 4.28 | 9.82 | 1.0 |
| Brazil | 1,079,576 | 2.79 | 5.80 | 368,317 | 1.26 | 1.94 | 5.2 |
| Germany | 972,615 | 2.52 | 11.79 | 787,936 | 2.69 | 9.58 | –2.7 |
| Canada | 725,606 | 1.88 | 22.46 | 557,340 | 1.90 | 16.90 | –0.5 |
| UK | 672,148 | 1.74 | 11.16 | 539,617 | 1.84 | 8.85 | –0.8 |
| Mexico | 627,825 | 1.62 | 6.09 | 471,459 | 1.61 | 4.48 | 6.9 |
| Indonesia | 625,677 | 1.62 | 2.85 | 397,143 | 1.35 | 1.77 | 16.4 |
| Australia | 601,444 | 1.56 | 29.49 | 374,045 | 1.28 | 17.75 | 2.7 |
| Iran | 598,479 | 1.55 | 8.66 | 495,987 | 1.69 | 6.98 | 16.2 |
| Italy | 571,378 | 1.48 | 9.75 | 456,428 | 1.56 | 7.69 | –2.5 |
| France | 542,980 | 1.40 | 8.92 | 371,757 | 1.27 | 6.00 | –5.2 |
| Republic of Korea | 535,836 | 1.39 | 11.13 | 503,321 | 1.72 | 10.39 | 8.7 |
| South Africa | 499,842 | 1.29 | 10.66 | 433,527 | 1.48 | 9.06 | 6.2 |
| Spain | 457,776 | 1.18 | 10.55 | 359,260 | 1.23 | 8.01 | 1.6 |
| Saudi Arabia | 439,516 | 1.14 | 19.01 | 402,450 | 1.37 | 16.66 | 9.6 |
| Ukraine | 427,297 | 1.11 | 9.07 | 317,537 | 1.08 | 6.83 | –2.8 |
| Other developed countries | 2,237,764 | 5.79 | 9.46 | 1,791,983 | 6.11 | 7.55 | 1.1 |
| Rest of Asia and Pacific | 3,527,583 | 9.13 | 3.51 | 2,460,617 | 8.39 | 2.37 | 7.3 |
| Rest of Latin America and the Caribbean | 1,329,867 | 3.44 | 5.04 | 749,694 | 2.56 | 2.77 | 10.0 |
| Rest of Africa | 1,659,120 | 4.29 | 1.90 | 699,867 | 2.39 | 0.77 | 4.1 |
| **World total** | **38,655,189** | **100.00** | **6.00** | **29,319,295** | **100.00** | **4.45** | **6.0** |

*Note:* The world totals include only emissions that have been accounted for in national inventories.

*Source:* a http://data.worldbank.org/indicator, last accessed 21 October 2010; b http://mdgs.un.org/unsd/mdg, last accessed 21 October 2010; see also Statistical Annex, Tables B.7 and B.8

pheric nitrous oxide levels were 18 per cent higher than pre-industrial levels, at 319 parts per billion. The gas has a lifetime in the atmosphere of 114 years, and over a 100-year period has a global warming potential that is 298 times greater than $CO_2$.

### ■ Halocarbons

Halocarbons – including chlorofluorocarbons (CFCs) and hydrochlorofluorocarbons (HCFCs) – are GHGs that are produced solely by human activities. CFCs were widely used as refrigerants before it was discovered that their presence in the atmosphere caused the depletion of the ozone layer. International regulations to protect the ozone layer – notably the Montreal Protocol of 1987 – have been successful in reducing their abundance and their contribution to global warming. However, the concentrations of other industrial fluorinated gases (hydrofluorocarbons, perfluorocarbons and sulphur hexafluoride) are relatively small but are increasing rapidly. Although these gases occur in much smaller concentrations than $CO_2$, methane and nitrous oxide, some of them have extremely long lifetimes and high global warming potentials, which means that they are important contributors to global warming. For example, HFC-23 (CHF3) has a lifetime of 270 years and a global warming potential over 100 years 14,800 times greater than $CO_2$.

### The causes of climate change

The main human sources of GHGs contributing to global warming are the dramatic rise in energy use, land-use changes and emissions from industrial activities (see Figure 1.4). Furthermore, between 1970 and 2004, changes in factors such as increased per capita income (up 77 per cent) and population growth (up 69 per cent) have favoured increases in GHG emissions. These have been, to a limited extent, offset by increases in efficiency and/or reductions in the carbon intensity of production and consumption; but the overall global trend has still been towards large increases in anthropogenic GHG emissions.

Not every country has contributed at the same level to global warming. In 2007, developed countries accounted for 18 per cent of the world's population and 47 per cent of global $CO_2$ emissions, while developing countries accounted for 82 per cent of the population and 53 per cent of $CO_2$ emissions.[20] Developing countries, therefore, generated only 25 per cent of the per capita emissions of developed countries. A select number of developed countries and major emerging economy nations are the main contributors to total $CO_2$ emissions (see Table 1.4). In fact, three developed countries (Australia, the US and Canada) have among the highest $CO_2$ emissions per capita, while some developing countries lead in the growth rate of $CO_2$ emissions (e.g. China and Brazil). These uneven contributions to the climate change problem are at the core of both international environmental justice issues and the challenges that the global community faces in finding effective and equitable solutions (see Chapter 2).

In this context, humanity is facing two main challenges that urban centres can to help address:

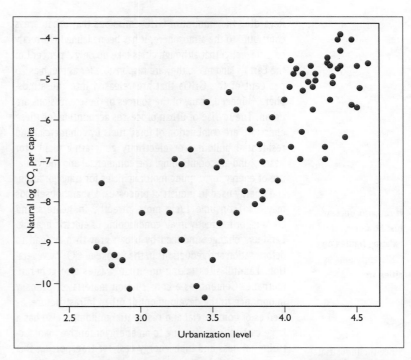

1 *There is a need to adapt*, at least to some amount of continued warming, because even if the concentrations of GHGs and aerosols are kept constant at year 2000 levels, 'a further warming of about 0.1°C per decade would be expected'.[21]

2 *There will also be a need to mitigate* – that is, to achieve development paths that bring about a peaking of emissions by 2015 and a stabilization of GHG concentrations in the atmosphere at about 445 to 490 parts per million by volume of $CO_2$ equivalents ($CO_2$eq) by the end of the century.[22] This path would keep global average temperature increases within 2°C to 2.4°C above pre-industrial levels, in keeping with the objective outlined in the United Nations Framework Convention on Climate Change, Article 2 (see Chapter 2).

**Figure 1.5**

**Relationships between urbanization levels and $CO_2$ emissions per capita**

*Source:* Romero Lankao et al, 2008

**Figure 1.6**

**Carbon intensity and economic development (2003)**

*Source:* Romero Lankao et al, 2008

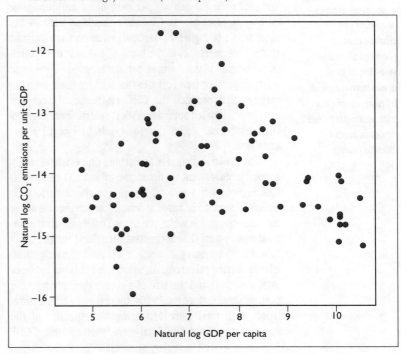

Regarding the amount of GHG emissions that urban areas contribute to the atmosphere, it has been claimed (correctly or incorrectly) that although cities take up only 2 per cent of the Earth's land mass, they are responsible for as much as 75 per cent of the GHGs that are released into the atmosphere.[23] Indeed, many of the sources of these emissions are urban. The myriad of urban processes accounting for these emissions are combustion of fossil fuels by commercial and residential buildings or electricity generating plants for heating and air conditioning, the commercial and individual use of energy for running motor vehicles for transportation, and energy used in industrial processes. Urban households may also consume fuels more directly, in heaters and cookers, or indirectly in air conditioning or electric heating. Land-use changes induced by urban growth may lead to deforestation and reductions in the uptake of $CO_2$ by vegetation. Landfill sites taking up urban wastes also generate methane. Cement, as a construction material of primary importance to the development of urban infrastructure, as well as of commercial and residential buildings, also has a large carbon footprint due to an energy-intensive manufacturing process and high energy cost for transporting this dense material. Lastly, many activities, such as agriculture, livestock production, mining, timber collection and lumber production, increase GHG emissions as direct emitters or reduce the uptake of these gases by vegetation. While these are often undertaken outside the boundaries of urban centres, they are aimed at satisfying urban needs for food, raw materials, forest products and construction materials.

As will be shown in Chapter 3, it remains unclear just how accurate existing figures on GHG emissions by cities are. Many different criteria have been used to measure these emissions, and the choice by researchers to use one or the other can greatly skew the final calculations on how large those contributions are.[24] For instance, if GHG emissions are allocated based on the generating activities within urban centres (the production-based approach), then these centres emit between 30 and 40 per cent of all anthropogenic GHGs. The proportion of GHGs that should be attributed to cities would be higher, however, if emissions were assigned to the consumers (i.e. to the home or business or organizational location of those whose demand for goods, services or waste disposal or travel creates the need for those goods or services that produce the GHG emissions). Under this consumption-based approach, cities' contribution to global GHG emissions would rise to almost half of all global emissions.

A dynamic, complex and strong link exists between economic development, urbanization and $CO_2$ emissions (see Figures 1.5 and 1.6). Urban contributions to $CO_2$ emissions seem to be based at least in part on the size of the national economy in which the urban centres are located (as measured by total GDP in constant US dollars) and the structure of that economy (i.e. whether it is predominantly industrial or service oriented). Although the relations between total emissions and the size of a country's economy have been weakening since the 1960s, there is still a strong correlation, with total emissions rising with the size of the economy (see Figure 1.6). Total energy used per unit of GDP

went down by 33 per cent between 1970 and 2004, yet the rate of improvement has not been enough to globally reduce GHG emissions, which are rising beyond the worst-case scenario and have already resulted in an Earth that is 0.8°C warmer on average than it was in pre-industrial times. Based on the significant roles that they play in their countries' economies, urban areas can be seen as playing a major role in this connection.[25]

Nevertheless, the relationship between levels of urban development, as measured by GDP and levels of GHG emissions, is not so straightforward. It is clear that differences in GHG emissions result from the peculiarities and weight of different sectors, as shown in the next section.

# FRAMEWORK FOR EXPLORING THE LINKAGES BETWEEN URBAN AREAS AND CLIMATE CHANGE

Reducing the contribution of cities to climate change, or mitigation, requires an adequate understanding of the drivers of urban GHG emissions, while effective adaptation must be based on a good understanding of what makes cities and their constituent socio-economic groups either vulnerable or resilient to climate change impacts. This section therefore focuses on the drivers of GHG emissions in urban areas and the concepts of vulnerability and resilience as frameworks for both analysis and for formulating mitigation and adaptation policy options.

## Drivers of urban contributions to GHG emissions

Since the industrial revolution, urban centres have concentrated industries, construction, transportation, households and other activities that release large quantities of GHGs. Other sources that occur both inside and outside cities, but serve urban development, include deforestation and other land-cover changes, agriculture, waste disposal, power generation, and refrigeration and air conditioning. Chapter 3 presents findings from a wide range of urban emissions inventories to show how the data on urban emissions varies from place to place, and how the figures on emissions vary depending on the approaches used (i.e. consumption- or production-based approaches). It is therefore important to have a framework for understanding the levels and drivers of emissions by different demographic and economic sectors, buildings and infrastructures within, or serving, urban areas.

The exploration of how urban centres contribute to climate change requires an understanding of how transportation, heating and cooling systems, industries and other urban activities and infrastructures act both as emitters and direct causes of climate change. They create two main categories of impacts on the carbon cycle and the climate system:

*It remains unclear just how accurate existing figures on GHG emissions by cities are*

*Since the industrial revolution, urban centres have concentrated industries, construction, transportation, households and other activities that release large quantities of GHGs*

*The exploration of how urban centres contribute to climate change requires an understanding of how ... urban activities and infrastructures act both as emitters and direct causes of climate change*

1   *Changes related to the emission of aerosols, GHGs and solid wastes.* GHGs are the main source of changes in the climate system. Not only do they change the dynamics of the carbon cycle, but together with aerosols they also generate changes in the Earth's radiation that induce climate change.[26] Wastes affect the growth, function and health of vegetation and of ecosystems in general.[27]

2   *Land-use related changes.* Urbanization is a process that changes the uses of land and by creating impervious surfaces, filling wetlands and fragmentation of ecosystems has disproportional impacts upon the carbon cycle. The built environment of urban areas is also a forcing function on the weather–climate system of urban centres because it is a source of heat and a poor storage system for water.

Both within and across cities, different populations, economic activities and infrastructures contribute at different levels to global warming. Some studies point to the fact that *gender inequities* exist both in energy use and GHG emissions and that the differences are related not only to wealth, but to behaviour and attitudes. For instance, women tend to buy efficient electric appliances, while men tend to undertake efforts to insulate their houses. Men tend to eat more meat, while women tend to eat more vegetables, fruits and dairy products. Men tend to use more private motorized transport than women, and to use larger, more fuel-consuming vehicles.[28]

Urban centres in developing countries have lower levels of emissions per capita than cities in developed countries.[29] Houston and Washington, DC (US), for instance, have carbon emissions that are about 9 to 18 times higher than those in São Paulo (Brazil), Delhi and Kolkata (India) (see Chapter 3). Yet, other wealthy cities such as Stockholm (Sweden) and Barcelona (Spain) have lower levels of emissions per capita than some South African cities. This is because several *interrelated* factors shape or determine the patterns of energy use and emissions by different populations and sectors.

The *climate and natural endowments* of an urban area are significant factors shaping its energy-use pattern. A city located in high latitudes, for instance, might consume more energy to heat its buildings and houses than one situated in the tropics; and conversely, an urban centre located in the tropics might consume more energy for air conditioning. Thus, climate change will affect energy consumption behaviour in many urban areas of the world.

Weather undoubtedly plays a role in cities' carbon footprints, but does not act alone. For instance, many relatively colder urban areas in the northeast of the US have larger residential carbon footprints because they rely on carbon-intensive home heating fuels such as fuel oil. Warm areas in the south, likewise, have large residential carbon footprints because they rely on carbon-intensive air conditioning.[30] The carbon intensity of the fuels used is, hence, another key factor. For instance, the carbon intensity of coal is almost two times higher than the carbon intensity of natural gas.

The *economic base* of a city is another important factor. In Beijing and Shanghai (China), industry contributes 43 and 64 per cent of the total emissions, respectively.[31] Industrial emissions of GHGs in cities elsewhere are much lower: 28.6 per cent in Mexico City, 7 per cent in London (UK), 9.7 per cent in São Paulo (Brazil), and 10 per cent in Tokyo (Japan) and New York (US). This reveals that many cities have already transitioned to service-based urban economies and, thus, have been able to reduce their carbon footprints. The differences reflect a shifting international pattern in the location of industrial activities – a pattern determined by differences in profitability, costs and environmental legislation among cities.[32] The current pattern reflects the fact that China has become the main manufacturer of commodities for the world, allowing developed countries to shift responsibility for their own GHG emissions in spite of the fact that their consumer-driven impact upon the market has created much of the need for a high industrial output in China. This international shifting of the location of industrial production calls for the use of consumption-based approaches, and not only production-based ones, in the measurement of emissions in order to have a true picture of responsibility for industrial emissions among and within countries and urban areas.[33]

*Affluence* has been repeatedly acknowledged as a significant driver of GHG emissions and other environmental impacts; but again it does not act alone – rather, together with such factors as technology, natural endowments and equity. According to ecological modernization theory, environmental problems such as climate change are addressed by development or modernization. A structural change, or shift, to less carbon-intensive societies occurs at the macroeconomic level through the development of new and less carbon-intensive technologies whose use is induced by market mechanisms.[34]

As an economy develops (modernizes), sectors such as agriculture and fisheries are replaced by manufacturing industries and, with further development, service industries. Ecological modernization theorists argue that economic growth within developing economies will follow a natural path, driven by economic forces and market dynamics, from higher to lower states of environmental stress. The environmental impacts of economic growth, thus, increase in the early stages of development, but stabilize and then decline as economies mature. The process is depicted by an inverted U-shape curve, also known as the Environmental Kuznets Curve. Indeed, the relation between national carbon intensity and level of economic development has changed from essentially linear in 1965 to essentially curvilinear in 1990.[35] The tendency to an essentially curvilinear relation was still valid for the year 2003 (see Figure 1.6). A linear relation means that a one unit increase in GDP essentially translates to a similar increase in emissions, while in a curvilinear relation a one unit increase relates to a smaller than one unit increase in emissions. However, at least part of this tendency might be understood in terms of the shifting of manufacturing activity to other areas due to economic, political and environmental factors, as illustrated in the example of China given above. Because developed countries'

Gender inequities exist both in energy use and GHG emissions

Urban centres in developing countries have lower levels of emissions per capita than cities in developed countries

The international shifting of the location of industrial production calls for the use of consumption-based approaches ... in the measurement of emissions ... among and within countries and urban areas

economies have become service based and because their industrial production has been relocated to some developing countries, GHGs emitted by their urban areas have decreased. However, their responsibility for that percentage of the GHGs emitted in the industrial manufacturing countries producing goods for them should be accounted to them as the consumers creating the need for the goods and not to the manufacturing country.[36] Some researchers have suggested that this change in the attribution of GHGs would alter the features of the curve.[37]

Affluence theory has empirical and political relevance for this Global Report for two reasons. While the 'environmental burdens of urban poverty primarily affect the poor living in the immediate locality', the environmental burdens of affluence, such as climate change, can affect both rich and poor people around the globe; but these also tend to fall disproportionately upon the poor.[38] The second reason, relevant to the debate around climate change impacts upon cities, follows from the fact that the very urban dwellers most at risk from local environmental degradation – the poor – seem also to be most at risk from floods, heat waves, storms and other climate-related threats.[39]

It can be misleading to concentrate on urban emissions per capita, as there are very large differentials within urban centres. Both gender and socio-economic equity is, therefore, a key dimension affecting GHG emissions by urban populations and activities. There is no adequate information to provide an accurate picture on the role of equity in determining different levels of emissions among demographic sectors of an urban area. Yet, some examples can be used to draw preliminary conclusions. According to a study on the per capita emissions footprints of single-person households in Germany, Norway, Greece and Sweden, on average men consumed between 6 per cent (Norway) and 39 per cent (Greece) more energy than women, and this gender difference is independent of income and age.[40] The per capita emissions of Dharavi, a predominantly low-income, high-density inner-city neighbourhood of

> The environmental burdens of affluence, such as climate change, can affect both rich and poor people around the globe; but ... tend to fall disproportionately upon the poor

> The size, growth, structure and density of population are key determinants of cities' GHG emissions

Mumbai (India), are a tiny fraction of the per capita emissions of high-income districts of Mumbai, where a high proportion of the population commutes to work by car.[41]

According to human ecologists, the size, growth, structure and density of *population* are key determinants of cities' GHG emissions and other environmental impacts.[42] A negative correlation exists between population density and atmospheric GHG emissions; for instance, a 1 per cent increase in the density of urban areas would relate to approximately 0.7 per cent decline in carbon monoxide ($CO$) pollution at the city level, with other factors held constant.[43] Spatially compact and mixed-use urban developments have significant benefits in terms of GHG emissions.[44] However, attention also needs to be given to other explaining factors, such as land-use patterns and the layout of the transportation system.[45] Furthermore, urban density poses a dilemma: while 'tailpipe emissions and fossil-fuel consumption are greatly increased with urban sprawl', levels of human exposure to emissions of other pollutants (e.g. nitrogen dioxide) might actually increase with density if no measures are undertaken to reduce atmospheric emissions.[46] The implications of urban form on climate change mitigation and adaptation are discussed in Chapters 5 and 6.

## Urban vulnerability and 'resilience'

As described above, urban settlements are already at risk from sea-level rise, droughts, heat waves, floods and other hazards that climate change is expected to aggravate. Yet, a focus on the *exposure* to these hazards alone is insufficient to understand climate change impacts upon urban centres, their populations and economic sectors. Attention to urban resilience, development, socio-economic and gender equity, and governance structures as key determinants of adaptive capacity and actual adaptation actions is also necessary. Many scholars and practitioners view resilience in the context of responses to hazards and recovery from disasters.[47] In this view:

- Cities can increase or reduce the impacts of such hazards as floods and heat waves as a result of their socio-environmental history. Urban activities invariably alter their environment, but two results are possible: environmental degradation and reduced resilience (see Box 1.3), or urban populations' growing ability to repair damage, sustain the environment and increase cities' resilience.[48]
- Urban populations and the different tiers of government responsible for their well-being are resilient if they are able to build capacity for learning and adaptation, and even capitalize on the learning opportunities that might be opened by a disaster. The urban populations of Dhaka and other human settlements of Bangladesh offer an example of this (see Box 1.4).

The significance of urban vulnerability and adaptive capacity to climate impacts can be analysed on at least two distinct levels: from the perspective of the city as a whole and the way in which it develops; and from the perspective of the city as it can be broken down to reveal its different socio-

---

**Box 1.3 Mexico City: Environmental degradation and vulnerability**

The water management system of Mexico City has developed features which do not allow it to cope with floods and droughts. It is overexploiting not only its water resources by between 19.1 and 22.2 cubic metres per second, but also the water of two providing basins (Lerma and Cutzamala). According to projections where no consideration is given to global warming, between 2005 and 2030 the population of Mexico City will increase by 17.5 per cent, while between 2007 and 2030 available water will diminish by 11.2 per cent. The situation might get worse if, as expected, climate change brings lower precipitation to this area. Those water users who already face recurrent shortages during the dry season, or when droughts hit Mexico City, will be especially affected. For example, 81.2 per cent of people affected by droughts during 1980 to 2006 live in Netzahualcoyotl, one of the poorer municipalities of the city.

This overexploitation of water resources creates two sources of vulnerability: first, problems of water availability (scarcity) that make water users (especially poor sectors already facing scarcity) vulnerable to the changes in the availability of water that are expected from climate change. Second, groundwater levels are continuously falling, which historically has caused subsidence (and continues to do so in some areas), thus undermining the foundations of buildings and urban infrastructure and increasing the vulnerability of these areas and the populations within them to such hazards as heavy earthquakes and rains.

*Source:* Romero Lankao, 2010

demographic groups' access to the determinants of adaptive capacity.

### ■ Urban development can bring increased vulnerability to climate hazards

The concentration, in urban centres, of people and their homes, infrastructure, industries and waste within a relatively small area can have two implications for the urban impacts of climate change and other stresses. On the one hand, urban areas can be dangerous places in which to live and work; their populations can be very vulnerable to extreme weather events or other hazards, with the potential to become disasters. For instance, the urban concentration of these elements can generate risk when residential and industrial areas lack space for evacuation and emergency vehicle access (as in the case of slums), when high-income populations are lured by low-lying coastal zones or green areas (as in California or Florida in the US, or Melbourne, Australia), or when lower-income groups, lacking the means to access safer land, settle on sites at risk from floods or landslides (as in Rio de Janeiro in Brazil, Mumbai in India and many urban centres in developing countries).

Urban settlements can increase the risk of 'concatenated hazards'.[49] This means that a primary hazard (heavy storm) leads to secondary hazard (e.g. floods creating contamination of water supplies, or landslides destroying houses and infrastructures). Industrialization, inadequate planning and poor design are key determinants of secondary or technological risks. As illustrated by Bogotá (Colombia), Buenos Aires (Argentina) and Santiago (Chile), the populations of many cities are already at risk from exposure to high levels of pollution, exceeding World Health Organization (WHO) standards in particulate matter and nitrogen dioxide concentration in the air.[50] It is possible that the impacts of climate hazards such as heat waves will overlap with pollution events and the urban heat-island effect, and compound one another, making urban disaster risk management even more complex.

On the other hand, the same concentration of people, infrastructure and economic activities in urban centres also means economies of scale for many of the measures that reduce risks from extreme weather events. These economies of scale might manifest themselves in a reduced per capita cost of better watershed management, warning systems and other measures to prevent and lessen the risks when a disaster threatens or occurs. Furthermore, when provided with policies focused on enhancing sustainability and moving from disaster response to disaster preparedness, urban settlements can increase their effectiveness at coping with climate hazards.

Exposure to current climate hazards is, for many cities, a result of historical location factors and a long development process. Many cities have developed without consideration of the risks that climate change will induce. Most large cities have been built on sites that were originally chosen for trade or military advantage (e.g. Shanghai, China; New York, US; Cartagena, Colombia; and Cape Town, South Africa). In the majority of cases, this has meant that they were located on the coasts or near the mouths of major

---

**Box 1.4 Capacity to learn and adapt in Bangladesh**

Bangladesh is situated in an area at risk from tropical storms, whose intensity and frequency have increased over the last years. A hurricane hit Bangladesh in 1991 killing at least 138,000 people and leaving as many as 10 million people homeless. Serious efforts have been undertaken, promoted by local and national governments and international organizations, to decrease the risk from tropical cyclones in the area. These efforts have included the development of an early warning system and the construction of public shelters to host evacuated people. These improvements were tested in 2007, when between 8 and 10 million Bangladeshis were exposed to Sidr, perhaps the strongest cyclone to hit the country since 1991. There was a 32-fold reduction in the death toll (i.e. 4234 people compared to 138,000) and Bangladesh's capacity for learning and adaptation was proven (see also Boxes 4.4 and 6.2).

*Source: Paul, 2009*

---

rivers where trade by sea with other coastal cities or by rivers with the interior hinterlands could best be accomplished. These urban centres then became the hubs of trade for their countries and, as such, greatly increased their wealth.

As this wealth continued to build, further development was fuelled and these areas became engines of economic growth for their countries, attracting more capital from private-sector investment and labour migration from rural areas and immigration from other countries. The movement to urban centres continues today and these areas have become magnets of industry and labour without regard to the many environmental risks that are endemic to these areas and the mounting hazards resulting from climate change.

### ■ Why are some sectors of the population more vulnerable?

Not all demographic segments of the urban population are equally affected by the hazards aggravated by climate change. The capacity of different urban populations to cope or adapt is influenced not only by age and gender, but also by one or a combination of some or many factors[51] (see Chapter 4). These factors include:

- Labour, education, health and the nutrition of the individuals (*human capital*). As a critical asset, labour is linked to investments in human capital. Health status determines people's capacity to work; education and skills determine the returns from their labour.
- The financial resources available to people (savings, supplies of credit – i.e. *financial capital*).
- The extent and quality of infrastructure, equipment and services (*physical capital*), some of which are owned by individuals (e.g. housing).
- Stocks of such environmentally provided assets as soil, land and atmosphere (*natural capital*). In urban areas, land for shelter is a critical productive asset.
- The quality and inclusiveness of governance structures and community organizations that provide or manage safety nets and other short- and longer-term responses, or *social capital* – an intangible asset defined as the rules, norms, obligations and reciprocity embedded in social relations and institutional arrangements.

> Urban centres also means economies of scale for many of the measures that reduce risks from extreme weather events

> When provided with policies focused on enhancing sustainability and moving from disaster response to disaster preparedness, urban settlements can increase their effectiveness at coping with climate hazards

Wealthy individuals and households have many of the requirements for higher adaptive capacity. They have more resources to reduce risks – that is, safer housing, more stable jobs, safer locations to live in, and better means of protecting their wealth (e.g. insurance of assets that are at risk). Wealthier groups often have more influence on public expenditures. In many urban areas, middle- and upper-income groups have been the main beneficiaries of government investment in such determinants of adaptive capacity as infrastructure and services. If government does not provide these, higher-income groups have the means to develop their own provisions for water, sanitation and electricity, or to move to private developments which provide them. Wealthier groups, therefore, have higher adaptive capacity.

Although systematic evidence of the gender implications of climate change at the city level both among wealthy and poor sectors and countries is still lacking,[52] some evidence points to the fact that gender gaps exist in access to such assets and options as credit, services, education, information, decision-making power and technology. For instance, in sub-Saharan Africa, 84 per cent of women's non-agricultural employment is informal (compared to 63 per cent of men's).[53] The informal sector is also important in capital and large cities, where more than half of all women are employed in the informal sector (except in South Africa and Namibia), although informal employment is actually higher in small cities and towns and rural areas.[54] Due to this situation, women do not have adequate livelihood options and can be particularly vulnerable to disasters. As illustrated by Hurricane Mitch in Honduras and floods in Dhaka, Bangladesh, disaster warnings often do not reach women or are not understood by women. Furthermore, in many instances, women cannot evacuate without the authorization of their husbands.[55]

Scattered evidence points to the fact that children are more at risk of being affected by the adverse impacts of climate change.[56] There are several reasons for this: they are in a stage of rapid development which can be severely interrupted by the stress of severe weather events and climate hazards. They are relatively more vulnerable to warm spells and heat waves, heavy precipitation, droughts and other climate hazards because of their immature organs and nervous systems, limited experience and behavioural characteristics. This can be intensified by poverty and the difficult choices that poor households make as they cope with challenging situations. However, it is also true that 'with adequate support and protection, children can be extraordinarily resilient' when faced with hazards and stresses.[57]

Very elderly men and women can also be at risk, as illustrated by the high elderly mortality rates in the heat waves that hit Chicago (US) in 1995 and Europe in 2003. Indeed, as illustrated by research in the cities of London and Norwich (UK), the elderly might feel, falsely, that heat waves do not pose a significant risk to them personally.[58] The elderly can also be limited in their capacity to move rapidly away from rising floodwaters by their isolation, their health conditions or their perceptions.

The urban poor tend to be highly vulnerable, especially in developing countries, and may also fall into other disadvantaged categories that increase vulnerability by also being women, very young or very old. Many poor populations face additional risks: they live in informal settlements, live on floodplains, unstable slopes, over river basins and in other highly risk-prone areas, or work within the informal economy. They are also constantly faced with the possibility that governments may forcibly move them off land sites deemed to be vulnerable to weather risks, or they may be moved simply because other actors want the land they occupy for more 'profitable' uses, but with the consequence that they are also moved away from their means of livelihood.[59]

Furthermore, poorer groups are most affected by the combination of greater exposure to a range of other possible urban hazards (e.g. poor sanitary conditions and lack of hazard-removing infrastructure such as drainage). They have less state provision to help them cope, along with less legal and insurance protection. Low-income groups also have far fewer possibilities to move to less dangerous sites. This should not, however, lead to the conclusion that the poor are merely passive recipients of the risks of climate change and other hazards. As illustrated by Cavity City in the Philippines, or the *Baan Mankong* ('secure tenure') programme in Thailand,[60] many poor groups have developed mechanisms to adapt. It just means that the structural issues referred to here pose severe limits to their coping mechanisms and create constraints upon their adaptation options.

# CONCLUDING REMARKS AND STRUCTURE OF THE GLOBAL REPORT

Urbanization and climate change are sources of both developmental and environmental challenges and opportunities. Industrialization and urbanization have been critical components of rapid economic growth and of technological changes that have contributed to improvements in the economy and the quality of life of many urban populations around the world. Both have also helped to decrease the carbon intensity and increase the efficiency of production and consumption. Yet, notwithstanding these socio-economic and technological achievements, poverty – which has increasingly been acquiring an urban face – remains a formidable challenge. 'The needs remain enormous, with the number of hungry people having passed the billion mark.'[61] Poverty alleviation thus remains the overarching priority, especially in developing countries.

Climate change, which is both a developmental and environmental issue, complicates the picture in several ways. The impacts of global GHG emissions are currently manifest in stronger and more frequent floods, droughts and heat waves, adversely affecting the industries, populations and governments of many urban centres. Therefore, urban populations and economic sectors are faced with two challenges: the need to adapt, at least to some amount of warming, and the urgency to mitigate the causes of global climate change.

---

**Higher-income groups ... have higher adaptive capacity**

**Women ... can be particularly vulnerable to disasters**

**Children are more at risk of being affected by the adverse impacts of climate change**

**Poorer groups are most affected by the combination of greater exposure to a range of other possible urban hazards (e.g. poor sanitary conditions and lack of hazard-removing infrastructure such as drainage)**

Urban centres of developed countries and wealthy sectors within cities of developing countries must play a vital role in reducing their carbon footprints. Their actions cannot be reduced to technological fixes aimed at increasing energy efficiency and reducing the carbon intensity of cars, fabrics, utilities and other devices. Because goods, services, waste disposal and transportation are aimed at satisfying urban markets, the responsibility for the emissions produced in their manufacture, production and energy expenditures needs to be allocated to urban consumers, even when these goods and services are generated outside urban boundaries. This has very profound implications and difficulties for creating real mitigation strategies. A call for a change in consumption patterns and lifestyles away from a focus on more and bigger is, clearly, fundamental.

Actions to induce changes in the factors shaping population density, urban form, lifestyles, equity and other components of urban development are equally fundamental for mitigation, adaptation and sustainable development. Transport strategies, for instance, need to be consistent with the spatial structures of cities.

Urban development can also be a source of resilience. Population densities can create the potential for city-scale changes in behaviour that can mitigate human impacts upon climate and create opportunities for adaptation to floods, heat waves and other climate hazards. Properly designed infrastructure developments can provide physical protection; well-designed communications and early warning systems; can help people to deal with disasters; and appropriate urban planning can help restrict the growth of populations and activities in risk-prone areas.

Those urban centres with populations lower than 500,000 people will be faced with great difficulties in coping with the impacts of climate change, given their relatively low management capacity. However, they can also take advantage of their relatively small size to redirect their future growth in more sustainable and resilient ways that reduce their emission levels to a desired minimum and enhance their resilience and ability to cope with climate hazards and other stresses.

This Global Report is organized into seven chapters. Chapter 2 focuses on the international climate change framework and the implications, opportunities and challenges that it offers for urban action. It describes the process by which climate change became an international regime: the Climate Convention; the main mechanisms, instruments and financing strategies of the Climate Convention; and the main positions of the parties to the Kyoto Protocol. Aimed at providing policy-makers with a navigational tool to better steer a course through the complex universe of climate policy and action, the chapter presents various components of the multilevel climate change governance elaborated upon throughout the report and describes the main actors, components and actions of climate governance at the international, supra-national (regional), national, and sub-national levels.

Chapter 3 examines the contribution of urban areas to climate change. It discusses the main protocols and methods for measuring GHG emissions and examines transportation, industry, buildings and other sources of GHG emissions in more detail. A summary of the scale of urban emissions and how they vary between countries at different stages of economic development is provided. The chapter illustrates how the total volume of emissions is strongly shaped by such factors as a city's geographic situation, demographic situation, urban form and density, and economic activities. It includes a discussion of both the main factors and underlying drivers influencing emissions.

Climate impacts and vulnerabilities are the main focus of Chapter 4. The chapter describes how climate change may exacerbate the physical, social and economic challenges that cities are currently experiencing. First describing the physical climate change hazards facing urban centres, it goes on to look at how the direct and indirect physical, economic and social impacts of these changes vary with disparities in existing vulnerabilities within and across cities, identifying specific urban populations, regions and cities that are particularly vulnerable to climate change and the reasons why this is so. The chapter ends with concluding remarks on the impact of climate change in cities and the lessons for policy.

Chapter 5 focuses on mitigation, one of the two main responses to climate change. It describes the mitigation policy responses and initiatives that are currently taking place in cities in the areas of urban planning and infrastructure development, transportation, the built environment and carbon sequestration. It examines how such strategies and measures have been undertaken through different modes and mechanisms of governing (e.g. provision, regulation, self-governing and enabling), and explores the factors shaping urban mitigation in institutional, economic, technical and political terms (e.g. individual and institutional leadership, knowledge and institutional capacity). Finally, the chapter provides a comparative analysis of emerging trends in mitigation responses.

Chapter 6 looks at adaptation to climate change from the fundamental position that because the international community has been unable to effectively respond to the challenge of reducing GHGs to a level that would avoid dangerous interference with the climate system, adaptation responses over the next decade will be critical. The chapter starts by defining urban adaptation and adaptive capacity, followed by a review of some existing coping and adaptation experiences by individuals, households, communities and urban governments, and then examines the relative roles and potential partnerships between stakeholders, and looks at some mechanisms for financing adaptation.

Chapter 7 summarizes the key findings and messages of the report, and proposes a set of integrating themes with respect to urban areas facing climate change challenges. The chapter first looks at the constraints and challenges to, and opportunities from, mitigation and adaptation actions, along with some of the linkages among drivers and vulnerabilities. It then goes on to highlight a variety of synergies and trade-offs between mitigation, adaptation and urban development. After briefly describing the current state of knowledge along with the gaps, uncertainties and challenges, the chapter provides a series of suggestions on future policy directions in terms of local, national and international principles and policies to support and enhance urban responses to climate change.

**Urban development can ... be a source of resilience**

**Properly designed infrastructure developments can provide physical protection ... and appropriate urban planning can help restrict the growth of populations and activities in risk-prone areas**

**Urban centres with populations lower than 500,000 people ... can ... take advantage of their relatively small size to redirect their future growth in more sustainable and resilient ways**

## NOTES

1   UN, 2010.
2   Due to increasing $CO_2$ concentrations.
3   The low-elevation coastal zone is the contiguous area along the coast that is less than 10m above sea level (IPCC, 2001b).
4   UN, 2010.
5   UN, 2010.
6   IPCC, 2007b; Satterthwaite et al, 2007b, 2009c.
7   UN, 2010.
8   UN, 2010. See also Statistical Annex, Table A.4.
9   See Chapter 3 and Tables 3.5, 3.11 and 3.12.
10  Dodman, 2009; Romero Lankao et al, 2008, 2009a.
11  McGranahan et al, 2005; Balk et al, 2009.
12  See UN-Habitat, 2007, p319.
13  Le Treut et al, 2007.
14  Ammann et al, 2007.
15  Sabine et al, 2004.
16  Sabine et al, 2004; Raupach et al, 2007.
17  Füssel, 2009.
18  The data in this section are derived from IPCC (2007d).
19  For a full list of gases to be assessed in national GHG inventories under the Kyoto Protocol, see the note in Table 3.1.
20  See Table 1.4; UN, 2010.
21  IPCC, 2007d, p12.
22  IPCC, 2007d.
23  Examples include the Clinton Foundation, Nicolas Stern, and Munich Insurance. See Satterthwaite, 2008a; Dodman, 2009; and Chapter 4.
24  Satterthwaite, 2008a; Dodman, 2009.
25  Zhang, 2010.
26  IPCC, 2007b.
27  Alberti and Hutyra, 2009.
28  Alber, 2010, p21.
29  Romero Lankao, 2007a; Satterthwaite, 2008a. See also Chapter 3.
30  Brown et al, 2008.
31  Ru et al, 2009.
32  Satterthwaite, 2007; Romero Lankao et al, 2005.
33  See Chapter 3.
34  Murphy, 2000; Gibbs, 2000.
35  Roberts and Grimes, 1997.
36  See Chapter 3.
37  Bin and Harris, 2006.
38  Satterthwaite, 1997a; McGranahan et al, 2001, p15;.
39  Parry et al, 2007a; Wilbanks et al, 2007; Satterthwaite et al, 2007b; Romero Lankao et al, 2008.
40  Räty and Carlsson-Kanyama, 2010.
41  Satterthwaite, 2008a.
42  Walker and Salt, 2006.
43  Romero Lankao et al, 2009a.
44  See Chapter 3.
45  Other determinants of transportation emissions are transit accessibility, pedestrian friendliness and local attitudes and preferences, which also influence driving behaviour. Handy et al, 2005.
46  Marshall et al, 2005, p284.
47  Vale and Campanella, 2005.
48  This can be done through two mechanisms: patterns of use that do not overexploit local resources and go beyond the carrying or absorbing capacity of local ecosystems; and the ability and power to import resources from, or export emissions to, surrounding and remote areas and to make up for the impact (i.e. avoid local overexploitation and pollution). See Turner et al, 2003.
49  Allan Lavell, cited in Satterthwaite et al, 2007b.
50  Romero Lankao et al, 2009b.
51  Moser, 2008; Moser and Satterthwaite, 2010; Harlan et al, 2006; Romero Lankao and Tribbia, 2009.
52  Alber, 2010.
53  UN-Habitat, 2008c; Alber, 2010 (citing WIEGO and Realizing Rights: The Ethical Globalization Initiative, 2009).
54  UN-Habitat, 2008c.
55  Alber, 2010.
56  Bartlett, 2008.
57  Bartlett, 2008, p1.
58  Wolf et al, 2010.
59  Satterthwaite et al, 2007b.
60  Satterthwaite et al, 2007b.
61  World Bank, 2009c, p1.

# 2

# CITIES AND THE INTERNATIONAL CLIMATE CHANGE FRAMEWORK

Responses to the climate change challenge are taking place within the context of an international framework that shapes related actions and decisions at all levels.[1] This framework is defined here as the spectrum of agreements, mechanisms, instruments and actors governing and driving climate change action globally. The overall structure of this framework is complex and multidimensional in that it is comprised of elements that are quite different and distinct in many of their functions and approaches, constituencies, scope and focus.[2] While international agreements negotiated by national governments such as the United Nations Framework Convention on Climate Change (UNFCCC) and its Kyoto Protocol remain crucial aspects of the framework, they are not the only mechanisms governing climate change action. Other layers of intervention have become equally important in implementing innovative climate change responses and policies, including those at the regional, sub-national and local levels.

Cities have a vital role to play in the implementation and achievement of commitments within the international climate change framework. They also stand to benefit from the opportunities created by this framework for local responses to climate change. Yet, local-level actors and authorities often lack an understanding of the nature and functioning of the various components of the international climate change framework and how they could utilize these to enhance their mitigation and adaptation strategies. For instance, many decision-makers operating at the city level lack a working knowledge of the opportunities and constraints associated with international financing options, including those established as part of the UNFCCC.[3] In view of this, the aim of this chapter is to highlight the key elements of the international climate change framework and its effects on interventions at the local level. It is also intended to frame discussions of climate change conditions, trends and policies in the rest of this Global Report.

The chapter starts by briefly describing the process by which climate change emerged as an issue of international concern culminating in the establishment of the UNFCCC as the key element of the international regime governing climate change issues. The core mechanisms, instruments and financing strategies of this Convention are then outlined. The Kyoto Protocol is also reviewed as the main

international treaty with legally binding emission reduction commitments. Subsequently, the key actors, components and actions of climate governance at the international, regional, national and sub-national levels are considered. Finally, the implications of the international climate change framework for local action at the city level are outlined.

## THE UNITED NATIONS FRAMEWORK CONVENTION ON CLIMATE CHANGE

Climate change issues have been discussed since the early 19th century (see Table 2.1), but only emerged as an international policy concern during the 1970s and 1980s when technological advances allowed scientists to state with more certainty that atmospheric concentrations of greenhouse gases (GHGs) were on the rise and that this could have profound ramifications for the Earth's climate. Between 1988 and 1990, national governments began to play a greater role in defining the climate change agenda, and the Intergovernmental Panel on Climate Change (IPCC) was established in 1988 to provide them with information on global warming trends through regular scientific assessments (see Box 2.1).

The process of formally negotiating an international climate change treaty started in December 1990, when the United Nations General Assembly created the Intergovernmental Negotiating Committee for a Framework Convention on Climate Change. In 1992, the committee adopted the United Nations Framework Convention on Climate Change (UNFCCC) at the United Nations Headquarters, New York. The UNFCCC, also known as the Climate Convention, entered into force in 1994 and had been ratified by 193 countries by October 2010.[4] The ultimate objective of the Convention is to stabilize global greenhouse gas (GHG) concentrations at a level that would prevent human interference with the climate system.[5] The Convention also aims to assist countries, especially developing ones, in their efforts to adapt to the effects of climate change.

The Convention's efforts to curb emissions are premised on some explicit and implicit norms which have

*Local-level actors and authorities often lack an understanding of the nature and functioning of the various components of the international climate change framework*

**Table 2.1**

**Major milestones in international climate change governance**

| | |
|---|---|
| 1827 | French scientist Jean-Baptiste Fourier is the first to consider the 'greenhouse effect' – the phenomenon whereby atmospheric gases trap solar energy, increasing the Earth's surface temperature. |
| 1896 | Swedish chemist Svante Arrhenius blames the burning of fossil fuels producing $CO_2$, the main greenhouse gas, for contributing to climate change. |
| 1950s | Global warming science grows with increasing information on the impacts of greenhouse gases upon the world's climate, together with the development and growth of environmental movements. |
| 1979 | First World Climate Conference in Geneva, Switzerland, calls on governments to forecast and prevent potential human-made changes in climate. |
| 1988 | The Intergovernmental Panel on Climate Change (IPCC) is established to produce regular scientific and technical assessments of climate change. |
| 1992 | The United Nations Framework Convention on Climate Change (UNFCCC) is adopted in New York, US, on 9 May 1992, and enters into force on 21 March 1994. |
| 1997 | The Kyoto Protocol to the Convention is adopted at COP-3 in Kyoto, Japan, and enters into force on 16 February 2005. |
| 2001 | The Marrakesh Accords, a set of detailed rules for the implementation of the Kyoto Protocol, is adopted during COP-7 in Marrakesh, Morocco. |
| 2007 | Negotiations for a new international treaty to take over from the Kyoto Protocol in 2012 begin in Bali, Indonesia, during COP-13. The Bali Road Map, a two-year process to finalize a binding agreement in 2009 during COP-15, is agreed upon. |
| 2009 | The main outcome of COP-15, the Copenhagen Accord, is a non-binding agreement which seeks to cap the global temperature rise and raise finances for climate change action in developing countries. |
| 2010 | The Cancún Agreements are adopted during COP-16 in Cancún, Mexico, containing a package of decisions on mitigation and adaptation targets, implementation and funding. |
| 2011 | COP-17, Durban, South Africa, 28 November–9 December 2011. |

*Sources:* Baumert et al, 2005; ICLEI et al, 2009; *New Scientist*, 2009

become fundamental to the international climate regime. Chief among these are the principle of 'common but differentiated responsibilities and respective capabilities' and the 'precautionary principle'.[6] The first recognizes historical differences in the contribution of developed and developing countries to climate change, as well as differences in their respective economic and technical capacity to tackle these problems.[7] In this regard, the Convention places the greatest responsibility for fighting climate change on developed countries, given their role in generating much of the GHG emissions in the past. The second implies that even in the absence of full scientific certainty, countries are obliged to

The IPCC's assessment process is a vital interface between science and policy and a crucial mechanism by which science informs policy-making

**Box 2.1 The Intergovernmental Panel on Climate Change**

The IPCC was created in 1988 by the World Meteorological Organization (WMO) and the United Nations Environment Programme (UNEP) in order to keep world governments informed of climate change issues. The IPCC's 194 member countries meet once a year during sessions also attended by numerous other institutions and observer organizations.

The United Nations General Assembly resolution 43/53 of 6 December 1988 states that the role of the IPCC is to 'provide internationally coordinated scientific assessments of the magnitude, timing and potential environmental and socio-economic impact of climate change and realistic response strategies'. The same resolution requested the WMO and UNEP to initiate a comprehensive review and subsequent development of recommendations with respect to the following vis-à-vis the IPCC:

- the state of knowledge of the science of climate and climatic change;
- programmes and studies on the social and economic impact of climate change, including global warming;
- possible response strategies to delay, limit or mitigate the impact of adverse climate change;
- the identification and possible strengthening of relevant existing international legal instruments having a bearing on climate; and
- elements for inclusion in a possible future international convention on climate.

The IPCC analyses scientific and socio-economic information on climate change and its impacts, and assesses options for mitigation and adaptation. It provides scientific, technological, and socio-economic findings to the Conference of the Parties (COP) to the UNFCCC. The IPCC's assessment process is a vital interface between science and policy and a crucial mechanism by which science informs policy-making. Accordingly, the IPCC has played a crucial role in establishing the importance of the climate change issue; providing an authoritative resolution of policy-relevant scientific questions; demonstrating the benefits and costs of various policy options; identifying new research directions; and providing technical solutions.

To date, the IPCC has prepared comprehensive scientific reports on climate change on a regular basis. The First Assessment Report of the IPCC (published in 1990) indicated that levels of human-made GHGs were increasing in the atmosphere and predicted that these would exacerbate global warming. It also illustrated the need for a political platform for countries to tackle the consequences of climate change, thereby playing a critical role in the creation of the UNFCCC. Both the Second (1995) and Third (2001) Assessment Reports implied stronger linkages between human activity and climate change, thereby strengthening efforts for the negotiation of the Kyoto Protocol. The Fourth (and latest) Assessment Report (2007) noted that the evidence for global warming is 'unequivocal' and forecasted warming of 1.8 °C to 4.0°C by 2100. The IPCC is currently working on the Fifth Assessment Report, which is due to be released in 2014.

In addition to the assessment reports, the IPCC has prepared numerous other reports, methodologies and guidelines to support countries in implementing their commitments.

*Sources:* IPCC, undated a, undated b; UN, 1988; Brasseur et al, 2007

anticipate, prevent or minimize the causes of climate change and mitigate its adverse effects.[8]

Countries ratifying the treaty are referred to as 'Parties to the Convention' and agree to develop national programmes to slow climate change. 'Annex I' countries include developed countries that were members of the Organisation for Economic Co-operation and Development (OECD) in 1992 and also countries with economies in transition. These countries are required to provide regular inventories of their GHG emissions using 1990 as the base year for these tabulations.[9] 'Annex II' countries consist of Annex I countries excluding countries with economies in transition. These parties are expected to support mitigation and adaptation activities in developing countries financially and through the transfer of technology. 'Non-Annex I' countries are developing countries and are given special consideration due to their limited capacity to respond to climate change.[10]

The main authority of the Convention is the 'Conference of the Parties' (COP), which is comprised of all parties and meets annually to assess 'progress made by Parties in meeting their commitments and in achieving the Convention's ultimate objectives'.[11] Sessions of the COP, of which 16 have taken place (by the end of 2010) since the Convention entered into force in 1994, serve as the main forums for negotiations between the parties and the adoption of key decisions and resolutions. This is particularly important since the Convention mostly contains general formulations that are deliberately ambiguous to accommodate the diverse positions of the parties. The COPs are also attended by a large number of observers, including intergovernmental, non-governmental and other civil society observers.[12]

The first Conference of the Parties (COP-1) took place in December 1995 in Berlin, Germany, and expressed concern about the ability of countries to meet their emissions targets and commitments. Through the Berlin Mandate adopted at this meeting, a committee was established to negotiate a protocol on climate change by 1997, including additional GHG emissions reduction commitments for developed countries for the post-2000 period.[13] By the time COP-2 took place in July 1996 in Geneva, Switzerland, consensus on the negotiation of a protocol was not yet in sight and preliminary national communications suggested that countries were unlikely to meet their emissions reduction targets (i.e. to return to their 1990 emissions levels by 2000).[14] However, the meeting endorsed the Second Assessment Report of the IPCC, and reaffirmed the need for legally binding 'quantified emission limitation reduction objectives'.[15] In 1997, the principles under the UNFCCC were finally translated into legally binding commitments through the Kyoto Protocol, which was adopted at COP-3 in Kyoto, Japan.[16]

In addition to its focus on emissions reduction, the UNFCCC also seeks to support adaptation activities in developing countries. Accordingly, in 2001, during COP-7 in Marrakesh, Morocco, three main funding mechanisms for adaptation were set up under the UNFCCC – namely, the Special Climate Change Fund, the Least Developed

---

**Box 2.2 Funding mechanisms of the UNFCCC**

The *Special Climate Change Fund* is intended to finance activities related to adaptation, technology transfer and capacity-building, energy, transport, industry, agriculture, forestry, waste management and economic diversification. By September 2009, voluntary contributions of around US$120 million had been pledged for the fund and 24 projects had been approved.[a]

The *Least Developed Countries Fund* aims to assist 48 least developed countries to prepare and implement National Adaptation Programmes of Action (NAPAs) through which they identify priority adaptation activities for funding.[b] The rationale for this fund lies in the recognition of the limited ability of such countries to adapt to the consequences of climate change.[c] By March 2010, the United Nations Framework Convention on Climate Change (UNFCCC) had received NAPAs from 44 countries.[d] As of 30 September 2009, US$180 million had been pledged for this fund through voluntary contributions and, by 2010, 84 projects had been approved.[e]

The *Adaptation Fund* was established to finance adaptation projects and programmes in developing countries that are especially vulnerable to climate change impacts.[f] It is to be funded from a 2 per cent levy on all Clean Development Mechanism (CDM) project activities (see Box 2.3). The fund only became operational in 2010 and by October 2010, projects had been approved in only four countries – namely, the Solomon Islands, Nicaragua, Senegal and Pakistan.[g] Although the fund is expected to have grown to US$500 million by 2012, this falls short of the estimated US$50 billion required annually for adaptation activities in developing countries.[h]

*Sources:* a Climate Fund Update, undated a; UNFCCC, undated f; GEF, undated; World Bank, 2009b; b UNFCCC, undated g; c UNFCCC, undated h; d UNFCCC, undated i; e Climate Fund Update, undated b; GEF, undated; World Bank, 2009b; f Climate Fund Update, undated c; UNFCCC, undated j; g AlertNet, 2010a, 2010b; h IIED, 2009

---

Countries Fund and the Adaptation Fund (see Box 2.2). These are administered by the Global Environment Facility (GEF), an international partnership between 182 countries, international institutions, non-governmental organizations (NGOs), and the private sector to address global environmental challenges. The GEF was established in 1991 as a pilot programme at the World Bank with UNEP and the United Nations Development Programme (UNDP) as implementing partners. During the United Nations Conference on Environment and Development (UNCED) in 1992, it was restructured to become a separate institution and the main entity managing the funding mechanisms of the UNFCCC.[17]

A key challenge for the UNFCCC is that its main goal is somewhat 'indeterminate'. In other words, although it conveys the long-term goal of reducing emissions, it cautiously avoids any quantitative expression of it.[18] This is partly because the climate domain is characterized by uncertainties regarding causes, impacts and relationships. Although the publication of the IPCC's Fourth Assessment Report in 2007 signalled that the scientific community has established with greater clarity that human activities are the main causal factors of the unprecedented changes in our climate system, climate science still faces challenges. For instance, it cannot currently help policy-makers to know, with absolute certainty, how much is too much (e.g. what is the point beyond which emissions are too high). Science also cannot objectively ascertain at what level human interference with climate becomes dangerous. Some form of value judgement is unavoidable. And value judgements are context specific, not only because climate impacts differ from place to place, but also because different people perceive the risks in diverse ways.[19]

A key challenge for the UNFCCC is that its main goal is somewhat 'indeterminate'

Furthermore, because many of the cause-and-effect relationships are long and potentially irreversible, they require planning that goes beyond the tenure and even the lifetime of most current decision-makers and stakeholders. Complex interdependencies exist between different policy areas within and beyond climate policy, and the international community may fail to put in place the unprecedented series of response mechanisms that are required.[20] The difficulties related to international climate change negotiations (i.e. stalled negotiations during most of the COPs followed by last-minute key decisions by some parties) further complicate the operationalization and implementation of the UNFCCC.

## THE KYOTO PROTOCOL

The Kyoto Protocol was adopted on 11 December 1997 in Kyoto, Japan, during COP-3, and entered into force on 16 February 2005. By the end of 2010, the protocol had been ratified by 191 countries.[21] While the protocol holds in common the objective and institutions of the UNFCCC, the two differ in that the protocol is a binding agreement which commits developed countries to stabilize their GHG emissions, while the Convention only encourages the same.[22] Key decisions and resolutions on the implementation of the Kyoto Protocol's provisions are taken during the Meeting of the Parties to the Kyoto Protocol (MOP), which

*The Kyoto Protocol ... is a binding agreement which commits developed countries to stabilize their GHG emissions*

occurs in conjunction with the meetings of the COP to the UNFCCC.[23] The rules for implementing the protocol were spelt out in the Marrakesh Accords adopted in 2001 at COP-7 in Marrakesh, Morocco.[24]

According to the protocol, developed countries commit to reduce their overall GHG emissions by at least 5 per cent below 1990 levels during the commitment period from 2008 to 2012.[25] They submit annual emission inventories and national reports at regular intervals and a compliance system is in place to assist countries to meet their targets. Some developed countries rejected the protocol but are developing alternative regulatory approaches.[26] Developing countries have also ratified the protocol but do not need to limit or reduce their emissions. In addition to reducing emissions, the Kyoto Protocol also seeks to assist vulnerable developing countries to adapt to the adverse effects of climate change, primarily through the Adaptation Fund (see Box 2.2). During COP-16 (in Cancún, Mexico) a decision on binding emissions targets for a 'second commitment period' (i.e. beyond 2012) was deferred to a future date.

Before its adoption, negotiations of the Kyoto Protocol were stalemated over two critical issues. *First*, developed countries were in disagreement regarding mitigation targets. The European Union (EU) supported a 15 per cent reduction in GHG emissions below 1990 levels; the US and Australia proposed lower targets; and Japan's position was somewhere in the middle. To deal with these differences, diverse emissions targets were set, ranging from a 10 per cent increase for Iceland to an 8 per cent reduction for Germany, Canada and other countries.[27] Rather than being based on what the scientific community would consider necessary to stabilize emissions at current levels, or reflecting the levels of reductions that countries could achieve, emissions targets were the outcome of tough bargaining in closed-door sessions between representatives of the US, the EU and Japan during the final hours of COP-3 in Kyoto, Japan.[28]

*Second*, the flexibility of implementation mechanisms was an issue of contention. While developing countries and the EU supported domestic action as the main means to achieve emissions reduction targets, the US and some industries (mostly from the energy sector) argued that developed countries could achieve their targets through emissions-abatement projects in other countries or through emissions trading. Thus, although countries are expected to meet their mitigation targets primarily through national programmes, the Kyoto Protocol enables them to cut their emissions through three flexible mechanisms – namely, the Clean Development Mechanism (CDM), joint implementation and emissions trading (see Box 2.3).

Despite already contributing to emissions reductions globally, the flexible mechanisms have also been criticized. For instance, CDM has been criticized for simply moving emission reduction activities and their socio-economic and environmental impacts to where it is cheapest to make them, which normally means a shift from developed to developing countries.[29] Also, the CDM is not necessarily able to deliver the promised development dividends to the host country.[30] Emissions trading has been critiqued for allowing developed countries to earn emissions reduction credits primarily

---

**Box 2.3 Flexible mechanisms under the Kyoto Protocol**

The three flexible mechanisms of the Kyoto Protocol are as follows:

1  *Emissions trading* allows developed countries that exceed their target emissions to offset them by buying 'credits' from countries that stay below their emission targets. Emission quotas were agreed with the intention of reducing overall emissions by developed countries by 5 per cent of the 1990 levels by the end of 2012. For the five-year compliance period from 2008 until 2012, countries that emit less than their quota will be able to sell emissions credits to countries that exceed their quota.[a] In 2010, the value of the global carbon market was estimated to be worth a staggering US$144 billion.[b]

2  The *Clean Development Mechanism (CDM)* – which has been operational since 2006 – enables emission reduction projects in developing countries to earn certified emission reduction credits, which can then be traded or sold. These credits can be purchased by developed countries to achieve a twofold purpose: to meet their own emissions reduction targets under the Kyoto Protocol and to assist other countries in achieving sustainable development through climate change mitigation.[c] CDMs have registered an astounding growth, with over 5000 projects in the pipeline as of August 2010.[d]

3  *Joint implementation* allows developed countries to invest in emissions reduction activities in other developed countries. A developed country can thus earn emission reduction units from an emission reduction or emission removal project in another developed country, which can be counted towards meeting its Kyoto target.[e] A total of 243 joint implementation projects were in the pipeline as of 1 November 2009.[f]

Transactions by parties to the Kyoto Protocol under the above three flexible mechanisms are tracked and recorded through an international transaction log.[g] The log monitors the compliance of transactions with the rules of the Kyoto Protocol and may reject entries where this is not the case. Between 1 November 2008 and 31 October 2009, a total of 225,119 transaction proposals were submitted to the international transaction log.

*Sources:* a UNFCCC, undated q; b World Bank, 2010b; c UNFCCC, undated m; d CD4CDM, undated; e UNFCCC, undated n; f Gilbertson and Reyes, 2009; g UNFCCC, undated l

through trading rather than through cutting their domestic emissions. It also encourages developed nations to avoid their obligation to develop pollution reduction innovations to enable developing countries to increase production while limiting pollution.[31]

In an effort to create a framework of action for the period after the end of the current commitment period of the Kyoto Protocol in 2012, the Bali Road Map was adopted in 2007 during COP-13 to finalize a binding agreement in 2009 during COP-15 in Copenhagen, Denmark. However (and as was the case with the negotiations within both the UNFCCC and the Kyoto Protocol, and despite two years of advance work initiated by the Bali Road Map), little progress was made during two weeks of negotiations in Copenhagen. With time running out, the US forged the Copenhagen Accord, a 'non-binding' agreement that all but a handful of parties accepted. While the Copenhagen Accord succeeded in forging agreement on the need to address climate change, it is viewed as a major compromise that emerged due to the failure of countries to agree on a binding agreement to govern emissions reduction in the post-Kyoto period.

In contrast, the latest meeting, COP-16 in 2010 (Cancún, Mexico) has been dubbed a 'beacon of hope' that has restored faith in international climate change negotiations. While an agreement for the post-Kyoto period was not reached, the adoption of the 'Cancún Agreements', a package of decisions on adaptation and mitigation targets, implementation and funding, has managed to ebb some of the pessimism that emerged following COP-15. In addition to encouraging countries to push their emissions reduction targets over and above the commitments within the Kyoto Protocol, the Cancún Agreements establish mechanisms such as the Green Climate Fund, the Cancún Adaptation Framework and the Climate Technology Centre and Network to strengthen climate change action.[32]

The next Conference of the Parties will take place in 2011 (28 November to 9 December in Durban, South Africa) and an attempt to forge a binding agreement for the post-2012 period will once again be made. However, it remains uncertain whether the international community will be able to reach a legally binding agreement to replace the Kyoto Protocol. The continued delay in reaching such an agreement is expected to have serious negative consequences for global emissions reduction efforts.[33]

Despite its significance as the main binding agreement between parties, the Kyoto Protocol has been criticized on a variety of grounds. Some argue that it imposes high burdens on developed countries, while others suggest that it provides ineffective incentives for participation and compliance. Yet others point out that it creates modest short-term climate benefits while failing to provide a long-term solution. Indeed, numerous alternatives to the protocol have been suggested to address these shortcomings.[34] The existence of a set of initiatives parallel to the Kyoto Protocol is a sign of the fragmented nature of the international climate change framework and has led to an extensive debate on how to continue the negotiation of future treaties. The majority of policy proposals still support a universal framework of climate governance, while other recent proposals implicitly

create the possibility of further institutional fragmentation of this framework (e.g. starting a bottom-up process in which countries would put on the table acceptable measures in line with national circumstances).[35]

# OTHER CLIMATE CHANGE ARRANGEMENTS

Although international climate change negotiations between national governments remain crucial, the last two decades have witnessed the multiplication of other regional, national and local (e.g. city) mechanisms and actors responding to the climate challenge. These include initiatives of multilateral and bilateral entities, sub-national tiers of government, grass-roots groups, private enterprises, NGOs and individuals. This section describes the role of these in curbing GHG emissions (mitigation) and in climate change adaptation. Furthermore, it examines the levels at which these actors operate and outlines some of the actions, initiatives and instruments that they have developed and implemented to date.

## International level

A number of actors are actively developing strategies for climate change adaptation and mitigation at the international level, including the United Nations, multilateral and bilateral agencies. These initiatives are mostly designed to support the implementation of the commitments of the Kyoto Protocol as the main international treaty for climate change. Although a multitude of international actors are currently active in responding to climate change, many of their strategies, programmes and actions have evolved in isolation from each other. The lack of a clear division of responsibilities between the numerous international actors has led in some cases to overlapping functions, conflicting mandates and blurred objectives, and in other cases to constructive collaboration.[36] In turn, this has implications for the extent to which city authorities are able to make use of international funds and programmes to implement local adaptation and mitigation initiatives.

### ■ The United Nations

The United Nations is one of the key climate change actors at the international level. In addition to its work through the UNFCCC and the IPCC described earlier, a number of its programmes and other entities are contributing to the global response to climate change. Since 2007, the UN has embarked on an initiative to ensure better coordination of its response to climate change. Towards this end, five focus areas were defined and convening UN entities identified for each focus area (see Table 2.2). Some additional cross-cutting areas were also identified, including climate science and knowledge and public awareness.[37] This approach is intended to minimize duplication of activities across various entities, thereby making the UN's work on climate change more effective and efficient.

UNEP is one of the organizations which has played a pivotal role in action on climate change, having jointly

It remains uncertain whether the international community will be able to reach a legally binding agreement to replace the Kyoto Protocol

The UN has embarked on an initiative to ensure better coordination of its response to climate change

**Table 2.2**

Focus areas for a coordinated United Nations response to climate change

| Focus area | Convening United Nations entities |
|---|---|
| Adaptation | High-level Committee on Programmes of the Chief Executive Board of the United Nations |
| Technology transfer | United Nations Industrial Development Organization (UNIDO) and United Nations Department for Economic and Social Affairs (UNDESA) |
| Reduction of emissions from deforestation and degradation (REDD) | United Nations Development Programme (UNDP), the Food and Agricultural Organization (FAO) and the United Nations Environment Programme (UNEP) |
| Financing mitigation and adaptation action | UNDP and the World Bank group |
| Capacity-building | UNDP and UNEP |

*Source:* UN, 2008

established the IPCC with the World Meteorological Organization (WMO) in 1988 and actively engaged with adaptation and mitigation efforts since then.[38] In addition to a wide range of activities on the urban environment, UNEP is also implementing climate change-related activities within the context of cities through its Campaign on Cities and Climate Change. This campaign aims to enable cities to fruitfully engage in the global climate debate and reduce their GHG emissions.[39]

The WMO – which is the UN specialized agency for weather, climate, hydrology and related environmental issues – has led the process of generating scientific evidence and knowledge on climate change trends and has been the principal provider of the information underlying the IPCC's assessment reports (see Box 2.1). The WMO has also been issuing 'annual statements on the status of the global climate' to document extreme weather events and provide a historical overview of climate variability.[40]

As the agency with a mandate to foster sustainable urbanization, the United Nations Human Settlements Programme (UN-Habitat) is well positioned to address climate change issues specifically within the urban context. In 2008, UN-Habitat launched its Cities and Climate Change Initiative to enhance the adaptive capacities and responsiveness of local governments in developing countries to climate change, as well as to support their efforts at reducing greenhouse gas emissions (see Box 2.4).

Different UN entities have also been collaborating in the area of climate change. A case in point is the joint establishment of the United Nations Collaborative Programme on Reducing Emissions from Deforestation and Forest Degradation in Developing Countries (UN-REDD) in 2008 by UNEP, UNDP and the Food and Agriculture Organization (FAO).[41] Furthermore, UN agencies frequently implement climate change activities jointly with a number of partners outside the UN system. One example is the collaboration between UN-Habitat, UNEP and the World Bank to establish the International Standard for Determining Greenhouse Gas Emissions for Cities, a common standard for measuring emissions from cities (see Box 2.5).[42]

The UN has also been playing a leading role in terms of disaster risk management, which is fundamental to climate change adaptation efforts. The International Strategy for Disaster Reduction (UNISDR), which was adopted in 2000, is a system of partnerships between local, national, regional and international organizations with the overall objective of supporting global disaster risk reduction. UNISDR functions as the United Nations focal point for the coordination of disaster reduction. It is also tasked to mobilize political and financial commitments to implement the Hyogo Framework for Action 2005–2015, the main international agreement which lays out principles and priorities for global disaster risk reduction action, though it is not legally binding.[43] The 'urban agenda' is receiving greater attention in the work of the UN on disaster issues, with the UNISDR launching a campaign on Making Cities Resilient: My City is Getting Ready in 2010 to urge mayors and local governments to commit to making their cities more resilient to disasters, including those related to climate change.[44]

On the whole, the UN has been performing a crucial role in steering and coordinating climate change action internationally. It has also been at the forefront of generating scientific knowledge on climate change to support international negotiations and evidence-based policy-making. The initiative to harmonize the work of various UN entities on

UN-Habitat is well positioned to address climate change issues specifically within the urban context

**Box 2.4 UN-Habitat's Cities and Climate Change Initiative**

Launched in 2008, the Cities and Climate Change Initiative (CCCI) seeks to promote collaboration between local governments and their associations and partners on climate change-related topics, enhance policy dialogue between local and national governments on addressing climate change, support local governments in addressing climate change impacts while reducing greenhouse gas (GHG) emissions, and foster awareness, education and capacity-building for the implementation of climate change policies and strategies.

CCCI initially helped four pilot cities in Asia, Africa and Latin America to carry out climate change assessments. These cities already are at risk of natural disasters. In Esmeraldas (Ecuador), for example, more than half the population live in areas at risk of floods or landslides, while in 2006 two typhoons hit Sorsogon City (the Philippines), destroying some 10,000 homes. Climate change will only exacerbate those vulnerabilities in the 21st century. CCCI currently plans to help those cities deepen their assessments in priority areas, develop climate change strategies and action plans, mainstream findings into ongoing planning processes, and build capacity. At the same time, CCCI has been expanding to include five new cities in Africa in 2009 (Bobo Dioulasso, Burkina Faso; Mombasa, Kenya; Walvis Bay, Namibia; Kigali, Rwanda; and Saint Louis, Senegal) and nine new cities in Asia and the Pacific in 2010 (Batticaloa and Negombo, Sri Lanka; Kathmandu, Nepal; Ulaanbaatar, Mongolia; Pekalongan, Indonesia; Port Moresby, Papua New Guinea; Lami City, Figi; Apia, Western Samoa; and Port Vila, Vanuatu).

CCCI also is developing capacity-building tools to help cities access carbon finance or to develop climate change plans, drawing on local experiences. Finally, CCCI is taking lessons that it has captured through its local-level work, and disseminating and applying them globally. For example, the recent experiences of Negombo (Sri Lanka) in determining a baseline for its GHG emissions are helping to inform the next iteration of the International Standard for Determining Greenhouse Gas Emissions for Cities (see Box 2.5).

*Source:* UN-Habitat, 2009b

climate change since 2007 is expected to further consolidate the organization's leading role in guiding the global response to climate change.

### ■ Other multilateral organizations

Other multilateral institutions are playing an increasingly important role in climate change adaptation and mitigation at various levels. For instance, although it was thought that climate considerations were marginal for multilateral development banks in the past, this has been changing in recent years.[45] The World Bank Group is one such actor that has been reinforcing its engagement with climate change issues (see Box 2.6). This includes working directly on climate change issues within the urban context. The World Bank Institute is implementing city-focused climate change activities specifically in four areas: South–South learning between cities; city-level networks and knowledge platforms; knowledge exchange and structured learning; and customized support to selected cities.[46] Furthermore, under its Carbon Finance Assist Programme, which aims to enhance the capacity of developing countries to engage fully with the flexible mechanisms of the Kyoto Protocol (see Box 2.3), the World Bank has further initiated a twinning initiative for climate change knowledge-sharing between cities and a Carbon Finance Capacity Building programme for emerging megacities.[47] This programme seeks to promote the role of carbon finance for sustainable urbanization and poverty reduction.[48] In addition, in 2009, the World Bank established a Mayors' Task Force on Urban Poverty and Climate Change during COP-15 in Copenhagen, Denmark, and intends to prepare a *Mayor's Handbook on Adaptation*.[49]

The regional development banks are also key multilateral actors responding to climate change. In 2007, the Asian Development Bank established the Clean Energy Financing Partnership Facility to enhance energy security and to abate climate change in developing member countries. Potential investments under this facility include those related to developing and promoting clean energy technologies, including for low-income groups. By 2010, the funds for this facility had reached US$44.7 million.[50] In 2009, the Inter-American Development Bank launched the Sustainable Energy and Climate Change Fund with a total annual contribution of US$20 million. The fund aims to support sustainable energy initiatives and innovations, as well as responses to climate change in Latin America and the Caribbean.[51] Elsewhere, the European Investment Bank, whose lending activities focus mainly on EU member states, has been a key player in supporting climate change responses through mitigation, adaptation, research, development and innovation, technology transfer and cooperation, and support for carbon markets.[52]

The OECD is another multilateral organization which has been working on climate change issues for almost three decades, particularly on economic and policy analysis. With respect to climate change issues in cities, the OECD aims to support climate-sensitive local and regional development policies. Accordingly, it has published a number of reports on this subject analysing the linkages between climate change and urban development.[53] The organization intends to

---

**Box 2.5 International Standard for Determining Greenhouse Gas Emissions for Cities**

Introduced in March 2010, the International Standard for Determining Greenhouse Gas Emissions for Cities seeks to establish a common standard for measuring emissions from cities. In addition to emissions generated within urban areas, the standard also measures emissions generated outside urban boundaries that are driven by urban-based activities. This includes the following:

• out-of-boundary emissions from the generation of electricity and district heating which are consumed in cities (including transmission and distribution losses);
• emissions from aviation and marine vessels carrying passengers or freight away from cities; and
• out-of-boundary emissions from waste that is generated in cities.

Rather than attributing the responsibility for emissions to local governments, the standard seeks to illustrate the extent to which the urban economy is carbon dependent. Accordingly, emissions from the generation of power for consumption in cities, from city-bound aviation and marine transport, and from waste generated in cities are included. Furthermore, standardized reporting will help cities to benchmark themselves.

*Source:* UNEP et al, 2010

---

continue its work on climate change in the urban context with a focus on the impacts of green growth and the effect of urban spatial form on GHG emissions.[54]

In sum, multilateral actors are playing an increasingly important role in supporting climate change responses. They have especially become a prominent source of financial and technical assistance for climate change action in developing countries.

---

**Box 2.6 Climate change initiatives at the World Bank**

Some of the major climate change activities at the World Bank during recent years include the following:

• In 2005, the Clean Energy Investment Framework was created to accelerate clean energy investments in developing countries. The framework functions as a collaborative endeavour between multilateral development banks and countries to identify investments needed to accelerate the transition to a low-carbon economy and support adaptation programmes.
• In 2008, a Strategic Framework was prepared to guide the World Bank's work on climate change issues with a focus on the following six action areas: supporting climate actions in country-led development processes; mobilizing additional finance; facilitating the development of market-based financing mechanisms; leveraging private-sector resources; supporting the development and deployment of new technologies; and enhancing policy research, knowledge and capacity-building.
• In 2008, the Climate Investment Fund was launched with pledges of US$10 billion from ten donor countries to fund the demonstration, deployment and transfer of low-carbon programmes to developing countries. There are two main funds under this initiative – namely, the Clean Technology Fund for activities related to the power sector, transport and energy efficiency; and the Strategic Climate Fund to support pilot approaches with the potential for scaling up. The latter focuses on key areas of relevance to climate change mitigation in cities, including energy efficiency in buildings and industry.

*Sources:* World Bank, undated b; UNCTAD, 2009; Climate Investment Funds, undated

## ■ Bilateral organizations

A number of bilateral initiatives to address climate change have emerged over the past few years, although less attention has been given to financial flows emanating from these initiatives.[55] For instance, one of the largest funds of this type is Japan's Cool Earth Partnership, established to support climate change mitigation and adaptation, as well as access to clean energy in developing countries for the period of 2009 to 2013. The fund is worth US$10 billion, with the bulk of it (80 per cent) allocated for activities related to the reduction of GHG emissions rather than adaptation. Another such fund is the UK's Environmental Transformation Fund – International Window, launched in 2008 to support development through environmental protection and climate change adaptation in developing countries. US$1.6 billion was made available for this fund. The International Climate Protection Initiative of Germany, launched in 2008, is a mechanism for financing climate change projects and is funded from the sale of emissions certificates. The focus is on developing, newly industrializing and transition countries. Since 2008, 181 projects worth a total of €354 million have been launched.

The EU, another major bilateral actor, works on climate change issues mainly through the Global Climate Change Alliance, an initiative launched in 2007 to support, through direct financial and technical assistance, adaptation and mitigation activities mainly in the least developed countries and the small island developing states. The alliance also seeks to strengthen dialogue between these countries and the EU on climate change issues in the context of international negotiations.[56] The EU earmarked an initial €90 million for the work of the alliance between 2008 and 2010.[57] The work of the alliance is organized around five priority areas – namely, adaptation; reducing emissions from deforestation and degradation; enhancing the participation of developing countries in CDMs; promoting disaster risk reduction; and mainstreaming climate change into poverty reduction strategies.[58]

While bilateral funds such as the ones described above are actively supporting climate change responses in developing countries, most are considered to be part of donors' official development assistance. Questions have arisen as to whether this is the best approach for bilateral assistance and whether traditional development aid agencies are best placed to dispense such funds. Furthermore, some of the funds are loans that need to be repaid by recipient countries rather than grants.[59]

## ■ Regional (supra-national) initiatives

Arrangements for climate change action have also been emerging at the regional level. One example is the Asia-Pacific Partnership on Clean Development and Climate. Launched in 2006, this is a partnership between seven major Asia-Pacific countries (Australia, Canada, China, India, Japan, the Republic of Korea and the US), all of which are among the world's top GHG-emitting countries. These countries are cooperating to respond to the challenge of increased demand for energy and the related problems of air pollution, energy security and climate change.[60] The partnership

differs from the UNFCCC and the Kyoto Protocol in its focus on voluntary approaches and technological cooperation, rather than on binding emissions targets.

Another example, the European Emissions Trading Scheme, became operational in 2005 and is the largest multinational GHG emissions trading scheme in the world, involving 25 countries. It is designed to assist countries to meet their emission reduction commitments under the Kyoto Protocol. The scheme limits the amount of $CO_2$ that can be emitted from large industrial facilities, such as power plants and carbon-intensive factories. It covers almost half (46 per cent) of the EU's $CO_2$ emissions. Countries are allowed to trade amongst themselves and in validated credits from developing countries through the CDM of the Kyoto Protocol.[61] The first phase of the scheme ran from 2005 to 2007, and the second runs from 2008 to 2012. Because all EU member states have ratified the Kyoto Protocol, the second phase of the scheme was designed to support the Kyoto mechanisms and compliance period. The scheme is expected to account for around two-thirds of the overall emissions reductions which the EU plans to achieve by 2020.[62] However, there has been some concern that the entirety of the emissions reductions required in the second phase could be met through various activities outside of the EU itself instead of through domestic reductions.[63]

## National level

The sustained attention of policy-makers, scholars and the media to climate policies at the international level has led them to focus less on other levels of intervention, such as the national level.[64] National governments have the primary responsibility for signing international agreements, curbing GHG emissions and responding to climate-related disasters. So far, their actions have focused mainly on mitigation efforts in a few energy-intensive sectors (e.g. energy, transportation and the built environment); but adaptation actions have recently gained growing attention.

Some countries such as the US and China have been relatively less supportive of international climate policies, but have established rather robust national climate change initiatives. Other countries such as the UK and Germany have been key promoters of climate policies and have introduced an array of policies to achieve long-term reductions. For instance, Germany has an integrated set of 'ecotaxes' to foster alternative energy development and to discourage fossil fuel consumption. The UK has designed a mixed set of regulatory and taxation mechanisms (e.g. a levy on carbon-based electricity generation) that supports energy-efficient and renewable energy programmes.

Yet, even climate champions such as the UK and Germany face challenges complying with their carbon reduction targets. For instance, by 2004 it was clear that the UK's Climate Change Programme, introduced in 2000 to meet the country's Kyoto target, would not achieve its mitigation targets because GHG emissions had been growing at 2 per cent annually from 2002.[65] A review of the programme was thus launched and a revised programme introduced in 2006. Furthermore, national mitigation strategies as well as adapta-

*A number of bilateral initiatives to address climate change have emerged ... although less attention has been given to financial flows emanating from these initiatives*

*Even climate champions such as the UK and Germany face challenges complying with their carbon reduction targets*

tion and disaster management plans often omit urban areas[66] and lack an in-depth understanding of the relevant social science necessary to achieve an integrated assessment of the linkages between climate change and development,[67] and to undertake that assessment in such a way that stakeholders participate effectively and meaningfully.

Developing countries still lag behind developed countries in terms of climate change action, although an increasing number are introducing national programmes of action on climate change. For instance, in 2008, India introduced its first National Action Plan on Climate Change outlining a number of core missions running through to 2017.[68] According to the plan, the country aims to dramatically increase the use of solar energy and enhance energy efficiency, including within the context of urban areas. In this respect, the plan aims to make 'habitat sustainable through improvements in energy efficiency in buildings, management of solid waste and modal shift to public transport'.[69] Mexico's Climate Change Programme aspires to achieve 50 per cent reductions in greenhouse gas emissions by 2050, while also seeking to reduce the vulnerability of human and natural systems to the effects of climate change during this period.[70] China's National Climate Change Programme states that 'China will achieve the target of about 20 per cent reduction of energy consumption per unit GDP by 2010, and consequently reduce $CO_2$ emissions'.[71] It also outlines a number of actions and targets to enhance adaptation to climate change, including through protecting ecosystem resources such as grasslands, forests and water reserves.[72]

Generally, there has been greater focus on mitigation than adaptation responses in developing countries, although the latter will be strengthened vis-à-vis the National Adaptation Programmes of Action (see Box 2.2). Furthermore, while developing programmes of action clearly demonstrate 'intent' to take action on the part of developing countries, numerous constraints may hinder the achievement of mitigation and adaptation targets, as elaborated upon in Chapters 5 and 6 of this Global Report.

## State/provincial level

National governments are not able to meet their international commitments for addressing mitigation and adaptation without localized action. This is not only because GHG emissions originate in activities and processes taking place at the sub-national level (e.g. states/provinces, municipalities and urban centres), but also because many impacts of climate change are locally felt. Already, sub-national governments at the state/provincial level are playing an increasingly important role in climate change mitigation and adaptation. For instance, local authorities in the Federal District of Mexico City have developed important efforts to curb its GHG emissions. One of these is the Mexico City government, which has prepared the Mexico City Climate Action Programme for the period of 2008 to 2012. The programme aims to reduce greenhouse gas emissions, as well as vulnerability to the impacts of climate change, while strengthening adaptation.[73] Policy networks, political leaders and research

groups have been critical in launching a climate agenda. Nevertheless, this has not been enough to push effective policies. Policy-making has been constrained by two sets of institutional factors: the problem of fragmentation in local governance and lack of institutional capacity.[74]

The US offers an example of the multiple interactions between state/province and national tiers of government.[75] In the absence of federal leadership, state (and local) government efforts have become a form of 'bottom-up governance' on climate change issues in the US. With its Global Warming Solutions Act of 2006, California was the first state in the US to introduce enforceable legislation to curb GHG emissions (see also Box 5.18). As per this bill, state-wide emissions are to be reduced to 1990 levels by the year 2020.[76] The State of Washington introduced a similar bill in 2008, and even went further to identify emissions limits up to 2050.[77]

A number of other initiatives across different US states have also emerged. For instance, the Regional Greenhouse Gas Initiative is a market-based initiative involving ten north-eastern and mid-Atlantic states to cap GHG emissions from the power sectors by 10 per cent by 2018.[78] Another similar initiative is the US Mayors Climate Protection Agreement, which has been signed by hundreds of mayors across the country. The agreement encourages mayors to work towards achieving the Kyoto Protocol targets through local action and to urge their state and the federal government to introduce policies for GHG emissions reductions.[79] The Urban Leaders Adaptation Initiative, whose partners represent nine US counties and cities (and the city of Toronto in Canada), aims to assess and project climate change impacts and support its partners in mitigation and adaptation activities.[80] The initiative is aimed at serving as a resource for local governments and as a means to empower local communities to develop and implement climate-resilient strategies.

## Local/city level

Although the Kyoto Protocol does not explicitly identify a role for cities and local governments in responding to climate change, city-level actors are actively participating in climate strategies, projects and programmes. These include local authorities, community-based organizations, the private sector, the academic sector and individuals. Local governments, for instance, have held municipal leadership summits parallel to the four COPs of 1993, 1995, 1997 and 2005. Since 2005, the 'local government and municipal authorities constituency' has operated as an observer in the UNFCCC negotiations.[81] Indeed, 'compared to national politicians, city leaders seem willing and able to take action to protect their cities against these threats and to help make a global difference'.[82]

Depending on their national contexts and histories, city authorities can have a considerable level of influence over both GHG emissions and adaptation to climate change, as elaborated upon in detail in Chapters 5 and 6 of this Global Report. In addition, they are increasingly becoming involved in international city networks, which represent a

Developing countries still lag behind developed countries in terms of climate change action

National governments are not able to meet their international commitments for addressing mitigation and adaptation without localized action

**Box 2.7 Major international city networks and initiatives on climate change**

*ICLEI* (Local Governments for Sustainability) was previously known as the International Council for Local Environmental Initiatives. Created in 1991, it is an association of more than 1200 local governments from 70 countries who are committed to sustainable development. ICLEI has worked with cities worldwide on climate change through its urban $CO_2$ Reduction Campaign, Green Fleets Campaign and its Cities for Climate Protection Campaign (CCP Campaign). Local governments participating in the CCP Campaign commit to undertake and complete five performance milestones, as detailed in Box 5.1.[a]

The Large Cities Climate Leadership Group, also known as the *C40* (and originally as the C20), was created in 2005 with the main goals of fostering action and cooperation on reducing GHG emissions, creating policies and alliances to accelerate the uptake of climate-friendly technologies. C40 is composed of cities from all world regions.[b]

The *Clinton Climate Initiative* was launched in 2005 by the William J. Clinton Foundation to create and advance solutions to the core issues driving climate change. In collaboration with governments and businesses around the world, the initiative focuses on three strategic programme areas: increasing energy efficiency in cities; catalysing the large-scale supply of clean energy; and working to stop deforestation. In 2006, the initiative became the delivery partner of the C40 to assist in the delivery of urban mitigation projects. The initiative launched the Climate Positive Development Program in 2009 to support 'climate positive' development in 17 urban locations across six continents. Nearly 1 million people are expected to live and work in these developments when they are complete.[c]

Founded in December 2005, the *World Mayors Council for Climate Change* has more than 50 members from all of the world and seeks to promote policies addressing climate change and its local impacts; to foster the international cooperation of municipal leaders on achieving relevant climate, biodiversity and Millennium Development Goals (MDGs); and to have a say in the design of effective multilateral mechanisms for global climate protection.[d]

*United Cities and Local Governments (UCLG)* represents and defends the interests of local governments globally. In 2009, more than 1000 cities in 95 countries were direct members of UCLG.[e] It is involved in the Partnership for Urban Risk Reduction, an ad hoc coalition of international organizations with the following objectives:

- promote worldwide awareness campaigns about risk reduction in regions regularly affected by natural disasters;
- build capacity at the local level to foresee and manage risks through the transfer of technical know-how to local actors and decision-makers; and
- set up a global platform for local authorities on disaster risk reduction.[f]

The *Climate Alliance* is an association of cities and municipalities in 17 European countries that have developed partnerships with indigenous rainforest communities. Since 1990, when it was founded, around 1500 cities, municipalities and districts together with more than 50 provinces have joined the alliance. NGOs and other organizations have also joined as associate members. Its aim is to preserve the global climate through a twofold mechanism: the reduction of GHG emissions by developed countries and the conservation of forests in developing countries. The hope is that the former will be achieved through an exchange of information on best practices and by providing recommendations, aids and tools for local climate change policies; while the latter will be achieved through the organization of campaigns and political initiatives on the conservation of the tropical rainforests and the defence of indigenous rights, and by raising awareness of the political situation and living conditions of the indigenous peoples in Amazonia.[g]

The *Asian Cities Climate Change Resilience Network* is an initiative of the Rockefeller Foundation in partnership with other entities such as academic, non-governmental, governmental, international, regional and national organizations.[h] The network seeks to catalyse attention, funding and action on building climate change resilience for poor and vulnerable people in Asian cities. In order to accomplish this, the network is in the process of testing and demonstrating a range of actions to build climate change resilience in India, Viet Nam, Thailand and Indonesia. Lessons from these interventions will be used to support climate change resilience-building in other urban areas of the region.

The *Covenant of Mayors* is a mechanism intended to encourage mayors of cities in EU countries to significantly reduce their GHGs. Accordingly, signatories to the covenant enter a formal commitment to go beyond the target to curb their $CO_2$ emissions by at least 20 per cent by 2020, as already set by the EU's Climate Action and Energy Package. About 2000 cities in 42 countries were signatories to the covenant by end of 2010. Within one year of signing the covenant, cities are expected to prepare a Sustainable Energy Action Plan indicating how they intend to meet their commitments.[i] Energy Cities, the European association of more than 1000 cities and towns, created in 1990, plays a leading role in the implementation of the covenant.[j]

*Sources:* a ICLEI, undated; b C40 Cities, undated; c Rosenzweig et al, 2010; a Clinton Foundation, undated; d World Mayors Council on Climate Change, undated; e Prasad et al, 2009; f United Cities and Local Governments, undated; g Climate Alliance, undated; h Rockefeller Foundation, 2010; i EU, undated; j Energy Cities, undated

**City authorities can have a considerable level of influence over both GHG emissions and adaptation to climate change**

form of multilevel environmental governance across national boundaries with the involvement of multiple governmental, private-sector, non-profit and other civil society stakeholders. International city networks – associations between cities at the international level – have been found to be important in developing the capacity of municipalities because 'they facilitate the exchange of information and experiences, provide access to expertise and external funding, and can provide political kudos to individuals and administrations seeking to promote climate action internally'.[83] In São Paulo, Brazil, for instance, participation in international municipal networks was seen as import for two key reasons. First, they provided the opportunity to 'join the international task force against climate change ... bypassing the nation-state with its lack of both binding international obligations and lack of national limits upon GHG emission'.[84] Second, such networks were an important source of personal motivation, offering individuals opportunities to engage with broader debates and keeping them 'passionate about the topic'.[85]

The number of these networks has been on the rise during recent years, as illustrated in Box 2.7. A number of the city networks for climate change have global membership, while others such as the Climate Alliance and the Asian Cities Climate Change Resilience Network have membership which is restricted to certain world regions. While some of the networks have been functional since the early 1990s, others have been launched only recently. In general terms, most city networks focus on climate change mitigation, although adaptation has been receiving greater attention during recent years.

National city networks have also been important in developing municipal capacity in countries where national governments have not taken action to address climate change – for example, the Partners for Climate Protection programme in Canada, ICLEI's CCP Australia programme and the US Mayors Climate Protection Agreement. Such networks have offered political support, additional funding (paradoxically often derived from national government) and a means of sharing information. In the case of the US Mayors Climate Protection Agreement, 'city representatives often cited a moral imperative to help other cities by sharing information on how best to address climate change … of city solidarity', while 'friendly competition to be the greenest city also served to further amplify engagement … engagement to address climate change … spread as cities promoted themselves (and were promoted by policy actors), competed with each other, and inspired other cities to go green'.[86]

However, networks have had an uneven impact, with evidence suggesting that they are more important in developing the capacity of those municipalities that are already leading responses to climate change, and that while the political support and knowledge transfer functions that such networks perform is valuable, 'in the absence of the financial and technological resources to execute programmes, the power of knowledge can be limited'.[87] In effect, networks appear to be most important for those with a degree of existing capacity to act, leading to a virtuous circle where additional resources and support can be accessed. However, for those without the capacity to access such networks in the first place, such initiatives may do little to build capacity to respond to climate change and, in effect, may serve to concentrate resources and attention on cities that are already leading the response to mitigating climate change.

In addition to city authorities, individuals, households and community-based organizations and other local actors have an important role to play in both international climate change negotiations and city-level mitigation and adaptation activities. These actors are recognized non-governmental constituencies in the UNFCCC negotiations and processes (see Box 2.8). As key emitters, the behaviour of these actors may directly result in the success or failure of mitigation efforts. Their actions may also be helpful in facilitating coping responses and in the integration of climate-risk reduction, in emergency responses to climate hazards and in development planning. Any efforts that local actors make to support mitigation, adaptation or emergency preparedness, however, first needs to be made possible by the existence of infrastructural support and regulative incentives. For instance, as illustrated by Dhaka, Bangladesh, and Lagos, Nigeria (see Boxes 6.1 and 6.2), if not supported by broader (governmental) policies and investments, the responses of local actors can merely reduce rather than prevent impacts. Manizales, Colombia, and Ilo, Peru, provide examples of how community-level actors can implement effective responses.[88] In these cities, community-based organizations have worked together with local authorities and the academic sector to become vehicles for more inclusive urban governance, and have implemented actions to prevent the spread of low-income populations in dangerous sites. Although these actions were not directly

---

**Box 2.8 Non-governmental constituencies of the UNFCCC**

Non-governmental organizations admitted as observers to the sessions of the United Nations Framework Convention on Climate Change (UNFCCC) have been grouped as follows:

- business and industry non-governmental organizations (BINGOs);
- environmental non-governmental organizations (ENGOs);
- farmers and agricultural non-governmental organizations;*
- indigenous peoples organizations (IPOs);
- local government and municipal authorities (LGMA);
- research and independent non-governmental organizations (RINGOs);
- trade union non-governmental organizations (TUNGOs);
- women and gender non-governmental organizations;*
- youth non-governmental organizations (YOUNGOs).*

A focal point is appointed for each constituency to:

- provide a conduit for the exchange of official information between their constituents and the UNFCCC secretariat;
- assist the UNFCCC secretariat in ensuring an effective participation appropriate to an intergovernmental meeting;
- coordinate observer interaction at sessions, including convening constituency meetings, organizing meetings with officials, providing names for the speakers list and representation at official functions;
- provide logistical support to their constituents during UNFCCC sessions; and
- assist the UNFCCC secretariat in realizing representative observer participation at workshops and other limited-access meetings.

*Note:* * Recognized on a provisional basis pending final decision on their status by COP-17 (28 November–9 December 2011).

*Sources:* UNFCCC, undated o, undated p

---

aimed at addressing climate change risks, such pro-poor and pro-development policies can enhance adaptive capacity and resilience to climate hazards.

Although they are a necessary component of successful climate change actions, grassroots actors should not be idealized. In some cases, their extensive involvement in these efforts can make things more difficult.[89] Sometimes, for instance, local associations are closely related to the state, or hold private or sectarian interests that distort local action. Bringing about change through grassroots efforts is perhaps most problematic in settlements within countries that have experienced strong centralized control. As documented in projects aimed at enhancing local capacity to respond to floods and other hazards in Guyana and Viet Nam,[90] the attempt by the international community to modify urban governance through funding community-sponsored development projects runs the danger that local elites or state agents will hijack the benefits of grassroots funding.

NGOs are also actively seeking to engage with climate change issues, as exemplified by the Climate Action Network, a network of around 500 NGOs working to promote climate change mitigation.[91] However, while NGOs are plentiful in large cities, they tend to be less common or even absent from smaller urban settlements. Where present, local NGOs are well placed to produce, accumulate and transfer climate change knowledge. As partners in develop-

> Networks … are more important in developing the capacity of those municipalities that are already leading responses to climate change

ment projects aimed at reducing emissions, capturing carbon and reducing risk, they are cost effective, increase transparency and accountability to beneficiaries, and strengthen inclusive governance. However, by increasing their accountability to upper levels of governance, NGOs can lose their flexibility and power to contest the decisions of governments and powerful interests. This can distance them from grassroots partners, reduce inclusiveness and horizontal accountability, and, thus, undermine climate change mitigation and adaptation efforts. On the other hand, when they can maintain independence, NGOs can enhance climate change policy efforts by providing a channel for feedback between the grassroots level and urban government or international civil society actors.[92]

In addition to the leading role of the IPCC in consolidating scientific knowledge to inform policy making, researchers around the world have been generating and disseminating climate change information including specifically in relation to cities. A case in point is the Urban Climate Change Research Network (UCCRN), an international group of researchers with 200 members from 60 cities globally.[93] UCCRN aims to provide climate change information and data specifically for urban decision makers and published its first assessment report on climate change and cities in 2011.[94] The Urbanization and Global Environmental Change project of the International Human Dimensions Programme, established in 2005, is another initiative researching the interactions between environmental change and urban processes.[95]

The private sector also has an important role to play in efforts aimed at curbing GHG emissions – for example, through producing more efficient vehicles and utilities, creating technologies for alternative energy, and constructing controlled wastewater treatment plants.[96] A growing number of private-sector companies are also considering how to mitigate emissions through transforming their own work practices. For instance, the Carbon Disclosure Project, established in 2000, has been reporting on GHG emissions from some of the world's largest companies. In 2010, this project collected data from 4700 of the world's largest corporations, on their GHG emissions, the risks and opportunities related to climate change they faced, and strategies for managing them. This process was supported by 534 investors with assets worth US$64 trillion.[97]

With regard to adaptation to climate change, the private sector has been subject to comparatively little attention, although it is playing a key role in defining investments in climate-proofing infrastructures, energy utilities and other urban sectors. Some specialized investment entities are already taking positions around climate-related risks via investments in reinsurance companies, in resource prices such as oil and gas with the potential to be affected by hurricanes, and through participation in alternative risk-transfer products (e.g. insurance-linked securities such as 'catastrophe bonds' and 'weather derivatives').[98] A key concern is that privatized actions in the area of adaptation may present a potential conflict of interest with the public good. The role of private security firms and privatized healthcare during emergency periods, for instance, requires greater study, with

*Partnerships between public, private, civil society and other actors are becoming critical in building urban capacity to respond to climate change*

*Mechanisms within the international climate change framework ... do not indicate a clear process by which urban areas and actors may participate*

potentially profound implications for governance in urban risk management and disaster response. Nevertheless, as recently emphasized by the executive director of the UNFCCC:

> *Traditional thinking would have us believe that adaptation is the exclusive ambit of the public sector. This is false on two levels: (1) business needs to adapt itself, and (2) adaptation holds investment opportunities for the private sector.*[99]

Indeed, urban capacity to address climate change is increasingly shaped by the presence of more formalized collaboration between public and private actors. Partnerships between public, private, civil society and other actors are becoming critical in building urban capacity to respond to climate change. For instance, in November 2010, R20 – Regions of Climate Action, an innovative coalition was launched to support clean technologies, climate resilient projects, green investment and also influence national and international policies. The coalition includes sub-national government members from developed and developing countries as well as organizations and individuals from the private sector, academia, national governments, international organizations and civil society.[100]

# THE POTENTIAL OF THE INTERNATIONAL CLIMATE CHANGE FRAMEWORK FOR LOCAL ACTION

This section briefly reviews the opportunities and challenges posed by the existing international governance framework for local action. It also discusses the existing mechanisms that urban areas could potentially take advantage of, what constraints exist to the use of these mechanisms by urban actors, and explains, briefly, some possible ways in which these constraints could be addressed.

A key factor constraining urban actors' use of mechanisms within the international climate change framework is the fact that these mechanisms are primarily addressed to national governments and do not indicate a clear process by which urban areas and actors may participate. The related international structure for climate change financing, in particular, has been described as 'diverse and complex, and not primarily designed for local governments'.[101] The funding mechanisms of the UNFCCC discussed earlier in this chapter (see Box 2.2) can be used to finance projects within urban areas, but they are only accessible for urban actors through their national governments. Even though national governments represent the interests of their urban populations in international discussions on allocating responsibility for climate change mitigation, and in developing international funding mechanisms and institutions to support adaptation, getting urban priorities moved up on national agendas can be problematic, at best. For instance, although National Adaptation Programmes of Action have been prepared for developing countries under the Least Developed Countries Fund, there have been few initiatives for adaptation at the sub-national level.[102]

Similarly, emissions trading currently takes place between countries and groups of countries or tends to target particular industries, thereby offering limited possibilities for actions at the urban level. For instance, the European Emissions Trading Scheme targets carbon-intensive factories and power plants by capping the amount of $CO_2$ that they emit. While some of these facilities are certainly located within urban areas, and it may be safe to assume that a large percentage of their output serves urban needs, local authorities are not generally in direct control of these activities. Of course, some exceptions exist, where, for instance, an urban centre owns a public utility, such as one for electricity generation.

In contrast, CDM offers significant potential for urban projects in developing countries in such sectors as transportation, waste and the building industry. Indeed, a recent study shows that it is one of the international financing mechanisms that city authorities are most aware of. However, urban CDM-based projects account for only 8.4 per cent of the total number of CDM projects registered with the UNFCCC in 2009. Most of these were related to solid waste, with only two projects related to transport. Furthermore, the majority of the urban CDM projects are concentrated in a few countries – namely, Brazil (36 per cent), China (14 per cent), Mexico (5 per cent) and India (2 per cent).[103]

A number of reasons have been identified for the small proportion of CDM projects being urban based. First, the responsibility for climate change action is perceived to lie with national rather than local governments. Second, city authorities are already overwhelmed by immediate local challenges and have difficulty justifying climate change-related projects and expenditures. Third, the financial resources required for climate change action (e.g. introducing energy-efficient technology and equipment) may be absent in developing countries. Fourth, the high transaction costs associated with project development and approval by authorities has been identified as an additional constraint.[104] Additional barriers to expanding the use of CDM in urban areas are considered in greater detail in Chapter 5 (see section on 'Financial resources').

The joint implementation mechanism is very similar to CDM, but it applies only between developed countries.[105] Since most of the joint implementation projects are in countries with economies in transition and emissions reduction activities are generally more expensive in these countries compared to similar activities in developing countries, the joint implementation mechanism has been used far less than the CDM.[106] The use of joint implementation by urban actors has, therefore, also been very limited.

A further major challenge for local authorities to take advantage of the international climate change framework to implement climate responses locally is that they are often overwhelmed by competing priorities. Besides coordinating policy efforts with organizations and actors at the national and state/provincial levels to address an array of non-climate-related developmental and environmental issues, they now need to deal with a multitude of issues centring on climate-related mitigation, adaptation, development, and disaster preparedness and response. While coping with a myriad of competing priorities within their boundaries, they also need to explore ways in which they can better connect to multiple levels of action and information on climate change, and know how their issues fit into the larger picture of regional, national and international climate change issues. In addition, mismatches exist between climate and local policy-making timeframes. Given the fact that many of the cause-and-effect relationships are long term and potentially irreversible, they require planning that goes beyond the tenure, the administrative power and even the lifetime of most current decision-makers and other stakeholders.

Despite the above challenges, local authorities can coordinate efforts with national and state/provincial authorities to make use of financial resources offered under the UNFCCC to invest in local mitigation initiatives which offer high mitigation potentials. These include investments in the areas of transport, energy generation, waste management and the like. Local urban authorities and actors can also take advantage of existing networks and organizations that focus specifically on enhancing local climate change action at the city level. Urban authorities could get support from the UNFCCC to finance adaptation projects, not only through their national governments, but also through their participation in various city networks. For instance, the Federation of Canadian municipalities is working with ICLEI through their Cities for Climate Protection Campaign (see Box 2.7). A total of 180 Canadian municipalities are engaged in assessing and reducing GHG emissions through the campaign.[107] Several initiatives also offer opportunities for urban authorities to learn from and share climate change best practices and lessons (see Box 2.7).

Urban authorities may also try to benefit from initiatives of multilateral and bilateral organizations seeking to enhance the capacity of developing countries to take part in and take advantage of international climate change discussions and the resulting instruments and mechanisms (see Box 2.7). For instance, The World Bank's Carbon Finance Assist Programme seeks to enhance the capacity of developing countries to engage with the flexible mechanisms of the Kyoto Protocol.[108] Similarly, the Climate Alliance aims to enhance the participation of developing countries in CDMs. UNEP's Campaign on Cities and Climate Change explicitly seek to support the engagement of cities in international climate change negotiations and forums.

Local authorities can also seek to harmonize climate change interventions with existing development interventions and concerns. For instance, mitigation can be integrated within local development concerns such as energy security and infrastructure provision. Adaptation measures can serve and be integrated not only within disaster risk reduction, but also within components of the development agenda such as land-use planning and access to water, sanitation and housing. For instance, existing coping actions such as community savings networks might be combined with insurance mechanisms sponsored by NGOs.

**A further major challenge for local authorities to take advantage of the international climate change framework...is that they are often overwhelmed by competing priorities**

**Urban authorities could get support from the UNFCCC to finance adaptation projects... through their participation in various city networks**

## CONCLUDING REMARKS

During the last few decades, climate change has gained importance as a major 21st-century challenge partly due to the consolidation of scientific evidence of the contribution of human activities to global warming. Knowledge – whether generated by scientific communities or brokered by the media, scientific entrepreneurs or NGOs at different levels (from the international to the local) – has been a fundamental factor shaping climate action at all levels. However, the move from knowledge to action is not straightforward. The political mechanisms by which individuals, groups, organizations and governments translate the scientific knowledge of climate change into concrete actions have played a critical role in this regard.

The UNFCCC and the Kyoto Protocol are the key elements of the overarching framework adopted by world governments to guide climate change responses globally. Although the adoption of the Kyoto Protocol was hailed as a significant milestone and has enabled substantial emissions reductions since, the failure to reach a legally binding agreement for the period after the end of the protocol's commitment period in 2012 is seen as a major failure of international climate change negotiations.

The UNFCCC and its Kyoto Protocol coexist with a multitude of parallel initiatives and frameworks operating at different sectors and spatial levels. Even if national governments are leading negotiations of climate change agreements at the international level, mitigation and adaptation activities are being implemented by numerous other actors at the regional, sub-national (e.g. state/provincial) and local levels. The sustained attention of policy-makers, the scientific community and the media to climate policies at the international level has mainly led them to overlook these other equally important levels of climate intervention.

Local action is indispensable for the realization of national climate change commitments agreed through international negotiations. Yet, the international framework described in this chapter presents both challenges and opportunities for climate change action at the local city level. Most of the mechanisms within the international climate change framework are addressed primarily to national governments and do not indicate a clear process by which local governments, stakeholders and actors may participate. Furthermore, local authorities can quickly be overwhelmed by competing priorities and therefore may not actively pursue opportunities offered by the international governance framework. Thus, in practice, mitigation and adaptation actions have been by-products of policies designed to address more pressing local problems or problems for which there is more pressure by interested parties. The overall complexity of the international climate change framework – as well as the multiplicity of related actors and mechanisms – may further prevent city authorities from benefiting from available opportunities. Also, administrative structures, party politics, political timetables, individual ambitions, inertias, and many other institutional and political constraints need to be overcome, thus requiring a broader-based institutional capacity for climate action. Its absence has deterred key mitigation and adaptation efforts. Yet, in some cases it has become another source of opportunity for state and local actors to fill a leadership gap.

Despite the challenges, the multilevel climate change framework briefly described in this chapter does offer opportunities for local action at the city level. While the proportions remain low, urban-based emissions reduction projects are being implemented through some of the mechanisms of the UNFCCC (e.g. the CDM). There is also great potential for expanding such projects given the role of urban areas in contributing to GHG emissions.[109] In addition, today, more urban authorities than ever before participate in international city networks for climate change adaptation and mitigation. These urban actors have developed a more aggressive approach, seeking to secure the economic competitiveness of their cities and to get a local voice in international negotiations and organizations.

The crux of the challenge is that actors of climate change at all levels – including governments, NGOs and civil society that are, more often than not, preoccupied with immediate and often localized interests and priorities – need to move within short timeframes to guarantee long-term and wide-ranging global interests, which can seem remote and unpredictable at best. The hope is that a wave of actions from local actors centring their work at the local level, where all the impacts of climate change will ultimately be felt, will join together to create the momentum to build broad-based support for mitigation and increase adaptive capacity in the areas and populations that are most vulnerable to the effects of climate change.

Local action is indispensable for the realization of national climate change commitments agreed through international negotiations

## NOTES

1   Biermann et al, 2009, p31.
2   Betsill and Bulkeley, 2007; Alber and Kern, 2008; Biermann et al, 2008.
3   ICLEI, 2010.
4   UNFCCC, undated a.
5   UNFCCC, undated b.
6   UN, 1992, Article 3. See also CISDL, 2002.
7   De Lucia and Reibstein, 2007.
8   UN, 1992.
9   UNFCCC, undated c.
10  UNFCCC, undated d.
11  UNFCCC, undated d.
12  UNFCCC, undated e.
13  UNFCCC, 1995.
14  UNFCCC, 1996.
15  Bodansky, 2001.
16  See section entitled 'The Kyoto Protocol' below.
17  GEF, undated.
18  Bodansky, 1993, cited in Gupta and van Asselt, 2006.
19  Gupta and van Asselt, 2006, p84.
20  Biermann et al, 2008.
21  UNFCCC, undated r.
22  UNFCCC, undated k.
23  See previous section on 'The United Nations Framework Convention on Climate Change'; UNFCCC, undated e.
24  UNFCCC, undated l.
25  UN, 1998.
26  After Australia's ratification of the protocol in 2007, the US is the only developed country not to have ratified the Kyoto Protocol.
27  UN, 1998.

28  Bulkeley and Betsill, 2003.
29  Gilbertson and Reyes, 2009.
30  Disch, 2010, p55.
31  Richman, 2003.
32  UNFCCC, 2010.
33  Beccherle and Tirole, 2010.
34  Bodansky, 2001; Aldy et al, 2003.
35  Biermann et al, 2008.
36  Held and Hervey, 2009.
37  UN, 2008.
38  UNEP, undated a.
39  UNEP, undated b.
40  WMO, 2007.
41  UN-REDD, undated.
42  UNEP et al, 2010.
43  UNISDR, undated a.
44  UNISDR, undated b.
45  Sohn et al, 2005.
46  World Bank, undated a.
47  World Bank, 2009a.
48  CFCB, undated.
49  World Bank, 2010d.
50  ADB, undated.
51  Inter-American Development Bank, 2007.
52  European Investment Bank, 2010.
53  For example, see OECD, 2010.
54  OECD, 2009.
55  Atteridge et al, 2009.
56  GCCA, undated a.
57  European Commission, 2009; GCCA, undated b.
58  GCCA, undated c.
59  Bird and Peskett, 2008.
60  Asia-Pacific Partnership on Clean Development and Climate, undated.
61  Parker, 2006. See also Box 2.2.

62  European Commission, 2009.
63  Gilbertson and Reyes, 2009, p42.
64  Rabe, 2007.
65  Department of Energy and Climate Change, undated.
66  Bulkeley and Betsill, 2003; Pelling, 2005; Satterthwaite et al, 2007a.
67  Satterthwaite et al, 2007a; Romero Lankao, 2007b; Moser, 2010.
68  Pew Center on Global Climate Change, 2008.
69  Pew Center on Global Climate Change, 2008.
70  Sandoval, 2009.
71  People's Republic of China, 2007.
72  People's Republic of China, 2007.
73  Secretaría del Medio Ambiente del Distrito Federal, 2008.
74  Romero Lankao, 2007b.
75  Rabe, 2007; Wheeler, 2008; see also Pew Center on Global Climate Change, undated.
76  California Solutions for Global Warming, undated.
77  Department of Ecology, State of Washington, undated.
78  Regional Greenhouse Gas Initiative, undated.
79  The United States Conference of Mayors, 2008.
80  Centre for Clean Air Policy, undated.
81  ICLEI, 2010, p15. See Box 2.8.
82  Rosenzweig et al, 2010.
83  Bulkeley et al, 2009, p26; see

    also Collier, 1997; Bulkeley and Betsill, 2003; Bulkeley and Kern, 2006; Granberg and Elander, 2007; Holgate, 2007; Kern and Bulkeley, 2009.
84  Setzer, 2009, p10
85  Setzer, 2009, p10.
86  Warden, 2009, p8.
87  Gore et al, 2009, p22.
88  See the section on 'Adaptation planning and local governance' in Chapter 6.
89  Pelling, 2005.
90  Pelling, 1998.
91  CAN, undated.
92  Pelling, 2005.
93  Rosenweig et al, 2010.
94  Rosenweig et al, 2011.
95  Rosenweig et al, 2010; IHDP, undated.
96  IPCC, 2007a.
97  PricewaterhouseCoopers, 2010.
98  Wilbanks et al, 2007.
99  Figueres, 2010.
100 Office of the Governor, California, US, undated.
101 ICLEI, 2010, p79.
102 ICLEI, 2010.
103 ICLEI, 2010.
104 World Bank, 2010a; Clapp et al, 2010.
105 See Box 2.3.
106 Gilbertson and Reyes, 2009.
107 Federation of Canadian Municipalities, 2009.
108 See the section on 'Other multilateral organizations' above.
109 See Chapter 3.

# THE CONTRIBUTION OF URBAN AREAS TO CLIMATE CHANGE

The allocation of responsibility for greenhouse gas (GHG) emissions – and, hence, climate change – is an important global policy debate. Indeed, the importance of responsibility is explicitly recognized in the United Nations Framework Convention on Climate Change (UNFCCC) allocation of 'common but differentiated responsibilities' among countries for addressing emissions.[1] As shown in Chapter 1, urban areas – which are now home to more than half of the world's population – clearly have an important role to play in facilitating reduced emissions; yet the contribution of urban areas to emissions is often unclear.

There are several reasons why it is important to consider the contribution of urban areas to climate change. *First*, a range of activities are associated with cities and their functioning that contribute to GHG emissions. Transportation, energy generation and industrial production within the territorial boundaries of towns and cities generate GHG emissions directly. Urban centres rely on inward flows of food, water and consumer goods that may result in GHG emissions from areas outside the city. In addition, individuals consume a range of goods and services that may have been produced locally or outside the urban area. An analysis of the relative impacts of these activities is an important step towards understanding the extent of the contribution of urban areas to climate change.

*Second*, measuring emissions from different cities provides a basis for comparisons to be made and the potential for inter-urban competition and cooperation. Climate-friendly development has the potential to attract external investment, and the growing importance of international urban networks[2] provides spaces for learning and knowledge sharing. Emissions measuring has recently been inserted into global policy debates. For example, the United Nations Environment Programme (UNEP), the United Nations Human Settlements Programme (UN-Habitat) and the World Bank launched an International Standard for Determining Greenhouse Gas Emissions for Cities at the World Urban Forum in Rio de Janeiro in March 2010.[3] This standard provides a common method for cities to calculate the amount of GHG emissions produced within their boundaries.

*Third*, an assessment of the contribution of cities to climate change is a vital first step in identifying potential solutions. The large and growing proportion of the Earth's population living in towns and cities, and the concentration of economic and industrial activities in these areas, means that they need to be at the forefront of mitigation. The establishment of emission baselines is necessary if effective mitigation benefits are to be identified and applied.

*Finally*, it is important to highlight the differences between *production*- and *consumption*-based analyses of GHG emissions. Most assessments of the contribution of cities – and countries – to climate change have focused on the emissions that are produced by activities taking place within given territorial boundaries. However, an alternative approach is to consider the emissions associated with the consumption patterns of individuals, recognizing that many agricultural and manufacturing activities that meet the needs of urban residents take place outside city boundaries, and often in other countries. This consumption-based approach provides an alternative framework for suggesting appropriate ways of reducing GHG emissions by focusing on consumer choices as potential drivers for mitigation.

The first two sections of this chapter explain the scientific and technical basis for measuring GHG emissions from urban areas. The third section presents findings from a wide range of urban emissions inventories to show how these vary from place to place, while the fourth section describes the factors influencing emissions at the urban level. Finally, the chapter examines different approaches to measuring GHG emissions and shows that the simple analyses that have been frequently used until now are no longer sufficient for addressing this urgent global challenge.

## MEASURING GREENHOUSE GAS EMISSIONS

In order to account for the contribution of urban areas to climate change, it is necessary to measure their emissions of GHGs.[4] This requires particular methodologies to account for the various activities and the volume of these gases that they produce. And in order to make meaningful comparisons over time, or between different places, there is a need for standardized protocols to be developed. According to the UNFCCC, inventories should meet the following five quality criteria:[5]

The allocation of responsibility for greenhouse gas ... emissions – and, hence, climate change – is an important global policy debate

Climate-friendly development has the potential to attract external investment

the ... International Standard for Determining Greenhouse Gas Emissions for Cities ... provides a common method for cities to calculate the amount of GHG emissions produced within their boundaries

1    *Transparency:* assumptions and methodologies should be clearly explained.
2    *Consistency:* the same methodology should be used for base and subsequent years.
3    *Comparability:* inventories should be comparable between different places.
4    *Completeness:* inventories should cover all relevant sources of emissions.
5    *Accuracy:* inventories should be neither over nor under true emissions.

The UNFCCC is the global instrument responsible for ensuring that countries measure national GHG emissions and set targets for their reduction

## International protocols for measuring greenhouse gas emissions

As noted in Chapter 2, the UNFCCC is the global instrument responsible for ensuring that countries measure national GHG emissions and set targets for their reduction. Under the Convention, national governments gather and share information on GHG emissions; launch national strategies to reduce emissions; and cooperate to prepare for adaptation to climate change impacts. A total of 36 developed countries have – under the Kyoto Protocol – accepted emission 'caps' that limit their total GHG emissions within a designated timeframe and are required to submit annual inventories of their national emissions, while other signatories to the Kyoto Protocol submit emission inventories in their periodic national communications.

National inventories are prepared according to a detailed set of criteria developed by the Intergovernmental Panel on Climate Change (IPCC): the *IPCC Guidelines for National Greenhouse Gas Inventories*.[6] This is a detailed series of five volumes prepared by the IPCC as a result of a request by the UNFCCC. It is intended to ensure that

countries are able to fulfil their commitments under the Kyoto Protocol and subsequent international agreements. These criteria recognize that figures will be estimates, but seek to ensure that these do not contain any biases that could have been identified and eliminated. There are also different tiers of estimation methods, which take into account varying availability of data between countries. The inventories provide measuring strategies and global warming potentials for the full range of GHGs, and include methodologies for estimating emissions in four key sectors. These are: energy; industrial processes and product use; agriculture, forestry and other land use; and waste (see Table 3.1).

In the case of urban areas, emissions from the use of fossil fuels, industrial processes and product use, and waste are of particular importance. Stationary combustion mainly relates to energy industries, manufacturing industries and construction; while mobile combustion includes transportation emissions from civil aviation, road transportation, railways and waterborne navigation (although only within national boundaries – fuel use associated with international maritime transportation is not included). These distinctions are important, as – taken as a whole – energy, transportation and buildings account for almost half of all global emissions (see Figures 1.4 and 3.1).

National GHG inventories are based on the assumption that a country is responsible for all emissions produced within its area of jurisdiction. As a pragmatic measure to facilitate national targets and reductions, this is likely to be the only enforceable strategy – as countries only have legislative power within their own national boundaries. However, it means that the patterns of consumption that drive emissions (notably in the energy and industry sectors) are often veiled. For example, many polluting and carbon-intensive manufacturing processes are no longer located in developed countries, but have been sited elsewhere in the world to take advantage of lower labour costs and less rigorous environmental enforcement, and this reduces emissions in developed countries.[7] This is an important issue when assessing the underlying factors influencing emissions, which is discussed later in this chapter.[8]

## Protocols for measuring corporate greenhouse gas emissions

As industries and corporations have become increasingly aware of the impact that their activities have upon the environment, they have increasingly engaged in conducting GHG inventories. This enables companies to develop effective strategies to manage and reduce GHG emissions, and to facilitate their participation in voluntary and mandatory emissions reductions programmes. The most frequently utilized of these is the Greenhouse Gas Protocol developed by the World Resources Institute (WRI) and the World Business Council for Sustainable Development (WBCSD). This protocol puts forward an accounting system that is based on relevance, completeness, consistency, transparency and accuracy, and provides a mechanism by which private-sector actors can contribute to the global goal of reducing GHG emissions.[9]

**Table 3.1**

**Sectors assessed for national GHG inventories**

| Sector | Sub-sectors |
|---|---|
| Energy | Stationary combustion |
| | Mobile combustion |
| | Fugitive emissions |
| | $CO_2$ transport, injection and geological storage |
| Industrial processes and product use | Mineral industry emissions |
| | Chemical industry emissions |
| | Metal industry emissions |
| | Non-energy products from fuels and solvent use |
| | Electronics industry emissions |
| | Emissions of fluorinated substitutes for ozone-depleting substances |
| | Other product manufacture and use |
| Agriculture, forestry and other land use | Forest land |
| | Cropland |
| | Grassland |
| | Wetlands |
| | Settlements |
| | Other land |
| | Emissions from livestock and manure management |
| | Nitrous oxide emissions from managed soils, and $CO_2$ emissions from lime and urea application |
| | Harvested wood products |
| Waste | Solid waste disposal |
| | Biological treatment of solid waste |
| | Incineration and open burning of waste |
| | Wastewater treatment and discharge |

Note: The GHGs to be assessed are $CO_2$, methane, nitrous oxide, hydrofluorocarbons, perfluorocarbons, sulphur hexafluoride, nitrogen trifluoride, trifluoromethyl sulphur pentafluoride, halogenated ethers and other halocarbons (see also Table 1.3).

*Source:* IPCC, 2006

| Sector | Definition | Examples |
|--------|-----------|----------|
| Scope 1: | Direct GHG emissions that occur from sources owned or controlled by the company. | Emissions from combustion in owned or controlled boilers, furnaces, vehicles; emissions from chemical production in process equipment. |
| Scope 2: | GHG emissions from the generation of purchased electricity consumed by the company. | Purchased electricity in which the emissions physically occur at the facility where electricity is generated. |
| Scope 3 (optional): | GHG emissions that are a consequence of the activities of the company, but occur from sources not owned or controlled by the company. | Extraction and production of purchased materials; transportation of purchased fuels; use of sold products and services. |

*Source:* WRI/WBCSD, undated

**Table 3.2**

**Emissions scopes for companies**

The protocol addresses the issues peculiar to the corporate sector of accounting for emissions from group companies, subsidiaries, affiliated companies, non-incorporated joint ventures or partnerships, fixed-asset investments and franchises. The concept of 'scope' was developed in this process,[10] which takes into account direct and indirect GHG emissions in a more effective manner (see Table 3.2). It also describes the processes by which GHG emissions can be identified and calculated, how these can be verified for a formal reporting process, and how targets can be set.

## Protocols for measuring local government greenhouse gas emissions

Urban authorities often function at two distinct levels. First, they function as *business enterprises* – owning or leasing buildings, operating vehicles, purchasing goods and carrying out various other activities. In this regard, urban authorities can assess their emissions as corporate entities, including the direct and indirect impacts of their work. Second, urban authorities function as *governments* – with varying levels of oversight for and influence on the activities taking place within the spatial area over which they have jurisdiction.

The most widely accepted methodology for measuring emissions within local government boundaries has been developed by Local Governments for Sustainability (ICLEI), which is an international association of more than 1000 local government authorities that have made a commitment to sustainable development. More than 700 of these are members of the Cities for Climate Protection Campaign (see Box 2.7), which aims to assist cities in adopting policies to reduce local GHG emissions, improve air quality, and improve urban liveability and sustainability. The campaign sets five milestones for participating authorities:

1. Conduct a baseline emissions inventory and forecast.
2. Adopt an emission reduction target for the forecast year.
3. Develop a local action plan.
4. Implement policies and measures.
5. Monitor and verify results.[11]

The first stage of this process requires an emissions inventory; yet no appropriate methodology existed when the campaign was devised. The IPCC methodology for countries does not provide specifications at the local authority level for discussing energy consumption, transportation or waste disposal; and the existing corporate accounting protocols do not cover the details of municipal operations such as street lighting, landfill emissions and emissions from wastewater treatment or other industrial activities.

ICLEI's International Local Government GHG Emissions Analysis Protocol[12] was developed in response to this need. This protocol takes into account both government and community sectors, using the main categories derived from the IPCC's guidelines on national inventories of GHG emissions from stationary combustion, mobile combustion, fugitive emissions,[13] product use, other land use and waste. The protocol organizes government emissions according to:

- buildings and facilities;
- electricity or district heating/cooling generation;
- vehicle fleet;
- street lighting and traffic signals;
- water and wastewater treatment, collection and distribution;
- waste;
- employee commute;
- others.

It also breaks down the 'macro-sectors' used by the IPCC methodology (shown in Table 3.1) into community sectors for analysis, as shown in Table 3.3. ICLEI's framework provides the basis for the calculation of most current city-wide GHG emissions inventories.[14] It also recognizes the concept of *scopes*[15] in both government and community sectors (see Table 3.4). However, these do not extend as far as the consumption-based approaches,[16] many of which are impractical given the financial and technical resources available to local authorities to conduct inventories of this type. A more detailed consumption-based analysis requires much more information relating to the embedded carbon content of consumer goods purchased by individuals.

The most widely accepted methodology for measuring emissions within local government boundaries has been developed by Local Governments for Sustainability (ICLEI)

**Table 3.3**

**ICLEI categorization of community sectors**

| Macro-sector (IPCC) | | Community sector (ICLEI) |
|---------------------|--|--------------------------|
| Energy | Stationary combustion | Residential |
| | | Commercial |
| | | Industrial |
| | Mobile combustion | Transportation |
| | Fugitive emissions | Other |
| Industrial processes and product use | | Other |
| Agriculture, forestry and other land use | | Agricultural emissions |
| | | Other |
| Waste | Solid waste disposal | Waste |
| | Biological treatment of solid waste | |
| | Incineration and open burning of waste | |
| | Wastewater treatment and discharge | |

*Source:* ICLEI, 2008

| | Definitions | Examples |
|---|---|---|
| **Government operations emissions:** | | |
| Scope 1: | Direct emission sources owned or operated by the local government. | A municipal vehicle powered by gasoline or a municipal generator powered by diesel fuel. |
| Scope 2: | Indirect emission sources limited to electricity, district heating, steam and cooling consumption. | Purchased electricity used by the local government, which is associated with the generation of GHG emissions at a power plant. |
| Scope 3: | All other indirect or embodied emissions over which the local government exerts significant control or influence. | Emissions resulting from contracted waste-hauling services. |
| **Community-scale emissions:** | | |
| Scope 1: | All direct emissions sources located within the geopolitical boundary of the local government. | Use of fuels such as heavy fuel oil, natural gas or propane used for heating. |
| Scope 2: | Indirect emissions that result as a consequence of activity within the jurisdiction's geopolitical boundary limited to electricity, district heating, steam and cooling consumption. | Purchased electricity used within the geopolitical boundaries of the jurisdiction associated with the generation of GHGs at the power plant. |
| Scope 3: | All other indirect and embodied emissions that occur as a result of activity within the geopolitical boundary. | Methane emissions from solid waste generated within the community which decomposes at landfills either inside or outside of the community's geopolitical boundary. |

*Source:* compiled from ICLEI, 2008

## New baseline inventories for urban emissions

Increased interest in the contribution of urban areas to GHG emissions, and a growing recognition of the importance of urban areas in addressing the causes of climate change, means that there have been increasing attempts to develop appropriate inventories to account for city-level emissions. Many of these now grapple with complex issues of production- and consumption-based measures for allocating emissions.[17] An important component of these inventories is the setting of baselines, which can then be used to set targets for emissions reductions in subsequent years. It is also possible that a widely accepted baseline methodology might form the basis for emissions trading schemes, or for urban areas as a whole to trade carbon credits either on the formal or the voluntary market.[18] Table 3.5 presents a World Bank compilation of GHG emission baselines for selected cities and countries.

In addition, the recently launched International Standard for Determining Greenhouse Gas Emissions for Cities[19] provides a common method for cities to calculate the amount of GHG emissions produced within their boundaries. This standard builds on and is consistent with the IPCC protocols for national governments, and provides a common format to facilitate compilation by local authorities. It is hoped that this will provide a standard that can be used by cities around the world.

## Boundary issues

Efforts to develop a standardized globally comparable methodology for GHG emissions at the local or municipal level are made more complicated by the wide range of boundary definitions used for these areas. In general, the smaller the scale, the greater the challenges posed by 'boundary problems' in which it is increasingly difficult to identify which emissions ought or ought not to be allocated to a particular place.[20]

The importance of boundary definitions is clear from studies of urban populations, where differences in how governments define city boundaries have direct effects of spatial structure. For instance, it has been shown how eight different lists of the world's 20 largest cities vary, with only nine cities appearing on all eight lists; and with four different areas competing for the first two ranks.[21] The population figures for some large cities are for people living within long-established city boundaries enclosing areas of only 20 to 200 square kilometres; whereas for others (particularly in China) this includes regions with many thousands of square kilometres and a significant rural population.[22] These complications – related to different definitions of *cities and urban areas*, and different conceptions of the spatial extent of these – are all equally relevant in relation to identifying GHG emissions from a particular urban area. Similarly, energy consumption in *urban areas* in the US can vary between 37 and 81 per cent depending on how these areas are defined and bounded in space.[23] Thus, even within a single country, the potential contribution of urban areas to climate change can vary by a factor of two depending on the spatial definition of these areas.[24]

# THE SOURCES OF GREENHOUSE GAS EMISSIONS

Towns and cities do not themselves emit GHGs. Rather, specific activities that take place within urban areas – and that are undertaken in different ways by people of different ages, genders and income groups – are the sources of these GHGs. Different activities or sectors emit different quantities of different gases – with diverse impacts upon climate change (see Figure 3.1). Some of these activities have been integral to the process of urbanization over the last 300 years, and some have the clear potential to reduce emissions to mitigate climate change.

The main sources of GHG emissions from urban areas are related to the consumption of fossil fuels: whether this is for electricity supply, transportation or industry. This section explores the main sources of GHG emissions from urban

*A ... consumption-based analysis requires much more information relating to the embedded carbon content of consumer goods purchased by individuals*

*Efforts to develop a ... methodology for GHG emissions at the local or municipal level are made more complicated by the wide range of boundary definitions used for these areas*

*Towns and cities do not themselves emit GHGs. Rather, specific activities that take place within urban areas ... are the sources of these GHGs*

| Country | Annual GHG emissions (tonnes of CO$_2$eq per capita) | Year | City | Annual GHG emissions (tonnes of CO$_2$eq per capita) | Year |
|---|---|---|---|---|---|
| Argentina | 7.64 | 2000 | Buenos Aires | 3.83 | |
| Australia | 25.75 | 2007 | Sydney | 0.88 | 2006 |
| Bangladesh | 0.37 | 1994 | Dhaka | 0.63 | |
| Belgium | 12.36 | 2007 | Brussels | 7.5 | 2005 |
| Brazil | 4.16 | 1994 | Rio de Janeiro | 2.1 | 1998 |
| | | | São Paulo | 1.4 | 2000 |
| Canada | 22.65 | 2007 | Calgary | 17.7 | 2003 |
| | | | Toronto (City of Toronto) | 9.5 | 2004 |
| | | | Toronto (Metropolitan Area) | 11.6 | 2005 |
| | | | Vancouver | 4.9 | 2006 |
| China | 3.40 | 1994 | Beijing | 10.1 | 2006 |
| | | | Shanghai | 11.7 | 2006 |
| | | | Tianjin | 11.1 | 2006 |
| | | | Chongqing | 3.7 | 2006 |
| Czech Republic | 14.59 | 2007 | Prague | 9.4 | 2005 |
| Finland | 14.81 | 2007 | Helsinki | 7.0 | 2005 |
| France | 8.68 | 2007 | Paris | 5.2 | 2005 |
| Germany | 11.62 | 2007 | Frankfurt | 13.7 | 2005 |
| | | | Hamburg | 9.7 | 2005 |
| | | | Stuttgart | 16.0 | 2005 |
| Greece | 11.78 | 2007 | Athens | 10.4 | 2005 |
| India | 1.33 | 1994 | Ahmedabad | 1.20 | |
| | | | Delhi | 1.50 | 2000 |
| | | | Kolkata | 1.10 | 2000 |
| Italy | 9.31 | 2007 | Bologna (Province) | 11.1 | 2005 |
| | | | Naples (Province) | 4.0 | 2005 |
| | | | Turin | 9.7 | 2005 |
| | | | Veneto (Province) | 10.0 | 2005 |
| Japan | 10.76 | 2007 | Tokyo | 4.89 | 2006 |
| Jordan | 4.04 | 2000 | Amman | 3.25 | 2008 |
| Mexico | 5.53 | 2002 | Mexico City (City) | 4.25 | 2007 |
| | | | Mexico City (Metropolitan Area) | 2.84 | 2007 |
| Nepal | 1.48 | 1994 | Kathmandu | 0.12 | |
| The Netherlands | 12.67 | 2007 | Rotterdam | 29.8 | 2005 |
| Norway | 11.69 | 2007 | Oslo | 3.5 | 2005 |
| Portugal | 7.71 | 2007 | Porto | 7.3 | 2005 |
| Republic of Korea | 11.46 | 2001 | Seoul | 4.1 | 2006 |
| Singapore | 7.86 | 1994 | | | |
| Slovenia | 10.27 | 2007 | Ljubljana | 9.5 | 2005 |
| South Africa | 9.92 | 1994 | Cape Town | 11.6 | 2005 |
| Spain | 9.86 | 2007 | Barcelona | 4.2 | 2006 |
| | | | Madrid | 6.9 | 2005 |
| Sri Lanka | 1.61 | 1995 | Colombo | 1.54 | |
| | | | Kurunegala | 9.63 | |
| Sweden | 7.15 | 2007 | Stockholm | 3.6 | 2005 |
| Switzerland | 6.79 | 2007 | Geneva | 7.8 | 2005 |
| Thailand | 3.76 | 1994 | Bangkok | 10.7 | 2005 |
| UK | 10.50 | 2007 | London (City of London) | 6.2 | 2006 |
| | | | London (Greater London Area) | 9.6 | 2003 |
| | | | Glasgow | 8.8 | 2004 |
| US | 23.59 | 2007 | Austin | 15.57 | 2005 |
| | | | Baltimore | 14.4 | 2007 |
| | | | Boston | 13.3 | |
| | | | Chicago | 12.0 | 2000 |
| | | | Dallas | 15.2 | |
| | | | Denver | 21.5 | 2005 |
| | | | Houston | 14.1 | |
| | | | Philadelphia | 11.1 | |
| | | | Juneau | 14.37 | 2007 |
| | | | Los Angeles | 13.0 | 2000 |
| | | | Menlo Park | 16.37 | 2005 |
| | | | Miami | 11.9 | |
| | | | Minneapolis | 18.34 | 2005 |
| | | | New York City | 10.5 | 2005 |
| | | | Portland, OR | 12.41 | 2005 |
| | | | San Diego | 11.4 | |
| | | | San Francisco | 10.1 | |
| | | | Seattle | 13.68 | 2005 |
| | | | Washington, DC | 19.70 | 2005 |

**Table 3.5**

**Representative GHG baselines for selected cities and countries**

*Note:* Sources of the data presented above and details on which emissions have been included in the baselines are specified in the original source.

*Source:* based on World Bank, undated c

**Figure 3.1**

**Global GHG emissions by sector and end use/activity**

*Notes:* All data are for 2000. All calculations are based on CO₂eq, using 100-year global warming potentials based on an IPCC total global estimate of 41,755 million tonnes CO₂eq. Land-use change includes both emissions and absorptions. Dotted lines represent flows of less than 0.1 per cent of total GHG emissions.

*Source:* World Resources Institute (http://cait.wri.org/figures/World-FlowChart.pdf)

**Figure 3.1**

**Global GHG emissions by sector and end use/activity**

*Notes:* All data are for 2000. All calculations are based on $CO_2$eq, using 100-year global warming potentials based on an IPCC total global estimate of 41,755 million tonnes $CO_2$eq. Land-use change includes both emissions and absorptions. Dotted lines represent flows of less than 0.1 per cent of total GHG emissions.

*Source:* World Resources Institute (http://cait.wri.org/figures/World-FlowChart.pdf)

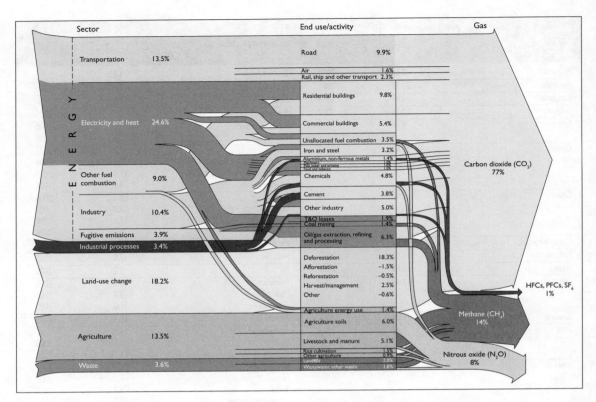

**Figure 3.2**

**World electricity generation by fuel type (1971–2008)**

*Note:* 'Other' includes geothermal, solar, wind, combustible renewables and waste, and heat.

*Source:* IEA, 2010, p24

areas, with a focus on energy supply (for electricity generation, transportation, commercial and residential buildings), industry, waste, agriculture, land-use change and forestry. It examines the activities that contribute to GHG emissions from these sectors, the types of gases that are generated and the importance of these for climate change. It also highlights the potential for mitigation in each of these sectors, setting the stage for detailed discussion in Chapter 5.

## Energy supply for electricity generation

Energy is perhaps the broadest possible category for assessing GHG emissions. The combustion of fossil fuels is the major source of these, and is used throughout the world for electricity generation, heating, cooling, cooking, transportation and industrial production. Energy is obtained from fossil fuels, biomass, nuclear power, hydroelectric generation and other renewable sources. Urban areas rely heavily on energy systems (shaped by the quantity of energy used), the energy structure (types of energy forms used) and the quality of the energy (its energetic and environmental characteristics). This section will thus focus on the use of energy for electri-city generation in urban areas, the different sources of energy and the implications for GHG emissions. Overall (in 2008), transport accounted for approximately 1.6 per cent of global electricity use; industry accounted for 41.7 per cent; while other sectors (agriculture, commercial and public services, residential, and non-specified other sectors) accounted for 56.7 per cent.[25] The particular aspects of electricity consumption by industry, and in commercial and residential buildings, will be examined later in this section.

Electricity functions as an 'energy carrier' – that is, an intermediate step between the original source of energy and the end user. In concentrated and densely populated urban areas, electricity is much more practical than the more direct use of fuels, particularly because large generation plants can be located hundreds of kilometres away and can feed electricity grids covering large areas. In 2008, a total of 20,181 terawatt hours (TWh) of electricity were produced around the world, most of which was generated from thermal energy (fossil fuels of oil, coal and gas) (see Figure 3.2 and Table 3.6). The world has a continued dependence on these GHG-generating fuels. Although the relative reliance on fossil fuels for electricity generation declined from 75.1 per cent in 1973 to 67.7 per cent in 2008, the total amount of energy produced from these sources grew from 4593TWh to 13,675TWh over the same period – recording an increase of 197.7 per cent.[26] The IPCC Fourth Assessment Report concluded that 'the global energy supply will continue to be dominated by fossil fuels for several decades'[27] and projected that without the introduction of effective new policy actions, energy-related GHG emissions are expected to rise by over 50 per cent between 2004 and 2030 (from 26.1 billion tonnes $CO_2$eq to between 37 and 40 billion tonnes $CO_2$eq).[28]

Electricity consumption varies significantly between urban areas around the world, although there is some clustering in the region of 4.5 to 7MWh per capita per annum for those connected to the grid: Bangkok (Thailand), Barcelona (Spain), Geneva (Switzerland), London (UK), Los Angeles (US), New York City (US) and Prague (Czech Republic) all fall within this range. In contrast, Cape Town (South Africa) has considerably lower per capita consumption (3.49MWh per capita per annum – perhaps because a large proportion of the population is not connected to electricity supply); while Toronto (Canada) and Denver (US) have considerably higher per capita consumption of 10.04 and 11.49MWh per capita per annum, respectively.[29]

The type of fuel used to generate electricity has a significant impact upon the volume of GHG emissions. Indeed, this is one of the most striking features influencing the emissions from different areas. Cities relying on nuclear or hydroelectric power generate substantially lower emissions than those that depend primarily on coal-fired power stations.[30]

In countries relying heavily on coal for electricity generation, electricity can be the single largest contributor to GHG emissions. A study of 15 South African cities indicated that electricity generation was responsible for more than 100 million tonnes of $CO_2$ emissions annually, or 66 per cent of the total – despite accounting for only 32 per cent of energy consumption.[31] However, average figures for electricity consumption – and the GHG emissions relating to these – are problematic because of industrial use of electricity. More detailed surveys are required to identify which members of society are utilizing large quantities of electricity and, thus, are responsible for the accompanying GHG emissions. In China, although the direct use of coal for energy has declined substantially during the last two decades, this remains the majority source for the generation of electricity.[32]

Among fossil fuels there are differences in the emissions generated for a given unit of electricity. Although coal is the world's most abundant fossil fuel and continues to be a vital resource in many countries, the typical efficiency of its conversion into electricity is about 35 per cent; and the burning of coal introduced approximately 9.2 billion tonnes of $CO_2$ into the atmosphere in 2005. The use of natural gas for electricity generation is growing rapidly (at an annual rate of 2.3 per cent during the 1990s), and contributes around 5.5 billion tonnes of $CO_2$ into the atmosphere each year. In contrast, the use of oil for direct generation of electricity has declined during recent years.[33] The operation of nuclear generation plants generates very low emissions of GHGs, although large indirect emissions are associated with the mining (and refining) of uranium and the building of nuclear plants.[34] This source of energy, however, has risks associated with its operation, storage of waste and the generation of

materials that can be used in nuclear weapons. For these reasons, public acceptance of nuclear power is limited in many countries.[35]

A variety of renewable energy systems can contribute to the security of energy supply and the reduction in GHG emissions (see Table 3.7).[36] However, there are still many challenges to be overcome in relation to the development of these technologies and in ensuring consistency in their generation capacity. For example, energy available from solar, wind and wave energy varies over time.

Many developing countries still face considerable challenges in expanding electricity networks to households – with low-income and female-headed households particularly seriously affected. In South Africa, for example, 64 per cent of the total population has no access to grid electricity. Although this is higher in rural areas, 16 per cent of households in cities did not use electricity for lighting.[37] Drawing on data for urban populations from a range of sources, it has been estimated that among the 117 developing countries for

| Type of fuel | Share of total energy generation (%) | |
|---|---|---|
| | 1973 (total generation: 6116TWh) | 2008 (total generation: 20,181TWh) |
| Thermal | 75.1 | 67.8 |
| Coal/peat | 38.3 | 41.0 |
| Oil | 24.7 | 5.5 |
| Gas | 12.1 | 21.3 |
| Nuclear | 3.3 | 13.5 |
| Hydro | 21.0 | 15.9 |
| Other (geothermal, solar, wind, combustible renewable and waste, heat) | 0.6 | 2.8 |

*Source:* IEA, 2010, p24

**Table 3.6**

**Electricity generation by energy source**

The type of fuel used to generate electricity has a significant impact upon the volume of GHG emissions

**Table 3.7**

**Categories of renewable energy technologies**

| Examples of technology | Stage of maturity |
|---|---|
| Large and small hydro<br>Woody biomass combustion<br>Geothermal<br>Landfill gas capture<br>Crystalline silicon photovoltaic solar-water heating<br>Onshore wind<br>Bio-ethanol from sugars and starch | Technologically mature with established markets in several countries. |
| Municipal solid waste to energy<br>Anaerobic digestion<br>Biodiesel<br>Co-firing of biomass<br>Concentrating solar dishes and troughs<br>Solar-assisted air conditioning<br>Mini- and micro-hydro<br>Offshore wind | Technologically mature but with relatively new and immature markets in a small number of countries. |
| Thin-film photovoltaic<br>Concentrating photovoltaic<br>Tidal range and currents<br>Wave power<br>Biomass gasification and pyrolysis<br>Bio-ethanol from ligno-cellulose<br>Solar thermal towers | Under technological development with demonstrations or small-scale commercial application, but approaching wider market introduction. |
| Organic and inorganic nanotechnology solar cells<br>Artificial photosynthesis<br>Biological hydrogen production involving biomass, algae and bacteria<br>Bio-refineries<br>Ocean thermal and saline gradients<br>Ocean currents | Still in technology research stages. |

*Source:* compiled from Sims et al, 2007

which data were available, there were 21 countries where more than half of urban households did not have access to electricity. Furthermore, there seems to be a strong correlation between the proportion of the urban population who have access to electricity and gross domestic product (GDP).[38] Increasing low-income urban residents' access to electricity generates important improvements to quality of life, including through reducing the use of potentially harmful or dangerous alternative fuels for heating, lighting and cooking. The provision of energy services – including electricity – is therefore an important component of alleviating poverty.

## Transport[39]

Globally, transportation is responsible for about 23 per cent of total energy-related GHG emissions[40] and 13 per cent of global GHG emissions (see Figure 1.4). In addition, transport activities increase as economies grow, and are expected to continue increasing in the decades ahead, especially with increasing levels of urbanization. Urban areas rely heavily on transportation networks of various kinds for both internal and external movements of goods and people. The proportion of journeys made by private as opposed to public transportation – particularly in larger cities – is an important factor influencing GHG emissions from an urban area. Urban areas, particularly in developed countries, often generate smaller amounts of GHG emissions per capita than rural areas due to the advantages of density. In the US, for example, per capita gasoline consumption is 12 per cent lower in urban counties than the national average.[41] Increases in public transportation use tend to reduce GHG emissions. A recent study suggested that a 1 per cent increase in public transportation would lead to a 0.48 per cent decrease in GHG emissions.[42]

Urban density is one of the most important factors influencing the amount of energy used in private passenger transport, and therefore has a significant effect on GHG emissions. Table 3.8 shows the ranking of ten cities on the basis of private passenger energy use, urban density and GHG emissions. With the exception of the Chinese cities, the most densely populated cities utilize less energy for private passenger transport and generally have lower GHG emissions per capita.

Access to public transport need not necessarily imply high density, as shown by the concepts of 'transit-oriented development' and 'transit villages' pioneered in California (US). These forms of development utilize moderate- to high-density housing within easy walking distance of major transit stops. However, this also requires careful planning of transit systems, the formation of community partnerships, detailed understanding of local real estate markets, and coordination among local, national and regional authorities. If successful, these developments can provide mobility choices, increase public safety, reduce the number of vehicle kilometres travelled (lowering annual household rates of driving by 20 to 40 per cent for those living, working and/or shopping near transit stations), reduce air pollution, reduce energy consumption, conserve resource lands and open space, reduce infrastructure costs, and contribute to more affordable housing.

Similar processes can be facilitated through the development of bus rapid transit systems in developing countries, which is discussed in more detail in Chapter 5. These are most efficient in servicing densely populated linear developments, which facilitate a large number of urban residents living within walking distances of the main trunk routes – generating an urban form often described as being shaped like a hand with fingers (where the 'palm' is the urban centre, and the 'fingers' the linear densely settled areas spreading out from the core).

Innovative thinking in relation to the planning of transportation infrastructure can therefore meet both environmental and social needs. Localized areas of relatively high densities are required to generate greater efficiencies in the usage of public transportation; but this can be consistent with meeting a variety of other demands from urban residents. Of course, the precise form that these transportation networks should take requires detailed local study. Overall, density is one of several factors that affects energy use – and, by extension, GHG emissions. However, addressing these issues requires ongoing analysis of urban *processes* rather than simply taking a snapshot of urban *form* at a particular moment in time.

A key component of GHG emissions from transportation is the number of vehicle kilometres travelled. The number of vehicle kilometres travelled is affected by several key aspects of urban design, including:

- density (higher number of people, jobs and/or dwelling units per unit area);
- diversity (greater mix of land uses);
- design (smaller block sizes, more sidewalk coverage, smaller street width);
- destination accessibility; and
- distance to transit.[43]

These 'five Ds' can be affected by the choices of planners and developers, and, in turn, will affect the travel choices of residents living in these areas. These aspects of urban design intersect with issues of personal choice and economic necessity – for example, there is some evidence from Sweden that women are more likely to use public transportation than men.[44] Chapter 5 shows the ways in which efficient urban

---

*Sidebar (left margin):*

The proportion of journeys made by private as opposed to public transportation ... is an important factor influencing GHG emissions from an urban area

Innovative thinking in relation to the planning of transportation infrastructure can ... meet both environmental and social needs

---

**Table 3.8**

Private passenger transport energy use, urban density and GHG emissions, selected cities

| Private passenger transport energy per person (ascending order) | Urban density (descending order) | GHG emissions per capita (ascending order) |
|---|---|---|
| Shanghai (China) | Seoul | São Paulo |
| Beijing (China) | Barcelona | Barcelona |
| Barcelona (Spain) | Shanghai | Seoul |
| Seoul (Republic of Korea) | Beijing | Tokyo |
| São Paulo (Brazil) | Tokyo | London |
| Tokyo (Japan) | São Paulo | Beijing |
| London (UK) | London | New York City |
| Toronto (Canada) | Toronto | Shanghai |
| New York (US) | New York | Toronto |
| Washington, DC (US) | Washington, DC | Washington, DC |

*Source:* compiled from Newman, 2006

design – including the use of brownfield developments – can help to minimize the distances that urban residents have to travel and, hence, reduce GHG emissions.

However, a variety of other factors also affect emissions from ground transportation, including the extent of private motor vehicle use, the quality of public transport, land-use planning and government policy. As shown in Table 3.9, there are significant variations between North American cities as a result of these factors. For example, Denver's per capita GHG emissions from ground transportation are four times greater than those of New York. Similarly, the high level of private motor vehicle dependence in Bangkok means that its per capita emissions from ground transportation are twice those of London, a much more affluent city but one with a more comprehensive public transportation system.

These variations can also be seen in the proportion of a city's GHG emissions that can be attributed to the transport sector. Shanghai and Beijing generate approximately 11 per cent of their emissions from the transportation sector, a figure dwarfed by their emissions from manufacturing.[45] However, the emissions from transport are increasing rapidly in Chinese cities, as shown by a recent study. The $CO_2$ emissions from transport in all 17 sample cities increased between 1993 and 2006. On average, the increase between 2002 and 2006 was 6 per cent per year and ranged from 2 to 22 per cent between the sample cities. The $CO_2$ emissions per capita from transport also increased in all cities and ranged between 0.5 and 1.4 tonnes per person in 2006, with Beijing being the highest.[46]

In London, New York and Washington, DC, transportation represents a significant contribution to the cities' emissions (22, 23 and 18 per cent, respectively); whereas in Barcelona (35 per cent), Toronto (36 per cent), Rio de Janeiro (30 per cent) and São Paulo (60 per cent), these figures are much higher.[47] However, the high figures in Rio de Janeiro and São Paulo are partially because these cities are strongly reliant on private motor vehicle transportation At the same time, it should be noted that London's emissions from transportation are lower than most developed country cities of similar size – as a result of high levels of public transport usage, strong investment in infrastructure and policies to promote alternatives to private motor vehicle use – while the extensive public transport system in New York City means that car ownership and usage levels are much lower than those in the US as a whole. The contribution of transportation to GHG emissions in Bangkok is described in Box 3.1.

Even when cars are chosen as the mode of transport, there are large variations in the GHG emissions produced by different sizes and types of vehicles. Within conventional private automobiles, there is a fourfold difference in their GHG emissions per kilometre. More efficient engines and greater use of diesel engines have the potential to reduce emissions. However, in the US, gains from engine efficiency have been offset as car weights and power have increased – meaning that overall fuel economy has hardly changed in the last 15 years.[48] Other factors that affect the contribution of urban transportation to GHG emissions include vehicle trip frequencies (starting a vehicle when it is cold uses more

energy and emits more $CO_2$ than starting the vehicle after it has warmed up) and vehicle operating speeds (motor vehicles with internal combustion engines are most efficient at an average speed of about 72km per hour). In spite of these issues, urban design that encourages less frequent car use will generate far greater benefits than the small losses associated with engine start-up; and roadway design that encourages higher speeds is likely to cause an increase in distances travelled by cars that will be far greater than any efficiency benefits.[49]

There is a strong association between rising income and car use in developing countries, meaning that economic growth in developing countries is very likely to result in increased car use and rising traffic congestion.[50] In addition, in many developing countries, the stock of motor vehicles is old and consists largely of second-hand and less efficient vehicles imported from developed countries. At the same time, the conversion of vehicles to use different fuels has the potential to reduce GHG emissions in many cities in developing countries. In Mumbai (India), for example, it has been estimated that the conversion of more than 3000 diesel-fuelled buses to compressed natural gas would result in a reduction of 14 per cent of total transport-related emissions.[51]

| City | Gasoline consumption (million litres) | Diesel consumption (million litres) | GHG emissions (tonnes $CO_2$eq per capita) |
|---|---|---|---|
| Denver (US) | 1234 | 197 | 6.07 |
| Los Angeles (US) | 14,751 | 3212 | 4.74 |
| Toronto (Canada) | 6691 | 2011 | 3.91 |
| Bangkok (Thailand) | 2741 | 2094 | 2.20 |
| Geneva (Switzerland) | 260 | 51 | 1.78 |
| New York City (US) | 4179 | 657 | 1.47 |
| Cape Town (South Africa) | 1249 | 724 | 1.39 |
| Prague (Czech Republic) | 357 | 281 | 1.39 |
| London (UK) | 1797 | 1238 | 1.18 |
| Barcelona (Spain) | 209 | 266 | 0.75 |

Source: Kennedy et al, 2009b

**Table 3.9**

**Ground transportation, fuel consumption and GHG emissions, selected cities**

There is a strong association between rising income and car use in developing countries, meaning that economic growth ... is very likely to result in increased car use

---

**Box 3.1 The contribution of transportation to GHG emissions in Bangkok, Thailand**

With approximately 7 million people, or more than 10 per cent of Thailand's population, Bangkok is not only the national capital, but also its focus for communications, its administrative centre and a major business hub for Southeast Asia. While the city's population has grown rapidly during the last 50 years, the population of the inner city has been declining as people have moved to suburban areas. Between 1978 and 2000, the population of the inner city declined from 3.25 million to 2.86 million, with a corresponding decrease in density from 15,270 to 11,090 people per square kilometre. Per capita emissions from Bangkok are at a similar level to many European and North American cities, with a figure of 7.1 tonnes per capita per annum.

Transportation is the single greatest source of Bangkok's GHG emissions, responsible for 23 million tonnes $CO_2$eq per year, or 38 per cent of the city's total. Electricity is the second largest contributor (33 per cent), followed by solid waste and wastewater (20 per cent). The contribution of this sector is growing rapidly: the number of vehicles registered in the city has soared from 600,000 in 1980 to 5.6 million in 2007, an almost tenfold increase. Indeed, since 2003, more than 500,000 additional vehicles have been registered in Bangkok each year.

Sources: Bangkok Metropolitan Administration, 2009; UN, 2010

The issue of emissions from transportation in developing countries is particularly important in countries where motor vehicle ownership is expanding rapidly. There are currently (2011) nearly 1.2 billion passenger vehicles worldwide. By 2050, this figure is projected to reach 2.6 billion – the majority of which will be found in developing countries.[52] The emissions associated with the increase in passenger vehicles can be reduced either through advances in fuel technology or by changes from one mode of transportation to another. The potential for such reductions is particularly strong in urban areas – with the advantage of relatively high densities of people, economic activities and cultural attractions. In Bogotá (Colombia), a bus rapid transit system (known as TransMilenio) – combined with car-restriction measures (including a path-breaking 'car-free' Sunday in which 120km of arterial roadways are closed to private motorized vehicles) and the development of new cycle-ways – has shown that an erosion of the relative importance of the public transport mode is not preordained.[53] Similar systems have also been proposed for several cities in Africa and Asia – for example, in Dar es Salaam.[54]

Current modes of urban transportation have many other adverse effects. According to the World Health Organization, more than 1.2 million people were killed in road traffic accidents in 2002. Projections indicate that this figure will increase by 65 per cent by 2020.[55] Reducing the reliance on private motor vehicles may help to reduce this figure. In addition, heavy reliance on personal transportation results in physical inactivity, urban air pollution, energy-related conflict and environmental degradation. Alternative modes of transport – particularly walking and cycling – can generate co-benefits, including improved human health through reducing obesity alongside reduced GHG emissions.[56] In addition, reducing the number of vehicles on the road can reduce local-source air pollution which is directly linked to mortality, cardiovascular diseases and respiratory illnesses, including asthma among young children.[57]

Perhaps the most notable omission from the above discussion is emissions from the aviation industry, which account for about 2 per cent of total anthropogenic GHG emissions.[58] These are not included within a country's national GHG inventory as a result of the lack of consensus as to where exactly these should be allocated. Should these be assigned to the country from which the aircraft takes off; the country in which the aircraft lands; the country in which the aircraft is registered; or the country of origin of the passengers? These issues are even more complex in the case of city emissions, as many of the passengers using major international airports situated in or close to major cities may be from elsewhere in the country, or may only be using these airports for transit purposes. The IPCC has estimated that aviation is responsible for around 3.5 per cent of human-induced climate change – the emissions from high-flying aircraft cause a greater amount of warming than the same volume of emissions would do at ground level – and that this is growing by approximately 2.1 per cent per year.[59]

Globally, shipping is responsible for about 10 per cent of transportation energy use;[60] but emissions from international maritime transportation are not included within national GHG inventories for similar reasons. Although international maritime transport is essential for servicing the needs of urban areas, the role of urban centres as ports and trans-shipment points means that allocating responsibility for the emissions associated with this is difficult, if not impossible.

The GHG inventories produced by London and New York offer an alternative set of figures that do take emissions from aviation into account. As a major air travel hub, London's airports handle 30 per cent of the passengers entering or departing the UK. If incorporated within the city's emissions inventory, aviation would be responsible for 34 per cent of London's emissions, and would raise total emissions from 44.3 million tonnes to 67 million tonnes for 2006.[61] In the case of New York (US), aviation would add 10.4 million tonnes per year to the city's emissions.[62] However, as is the case with industrial emissions, allocating the responsibility for all aviation-based emissions to a city's inventory is misleading – large city airports provide a service not only to individuals from elsewhere in the same country, but also from abroad.

## Commercial and residential buildings

GHG emissions from commercial and residential buildings are closely associated with emissions from electricity use, space heating and cooling. When combined, the IPCC estimates global emissions from residential and commercial buildings at 10.6 billion tonnes of $CO_2$eq per year,[63] or 8 per cent of global GHG emissions (see Figure 1.4). Commercial and residential buildings are responsible for both direct emissions (onsite combustion of fuels), indirect emissions (from public electricity use for street lighting and other activities and district heat consumption) and emissions associated with embodied energy (e.g. in the materials used for their construction). Emissions are affected by the need for heating and cooling, and by the behaviour of building occupants.

The type of fuel used for heating and cooling also determines the amount of GHGs emitted. Although Prague in the Czech Republic uses less energy per capita for heating than New York City (US), its emissions from heating are higher due to its reliance on coal. To a lesser extent, the emissions of Cape Town (South Africa) and Geneva (Switzerland) are also slightly higher than other comparable cities due to the predominance of oil instead of natural gas for heating.[64]

Data are available for the US on the direct final consumption of fossil fuels in buildings and industry.[65] However, it is not possible to separate the consumption of fuel between residential, commercial and industrial uses. Natural gas and fuel oil are the primary sources of energy for heating these buildings, while electricity is the main source of energy for cooling. Consequently, urban areas in warmer climate zones (where cooling, rather than heating, is required) tend to have lower direct final consumption of fossil fuels, whereas the inverse is the case in cooler climate zones. However, there are also regional differences in fuel composition. The US Northeast relies more on fuel oil,

---

**There are currently ... nearly 1.2 billion passenger vehicles worldwide. By 2050, this figure is projected to reach 2.6 billion – the majority of which will be found in developing countries.**

**GHG emissions from commercial and residential buildings are closely associated with emissions from electricity use, space heating and cooling**

**The type of fuel used for heating and cooling also determines the amount of GHGs emitted**

which emits a greater volume of $CO_2$ per unit of energy, while the US north Midwest relies more on natural gas.

The average size of new single-family homes in the US has expanded during recent years. Larger houses occupy more land, require greater amounts of material for construction, and consume more energy for heating and cooling. Larger homes are typically associated with higher incomes and higher energy consumption (see Table 3.10), and are found more often in sprawling counties.[66] Overall patterns of residential density also affect GHG emissions: the most densely settled area of the US (the Northeast, with 873 persons per square kilometre) has the lowest per capita annual energy consumption (283 million kilojoules), $CO_2$ emissions (15.0 tonnes) and vehicle distance travelled (13,298km) in the country.[67]

In the UK, residential buildings are responsible for 26 per cent of all $CO_2$ emissions, commercial and public buildings for 13 per cent, and industrial buildings for 5 per cent.[68] Most energy consumption (84 per cent) in residential buildings is from the heating of space and water. As the primary fuel for this is natural gas (which has a lower carbon content than electricity), this is only responsible for 74 per cent of $CO_2$ emissions. In non-residential buildings, heating (37 per cent) and lighting (26 per cent) are the main sources of $CO_2$ emissions.

Estimates for residential energy use per unit area for the UK and the US are 228 and 138kWh per square metre per year, respectively.[69] Since the average dwelling size is much greater in the US (200 compared to 87 square metres), the US uses more energy per household. For non-residential buildings, energy use in the UK, the US and India are 262, 287 and 189kWh per square metre per year, respectively.

In China, energy consumption of buildings accounts for 27.6 per cent of national energy consumption and contributes 25 per cent of national GHG emissions. In total, the country has 40 billion square metres of built space, of which only 320 million square metres can be identified as 'energy-saving buildings'. The Energy Efficiency in Building and Construction programme aims to ensure that, by 2010, new urban buildings will reach a new design standard of reducing energy consumption by 50 per cent.[70] At a smaller scale, various initiatives are being implemented in other middle-income countries such as South Africa, including the development of low-cost, low-energy houses for the urban poor. These employ simple technologies such as north-facing orientation and roof overhangs to maintain comfortable indoor temperatures without the use of heating or cooling equipment.[71]

The growth in India's population, coupled with the rapidly increasing consumer expectations of its middle class, means that emissions from that country's building stock are becoming increasingly important. During 2005/2006, only about 1 per cent of India's urban households used electricity for cooking, 59 per cent used natural gas or liquefied petroleum gas, 8 per cent used kerosene, 4 per cent used coal or lignite, and the rest, about 27 per cent, used firewood, charcoal, biomass or other energy sources.[72] Approximately one third of India's energy is supplied by large hydropower plants with limited GHG emissions. In the Indian residential

| Household income level (US$) | Household area (m²) | Energy consumption of household (million kilojoules) |
|---|---|---|
| $15,000–$19,999 | 139.4 | 85 |
| $30,000–$39,999 | 157.9 | 92 |
| $75,000–$99,999 | 250.8 | 119 |
| $100,000 or more | 315.9 | 143 |

*Source:* based on Markham, 2009, p12

**Table 3.10**

**Energy consumption by income level and dwelling size in the US (2008)**

sector, fans account for 34 per cent of total electricity consumption, lighting for 28 per cent and refrigeration for 13 per cent.[73] Of particular relevance for buildings in India is the large amount of electricity used by heating, ventilation and air-conditioning systems. In buildings without this facility, lighting is the main component of energy consumption, whereas in buildings with heating, ventilation and air-conditioning systems, these account for 40 to 50 per cent of consumption. However, there are large disparities between income groups: low-income groups in urban areas rely heavily on kerosene and liquefied petroleum gas for energy alongside electricity, and a significant proportion still use firewood or other biomass fuels. India's commercial sector uses electricity primarily for lighting (60 per cent of consumption) and heating, ventilation and air-conditioning systems (32 per cent).[74]

## Industry[75]

Globally, 19 per cent of GHG emissions are associated with industrial activities (see Figure 1.4). Although most cities in North America and Europe emerged and developed as a result of industrial activities, and still require industries to provide jobs and revenue, these same activities generate pollution. However, during recent decades the global pattern of industrial activities has shifted, in part due to transnational corporations seeking lower wages and higher profitability, and in part due to the increasing success of companies and corporations from China, India, Brazil and elsewhere in competing on the world market. Differences in environmental legislation have also transformed the geography of industrial location. It has been noted that 'when [cities] are able, they will get rid of polluting industries, pushing them away from city centers to suburbs or to other cities'.[76]

Many industrial activities are energy intensive in their operation. These include the manufacture of iron and steel, non-ferrous metals, chemicals and fertilizer, petroleum refining, cement, and pulp and paper. There are differences in industrial emissions according to location and according to the size of the industry. In developing countries, some facilities are new and incorporate the latest technology. Yet, small- and medium-sized enterprises may not have the economic or technical capacity to install the latest energy-saving equipment. The industrial processes mentioned above are responsible for direct GHG emissions. In Los Angeles (US), Prague (Czech Republic), and Toronto (Canada) these add 0.22, 0.43 and 0.57 tonnes of $CO_2$eq per capita per annum, respectively.[77]

Larger houses ... consume more energy for heating and cooling

Differences in environmental legislation have also transformed the geography of industrial location

A study of 15 South African cities shows that, on average, manufacturing caused half of all GHG emissions. However, in towns characterized as 'heavy industrial', manufacturing's share rose to 89 per cent, and residential share declined to 4 per cent. This accounts for the extraordinarily high per capita emission in some of these industrial towns. Saldanha Bay (a large port and industrial centre in Western Cape Province) and uMhlatuze (a port in KwaZulu-Natal that includes two aluminium smelters and a fertilizer plant) have per capita emissions of 50 and 47 tonnes of $CO_2$ emissions per annum, respectively (see Table 3.13). However, 93 per cent of emissions from Saldanha Bay and 95 per cent of emissions from uMhlatuze are generated by industry and commerce, with households being responsible for only 1.9 per cent and 1.5 per cent of emissions, respectively.[78]

Similar patterns exist in China, where the ratio of urban to rural per capita commercial energy usage is 6.8.[79] This reflects a situation in which the industrial sector has tended to dominate urban $CO_2$ emissions, although this has been declining in some cases. Between 1985 and 2006, the share of the industrial sector in total $CO_2$ emissions from Beijing declined from 65 to 43 per cent; and in Shanghai from 75 to 64 per cent. However, rather than representing an absolute decline, this reflects the growing significance of the transportation sector. $CO_2$ emissions from transportation rose sevenfold in Beijing and eightfold in Shanghai over the same time period. Another set of estimates suggests that between 1990 and 2005, 90 per cent of Shanghai's energy was consumed by the industrial sector and only 10 per cent by 'family life', although these figures appear to exclude energy consumption by transportation.[80]

## Waste

Emissions from waste represent about 3 per cent of total emissions (see Figure 1.4). Despite being only a small contributor to global emissions, rates of waste generation have been increasing during recent years, particularly in developing countries that have been experiencing increasing affluence. Although waste generation is linked to population, affluence and urbanization,[81] emissions from waste may be lower in more affluent cities, as urban areas have the potential to greatly reduce – or even eliminate – emissions from waste. The concentration of people and activities in urban areas means that waste can be collected efficiently, and methane emissions from landfills can be captured and flared or used to generate electricity. Although many developing countries lack the technology for methane recuperation, landfill gas capture projects funded through the Clean Development Mechanism (CDM),[82] which are increasingly being implemented even in the least developed countries (e.g. at the Mtoni dumpsite in Dar es Salaam, Tanzania), offer numerous potentials.

As a result of some of the above-mentioned initiatives, GHG emissions from waste are relatively low in many urban areas in developed countries. Per capita emissions from waste in Barcelona (Spain), Geneva (Switzerland), London (UK), Los Angeles (US), New York (US), Prague (Czech Republic) and Toronto (Canada) are all less than 0.5 tonnes $CO_2$eq per capita per annum; while those from Bangkok (Thailand) and Cape Town (South Africa) are 1.2 and 1.8 tonnes $CO_2$eq per capita per annum, respectively.[83] Indeed, New York has negative net emissions as a result of the capture of methane from managed landfills, while in São Paulo (Brazil), Barcelona (Spain) and Rio de Janeiro (Brazil) solid waste is responsible for 23.6, 24 and 36.5 per cent of urban emissions, respectively.[84] These variations are likely to be due not only to different patterns of consumption and waste generation, but also to differences in the management of waste and differences in accounting mechanisms – variations that are almost impossible to assess in the absence of a standardized urban framework for conducting emissions inventories.

## Agriculture, land-use change and forestry

At a global level, 14 per cent of GHG emissions can be allocated to activities related to agriculture and 17 per cent to forestry (see Figure 1.4). Although these are often thought of as being rural activities, urban agriculture can be an important component of local economies and food supply systems, and many urban areas do include parks and forests. The broader ecological impacts of cities have been increasingly studied during recent years,[85] based on the recognition that the expansion of built-up areas transforms rural landscapes; that city-based enterprises, households and institutions place demands on forests, farmlands and watersheds outside urban boundaries; and that solid, liquid and airborne wastes are transferred out of the city to the surrounding region or further afield.[86]

Specifically in relation to climate change, urban areas can shape emissions from agriculture, land-use change and forestry in two major ways. First, the process of urbanization can involve direct changes in land use, as formerly agricultural land becomes incorporated within built-up areas. The world's urban population multiplied tenfold during the 20th century, meaning that more land was covered by urban development. Equally, global urban trends towards suburbanization mean that cities are sprawling and encroaching on land that may previously have been covered with vegetation – thereby reducing its potential to absorb $CO_2$. Second, the consumption patterns of increasingly wealthy urban residents can shape the type of agricultural activities that is taking place. For example, the growing consumption of meat products is associated with emissions of methane from livestock rearing.

---

**Margin notes:**

Although waste generation is linked to population, affluence and urbanization, emissions from waste may be lower in more affluent cities, as urban areas have the potential to greatly reduce – or even eliminate – emissions from waste

GHG emissions from waste are relatively low in many urban areas in developed countries

Global urban trends towards suburbanization mean that cities are sprawling and encroaching on land that may previously have been covered with vegetation – thereby reducing its potential to absorb $CO_2$

# THE SCALE OF URBAN EMISSIONS

It is impossible to make definitive statements about the scale of urban emissions. There is no globally accepted standard for assessing the scope of urban GHG emissions – and even if there was, the vast majority of the world's urban centres have not conducted an inventory of this type. Current administrative differences in the extent, frequency and thoroughness of inventories contribute to substantial variations in the scope, accuracy and comparability of these. Perhaps the two most substantial differences between inventories are related to boundary issues and scope issues. These issues are discussed in greater detail in the next two sections.[87] This section presents the results from a range of previously conducted GHG inventories, assesses their findings, and identifies common themes across developed and developing countries.

## Global patterns of emissions

The economic activities, behavioural patterns and GHG emissions from urban areas are shaped by the overall economic, political and social circumstances of the countries in which they are located. At a global level, there are striking differences in GHG emissions between regions and countries. The 18 per cent of the world's population living in developed countries account for 47 per cent of global $CO_2$ emissions, while the 82 per cent of the world's population living in developing countries account for the remaining 53 per cent. The US and Canada alone account for 19.4 per cent of global GHG emissions, while South Asia accounts for 13.1 per cent and Africa just 7.8 per cent.[88] Even greater differences can be seen if individual countries are compared: per capita $CO_2$eq emissions vary from less than 1 tonne per year for Bangladesh and Burkina Faso to more than 20 tonnes for Canada, the US and Australia.[89] These variations among countries are shown in greater detail in Figure 3.3.

In addition, global growth in GHG emissions has not been distributed evenly between countries. Between 1980

Per capita $CO_2$eq emissions vary from less than 1 tonne per year for Bangladesh and Burkina Faso to more than 20 tonnes for Canada, the US and Australia

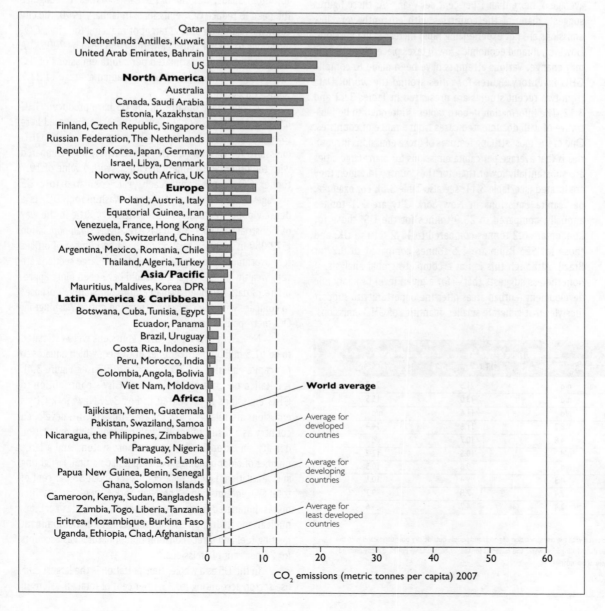

**Figure 3.3**

**$CO_2$ emissions per capita in selected countries and world regions (2007)**

*Sources:* based on http://mdgs.un.org/unsd/mdg (last accessed 21 October 2010); and UN, 2010

| City | GHG emissions per capita (tonnes of $CO_2$eq) (year of study in brackets) | National emissions per capita (tonnes of $CO_2$eq) (year of study in brackets) |
|---|---|---|
| Washington, DC (US) | 19.7  (2005) | 23.9  (2004) |
| Glasgow (UK) | 8.4  (2004) | 11.2  (2004) |
| Toronto (Canada) | 8.2  (2001) | 23.7  (2004) |
| Shanghai (China) | 8.1  (1998) | 3.4  (1994) |
| New York City (US) | 7.1  (2005) | 23.9  (2004) |
| Beijing (China) | 6.9  (1998) | 3.4  (1994) |
| London (UK) | 6.2  (2006) | 11.2  (2004) |
| Tokyo (Japan) | 4.8  (1998) | 10.6  (2004) |
| Seoul (Republic of Korea) | 3.8  (1998) | 6.7  (1990) |
| Barcelona (Spain) | 3.4  (1996) | 10.0  (2004) |
| Rio de Janeiro (Brazil) | 2.3  (1998) | 8.2  (1994) |
| São Paulo (Brazil) | 1.5  (2003) | 8.2  (1994) |

*Source:* Dodman, 2009

**Table 3.11**

**Comparisons of city and national GHG emissions, selected cities**

**A wide range of cities in the US have produced GHG inventories**

**Table 3.12**

**GHG emissions inventories, selected cities**

| City | Emissions per capita (tonnes of $CO_2$eq per year) | | |
|---|---|---|---|
| | Emissions within city | Direct emissions | Life-cycle emissions[a] |
| Denver (US) | n/d | 21.5 | 24.3 |
| Los Angeles (US) | n/d | 13.0 | 15.5 |
| Cape Town (South Africa) | n/d | 11.6 | n/d |
| Toronto (Canada) | 8.2 | 11.6 | 14.4 |
| Bangkok (Thailand) | 4.8 | 10.7 | n/d |
| New York City (US) | n/d | 10.5 | 12.2 |
| London (UK) | n/d | 9.6 | 10.5 |
| Prague (Czech Republic) | 4.3 | 9.4 | 10.1 |
| Geneva (Switzerland) | 7.4 | 7.8 | 8.7 |
| Barcelona (Spain) | 2.4 | 4.2 | 4.6 |

*Notes:* n/d = not determined.

a Life-cycle emissions are associated with the transportation of goods and people outside city boundaries, and with the production of key urban materials, including food, water and materials for shelter, but that may not be directly emitted from within the cities' geographical boundaries.

*Source:* Kennedy et al, 2009b

and 2005, $CO_2$ emissions increased by more than 5 per cent per year in the Republic of Korea, China and Thailand. Yet, over the same period, emissions from Chad, the Democratic Republic of Congo, Liberia and Zambia declined by an average of more than 1 per cent per year.[90] As these figures suggest, many of the countries with currently very low emissions are not experiencing rapid increases in emissions. However, if rapid economic growth takes place, this situation may change. Various attempts have been made to compile GHG inventory figures for cities around the world. Data from two recent studies are presented in Tables 3.11 and 3.12. The information in both tables is intended to be illustrative of well-documented cases from a range of countries. One of the most striking features of these emissions inventories is that average per capita emissions for many large cities are substantially lower than for the country in which they are located (see Table 3.11; see also Table 3.5). For example, per capita emissions in New York City are 7.1 tonnes annually, compared to 23.9 tonnes for the US; those for London are 6.2 tonnes compared to 11.2 for the UK; and those for São Paulo are 1.5 tonnes compared to 8.2 for Brazil. Although this is not a complete global analysis, it nonetheless suggests that – for a given level of economic development – urban areas offer the opportunity to support lifestyles that generate smaller quantities of GHG emissions.

## Urban emissions in developed countries

Many urban areas in developed countries have their origins (or their change in scale) in the industrial revolution and the rapid expansion of manufacturing during the 18th and 19th centuries. The industrial hubs of northern England, the Rhine-Ruhr Valley in Germany and the north-eastern seaboard of the US all developed in close proximity to heavy raw materials such as coal and iron ore. However, since the middle of the 20th century, the economies of these regions have shifted away from secondary industry into tertiary and quaternary industries. As will be seen from the examples discussed below, this means that their emissions from the manufacture of products are relatively low. At the same time, these urban areas have become centres of wealth and consumption. The lifestyles of their residents – particularly related to consumption and travel – generate a large carbon footprint; yet this is seldom accounted for in emissions inventories.

### ■ Urban emissions in North America

Toronto, Canada, was one of the earliest cities to recognize the need to reduce $CO_2$ emissions. In January 1990, the city council declared an official target of reducing the city's $CO_2$ emissions to 20 per cent below the 1988 level by 2005.[91] A more recent survey estimated per capita emissions of 8.2 tonnes in 2001, compared to a Canadian average of 23.7 tonnes (in 2004).[92]

A wide range of cities in the US have produced GHG inventories, although only a few of these are discussed here to highlight particular issues. The overall per capita GHG emissions for Washington, DC, are relatively high compared with the other North American cities – with a value of 19.7 tonnes $CO_2$eq per capita each year, compared to a US average of 23.9 tonnes. Although Washington, DC, is a densely populated urban centre, with very little in the way of industrial activities, it also has a relatively small population (572,059 in 2000) in relation to the large number of offices for government and related functions, and large sections are very wealthy. In this regard, it would have been more appropriate (if data had been available) for comparative purposes to compare with the emissions from the entire Washington, DC, metropolitan area.[93]

New York City's total GHG emissions were estimated to be 61.5 million metric tonnes of $CO_2$eq, which equates to per capita emissions of 7.1 tonnes in 2007 (see Figure 3.4). A detailed description of New York City's contribution to GHG emissions is presented in Box 3.2. New York City's emissions are relatively low for a wealthy city in a developed country as a result of small dwelling sizes, high population density, an extensive public transport system, and a large number of older buildings that emphasize natural daylighting and ventilation. Electricity accounts for about 38 per cent of total $CO_2$eq emissions. The New York City electricity fuel mix is dominated by natural gas, but also includes coal, oil, nuclear and hydropower. Natural gas is also the dominant heating fuel and direct consumption of natural gas accounts for 24 per cent of emissions.[94]

In the US as a whole, transportation is the largest end-use sector, accounting for 33.1 per cent of total emissions in

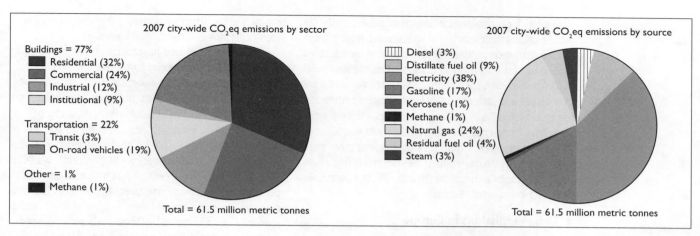

**2007 city-wide CO₂eq emissions by sector**

Buildings = 77%
- ■ Residential (32%)
- Commercial (24%)
- Industrial (12%)
- Institutional (9%)

Transportation = 22%
- Transit (3%)
- On-road vehicles (19%)

Other = 1%
- ■ Methane (1%)

Total = 61.5 million metric tonnes

**2007 city-wide CO₂eq emissions by source**

- Diesel (3%)
- Distillate fuel oil (9%)
- Electricity (38%)
- Gasoline (17%)
- Kerosene (1%)
- ■ Methane (1%)
- Natural gas (24%)
- Residual fuel oil (4%)
- Steam (3%)

Total = 61.5 million metric tonnes

the country.[95] Emissions from transportation in New York City were 13 million metric tonnes of $CO_2$eq during the fiscal year of 2007, or 22 per cent of the city's total $CO_2$eq emissions, which was well below the average for the US.[96] Although the city's official inventory did not include GHG emissions from aviation and shipping, emissions from aviation have been estimated at 10.4 million metric tonnes, while emissions from transportation of freight by water have been estimated at 6.2 million metric tonnes.[97] Residential and commercial buildings each account for a larger share of emissions than transportation, and, overall, buildings account for 77 per cent of GHG emissions (although this does include industry) (see Figure 3.4). The share of building-sector emissions in New York City is larger than in the US as a whole both because of reduced transportation-sector emissions resulting from an effective public transit network and because of the city's large and energy-intensive commercial sector. Since buildings account for the majority of emissions, this sector is likely to be the key focal point for emissions reduction policies in New York City.

As noted in Box 3.2, New York City has set a target of a 30 per cent reduction in government operations emissions by 2030.[98] The GHGs emitted as a result of government operations were 4.3 million metric tonnes $CO_2$eq in 2007, representing approximately 7 per cent of the city-wide total. Similar to the city-wide results, buildings account for the vast majority of city government emissions (63 per cent). The city's municipal vehicle fleet accounts for 8 per cent of the total.[99]

New York City's 2007 city-wide inventory showed that $CO_2$eq emissions decreased by 2.5 per cent between 2005 and 2007. Although energy consumption increased between these two years, the carbon intensity of the electricity supply decreased when two new efficient power plants were introduced in 2006, displacing electricity generated from less efficient plants with higher $CO_2$eq coefficients. This change alone reduced emissions by approximately 3.2 million metric tonnes (5 per cent reduction). Milder winter and summer weather conditions in 2007 compared with 2005 also contributed to the reduction. 'Heating degree days' and 'cooling degree days', which reflect the demand for energy needed to heat or cool a home or business, decreased by 0.6 and 17.7 per cent, respectively, from 2005 to 2007.[100]

A very different picture emerges from the emissions inventory of Aspen, Colorado (US).[101] The inventory gives an overall figure of 50 tonnes $CO_2$eq per capita per annum, but also shows how various methods for calculation can lead to different results. Aspen is a major tourist destination, and the figure includes a calculation to allocate emissions to this group of temporary residents. If this is removed, then per capita emissions for the *resident* population rise to 102.5 tonnes $CO_2$eq. The overall figure includes air transportation – and if this is removed, as well as tourist driving and commuting to locations outside Aspen, then the equivalent figure falls to 40.3 tonnes $CO_2$eq. Whichever figure is used is considerably higher than the US national average, and this indicates the high emissions from relatively small towns in predominantly rural areas. Aspen is a very wealthy town with a high reliance on the tourism industry (which may be seen as energy intensive in its own right), and requires substantial heating because of its mountainous location.

A report by the Brookings Institution examines the per capita carbon footprints of individuals (rather than comprehensive GHG inventories) from the 100 largest metropolitan areas in the US from 2000 to 2005.[102] The

**Figure 3.4**

**GHG emissions inventory, New York City, US**

*Source:* City of New York, 2009

New York City's 2007 city-wide inventory showed that CO₂eq emissions decreased by 2.5 per cent between 2005 and 2007

---

**Box 3.2 Contribution to GHG emissions, New York City, US**

New York is the largest city in the US and a global centre of commerce and culture. The city itself has a population of 8.25 million and forms the core of the New York Metropolitan Area, with a population of 18.8 million. The city produces approximately 8 per cent of total US gross domestic product (GDP) and is a leading financial centre. In general, the city's total emissions are high, but per capita emissions are low in comparison to other urban areas in the US.

The city's greenhouse gas (GHG) emissions are dominated by energy-related activities: more than two-thirds of the city's emissions are associated with electricity and fuel consumption in residential, commercial and institutional buildings. A further 22 per cent is associated with transportation – this is low by US standards, as the city has the highest rate of commuting by public transport in the country. Three-quarters of the methane produced at landfills and wastewater treatment plants is captured, so this represents a very small source of emissions.

The City of New York has been completing GHG emissions inventories since 2007 and has passed a law requiring annual updates to this. This is associated with PlaNYC: the mayor's comprehensive sustainability plan for the city's future, which has set goals including reducing GHG emissions by 30 per cent by 2030. Measures to achieve this are associated with upgrading the local power supply (by replacing inefficient power plants with more advanced technologies), reducing energy consumption (by imposing more aggressive energy codes for new buildings and promoting energy efficiency in existing buildings) and reducing transport-related emissions (through the expanded use of public transportation).

*Sources: Parshall et al, 2009, 2010*

report concludes that despite housing two-thirds of the US population and three-quarters of its economic activity, these metropolitan areas emitted just 56 per cent of the country's GHG emissions from highway transportation and residential buildings. However, the footprints that were assessed were only partial, and included only highway transportation and energy consumption in residential buildings – and not emissions from commercial buildings, industry or non-highway transportation. Even within these parameters, there were large differences between metropolitan areas: from 1.36 tonnes per capita in Honolulu (Hawaii, US) to 3.46 tonnes per capita in Lexington (Kentucky, US).

### ■ Urban emissions in Europe

In comparison to North American cities, the contribution of urban areas in Europe to climate change is relatively low. This is as a result of several factors. European urban areas tend to be more compact. They tend to have lower car ownership and car usage rates, smaller, more fuel-efficient cars, reducing emissions from private transportation. They tend to have more effective public transportation networks, which are deemed socially acceptable to a broader range of individuals. Furthermore, urban areas in Europe have higher levels of densification and lower levels of sprawling in relation to North American cities.

The overall $CO_2$ emissions of London in 2006 were 44.3 million tonnes – representing 8 per cent of the UK's total emissions, and a slight decline from the 45.1 million tonnes produced in 1990 despite a rise in population of 0.7 million people during the same time period.[103] This reduction can be attributed to the halving of industrial emissions, as industrial activity declined or has relocated to other parts of the UK or overseas. The per capita emissions from London are the lowest of any region in the UK, and at 6.2 tonnes per capita, in 2006, were just over half of the national average of 11.2 tonnes per capita (see Table 3.11). Per capita emissions from Glasgow at 8.4 tonnes per capita in 2004 are higher than those for London; but this may also reflect the fact that the analysis covered the entire area of Glasgow and the Clyde Valley, an area comprised of eight local authorities and covering an area of 3405 square kilometres. This area also emits a higher than average quantity of agricultural emissions due to a proportionally larger dairy farming sector in the area.

Barcelona, the second largest city in Spain, had a population of 1.6 million people within its administrative unit in 1996. Over the period of 1987 to 1996, the total emissions for the city grew from 4.4 million tonnes to 5.1 million tonnes. There was, however, a decline from 5.3 million tonnes to 4.9 million tonnes between 1992 and 1995. Part of this decline can be attributed to a decline in population. Indeed, between 1987 and 1995, the population of Barcelona shrank from 1.7 million to 1.5 million. Barcelona's relatively low level of per capita emissions can be attributed to several factors. These include: the city's economy is primarily service based rather than manufacturing based; 90 per cent of the city's electricity is generated by nuclear and hydro energy; the city's mild climate and the rarity of household air-conditioning systems; and the

compact urban structure, in which many residents live in apartments rather than individual houses.[104]

Given the reduced importance of manufacturing in some cities, some inventories have been produced that only take into account emissions associated with electricity, transportation and waste generation. In Oeiras municipality (part of the metropolitan area of Lisbon, Portugal, with a population of 160,000), electricity accounted for 75 per cent of municipal emissions in 2003. Other sources of emissions were gaseous fuel consumption (11 per cent), waste in sanitary landfills (8 per cent), liquid fuels consumption (5 per cent) and wastewater treatment (1 per cent). The total municipal emissions in 2003 were 525,550 tonnes – or 3.3 tonnes $CO_2$eq per capita.[105] Liquid fuels serve as a proxy for transportation. However, this only provides information on the amount of fuel purchased, and not on the vehicle kilometres driven within the municipality. In addition, it fails to resolve important issues related to the allocation of emissions from motor vehicles – whether these should fall in the location of residence of the driver, the origin or destination of any given journey, or some combination of the above.

Emissions inventories for Geneva, Switzerland, show annual within-city emissions of 7.4 tonnes $CO_2$eq per capita, and for Prague, Czech Republic, show annual within-city emissions of 4.3 tonnes $CO_2$eq per capita. If 'life-cycle' emissions are taken into account, these figures rise to 8.7 and 10.1 tonnes $CO_2$eq per capita, respectively (see Table 3.12).

## Urban emissions in developing countries

Very few detailed emissions inventories have been produced by cities in developing countries. Cities in these countries are frequently economic centres that contribute significantly to the gross national product (GNP), and act as economic, political, social and cultural centres. Consequently, these cities are centres of consumption and wealth, with likely consequences including higher per capita GHG emissions than surrounding areas.

As manufacturing has declined in importance in developed countries, it has expanded rapidly in some developing countries. Countries such as Brazil, China, India and South Africa – encouraged by economic and geopolitical changes – are now centres for global manufacturing. The relatively cheap and plentiful supply of labour, and the increased ease of transporting raw materials and finished products, means that these countries can compete effectively on the world market. Yet, this competition is not without costs, some of which are associated with local environmental degradation and others with the emission of GHGs. Current global protocols for measuring emissions, and setting targets for their reduction, solely address the location of production of these GHGs, which means that some developing countries with prospering manufacturing industries appear to be major emitters.

Some developing countries play an increasingly important role in contributing to global GHG emissions. China has recently overtaken the US as the world's leading total emitter of GHGs, although its per capita emissions are

---

*Sidebar notes (left margin):*

In comparison to North American cities, the contribution of urban areas in Europe to climate change is relatively low

Very few detailed emissions inventories have been produced by cities in developing countries

China has recently overtaken the US as the world's leading total emitter of GHGs, although its per capita emissions are significantly lower

### Box 3.3 Contribution to GHG emissions, São Paulo, Brazil

The São Paulo Metropolitan Region has a population of 18 million and is the largest urban area in Brazil. The city is a major driving force for the national economy, with a gross domestic product (GDP) of US$83 billion in 2003. The service industry is the main driver, accounting for 62.5 per cent of GDP. This is followed by the industrial sector, which accounts for 20.6 per cent.

A comprehensive greenhouse gas (GHG) emissions inventory was conducted in 2005. It shows that energy use accounts for more than three-quarters of the city's emissions (see figure below). Approximately two-thirds of this was associated with diesel and gasoline, and 11 per cent with electricity generation. However, the contribution of urban transportation to GHG emissions is still relatively low as a result of the mandatory blend of ethanol (23 per cent) and gasoline (77 per cent) used in most of the private fleet. Similarly, the contribution of the electricity generation sector is low as the city relies heavily on hydroelectric generation. Solid waste disposal accounted for almost one quarter of the city's emissions, or 3.7 million tonnes of $CO_2$eq. However, Clean Development Mechanism (CDM) projects at the Bandeirantes and São João landfills will prevent the generation of 11 million tonnes of $CO_2$eq by 2012 – almost removing the contribution of solid waste to the city's emissions.

Per capita emissions from the city are low, at about 1.5 tonnes $CO_2$eq per year (in 2003), compared to a national average for Brazil of 8.2 tonnes (1994 figure). Despite this, the growing importance of reducing global GHG emissions means that cities in middle-income countries will increasingly need to identify their emissions reduction potential and act on this.

It is important to note that although the city of São Paulo accounts for 6.8 per cent of the population of Brazil, its GHG emissions are relatively small. This is because Brazil is a large emitter of GHGs from agriculture, land-use change and forestry. In the case of deforestation, due to high rates, emissions account for 63.1 per cent of total national emissions of $CO_2$ and methane. The agriculture sector as a whole is responsible for 16.5 per cent of the same gases, mainly because of the size of the national herd. In the case of the extremely urbanized city of São Paulo, these emissions are insignificant.

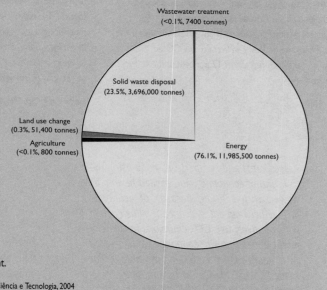

*Sources:* Dubeux and La Rovere, 2010; La Rovere et al, 2005; Ministério da Ciência e Tecnologia, 2004

significantly lower.[106] Brazil, China, India and South Africa together form the 'BASIC' group of countries that – although not part of the legally binding framework to reduce emissions – recognize that their substantial emissions compel them to take a more progressive role in international climate negotiations. Specifically, these countries are at the forefront of the development of 'nationally appropriate mitigation actions'[107] and the reduction in carbon intensity of their industries, rather than absolute reductions in emissions.

Box 3.3 provides an overview of the GHG emissions from São Paulo, Brazil. As indicated, total annual emissions from São Paulo are 15.7 million tonnes, or 1.5 tonnes per capita. However, there is the potential for the transportation and solid waste sectors in the city to engage in GHG mitigation activities – for example, through CDM projects such as the landfill gas to energy project at Bandeirantes.

GHG emissions have also been calculated for Mexico City, although the figures generated vary widely: from a total of 34.9 million tonnes $CO_2$eq in 1996 to 60.0 million tonnes in 2000, and 62.6 million tonnes in 2004. This variation is a result of the scarcity and inconsistency of official inventories, and methodological issues related to the inclusion and exclusion of emissions from solid waste and aviation.

However, even at the higher level this equates to per capita emissions of 3.6 tonnes per year, lower than the national figure of 4.6 tonnes.[108] Box 3.4 provides further insights on GHG emissions for Mexico City.

In South Africa, $CO_2$ emissions have been calculated from six large metropolitan areas ('metros'), four industrial towns/cities and five non-industrial towns/cities (see Table 3.13). The average annual per capita emissions from all these urban areas is 8.1 tonnes of $CO_2$. This is slightly lower than the national average of 8.9 tonnes, but considerably higher than the African average of 1.1 tonnes. When broken down according to type of town, 'non-industrial towns and cities' have average per capita emissions of 3.4 tonnes; 'metros', 6.5 tonnes; and 'industrial towns and cities', 26.3 tonnes. The averages for individual urban areas range from 1.7 tonnes for King Sabata to a massive 49.5 tonnes for Saldanha Bay. The main sources of these emissions also vary substantially. In industrial towns, manufacturing accounts for as much as 89 per cent of the emissions, while in non-industrial towns, this figure was only 36 per cent. This is reflected in the $CO_2$ emissions per economic unit of value added, which shows that this is much higher in industrial centres than in service-oriented cities. Thus:

*Brazil, China, India and South Africa ... are at the forefront of the development of 'nationally appropriate mitigation actions' and the reduction in carbon intensity of their industries, rather than absolute reductions in emissions*

**Box 3.4 GHG emissions and climate change in Mexico City**

Mexico City has one of the highest levels of air pollution in the world. With an estimated population of 20.1 million people, 3.75 million cars and 35,000 industries, the Mexico City Metropolitan area is a major emitter of greenhouse gases (GHGs). In 2007, its emission was estimated at 60 million tonnes of $CO_2$eq, accounting for 9.1 per cent of national emissions. Of these, 37 million metric tonnes of $CO_2$eq were produced within Mexico City, with transportation, industry, housing and solid waste management serving as the major contributors (43, 22, 13 and 11 per cent, respectively). About 88 per cent of all GHG emissions in Mexico City are attributed to energy consumption in the form of fossil fuels and electricity used in transportation, industry, trade, housing or services.

The Mexico City government recognizes that climate change is now the most serious threat to the ecosystems of the city, with unquestionable socio-economic consequences for the population. Consequently, the city has developed the Mexico City Climate Action Programme 2008–2012, with a total of 26 GHG mitigation actions. If implemented, they will reduce the $CO_2$eq emissions by 4.4 million tonnes a year, which represents 12 per cent of the annual GHG emissions in Mexico City.

*Sources:* Delgado, 2008; Casaubon et al, 2008

**Box 3.5 Contribution to GHG emissions, Cape Town, South Africa**

With a population of about 3.2 million people, the City of Cape Town accounts for 5 per cent of South Africa's total energy consumption and 5.2 per cent of the national electricity consumption. Greenhouse gas (GHG) emission from fuel types used in the city has been significant, with electricity accounting for 69 per cent; petrol, 17 per cent; and diesel, 9 per cent. Other fuels (paraffin, liquid petroleum gas, coal, heavy furnace oil and wood) account for the rest.

The city's energy and climate strategy identifies industry, transport and residential as the major sectors contributing to GHG emissions both within the City of Cape Town and in the entire Western Cape region. As a result, the City of Cape Town has an average annual emission of 6.4 tonnes of $CO_2$ per person (compared to Western Europe at 4.5 tonnes, and the rest of Africa at 0.6 tonnes). The emissions cause visible pollution ('brown haze'). In 2003, air quality monitoring stations recorded 162 days of poor air quality. An earlier *Cape Town Brown Haze Study*, conducted in 1997, attributed 65 per cent of the brown haze to vehicular emission, of which 49 per cent is caused by diesel vehicles and 16 per cent by petrol vehicles.

*Sources:* City of Cape Town, 2006, 2007

*Assessments and comparisons done using the $CO_2$/unit of economic value created have to be viewed with caution. Because of economic linkage between cities, economic value created with relatively low $CO_2$ emissions in one city might depend on the economic value created with much higher emissions in another. For example, while the City of Johannesburg measures 9.6 tonnes $CO_2$/R100,000 [South African rand] compared with 133.6 tonnes $CO_2$/R100,000 for Sedibeng, the City of Johannesburg derives components of its economic value creation through provision of low-energy-intensity services to high-energy-intensity industries in Sedibeng ... the cities cannot be seen in isolation.[109]*

Further assessment of Cape Town's GHG emissions shows that electricity use is responsible for 69 per cent of emissions (see Box 3.5). This high level of emissions is related to the fact that 95 per cent of South Africa's electricity generation is coal fired, with high levels of emissions for given quantities of energy. By sector, transport accounts for 54 per cent of energy use, followed by commerce and indus-

| Town/city | Population | Total $CO_2$ emissions (tonnes) | $CO_2$ per capita (tonnes) | $CO_2$ per 100,000 South African rand value added* |
|---|---|---|---|---|
| City of Cape Town | 3,069,404 | 19,736,885 | 6.4 | 13.7 |
| City of Johannesburg | 3,585,545 | 19,944,863 | 5.6 | 9.6 |
| City of Tshwane | 1,678,806 | 13,537,109 | 8.1 | 12.7 |
| Ekurhuleni | 2,761,253 | 22,917,257 | 8.3 | 24.9 |
| eThekwini | 3,269,641 | 18,405,182 | 5.6 | 15.6 |
| Nelson Mandela | 1,013,883 | 4,754,204 | 4.7 | 13.8 |
| **Total for 'metros'** | **15,378,532** | **99,295,500** | **6.5** | **14.1** |
| uMsunduzi | 562,373 | 3,543,806 | 6.3 | N/A |
| Saldanha Bay | 79,315 | 3,923,771 | 49.5 | 30.2 |
| Sedibeng | 883,772 | 25,257,942 | 28.6 | 133.6 |
| uMhlatuze | 360,002 | 16,816,074 | 46.7 | 140.1 |
| **Total for 'industrial towns/cities'** | **1,885,462** | **49,541,593** | **26.3** | **123.1** |
| Buffalo City | 702,671 | 2,449,144 | 3.5 | 106.9 |
| King Sabata | 421,233 | 713,526 | 1.7 | N/A |
| Mangaung | 662,063 | 2,495,297 | 3.8 | 16.7 |
| Potchefstroom | 129,075 | 634,580 | 4.9 | 15.4 |
| Sol Plaatje | 196,846 | 882,234 | 4.5 | 13.9 |
| **Total for 'non-industrial towns/cities'** | **2,111,888** | **7,174,781** | **3.4** | **15.7** |
| **Total for towns/cities reviewed** | **19,375,882** | **156,011,874** | **8.1** | **13.0** |
| **Total for South Africa** | **46,586,607** | **391,327,499** | **8.4** | **N/A** |

**Table 3.13**

**$CO_2$ emissions from South African urban areas (2004)**

*Notes:* * In 2004, US$1 was worth an average of 6.5 South African rand. N/A = not available.

*Source:* Sustainable Energy Africa, 2006

try (29 per cent), residential (15 per cent) and local authority (2 per cent).[110]

It needs to be emphasized that the responsibility for these emissions is not distributed evenly throughout the urban population. While more affluent urban residents consume resources – including fuel for heating or cooling, petrol or diesel for transportation, food items and consumer goods with high levels of 'embedded carbon' – in similar quantities to urban residents in developed countries, poorer urban residents use very little of these resources. Although this is an issue in all urban centres, it is particularly evident in highly unequal societies.

A recent study in India showed that the average GHG emissions of the wealthiest 1 per cent of the Indian population are 4.52 tonnes $CO_2$eq per annum, or more than four times as much as the 1.11 tonnes $CO_2$eq per annum generated by the poorest 38 per cent of the population.[111] A significant proportion of urban residents in low-income countries have very low levels of GHG emissions because of their limited use of fossil fuels, limited use of electricity, and limited consumption of goods and services that require GHG emissions for their production and transportation. Indeed, many low-income urban dwellers whose livelihoods are based on reclaiming, reusing or recycling waste may actually generate negative emissions through these activities.[112]

Various efforts are being made to reduce greenhouse gas emissions in urban areas in low-income countries, frequently in association with broader goals of improving air quality and implementing more effective solid waste management. In Dhaka (Bangladesh), motorized rickshaws have been banned from using two-stroke petrol engines since September 2002 and have been replaced by engines using compressed natural gas.[113] Although introduced more

as a measure to improve local air quality, compressed natural gas engines can reduce lifetime greenhouse gas emissions by 21 to 26 per cent.[114] Similarly, a range of interventions developed for the Dhaka Metropolitan Development Planning Support System – including environmental and physical quality, urban and infrastructure development, social and economic development, governance, and educational and scientific development – have relevance to climate change mitigation efforts.[115]

## Estimating the global-level urban emissions

Any blanket statements about the total contribution of urban areas or cities to GHG emissions need to be treated with caution. There is no globally accepted definition of an *urban area* or *city*, and there are no globally accepted standards for recording emissions from sub-national areas. In addition, there is little clarity on the relative allocation of responsibility from production- or consumption-based approaches. This is made particularly clear in the South African cases presented above, as well as in the comparison between Japanese and Chinese cities in which vastly different proportions of emissions can be attributed to the manufacturing sector, which produces goods for consumption in many other locations around the world.

At one extreme, it could be argued that urban areas make no contribution to GHG emissions: that the economic and social benefits, and environmental costs, associated with their commercial, industrial and manufacturing sectors are distributed more widely to individuals throughout countries, regions and around the world. In this regard, individuals or sectors could be deemed 'responsible' for certain levels of emissions. At the other extreme, it could be argued that all

A significant proportion of urban residents in low-income countries have very low levels of GHG emissions

Any blanket statements about the total contribution of urban areas or cities to GHG emissions need to be treated with caution

| Sector | Percentage of global GHG emissions | Justification for estimating the proportion of GHGs from cities, from the perspective of the location of activities that produced them | Percentage of GHGs allocated to cities |
|---|---|---|---|
| Energy supply[a] | 25.9 | A high proportion of fossil fuel power stations are not in cities, especially the largest cities.<br>One third to one half of emissions from city-based power stations. | 8.6–13.0 |
| Industry | 19.4 | A large proportion of heavy industry (which accounts for most GHGs from industry) is not located in cities, including many cement factories, oil refineries, pulp and paper mills, metal smelters.<br>Two-fifths to three-fifths of emissions in cities. | 7.8–11.6 |
| Forestry[b] | 17.4 | No emissions assigned to cities. | 0 |
| Agriculture | 13.5 | Some large cities have considerable agricultural output, but mostly because of extended boundaries encompassing rural areas.<br>No emissions assigned to cities. | 0 |
| Transport | 13.1 | Private use of motor vehicles a large part of this. Should commuting by car by those living outside cities be assigned to cities? Should city dwellers driving outside city boundaries be assigned to their city?<br>60 to 70 per cent of emissions assigned to cities. | 7.9–9.2 |
| Residential and commercial buildings | 7.9 | Large sections of middle- and high-income groups in developed countries live outside cities – and a significant and increasing proportion of commercial buildings are located outside cities.<br>60 to 70 per cent of emissions assigned to cities. | 4.7–5.5 |
| Waste and wastewater | 2.8 | More than half of this is landfill methane; but a proportion of this would be released outside urban boundaries from waste generated inside cities.<br>54 per cent of emissions assigned to cities. | 1.5 |
| Total[c] | 100 | | 30.5–40.8 |

*Notes:* a A large part of this is from fossil fuel power stations. Excludes refineries, coke ovens, etc., which are included under industry.
b Land use and land-use changes.
c Total emissions for the GHGs covered by the Kyoto Protocol amounts to 49 billion tonnes of $CO_2$eq.

*Sources:* based on Barker et al, 2007; Satterthwaite, 2008a, p544

**Table 3.14**

**Cities' contribution to global anthropogenic GHG emissions, by sector**

| Cities' share in ... | 2006 (%) | 2030 (%) |
|---|---|---|
| Global energy consumption | 67 | 73 |
| Global energy-related $CO_2$ emissions | 71 | 76 |
| Global anthropogenic GHG emissions | 40–70 | 43–70 |

Source: Walraven, 2009

**Table 3.15**

The contribution of urban areas to various aspects of climate change

human activities other than those directly associated with rural land-use change and agriculture are urban in their character. In this situation, all non-forestry and agriculture emissions could be allocated to urban areas – which constitute 69 per cent of total global emissions according to the IPCC. Neither of these perspectives is valid – but they do highlight that 'drawing the line' as to exactly how urban areas 'contribute' to climate change is a highly subjective process.

From the perspective of the location of production, it has been suggested that *cities* (in this case, excluding smaller towns and other small urban settlements) probably emit between 30 and 40 per cent of all anthropogenic GHG emissions (see Table 3.14). This is based on an analysis of the contribution of different sectors to global emissions, and an assessment of the proportion of each of these that is associated with cities.[116] Table 3.15 provides higher estimates of the contribution of cities to global GHG emissions,[117] although, in this case, cities refer to all urban areas, including towns and other small urban settlements. It has been suggested that various overestimates of urban contributions to climate change are related to an Organisation for Economic Co-operation and Development (OECD) report, which states that 80 per cent of global energy use is linked to cities.[118] Consequently, a production-based figure of 40 to 70 per cent, and a consumption-based figure of 60 to 70 per cent has been presented.[119] The differences between consumption- and production-based approaches will be discussed in greater detail below.[120]

The climatic situation of any given urban area affects the energy demands for heating and cooling

# FACTORS INFLUENCING EMISSIONS

As the previous section showed, the contribution of urban areas to GHG emissions in different countries – and even of different urban areas within the same country – varies greatly. This is due to a variety of factors, including differences in the sources of emissions.[121] This section examines the main factors that affect the sectors generating GHGs – namely, geographic situation, demographic variation, urban form and the types of economic activities. None of these factors operate in isolation, and it is perhaps more appropriate to conceptualize the urban system as a whole – recognizing that any urban area is intricately linked with rural areas, urban areas within the same country, and has international linkages. The discussion also looks at the politics of measuring GHG emissions by discussing alternative approaches, such as ecological versus carbon footprints, production-based versus consumption-based approaches, and individual versus urban drivers of emissions.

The geographical location in relation to natural resources influences the fuels that are used for energy generation and, hence, the levels of GHG emissions

## Geographic situation

Various aspects of geography affect the contribution of urban areas to climate change. These can be broadly categorized as climatic situation, altitude and location in relation to natural resources. The climatic situation of any given urban area affects the energy demands for heating and cooling. High-latitude locations have longer hours of darkness in the winter, requiring additional energy consumption for lighting. High-latitude locations are also colder in winter, with additional heating requirements. Both space and water require heating, and in many countries (e.g. the UK), the heating of water is a major consumer of energy in residential households. Heating requirements are usually met through the direct burning of an energy source such as coal, oil or natural gas. In contrast, space cooling through air conditioning is normally powered by electricity. Urban areas in warmer locations therefore have an emissions profile strongly influenced by the energy source used to generate this.

In Spain, electricity demand shows an increasing trend that can be linked to demographic, social and economic factors. Within this trend, variations in consumption can be seen as a result of particular weather conditions. In the winter months, colder than normal days are associated with increased electricity consumption for heating, whereas in the summer months, warmer than normal days are associated with this for cooling.[122]

In the US, the consumption of fuel oil and natural gas by households is determined primarily by climate. There is a very strong negative correlation between home heating-related emissions and lower temperatures in January – a factor that is, itself, determined by geographical location.[123] In contrast, many locations in the US with particularly hot summers (higher temperatures in July) have higher electricity consumption associated with space cooling. Solely taking these issues into account, it has been noted that areas with moderate temperatures have lower emissions and lower associated expenditure on energy. Although comparable studies have not been undertaken elsewhere, it is likely that this pattern is replicated on a global level, with areas experiencing very hot or cold climatic conditions requiring a greater use of energy for the cooling or heating of residential and commercial buildings.

The geographical location in relation to natural resources influences the fuels that are used for energy generation and, hence, the levels of GHG emissions. This is a factor of transportation costs: where a more efficient source of fuel is available in close proximity to the city or town, it can be used more economically. For example, urban areas that are able to draw on nearby sources of natural gas will emit a smaller volume of GHGs for a given amount of energy than areas that rely on coal for energy. China's continued reliance on coal, which provides 70 per cent of its total energy requirements, is largely due to the abundance of this resource – China has the world's third largest coal reserves and is the largest coal producer in the world. In contrast, countries with larger reserves of natural gas tend to be more reliant on this cleaner source of energy. For example, the UK has increased the proportion of its energy from natural gas from 20 to 34 per cent between 1980 and 2003.[124]

The potential for using renewable sources of energy – and the reductions in GHG emissions associated with this – are also affected by locational factors. Some renewable energy is entirely reliant on natural resources – for example, the availability of large rivers for hydroelectric generation: Rio de Janeiro and São Paulo (in Brazil) have low levels of emissions from electricity generation for this reason. In other locations, particularly in smaller or more arid regions, large-scale hydroelectric generation is not a viable source of energy. Wind, geothermal, tidal and wave energy all rely on natural phenomena existing in particular locations, although solar photovoltaic and solar thermal energy are less tied to specific locations.[125]

Geographical location affects the type of economic activities taking place in urban areas. Similarly, these can both be environmental and social factors of location. Historically, heavy industry has been located close to bulky raw materials, particularly coal and iron ore. The new international division of labour, coupled with increasing reductions in cross-border tariffs, means that spatial variation in the cost of labour are increasingly important in determining the location of manufacturing activities.

The geographic level at which decisions on energy are made can also influence emissions from urban areas. Decisions to construct and operate nuclear power plants are made at the country level, rather than at the city level. Nuclear power generation – assessed purely from a GHG emissions perspective – enables reduced climate impacts from a given amount of energy; yet urban areas themselves are unable to make decisions of this nature. National policy decisions in France and Spain are one reason why city emissions in these countries are lower than those of cities in the UK.

## Demographic situation

The relationship between population growth and GHG emissions is complicated, and varies according to the level of analysis. As can be seen from the wide variations in GHG emissions from countries around the world, population size in itself is not a major driver of global warming. At a global level, the areas experiencing the highest rates of population growth are areas with currently low levels of per capita emissions. As Table 3.16 shows, the developing countries with the highest rates of population growth have had much lower rates of growth in their $CO_2$ emissions than developed countries with much lower rates of population growth. While larger urban populations do lead to increased total sulphur dioxide emissions and transportation energy use,[126] there is no evidence to suggest that this results in higher per capita emissions. Table 3.16 also shows that the increase in $CO_2$ emissions in high-income countries has been significantly higher than in the upper-middle income countries, despite relatively similar rates of population growth.

The demographic composition of a society has a wide range of effects on consumption behaviour and GHG emissions. In some urban areas, changing age structures will affect GHG emissions associated with energy use. Population aging in the US has been shown to cause reductions in

| Income category in 2005 | 1950–1980 | | 1980–2005 | |
|---|---|---|---|---|
| | Growth in population (%) | Growth in CO₂ emissions (%) | Growth in population (%) | Growth in CO₂ emissions (%) |
| Low-income countries | 36.0 | 5.6 | 52.1 | 12.8 |
| Lower-middle income countries | 47.1 | 39.7 | 35.7 | 53.2 |
| Upper-middle income countries | 5.7 | 9.6 | 5.0 | 5.0 |
| High-income countries | 11.2 | 45.1 | 7.2 | 29.1 |

*Source:* Satterthwaite, 2009, p558

**Table 3.16**

**CO₂ emissions, population growth and national income**

labour income and changes in consumption patterns – both of which result in lower GHG emissions. Depending on the factors taken into consideration, these savings can range from 15 to 40 per cent of the emissions that would otherwise be expected. Conversely, in China, estimates of GHG emissions that incorporate aging suggest that this will initially lead to higher emissions (as the proportion of the population in the labour force and, hence, their power to consume increases), followed by reduced emissions (as this proportion declines).[127]

Another demographic change that may affect patterns of GHG emissions from urban areas is the increasing trend towards smaller households. In many developed countries and some developing countries such as China, the average household size has declined – meaning that the number of households has expanded more rapidly than the total population size. Economies of scale are therefore reduced, with the result that the per capita energy consumption of smaller households is significantly higher than that of larger households.[128]

Paradoxically, the slowing of population growth may result in increased emissions, as lower population growth and smaller household sizes may be associated with a rise in the number of separate households and increased disposable income to be spent on consumption. The decline in fertility in Brazil has been associated with a rapid rise in the number of households. The total fertility rate declined from six births per woman during the mid 1960s to replacement level in the mid 2000s. While the population grew at an annual rate of 1.4 per cent between 1996 and 2006, the actual number of households increased at an annual rate of 3.2 per cent. Over the same time period, the number of 'double income no kids' households almost doubled (from 1.1 million to 2.1 million). These households have relatively high incomes and the ability to consume larger quantities of goods and services with associated consequences for their total GHG emissions.[129]

The relationship between population size, population composition and the contribution of urban areas to climate change is therefore complex. As cities grow, they concentrate the demand for fresh water and other natural resources, and concentrate the production of pollutants and GHGs. Cities with considerable local ecological impact – such as Solapur, India, with 1.1 million people but low average consumption – may have much smaller global impacts than similar-sized cities (such as Perth, Australia, or Portland, US). These latter cities have moderated their local ecological impact by importing most of the goods that they consume – potentially producing far larger global impacts.[130]

**Per capita energy consumption of smaller households is significantly higher than that of larger households**

**Paradoxically, the slowing of population growth may result in increased emissions, as lower population growth and smaller households sizes may be associated with a rise in the number of separate households and increased disposable income to be spent on consumption**

Taking this into account, it is not the absolute number of people who live in urban areas that affects the contribution of these areas to climate change. Rather, it is the way in which these areas are managed, and the choices that are made by the urban residents living there that have the greatest effects.

## Urban form and density[131]

Urban form and density are associated with a range of social and environmental consequences. On the one hand, the extremely high densities of many cities in developing countries – particularly in informal settlements and other slums – result in increased health risks, and high levels of vulnerability to climate change and extreme events. At the other extreme, the extremely low densities of many suburban areas in North America are associated with high levels of household energy consumption as a result of sprawling buildings and extensive car usage.

Urban sprawl refers to the increasing geographical spread of urban areas into areas that were not previously built up.[132] In many developing countries, the related process of peri-urbanization is increasingly taking place.[133] These interfaces are affected by some of the most serious problems of urbanization, including intense pressures on resources, slum formation, lack of adequate services such as water and sanitation, poor planning, and degradation of farmland. They are of particular significance in developing countries, where planning regulations may be weak or weakly enforced, and result in areas with complex patterns of land tenure and land use.

At its simplest, urban density is measured by dividing a given population by its area. In the case of urban areas, the widely varying definitions of the spatial extent of these areas lead to a great deal of difficulty in generating comparable statistics for different towns and cities. Dividing the population of a metropolitan area by the administrative area contained within its official boundaries is a highly unreliable measure – particularly for comparisons – because the density will vary according to the definition of these boundaries.[134] In addition, standard measures of density are calculated over an entire land area, without taking into account the levels of

connectivity. In this regard, the gradual transformation of the urban form of Curitiba (Brazil) from a predominantly radial-circular form to a more linear pattern of development has reduced the city's overall density, yet has facilitated the development of a more rapid and effective public transportation system and various other social and environmental benefits.[135]

In general and at a global level, there is strong evidence that urban densities have been declining over the past two centuries.[136] Perhaps the most detailed and compelling assessment of this phenomenon is provided by a World Bank report that records the decline in the average density of developed country cities from 3545 to 2835 people per square kilometre between 1990 and 2000. During the same period, the average urban population density in developing countries declined from 9560 to 8050 people per square kilometre.[137] The reduction in urban densities is likely to continue into the future. It is estimated that the total population of cities in developing countries will double between 2000 and 2030; but their built-up areas will triple from approximately 200,000 square kilometres to approximately 600,000 square kilometres. During the same period, the population of cities in developed countries is projected to increase by approximately 20 per cent, while their built-up areas will increase by 2.5 times: from approximately 200,000 square kilometres to approximately 500,000 square kilometres. These agglomerated figures for developed and developing countries conceal a great deal of regional variation (see Table 3.17). Southeast Asian cities were almost four times as densely populated as European cities, and almost eight times as densely populated as those in North America and Australasia. When disaggregated by national income levels, Table 3.17 shows that cities in low-income countries are more than four times as densely populated as cities in high-income countries.

In summary, spatially compact and mixed-use urban developments have several benefits in terms of GHG emissions:

- Reduced costs for heating and cooling resulting from smaller homes and shared walls in multi-unit dwellings (see above).
- The use of energy systems covering a broader area (e.g. district) for cooling, heating and power generation, as well as lesser line losses related to electricity transmission and distribution. The use of micro-grids to meet local requirements of electricity can create efficiencies in storage and distribution.[138]
- Reduced average daily vehicle kilometres travelled in freight deliveries and by private motor vehicles per capita. Population density increases accessibility to such destinations as stores, employment centres and theatres.[139] It has been found that with all other variables constant except density, a 'household in a neighbourhood with 1000 fewer units per square mile drives almost 1200 miles more and consumes 65 more gallons of fuel per year' over a household in a higher density neighbourhood.[140]

**Sidenotes (left margin):**

There is strong evidence that urban densities have been declining over the past two centuries

Spatially compact and mixed-use urban developments have several benefits in terms of GHG emissions

**Table 3.17**

Average population density of cities' built-up areas

| Region | Persons per square kilometre | |
| --- | --- | --- |
| | 1990 | 2000 |
| Developed countries | 3545 | 2835 |
| Europe | 5270 | 4345 |
| Other developed countries | 2790 | 2300 |
| Developing countries | 9560 | 8050 |
| East Asia and the Pacific | 15,380 | 9350 |
| Latin America and the Caribbean | 6955 | 6785 |
| Northern Africa | 10,010 | 9250 |
| South and Central Asia | 17,980 | 13,720 |
| Southeast Asia | 25,360 | 16,495 |
| Sub-Saharan Africa | 9470 | 6630 |
| Western Asia | 6410 | 5820 |
| High-income countries | 3565 | 2855 |
| Upper-middle income countries | 6370 | 5930 |
| Lower-middle income countries | 12,245 | 8820 |
| Low-income countries | 15,340 | 11,850 |

*Source:* adapted from Angel et al, 2005

## ■ Urban density and greenhouse gas emissions

Urban form and urban spatial organization can have a wide variety of implications for a city's GHG emissions. The high concentrations of people and economic activities in urban areas can lead to economies of scale, proximity and agglomeration – all of which can have a positive impact upon energy use and associated emissions, while the proximity of homes and businesses can encourage walking, cycling and the use of mass transport in place of private motor vehicles.[141] Some researchers suggest that each doubling of average neighbourhood density is associated with a decrease of 20 to 40 per cent in per-household vehicle use with a corresponding decline in emissions.[142] An influential paper published in 1989 suggested that gasoline use per capita declines with urban density, although this relationship weakens once GDP per capita is brought into consideration.[143] It has also been argued that 'by the middle of the century the combination of green buildings and smart growth could deliver the deeper reductions that many believe are needed to mitigate climate change'.[144]

A recent study of GHG emissions in Toronto (Canada) deals with the issue of density explicitly. The study depicts both the overall patterns of GHG emissions and examines how these vary spatially throughout the Toronto Census Metropolitan Area: as the distance from the central core increases, private motor vehicle emissions begin to dominate the total emissions.[145] This pattern is supported by an earlier study, which found that low-density suburban development in Toronto is 2 to 2.5 times more energy and GHG intensive than high-density urban core development on a per capita basis.[146]

Density may also affect household energy consumption. More compact housing uses less energy for heating. For example, households in the US living in single-family detached housing consume 35 per cent more energy for heating and 21 per cent more for cooling than comparable households in other forms of housing. In addition, dense urban areas generate a more intense urban heat-island effect[147] that raises air temperature in a typical US city by 1°C to 3°C over the surrounding rural area. This increases the number of 'cooling days' and decreases the number of 'heating days', with the latter tending to have a greater effect on energy consumption. Consequently, residential buildings in dense urban areas tend to consume lower levels of energy.[148]

Any assessment of changing density and changing emissions needs to take multiple factors into account. It is necessary to assess the GHG emissions of different types of urban development both between different cities and within the same city. While it appears that decreasing urban density may be implicated in increasing GHG emissions, the data are affected by a variety of other variables, including overall income levels. For example, cities in South Asia are not only more densely settled than cities in North America, but also have much lower GHG emissions. The difference in the latter is due much more to income and consumption patterns than to variations in income levels. For example, London's annual emissions declined from 45.1 to 44.3 million tonnes between 1990 and 2006, despite the population growing by 0.7 million people, the built-up area increasing from 1573 to 1855 square kilometres, and urban density decreasing from 6314 to 5405 persons per square kilometre.[149] In this particular situation, per capita GHG emissions appeared to decline at the same time as urban density declined. While it may appear that the decreased density did not influence GHG emissions, in fact the decline in emissions can be attributed to the halving of industrial emissions, as industrial activity has relocated to other parts of the UK or overseas.

Dense urban settlements can therefore be seen to enable lifestyles that reduce per capita GHG emissions through the concentration of services that reduces the need to travel large distances, the better provision of public transportation networks, and the constraints on the size of residential dwellings imposed by the scarcity and high cost of land. Yet, conscious strategies to increase urban density may or may not have a positive influence on GHG emissions and other environmental impacts. Many of the world's most densely populated cities in South, Central and Southeast Asia suffer severely from overcrowding, and reducing urban density will meet a great many broader social, environmental and developmental needs. However, people do often wish to stay in the same location, and improvements can be achieved through upgrading. High urban densities can cause localized climatic effects, such as increased local temperatures.[150] In addition, a variety of vulnerabilities to climate change are also exacerbated by density. Coastal location, exposure to the urban heat-island effect, high levels of outdoor and indoor air pollution, and poor sanitation are associated with areas of high population density in developing country cities.[151] However, these also provide clear opportunities for simultaneously improving health and cutting GHG emissions through policies related to transport systems, urban planning, building regulations and household energy supply.[152]

However, it should be noted that density is just one of a variety of factors that influences the sustainability of urban form. It has been argued that compactness alone is neither a necessary nor sufficient condition for sustainability,[153] and at least seven design concepts for a sustainable urban form have been identified – namely, compactness, sustainable transport, density, mixed land uses, diversity, passive solar design, and greening.[154] Based on these criteria, the 'compact city' model is identified as being most sustainable, followed by the 'eco-city', 'neo-traditional development' and 'urban containment' – although this classification and ranking is based on reviews of literature rather than empirical research. A more complex relationship between land use and GHG emissions involves a model that also takes into account landscape impacts (deforestation, carbon sequestration by soils and plants, urban heat island), infrastructure impacts, transportation-related emissions, waste management-related emissions, electric transmission and distribution losses, and buildings (residential and commercial). There are complex relationships between these factors – for example, denser residential areas may have lower levels of car use, but simultaneously present fewer options for carbon sequestration.[155]

The high concentrations of people and economic activities in urban areas can lead to economies of scale, proximity and agglomeration – all of which can have a positive impact upon energy use and associated emissions

Dense urban settlements ... enable lifestyles that reduce per capita GHG emissions

Conscious strategies to increase urban density may or may not have a positive influence on GHG emissions and other environmental impacts

Although the relationship between urban density and GHG emissions is complex, there are certain directions that can be identified that are of relevance for urban policy. These do not amount to wholesale recommendations in favour of densification, but rather look at strategically assessing population distributions in a manner that contributes to broader goals of climate change mitigation. Encouraging densification at an aggregate level – for example, within administrative boundaries – risks neglecting the important environmental and social roles played by gardens and open spaces. It is also worth considering the different housing needs for individuals at different life stages, and reconsidering the notion of 'housing for life' that has been prevalent in many national housing policies. In this regard, dense settlement patterns may meet the needs of certain groups within society, but not others.

In general, density provides the potential for access to and greater use of public transport and of walking and cycling – where urban space is designed to meet the needs of users. A study of London shows a 'positive link between higher density areas and levels of public transport access across London, which is reflected in the decisions that people make about how to get to work'.[156] It further concludes that 'on balance, people will use public transport where it is available, especially in high density, centrally located areas'. People appear willing to 'trade off' more space in their home for other qualities of a residential area, including personal and property safety, the upkeep of the area, and proximity to shops and amenities.

Localized areas of relatively high densities are required to generate greater efficiencies in the usage of public transportation; but this can be consistent with meeting a variety of other demands from urban residents. Of course, the precise form that these transportation networks – and other urban networks for supplying electricity, water, etc. – should take requires detailed local study. Overall, density is one of several factors that affects energy use (and, by extension, GHG emissions) from towns and cities. Addressing these issues requires ongoing analysis of urban *processes* rather than simply taking a snapshot of urban *form* at a particular moment in time.

The uses of land and spatial distribution of population densities within an urban area define its *structure* or *form*. Urban spatial structures play a major role in determining not only population densities, but also the transportation mode (e.g. the relative importance of public versus private modes) and with it cities' levels of energy use and GHG emissions. While urban structures do evolve with time, driven by changes in the localization of economic activities, real estate developments and population, their evolution is slow and can seldom be shaped by design. The larger the city, the less it is amenable to change its urban structure.

Four urban structures or forms can be distinguished.[157] In the first, *mono-centric*, represented by such cities as New York (US), London (UK), Mumbai (India) and Singapore, most economic activities, jobs and amenities are concentrated in the central business district (CBD). Here authorities should focus on promoting public transport as the most convenient transport mode, for most commuters

travel from the suburbs to the CBD. In the second, *polycentric*, exemplified by such cities as Huston (US), Atlanta (US) and Rio de Janeiro (Brazil), few jobs and amenities are located in the centre and most trips are from suburb to suburb. A very large number of possible travel routes exists, but with few passengers per route. Therefore public transport is difficult and expensive to operate and individual means of transportation or collective taxis are and should be promoted as the more convenient transportation options for users. The third one, *composite (or multiple-nuclei) model*, is the most common type of urban spatial structure containing a dominant centre together with a large number of jobs located in the suburbs. Most trips from the suburbs to the CBD are made and should be promoted by public transport, while trips from suburb to suburb are made with individual cars, motorcycles, collective taxis or minibuses. The fourth, also called *urban village model*, does not exist in the real world, but can be found only in urban master plans. In this model, urban areas contain many business centres, commuters travel only to the centre which is the closest to their residence and have more opportunities to walk or bicycle to work. It is an ideal because it requires less transportation and roads, thus, in theory, dramatically reducing distances travelled, energy used and, as a consequence, emissions of GHGs and other pollutants. However it is not feasible, as 'it implies a systematic fragmentation of labour markets which would be economically unsustainable in the real world'.[158]

## The urban economy

The types of economic activities that take place within urban areas also influence GHG emissions. Extractive activities (such as mining and lumbering) and energy-intensive manufacturing are obviously associated with higher levels of emissions – especially when the energy for these is supplied from fossil fuels. However, there are fewer of these activities in many cities in developed countries, as lower transportation costs and the lower cost of labour elsewhere have encouraged industries to relocate elsewhere. In London, for example, industrial emissions halved between 1990 and 2006, as industrial activity has relocated to other parts of the UK or overseas.[159]

Yet, all urban areas rely on a wide range of manufactured goods (produced within the urban area or elsewhere), and manufacturing areas similarly rely on the services provided by certain urban centres. This relationship can exist within countries. In South Africa (and as noted above), the industrial town of Sedibeng (population of 880,000 and annual per capita emissions of 28.6 tonnes $CO_2$eq) is linked with the services provided by the City of Johannesburg (population 3.6 million and annual per capita emissions of 5.6 tonnes $CO_2$eq) (see Table 3.13). As described above, this process exists across national boundaries with many of the world's cities acting as centres for the trading of commodities and consumption of manufactured goods, while generating few emissions from within their own boundaries. With this in mind, the next section examines alternative approaches on how to measure the emissions from urban areas.

*In general, density provides the potential for access to and greater use of public transport and of walking and cycling*

*Urban spatial structures play a major role in determining ... the transportation mode ... and with it cities' levels of energy use and GHG emissions.*

*The types of economic activities that take place within urban areas ... influence GHG emissions*

The influence of the urban economy on patterns of emissions can be seen in the large variations in the proportion of a city's GHG emissions that can be attributed to the industrial sector.[160] Industrial activities in many rapidly industrializing developing countries (such as China) are responsible for a large proportion of urban GHG emissions. Indeed, while 12 per cent of Chinese emissions were due to the production of exports in 1987, this figure had increased to 21 per cent in 2002 and 33 per cent (equivalent to 6 per cent of total global $CO_2$ emissions) in 2005.[161] A recent paper on this issue describes the situation as follows:

> ... many of the countries in the western world have dodged their own carbon dioxide emissions by exporting their manufacturing to ... China. Next time you buy something with 'Made in China' stamped on it, ask yourself who was responsible for the emissions that created it.[162]

In contrast, GHG emissions from the industrial sector in cities elsewhere are much lower, generally reflecting a transition to service-based urban economies. Industrial activities account for just 0.04 per cent in Washington, DC (US) (largely because of the narrow spatial definition of the District of Columbia); 7 per cent in London (UK); 9.7 per cent in São Paulo (Brazil); and 10 per cent in Tokyo (Japan) and New York City (compared to 29 per cent for the US as a whole). The declining importance of industry in causing emissions is evident in several cities. In Rio de Janeiro, the industrial sector's proportion of emissions declined from 12 per cent in 1990 to 6.2 per cent in 1998; and in Tokyo, it declined from 30 to 10 per cent during the last three decades.[163]

## The politics of measuring emissions

There are striking differences in the contribution of different urban areas to climate change. Measured purely in terms of direct emissions per person from a given urban area, these may vary by a factor of 100 or more. As noted earlier in this chapter, different 'scopes' of emissions may be taken into account[164] (see Tables 3.2 and 3.4). In practice, GHG emission inventories from urban areas that include Scope 3 emissions are very rare. And the extent to which these Scope 3 emissions (i.e. indirect or embodied emissions) are included is very arbitrary and there is no agreement as to a comparable framework to compare emissions of this type between urban areas. If Scope 3 or embodied emissions are included, it is likely that the per capita emission of GHGs allocated to a city will increase significantly – particularly if the city is large, well-developed and with a predominance of service and commercial activities.[165] In addition, it is almost impossible to compile a comprehensive inventory of Scope 3 emissions that takes into account all the consumption of the individuals living in an urban area. In other words, *'emissions can be attributed either to the spatial location of actual release or to the spatial location that generated activity that led to the actual release'*.[166] A detailed Scope 3 inventory should also subtract the embodied energy in goods made in that city and subsequently exported.

The data presented in this chapter show that urban areas with a heavy concentration of industrial and manufacturing activities have high levels of GHG emissions. They also show that wealthier urban areas have high emissions – although these may be lower than non-urban but equally wealthy areas. The per capita emissions of GHGs by individuals including those caused by the goods they consume and wastes they generate vary by a factor of more than 1000 depending on the circumstances into which they were born and their life chances and personal choices. Obviously, their lifetime contribution is also influenced by how long they live. Poorer groups with low annual per capita emissions often have life expectancies of 20 to 40 years less than high-income groups. Unsustainable levels of consumption, which drive the processes of production, are therefore crucial to understanding the contribution of urban areas to climate change. This section thus discusses alternative approaches on how to calculate the contribution of urban areas to climate change, thereby helping to provide a framework for understanding and addressing the root causes of GHG emissions.

As noted above, urban areas in different countries, and even within the same country, have different emissions profiles according to environmental, economic, social, political and legal differences over space and across national boundaries. This influences the balance of *production* and *consumption* of GHGs, as many of the most highly emitting activities have been displaced to rapidly industrializing developing countries. The Kyoto Protocol – and its likely successor treaty – also creates incentives for developed countries[167] to reduce the emissions from within their national boundaries, which may create perverse incentives for raised levels of emissions in developing countries which are not subject to these constraints. However, the principle of 'common but differentiated responsibilities'[168] adopted in the negotiations ought to prevent this from happening. Similarly, as the concept of 'nationally appropriate mitigation actions'[169] becomes adopted at the local level, positive incentives may be created to encourage urban areas to reduce their emissions in contextually appropriate ways.

In particular, political forces and the policy environment – at the global, national and local levels – can be a strong underlying factor in shaping GHG emissions. At the global level, the establishment of national targets for developing countries[170] is an important factor driving reductions in emissions. The implementation of CDM projects – in which emissions reductions in developing countries are supported by developed country actors – can also shape emissions patterns. Where local and national governments in developing countries support CDM activities,[171] this can have a substantial local impact upon local emissions. At the same time, local governments can shape city emissions through several different pathways: through undertaking emissions reductions activities in their own activities (e.g. local authority buildings and vehicle fleets); through changing the legislative environment (e.g. through increased taxation on highly polluting industries or tax incentives encouraging the use of low-carbon technology); and through encouraging behavioural change among citizens (e.g. through mass awareness or educational programmes).[172]

### ■ Ecological footprints versus carbon footprints

One useful approach for calculating GHG emissions from urban areas is to consider ecological footprints. The ecological footprint is a concept that measures the area of the Earth's surface required to provide the consumption needs of an individual, urban area or country. The concept of the ecological footprint recognizes that larger areas of land are required to sustain life inside urban areas than are contained within municipal boundaries of the built-up area – with this area being much larger for wealthy cities.[173] Most cities and regions depend on resources and ecological services – including food, water and the absorption of pollutants – from outside their boundaries; many depend on those from distant ecosystems; and the environmental consequences of urban activities are thus felt globally or in distant regions. Ecological footprint analysis has been used in recent years to develop a related concept: the carbon footprint (see Box 1.1). A full carbon footprint therefore takes into account all the emissions included in 'Scope 1', 'Scope 2' and 'Scope 3', but places a greater emphasis on the indirect emissions from products and services that are consumed but not directly controlled.

Although often used interchangeably, the implications of an emissions inventory and a carbon footprint can therefore be quite different. The *emissions inventory* is derived from the UNFCCC model of national inventories that accounts for GHG emissions produced within a geographically defined boundary. In contrast, the *carbon footprint* is derived from the concept of the ecological footprint, and is focused on the GHG emissions associated with the consumption of goods and services. The origin of GHG emissions is therefore better understood through the use of consumption-based assessments. In relation to ecological footprints, it has been concluded that 'wealthy nations appropriate more than their fair share of the planet's carrying capacity'.[174] Similarly, the use of a consumption-based analysis of emissions, derived from a carbon footprint approach, will help to make it clearer which countries, urban areas and individuals are responsible for more than their fair contribution to global climate change.

### ■ Production-based versus consumption-based approaches

The use of a production-based approach to assessing the contribution of urban areas to GHG emissions can lead to perverse and negative effects. Urban areas will be able to *reduce* their emissions through creating disincentives for *dirty* economic activities that generate high levels of GHGs (e.g. heavy industry), and incentives for *clean* economic activities that generate much smaller emissions (e.g. high-tech industries). This situation can already be seen: many polluting and carbon-intensive manufacturing processes are no longer located in Europe or North America, but have been sited elsewhere in the world to take advantage of lower

labour costs and less rigorous environmental enforcement. Since developing countries are not required to reduce emissions under the UNFCCC, a process of 'carbon leakage' can take place, where emissions are moved rather than reduced.[175] Yet, climate change is a global phenomenon: emissions of a given quantity of $CO_2$ have the same effect on the global climate wherever in the world these are released. From the perspective of global climate change, the consequences are the same irrespective of whether an industry is located within the urban areas or rural areas of a developed or a developing country. The underlying drivers for these emissions are the demands of consumers who desire particular products. Thus, an assessment of the contribution of urban areas to climate change needs to reflect the location of the people making these demands.

At a national level, input–output analyses have been used to show national average per capita carbon footprints. These take into account construction, shelter, food, clothing, mobility, manufactured products, services and trade. National average per capita footprints vary from approximately 1 tonne $CO_2$eq in many African countries to approximately 30 tonnes $CO_2$eq in Luxembourg and the US. The proportion of these emissions attributed to internal consumption varies greatly: with low figures in small city-states and countries with low levels of imports (e.g. 17 per cent in Hong Kong and 36 per cent in Singapore) and high figures in major industrial and manufacturing countries (e.g. 94 per cent in China and 95 per cent in India) and in countries with low levels of imports as a result of poverty (e.g. 90 per cent in Madagascar and Tanzania).[176] This method for measuring *responsibility* shows clearly that the countries with high levels of consumption and imports are responsible for much greater volumes of GHG emissions than a production-based approach would indicate.

The use of a production-based system to assess the contribution of urban areas to climate change diverts attention and blame from the high-consumption lifestyles that drive unsustainable levels of GHG emissions. This system fails to identify the areas in which interventions are required to reduce emissions by focusing attention on only one part of multiple complex commodity chains. In addition, analysing emissions at a city level generates a variety of logistical problems. For instance, there are large information gaps (particularly in developing countries); different information is available at different geographic levels; and political boundaries of cities may change over time and often include both rural and urban populations (as is the case for Beijing and Shanghai in China).[177]

Production-based emissions methodologies therefore distort the responsibility of different cities for generating GHGs. Different types of cities will be affected in different ways by this approach: 'in service-oriented cities, consumption-related emissions are more important than those produced by production'.[178] Consequently, the responsibility of successful production-oriented centres such as Beijing and Shanghai is exaggerated, while that of wealthy service-oriented cities including many cities in North America and Europe is underemphasized. The fact that Beijing and Shanghai have per capita emissions of more than twice the

---

*The use of a consumption-based analysis of emissions ... will help to make it clearer which countries, urban areas and individuals are responsible for more than their fair contribution to global climate change*

*From the perspective of global climate change, the consequences are the same irrespective of whether an industry is located within the urban areas or rural areas of a developed or a developing country*

*Production-based emissions methodologies ... distort the responsibility of different cities for generating GHGs*

Chinese average therefore reflects not only the relative affluence of these cities (and the spatially uneven incorporation of different parts of China into global economic networks), but also the role that they play in manufacturing consumer products that are used elsewhere in China and throughout the world.

In contrast, a consumption-based approach attempts to address the origin of emissions in a more comprehensive manner. This type of accounting system would result in a lower level of GHG emissions to developing countries (with a likely substantial reduction in the GHG emissions allocated to China and Chinese cities), and should – in theory – influence consumers in developed countries to assume responsibility for choosing the best strategies and policies to reduce emissions.[179] Consumption-based mechanisms inherently have greater degrees of uncertainty (as there are many more systems to be incorporated in the final calculation); but they do provide considerable insight into climate policy and mitigation, and should probably be used at least as a complementary indicator to help analyse and inform climate policy.[180] They respond to broader concerns about sustainability by ensuring that as well as improving environmental performance within city boundaries, there is a reduction in the transfer of environmental costs to other people, distant places or future times.[181]

A consumption-based approach can also be benchmarked against global needs to limit GHG emissions to prevent dangerous climate change. The best available estimates suggest that annual global GHG emissions need to be reduced from approximately 50 billion tonnes to 20 billion tonnes $CO_2$eq per year by 2050. With an estimated global population of 9 billion in 2050, this means that individual carbon footprints around the world will have to be at an average of less than 2.2 tonnes per year. In particular, it must be recognized that different locations have access to different sources of energy: and a *fair* allocation of emissions should not mean that individuals or urban areas located in geographical proximity to abundant geothermal or hydroelectric sources of energy are able to *cash in* these spatial advantages to produce greater emissions from other activities. Climate change is, indeed, a global challenge, and needs to be addressed with global solutions.

Recent research has highlighted the role of urban food consumption in generating GHG emissions. The processes involved in the production and distribution of food for urban consumption use significant amounts of energy and also generate substantial GHG emissions. Fruits and vegetables consumed in developed countries often travel between 2500km and 4000km from farm to store.[182] In North America, for example, the average food product in the supermarket has travelled 2100km before ending up on a shelf and the food system accounts for some 15 to 20 per cent of the energy consumption in the US.[183] Research has also shown that a basic diet composed of imported ingredients can use four times the energy and produce four times the GHG emission than an equivalent diet with local ingredients. The potential for localized urban food production and consumption in promoting energy efficiency and reducing GHG emissions is clear.

However, this needs to be seen alongside the development benefits that the export of agricultural products can bring to developing countries. Air-freighted products are frequently seen as major problems in the emission of GHGs; yet the issues are much more complex. In the UK, fresh produce air-freighted from Africa is responsible for less than 0.1 per cent of national emissions – and the emissions from sub-Saharan African countries are minuscule in the first place. At the same time, more than 1 million African livelihoods are supported by growing this produce.[184] In addition, some agricultural practices that reduce 'food kilometres' – such as using greenhouses to grow tropical crops in temperate latitudes – can have a larger impact upon emissions than the distance travelled.

Distinct challenges are associated with consumption-based approaches to measuring the contribution of individuals and urban areas to climate change; yet these can provide considerable insights into climate policy and mitigation.[185] In practice, both production- and consumption-based approaches will continue to be required. Table 3.18 shows the main driving forces for GHG emissions from both perspectives. In many sectors – particularly in energy, transport, and residential and commercial buildings – interventions to address production-related emissions are similar to those addressing consumption-related emissions. In terms of industry, however, a consumption-based approach places an added emphasis on the global dimensions of emissions – spreading the net wider in terms of where the impacts of individual activities are actually felt. Addressing emissions from a consumption-based approach is therefore much more about reducing emissions rather than merely shifting them elsewhere.

Consumption-based approaches help to ensure that the allocation of responsibility for GHG emissions simultaneously addresses concerns of climate, environmental and gender justice. Of course, global action is required to reduce the risks of climate change – yet the burden for meeting this goal should not fall on individuals or urban areas that have little responsibility for it.[186] Rather, a consumption-based analysis ensures that the responsibility for addressing this problem lies with the individuals, urban areas and countries who have the greatest responsibility for causing it. Similarly, production-based inventories mask the gendered nature of individual energy-use patterns.[187] Indeed, one study suggests that more men than women own cars in Sweden, and concludes that 'if women's consumption levels were to be the norm, both emissions and climate change would be significantly less than today'.[188]

### ■ Individual versus urban drivers of emissions

The preceding discussion makes it clear that consumption is perhaps the most important driver of GHG emissions. In this regard, individuals can be seen as the basic unit affecting emissions. It is the consumption choices and behaviour of individuals that ultimately lead to the use of energy and the production of GHGs. However, it needs to be stressed that the choices that individuals make are shaped by structural forces in the areas in which they live. For example, individuals living in urban areas with effective integrated public

**Consumption-based mechanisms ... respond to broader concerns about sustainability by ensuring that ... there is a reduction in the transfer of environmental costs to other people, distant places or future times**

**Climate change is ... a global challenge, and needs to be addressed with global solutions**

**Consumption-based approaches help to ensure that the allocation of responsibility for GHG emissions simultaneously addresses concerns of climate, environmental and gender justice**

| Sector | What drives growing GHG emissions in urban areas? | |
| | Production perspective | Consumption perspective |
| --- | --- | --- |
| Energy supply: | A large proportion comes from fossil fuel power stations – hence, a growth in electricity provision from high GHG-emitting sources. Many large fossil fuel power stations are located outside urban areas; but the GHG emissions from the electricity used in urban areas are usually allocated to these urban areas. | GHGs from energy supply now assigned to consumers of energy supplies/electricity, so growth in GHG emissions is driven by increasing energy use; consumers are also allocated the GHGs from the energy used to make and deliver the goods and services that they consume. |
| Industry: | Growing levels of production; growing energy intensity in what is produced; importance of industries producing goods whose fabrication entails large GHG emissions (e.g. motor vehicles). | GHGs from industries and from producing the material inputs that they draw on no longer allocated to the enterprises that produce them, but rather to the final consumers of the products, so again GHG growth driven by increased consumption. |
| Forestry and agriculture: | Many urban centres have considerable agricultural output and/or forested areas, but mostly because of extended boundaries that encompass rural areas; from the production perspective, GHGs generated by deforestation and agriculture are assigned to rural areas. | GHGs from these no longer allocated to rural areas (where they are produced), but rather to the consumers of their products (many or most in urban areas); note how energy intensive most commercial agriculture has become; also the high GHG implications for preferred diets among high-income groups (including imported goods, high meat consumption, etc.). |
| Transport: | Growing use of private motor vehicles; increases in average fuel consumption of private motor vehicles; increased air travel (although this may not be allocated to urban areas). | As in the production perspective; GHG emissions from fuel use by people travelling outside the urban area they live in are allocated to them (thus, includes air travel); also concern for GHG emissions arising from investment in transport infrastructure. |
| Residential/ commercial buildings: | Growth in the use of fossil fuels and/or growth in electricity use from fossil fuels for space heating and/or cooling, lighting and domestic appliances. | As in the production perspective, but with the addition of GHG emissions arising from construction and building maintenance (including the materials used to do so). |
| Waste and wastewater: | Growing volumes of solid and liquid wastes and of more energy-intensive waste. | Large and often growing volumes of solid and liquid wastes with GHGs; these are allocated to the consumers who generated the waste, not to the waste or waste dump. |
| Public sector and governance: | N/A | Conventional focus of urban governments on attracting new investment, allowing urban sprawl and heavy investment in roads, with little concern for promoting energy efficiency and low GHG emissions. |

*Notes:* For a discussion of mitigation action based on these two perspectives, see Table 5.11. N/A = not available.

*Source:* based on Satterthwaite, 2009, pp548–549

transportation systems or safe, well-maintained bicycle pathways will be much more able to reduce distances travelled by car. As an increasing proportion of the world's population lives in urban areas, the choices that are made in relation to investments in urban infrastructure will have a growing role in determining future GHG emissions.

In developed countries, high levels of wealth and disposable income lead to generally high levels of consumption and, hence, GHG emissions, as is evident from some of the national inventories discussed earlier in this chapter.[189] As was also discussed, the economies of scale and the advantages of density mean that urban residents in these countries tend to generate fewer GHG emissions than the national average – at least from a production perspective. Individual drivers of emissions in urban areas in developed countries are still – obviously – related to personal consumption habits. Yet, at the same time, the consumption patterns of wealthy residents in developing countries also drive up national GHG emissions. It should also be noted that in developing countries, average incomes in urban areas are often substantially higher than in rural areas. Individuals and households with higher incomes – and greater consumption – are therefore likely to be concentrated in these urban areas.

The behaviour of urban residents is shaped by cultural and social contexts. These factors can drive individual choices that affect emissions, including the choice of car, decisions about transportation modes and the ways in which energy is used at home (switching off lights, managing heating and cooling), all of which can make a difference to

urban emissions.[190] More broadly, the values placed on leading more self-sufficient lives can affect a wide range of consumption decisions and, therefore, emissions generation.

These contexts include gender roles and expectations. On average, women tend to contribute less to climate change as an outcome of consumption patterns, social roles and pro-environmental behaviour.[191] In general, people living in poverty tend to contribute less to climate change, and more women than men in almost all societies live in poverty. Prescribed gender roles mean that women tend to participate in different activities than men, and frequently travel less for business purposes. In addition, there is evidence that women in developed countries are more likely to consider the environmental impact of purchasing decisions. However, this analysis needs to be tempered by the fact that for various activities – particularly household services such as heating and food preparation – it is impossible to disaggregate the relative contribution of different members within the same household. Yet, despite their lower contribution to climate change, women are more likely to be affected by its impacts.[192]

However, these individual choices need to be seen in the context of the provision (or lack) of particular forms of infrastructure that can lead to marked differences in urban emissions. The same per capita electricity consumption can give widely diverging per capita GHG emissions based on the energy pathways adopted at the urban and national level. Among the factors leading to Tokyo's lower emissions than those in Beijing and Shanghai are its efficient urban infrastructure, greater reliance on lower-emitting sources of energy generation, and more efficient end-use technology, as well as the different types of industrial activities taking place.[193] Independently of the level of affluence, a well-managed city with a good public transportation system, whose population has access to water and sanitation, to adequate health services and to a good quality of life, is likely to have fewer problems in dealing with a wide range of environmental challenges – including climate change – than a city that is poorly managed.

Most US cities have three to five times the gasoline use per person of most European cities; yet it is difficult to see that Detroit has five times the quality of life of Copenhagen or Amsterdam. Indeed, wealthy, prosperous and desirable cities can have relatively low levels of fuel consumption per person.[194] Most European cities have high-density centres where walking and bicycling are efficient and pleasant modes of transport and public transportation is often well-planned and effective. In this regard, well-planned and well-governed cities are central to delinking high living standards and high quality of life from high consumption and high GHG emissions.[195]

However, it should be remembered that cities and towns also contain areas that have high concentrations of poverty and vulnerability, and many residents of these areas will have extremely low emissions. For low-income house-holds in most developing countries, recent demographic and health surveys show that fuel use is still dominated by charcoal, firewood or organic waste. Where access to this is commercialized – as is the case in many urban centres – total fuel use among low-income residents will be low because of its high cost. If urban households are so constrained in their income levels that families can afford only one meal a day, their consumption will be generating only miniscule amounts of GHGs. Moreover, low-income urban households use transport modes (walking, bicycling or public transporta-tion) with no or low emissions – most of which is used to more than full capacity.[196]

## CONCLUDING REMARKS AND LESSONS FOR POLICY

Activities taking place in urban areas – including the actions of individual urban residents – generate a range of GHGs that contribute to climate change. Assessing the contribu-tion of cities to climate change is an important step in developing locally appropriate mitigation actions. The recent launch of a standard methodology that can be used by urban areas around the world to produce comparable data is an important component of this. Yet, as has been shown in this chapter, assessing the contribution of cities to climate change is not a straightforward process, and there has been substantial debate about the proportion of global emissions that can or should be attributed to urban areas. This analysis leads to several key messages: the need for a better under-standing of the nature of emissions from cities; the considerable differences in GHG emissions between cities and the wide range of factors contributing to this; the substantial differences in responsibility for GHG emissions

from different groups of people within cities; and the impor-tance of examining the underlying drivers of emissions.

There has been an increasing debate about the proportion of global emissions that can or should be attrib-uted to urban areas. This is partially due to the absence of a standardized methodology that has been globally agreed as representing the emissions for which a city is 'responsible'. But it is also compounded by variations in the definition of 'urban areas' between different countries, the ways in which urban boundaries are defined, and the quality of data avail-able. The main sectors for which emissions can be assessed – energy for electricity, transportation, commercial and residential buildings, industry, waste, agriculture, land-use change and forestry – are all relevant to urban areas, which rely on goods, services and processes taking place both inside and outside their boundaries.

There are large differences in GHG emissions between countries and cities around the world, with per capita emissions varying by a factor of 100 or more between the lowest- and highest-emitting countries. There are a variety of factors influencing the total and per capita emissions of a city, including geographic situation (which influences the amount of energy required for heating and lighting), demographic situation (related to both total population and household size), urban form and density (sprawling cities tend to have higher per capita emissions than more compact ones), and the urban economy (the types of activities that take place, and whether these emit large quantities of GHGs).

However, there are more fundamental underlying factors affecting emissions, primarily related to the wealth and consumption patterns of urban residents. If consump-tion is taken into account, the emissions from wealthy cities increase substantially, while those from manufacturing cities in developing countries decline. A consumption-based approach has substantial value when considering global emissions reductions – as it removes the incentive simply to 'move' the location of production to countries that are not bound by specific carbon emissions reduction targets. These underlying drivers are inevitably complex, contextually specific, and contingent on a wide range of structural, social, economic and political variables. Reducing urban emissions requires recognizing this complexity and addressing it accordingly.[197]

In turn, these key findings generate several messages for policy at the global, national and local levels. The impor-tance of cities in directly and indirectly generating GHG emissions indicates that there should be a more central role for sub-national and urban governments in global responses to climate change. Several global networks of cities have been formed with the intention of reducing GHG emissions, sharing knowledge and engaging in advocacy within the UNFCCC. However, there are limited pathways for cities to engage directly in global climate change policy or to receive financing for mitigation activities. Addressing this will require changes in both global and national policy, as national governments will need to recognize the need for cities to act in this arena and provide appropriate legislative frameworks.

Most US cities have three to five times the gasoline use per person of most European cities; yet it is difficult to see that Detroit has five times the quality of life of Copenhagen or Amsterdam

The importance of cities in directly and indirectly generating GHG emissions indicates that there should be a more central role for sub-national and urban governments in global responses to climate change

A combination of
regulations ... and
incentives ... can
help to encourage
businesses in cities
to operate in a way
that reduces their
contribution to
climate change

Addressing the
challenge of climate
change in cities will
require citizens,
civil society, the
private sector, local
and national
governments, and
international
organizations to
work together in
partnerships

The assessment of the contribution of cities to climate change provided in this chapter also highlights some of the most important areas for responses by city authorities. First, it has been shown that activities undertaken directly by cities can be substantial producers of GHGs. City authorities are often responsible for operating large fleets of vehicles, large numbers of buildings and facilities such as waste disposal sites. These can all produce large amounts of GHGs, yet can also be modified to reduce their contributions.

Second, urban form and the urban economy have been shown to be key factors influencing emissions at the city level. Through their responsibilities for land-use planning and attracting investment, city authorities can help to shape the policy environment within which a range of other stakeholders act. Encouraging relatively dense urban settlements can reduce distances travelled by urban residents and can make public transportation a more appealing prospect. A combination of regulations (e.g. in relation to commercial and industrial energy standards) and incentives

(e.g. to support buildings with 'green roofs' or passive solar heating) can help to encourage businesses in cities to operate in a way that reduces their contribution to climate change.

Finally, local levels of government are appropriately positioned to engage directly with citizens to shape behaviour. This chapter shows the importance of individual consumption patterns in contributing to GHG emissions from both within and outside city boundaries. City authorities and civil society can help to generate awareness of the implications of consumption decisions, and can encourage individual urban residents to act in a less carbon-intensive manner. Addressing the challenge of climate change in cities will require citizens, civil society, the private sector, local and national governments, and international organizations to work together in partnerships. Local authorities are located at a crucial nexus for engaging with these different groups and playing a leading role in reducing the contribution of cities to climate change.

# NOTES

1    UNFCCC, Article 3. See also Chapter 2.
2    A brief overview of such networks is provided in Chapter 2.
3    UNEP et al, 2010.
4    For a discussion of the characteristics of the main GHGs, see Chapter 1.
5    UNFCCC, 2004.
6    IPCC, 2006.
7    Dodman, 2009.
8    See the section on 'Factors influencing emissions'.
9    WRI/WBCSD, undated. The procedures of this protocol have been adopted by a wide range of private-sector companies (for more details, see www.ghgprotocol.org).
10   Scope 1 emissions represent direct emissions from within a given geographical area; Scope 2 emissions are those associated with electricity, heating and cooling; and Scope 3 emissions include those that are indirect or embodied.
11   See also http://webapps01.un.org/dsd/partnerships/public/partnerships/1670.html.
12   ICLEI, 2008.
13   Fugitive emissions are 'intentional or unintentional release of ... [GHGs, which] may occur during the extraction, processing and delivery of fossil fuels to the point of final use' (IPCC, 2006, p4.6).
14   As discussed below in the section on 'The scale of urban emissions'.
15   See note 10.
16   As discussed below in the section on 'The politics of measuring emissions'.
17   These are discussed in more detail below in the section on

'The politics of measuring emissions'.
18   Any new baseline inventory is likely to include aspects of both WRI/WBCSD's Corporate Accounting and Reporting Standard and ICLEI's International Local Government GHG Emissions Analysis Protocol. This will thus take into account the direct emissions from a city, as well as a selected component of cross-boundary emissions included within the WRI/WBCSD concepts of Scope 2 and Scope 3 (Kennedy et al, 2009a).
19   UNEP et al, 2010.
20   Kates et al, 1998.
21   Forstall et al, 2009.
22   Satterthwaite, 2007.
23   Parshall et al, 2009, 2010.
24   These issues are discussed further in the sections below on 'The scale of urban emissions' and 'Factors influencing emissions'.
25   IEA, 2010, p35.
26   IEA, 2010, pp24–25.
27   IPCC, 2007e.
28   Sims et al, 2007.
29   Kennedy et al, 2009b.
30   World Nuclear Association, 2010.
31   Sustainable Energy Africa, 2006.
32   Dhakal, 2009.
33   Sims et al, 2007.
34   Nuclear power plants have very large levels of embedded energy in the building materials and plant construction and decommissioning, as well as in the processing and storing of radioactive waste – much of which comes from fossil fuels.
35   Although this may, in part, be

changing now – at least in some countries – in response to the need to reduce the dependence of electricity generation on fossil fuels in light of the increasing concern over climate change.
36   These issues are discussed in more detail in Chapter 5.
37   Sustainable Energy Africa, 2006.
38   Satterthwaite and Sverdlik (2009), citing data from Legros et al (2009).
39   This section draws extensively on Dodman (2009).
40   Barker et al, 2007.
41   Parshall et al, 2009.
42   Romero Lankao et al, 2009a.
43   Ewing et al, 2008.
44   Johnsson-Latham, 2007.
45   Compiled from data presented in Newman (2006) and Dodman (2009).
46   Darido et al, 2009.
47   Compiled from data presented in Newman (2006) and Dodman (2009).
48   Ewing et al, 2008.
49   Ewing et al, 2008.
50   Kutzbach, 2009.
51   Takeuchi et al, 2007.
52   Wright and Fulton, 2005, Figure 1.
53   Wright and Fulton, 2005.
54   Unpublished document, Dar es Salaam City Corporation.
55   WHO, 2004. See also UN-Habitat 2007, Chapter 9.
56   Woodcock et al, 2007.
57   Bloomberg and Aggarwala, 2008.
58   Kahn Ribeiro et al, 2007.
59   Kahn Ribeiro et al, 2007.
60   Kahn Ribeiro et al, 2007.
61   Mayor of London, 2007.
62   City of New York, 2009.
63   Barker et al, 2007.

64   Kennedy et al, 2009b.
65   Parshall et al, 2009.
66   Ewing et al, 2008.
67   Markham, 2009.
68   Gupta and Chandiwala, 2009, p4. The rest of the emissions were from transport (33 per cent), industrial processes (22 per cent) and agriculture (1 per cent).
69   Gupta and Chandiwala, 2009.
70   Yuping, 2009.
71   Sykes, 2009.
72   See www.statcompiler.com, last accessed 12 October 2010.
73   Gupta and Chandiwala, 2009, p11.
74   Gupta and Chandiwala, 2009, p11.
75   This section draws extensively on Dodman, 2009.
76   Bai, 2007
77   Kennedy et al, 2009b.
78   Sustainable Energy Africa, 2006.
79   Dhakal, 2009.
80   Ru et al, 2009.
81   Barker et al, 2007.
82   See Chapter 2.
83   Kennedy et al, 2009b.
84   Dodman, 2009.
85   See below in the section on 'The urban economy'.
86   Hardoy et al, 2001.
87   See section below on 'Factors influencing emissions'.
88   Rogner et al, 2007.
89   UN, undated.
90   Satterthwaite, 2009.
91   Harvey, 1993.
92   VandeWeghe and Kennedy, 2007.
93   That is, including counties such as Arlington and Alexandria which are located in the neighbouring state of Virginia.
94   City of New York, 2009.

95  United States Department of Energy, 2008.
96  City of New York, 2009.
97  City of New York, 2007.
98  City of New York, 2007.
99  City of New York, 2009.
100 City of New York, 2009.
101 Heede, 2006.
102 Brown et al, 2008.
103 Mayor of London, 2007.
104 Baldasano et al, 1999.
105 Gomes et al, 2008.
106 See Table 1.4.
107 The term 'nationally appropriate mitigation actions' was first used in the Bali Action Plan, the main outcome of COP-13, and recognizes that different countries may take different nationally appropriate action on the basis of equity and in accordance with the principle of 'common but differentiated responsibilities and respective capabilities' (see Chapter 2).
108 Patricia Romero Lankao, pers comm., 2009.
109 Sustainable Energy Africa, 2006, p83.
110 PADECO, 2009a.
111 Ananthapadmanabhan et al, 2007. The two categories refer to the 10 million people (1 per cent of the population) in India who earn more than 30,000 rupees (approximately US$700) per month, and the 432 million people (38 per cent of the population) who earn less than 3000 rupees (approximately US$23) per month.
112 Satterthwaite, 2009.
113 PADECO, 2009b.

114 Wang and Huang, 1999.
115 Roy, 2009.
116 Satterthwaite, 2008a.
117 Walraven, 2009.
118 OECD, 1995.
119 Walraven, 2009.
120 See the section on 'The politics of measuring emissions'.
121 See the section on 'The sources of greenhouse gas emissions'.
122 Valor et al, 2001.
123 Glaeser and Kahn, 2008.
124 Energy Information Administration, undated.
125 REN21, 2009.
126 Romero Lankao et al, 2009a.
127 Dalton et al, 2008.
128 Jiang and Hardee, 2009.
129 Martine, 2009.
130 Newman, 2006.
131 This section draws significantly on Dodman, 2009.
132 Gottdiener and Budd, 2005.
133 McGregor et al, 2006.
134 Angel et al, 2005.
135 Rabinovitch, 1992.
136 UNFPA, 2007.
137 Angel et al, 2005.
138 Brown et al, 2008, pp11–12.
139 Newman and Kenworthy, 1999.
140 Brown et al, 2008, p12.
141 Satterthwaite, 1999.
142 Gottdiener and Budd, 2005.
143 Newman and Kenworthy, 1989.
144 Brown and Southworth, 2008.
145 VandeWeghe and Kennedy, 2007.
146 Norman et al, 2006.
147 'The "urban heat island" effect is caused by day time heat being retained by the fabric of the buildings and by a reduction in cooling vegetation... In

tropical cities, the mean monthly urban heat island intensities can reach 10°C by the end of the night, especially during the dry season' (Kovats and Akhtar, 2008, p165).
148 Ewing et al, 2008.
149 Mayor of London, 2007.
150 Coutts et al, 2008.
151 Campbell-Lendrum and Corvalan, 2007.
152 The impacts of climate change on cities are discussed further in Chapter 4.
153 Neuman, 2005.
154 Jabareen, 2006.
155 Andrews, 2008.
156 Burdett et al, 2005, p4.
157 Bertaud et al, 2009.
158 Bertaud et al, 2009, p29. At least two reasons help to explain this: companies do not hire based upon who lives within their business areas, and economic realities prevent people from restricting their job searches to only those businesses that are within walking or biking distances from their homes.
159 Mayor of London, 2007.
160 See the section on 'Industry' above.
161 Weber et al, 2008.
162 Walker and King, 2008.
163 Compiled from data presented in Newman (2006) and Dodman (2009).
164 See note 10.
165 Dhakal, 2008.
166 VandeWeghe and Kennedy, 2007.
167 Referred to as 'Annex I countries' in the Kyoto

Protocol.
168 See Chapter 2.
169 See note 110.
170 Referred to as 'non-Annex I countries' in the Kyoto Protocol.
171 As seen in the case of São Paulo, Brazil (see Box 3.3; and Dubeux and La Rovere, 2010).
172 This is discussed in more detail in Chapter 5.
173 Rees, 1992; Rees and Wackernagel, 1998; Wackernagel et al, 2006; Girardet, 1998.
174 Rees, 1992, p121.
175 Hertwich and Peters, 2009.
176 Hertwich and Peters, 2009.
177 Dhakal, 2004.
178 Bai, 2007, p2.
179 Bastianoni et al, 2004.
180 Peters, 2008.
181 Satterthwaite, 1997b.
182 Halweil, 2002; Murray, 2005.
183 Hendrickson, undated.
184 Garside et al, 2007.
185 Peters, 2008.
186 Adger, 2001.
187 Terry, 2009.
188 Johnsson-Latham, 2007.
189 See the section on 'The scale of urban emissions'.
190 Dhakal, 2008.
191 Women's Environment Network, 2010.
192 Patt et al, 2009. See also Chapters 4 and 6.
193 Dhakal, 2008.
194 Newman, 2006.
195 Satterthwaite, 2008a.
196 Satterthwaite, 2009.
197 See Chapter 5.

# CHAPTER 4

# THE IMPACTS OF CLIMATE CHANGE UPON URBAN AREAS

Climate change impacts are now well documented and technological advancement has led to a clearer understanding of future risks and impacts. With increasing urbanization, understanding the impacts of climate change upon the urban environment will become ever more important. Evidence is mounting that climate change presents unique challenges for urban areas and their growing populations. Where urban areas grow rapidly without regard to current and future resource demands and climate change, large numbers of people and their assets can find themselves vulnerable to a range of disruptive and damaging risks.

These impacts extend far beyond the physical risks posed by climate change, such as sea-level rise and extreme weather events. Cities could face difficulties in providing even the most basic services to their inhabitants as a result of climate change. Climate change may affect water supply, ecosystem goods and services, energy provision, industry and services in cities around the world. It can disrupt local economies and strip populations of their assets and livelihoods, in some cases leading to mass migration. Such impacts are unlikely to be evenly spread among regions and cities, across sectors of the economy or among socio-economic groups. Instead, impacts tend to reinforce existing inequalities; as a result, climate change can disrupt the social fabric of cities and exacerbate poverty.

Although there is a burgeoning literature documenting climate change impacts in various cities, there are few comprehensive studies that evaluate the wider implications for cities across the globe. The purpose of this chapter is to identify and discuss the impact of climate change on cities, where an 'impact' is defined as a specific effect on natural or human systems, either positive or negative, that results from exposure to climate change.[1] The first section describes the physical climate change risks faced by cities and the extent of their variation across cities. 'Risk' is defined here as the combination of the magnitude of the impact with the probability of its occurrence.[2] The direct and indirect physical, economic, social and health impacts of these changes in cities are then reviewed in the context of existing vulnerabilities. Accordingly, impacts upon urban physical infrastructure, economies, public health and security are discussed, keeping in mind the differential impact of climate change upon specific vulnerable groups. The chapter then identifies key indicators of vulnerability to climate change for urban residents and cities themselves. Finally, the last section offers some conclusions and lessons for policy.

## CLIMATE CHANGE RISKS FACING URBAN AREAS

Atmospheric and oceanic warming as a result of human activities has been observed over the past several decades.[3] Climate research has illuminated the link between global warming and the alteration of the Earth's water cycle, which has led to changes in precipitation frequency and intensity, cyclone activity, glacial melt and sea-level rise. These physical changes, and the associated responses of ecosystems and economies, have discernible implications for cities worldwide, although these implications are characterized by wide geographical variation. Many of these changes assume a gradual building of climate impacts and are becoming a reality already; however, a not yet fully explored implication relates to the possible effects of abrupt climate change events (see Table 4.1).

This section describes the observed and predicted trends and geographical variations in physical climate change risks that confront urban settlements, including sea-level rise, tropical cyclones, heavy precipitation events, extreme heat events and drought. The local conditions generated by cities as a result of heat-island effects are also discussed, underscoring the exacerbated risks and unique challenges faced by the urban environment. The discussion in this chapter has been restricted to risks that have direct and indirect impacts upon urban settlements, and can be addressed through local planning and governance.

### Sea-level rise

Sea-level rise refers to the increase in the mean level of the oceans.[4] Average sea levels have been rising around the world during recent decades, but with significant regional variation. The average rate of rise accelerated from 1.8mm per year between 1961 and 2003 to 3.1mm per year between 1993 and 2003.[5] Sea-level rise has occurred fastest in the central Pacific region away from the Equator, the

*Climate change presents unique challenges for urban areas and their growing populations*

**Table 4.1**

Projected impacts
upon urban areas of
changes in extreme
weather and climate
events

| Climate phenomena | Likelihood | Major projected impacts |
|---|---|---|
| Fewer cold days and nights | Virtually certain | Reduced energy demand for heating |
| Warmer and more frequent hot days and nights over most land areas | Virtually certain | Increased demand for cooling |
| Warmer temperatures | Virtually certain | Reduced disruption to transport due to snow, and ice effects on winter tourism<br>Changes in permafrost, damage to buildings and infrastructures |
| Warm spells/heat waves: frequency increases over most land areas | Very likely | Reduction in quality of life for people in warm areas without air conditioning; impacts upon elderly, very young and poor, including significant loss of human life<br>Increases in energy usage for air conditioning |
| Heavy precipitation events: frequency increases over most areas | Very likely | Disruption of settlements, commerce, transport and societies due to flooding<br>Significant loss of human life, injuries; loss of, and damage to, property and infrastructure<br>Potential for use of rainwater in hydropower generation increased in many areas |
| Areas affected by drought increase | Likely | Water shortages for households, industries and services<br>Reduced hydropower generation potentials<br>Potential for population migration |
| Intense tropical cyclone activity increases | Likely | Disruption of settlements by flood and high winds<br>Disruption of public water supply<br>Withdrawal of risk coverage in vulnerable areas by private insurers (at least in developed countries)<br>Significant loss of human life, injuries; loss of, and damage to, property<br>Potential for population migration |
| Increased incidence of extreme high sea level (excludes tsunamis) | Likely | Costs of coastal protection and costs of land-use relocation increase<br>Decreased freshwater availability due to saltwater intrusion<br>Significant loss of human life, injuries; loss of, and damage to, property and infrastructure<br>Potential for movement of population |

northeast Indian Ocean and in the North Atlantic along the coast of the US. The Equatorial western Pacific, central Indian Ocean and Australia's northwest coast have experienced the lowest rates of rise.[6] The Intergovernmental Panel on Climate Change (IPCC) predicts that global sea levels will continue to rise anywhere from 0.18 to 0.59m above 1980 to 1990 levels by the end of the 21st century.[7]

Thermal expansion, or the increase in volume of ocean water as it warms, is considered to be the leading cause of sea-level rise; but melting ice sheets may become ever more important in the future.[8] An additional factor contributing to rising sea levels is melting ice from glaciers and land masses such as Greenland and Antarctica. Since 1978, the total area of Arctic sea ice has declined by an average of 2.7 per cent each decade.[9] Satellite surveys over West Antarctica show glacial melting consistent with a rate of sea-level rise of 0.2mm per year and indicate that melting has accelerated during the early 2000s compared to the late 1990s.[10] Estimates of sea level rise due to ice loss from Antarctica and Greenland from 1993 to 2003 are about 0.21mm per year for both; but loss of these sheets in the future, even partially, could greatly alter the projections of sea-level rise.[11]

Studies of past warming events suggest that the Antarctic and Greenland ice sheets melt rapidly in response to warming, and could contribute to sea-level rise exceeding 1m per century.[12] Given that some physical processes of glacial ice melt are not yet well understood by climate scientists, it has been difficult to provide an estimate of the upper bounds of sea-level rise.[13] When considering what is known about glacial ice dynamics in tandem with the record of past ice-sheet melting, however, it is possible that the rate of future melting and related sea-level rise could be faster than widely thought. There may be temperature thresholds or 'tipping points' that accelerate melting to rates not yet experienced in modern times.

Sea-level rise and its
associated impacts
will, by the 2080s,
affect five times as
many coastal
residents as they did
in 1990

The direct effects of sea-level rise include increased storm flooding and damage, inundation, coastal erosion, increased salinity in estuaries and coastal aquifers, rising coastal water tables and obstructed drainage. However, a great many indirect impacts are also probable (e.g. changes in the functions of coastal ecosystems and in the distribution of bottom sediments). Since ecosystems such as wetlands, mangrove swamps and coral reefs form natural protections for coastal areas, changes to or loss of these ecosystems will compound the dangers faced by urban coastal areas.

Sea-level rise is a serious concern for coastal cities as rising water levels and storm surges can cause property damage, displacement of residents, disruption of transportation and wetland loss. This is especially so in the low-elevation coastal zone which, as indicated in Chapter 1, refers to the continuous area along coasts that is less than 10m above sea level. It is predicted that sea-level rise and its associated impacts will, by the 2080s, affect five times as many coastal residents as they did in 1990.[14] In coastal North African cities, a 1°C to 2°C increase in temperature could lead to sea-level rise exposing 6 to 25 million residents to flooding. Sea-level rise projections from 2030 to 2050 indicate that Egyptian cities in the Nile Delta will be severely affected, including Port Said, Alexandria, Rosetta and Damietta.[15] Low-lying coastal cities such as Copenhagen (Denmark), which lies at only 45m above sea level, will be especially vulnerable to sea-level rise. Many small island communities in the South Pacific are also highly vulnerable to rising sea levels. In fact, there is concern that sea-level rise and flooding will occur to such an extent that some Pacific islands will be completely submerged and entire communities displaced.[16]

The impacts of sea-level rise will continue to be felt globally even if greenhouse gas (GHG) emissions are drastically reduced given the time-lag between rising atmospheric and oceanic temperatures and the resulting sea-level rise.

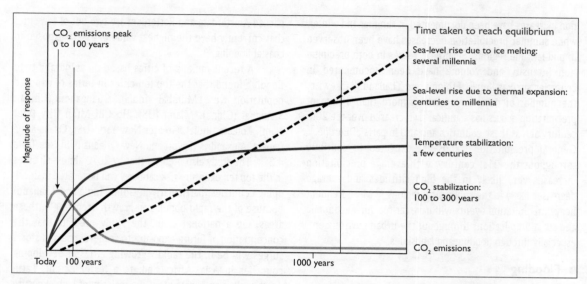

**Figure 4.1**

**Relationship between CO₂ emissions reduction, temperature stabilization and sea-level rise**

*Note:* After CO$_2$ emissions are reduced and atmospheric concentrations stabilize, surface air temperature continues to rise slowly for a century or more. Thermal expansion of the ocean continues long after CO$_2$ emissions have been reduced, and melting of ice sheets continues to contribute to sea-level rise for many centuries. This figure is a generic illustration for stabilization at any level between 450 and 1000 parts per million, and therefore has no units on the response axis. Responses to stabilization trajectories in this range show broadly similar time courses; but the impacts become progressively larger at higher concentrations of CO$_2$.

*Source:* IPCC, 2001a, p17

Regardless of future emission levels, past emissions have set sea-level rise on a trajectory that will not stabilize for millennia. Figure 4.1 presents a theoretical picture of this phenomenon: even if CO$_2$ emissions are reduced and atmospheric concentrations stabilized, global air temperature continues to rise for centuries and sea-level rise continues for millennia. Although emissions mitigation may prevent ever-worsening effects, the Earth is already 'locked into' a certain extent of climate change.[17]

## Tropical cyclones

Tropical cyclones are weather systems associated with thunderstorms and strong winds that are characterized by their wind circulation patterns and a well-defined centre.[18] These systems are so named because they originate near the Equator. Similar systems that originate in the mid-latitudes[19] are referred to as 'extra-tropical cyclones'. Both of these result in waves and storm surges (i.e. temporary offshore rise of water) that can damage property and threaten the safety of individuals in the affected area. Cyclones are classified as 'storms' when sustained wind speeds reach between 63km to 118km per hour, while a hurricane is a tropical cyclone with sustained wind speeds exceeding 118km per hour.[20]

Globally, tropical cyclones and extra-tropical storms have been increasing in intensity since the 1970s as measured by their wind speed and other indices of a storm's destructive power. With the exception of the South Pacific Ocean, all tropical cyclone basins show increases in wind speed, wind strength and storm duration, with the greatest increases in the North Atlantic and northern Indian oceans.[21] Although tropical storms have not increased in frequency, extreme extra-tropical storms have increased in number in the Northern Hemisphere since 1950.[22]

Accumulating evidence also suggests that the strongest storms are getting stronger around the world. The maximum wind speeds for satellite-observed cyclones between 1981 and 2006 show the increasing occurrence of cyclone wind speeds greater than the median. While the number of low-intensity hurricanes (category 1) has remained approximately constant, they occur less often as a percentage of the total number of hurricanes. On the other hand, hurricanes in the strongest categories (4 and 5) have almost doubled in number and in proportion (from around 20 per cent to around 35 per cent during the same period). These changes have been observed in all of the world's ocean basins.[23]

Although the relationship between temperature and formation of storm systems is not completely understood, increased temperature does correlate with increased occurrence of tropical cyclones and extra-tropical storms.[24] Rising sea surface temperatures change the Earth's water cycle, disrupting ocean currents and altering precipitation patterns, which may lead, in part, to the increases in storm intensity observed over the past several decades.[25] With global warming, potential intensity (i.e. the upper bound of cyclone intensity) is predicted to increase in most regions of tropical cyclone activity.[26]

The implications of increased cyclone activity and intensity are far reaching for cities. Power outages during storms disrupt transportation, economic activity and supply of potable water. Physical destruction caused by storms is often extremely expensive to repair and results in fatalities and injuries to humans and wildlife. Furthermore, inundation of water during storms can contaminate water supplies with saltwater, chemicals and waterborne diseases.

## Heavy precipitation events

Heavy precipitation events are defined as the percentage of days with precipitation that exceeds some fixed or regional threshold compared to an average 'reference period of precipitation from 1961–1990'.[27] On average, observations indicate that heavy one-day and heavy multi-day precipitation events have increased globally throughout the 20th century and these trends are very likely to continue throughout the 21st century.[28] Deviations from average weather patterns have been observed globally, with an increase in the frequency of heavy precipitation events in most areas of the world.[29]

Precipitation changes have been variable at the regional level. In the tropics, eastern North America, Northern Europe, and Northern and Central Asia, precipita-

*Tropical cyclones and extra-tropical storms have been increasing in intensity since the 1970s*

tion increases have been documented in summer and winter, while summer precipitation decreases have been observed in mid-latitude regions. Severe decreases in both precipitation intensity and volume have been documented in countries such as Kenya, Ethiopia and Thailand.[30] Likewise, the number of days during which more than 10mm of precipitation occurs has significantly increased over the 20th century across these countries and also in parts of Europe.[31] General precipitation trends are expected to continue throughout the 21st century, with average precipitation increases very likely in the high latitudes and average decreases likely in the subtropical regions.[32] More frequent heavy precipitation events will have far-reaching economic and social implications throughout the urban environment, especially through flooding and landslides.

*The concentration of future exposure to sea-level rise... will be in the rapidly growing cities of developing countries in Asia, Africa and, to a lesser extent, Latin America*

### ■ Flooding

Floods are among the most costly and damaging disasters posing a critical problem to city planners as they increase in frequency and severity. The frequency and severity of flooding has generally increased during the last decade (compared to 1950–1980 flood data), along with the frequency of floods that exceed levels that only typically occur once every 100 years. Although there is variation in regional predictions, it is generally accepted that both trends will continue, especially in Asia, Africa and Latin America. Flood risk is also projected to increase throughout Europe, particularly in eastern and northern regions and along the Atlantic coast. Assessments of vulnerability in Germany show that seaport cities Bremen and Hamburg may experience increased probabilities of flood risk from storms as climate change progresses, exposing billions of dollars of economic capital to potential damage.[33] The Netherlands is one of the most exposed countries in Europe, with nearly one third of the country located below average sea level in 2008.[34] Densely packed Amsterdam and Rotterdam are two out of ten cities that currently have the highest value of assets exposed to coastal flooding.[35]

A recent ranking of cities based on vulnerability to flooding found that the top ten cities in terms of exposed population were Mumbai (India), Guangzhou (China), Shanghai (China), Miami (US), Ho Chi Minh City (Viet Nam), Kolkata (India), Greater New York (US), Osaka-Kobe (Japan), Alexandria (Egypt) and New Orleans (US) (see Table 4.2).[36] The study also predicts that by 2070 almost all cities in the top ten exposure risk category will be located in developing countries (particularly in China, India and Thailand) because of the rapid population growth occurring in these areas. On a national scale, the study predicts that the concentration of future exposure to sea-level rise and storm surges will be in the rapidly growing cities of developing countries in Asia, Africa and, to a lesser extent, Latin America. It is anticipated that the majority of high-exposure coastal land area (90 per cent) will be located in only eight countries: China, US, India, Japan, The Netherlands, Thailand, Viet Nam and Bangladesh.

In addition to the evident structural damage and loss of life that they cause, floods can short-circuit transformers and disrupt energy transmission and distribution; paralyse transportation; contaminate clean water supplies and treatment facilities; mobilize trash, debris and pollutants; and accelerate the spread of waterborne diseases.[37] Poorly planned informal settlements are especially vulnerable to the impacts of flooding, as illustrated by the case of Mexico City where flash flooding has increased dramatically over the past decade (see Box 4.1).

### ■ Landslides

A landslide refers to a mass of material (e.g. rock, earth or debris) that slips down a slope by gravity. The movement is often rapid and assisted by water when the material is saturated.[38] Vegetative cover, precipitation patterns, slope angle, slope stability and slope-forming material all influence the vulnerability of an area to landslides.[39] Furthermore, the spatial distribution of landslides suggests a correlation between rapid land-use change and areas affected by landslides and mudflows.[40] Urban expansion and the clearing of vegetation for building and road construction can lead to soil erosion and weathering, and thereafter to loss of soil stability and the increased likelihood of landslides. Clearing vegetation interferes with the capacity for absorption of rainfall, which results in runoff and gully erosion. Also, as settlements develop, vegetation is replaced with paved or hard pack areas and rainwater is channelled through preferential flow channels instead of natural pathways, which increases the water's erosive power.

The risk from landslides is also likely to increase as urban development continues on marginal and dangerous lands. With rapid urbanization, populations, especially the urban poor, increasingly settle in areas that are prone to hazardous landslides and are unsuited for residential development.[41] City growth, chronic poverty, urban land speculation, insecure tenure, inadequate urban infrastructure investment, and poor urban planning policies contribute to continued development in vulnerable areas.[42]

---

**Box 4.1 Increased incidence of flash flooding in Mexico City**

The greater metropolitan area of Mexico City is one of the largest and most densely populated urban settlements in the world, containing an estimated 19.5 million residents at a population density of 3584 persons per square kilometre in 2010. The city and its residents have become increasingly vulnerable to flooding and related impacts of climate change over the past century. Annual rainfall in Mexico City increased from 600mm during the early 20th century to over 900mm towards the end of the century. Likewise, the annual incidence of flash flooding has increased from one to two annual floods, to six to seven annual floods over the same time period. On 2 August 2006, for example, a rainfall of 50.4mm in only 36 minutes caused severe flooding in the southern and western parts of the city. The incidence of flash flooding is expected to continue to rise due to climate change-related increases in the frequency of heavy precipitation.

Higher precipitation is associated with an increased frequency of flash flooding, which encompasses a wide range of conditions that threaten life and property, including submerged roads, overflowing rivers and mud- or rock-slides. Flooding damage including injury, death, property loss and water contamination are exacerbated by the infrastructure and development patterns in Mexico City. Informal settlements are often located in areas prone to flooding and landslides and, thus, particularly vulnerable. Inadequate drainage in these areas results in the accumulation of trash and debris that poses serious hazards to human health when flooding occurs. Poorly maintained and aging water drainage and sanitation systems throughout the city worsen the impacts of heavy rains and flash flooding, and make it more difficult for communities to recover.

*Source:* Ibarrarán, 2011

Faulty construction methods and missing or inadequate infrastructure design prevalent throughout informal settlements further contribute to slope degradation, increasing the risks of landslides. Construction practices, such as cut and fill, which move soil from one part of a site to another, increase the risk of a landslide, which, in turn, has been shown to weaken slope stability and increase the likelihood of a further landslide.[43] Recent estimates suggest that 32.7 per cent of the world's population live in slums,[44] which are often situated in marginal and dangerous areas (i.e. steep slopes, floodplains and industrial areas).[45] In cities such as Dhaka (Bangladesh), residents of informal settlements inhabit slopes surrounding the urban core, putting themselves at risk from flash floods and landslides.[46] Similarly, in Mexico City, landslides often adversely affect slum residents.[47] However, wealthy urban residents also occupy areas vulnerable to landslides primarily for aesthetic reasons, as illustrated by the case of Los Angeles, US.[48]

An increasing frequency of landslides will have a variety of direct and indirect impacts in urban areas. Damage to infrastructure can be substantial, resulting in high maintenance and repair costs. Indirect impacts as a result of this damage, such as constrained movement of goods and services, drive costs higher.

## Extreme heat events

Heat waves are typically defined as extended periods of hotter than average temperatures, although the precise timing and temperature differential varies regionally.[49] The lack of specificity in the definition of an extreme heat event or heat wave is due to the importance of local acclimatization to climate, which varies geographically. Previous research shows that populations in different locations have varying abilities to deal with temperature extremes. For example, studies in Phoenix (US) have found no statistically significant relationship between mortality rates and high temperatures below 43°C, while in Boston (US), an increase in the rates of mortality is observed at 32°C.[50] Several explanations exist for this phenomenon in Boston, including behavioural factors. Extremely high temperatures occur infrequently and, as a result, residents do not have the proper level of preparedness for heat waves. Also, Boston has extremely cold winters so a large percentage of homes are built from heat-retaining red brick and few homes have central air conditioning.[51] Consequently, during extreme heat events, ambient air temperature inside Boston homes can be dangerously high.

As a result of climate change, extreme heat events are predicted to become more frequent, intense and longer lasting over most land areas (see Box 4.2).[52] Some of the regions where more severe heat waves are expected in the future, due to increasing concentrations of atmospheric GHGs, include North America (particularly in the southern and north-western parts of the US) and Europe.[53]

Communities dependent upon glacial melt water also stand to be negatively affected by changes in the distribution of extreme heat. As air and ocean temperatures rise and the increasing frequency of heat waves changes stream flows, glaciers around the world will continue to shrink, threaten-

---

**Box 4.2 Extreme heat event trends in the US and Europe**

Around the world, extreme heat events are predicted to become more intense, more frequent and longer lasting. In general, the increasing frequency of extreme heat events is likely to affect cities in colder regions because of a lower saturation of cooling technologies, heat retention design of the existing building stock and cultures underprepared for extreme heat events:[a]

- On average in Chicago, 1.09 to 2.14 heat waves occur per year, whereas by 2080 to 2099 the region could see heat wave frequency increasing to 1.65 to 2.44 per year. Also, the duration of heat waves may increase from 5.39 to 8.85 days today, to 8.47 to 9.24 days by the same time.[b]

- Today, Paris averages 1.18 to 2.17 heat waves per year, which is expected to increase to 1.70 to 2.38 per year by the end of the century. The average duration of a heat wave is expected to increase from 8.33 to 12.69 days, to 11.39 to 17.04 days within this timeframe.[b]

- In the north-eastern US, cities typically experience 10 to 15 days with temperatures above 32°C and 1 or 2 days with temperatures above 38°C. However, by the end of the century, cities such as Philadelphia, Boston and New York can expect between 30 and 60 days each year with temperatures over 32°C, and between 3 and 9 days with temperatures over 38°C, depending on the emissions scenario.[c]

- In some parts of Switzerland, the average monthly temperatures were as much as 6°C above monthly averages in June and August 2003, when Europe experienced a major heat wave. It is likely that future climate conditions will resemble the summer of 2003 more than current conditions. Basel (Switzerland) could experience as many as 40 days above 30°C, as compared to 8 days today.[d]

*Sources:* a Basu and Samet, 2002; b Meehl and Tebaldi, 2004; c UCS, 2006; d Beniston and Diaz, 2004

---

ing the one sixth of the world's population dependent upon glacial melt water.[54] In a number of South American countries with communities dependent upon glacial melt water, water stress[55] could increase as small glaciers disappear due to warmer temperatures and less snowfall. Changes in precipitation and the rapid loss of glacial mass in this region will significantly affect water availability for cities across the region – for example, Quito (Ecuador), Lima (Peru) and Bogotá (Colombia) – both for human consumption and electricity generation. Communities dependent upon glacial melt water in China and Pakistan could also be negatively affected by shrinking glaciers.[56]

While physical climate changes can impact upon both rural and urban areas, urban settlements generate unique local conditions that interact with heat events. Compared to rural areas, cities tend to have higher air and surface temperatures due to the urban heat-island effect: the tendency of cities to retain heat more than their surrounding rural areas.[57] For the average developed country city of 1 million people, this phenomenon can cause air temperatures that are 1°C to 3°C higher than the city's surrounding area. At night, when urban heat-island effects are strongest, temperature differences can reach 12°C.[58] By increasing temperatures, urban heat-island effects can aggravate the heat-related negative implications of climate change and impose costly energy demands on urban systems as they attempt to adapt to higher temperatures.[59] The degree of these effects is not uniform across cities. The physical layout of a city, its population size and density, and structural features of the built environment all influence the strength of the urban heat-island effect. For example, the tendency

*Urban settlements generate unique local conditions that interact with heat events*

for French, Italian and Spanish cities to have stronger heat-island effects has been linked to their compactness and limited area of green space compared to other European cities.[60]

Extreme heat events negatively impact upon human health and social stability, increase energy demand and affect water supply. The costs of water treatment are likely to increase as high temperatures increase water demand. At the same time, water quality could decline as water pollution becomes increasingly concentrated.[61] Heat waves are more likely to impact upon vulnerable populations, including the elderly, very young, individuals with pre-existing health conditions and the urban poor. The urban poor in developed countries are especially at an increased risk from extreme heat events because of their low adaptive capacity.[62]

### Drought

The amount of land area under extreme drought conditions is expected to increase further in the future as a result of changes in precipitation

Drought can be defined as a phenomenon in which precipitation is significantly below normal levels, which leads to hydrological imbalances that negatively affect land resources and production systems. It can refer to moisture deficits in the topmost metre of soil (i.e. agricultural drought), prolonged deficits of precipitation (i.e. meteorological drought), below-normal water levels in a body of water (i.e. hydrological drought) or any combination of these.[63] Droughts can result from a number of different factors. In the western parts of the US, drought conditions have emerged largely as a result of decreases in snow pack, while areas in Australia and Europe have seen drought conditions due to extremely high temperatures associated with heat waves.[64] In Asia, increasing frequencies of droughts are likely to result from increasing temperatures.[65]

The IPCC concluded that not only have droughts become more common in the tropics and subtropics since 1970 but, more likely than not, humans have contributed to this trend.[66] Since the 1950s, significant drying trends have been observed across the Northern Hemisphere in portions of Eurasia, Northern Africa, Canada and Alaska. The Southern Hemisphere over the same time period has experienced slight drying trends. During the last century, mean precipitation in all four seasons of the year has tended to decrease in all of the world's main arid and semi-arid regions: northern Chile, the Brazilian northeast and northern Mexico, West Africa and Ethiopia, the drier parts of Southern Africa, and western China.[67] In Yemen, the capital city Sanaa is expected to run out of water by the year 2020, spurring mass migration and potential conflicts.[68]

Substantial damage to residential and commercial structures is expected with the increasing occurrence of climate change-related hazards and disasters

The amount of land area under extreme drought conditions[69] is expected to increase further in the future as a result of changes in precipitation.[70] Currently, as much as 1 per cent of all land area is considered as being under extreme drought conditions.[71] By 2100, this could increase to as much as 30 per cent.[72] Drying is likely to occur in continental interiors during summer periods, especially in the subtropics, low and mid-latitudes.[73] More intense and multi-annual droughts have occurred in sub-humid regions, including Australia, western US and southern Canada.[74] In

Africa, one third of all people already live in drought-prone areas and, by 2050, as many as 350 to 600 million could be affected by drought.

Drought affects urban areas in numerous ways. It can compromise water quality and increase the operating costs of water systems while reducing their reliability.[75] Water stress is likely to increase as a result of changes in precipitation and the consequent decline in water supply and quality and increased demand for water.

## IMPACTS UPON PHYSICAL INFRASTRUCTURE

This section describes the physical damage caused by climate change and their implications for urban areas. Climate change has direct effects on the physical infrastructure of a city – its network of buildings, roads, drainage and energy systems – which, in turn, affects the welfare and livelihoods of its residents. The severe weather events and related hazards outlined above can decimate roads, homes and places of business. These impacts will be particularly severe in low-elevation coastal zones where many of the world's largest cities are located. Although they account for only 2 per cent of the world's total land area, approximately 13 per cent of the world's urban population live in these zones.[76]

### Residential and commercial structures

Substantial damage to residential and commercial structures is expected with the increasing occurrence of climate change-related hazards and disasters. In this regard, flooding is one of the most costly and destructive natural hazards, and, as indicated earlier, one that is likely to increase in many regions of the world as precipitation intensity increases. In the absence of adaptive infrastructure changes, vast increases in spending on flood damage in cities are expected due to climate change.[77] In Boston (US), for example, river flooding could cause up to US$57 billion in damage by 2100 without adaptive measures, an estimated US$26 billion greater cost than would occur in the absence of climate change. Many of the homes likely to be affected are low-value houses that may not be insured against predicted damage. As in other areas of the world, the distributional nature of these impacts remains a challenge.[78]

The terms '100-year flood' and '500-year flood' are sometimes used to describe the flood risk to residents living in particular areas. These terms refer to the probability with which the flood occurs. For example, if there is a 1/100 chance of a given city experiencing a flood at a rate of 425 cubic metres per second, this level of flooding will occur, on average, once every 100 years. Likewise, a flood rate that occurs with a probability of 1/500 is referred to as a 500-year flood.[79] The terms 100-year floodplain and 500-year floodplain refer to the geographic areas that are affected during 100-year and 500-year floods, respectively.

Today, around 40 million people live in a 100-year floodplain. By 2070, the population living at this risk level

| Ranking by population exposure | Ranking by value of property and infrastructure assets exposure |
|---|---|
| Kolkata (India) | Miami (US) |
| Mumbai (India) | Guangzhou (China) |
| Dhaka (Bangladesh) | New York (US) |
| Guangzhou (China) | Kolkata (India) |
| Ho Chi Minh City (Viet Nam) | Shanghai (China) |
| Shanghai (China) | Mumbai (India) |
| Bangkok (Thailand) | Tianjin (China) |
| Rangoon (Myanmar) | Tokyo (Japan) |
| Miami (US) | Hong Kong (China) |
| Hai Phong (Viet Nam) | Bangkok (Thailand) |

*Source:* Nicholls et al, 2008

**Table 4.2**

**Exposure to floods in cities**

could rise to 150 million people. The estimated financial impact of a 100-year flood would also rise from US$3 trillion in 1999 to US$38 trillion by this time. Miami (US) is the most exposed city today and will remain so in 2070, with exposed assets rising from approximately US$400 billion today to over US$3.5 trillion. Over the coming decades, the unprecedented growth and development of Asian megacities will be a key factor driving the increase in coastal flood risk globally. By 2070, eight of the most exposed cities will be in Asia (see Table 4.2).

Damage to residential and commercial structures is not limited to large-scale disasters. Slow-onset climate-change physical risks such as sea-level rise can also affect the built environment in a number of ways. Coastal erosion is likely to affect cities around the world particularly in the mega-deltas of South, East and Southeast Asia, Europe and the North American Atlantic coast.[80] In the US, a 0.3m sea-level rise[81] would erode approximately 15m to 30m of shoreline in New Jersey and Maryland, 30m to 60m in South Carolina and 60m to 120m in California.[82] Parts of Louisiana and Mississippi along the Gulf Coast of the US are physically susceptible to loss of land from the combined effects of erosion and sea-level rise, whereas some areas of Florida and Texas are susceptible due to social and economic factors of vulnerability.[83] Coastal erosion and saltwater intrusion can ruin buildings and render some areas of land uninhabitable, which is a particular problem for coastal cities that rely on tourism as a major part of their economies. Mombasa (Kenya), for instance, could lose approximately 17 per cent of its land from a 0.3m rise in sea level, causing the loss of hotels, cultural monuments and beaches that draw tourists.[84]

Subsidence, or the downward shift of the Earth's surface, is another 'slow-onset' factor that poses a risk to residential and commercial structures in cities. Subsidence can be caused or exacerbated by overexploitation of groundwater resources during hot, dry periods which are likely to occur more frequently with climate change. Subsidence can be as rapid as 1m per decade, resulting in significant damage to pipelines, building foundations and other infrastructure.[85] In England, increased subsidence caused by drier, hotter summers led to significantly greater homeowner insurance claims throughout the late 1990s.[86] Subsidence has been noted in several megacities throughout the world, including Tokyo (Japan), Dhaka (Bangladesh), Jakarta (Indonesia),

Kolkata (India), Metro Manila (the Philippines), Shanghai (China), Los Angeles (US), Osaka (Japan) and Bangkok (Thailand).[87] During the late 1980s, Tianjin (China) experienced as much as 11cm of subsidence per year.[88] Portions of the Osaka-Tokyo metropolitan region would be under water as a result of subsidence if it were not for coastal defences and extensive flood-control systems.[89]

Accumulating damage to residential and commercial buildings due to sun exposure and low-intensity wind and precipitation may increase in some areas of the world as regional weather patterns change. In London (UK), more frequent heavy rains and higher peak wind speeds (predicted for the 2050s and 2080s) are expected to damage buildings, particularly those that are aging. Wind and rain damage can cause hazards to people in the vicinity of affected buildings and may lead to additional economic losses if commercial buildings need to close for repairs.[90] Studies on New Zealand indicate that commercial buildings throughout the country will experience increased damage in the face of wind damage, coastal flooding and extreme temperatures.[91] In the Arctic region, human settlements are expected to face serious challenges with the melting of permafrost, which is essential for the stability of buildings and infrastructure.[92]

## Transportation systems

Climate change impacts frequently disrupt transportation systems through weather conditions that have immediate consequences for travel and damage, causing lasting service interruptions. In coastal cities, in particular, sea-level rise can inundate highways and cause erosion of road bases and bridge supports. For example, along the Gulf Coast of the US an estimated 3862km of roadway and nearly 402km of rail tracks may become permanently submerged during the next 50 to 100 years due to the combined impacts of subsidence and sea-level rise. Total economic impacts resulting from this loss could reach hundreds of billions of dollars when considering the commercial and industrial activities that take place in the gulf's many seaports, highways and railroads.[93] For instance, weather-related highway accidents translate into annual losses of at least US$1 billion annually in Canada, while more than one quarter of air travel delays in the US are weather related.[94] In India, landslides in July 2000 resulted in 14 days without train service, leading to estimated losses of US$2.2 million.[95]

*Unprecedented growth and development of Asian megacities will be a key factor driving the increase in coastal flood risk globally*

*Climate change impacts frequently disrupt transportation systems*

Heavy precipitation and its effects in the form of flooding and landslides can cause lasting damage to transportation infrastructure, such as highways, seaports and bridges. In 1993, flood damage to transport systems in Midwestern US resulted in major traffic disruption from Missouri to Chicago for nearly six weeks.[96] Delays in public transportation including rail and air services often occur during heavy rains and storms. A study of the Konkan railway network in western India that facilitates trade and energy services between Mumbai and Mangalore revealed that 20 per cent of major repairs were due to climatic factors. Each year, US$1.1 million is spent to reduce the number of locations on the network that are vulnerable to heavy rains.[97] Heavy rains also affect the long-term functional capacity of airport runways, which will lead to the need for increased maintenance considerations in those areas where precipitation is likely to increase.

Increasingly higher temperatures, particularly long periods of drought and higher daily temperatures, compromise the integrity of paved roadways and necessitate more frequent repairs. For instance, by 2080, road buckling, rutting and speed restrictions are anticipated to increase in London (UK) as average temperature increases melt asphalt and accentuate subsidence.[98] Extreme heat also leads to joint expansion on bridges and rail deformation, which require costly maintenance and, in worst-case scenarios, could cause major accidents. Drier conditions can further cause lower water levels in rivers and interrupt trade and transportation via inland water routes.

Besides potentially endangering lives, the destruction or damage of transportation systems and lasting service disruptions greatly affect nearly all aspects of urban life. Disruptions in public transportation can limit the ability of residents to get to work, leading to declines in economic productivity. By 2100, as a result of increases in climatic change-related delays, motorists in Boston (US) could spend 80 per cent more time on roadways and 82 per cent more trips could be cancelled.[99] Interruptions in the transport of fuel for energy production can also lead to service disruptions in the electricity sector.

### Energy systems

By their very nature, cities are centres of high demand for energy and related resources. Climate change is likely to affect both energy demand and supply. The combination of urban population growth, changing local weather conditions, urban heat-island impacts and economic growth has the potential to substantially increase demand for energy (see Box 4.3). Although the relationship between energy demand and local weather fluctuations has long been confirmed, relatively few studies have taken on the task of examining how longer-term climate changes affect the energy sector.

Energy demand increases will depend upon regional climate differences. Higher winter temperatures can lead to decreased heating use, while increased summer temperatures can lead to increased need for cooling. In turn, greater use of air conditioning due to rising temperatures can worsen the urban heat-island effect and further increase the cooling demand in urban areas.[100] Studies indicate that there is high regional variation in energy demand sensitivity to climate change even in similar climates. In the US, for example, neighbouring states Florida and Louisiana have different patterns of industrial and residential energy use.[101] Likewise, an assessment of several US regions reveals unique demand sensitivities among four cities (Seattle, Minneapolis, Phoenix and Shreveport) and different directions of demand change between states with differing average local weather conditions.[102] The use of aggregate data, however, may be misleading because even if there is no net increase in regional demand, great increases in local demand for cooling may still require infrastructure investment, reconsideration of energy portfolios and energy conservation mechanisms.

Climate change will also affect energy generation and distribution. Across Africa, hydroelectric power generation is likely to be restricted with the more frequent occurrence of drought periods. For instance, climate change simulations suggest that the planned Batoka Gorge Hydroelectric Project on the Zambezi River, a joint project between Zambia and Zimbabwe, will be negatively affected if the mean monthly river flow significantly declines.[103] However, the worldwide impacts of climate change upon hydroelectricity production are variable. For example, electricity output from hydroelectric projects in Scandinavia and northern Russia is predicted to increase due to trends in future precipitation patterns and temperature.[104]

Reduced stream flows due to climate change may further reduce the availability of cooling water for thermal and nuclear power plants.[105] In Europe, the 2003 heat wave was accompanied by annual rainfall deficits of as much as 300mm.[106] Drought conditions had impacts upon power generation and several power plants were unable to physically or legally divert water because of extremely low stream flow, resulting in reductions in power generation. For instance, nuclear power plants in parts of France were forced to shut down as stream levels became too low or water temperatures exceeded environmental standards. Six nuclear reactors, as well as a number of conventional power plants, were given exemptions to continue operating in spite of exceeding legal limits.[107] In terms of energy distribution, electricity transmission infrastructure may become increasingly vulnerable to damage and interference as storms and flooding become more frequent and intense.[108]

> **Destruction or damage of transportation systems and lasting service disruptions greatly affect nearly all aspects of urban life**

> **Climate change is likely to affect both energy demand and supply**

---

**Box 4.3 Global changes in energy demand**

- Between 2010 and 2055 in the US, energy demand may increase capacity requirements for the electricity sector by 14 to 23 per cent over demand trends in the absence of climate change.[a]
- Daily peak energy loads for New York City (US) could increase by 7 to 13 per cent by the 2020s, 8 to 15 per cent by the 2050s and 11 to 17 per cent by the 2080s.[b]
- By 2080, a 30 per cent increase in energy demand is forecast for Athens (Greece), largely as a result of increasing air-conditioning use.[c]
- In Toronto (Canada), a 3°C increase in temperatures would result in increases in peak electricity demand of 7 per cent and an increase of 22 per cent in the variability of peak demand.[d]

*Sources:* a Linder, 1990; b Rosenzweig and Solecki, 2001; c Giannakopoulos and Psiloglou, 2006; d Colombo et al, 1999

## Water and sanitation systems

The availability, treatment and distribution of water could be affected by climate change as temperatures increase and precipitation patterns change.[109] On the one hand, climate change is expected to compromise water supplies, particularly in areas where water stress is expected to increase. In developing regions such as Africa, water stress is expected to increase as a result of population growth and is likely to be exacerbated by climate change. However, the impacts will not be uniform across the continent as populations in the north and south are expected to experience increases in water stress, while those in the east and west are likely to see a reduction in water stress.[110]

Water supplies can be reduced or increased through changes in precipitation patterns, reductions in river flows, falling groundwater tables and, in coastal areas, saline intrusion in rivers and groundwater.[111] For example, detected declines in glacier volumes in parts of Asia and Latin America are already reducing river flows at key times of the year. For cities located in the Andean valleys and in the Himalaya-Hindu-Kush region, this has substantial impacts upon water flows and affects multiple human uses of water in these areas, including reducing hydroelectric power generation.[112] The expected changes in runoff and water availability are, however, projected to be regionally differentiated by 2050: increases by 10 to 20 per cent at higher latitudes and in some areas in the wet tropics (e.g. populous areas in tropical East and Southeast Asia), and decreases by 10 to 30 per cent over areas in the mid-latitudes and dry tropics, some of which are currently water stressed.

On the other hand, with rising temperatures, more frequent extreme heat events and population growth in the future, demand for water in cities is expected to increase.[113] Many areas of the world have been getting drier across all seasons; if the trend continues, water resource limitations will become more severe.[114] By 2030, summer water use in Washington, DC (US), is expected to increase between 13 and 19 per cent relative to an increase from 1990 levels without climate change.[115] In Cape Town (South Africa), water demand is simultaneously projected to increase with temperature increases.[116] In Nagoya (Japan), temperature increases could induce an increase of 10 per cent in water use.[117] In Latin America, 12 to 81 million residents could experience increased water stress by the 2020s. By the 2050s, this number could rise to 79 to 178 million.[118] Stream flow supplying Melbourne (Australia) is likely to decline by 3 to 11 per cent by 2020 and 7 to 20 per cent by 2050, compared to 1961 to 1990 averages, thereby affecting water supplies. Concerns about drought and water demand increases have also been raised in other cities such as Auckland, Adelaide, Canberra, Perth, Brisbane and Sydney.[119]

Climate change-related changes in precipitation and sea levels can also affect the quality and treatment of water in cities. Saltwater intrusion can occur more frequently in communities experiencing sea-level rise and contaminate ground and surface water, thus reducing the supply of potable water and spreading harmful pollutants throughout urban water systems. Cases of saltwater intrusion due to sea-level rise have already been documented among most coastal cities across diverse environments, including eastern US (e.g. New Orleans), Latin America (e.g. Buenos Aires, Argentina), as well as both in the Yangtze River Delta in China and the deltas of Viet Nam.[120] Reduced precipitation and, thus, water supply can also cause saltwater intrusion. The city of Kochi (India) is located at 2m above sea level and is compromised by a network of rivers and canals. Saltwater intrusion into these rivers is worsened during hot, dry periods when evaporation increases the concentration of salts in the water, leading to economic losses and drinking water shortages.[121]

Furthermore, excess heat from buildings and roads due to the urban heat-island effect can be transferred to storm water, thereby increasing the temperature of water that is released into streams, rivers, ponds and lakes. Higher water temperatures, in conjunction with increased precipitation intensity and low flows, are predicted to exacerbate water pollution, including through thermal pollution, which can promote algal blooms and increase bacterial and fungal content.[122] Once they have been contaminated, it is, in most cases, expensive to clean drinking water supplies.

Water supply infrastructure is capable of adapting to small changes in mean temperatures and precipitation amounts as water systems have been designed with spare capacity for future growth.[123] Still, many systems will need improvements, such as building new reservoirs or extension of water intake pipes to handle increasingly variable precipitation levels. In addition, water supply infrastructure is vulnerable to damage from extreme climate events such as floods and storm surges, especially if it is adjacent to rivers.[124] In New York City (US), pumping stations and water treatment facilities, including intake and outflow sites, are vulnerable to storm surges.[125] Damage to water supply infrastructure, especially if electronics are damaged, can take weeks to fix and can cost as much to repair as their initial construction costs, as was the case during flooding in Mozambique during 2000.[126]

Climate change-related disasters can also affect sanitation systems in urban areas which already face serious challenges, especially in developing countries. Although access to improved water supply and sanitation has been increasing since 1990 in many areas of the world, there are still large proportions of the population living in unsanitary conditions.[127] In 2006, 38 per cent of the world's population and nearly half of the developing world's population lacked access to improved sanitation facilities, including flush toilets, pit latrines or composting toilets.[128] Access to sanitation infrastructure and services is likely to decline further due to climate change-related risks, as in the case of Hurricane Mitch, which destroyed 20,000 latrines in 1998.[129]

## ECONOMIC IMPACTS

The increasing frequency and intensity of extreme climatic events and slow-onset changes will increase the vulnerability of urban economic assets and, subsequently, the costs of doing business.[130] Studies suggest that developing countries

**The availability, treatment and distribution of water could be affected by climate change**

**Climate change-related disasters can also affect sanitation systems in urban areas... especially in developing countries**

typically suffer low economic losses but high human losses as a result of climate change-related risks, while developed countries suffer high economic costs and low human losses. However, recent events show that developed countries can suffer high human costs as well, especially amongst the urban poor. Also, when economic impacts are expressed as a share of the value of total assets or gross domestic product (GDP), economic costs incurred by developing countries may also be high and can result in increasing fiscal imbalances and current account deficits due to increased borrowing and spending to finance recovery.[131] This section explores the economic impacts of climate change within urban areas, including those related to economic sectors, ecosystem services and livelihoods.

## Sectoral economic impacts

Climate change will affect a broad range of economic activities, including trade, manufacturing, transport, energy supply and demand, mining, construction and related informal production activities, communications, real estate, and business activities.[132] Box 4.4 describes the cross-sectoral economic impacts of tropical cyclones in Dhaka (Bangladesh).

This section describes climate change impacts upon economic sectors – namely, retail and commercial services, industry, tourism and insurance – as these tend to operate in and around cities. Industrial infrastructure in coastal cities is particularly vulnerable to sea-level rise and coastal storms. The effects of climate change on tourism are also considered

*The increasing frequency and intensity of extreme climatic events and slow-onset changes will increase the vulnerability of urban economic assets*

as it can be a part of the urban economy directly or dependent on it for services, including travel (e.g. airports, seaports, etc.) and supplies. Furthermore, climate impacts upon the tourism industry can induce migration from rural to urban areas, thus increasing the demand for goods and services within urban areas.[133]

### ■ Industry and commerce

Industrial activities can bear potentially high direct and indirect costs from climate change and extreme climate events. Whether industries are located in the heart of urban areas or in adjacent suburban or rural areas, they provide services and resources that are vital to city function. Damage to industries due to climate events thus has direct and indirect impacts upon cities and their residents.

The direct effects of climate change and extreme climate events on industry include damage to buildings, infrastructure and other assets. These effects are especially severe where industrial facilities are located in vulnerable areas, such as coastal zones and floodplains. For example, sea-level rise in coastal cities such as New Orleans (US) will potentially necessitate the relocation of refineries, natural gas plants and facilities, as well as supporting industries to less at-risk areas or further inland, at a substantial cost (see Box 4.5).[134] The indirect impacts of climate change upon industry include those resulting from delays and cancellations due to climate effects on transportation, communications and power infrastructure.[135]

Similarly, retail and commercial services are vulnerable because of supply chain, network and transportation disruptions, and changes in consumption patterns.[136] An increasing likelihood of flooding, coastal erosion and other extreme events will stress and damage transport infrastructure, as indicated earlier in this chapter, disrupt retail and commercial services, and subsequently increase the costs of doing business.[137] For example, in 2001, the Great Lakes–St Lawrence region of Canada experienced drought conditions that lowered river levels to such an extent that it slowed river traffic, which partially explains the reduction in volume of goods shipped through the Great Lakes that year.[138] Similarly, the 2003 heat wave and drought across Europe resulted in record low river levels, which negatively affected the transportation of goods along inland waterways.[139]

Changes in the regulatory environment, including climate change mitigation policies (e.g. carbon tax and emissions targets) could potentially raise the costs of business for industries, especially if they are energy intensive.[140] For instance, the iron and steel industry is heavily dependent upon burning fossil fuels, with 15 to 20 per cent of the production costs going towards energy. In the US, the pulp and paper industry is the second most energy-consuming industry.[141]

Industries dependent upon climate-sensitive inputs are also likely to experience changes in the reliability, availability and cost of major inputs as a result of changes in climate and climate mitigation policies. For instance, industries dependent upon timber and agricultural inputs rely on an increasingly fragile resource because of changes in the

---

**Box 4.4 Cross-sectoral impacts of tropical cyclones: The case of Dhaka, Bangladesh**

Given that the majority of its land area is less than 6m above sea level, the population of Bangladesh and its assets are highly vulnerable to the impacts of tropical cyclones. Rising sea levels and increased prevalence of cyclones have been documented over the past decade, along with increased frequency and intensity of sudden and severe floods. Between 1991 and 2000, the country experienced 93 major disasters, resulting in nearly 200,000 deaths and costing US$5.9 billion in damage to agriculture and infrastructure.

Storm surges cause massive damage to the city of Dhaka, which has experienced four major floods in the past two decades, including one that submerged 85 per cent of the city. In addition to endangering lives, these events have multi-sectoral impacts that cause lasting damage to the economic and social fabric of the city. Disruption of activity in textiles, timber, food and agro-based industries results in massive economic loss. In 1998, it was calculated that total industry loss was more than US$66 million. All utility services essentially cease during flooding events, and structural damage can cause lasting disruption of utility services such as water supply, sanitation, waste and sewage management, telecommunications, and electricity and gas supplies.

The city's adaptation efforts have been aimed at mitigating the impacts of extreme flooding events by expanding the Integrated Flood Protection Project, a programme funded by the Asian Development Bank to improve flood protection structures, drainage and sanitation, and to resettle residents of slums into safer areas. Improving the drainage system and reinforcing the water system is a priority since the city's water has become contaminated in the past, and acute drinking water crisis has been a major problem in post-flood efforts. There is also an initiative to involve non-governmental organizations (NGOs), the business community and community-based organizations to enhance aid in relief, recovery and rehabilitation programmes.

*Source: Vaidya, 2010*

---

**Box 4.5 Economic impacts of Hurricane Katrina, US**

The city of New Orleans is located on vulnerable lands at the mouth of the Mississippi River on the Gulf of Mexico. Due to its proximity to the Mississippi and the gulf, the area has strategic economic importance for the petrochemical industry, as well as international trade. New Orleans's longstanding infrastructure and population centres have become increasingly at risk from climate events; coastal defences and other land areas are subsiding as a result of groundwater withdrawal, man-made changes to the flow of the Mississippi River prevents silting and the build-up of new land, and the below sea-level elevation of much of the city requires continuous pumping of water.

In 2005, Hurricane Katrina caused extensive damage to physical infrastructure and the economies of the gulf coast region. The economic losses were in the hundreds of billions of US dollars. An estimated 1.75 million property claims were filed, totalling more than US$40 billion. Over 250,000 claims were filed as a result of flood damage, which would have bankrupted the National Flood Insurance Program were it not given the right to borrow an additional US$20.8 billion.

In the Gulf of Mexico, over 2100 oil and natural gas platforms and 15,000 miles (24,140km) of pipeline were affected. A total of 115 platforms were lost, with 52 suffering heavy damage; 90 per cent of total Gulf of Mexico oil production and 80 per cent of natural gas were idled, with lost production equalling over 28 per cent of annual production. The damage to the petrochemical corridor, which produces half of the US supply of gasoline, caused disruptions in economic markets worldwide, resulting in the largest spike in oil and gas prices since the Organization of the Petroleum Exporting Countries (OPEC) embargo of 1973. In the first two months following Hurricane Katrina, over 390,000 people lost their jobs, with over half coming from low-wage earning jobs. As of 2006, only 10 per cent of businesses in New Orleans had returned and reopened.

Before Hurricanes Katrina and Rita, which also happened in 2005, the port of New Orleans was the fourth largest port in the world in terms of transported tonnage. However, as a result of the damage from hurricanes, port operations were halted for a period of time, which forced a realignment of shipping destinations and functions that, because of the high cost of realignment, could become permanent.

*Sources:* Petterson et al, 2006; Wilbanks et al, 2007

---

incidence of pests and diseases. Climate change holds the potential to shift the habitat of economically important tree and crop species, as well as changing the behaviour and distribution of pests.[142] The changing distribution of climate-sensitive inputs could result in increasing costs to industry, as industrial plants and their raw material inputs become geographically separated.

### ■ Tourism and recreation

The tourism industry is highly dependent upon reliable transportation infrastructure, including airports, ports and roadways. Climate change has the potential to not only shift regional temperature distributions, but also increase the incidence of severe weather events, which would increase transportation delays and cancellations. Since recreational activities and tourism are often major sources of revenue for urban areas, when climate change impacts affect these activities, local urban economies will incur monetary and job losses (see Box 4.6).

Tourism in cities of high-latitude countries could benefit from a pole-ward shift[143] of warmer conditions, increasing the area available for tourism activities.[144] However, winter activities (i.e. skiing and snowmobiling) are likely to become increasingly vulnerable because of climate change-related declines in natural snowfall leading to fewer days of snow cover.[145] Across much of the north-eastern region of the US, climate change will result in fewer days of natural snow cover and, in spite of snow-making technology, ski areas will experience a decline in the length of their seasons. To continue operating will necessarily mean an increase in costs because manufacturing snow is both water and energy intensive. Climate change will further result in declining season length as reliable snow is pushed into upper latitudes and higher elevations. As a result, the average

distance travelled to winter resources such as ski mountains is expected to increase dramatically.[146] The weakening of the ski industry would also affect related support industries such as hotels, restaurants and ski shops. The decline of winter recreational opportunities can thus result in great economic losses for those regions with economies heavily reliant on skiing and snowboarding.

The summer tourism industry across the temperate zone is thought to be resilient to increases in average temperatures because of the expectations of warm temperatures, as well as the availability of air conditioning.[147] However, changes in the frequency and intensity of extreme

> Industries dependent upon climate-sensitive inputs are also likely to experience changes in the reliability, availability and cost of major inputs

---

**Box 4.6 Climate change impacts upon the tourism industry**

- The annual number of tourists visiting Canada and Russia is estimated to increase by 30 per cent as a result of a 1°C rise in temperatures.[a]
- The cost of climate change for Switzerland is estimated to be US$1.4 to $1.9 billion by 2050 – of this amount, US$1.1 billion is from tourism alone. In Switzerland, 85 per cent of ski areas are considered as snow-reliable today; however, under climate change scenarios, only 44 per cent will remain snow reliable in the future. A number of communities in Switzerland are heavily dependent upon winter tourism as it provides a significant portion of their income.[b]
- The Norwegian ski industry could be negatively affected by climate change as summertime ski destinations are expected to experience more rainy weather during the summer months.[c]
- For Australia, a 3°C to 4°C increase in temperatures would cause catastrophic mortality to a large percentage of the coral species that make up the Great Barrier Reef. Even with a 1°C to 2°C increase in temperatures, between 58 and 81 per cent of the coral would be bleached every year. And because of the importance of the reef to Australian tourism, a US$32 billion industry, declines in reef health would negatively impact the tourism industry.[d]

*Sources* a Hamilton et al, 2005; b Elsasser and Bürki, 2002; c O'Brien et al, 2004; d Preston and Jones, 2006

weather events could negatively affect the perception of safety in these locations, environmental quality and tourism infrastructure reliability. For example, in Southern Europe, along the Mediterranean coast, increasing water scarcity as a result of climate change could negatively impact upon the tourism industry.[148]

Coastal areas, including those in cities, have often been extensively developed for tourism, leaving substantial investments in buildings and infrastructure at risk from extreme climatic events, which would significantly affect the economies of small island states.[149] Erosion as a result of coastal storms can cause beaches to recede by as much as 5m, but can recover quickly through natural sand deposition. If sea-level rise is accompanied by stronger and more frequent coastal storms, the costs of maintaining shore space for tourist activities could rise and reduce beach tourism for cities.[150] The city of Rio de Janeiro (Brazil), for instance, popular, among other reasons, for its beaches, is vulnerable to sea-level rise and increased erosion. In the capital city of Estonia, Tallinn, beachside resorts are particularly vulnerable to sea-level rise and storm surges, which could lead to more erosion of the beaches and negatively affect tourism.[151] Furthermore, extreme climate events can damage many reefs and coastal ecosystems, resulting in declining tourism.[152] A 1°C rise in temperatures would result in more frequent coral bleaching, with the coral recovering slowly, while a 2°C rise in temperatures would result in the annual bleaching of coral in many areas that might never recover.[153]

Tourism is an essential component of the local economy of many island countries in the Caribbean and elsewhere. In the eastern Caribbean, tourism accounts for between 25 and 35 per cent of the regional economy, one quarter of foreign exchange earnings and one fifth of all jobs. Each year, the region receives approximately 20 million tourists. The economic dependence upon tourism has led to intensive development and siting of infrastructure (i.e.

*Climate change could result in increasing demand for insurance while reducing insurability*

hotels, roads, etc.) that tend to be densely packed along coastlines. As a result of changes in sea level and wave action, the islands of the eastern Caribbean will experience submergence of low-lying areas, including population centres, erosion of soft shores, increasing salinity of estuaries and aquifers, and more severe coastal flooding and storm damage.[154]

### ■ Insurance

The insurance industry is vulnerable to climate change, particularly extreme climate events that can affect a large area.[155] Storms and flooding can cause significant amounts of damage and are often responsible for a large percentage of total losses, as illustrated in Box 4.7.[156]

Climate change could result in increasing demand for insurance while reducing insurability. Insurance industry catastrophe models forecast that annual insured claims and losses are likely to significantly increase over the next century as a result of the increasing intensity and frequency of extreme storms. The distribution of claims is unlikely to be even as construction quality, property values and insurance coverage vary widely worldwide. In response, the insurance industry could adapt by raising the cost of insurance through measures such as increasing premiums, restricting coverage, etc.[157] Indeed, the costs of insurance coverage are expected to increase significantly if infrequent but catastrophic events become more common in the future.

In addition, the uncertainty surrounding the probability of high-loss events in the future is likely to place upward pressure on insurance premiums.[158] The implications of this will be harshest on low- (and possibly middle-) income households in developed countries if they are no longer able to afford insurance to recover from climate change-related events. It is already the case in the insurance industry that individuals tend to be underinsured, especially against events with low probabilities of occurrence. Studies have

---

**Box 4.7 Impacts of climate change upon the insurance industry**

- In 1992, Hurricane Andrew hit southern Florida (US) and resulted in over US$45 billion in damage (2005 dollars). In the aftermath, 12 insurance companies dissolved.[a]
- The average annual damage from hurricanes in the US is estimated to increase by US$8 billion (2005 dollars) due to intensification, assuming a scenario in which $CO_2$ levels double.[b]
- By the 2080s, a severe hurricane season in the US would increase annual insured damage by 75 per cent, while in Japan, insured damage would increase by 65 per cent.[c]
- Insured damage in Europe are estimated to increase by 5 per cent as a result of extreme storms, with the costs of a 100-year storm doubling from US$25 billion to US$50 billion by the 2080s.[c]
- Miami (US) has over US$900 billion of capital stock at risk from severe coastal storms, and London (UK) has at least US$220 billion of assets located on a floodplain.[d]
- The gross regional product of the New York City region (US) is estimated to be nearly US$1 trillion annually and losses from a single large event could be in the range of 0.5 to 25 per cent, or as much as US$250 billion.[e]
- The full macroeconomic costs of Hurricane Katrina of 2005 are estimated at US$130 billion, while the gross state product for Louisiana (US) in the same year was US$168 billion.[f]
- In Russia, insurance costs along the Lena River have increased during recent years as a result of more frequent and severe flooding.[g]
- By 2100, flooding could cause over US$94 billion in property damage in metropolitan Boston (US) if no adaptive actions are taken, with homeowners on 100-year and 500-year floodplains sustaining an average of US$7000 to US$18,000 in flood damage per household.[h]

*Sources:* a Wilbanks et al, 2007, p369; b Nordhaus, 2006; c Hunt and Watkiss, 2007, p21; d Stern, 2006, p14; e Jacob et al, 2000; f Stern, 2006, p11; g Perelet et al, 2007; h Kirshen et al, 2006

| Impact of urbanization | Effects on ecosystem | Effects on ecosystem service |
|---|---|---|
| Reduced permeability of surfaces | Reduction of biodiversity<br>Surface and groundwater pollution<br>Alteration of surface and groundwater channels | Reduced capacity for natural pollutant filtration |
| 'Patchy' land-use patterns that fragment the landscape and spread into natural environments such as forests | Reduction of biodiversity<br>Loss of trees and soil | Reduction in $CO_2$ retention of nearby land<br>Reduction in local oxygen supply |
| Excess emissions of nutrients (e.g. nitrogen, phosphorus), sediments, metals and other wastes into waterways | Mass death of aquatic species | Reduction of food sources and other economic activity (e.g. recreation, tourism) |
| Development on wetlands | Loss of wetland area<br>Loss of biodiversity | Reduced capacity for natural pollutant filtration<br>Reduction in local oxygen supply<br>Reduction of natural storm buffer |

*Impacts of urbanization upon ecosystem services*

found that, despite favourable premiums, individuals often fail to purchase insurance for low-probability but high-loss events in part because of the costs associated with finding a policy.[159]

Insurance coverage can vary widely within and among developed and developing countries, as there tends to be a correlation between economic growth and insurance coverage.[160] While it is expected that insurance coverage will increase with economic development in many developing countries, at-risk infrastructure and buildings – including government-owned properties – compound their vulnerability by not having insurance coverage.[161] As much as 29 per cent of total property losses are covered by some form of insurance in developed countries.[162] In developing countries, however, only about 1 per cent of total losses are insured.[163]

Private insurers in developed countries will often not provide insurance or will restrict it in areas that have suffered significant past losses from floods, which then necessitates government involvement in order to provide flood insurance.[164] The risk of loss then falls upon government programmes and individual homeowners because insurance out-payments rarely cover the entire cost of reconstruction.[165] Furthermore, it would appear that government programmes are increasingly vulnerable to climate change as a result of an increasing frequency and intensity of extreme climate events. For example, Hurricane Katrina damage in New Orleans (US) and the surrounding region almost bankrupted the National Flood Insurance Program.[166] Populations worldwide are growing within coastal areas and growth is expected to increase rapidly, suggesting an increase in the vulnerability of property but also insurance providers, including government programmes.[167]

In some places, the availability of insurance in coastal and other vulnerable areas fails to discourage development in areas at risk of flooding from coastal storms.[168] In the eastern parts of the Caribbean, for instance, building quality and location are not typically factored into insurance availability or cost. Due to missing incentives to mitigate the impacts of extreme climatic events and given that only a small percentage of the risk is retained by local insurance companies, buildings are often ill-prepared targets for extreme weather events or climate change. Instead, insurance companies are encouraged by the system to underwrite as many policies as possible, regardless of their soundness.[169]

## Ecosystem services

Natural environmental processes provide benefits that are vital to city function and human health. These ecosystem services include oxygen production, carbon storage, natural filtering of toxins and pollutants, and protection of coastal societies from flooding and wind during storms. Human activities (e.g. development, pollution and wetland destruction) can harm such ecosystem services. Increasing urbanization places greater demand on natural resources and imposes significant changes on the environmental processes that drive the benefits that societies derive from ecosystem services.[170] Table 4.3 illustrates some of these changes and their effects on ecosystem services.

The Millennium Ecosystem Assessment[171] indicates that climate change has been identified as a key factor behind the accelerated loss and degradation of ecosystem services. The assessment found that approximately 60 per cent of the ecosystem services evaluated were being degraded or used unsustainably.[172] Wetland health may be particularly threatened in the coming decades as the combined impacts of landscape modification and sea-level rise cause the Earth's deltas to sink below oceans levels.[173]

Loss of ecosystem services, besides potentially affecting food provision and human health, can significantly reduce the revenue of cities. In Durban (South Africa), for example, the replacement value of the ecosystem services (e.g. water provision, flood prevention) within the city's network of open space was estimated at US$418 million per year in a study published in 2003.[174] This was approximately 38 per cent of the city's total capital and operating budget at that time, illustrating the financial consequences of losing access to these services. A further significant point is that it is the poorest and most vulnerable people/communities who are most directly reliant on these services in order to meet their basic needs. They therefore stand to lose the most from the damage of ecosystems goods and services under projected climate change conditions.

## Livelihood impacts

Extreme climate events can disrupt the ability of individuals and households in urban areas to sustain livelihoods.[175] Climate change-related disasters destroy livelihood assets or the means of production available to individuals, households or groups. These include stocks of natural resources (natural

*Populations worldwide are growing within coastal areas ... suggesting an increase in the vulnerability of property but also insurance providers*

*Climate change has been identified as a key factor behind the accelerated loss and degradation of ecosystems services*

capital), social relationships (socio-political capital), skills and health (human capital), infrastructure (physical capital) and financial resources (financial capital), which are necessary to sustain a livelihood. By affecting such assets, climate change-related events can pose a serious threat to urban livelihoods.

The effects of climate change on livelihoods will also depend on their geographical location and, thus, exposure to the physical risks associated with climate change. Livelihood activities located in low-elevation coastal zones, for instance, will be vulnerable to the impacts of sea-level rise and cyclones. Livelihood impacts will also vary from one context to another depending on the vulnerability of existing assets and opportunities. For instance, the livelihoods of the urban poor are likely to be most at risk from climate change effects since their assets and livelihoods are already meagre and unreliable. In particular, individuals living in informal settlements are likely to have meagre savings and any disruption to their livelihood directly affects their ability to buy food and pay bills, including for their children's education and healthcare. Livelihood activities of the urban poor are also more severely affected by climate events than other social groups because of their presence in at-risk zones. For instance, flooding makes it difficult for residents of informal settlements to conduct small-scale commerce, petty trading and artisan trades, and thus can leave them undernourished for days while the area and local economy recovers. A one-day rain event in Maputo (Mozambique) might result in floods that linger for three days, and if rains persist, floodwaters might rise as much as 1m and take a month to recede.[176]

Where livelihoods are dependent on climate-sensitive inputs, the impacts of extreme as well as slow-onset climate changes will be further accentuated. This is so in the case of agriculture and tourism sectors of the economy. Flooding associated with sea-level rise has reduced the level of tourism in Venice (Italy), resulting in fewer jobs and economic losses for the city. The city's productivity is largely tied to its aquaculture industry and tourism. By 2030, flooding and sea-level rise are projected to cost the city €35 to €42 million in decreased tourism levels and €10 to €17 million in aquaculture revenues.[177] Studies have revealed that tourists are unlikely to return to vacation spots in some islands such as Bonaire and Barbados if coral bleaching (which has been linked to warming waters) occurs, resulting in loss of fish and coral species.[178]

The agricultural sector is also vulnerable to climate variability; thus, individuals dependent on it for their livelihoods are at risk. Low-lying areas in Southeast Asia are particularly vulnerable to coastal erosion and flooding, which is likely to result in loss of cultivated land and fishery nurseries. In parts of Africa, livelihoods and national GDP are highly dependent upon the agricultural sector, which accounts for as much as 70 per cent of national GDP in some countries.[179] For urban centres, distant impacts upon tourism and agriculture can potentially result in increased migration from rural areas, which creates more demand for infrastructure and services, though this phenomenon is not well understood.[180]

**Extreme climate events can disrupt the ability of individuals and households in urban areas to sustain livelihoods**

**Climate changes cause local weather conditions ... that affect public health in urban areas**

# PUBLIC HEALTH IMPACTS

Climate changes cause local weather conditions – including extreme heat and severe weather events – that affect public health in urban areas. This section describes these key health issues, focusing on impacts related to extreme temperatures, disasters, epidemics, health services and psychological illnesses. It also considers how poverty acts as a compounding factor which exacerbates the health impacts of climate change.

Climate change can lead to extended periods of heat (i.e. heat waves) and drought. More heat waves have the potential to increase the incidence of heat stress and heat-related mortality.[181] Higher than average night-time temperatures compound heat stress by eliminating the typical period during which the human body can recover from heat stress accrued throughout the day.[182] In particular, several consecutive nights with temperatures above normal can negatively impact upon health, leading to heat-related illness and mortality.[183] For example, the heat wave of 2003 across much of Europe is believed to have caused the death of over 20,000 people. It was the warmest summer since 1540 and could become the norm by the end of the 21st century.[184] Sustained high temperatures in France raised mortality by an estimated 140 per cent compared to historical averages, while over 2000 excess deaths reportedly occurred in England and Wales.[185] In the US, high temperatures result in an average of 400 deaths and many more hospitalizations each year.[186] Projections of climate change impacts in New York City (US) further show significant increases in respiratory-related diseases and hospitalization.[187]

With more individuals moving to urban locations, higher temperatures and a rapidly aging society, the threat of heat-related mortality will become more severe in future.[188] Urban residents are especially at a higher risk of heat-related mortality as a result of the urban heat-island effect.[189] However, death from heat is significantly underreported, as widely accepted criteria in determining heat-related death do not exist. Often, a pre-existing condition is listed as the cause of death, while the role of environmental factors is not considered.[190]

Catastrophic events have both immediate and lasting impacts upon public health. For example, of 238 natural catastrophes occurring from 1950 to 2007, 66 per cent were climate related, most of which involved storms or flooding.[191] Recent flooding in Manila (the Philippines) and surrounding areas affected an estimated 1.9 million people and killed at least 240. Torrential downpours in cities and towns across north-eastern Brazil in 2010 caused floods that rendered at least 120,000 people homeless and killed at least 41 others.[192] As the intensity and frequency of precipitation increases, ever more urban residents will be at risk of injury and property loss.

Increasing intensity of storms and frequency of severe storms threaten to further impact upon urban areas and the health of their residents, as illustrated by recent floods in Pakistan which killed 1100 people.[193] Beyond causing immediate death and injuries, floods and storms can cause long-term damage to facilities that provide health-related services. Power outages can disrupt hospital services, as occurred in Dresden (Germany) in 2002 when floods from the River Elbe affected four out of the six major hospitals in the region.[194] Likewise, clean water provision can be compromised if treatment facilities are structurally damaged or lack power.

Physical climate changes, including temperature, precipitation, humidity and sea-level rise, can alter the range, life cycle and rate of transmission of certain infectious diseases. As indicated earlier, flooding can introduce contaminants and diseases into water supplies, which has been linked to increased incidence of diarrhoeal and respiratory illnesses in both developed and developing countries.[195] Psychological illnesses sometimes also increase following storms and other disasters. Post-traumatic stress disorder, anxiety, grief and depression are commonly observed among individuals following hurricanes and other disasters.[196] Declining local air quality is a further consequence of climate change which threatens health. The photochemical reactions of pollutants in the air which cause smog will intensify as temperatures rise. For example, in Los Angeles, California (US), a 1°C increase in temperatures above 22°C results in an increase in the incidence of smog by 5 per cent.[197]

While the complex relationship of disease incidence with both environmental and demographic factors makes the identification of cause–effect relationships difficult, it is likely that climate change will increase the global disease burden. The World Health Organization attributes at least 150,000 annual deaths to diseases associated with climate change that has occurred since the 1970s, and estimates that death rates from climate-induced disease risk may double by 2030.[198] Malaria may pose a particular problem for populations in developing countries, including those in sub-Saharan Africa. In contrast, precipitation decreases in some parts of Central America and the Amazon region may reduce the rate of malaria transmission.[199] Climate change is also likely to affect the transmission of a number of other diseases, including dengue fever, rodent-borne diseases and diarrhoeal illnesses.[200]

As discussed in a subsequent section of this chapter, diseases can weaken the defences of communities at large and of certain subgroups of a population in particular (e.g. low-income groups). Health impacts, both immediate and long term, tend to hit the poorest urban residents the hardest in part because they often lack mobility, resources and insurance. These residents also typically occupy the highest-risk areas of cities. These and other distributional impacts are discussed in the next section of this chapter.

# SOCIAL IMPACTS

The degree to which human settlements are vulnerable to climate change depends not only on the nature and magnitude of physical changes, but also on the socio-economic characteristics of each city. Cities that experience the same category of hurricane, for example, may incur very different mortality levels and economic losses based on relative wealth and infrastructure. Within cities, too, different population groups are differentially affected by the same weather events and climatic conditions. Climate change differentially impacts upon groups of individuals, such as marginalized minorities, women and men, young and old. These impacts have, until recently, received relatively little attention compared to other distributional issues.

The distributional effects of climate change in urban areas within the context of existing vulnerabilities are reviewed below. In doing so, it is critical to acknowledge and confront compounding vulnerabilities for specific groups in urban areas. Individuals, households and communities who fall into more than one category of vulnerability can find the deck dramatically stacked against them in terms of their ability to prepare for and respond to the varied impacts that they already face and will face in the future. Climate change impacts magnify gender and racial inequalities, often affecting poor minorities and poor women more than any other groups. These impacts often exacerbate poverty as individuals lose their livelihoods and possessions. Sickness and injury, two of the most important factors attributable to increasing poverty, affect the poor more than other groups.[201] A vicious cycle then develops whereby marginalized groups bear the greatest burdens of climate change, thus preventing them from escaping poverty and leaving them continuously vulnerable to further change. Urban planners and policy-makers are thus often charged with confronting multiple social issues at once. Understanding the nature of group-specific climate change dynamics can enable decision-making that seeks to break this cycle – for instance, by promoting inclusion of typically marginalized groups in planning, anticipating the unique needs of groups during disasters and preparing accordingly.

## Poverty

Climate change is considered a distributional phenomenon because it differentially impacts upon individuals and groups based on wealth and access to resources. In general, low-income households in both developed and developing countries are most vulnerable to climate change impacts primarily due to the scale and nature of the assets that they possess or can draw upon (see Box 4.8). The interactions between climate change and income do not affect developing countries alone. There are many examples of poor communities in developed countries faring worse than the wealthier groups during the same disaster. During Hurricane Katrina in New Orleans (US), residents without cars and financial resources to evacuate were left behind. Some of the hardest hit low-lying neighbourhoods were also the poorest, leaving those with few resources to bear most of the devastation.[202]

It has been suggested that the assessment of vulnerability to climate change impacts and the ways in which it is socially distributed can perhaps best be understood by considering six key questions:[203]

1  Who lives or works in the locations most exposed to hazards related to the direct or indirect impacts of climate change (such as on sites at risk of flooding or landslides)?

---

*Climate change will increase the global disease burden*

*Climate change differentially impacts upon groups of individuals, such as marginalized minorities, women and men, young and old*

2    Who lives or works in locations lacking the infrastructure that reduces risk (e.g. drains that reduce flood risk)?
3    Who lacks information, capacity and opportunities to take immediate short-term measures to limit impacts (e.g. to move family and assets before a disaster event)?
4    Whose homes and neighbourhoods face the greatest risks when impacts occur (e.g. because of poorer quality buildings that provide less protection for inhabitants and their physical assets)?
5    Who is least able to cope with the impacts (including illness, injury, loss of property and loss of income)?
6    Who is least able to avoid impacts (e.g. by building better homes, agitating for improved infrastructure or moving to a safer place)?

**The urban residents most vulnerable to climate change are the poor slum and squatter settlement dwellers**

A large proportion of the urban population in developing countries live on sites ill suited to housing – for instance, floodplains or mountain slopes or areas prone to flooding or affected by seasonal storms, sea surges or other weather-related risks.[204] Most such sites are occupied by low-income households because other 'safer' sites are beyond their means. There is also the growing proportion of the world's urban population living in the low-elevation coastal zone[205] – and many studies of particular coastal cities show that most of those most at risk are low-income groups.[206]

**Access to insurance is generally more inclusive in cities in developed countries compared to those in developing countries**

With regard to who lacks information, capacity and opportunity to take immediate short-term measures to limit impacts, the devastation caused in so many low-income settlements by extreme weather is not necessarily a matter of a lack of knowledge or capacities on the part of their residents, although this may be the case for some new arrivals.[207] Even if they know of an approaching storm that may threaten their homes, the residents of informal settlements are often reluctant to move even when advised to do so – for instance, for fear of losing valuables to looters, uncertainty about provisioning for their needs in the places they move to and the worry of not being allowed back if their house and settlement are damaged. For instance, in Santa Fe (Argentina), large-scale floods affecting large

sections of the population have become common – but many of those living in informal settlements at high risk from flooding did not want to move because they had no confidence in the police that they would stop looting and were worried that because they had no legal tenure, they might not be allowed back.[208]

In terms of whose homes and neighbourhoods face greatest risks when impacts occur, studies of disaster impacts from extreme weather in urban areas suggest the majority of those who are killed or seriously injured and who lose most or all their assets are from low-income groups.[209] Many disasters only affect the inhabitants of particular informal settlements and other slums, and most such disasters are not registered in national or international records of disasters.[210] The reasons why most of the inhabitants of informal settlements are so much at risk is obvious: poor-quality housing with inadequate foundations, high levels of overcrowding, lack of infrastructure, etc. Most low-income groups live in housing without air conditioning or adequate insulation, and during heat waves, the very young, older persons and people in poor health are particularly at risk.[211] For instance, in regard to urban centres in India:

> [T]he urban residents most vulnerable to climate change are the poor slum and squatter settlement dwellers and ... they are multiply challenged by even small events that impact their livelihoods, income, property, assets and sometimes their lives. Because of systematic exclusion from the formal economy of the city – basic services and entitlements and the impossibly high entry barrier into legal land and housing markets – most poor people live in hazardous sites and are exposed to multiple environmental health risks via poor sanitation and water supply, little or no drainage and solid waste services, air and water pollution and the recurrent threat of being evicted.[212]

Financial shocks from damage lasts months, even years, after a disaster occurs. Thus, the extent to which the population of a given city is protected by insurance will in large part determine the impact of disasters. Access to insurance is generally more inclusive in cities in developed countries compared to those in developing countries, where poor households typically lack access altogether.[213] Still, low-income households in developed countries can be excluded from insurance where public coverage is inadequate and the costs of private insurance are prohibitively high. Unlike their wealthy counterparts, low-income households often lack the resources to mitigate damages after they occur – for instance, through healthcare, structural repair, communication, food and water.[214] In the absence of adequate recovery assistance, the poor often sacrifice nutrition, children's education or any remaining assets to meet their basic needs, thereby further limiting their chance of recovery and escape from poverty.[215]

Evidently, climate change disproportionately affects lower-income groups in both developed and developing countries. Although these distributional impacts are far from being adequately addressed on international or national levels, the nexus between poverty and climate change has steadily worked its way into the climate change discourse, emerging in focus groups, meetings and reports by many international organizations, including the Organisation for Economic Co-operation and Development (OECD) and the World Bank.[216]

## Gender

In most urban centres, there are significant differences between women and men in terms of their exposure to climate-related hazards, and their capacity to avoid, cope with or adapt to them.[217] This is because men and women differ in their livelihoods, familial roles, production and consumption patterns and other behaviours, perceptions of risk, and are in some cases treated differently with respect to planning and relief efforts during and after disasters (see Table 4.4).

In general, women, especially poor women, are more likely than men to suffer injuries or death when a natural disaster occurs, with more severe disasters correlating with wider gaps in relative risk. Poor women have been found to be more exposed to direct harm from flooding or hurricanes compared to other socio-economic groups.[218] In 1991, a cyclone in Bangladesh killed five times as many females as males.[219] Females comprised more than three-quarters of the deaths in four Indonesian villages hit by the 2005 tsunami, while in the village of Kuala Cangkoy, where the worst devastation occurred, females accounted for 80 per cent of the deaths.[220] Gendered impacts are evident in rich countries as well, particularly in poor communities. For example, in the French heat wave of 2003, about 70 per cent of fatalities were women, although this number may be artificially high since there are more women than men in older age groups.[221]

To some extent, the higher death rates for women in disasters can be explained by the fact that women comprise the majority of the world's poor population, who face vulnerability factors as previously discussed. This statistic,

however, can obscure the many other important factors that place women at greater risk than men. In developing countries, women often experience unequal access to resources, credit, insurance, services and information. Women's socio-cultural roles and typical care-giving responsibilities often prevent them from migrating and seeking shelter before and after disaster events. In some cases, women may not be allowed to travel alone and may be prevented from learning skills that could aid their survival during a disaster. Also, women's lower economic status increases their vulnerability in the event of a disaster occurring. When homes are destroyed or damaged, this often affects women's incomes more than men's as they often engage in income-generating activities from home and therefore lose income when homes are destroyed.[222] Where access to resources and the social status of women are nearly equal to that of men, the mortality difference between the sexes is much smaller or, in fact, negligible, compared to societies with wide gender inequalities.[223]

The method by which aid is distributed following disasters further contributes to gendered vulnerability. In both developing and developed countries, women may have limited capacity to secure relief aid, whether due to formal assistance policies or cultural norms.[224] In Bangladesh, for example, women have traditionally had difficulty receiving relief aid after disasters because it was difficult for them to wait in long lines at recovery centres when they needed to care for children at home. Expanded recovery systems that provide door-to-door service are helping to address this issue.[225]

Households that are headed by women do not always receive the assistance they need when disaster relief is tailored to reintegrate men into the workforce or when it privileges male-headed households for relief aid.[226] For example, relief checks following Hurricane Andrew in Miami (US) were distributed to men as traditional heads of household, ignoring the reality that many families were then headed by women.[227] If men leave their families, as frequently occurs following a natural disaster, women are rendered ineligible for public assistance or may go unrecognized by the system.

> Women, especially poor women, are more likely than men to suffer injuries or death when a natural disaster occurs

> Women's lower economic status increases their vulnerability in the event of a disaster occurring

| Aspect of vulnerability | Contribution to urban vulnerability | Contribution to climate vulnerability |
|---|---|---|
| Gendered division of labour and 'poverty of time' | Women have prime responsibility for 'reproductive' labour; lack of time to engage in 'productive' labour | Limited financial assets to build resilience and to cope with disaster events |
| Gender-ascribed social responsibilities | Women have prime responsibility for 'reproductive' labour; lack of time to engage in 'productive' labour | Additional domestic responsibilities when access to food, water and sanitation are disrupted; additional time required to care for young, sick and elderly |
| Cultural expectations of gender norms | Constraints on women's mobility and involvement in certain activities | Higher mortality from disaster events due to lack of skills and knowledge |
| Unequal entitlements to land and property | Limited access to productive resources | Limited ability to invest in more resilient land or shelter |
| Higher representation of women in informal sectors | Lower wages and lack of financial security | Damage to homes and neighbourhoods affects women's incomes more severely as income-earning activities are often undertaken at home |
| Safety and security in public spaces | Limited freedom to use public space | Particular problem in temporary accommodation/ relocation sites; high rates of sexual abuse and violence |
| Limited engagement of women in planning processes | Urban plans fail to meet particular needs of women and children | Climate adaptation plans fail to meet needs of women and children; failure to incorporate women's perspectives may result in higher levels of risk being accepted |

*Source: IFRC, 2010*

**Table 4.4**

**Gender and climate vulnerability**

Likewise, trauma programmes are often not tailored to the specific, and, at times, unique needs of men and women. In the aftermath of storms, women often disproportionately experience sexual or domestic violence.[228] Disaster relief programmes are sometimes inadequate to meet the medical needs of women, especially those related to reproductive and psychological health. Women were 2.7 times more likely than men to exhibit clinical symptoms of post-traumatic stress disorder following Hurricane Katrina, and many went untreated for years after the event because of limited access to public assistance programmes and lack of health insurance.[229] In some cases, men may take greater risks following natural disasters and may not receive treatment for trauma because of gender roles and stereotypes.[230] The psychological needs of men may also be overlooked in disaster programmes; for example, men were not offered counselling following flooding of the Koshi River in 2008 that affected Bihar, India and Nepal.[231]

Restrictions on women's livelihood also increase their vulnerability to climate change in both developing and developed countries. Women sometimes have less access to education compared to men and tend to earn lower wages than men as well, especially in developing countries (even for the same work). Similarly, in developed countries, gender differences in employment opportunities and pay are one of the greatest contributors to increased poverty rates among women.[232] In many developing countries, marriage customs may prevent women from working at all outside the home, and may remove women from social networks and extended family.[233] Women's ability to contribute to their own welfare and garner resources and investments that could help them recover from disasters is thus limited.

Furthermore, women are often excluded from planning processes and discussions about climate change. As a result, the perspectives and needs of women are insufficiently incorporated within processes and mechanisms to address climate change, if they are included at all. There is little evidence of specific efforts to target women in adaptation activities funded by bilateral and multilateral programmes. In excluding them from planning processes, an opportunity is missed to gain the unique knowledge that women possess regarding mitigation strategies, natural resource use, and adaptation and coping strategies following disasters. For instance, as primary caregivers, women could provide vital information about storing and protecting food and valuables during a disaster, educating children about survival strategies, and reinforcing structures before and after a severe weather event.[234]

## Age

Young children are particularly vulnerable to climate change impacts, in part because of their physiological immaturity (see Box 4.9). Due to their limited cognitive ability and behavioural experiences compared to adults, children are less equipped to handle disaster risks. They are more susceptible to diarrhoeal diseases and malaria – which, as mentioned earlier, are anticipated to increase with climate change in many regions. Furthermore, physical health damage can be more severe and long lasting in children than adults because their bodies and organs are still developing,[235] and higher metabolism in children makes their need for constant sustenance more pressing than it is for adults. Food and water scarcity thus has particularly rapid and serious consequences for children living in poverty.[236]

Children have limited ability to care for their basic needs and take actions to adjust their physical conditions to cope with external conditions. Adults are responsible for these and other needs of children, including providing information. In the absence of adult support, these and other issues – including reduced ability to communicate effectively and highly restricted mobility – leave children especially vulnerable to climate change impacts.

For some children in some places, the added challenges brought by climate change (including higher risks from under-nutrition, intestinal parasites, diarrhoeal diseases or malaria) could erode their opportunities for learning and growth – for instance, through lower cognitive capacity and performance. Learning is also dependent on supportive social and physical environments and the opportunities to master new skills. Disasters often result in the interruption of formal schooling for months at a time, and children are more likely to be withdrawn from school when households face shocks.

Levels of psychological vulnerability and resilience depend on children's health and internal strengths, as well as household dynamics and levels of social support. Poverty and social status can have an important role in this regard. The losses, hardships and uncertainties surrounding stressful events can have high costs for children. Increased levels of irritability, withdrawal and family conflict are not unusual after disasters. High stress for adults can have serious implications for children, contributing to higher levels of neglect. Increased rates of child abuse have long been associated with such factors as parental depression, increased poverty, loss of property or a breakdown in social support.

Displacement and life in emergency or transitional housing after disasters or evictions have been noted in many contexts to lead to an erosion of the social controls that normally regulate behaviour within households and communities. Overcrowding, chaotic conditions, lack of privacy and the collapse of regular routines can contribute to anger, frustration and violence. Adolescent girls especially report sexual harassment and abuse. The synergistic and cumulative effects of such physical and social stressors can affect children's development on all fronts. As the numbers of displaced people grow, these dysfunctional environments are likely to become the setting within which more and more children spend their early years.

Even less extreme events can create havoc in families' lives, deepening the level of poverty. When times are hard, children can become an asset that is drawn on to maintain the stability of the household. Children may be pulled from school to work or take care of siblings. Some children may be considered more 'expendable' than others. Many of the young prostitutes in Bombay (India) are from poor rural villages in Nepal, where inadequate crop yields lead families to sacrifice one child so that others may survive.

**Women are often excluded from planning processes and discussions about climate change**

**Young children are particularly vulnerable to climate change impacts, in part because of their physiological immaturity**

---

**Box 4.9 Climate change risks for children**

Drawing on studies on children and their vulnerabilities, it is possible to highlight the following risks associated with climate change that have clear impacts upon child health and survival:

- *Mortality in extreme events:* in most developing countries, the loss of life is disproportionately high among children – especially among the poor – during such extreme events as flooding, high winds and landslides. Children are 14 to 44 per cent more likely to die due to environmental variables, including extremes in temperature, flooding and severe weather events, than the total population at large.[a] For example, drowning incidents in floods are particularly high for children in Kampala (Uganda).[b] A study of flood-related mortalities in Nepal found that the death rate for children aged two to nine was more than double that of adults; and pre-school girls were five times more likely to die than adult men.[c] The average death rate for children was twice that of adults in flooding in Nepal, with poor children suffering the highest death rates.[d]

- *Water and sanitation-related illnesses:* children under five are the main victims (80 per cent globally) of sanitation-related illnesses (primarily diarrhoeal diseases)[e] because of their less developed immunity systems and because their behaviour can bring them into contact with pathogens. This also results in higher levels of malnutrition and increased vulnerability to other illnesses. Droughts, heavy or prolonged rains, flooding and conditions after disasters – as well as climate change-related constraints on freshwater supplies in many urban centres – all intensify the risks, which are already very high in informal settlements or other areas with concentrations of low-income groups.

- *Malaria, dengue and other tropical diseases:* warmer average temperatures are expanding the areas where many tropical diseases can occur, with children most often the victims. In many locations, the most threatening tropical disease is malaria. Up to 50 per cent of the world's population is now considered to be at risk. In Africa, 65 per cent of mortality is among children under five.[f] Malaria also increases the severity of other diseases, more than doubling overall mortality for young children. Climate change is also accelerating the comeback of dengue fever in many countries in the Americas.[g]

- *Heat stress:* young children, along with older persons, are at highest risk from heat stress. Research in São Paulo (Brazil) found that for every degree increase above 20°C, there was a 2.6 per cent increase in overall mortality in children under 15 (same as for those over 65).[h] Risks for younger children are higher. Those in poor urban areas may be at highest risk because of the urban heat-island effect, high levels of congestion and little open space and vegetation.[i]

- *Malnutrition:* malnutrition results from food shortages (e.g. as a result of reduced rainfall, other changes affecting agriculture, or interruptions in supplies during sudden acute events) and is also closely tied to unsanitary conditions and to children's general state of health. If children are already undernourished, they are less likely to withstand the stress of an extreme event. Malnutrition increases children's vulnerability on every front and can result in long-term physical and mental stunting.

- *Injury:* after extreme events, injury rates go up. Children, because of their size and developmental immaturity, are particularly susceptible and are more likely to experience serious and long-term effects (from burns, broken bones, head injuries, etc.) because of their size and physiological immaturity.[j]

- *Quality of care:* as conditions become more challenging to health, so do the burdens faced by caregivers. These problems are seldom faced one at a time – risk factors generally exist in clusters. Overstretched and exhausted caregivers are more likely to leave children unsupervised and to cut corners in all the chores that are necessary for healthy living.

*Sources:* a Bartlett, 2008; b Mabasi, 2009, p5; c Pradhan et al, 2007; d UN, 2007; e Murray and Lopez, 1996; f Breman et al, 2004; g World Bank, 2009c; h Gouveia et al, 2003; i Kovats and Akhtar, 2008; j Berger and Mohan, 1996

---

Older persons share similar physical and social vulnerabilities with children. Pre-existing illnesses and physical ailments limit their mobility and coping capacity, which may prevent them from evacuating or seeking shelter in emergency situations. Since their bodies adjust more slowly to physical conditions than younger populations, they may not perceive excessive heat quickly enough to prevent heat stroke. Empirical evidence indicates that the elderly display disproportionately higher injury rates after natural disasters and higher rates of heat wave mortality.[237] A recent study on climate change impacts in Oceania found that 1100 people aged over 65 die each year in ten Australian and two New Zealand cities as a result of heat waves.[238]

While adaptive measures do exist to help the elderly combat their physical vulnerability, these mechanisms are often solely accessible to wealthy populations. The elderly are more likely than younger people to require assisted transportation out of a dangerous situation; but poor individuals may not be able to afford private transportation. Lack of personal contacts and distrust of strangers decreases their access to volunteered assistance, and they are also less likely to accept financial assistance from public recovery and aid programmes.[239]

The vulnerability of the elderly is, thus, like other groups, dependent upon their economic status. Yet, all else being equal, the elderly show disproportionate rates of poverty, for the most part because they no longer maintain a source of income and tend to have low endowments of assets.[240] While rates and magnitude of poverty are greater in developing countries, poor older persons in developed countries are more likely to live alone and be socially isolated.

## Ethnic and other minorities (including indigenous groups)

Racial and ethnic minorities also exhibit increased vulnerabilities to climate change in both developed and developing

The elderly display disproportionately higher injury rates after natural disasters and higher rates of heat wave mortality countries. Discriminatory practices often segregate groups of minorities into the highest-risk neighbourhoods, usually without access to insurance and loans as security against climate change impacts. The majority of flooding victims in Bihar (India) in 2007 were 'untouchable' low-caste groups who resided in floodplains and areas prone to landslides.[241] The most vulnerable low-lying communities in New Orleans (US) are comprised mostly of African-Americans. This group suffered the relatively most severe losses of life and assets during Hurricane Katrina.[242]

In both developed and developing countries, provision of government assistance following disasters is often less accessible to racial and ethnic minorities. Aid workers may not be properly educated regarding cultural norms, or important information regarding assistance may not be available in the right language.[243] Since aid is sometimes structured around the household as a single family unit, some ethnic minorities may not receive as much assistance as majority groups. For instance, in the US, the Federal Emergency Management Agency's assistance to Haitian residents of Florida following tropical storm damage has been found to be insufficient because several families tend to occupy a single household.[244] Outright exclusion of certain groups from disaster relief occurred during recent disasters in South Asia, including during the flooding of the Koshi River in 2008 in Bihar, India and Nepal, the 2005 Kashmir earthquake and the 2004 Indian Ocean tsunami. Assessments of the relief and reconstruction efforts following these disasters have revealed discriminatory practices and human rights abuses against women, the poor, indigenous groups and the disabled.[245] Furthermore, the knowledge of government views towards minority groups or previous examples of discrimination in some cases discourages minorities from seeking assistance.[246]

Similarly, indigenous peoples in many areas have historically faced factors that can increase their relative vulnerability. Alienation from decision-making, education, healthcare and information regarding assistance and relief programmes are common among indigenous peoples. Moreover, indigenous peoples often lack security of land tenure and legally recognized property rights, which can force them to settle in hazardous areas if they are removed from their land.[247] Lack of legal property can also limit the ability of indigenous peoples to adapt to climate change, particularly when, for example, their adaptation strategies involve seasonal migration due to drought. If their traditional means of adapting are restricted by denial to move into new areas, they may not be able to cope with changing climatic conditions.[248]

## DISPLACEMENT AND FORCED MIGRATION

Millions of people move each year, with over 5 million crossing international borders into developed countries and even greater numbers moving into or within developing countries.[249] The reasons why people move are complex and interrelated, and there is evidence that poor environmental conditions can contribute to the decision of groups or individuals to move. As the world's climate changes, resulting environmental degradation, drought and sea-level rise may lead to the permanent displacement of people and, consequently, increased internal and international migration. The term migrant does not imply that movement was forced, but refers to a person who has changed place of residence either by moving across international borders (international migrant) or moving within one's country of origin (internal migrant).[250] This section describes the observed role of the environment in migration, projections for future migration as a result of climate change, and the consequences of migration.

Migration has been documented around the world both as a response to sudden-onset natural disasters and slow-onset changing environmental conditions. In 2008, an estimated 20 million individuals were displaced due to sudden-onset natural disasters alone.[251] Flooding and severe storms have been linked to migration in the Philippines, Pakistan, China and the Democratic People's Republic of Korea.[252] Decades of drought and land degradation have contributed to the relocation of nearly 8 million inhabitants of north-eastern Brazil to the central and southern regions of the country since the 1960s.[253] In Ghana, studies have found evidence of drought-induced internal migration from north-western to central and southern regions. Northern Ghanaians relocate to Ghana's middle regions because of the combination of poor agro-ecological conditions at home and easy access to fertile lands in the more humid south. As such, 30.8 per cent of people born in north-western Ghana now live elsewhere.[254]

Still, figures for environment-related migration are contentious because it is difficult to ascribe a single cause to most migration events. Evaluation of historical migration events, both permanent and temporary, suggests that environmental decline can serve as an important 'push' factor in generating movement; but it is not typically the sole causative agent of migration. As a further complicating issue, environmental degradation itself may occur not only due to climate change impacts, but as a side effect of war, political instability, overpopulation or widespread poverty. Changing environmental conditions may exacerbate longstanding problems such as conflict or food shortages. Many factors that can be implicated in migration are difficult to entangle, and it is impossible to ascribe blame to a single starting factor.

The response of any particular community to environmental change depends on a variety of socio-economic and historical considerations. In the least developed countries where rural economic activities are disrupted by environmental conditions (e.g. drought), migration is usually temporary and internal.[255] If societies are able to adapt to slow-onset changes, they may only migrate seasonally or individuals may leave temporarily and send back resources to their remaining family. While sudden disasters often force people to move quickly to a safe location, the poor do not often have the resources to move, and loss of resources during disasters may only make it less likely that low-income households will eventually relocate.[256]

*Racial and ethnic minorities also exhibit increased vulnerabilities to climate change in both developed and developing countries*

*As the world's climate changes, resulting environmental degradation, drought and sea-level rise may lead to ... increased internal and international migration*

Projections for future climate change-related displacement average 200 million migrants by 2050; yet estimates depend greatly on the degree of climate change and how abruptly change occurs.[257] Despite difficulties in predicting global migration patterns, there are areas that may be particularly affected because of their vulnerability to risk factors. Populations located at low elevations are vulnerable to climate-induced migration, especially in areas where other vulnerability factors exist (e.g. overcrowding). Small island states, including the Bahamas, Marshall Islands and Kiribati, are located entirely below 3m or 4m above sea level, so their populations here may have to relocate entirely as sea-level rise and coastal subsidence continue.[258]

It is difficult to establish where displaced residents are likely to relocate to. In most historical cases, displaced residents move to other regions of their native country. Rural to urban migration has been a major component of urbanization across Africa and in Asia, though it should not be an assumed response to environmental degradation around the world. In regions with strong agriculture sectors, migrants may move from one rural area to another rather than from rural areas to cities. In rapidly urbanizing countries (e.g. in countries throughout the Latin American and Caribbean region), migration from one city to another is common. Rural to urban migration typically happens where economic growth is occurring or there is an expansive manufacturing or service economy.[259] However, it is also reasonable to expect that some international migration may occur where inland relocation is impossible or where cultural and historic relationships exist between countries.

Depending on the scale and nature of these events, migration can result in social disruption or conflict, especially if migratory events bring into contact peoples with pre-existing social or cultural tensions. New arrivals to cities may also be seen as competition for jobs or resources, generating distrust and possibly leading to conflict. Social disruption is particularly likely in developing countries where cities may be less able to absorb new residents. In addition, political instability common to many developing countries can at best fail to mitigate conflicts and can, at worst, facilitate them.[260]

Forced migrants can also find themselves vulnerable to a range of risks, climate related and otherwise. They often face threats to their health and personal security and, in some areas of the world, are at danger of human trafficking and sexual exploitation.[261] The nature and magnitude of these implications will depend on the location of events, number of people involved, and the time-scale over which migration occurs and preparations have been made.

A growing body of evidence suggests that threats to livelihoods, immigration and resource scarcity can become sources of violent conflict, and indicates that climate change can directly and indirectly influence these trends.[262] Indeed, the United Nations Security Council acknowledges climate change as a threat to human security as resource scarcity, water stress and migration potentially lead to competition and conflict.[263] Still, there is much uncertainty about the specific causal links between climate change, human insecurity and the risk of violent conflict. More research regarding conflict, especially at the regional level, would be useful in identifying where policy intervention may be necessary now and in future.[264]

# IDENTIFYING CITIES VULNERABLE TO CLIMATE CHANGE

The concept of vulnerability in relation to climate change is also applicable to larger systems such as cities or city-regions or to resources and ecosystem services. This section of the chapter describes the key indicators of vulnerability in urban areas with regard to risk of exposure and adaptive capacity. Cities may not only serve as sources of particular vulnerability to climate change, but also as centres of concentration for resources, novel ideas and capacity for technological innovation. In this sense, although cities face interacting risk factors from climate change, they may have the ability to respond to climate change while providing tools and lessons for others.

## Urbanization

As indicated in Chapter 1, levels of urbanization are increasing worldwide. Population growth in urban centres has the potential to significantly exacerbate climate change impacts. Increasing population means greater demand for resources – including energy, food and water – and greater volumes of waste products. Thus, for those regions of the world where resource scarcity is an existing problem, urbanization can be a significant vulnerability factor. Population growth can also cause stronger urban heat-island impacts, which can be a particular challenge for small compact cities such as those typical of Southern Europe.

Where population growth occurs rapidly, demand for housing, infrastructure and services can grow much faster than supply. This can force development in hazardous areas or with inadequate construction materials and techniques. In many cases in developing countries, urban slum expansion results in part because population growth outpaces the construction of adequate affordable housing. Unplanned population growth can also result in sprawling urban settlements that encroach upon natural flood and storm buffers.

Rates of urbanization are higher in developing countries, which are less prepared than developed countries to deal with the resulting impacts. For these regions of the world, population growth can act as an acute threat multiplier, concentrating residents in high-risk areas without infrastructure or services, and accelerating environmental degradation. As cities continue to rapidly urbanize surrounding areas, they typically increase their exposure to climate events as development patterns expand into areas that are more vulnerable to climate change and extreme climatic events.[265]

Urban areas face a dichotomy with regard to their vulnerability and resilience to climate change. On the one hand, larger cities are more likely than other regions to be affected by climate events because of their larger size and populations.[266] On the other hand, larger cities tend to have

*Projections for future climate change-related displacement average 200 million migrants by 2050*

*Population growth in urban centres has the potential to significantly exacerbate climate change impacts*

a significant accumulation of human and financial capital that allows them to plan and respond more effectively to extreme events, as cities can draw on talents and financial resources from around the world to aid in rescue and recovery.[267] Also, adaptation can be expensive and, as a result, larger cities tend to be better protected, both by engineering works and early warning systems.[268] However, this generalization does not necessarily hold true in developing cities, where prevalence of slums, inadequate governance and limited resources can reduce resilience.[269]

Despite the inherent issues associated with growing urban populations, most problems can be mitigated with urban planning that diverts growth away from highly hazardous areas, enforces energy and water efficiency standards for buildings, and minimizes urban heat effects. Thus, the extent to which urbanization acts as an additional source of vulnerability often depends on the integration of future population projections within land-use and infrastructure planning at the city level.

## Economic development

Climate change impacts are not experienced in the same way by cities in developing and developed countries. Risk is skewed towards developing countries such that more people are at risk of being affected by a natural disaster in a developing country compared to a similar disaster in a developed country.[270] Lack of economic strength, as is the case in many developing country cities, exacerbates vulnerability by limiting the ability to minimize and adapt to the impacts of climate-related hazards. Studies have linked the size of a city's coastal population and economy (e.g. GDP and GDP per capita) to its vulnerability to sea-level rise.[271] Other issues underlying the risk differential between cities in developing and developed countries include the integrity of infrastructure and urban planning, or lack thereof; the availability of resources and information; levels of risk awareness; presence of disease and malnourishment; and dependence on natural resources.

Developing country cities often lack risk management plans, early warning systems and the ability or foresight to move residents to safer locations when disasters are inevitable. Their local authorities do not have the capacity to respond to natural disasters, and if laws or plans do exist for disaster response, they are rendered ineffective from lack of human or financial capital to enact them. For instance, the capacity of local authorities in developing countries to minimize the effects of flooding is restricted compared to developed countries, including through physical protection such as complex and modern water treatment and catchment systems, flood barriers and other risk buffers. Moreover, due to the unequal distribution or general lack of resources, political instability and corruption, many developing country cities lack the network of governmental and non-governmental institutions that aid recovery efforts in wealthier countries.[272] As a result, developing countries can experience great physical damage during flooding or severe weather events and often have difficulty rebuilding their infrastructure and economy. Furthermore, a recent study

concludes that the National Adaptation Programmes of Action (NAPAs) which are intended to guide adaptation responses in least developed countries and small island states are inadequate to protect public health from climate change impacts.[273]

The diversity of local sources of income is a further important facet of the magnitude of climate change impacts in cities. Where cities are reliant upon few industries for the majority of local economic productivity, they can be seriously affected if those activities are affected by climate change both due to short-term monetary losses and longer-term economic decline. In those areas with low economic diversity, loss of a single industry leaves few other options for workers who lose their jobs. In Venice (Italy), for example, flood impacts on tourism and aquaculture leave the city's future uncertain.

An additional vulnerability factor for cities is the degree of disparity between high- and low-income groups. In both developing and developed country cities, the poorest are typically the hardest hit by natural disasters and least able to cope with a range of climate change impacts. Those cities with great income inequality and large populations of residents living in poverty have inherently high vulnerability.

Some developing country cities may be unable to prepare for climate change or to cope with climate change because they are hampered by outbreaks of disease or chronic malnourishment. Unhealthy populations have reduced mobility and may be especially sensitive to water and food shortages. Prevalence of HIV (human immunodeficiency virus) and AIDS (acquired immune deficiency syndrome), for example, have been cited among the primary reasons that the population of Malawi has been increasingly vulnerable to the effects of regional drought.[274] The impacts of disease do not end with infected individuals, but rather weaken the defences of entire communities. As a greater proportion of the population becomes sick, food and economic productivity declines and contributes to higher rates of poverty and malnourishment.[275] It is clear that similar effects could occur not only in regions with AIDS epidemics, but those experiencing outbreaks of plague, flu and other infectious disease.

## Physical exposure

The level of vulnerability of an urban area to climate change risks depends, in part, on how much of the city's population and economic assets are located in high-risk areas (i.e. exposure). In many cases, exposure level will be a function of the location of the city itself. Many of the world's largest cities are located in areas vulnerable to climate events, such as low-lying coastal areas. Though low-elevation coastal zones account for only 2 per cent of the world's total land area, this area accounts for approximately 13 per cent of the world's urban population.[276] Coastal cities in this zone have high levels of exposure – both of population and assets – to sea-level rise, storm surges and flooding simply as a function of being so near the ocean.

Exposure can also be linked to land-use planning within the city, including continued development in known

Climate change impacts are not experienced in the same way by cities in developing and developed countries

The level of vulnerability of an urban area to climate change risks depends ... on how much of the city's population and economic assets are located in high-risk areas

hazardous zones, and the destruction of natural protective areas.[277] Coastal communities who encroach onto wetlands, sand dunes and forested areas increase the likelihood of flooding, together with all its associated impacts upon housing structures, transportation networks and water quality.[278]

Weak structural defence mechanisms and oversight of building codes further increase the vulnerability of cities in high-risk areas. Sea walls, levees, dykes and water pumps can reduce the chances and intensity of flooding from storm surges and heavy precipitation, while reinforcements on housing and transportation systems can limit damage when flooding does occur. Those cities with inadequate, aging structural defences and infrastructure in need of repairs or upgrades are often highly vulnerable to climate change risks. The system of structural defences throughout cities in Japan, for example, has resulted in fewer cyclone damages than in cities in the Philippines, even though exposure risk in Japan is generally higher.[279]

In particular, the physical infrastructure of slums increases the vulnerability of residents to climate change impacts. In 2010, nearly 32.7 per cent of the urban population in developing countries lived in slums,[280] which are especially vulnerable to climate change. The very defining characteristics of slums – namely, structures of substandard quality, lack of basic services, overcrowding and social exclusion – clearly suggest that residents are particularly vulnerable to climate change impacts.[281] Disaster risk is often high for slums because construction occurs in particularly hazardous areas, including steep slopes or in floodplains. In Nairobi (Kenya), for example, poor urban planning has resulted in residential and commercial development in floodplains that restricts water flow and increases the likelihood of flooding.[282] The lack of adequate drainage systems leaves such settlements open to rapid flash floods, as in the case of those that occurred near Caracas (Venezuela) in 1999 and Mumbai (India) in July 2005.[283] In Mozambique, politicized land distribution systems and high pricing forces urban residents to live in unregulated slums and informal settlements with inadequate drainage. As a result, severe flooding in 2000 disproportionately affected the urban poor living in a number of urban locations.[284] Box 4.10 further illustrates the challenge of flooding in the slums of Kampala (Uganda).

## Urban governance and planning

The ability of urban centres to prepare for and respond to climate change is linked in large part to the quality of local governance and the strength of the institutional networks available to provide assistance to residents, as elaborated upon in greater detail in Chapters 5 and 6. Urban governance and planning can improve resilience to climate change impacts through targeted financing of adaptation, broad institutional strengthening and minimizing the drivers of vulnerability.[285] Urban areas with weak governance systems – as a result of political instability, exclusion of climate change from the political agenda or lack of governmental resources – are highly vulnerable to climate change impacts.

**Box 4.10 Vulnerability of slums to climate change: The case of Kampala, Uganda**

Kampala, the capital city of Uganda, has been experiencing rapid urbanization and slum expansion. Currently, over 50 per cent of the urban population live in informal settlements characterized by poor sanitary conditions, infrastructure deficiencies and lack of waste disposal services.

In these areas, even relatively small amounts of rain can cause flooding. The natural drainage capability of the land has been impaired because of the extensive amounts of construction, complex roadways and collection of trash and debris. Runoff is therefore six times that which would occur in a natural environment, leading to hazardous conditions during rains. Flood-related accidents result in deaths of slum residents each year, many of them children. Sewers are available to only a small proportion of the population, so flooding carries faeces and spreads diarrhoeal diseases such as cholera.

Increasing variability of rainfall and more intense storms have compounded the problems that already exist in the slums of Kampala. Climate change is likely to increase the incidence of flooding and accelerate the spread of diseases, including malaria and waterborne diseases. Climate change here has the potential to worsen poverty, especially among poor women, who have limited, if any, access to credit or property compared to men, and who are often excluded from decision-making processes.

*Source:* Mabasi, 2009

In many cities throughout developing countries, populations continue to grow in the absence of effective urban planning, resulting in living conditions that exacerbate climate change impacts, and development in vulnerable areas such as coastal zones at risk from sea-level rise, flooding and coastal storms. Similarly, weak building codes and standards (or lack of enforcement) increase the vulnerability of individual households and entire communities.[286]

Civil society institutions – including community-based organizations, NGOs, faith-based organizations and organizations for minorities and women – can mitigate vulnerability by helping populations cope with and adapt to change. These may be especially powerful resources for underrepresented minorities, women and indigenous peoples whose unique needs are often overlooked even where climate change is a focus of political institutions. Cities where these resources are unavailable or discouraged may be particularly vulnerable to change.

## Disaster preparedness

Natural and human-made disasters have been on the rise worldwide since the 1950s, coinciding with the rise in world urban population (see Figure 4.2).[287] As climate change continues to occur, disasters such as landslides, floods, windstorms and extreme temperatures may occur with greater frequency and intensity. Urban vulnerability to climate change will therefore depend upon disaster preparedness, defined by the International Strategy for Disaster Reduction Secretariat as 'activities and measures taken in advance to ensure effective response to the impact of hazards, including the issuance of timely and effective early warnings and the temporary evacuation of people and property from threatened locations'.[288]

Disaster preparedness may be linked to governance

*The physical infrastructure of slums increases the vulnerability of residents to climate change impacts*

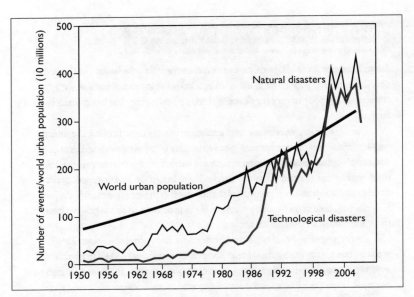

**Figure 4.2**

**World population and recorded natural and technological disasters (e.g. industrial and transport accidents) (1950–2005)**

*Source:* UN-Habitat, 2007, p170

Urban vulnerability to climate change will therefore depend upon disaster preparedness

and institutional capacity and the availability of information to residents; but it is not necessarily the case that poorer countries or cities will always be less prepared. For example, despite being a relatively poor country, Cuba has implemented effective disaster preparation mechanisms. On the other hand, although the US is a relatively rich country, it has sometimes proven to be ill prepared for disasters; for instance, emergency response was inadequate both before and after Hurricane Katrina struck the city of New Orleans.

# CONCLUDING REMARKS AND LESSONS FOR POLICY

Climate change impacts have real implications in the urban environment, many of which will continue to exacerbate existing vulnerabilities and social issues in the future. While local climate change risks, vulnerabilities and adaptive capacity vary across cities, the global review undertaken in this chapter reveals several key common themes.

First, climate change impacts may have compounding effects across many sectors of city life. The specific nature of climate change risks is heterogeneous around the world; but these risks have compounding effects in nearly any context. For example, extremely high temperatures have direct impacts upon human health, placing individuals at risk of heat-related illness or mortality. At the same time, increasing temperatures in certain locations increase demand for energy, which can reinforce climate change by increasing greenhouse gas emissions and exacerbating the urban heat-island effect. Cities have inherent properties that can interact with climate change effects – including rapid population growth, high population density, urban heat-island impacts and the presence of poverty – such that impacts that may appear minor when considered individually may have serious effects when considered together in local context.

Second, climate change does not affect everyone within a city in the same way: gender, age, race and wealth have implications for the vulnerability of individuals and groups. Racial and ethnic minorities, indigenous peoples, poor populations and socially isolated individuals are highly

vulnerable to climate change impacts. The poor are often least able to cope and adapt to climate change impacts because they have relatively few resources and tend to be located in the most hazardous areas. Indigenous peoples, minorities and women may be explicitly or tacitly removed from decision-making processes and, in some cases, have limited access to insurance, information and resources. As a result, these groups are both less prepared for physical hazards and less able to adapt. These effects tend to be particularly pronounced in cities in developing countries compared to developed countries, but are evident worldwide.

Third, planning within cities – including siting of residential areas, businesses and transportation infrastructure – often proceeds based on historic climate data, increasing the risk of various sectors to changing conditions. Because of low land prices and less resistance from residents, infrastructure (including ports, water sanitation facilities, power plants, roads and airports) tends to be constructed in vulnerable areas. These assets are long lived and will therefore be subjected to changing conditions such as sea-level rise, more variable precipitation and increased intensity of storms. Failure to adjust zoning and building codes and standards with an eye to the future may limit the prospect of infrastructure adaptation and place lives and assets at risk. Likewise, failure to consider the impacts of rising populations in city planning leads to conditions that exacerbate the vulnerability of residents to climate change, as illustrated by the case studies reviewed in this chapter. These conditions include water and other natural resource scarcity, environmental degradation and development of urban slums.

Fourth, climate change impacts can be long lasting and propagate worldwide. When disasters related to climate change occur, focus on the affected areas tends to be limited to a short period of time following the event. Yet, experience reveals that the social and economic impacts of these disasters can extend for months or years. Damage to transportation infrastructure can interfere with a city's ability to recover from extreme climate events. Lack of insurance coverage can make it very difficult for individuals to cope in the aftermath of disasters, particularly among the poor, who may not have savings or assets to use to repair damage to their homes or to purchase the necessities of recovery. Moreover, cities around the world, in particular large cities, are interconnected by capital and labour markets. Extreme climate events that result in economic losses in urban areas or interruption of trade routes can thus result in long-lasting rebounding global impacts.

Fifth, limitations on governance and planning increase the vulnerability of cities, especially in developing countries, to climate change. Poor planning resulting from scarce resources, limited information and/or political corruption limit the ability of cities to prepare for climate change as well as to recover when climate-related impacts occur. In developing countries, in particular, poor planning has encouraged the development of slums and informal settlements that are prone to damage from climate-related impacts. Slum expansion can be difficult to control because these settlements sometimes develop outside of the jurisdiction of local government. In both developed and developing countries, inadequate preparation for climate-related disasters has led to great losses of life and assets when individuals were not evacuated before a disaster or rapidly attended to afterwards.

Taken together, the themes discussed above suggest that the direct and indirect impacts of climate change will continue to threaten the very social and economic fabric of cities. International, national and local governments and institutions can benefit from the growing body of research on climate change impacts by adjusting their policy approaches with a mind to the future.

The many examples of climate change impacts reviewed in this chapter highlight the context-specific nature of impacts. Accordingly, policies ought to be designed to address local physical impacts and vulnerabilities to the greatest extent possible. This does not, however, preclude the importance of national government and international collaboration on the global challenge of climate change. In fact, security issues, migration and resource scarcity will often raise issues that cross local and national boundaries.

Likewise, policies and interventions should be devel-

oped with attention to the social and economic characteristics of resident populations in order to reduce, rather than reinforce, inequalities. Care should be taken to identify who bears the greatest burden of climate change in a given area and to develop policies with the goal of minimizing this burden. Increasing the participation of groups who have been typically marginalized – whether indigenous groups, low-income groups, women and/or racial minorities – can help to both reduce the distributional impacts of climate change and broaden the knowledge base used to tackle climate change.

Perhaps the most important lesson for policy-makers is that climate change should no longer be considered a solely environmental challenge, addressed in isolation from other social and economic issues. Climate change in urban areas interferes with a wide range of existing and emerging policy challenges, among them poverty eradication, water sanitation, scarcity of food and water, and population growth. When climate change is embraced as an integral part of these challenges, solutions can be designed to more adequately reflect and address its myriad impacts upon cities.

**The direct and indirect impacts of climate change will continue to threaten the very social and economic fabric of cities**

# NOTES

1   IPCC, 2007c, Annex I, p82.
2   Schneider et al, 2007, p781.
3   As illustrated in Chapter 1.
4   IPCC, 2001a.
5   IPCC, 2007b, p30.
6   Church et al, 2004.
7   IPCC, 2007b, p45.
8   Thomas et al, 2004.
9   IPCC, 2007b, p30.
10  Thomas et al, 2004.
11  IPCC, 2007b, p28.
12  Scambos et al, 2004; Overpeck et al, 2006.
13  IPCC, 2007b, p45.
14  Nicholls et al, 1999.
15  Bigio, 2009.
16  Adams, 2007.
17  Ruth and Gasper, 2008.
18  Definition from the National Weather Service *Glossary of National Hurricane Center Terms*, National Weather Service (undated).
19  Areas of the Earth between the tropics and polar regions: http://en.wikipedia.org/wiki/Mid-latitudes.
20  In this section of the chapter, the terms 'tropical cyclone,' 'tropical storm' and 'hurricane' are used interchangeably. Definitions from the National Weather Service *Glossary of National Hurricane Center Terms*, National Weather Service (undated).
21  Emanuel, 2005; Elsner et al, 2008.
22  IPCC, 2007b.
23  Webster et al, 2005.
24  IPCC, 2007c.
25  Donnelly and Woodruff, 2007.
26  Vecchi and Soden, 2007.
27  IPCC, 2007b.

28  Easterling et al, 2004.
29  IPCC, 2007b, p30.
30  Easterling et al, 2004.
31  Frich et al, 2002.
32  IPCC, 2007b, p30.
33  Sterr, 2008.
34  VanKoningsveld et al, 2008.
35  Nicholls et al, 2008.
36  Nicholls et al, 2008.
37  Ruth and Rong, 2006.
38  IPCC, 2007b, p877.
39  Smyth and Royle, 2000.
40  Smyth and Royle, 2000.
41  Cross, 2001.
42  Smyth and Royle, 2000.
43  Smyth and Royle, 2000.
44  UN-Habitat, 2010.
45  UN-Habitat, 2003, 2009a.
46  Rashid, 2000.
47  Ibarrarán, 2011.
48  Smyth and Royle, 2000; Cross, 2001.
49  Robinson, 2001.
50  Kalkstein and Davies, 1989; Ruth et al, 2006.
51  Smyth and Royle, 2000.
52  IPCC, 2007b, p33.
53  Meehl and Tebaldi, 2004.
54  Stern, 2006, p63; IPCC, 2007b, p53.
55  A country is water stressed if water supply acts as a constraint on development or if withdrawals exceed 20 per cent of the renewable water supply (Wilbanks et al, 2007).
56  Bates et al, 2008, p43.
57  Oke, 1982.
58  Akbari, 2005.
59  Akbari, 2005.
60  Meehl and Tebaldi, 2004; Schwartz and Seppelt, 2009.
61  IPCC, 2007b, p53.
62  IPCC, 2007b, p53; see section

on 'Poverty' later in this chapter.
63  IPCC, 2007b; Bates et al, 2008, p38.
64  Smyth and Royle, 2000.
65  Bates et al, 2008, p85.
66  Bates et al, 2008, p38.
67  Folland et al, cited in Wilbanks et al, 2007.
68  *The Sunday Times*, 2009.
69  Symptoms of extreme drought are widespread water shortages or restrictions. National Drought Mitigation Center, 2010.
70  Bates et al, 2008, p3.
71  Bates et al, 2008, p26.
72  Burke et al, 2006.
73  Bates et al, 2008, p3.
74  Bates et al, 2008, p38.
75  Bates et al, 2008, pp3, 43.
76  McGranahan et al, 2007.
77  Choi and Fisher, 2003; Hall et al, 2005; Kirshen et al, 2006.
78  Kirshen et al, 2006.
79  Definitions from United States Geological Survey (undated): http://ga.water.usgs.gov/edu/100yearflood.html.
80  IPCC, 2007c, p48.
81  This is a mid-range estimate of sea-level rise at the end of the 21st century from IPCC, 2007c, p45.
82  Ruth and Rong, 2006.
83  Boruff et al, 2005 (see section on 'Social impacts' later in this chapter).
84  Awuor et al, 2008.
85  Klein et al, 2003.
86  Graves and Phillipson, 2000.
87  Klein et al, 2003.
88  Klein et al, 2003.
89  Klein et al, 2003.

90  Sanders and Phillipson, 2003.
91  Camilleri et al, 2001.
92  UN-Habitat, undated.
93  Transportation Research Board, 2008, p62.
94  Andrey and Mills, 2003, cited in Wilbanks et al, 2007.
95  Shukla et al, 2005.
96  Transportation Research Board, 2008, p64.
97  Shukla and Sharma, undated.
98  Darch, 2006.
99  Kirshen et al, 2006.
100 Hunt and Watkiss, 2007.
101 Sailor, 2001.
102 Scott et al, 1994.
103 Harrison and Whittington, 2002.
104 Lehner et al, 2005.
105 EEA, 2005.
106 IPCC, 2007f, p562.
107 de Bono et al, 2004.
108 IPCC, 2007f, p362.
109 Ruth and Gasper, 2008.
110 IPCC, 2007f, p445.
111 Wilbanks et al, 2007.
112 Vergara, 2005; Magrin et al, 2007; Füssel, 2009.
113 Ruth and Gasper, 2008.
114 Rhodes, 1999.
115 Boland, 1997.
116 Bates et al, 2008, p79.
117 Hunt and Watkiss, 2007, p27.
118 Wilbanks et al, 2007.
119 Bates et al, 2008, p92.
120 IPCC, 2001a, p57; de Sherbinin et al, 2007.
121 Tanner et al, 2009.
122 Environment Canada, 2001; Kumagai et al, 2003; Hall et al, 2005.
123 Rosenzweig and Solecki, 2001; Wilbanks et al, 2007, p370.
124 Wilbanks et al, 2007, p372.

125  Rosenzweig and Solecki, 2001.
126  World Bank, 2000; Wilbanks et al, 2007, p371.
127  UN-Habitat, 2009a, p230.
128  UN-Habitat, 2009a, p230. UN-Habitat defines 'improved drinking water coverage' by the percentage of people having access to improved drinking water technologies such as piped water and protected wells. 'Improved sanitation facilities' are more likely to separate human excreta from human contact (UN-Habitat, 2009a, p224).
129  Fricas and Martz, 2007.
130  Kirshen et al, 2006.
131  Petterson et al, 2006; Wilbanks et al, 2007, p376.
132  Wilbanks et al, 2007, p366.
133  O'Brien et al, 2004; Adger et al, 2005; Kirshen et al, 2006; Wilbanks et al, 2007, p362.
134  Stern, 2006, p17.
135  Kirshen et al, 2006.
136  Wilbanks et al, 2007, p368.
137  Kirshen et al, 2006.
138  Wheaton et al, 2005.
139  Bates et al, 2008, p75.
140  Ruth et al, 2004; Wilbanks et al, 2007, p368.
141  Ruth et al, 2004.
142  UCS, 2008.
143  Defined as towards or in the direction of a pole of the Earth (Merriam-Webster, undated).
144  Agnew and Viner, 2001; Gomez Martin, 2005; Perelet et al, 2007.
145  Elsasser and Bürki, 2002; Scott et al, 2007.
146  Scott et al, 2007.
147  Wilbanks et al, 2007, p368.
148  Hunt and Watkiss, 2007, p28.
149  Lewsey et al, 2004; Wilbanks et al, 2007, p368.
150  de Sherbinin et al, 2007.
151  Kont et al, 2003.
152  Adger et al, 2005.
153  Donner et al, 2005.
154  Lewsey et al, 2004.
155  O'Brien et al, 2004; Petterson et al, 2006; Stern, 2006, p10.
156  Stern, 2006, p78.
157  Dlugolecki, 2001; ABI, 2005; IPCC, 2007f, p557, p723.
158  Mills, 2005.
159  Kunreuther et al, 2001; IPCC, 2007f, p734.
160  Petterson et al, 2006; Wilbanks et al, 2007, p369.

161  Enz, 2000; Lewsey et al, 2004; Wilbanks et al, 2007, p369.
162  Defined here as countries with median per capita incomes above US$9361 (Freeman and Warner, 2001).
163  Freeman and Warner, 2001.
164  Wilbanks et al, 2007, p369.
165  Petterson et al, 2006.
166  Wilbanks et al, 2007, p369.
167  Wilbanks et al, 2007, p371.
168  Petterson et al, 2006.
169  Lewsey et al, 2004.
170  Grimm et al, 2008.
171  The Millennium Ecosystem Assessment was a global effort initiated in 2001 'to assess the consequences of ecosystem change for human well-being and the scientific basis for action needed to enhance the conservation and sustainable use of those systems and their contribution to human well-being' (ICSU et al, 2008).
172  Millennium Ecosystem Assessment, 2005.
173  Syvitski et al, 2009.
174  Environmental Management Department, 2003.
175  IPCC, 2007f, p362.
176  Douglas et al, 2008.
177  Sgobbi and Carraro, 2008.
178  Uyarra et al, 2005.
179  Mendelsohn et al, 2000.
180  McLeman and Smit, 2005.
181  Beniston and Diaz, 2004.
182  Beniston and Diaz, 2004.
183  Basu and Samet, 2002.
184  Beniston and Diaz, 2004.
185  Haines et al, 2006.
186  Basu and Samet, 2002.
187  Rosenzweig and Solecki, 2001.
188  Basu and Samet, 2002.
189  Lee, 1980.
190  Wolfe et al, 2001; Basu and Samet, 2002.
191  Costello et al, 2009.
192  BBC News, 2010a.
193  As of 1 August 2010; BBC News, 2010b.
194  Meusel and Kirch, 2005.
195  Ahern et al, 2005.
196  Silove and Steel, 2006.
197  Akbari, 2005.
198  Patz et al, 2005.
199  Tanser et al, 2003.
200  McMichael et al, 2003.
201  Bartlett, 2008.
202  UN-Habitat, 2006.
203  Hardoy and Pandiella, 2009.
204  Hardoy et al, 1992, 2001.

205  McGranahan et al, 2007.
206  See Awuor et al (2008) for Mombasa (Kenya); Revi (2008) for cities in India; Alam and Rabbani (2007) for Dhaka (Bangladesh); and Dossou and Glehouenou-Dossou (2007) for Cotonou (Benin); also Adelekan (2010) for Lagos.
207  See Nchito (2007) for Lusaka (Zambia); and de Sherbinin et al (2007) for Rio de Janeiro (Brazil).
208  Hardoy and Pandiella, 2009.
209  Satterthwaite et al, 2007a; UN, 2009. However, note that it is the disasters in the developed countries that generally have the highest economic costs (at least in absolute terms).
210  Bull-Kamanga et al, 2003; UN-Habitat, 2007; UN, 2009.
211  Bartlett, 2008.
212  Revi, 2008, p219.
213  UN, 2007, p80.
214  Adger, 1999, 2000.
215  UNDP, 2007, p74.
216  See, for example, African Development Bank et al (2003).
217  Alber, 2010.
218  Neumayer and Plümper, 2007.
219  UNDP, 2007, p77.
220  Oxfam, 2005.
221  Toulemon and Barbieri, 2008.
222  Bartlett, 2008.
223  Neumayer and Plümper, 2007.
224  Enarson, 2000.
225  Enarson, 2000.
226  Enarson and Phillips, 2008.
227  Enarson, 2000.
228  WEDO, 2008, p55.
229  Overstreet and Burch, 2009.
230  Enarson, 2000.
231  Brookings Institution, 2009.
232  Ruth and Ibarrarán, 2009.
233  Schroeder, 1987.
234  WEDO, 2008, p55.
235  Bartlett, 2008.
236  Ruth and Ibarrarán, 2009, p61.
237  Bartlett, 2008.
238  McMichael et al, 2003.
239  Langer, 2004.
240  Ruth and Ibarrarán, 2009, p61.
241  Fothergill et al, 1999.
242  UN-Habitat, 2006.
243  Ruth and Ibarrarán, 2009.
244  Fothergill et al, 1999.
245  Brookings Institution, 2009.
246  Langer, 2004.
247  UN-Habitat and OHCHR, 2010.

248  Macchi, 2008, p19.
249  UNDP, 2009, p9.
250  UNDP, 2009, p15.
251  OCHA and IDMC, 2009.
252  Reuveny, 2007.
253  Alston et al, 2001.
254  Rain et al, 2011.
255  Raleigh et al, 2008.
256  UNDP, 2009, p45.
257  See Myers, 1997. See also Stern Review Team, 2006. This estimate is tentative, and Myers himself has acknowledged that the figure is based upon 'heroic extrapolation' (see Brown, 2007, p6).
258  Myers, 2005.
259  Tacoli, 2009.
260  Reuveny, 2007.
261  UNDP, 2007, p24.
262  Kumssa and Jones, 2010.
263  At the 5663rd Meeting of the Council in 2007, representatives from across the world echoed the belief that climate change issues could have real national and international implications, and that these issues ought to be addressed in a global forum (UN, 2007).
264  Gulden, 2009, p187.
265  UN-Habitat, 2007.
266  Cross, 2001.
267  Klein et al, 2003.
268  Cross, 2001.
269  Klein et al, 2003.
270  UNDP, 2007.
271  Nicholls and Tol, 2007.
272  Ruth and Ibarrarán, 2009.
273  WHO, 2010.
274  Benson and Clay, 2004.
275  UNDP, 2007, p93.
276  Romero Lankao, 2009.
277  Romero Lankao, 2009.
278  Ruth and Gasper, 2008.
279  UN, 2007, p80.
280  UN-Habitat, 2010, p32.
281  UN-Habitat, 2003, p13.
282  Douglas et al, 2008.
283  Cambell-Lendrum and Corvalan, 2007.
284  UN-Habitat, 2007, p170.
285  Tanner et al, 2009.
286  Smyth and Royal, 2000.
287  See UN Habitat, 2007, on the trends of natural and human-made disasters in cities.
288  ISDR Terminology: www.unisdr.org/eng/library/lib-terminology-eng home.htm, last accessed 1 November 2010.

# 5

# CLIMATE CHANGE MITIGATION RESPONSES IN URBAN AREAS

Mitigation – the reduction of greenhouse gas (GHG) emissions and their capture and storage – has been at the heart of policy responses to climate change over the past two decades. At the international level, the 1992 United Nations Framework Convention on Climate Change (UNFCCC) has as its core objective the 'stabilization of greenhouse gas concentrations in the atmosphere at a level that would prevent dangerous anthropogenic interference with the climate system'.[1] Subsequent agreements, including the 1997 Kyoto Protocol and the 2009 Copenhagen Accord, have developed targets and timetables for the international community to reduce GHG emissions.[2] Many national governments have made commitments which go beyond the rather modest goals that have so far been agreed internationally. However, achieving these international and national ambitions is dependent on the implementation of policies and measures to reduce or capture GHG emissions on the ground. Cities are therefore critical places for achieving mitigation. As Chapter 3 has shown, a significant proportion of GHG emissions arise from activities undertaken in urban areas.[3] Cities represent concentrations of population and economic activities, with growing demands for energy for domestic services such as heating, cooling and lighting, as well as commercial buildings, industrial processes, telecommunications systems, the provision of water, the production of waste, leisure activities, travel and so on. Cities can therefore be seen as part of the problem of climate change and reducing GHG emissions in cities is a key policy challenge (see Table 5.1).

However, cities can also be seen as part of the solution to addressing climate change (see Table 5.1), both in terms of the role of urban governments and because of the potential for private-sector and civil society actors to respond to climate change at the urban level. Municipal authorities are potentially important actors in tackling the challenge of mitigation for three reasons. First, they have jurisdictional responsibility for key processes – land-use planning, transportation, waste collection and disposal, and energy consumption and generation – which shape GHG emissions. Second, the concentration of people/business in urban areas means that solutions (e.g. mass transit or requirements for energy savings in offices) are feasible. In other words, cities can act as laboratories where solutions for addressing climate change can be tried and tested. Third, municipal governments also provide a key interface for engagement with stakeholders in the private sector and civil society. It is increasingly clear that non-governmental actors have a significant role in addressing climate change at the urban level. Private-sector organizations and civil society groups are now involved in a range of measures (e.g. promoting behavioural change and reducing energy use in commercial buildings) independently of local and national governments.

Over the past two decades, cities have provided a crucial arena within which the challenges of climate change

Mitigation – the reduction of greenhouse gas (GHG) emissions and their capture and storage – has been at the heart of policy responses to climate change over the past two decades

Municipal authorities are ... important actors in tackling the challenge of mitigation

| Part of the problem | Part of the solution |
| --- | --- |
| • In 2010, half of the world's population lived in cities.[a]<br>• Between 2010 and 2020, 95% of the global population growth (766 million) will be urban residents (690 million), and the bulk of these (632 million) will be added to the urban population of developing countries.[a]<br>• Between 2000 and 2010, the number of slum dwellers in developing countries increased from 767 million to 828 million. The figure might reach 889 million by 2020.[b]<br>• Cities represent concentrations of economic and social activities that produce GHG emissions.[c]<br>• Cities and towns produce between 40 and 70 per cent of global anthropogenic GHG emissions.[c]<br>• By 2030, over 80 per cent of the increase in global annual energy demand above 2006 levels will come from cities in developing countries.[d] | • Municipal authorities have responsibility for many processes that affect GHG emissions at the local level.<br>• Municipalities can act as a 'laboratory' for testing innovative approaches.<br>• Municipal authorities can act in partnership with private-sector and civil society actors.<br>• Cities represent high concentrations of private-sector actors with growing commitment to act on climate change.<br>• Cities provide arenas within which civil society is mobilizing to address climate change. |

*Sources:* a UN, 2010; b UN-Habitat, 2010; c see Chapter 3; d IEA, 2008, 2009

**Table 5.1**

Cities and the mitigation of climate change

mitigation are being addressed. During the 1990s, these responses were primarily concentrated in developed countries and undertaken through three international municipal networks: Local Governments for Sustainability's (ICLEI's) Cities for Climate Protection Campaign (CCP), the Climate Alliance and Energie-Cités.[4] During the 2000s, the cities involved in responding to climate change have grown in number and now include cities in the developing world, in part facilitated by the emergence of new international initiatives such as the Cities Climate Leadership Group (C40), as well as the continuing work of more established networks.[5] Despite this recent growth in interest, and in their potential significance in responding to climate change, the understanding of how and why cities are responding to climate change remains limited, particularly in developing countries. Studies of the responses to the issue of climate change mitigation in cities rely heavily on individual case studies from 'pioneering' cities in developed countries,[6] with some notable exceptions.[7] This body of research suggests that the response of cities to the challenges of mitigation has been fragmented,[8] that significant gaps exist between the rhetoric of addressing climate change and the realities of action on the ground,[9] and that the possibilities, and responsibilities, for acting to reduce GHG emissions vary significantly between cities.[10] In short, attempts to mitigate climate change in cities have been far from straightforward.

Given that cities lie at the heart of the contemporary neo-liberal political-economic model, this is not surprising. Cities are pivotal sites in the 'metabolism' of natural resources and the consequent production of GHG emissions, upon which this model of development rests:[11]

> ... cities have extended their ecological hinterland by importing natural resources or resource-based infrastructure services, like electricity ... from afar, but also by using ecosystems far beyond the urban bioregion as sinks for their emissions. The patterns of modern urbanization have thus been highly dependent on the functioning of the networks driving material flows in and throughout the city.[12]

This pattern of urbanization and its environmental consequences has been uneven. While cities in the developed world have historically been the source of the bulk of urban GHG emissions, as the location of production of goods and services shifts to cities in developing countries so do environmental burdens. At the same time, as the consumption of energy-intensive goods and services increases amongst affluent sectors of urban societies in developing countries, so too will GHG emissions. However, the levels of GHG emissions from poor urban populations remain negligible, suggesting that urban efforts to mitigate climate change need to be targeted at cities where there is both a responsibility and a capacity to act. Furthermore, climate change will deepen a range of existing inequalities; thus, discussions of climate change mitigation in cities need to include broader concerns about the vulnerability of different social groups. Specifically, the gender dimension of climate change mitigation,

and the potential for women to contribute to climate change mitigation strategies, has not yet been fully acknowledged.[13]

The result is a complex geography of urban GHG emissions,[14] where responsibility for action, and the capacity to act, rest with affluent urban societies, but where the brunt of the future impacts of climate change will be borne by vulnerable urban populations.[15] In this context, building an understanding of how cities in developed countries are responding to the challenge of climate change mitigation – beyond the small number of case studies currently available – is a critical task. At the same time, there is a need to understand how climate change mitigation is being addressed in the world's megacities, which because of their sheer size are potentially critical sites of current and future GHG production, as well as the small urban centres within which the bulk of population growth and energy demand over the next few decades is forecast to occur.[16] In Asia and Latin America, recent industrialization and the growth of affluent urban communities suggests that climate change mitigation may be an increasingly pressing challenge.

This chapter seeks to address these knowledge gaps by providing a review of urban responses to climate change in a comparative context. It focuses on the responses of so-called 'global' cities (those regarded as having particular strategic economic and/or political importance)[17] and megacities (those with a population of more than 10 million people). These cities are critical to the urban mitigation of climate change both because of their current and potential contribution to GHG emissions and their wider economic and political influence.[18] First, the chapter considers the policy responses and initiatives that are emerging in cities. Second, it examines how such strategies and measures have been undertaken through different modes and mechanisms for governing climate change in the city. Third, the chapter assesses the opportunities and constraints that cities have encountered in institutional, economic, technical and political terms, before, fourth, providing a comparative analysis of emerging trends in urban responses to climate change. Finally, the chapter offers some concluding comments and lessons for policy.

# RESPONSES TO CLIMATE CHANGE MITIGATION IN URBAN AREAS

Over the past two decades, municipal authorities have engaged in the development of urban climate change policies as well as initiatives and schemes to reduce GHG emissions in the city. More recently, a range of other actors – including non-governmental organizations (NGOs), donor agencies and private corporations – have also become involved in urban climate change mitigation initiatives. This section reviews different policy approaches that municipalities have developed for dealing with climate change mitigation before considering the strategies and measures that have been adopted by both public and private actors in five key sectors: urban development and design; built

---

**Over the past two decades, cities have provided a crucial arena within which the challenges of climate change mitigation are being addressed**

**Responsibility for action, and the capacity to act, rest with affluent urban societies, but ... the brunt of the future impacts of climate change will be borne by vulnerable urban populations**

environment; urban infrastructures; transport; and carbon sequestration.

## Municipal policy approaches

The policy approaches adopted by municipal governments to address the mitigation of climate change in urban areas vary considerably in terms of the sources of GHG emissions that are targeted – whether these are from the municipalities' own activities or from across the urban community – and whether they are undertaken on a strategic or ad hoc basis (see Table 5.2). In each case, a variety of mechanisms for developing and implementing climate change mitigation measures have been used.[19]

Municipalities have undertaken ad hoc measures to reduce GHG emissions from their own operations, often on a *reactive* basis – for example, in response to a particular funding opportunity or the initiative of an individual (see Table 5.2). Municipal authorities have also been *opportune* in developing one-off schemes or projects at the community scale, often in collaboration with other partners. Such ad hoc approaches are popular and 'numerous cities, which have adopted GHG reduction targets ... prefer to implement ... measures on a case by case basis'.[20] The wide range and significant number of such ad hoc responses suggest that given the right financial and political conditions, municipal governments have been more than able to respond positively to the challenges of mitigating climate change.

Strategic approaches, in contrast, have usually been developed where there has been access to secure funding, new institutional structures – such as a central unit for addressing climate change – and strong political support for action. These can either involve setting out a programme of goals and measures through which municipalities seek to reduce their own GHG emissions over the medium to long term (a *managerial* approach), or a *comprehensive* approach, developed by only a few municipalities, involving target setting, planning and the development of initiatives at the community level.[21] Such strategic approaches were first promoted by ICLEI's CCP Milestone programme established during the mid 1990s (see Box 5.1). A similar approach has also been adopted by the Climate Alliance in its Climate Compass initiative (see Box 5.2). Evidence suggests that some substantial reductions in GHG emissions have been achieved by these means. For example, in 2006, 546 local governments in 27 countries were members of the CCP campaign, accounting for 20 per cent of global GHG emissions. Estimates suggest that the annual emission reduction by these cities was 60 million tonnes of $CO_2$eq, which amounts to a 3 per cent annual reduction among the participants and 0.6 per cent globally.[22] However, while those municipalities that have focused on their own operations have made substantial progress against their targets, achieving such goals beyond the confines of the municipality itself has been both more difficult to monitor and more challenging to implement.

Despite differences in the approaches that municipalities have adopted to the formation and implementation of climate policy, research suggests that attention has primarily

| | Ad hoc | Strategic |
|---|---|---|
| **Municipality** | Reactive | Managerial |
| **Community** | Opportune | Comprehensive |

**Table 5.2**

Typology of policy response to climate mitigation in the urban arena

### Box 5.1 Strategic approaches to urban climate change policy: The CCP Milestone Methodology

- *Milestone 1:* establish an inventory and forecast for key sources of GHG emissions in the corporate (municipal) and community areas, and conduct a resilience assessment to determine the vulnerable areas based on expected changes in the climate.
- *Milestone 2:* set targets for emissions reduction and identify relevant adaptation strategies.
- *Milestone 3:* develop and adopt a short- to long-term local action plan to reduce emissions and improve community resilience, addressing strategies and actions for both mitigation and adaptation.
- *Milestone 4:* implement the local action plan and all the measures presented therein.
- *Milestone 5:* monitor and report on GHG emissions and the implementation of actions and measures.

*Source:* www.iclei.org/index.php?id=810, last accessed 18 October 2010; see also Box 2.7

### Box 5.2 Strategic approaches to urban climate change policy: The Climate Alliance's Climate Compass

*Module 1 – Initiation:*
- informing relevant departments of the administration;
- clarifying needs and expectations;
- raising awareness of local climate change policies;

*Module 2 – Inventory:*
- analysing the setting;
- surveying previous priorities and activities;
- characterizing the initial conditions;

*Module 3 – Institutionalization:*
- building organizational structures;
- assigning responsibilities and nominating persons in charge;
- forming a Climate Compass working group;

*Module 4 – Climate action programme:*
- defining targets;
- selecting priority measures;
- formulating strategic resolutions (on criteria, standards, etc.);
- agreeing the mid- and long-term climate strategy;

*Module 5 – Monitoring and reporting:*
- developing indicators;
- collecting data for $CO_2$ monitoring;
- preparatory work for future reporting.

*Source:* www.climate-compass.net/_modules.html, last accessed 18 October 2010

been focused on initiatives in the energy sector, and in particular on improving energy efficiency.[23] Energy efficiency is a particularly potent issue as it can 'advance diverse (and often divergent) goals in tandem',[24] serving to translate various interests into those concerning climate

Various strategies of land-use planning ... have been used in order to limit urban expansion, reduce the need to travel and increase the energy efficiency of the urban built form

In developing countries, there are few initiatives to explicitly mitigate climate change through urban design and development

In developed countries, private developers and community groups have led new urban development, brownfield regeneration and neighbourhood renewal projects which seek to address climate change

change and effectively forging new partnerships. While energy efficiency still dominates many municipal responses to mitigating climate change, the growing diversity of those cities involved in mitigating climate change together with the range of private-sector and civil society actors becoming involved with this policy agenda has led to a growing array of projects and measures being adopted.

Nonetheless, it is possible to identify five key sectors in which urban responses to mitigating climate change have been concentrated: urban form and structure; built environment; urban infrastructures; transport; and carbon sequestration. Reviewing the evidence across these sectors, the following sections examine the range of activities being undertaken by municipal authorities and other actors in the city to reduce GHG emissions and the strengths and weaknesses of the initiatives that have been undertaken.

## Urban development and design

The use of energy within a city, and the associated production of GHG emissions, is dependent on both the form of urban development (i.e. its location and density) and its design.[25] As urbanization continues apace, one of the critical challenges is managing the process of urban development and, in particular, the twin challenges of urban sprawl and the growth of informal urban settlements (see Box 5.3).[26] Urban sprawl is an increasing challenge for cities in developed and developing countries. As the distances between home, work, education and leisure activities increase, so often does the reliance on private motorized transport. In some cities sprawl has meant the development of middle-class urban fringe districts where dwelling sizes tend to increase, leading to an increase in per capita GHG emissions. In other cities, sprawl is fuelled by the growth of informal settlements. Between 2000 and 2010 the number of slum dwellers in developing countries increased from 768 millions to 828 million, and estimates suggest that the number of slum dwellers will increase to 889 million by 2020.[27] Slum populations lack adequate access to reliable

and affordable energy supplies and shelter, meaning that, in parallel with the other significant challenges that such settlements pose for sustainability and well-being, many households are unable to heat or cool their dwellings effectively and experience fuel poverty.

In seeking to address these challenges, various strategies of land-use planning, including land-use zoning, master-planning, urban densification, mixed-use development and urban design standards, have been used in order to limit urban expansion, reduce the need to travel and increase the energy efficiency of the urban built form.[28] Such approaches can be deployed at a range of locations within the city and at different scales (see Table 5.3). Overall, research suggests that large-scale schemes, including large regeneration projects, projects to prevent urban expansion and the reuse of derelict land appear to be a more common response for mitigating climate change than small regeneration projects. Most such projects are undertaken in developed countries. In developing countries, there are few initiatives to explicitly mitigate climate change through urban design and development, and where they do exist, the local governments' capacity to implement such measures is often limited.

Most often, these projects are led by municipal authorities through the use of planning regulations and planning guidance. This is the case, for example, of the principles of 'compact city planning'[29] incorporated within the municipal ordinances of cities such as São Paulo (Brazil) and Cape Town (South Africa),[30] although, in practice, it is not clear that such principles can actually be implemented in an effective way. These principles advocate a combination of planning measures to combine high-density development and mixed land-use principles to prevent urban sprawl and reduce the dependence on motorized transport, while focusing on the integration of green areas in the city. Although this principle may appear to be linked with more sustainable urban form models, research in developed country cities[31] suggests that the effectiveness of the compact city model in reducing GHG emissions depends on the lifestyle and space demands of the city inhabitants.

Alongside initiatives undertaken by municipal authorities, particularly in developed countries, private developers and community groups have led new urban development, brownfield regeneration and neighbourhood renewal projects which seek to address climate change specifically, such as the Onion Flats in Philadelphia (US), the Green Building in Manchester (UK), the A101 neighbourhood in Moscow (Russia) and the project T-Zed in Bangalore (India).[32] The combination of sustainability and climate mitigation objectives with business interests has led to the development of large-scale flagship urban developments that may bring together local and international partners to advance economic interests alongside environmental ones. One famous example from China was the proposed eco-city Dongtan, in Shanghai's 'last piece of pristine land' in Chongming Island. The developer, Shanghai Industrial Investment Corporation, contracted Arup, the international professional services firm in 2005 to design a master plan for Dongtan as an 'experiment' to showcase a national model for sustainability, energy efficiency and environmental awareness.[33] Some

---

**Box 5.3 Urban development challenges for mitigating climate change: Thailand and Canada**

In Chiang Mai (Thailand), research found that urban and commercial development coupled with growing economic prosperity has led to a surge in personal vehicle usage, related to both work commuting and leisure. The number of registered passenger cars and motorcycles increased more than 20-fold between 1970 and 2000, while the population only doubled, with a significant impact on greenhouse gas (GHG) emissions.[a]

Few Canadian cities appear to be prioritizing climate change-related action in land-use planning. While most cities do not acknowledge the emission reduction benefits of growth management and increased density, Calgary, Vancouver and Toronto are making explicit connections between land use and emissions. Yet, even in these three cities – which are leading climate change action in Canada – few specific initiatives address these connections. Research has attributed this to two main reasons: first, cities depend on provinces to review land-use planning policies, and this relationship may act in delaying or even discouraging action in this area; and second, actions required may be extremely divisive, openly challenging the traditional preference for suburban development in Canada.[b]

*Sources:* a Lebel et al, 2007, p101; b Mackie, 2005; Gore et al, 2009, p11

commentators, however, cast doubt on whether the Dongtan plans will ever be realized.[34] Whether the reason is the lack of leadership,[35] the conflicts of interest between local developers and international partners[36] or the permissive policies of local authorities,[37] many have criticized the project as not offering real solutions to address climate change.[38]

Furthermore, even where individual developments may be successful, the logic of developing greenfield urban fringe sites as a means of addressing climate change mitigation can be questioned, both in terms of their overall carbon footprint and, because of their exclusive nature, their potential for exacerbating social inequalities. Despite these criticisms, the trend for developing new 'eco-cities' shows little sign of abating in developed and developing countries alike. For example, the Clinton Climate Initiative has recently launched the Climate Positive programme, focusing on large-scale developments in 17 cities on six continents which are aiming to become carbon neutral.[39] A contrasting trend is the proliferation of initiatives, primarily in developed countries for the regeneration of brownfield land and neighbourhood renewal, which combine social and environmental justice objectives (see Box 5.4).

While municipal authorities can be crucial to the development of these projects, grassroots civil society organizations are also important. In the US, the Tent City project in Boston, the Plaza Apartments in San Francisco and the Intervale Green and Louis Nine House in New York[40] are all associated with civil society actors, sometimes led by NGOs, and have sought to promote carbon-saving technologies as suitable cheap alternatives for providing energy to low-income residents.

The confluence of a variety of interests and material circumstances in initiatives to mitigate climate change through urban design and development makes them complex and difficult to manage. The development and implementation of 'low-carbon' planning principles by municipal governments may encounter political opposition, lack enforceability, and have limited impacts upon the behaviour of individuals who live and work in the city. Furthermore, such principles may be socially divisive, reinforcing patterns of inequality in the city by creating enclaves of 'sustainable' living while failing to address the basic needs of the majority of urban citizens. Moreover, gender concerns have not been fully integrated within climate change policies and planning.[41]

In terms of low-carbon urban development projects, the circumstances that lead to their inception may change rapidly, thus challenging their feasibility, as was the case in the Dongtan project in Shanghai (China). One means of ensuring the long-term feasibility of such schemes is to take other issues of social and environmental justice into consideration, either through public consultation or through the participation of a range of stakeholders in the design and management of the project. Current examples suggest that small-scale developments which aim to simultaneously address environmental and social issues (e.g. homelessness, poverty, etc.) are more likely to find support from civil society groups, who in turn can facilitate their implementation. This, however, does not dismiss the idea that visionary

| Type of scheme | Description |
|---|---|
| Urban expansion, informal settlements or suburban development: | Application of land-use planning and design policies to limit energy use in the expanding areas of existing cities. |
| New urban development: | Application of land-use planning and design policies to limit energy use in new urban areas. |
| Reuse of brownfield land: | Urban development on old industrial or other derelict areas of the city to encourage densification, mixed-use development and reduce energy use in the city. |
| Neighbourhood and small-scale urban renewal: | Schemes which seek to renew existing housing stock and redevelop urban layout and design at a neighbourhood or street scale in order to reduce energy use in the city. |

cutting-edge projects may be able to provide best practice examples to challenge current socio-technical barriers, but suggests that the focus needs to change towards the development of projects that can address global demands for climate change mitigation and local demands for quality of life.

**Table 5.3**

**Climate change mitigation through urban development and design**

## Built environment

The design and use of the built environment is a critical arena for climate change mitigation because 'the building sector consumes roughly one-third of the final energy used in most countries, and it absorbs an even more significant share of electricity'.[42] The built environment includes public (e.g. government offices, hospitals, schools) domestic (housing) and commercial/industrial (e.g. offices, factories) buildings, with the latter increasingly recognized as important in driving peak demand and significant sources of GHG emissions in cities in developing countries.[43] The use of energy within the built environment is the result of complex interactions among building materials, design, the systems used

*The design and use of the built environment is a critical arena for climate change mitigation*

---

**Box 5.4 Sustainable living and brownfield development, Stockholm, Sweden**

Hammarby Sjöstad, Stockholm's largest new urban development project, is a model for closed-loop sustainable urban development. Their strategy is outlined in the Hammarby Model, an eco-cycle that optimizes resource use and minimizes waste in order to meet a wide range of sustainability targets in the areas of land use, energy, water, transport, building materials and socio-economic indicators. The new district, expected to house 25,000 inhabitants, is built on 200ha of industrial and harbour brownfield land in southern Stockholm, using tested eco-friendly building materials.

The district has its own recycling model, an underground vacuum-based system, which reduces waste and its associated collection costs (by 40 per cent overall, and 90 per cent for non-recyclable waste). Rainwater harvesting and the diversion of storm water from the sewerage system to be reused for heating, cooling and power generation help to offset demands for both water and power. The Hammarby district achieves 100 per cent renewable energy in its district heating network and transport use by making use of heat recovery from waste incineration, and biogas from the digestion of organic waste and sludge for household and transport use. Rooftop solar panels are also widely employed.

Suggested reasons for the successful realization of the Hammarby project (due to be completed in 2015) include acknowledgement of Stockholm's strong leadership in sustainable development planning; the implementation of innovative policies; high stakeholder involvement and commitment; and the successful coordination between and within the municipality and the Swedish national government.

*Source: Hammarby Sjöstad, 2010*

| Type of scheme | Description |
|---|---|
| Energy-efficient materials: | The use of energy-efficient materials in the construction of the built environment. |
| Energy-efficient design: | The use of energy- and water-efficient design principles, such as 'passive' heating and cooling. |
| Building-integrated alternative energy supply: | The use of renewable and low-carbon energy technologies to provide energy to individual buildings. |
| Building-integrated alternative water supply: | The use of off-grid water supply and processing techniques which reduce energy use in the production and heating of clean water. |
| New-build energy and water-efficient technologies: | The use of energy- and water-efficient devices in the construction and development of new buildings. |
| Retrofitting energy- and water-efficient technologies: | The use of energy- and water-efficient devices in the renovation of existing buildings. |
| Energy- and water-efficient appliances: | The use of efficient appliances within the built environment. |
| Demand-reduction measures: | Measures aimed at reducing the demand for energy and water within the built environment. |

**Table 5.4**

**Climate change mitigation in the built environment**

**Policy approaches for reducing GHG emissions from the built environment have primarily focused on issues of energy efficiency**

to provide buildings with energy and water, and the ways in which buildings are used on a daily basis.[44] Gender differences may play an important role in how energy within the household is used.[45]

Policy approaches for reducing GHG emissions from the built environment have primarily focused on issues of energy efficiency, with approaches grouped into 'three categories: economic incentives (e.g. taxes, energy pricing); regulatory requirements (e.g. codes or standards); or informational programmes (e.g. energy awareness campaigns, energy audits)'.[46] More recently, there has been a growth in voluntary rating systems (e.g. Energy Star in the US and the Carbon Trust Standard in the UK) and in the involvement of private actors (e.g. the C40 and Clinton Climate Initiative) in schemes to reduce energy use, which has led to an increase in expectations concerning energy efficiency in the

(commercial) built environment. This combination of financial, regulatory, education-based and voluntary mechanisms[47] has led to an explosion in the range of schemes deployed to address energy use in the built environment, which has also been assisted by the development of micro-generation technologies and new building materials (see Table 5.4).

Despite the potential range of initiatives that could be undertaken, measures in the built environment sector tend to focus on energy-efficient technologies, alternative energy supply technologies and demand-reduction practices. Existing evidence suggests that initiatives in the built environment sector have primarily been located in cities in developed countries.[48] In particular, efforts have been concentrated on retrofitting existing buildings, those which are municipally owned and in the residential sector, with energy-efficient technologies – for example, in the European cities of Vienna (Austria), Stockholm (Sweden), London (UK), Munich (Germany) and Rotterdam (The Netherlands) (see also Box 5.5). National governments in developed countries have also intervened in implementing retrofitting programmes at the local level. For example, the US Department of Energy has led the Weatherization Assistance Program, which since 1999 has sought to increase the energy efficiency of low-income households while ensuring their safety in New York and other US cities.[49]

Research also suggests that in developed countries, many successful projects are led by grassroots organizations and housing co-operatives, such as the case in Tel Aviv (Israel), where a group of house buyers announced in 2009 the launch of the first Tel Aviv ecological housing project.[50] This suggests that innovative forms of social organization are emerging to coordinate and lead initiatives to address climate change in the built environment, with significant potential for addressing issues of social and environmental justice. Private developers may also have a strong role in promoting and implementing sustainable technologies. However, dealing with existing building stock poses problems in terms of conservation of heritage and dealing with demolition materials: in the UK, in order to achieve the existing targets for GHGs emission reductions, higher rates of demolition are advocated if sustainable technologies alone cannot meet the heating needs of insufficiently insulated housing.[51]

Despite the focus on measures to address climate change in the built environment, few cities in developed countries have sought to develop energy-efficient building materials or to address issues of the sustainable supply and use of water. However, when the intention is to establish best practice examples or to showcase new technologies, projects often include a range of different measures, including novel materials as well as low-carbon energy and water systems and passive designs. The ability of these measures to make significant gains in emissions reductions will depend on the current building standards, which vary greatly from city to city. Universities, architectural practices and engineering firms have been important sources of innovation, leading pilot projects designed to showcase a range of technologies.[52] The use of energy-efficient materials is not

---

**Box 5.5 Retrofitting domestic, public and commercial buildings in the UK and the US**

- *London (UK):* the Carbon 60 project followed the commitment of the Sandford Housing Co-operative to reduce greenhouse gas (GHG) emissions by 60 per cent. The combined financial support from private energy companies, the UK government and the rent increases within the co-operative made possible the retrofitting of 14 houses with wood pellet boilers and solar water heating.[a]
- *Birmingham (UK):* the Summerfield eco-housing project in Birmingham (supported by Birmingham City Council and Urban Living and the Family Housing Association) developed a demonstration project in a Victorian house featuring solar photovoltaic panels; grey water recycling and air source heat pumps; sunpipes; high-performance insulation made from recycled paper, denim and sheep's wool; and kitchens made from recycled materials.[b]
- *Manchester (UK):* the Cooperative Insurance Services 'Tower' was built in 1962 and is the tallest office building in the UK outside of London. In 2004, the Cooperative Financial Services started a UK£5.5 million project to retrofit photovoltaic technology, funded by the Northwest Regional Development Agency.[c]
- *Philadelphia (US):* the Friends Center Building Project, initiated in 2006, involves the retrofitting of an 1856 building with sustainable technologies. The project integrates recycled materials, recycled construction waste, white roof, and windows with spectrally selective glass, alongside sustainable and renewable technologies (e.g. geothermal exchange; solar array; wind power; storm water capture and reuse) and green building design with natural light.[d]

*Sources:* a Sanford Housing Co-operative, undated; b Office of the Deputy Prime Minister, 2003; c Energy Planning Knowledge Base, undated; d www.friendscentercorp.org, last accessed 18 October 2010

only possible in individual household projects, but can also be promoted as a strategy for commercial projects or, more widely, to encourage social and environmental sustainability (see Box 5.6).

In cities in developing countries there has been less emphasis on retrofitting residential buildings or on reducing demand for energy and water use. However, initiatives have been established to install energy-efficient appliances in municipal buildings in several cities, including Mexico City and Cape Town (South Africa),[53] and to reduce energy use in commercial buildings, especially in cities in Asia.[54] In addition, the use of energy-efficient materials has been an important means through which municipal governments and other actors have sought to address GHG emissions reductions and the provision of low-cost housing to low-income groups. South American cities such as Buenos Aires (Argentina) and Rio de Janeiro (Brazil) have piloted the use of energy-efficient and low-cost materials to deliver sustainable houses in low-income areas. In June 2009, the Argentinian Ministry of Infrastructure signed a contract with the Housing Institute, the National University of La Plata and the National Institute of Industrial Technology to start a pilot project to deliver social housing 'bioclimatic houses' in Buenos Aires.[55]

In addition to measures to improve energy efficiency and reduce demand, cities are also experimenting with alternative forms of renewable and low-carbon energy supply. In the built environment, initiatives have primarily focused on the use of solar water heaters, relatively simple devices used to heat water using sunlight,[56] rather than other autonomous energy supply devices, such as photovoltaic cells, wind power or biomass technologies. Some cities – such as Barcelona (Spain), São Paulo (Brazil) and Buenos Aires (Argentina) – enforce the adoption of solar water heaters in municipal ordinances. In China, given its leading position in the manufacture of domestic solar water heaters, there is potential for this technology to be widely adopted. The main barrier for the adoption of solar water heaters is their large initial installation cost; but given that solar water heaters have a longer lifetime, the overall costs of solar water heaters may be considerably lower.[57] A study of a project to install 200,000 solar water heaters in the quickly urbanizing and industrializing district of Yinzhou (China) concluded that such a project could have significant benefits (see Table 5.5). In addition to their climate change benefits, the decentralization of energy provision is often seen as a way of addressing the energy needs of those actors who do not have access to a reliable supply of energy. From a gender perspective, low-carbon options for cooking, such as biogas digesters and solar cookers, may facilitate women's access to energy as long as they are adapted to the local context and compatible with women's daily routines and workloads.[58]

Within the built environment, the potential for mitigation gains from reducing the demand for energy is also significant. Municipal governments, private-sector companies and civil society groups have undertaken a wide range of initiatives aimed at changing the ways in which their own employees and urban citizens use energy. To date, these efforts have not taken issues of gender into account.[59] This

could be a critical omission as women are often thought to have a greater share of decision-making within the household. For example, in Organisation for Economic Co-operation and Development (OECD) countries, women make over 80 per cent of consumer decisions in the household[60] and thus may determine the sustainable consumption decisions within the home. In general, women appear to be more prepared for behavioural changes than men, as men tend to rely most on technological solutions. For example, women tend to place more emphasis on eco-labelled food, recycling and energy efficiency than men.[61] This suggests that women-oriented sustainable consumption policies could work as a tool that municipalities and other urban actors could use to reduce GHG emissions from households.

Among the approaches to mitigate GHG emissions in the built environment applied over the last two decades, the emphasis has been on energy efficiency measures – both in terms of technologies and initiatives to reduce demand – with far fewer projects to reduce GHG emissions through alternative forms of energy supply and limited evidence of other initiatives targeting resource use. Initial climate

*Cities are also experimenting with alternative forms of renewable and low-carbon energy supply*

---

A project developed by the Habitat Kyrgyzstan Foundation has provided more than 48 affordable environmentally sustainable homes for low-income families using a traditional cane reed and clay construction technology. Heating is provided by an innovative coiled-circuit under-floor heating system. The houses meet local building regulations, but allow families to save up to 40 per cent of the construction costs compared with conventional brick housing. The use of volunteer labour further reduces the cost of the houses, and low-cost housing loans help to ensure affordability.

The use of traditional building methods and locally available materials relies on the revival of a traditional cost-effective building technology commonly used during the 19th century, but replaced during the 20th century by brick building. The Habitat Kyrgyzstan Foundation has adapted the traditional cane reed construction method to include a timber frame with cane reed and clay wall sections, improving insulation without compromising comfort.

*Source:* www.worldhabitatawards.org, last accessed 18 October 2010

---

**Table 5.5**

**Costs and benefits of a project to install 200,000 solar water heaters in the residential sector in Yinzhou, China**

| Benefits | |
|---|---|
| Climate benefits: | Abatement of 88,900 tonnes of $CO_2$eq per annum, 1.3 million tonnes over 15 years. |
| Other environment effects: | Reduction of sulphur dioxide, nitrogen oxides, other air pollutants and wastes. |
| Economic and social effects: | Potential health improvement. Low-cost water heating supply. |
| **Costs** | |
| Subsidy:[a] | US$1.28 million. |
| Estimated gross financial costs:[b] | 400 million RMB (US$48 million). |
| Administrative, institutional and political considerations: | Transaction costs likely to reach US$2 per heater to meet the need for advertisements and a good distribution system (US$0.4 million). |
| The cost of certified emission reductions: | Approximately US$1.3 per tonne $CO_2$eq. |

*Notes:* a Subsidy is calculated as the amount necessary to cover the cost differential between the solar water heater and the electric water heater for the first five years, including the electricity cost reduction.
b Gross financial costs here refer to the total cost of initial purchase and installation. The price of residential electricity adopted here is 0.65 RMB (US$0.08) per kilowatt hour (kWh).

*Source:* adapted from Zhao and Michaelowa, 2006

change action within the built environment has focused on easy gains from energy and water efficiency, and the adaptation of existing technologies.[62] The combination of regulation, civil society action and the inclusion of sustainable building principles could have a big impact upon incorporating climate change mitigation technologies and principles within new buildings. Yet, there are obstacles to retrofitting existing buildings, such as inadequate returns on investment, difficulties in dealing with existing stock, lack of financial incentives and regulatory constraints, dependency upon occupancy cycles, and general lack of information about available technological solutions. The combination of social and environmental benefits that energy efficiency can generate is particularly relevant in developing countries, where environmental measures may tackle other social problems such as 'fuel poverty'. However, there is a case to look beyond these measures as their impact may be reduced by the 'rebound effect' – that is, the tendency to use efficiency gain to increase consumption.[63] In this context, energy efficiency measures need to be coupled with those to develop low-carbon renewable energy sources and the reduction of energy demand.

## Urban infrastructure

Urban infrastructure – in particular, energy (electricity and gas) networks, and water and sanitation systems – is critical in shaping the current and future trajectories of GHG emissions. The type of energy supply, the carbon intensity of providing water, sanitation and waste services, and the release of methane from landfill sites are important, though often hidden, components of GHG emissions at the local level. Infrastructure systems frequently lie outside the direct control of municipal governments, are intertwined with power struggles over the rights of those living in informal settlements,[64] and require significant resources and long-term planning. The significant upfront costs of renewing or replacing existing infrastructures, or of providing such systems to the expanding areas of cities, means that investment in infrastructure is often delayed in favour of more pressing immediate concerns. Furthermore, although urban infrastructure systems are often regarded as gender neutral, men and women are affected differently by water, waste and energy policies as their work and community roles differ. For example, while women are often responsible for ensuring energy supply at the household level, they may be excluded from technical work on the energy systems, often regarded

as a male domain.[65] Equally, women's safety and security may be more dependent on adequate infrastructure systems, such as the provision of adequate lighting and sanitation facilities.[66] Urban infrastructure systems therefore pose unique and complex challenges for mitigating climate change.

At the same time, the very nature of urban infrastructure systems is changing significantly. In developed countries, research has documented the demise of the nationally integrated, 'modern' homogeneous utility networks in the face of processes of market liberalization, privatization, neoliberal political ideologies, shifts in urban planning, new technologies and new practices of consumption, leading to the 'splintering' of urban infrastructure systems.[67] Similar processes, although often less apparent, are taking place in cities in developing countries. Thus, across a diverse range of cities, a sense of social, political and technical dynamism and instability now characterizes the provision of basic services and infrastructure development. Within this context, mitigating climate change is becoming an important issue, but one that competes for attention with other pressures for energy security and affordability, and the provision of basic services. Nonetheless, municipal authorities – together with other government, private and civil society actors – have undertaken a range of schemes in order to reduce GHG emissions through the refurbishment and development of urban infrastructure systems (see Table 5.6).

Of the three infrastructure areas considered in this section, research suggests that initiatives to explicitly address climate change have been concentrated in the energy and energy-from-waste domains and on the provision of new forms of energy supply, with fewer initiatives to address the carbon intensity of the provision of water, sanitation and waste services or to reduce demand. In some cities in developing countries, the Clean Development Mechanism (CDM)[68] has been an important driver of infrastructure projects, particularly landfill gas capture (see discussion below). Issues of energy security have also been important drivers for the development of low-carbon energy supply systems in developing countries and initiatives for demand reduction in some Latin American and African cities. In India, cities such as Chennai have been successful in promoting rainwater harvesting as a form of water conservation. In Latin America, concerns for water security are also leading to the development of initiatives with benefits for mitigating climate change. While urban infrastructure initiatives are often led by municipal governments or urban utility authorities, regional and national governments, international agencies and private companies are frequently involved because of the multilevel nature of such systems.

In terms of energy systems, three different approaches for developing low-carbon forms of urban energy supply can be identified. *First*, many municipalities have sought to reduce the carbon footprint of existing supply networks. An increasingly common initiative, found in cities such as Melbourne (Australia), Beijing (China) and Jogjakarta (Indonesia), has been to retrofit street-lighting systems with energy-efficient bulbs. Cities, particularly in Europe, have also sought to develop existing district heating and combined

**Municipal authorities ... have undertaken a range of schemes in order to reduce GHG emissions through the refurbishment and development of urban infrastructure systems**

**In some cities in developing countries, the Clean Development Mechanism (CDM) has been an important driver of infrastructure projects**

**Climate change mitigation and urban infrastructures**

| Type of scheme | Description |
| --- | --- |
| Alternative energy supply: | Development of renewable energy or low-carbon energy supply systems at the city scale. |
| Landfill gas capture: | Use of gas produced by landfill sites for energy provision. |
| Alternative water supply: | Use of alternative forms of water supply, storage and processing to reduce energy use at city scale. |
| Collection of waste for recycling or reuse: | Development of alternative collection systems and ways of using waste to reduce methane produced at landfill sites. |
| Energy and water efficiency/conservation: | Enhancing the efficiency of existing infrastructure systems or development of new efficient systems. |
| Demand reduction: | Schemes to reduce demand for energy and water use, and for the collection of waste. |

heat and power (CHP) plants. In Germany, Berlin is home to Western Europe's largest urban heat network, with over 1500 kilometres of pipes and over 280 district-level CHP plants stretching across the city, delivering low-carbon energy to a wide variety of consumers.[69]

A *second* approach has been for municipalities to purchase renewable energy, either for their own buildings and operations, or as a means of offering consumers access to green energy at a reduced cost. This approach is often facilitated by purchase agreements between the municipality and a private supplier of low-carbon or renewable energy, such as in the case between Cape Town and the Darling Wind Farm in South Africa, or the commitment of the City of Sydney (Australia) to achieve the supply of 100 per cent of the city's energy from renewable sources by using a system to accredit private energy companies.[70]

A *third* approach has been to develop new low-carbon and renewable energy systems within cities. In these initiatives, climate change mitigation is often expressed as a secondary objective in relation to ensuring energy security. This is the case in the growing interest of Latin American cities such as Quito (Ecuador), Bogotá (Colombia) and Rio de Janeiro (Brazil) in sources of energy that may reduce their dependence on oil by promoting the use of natural gas in households. In Cape Town (South Africa), the company Eskom, backed by the national government, has proposed the construction of a nuclear plant to meet the twin objectives of guaranteeing the energy safety of the region and reducing the city's carbon emissions; but the project has found considerable local opposition from both the City of Cape Town and diverse stakeholder groups, mirroring global debates about the role of nuclear power in climate change mitigation.[71] In China, the Beijing municipal government has accelerated the development of clean energy sources, including geothermal resources, biomass and wind power. In addition to the 118 plants already in operation by the end of 1998, 174 new geothermal wells were constructed between 1999 and 2006. Beijing now consumes about 8.8 million cubic metres of geothermal water each year, reducing $CO_2$ emissions by 850,000 tonnes during the period of 2001 to 2006.[72] Beijing is also increasingly looking at wind power and biomass generation, and the city had planned to increase the energy share of these renewable energy sources to 4 per cent by 2010. The Guanting wind farm, located on the southern bank of the Guanting Reservoir, is Beijing's first wind power generation station, with 33 wind turbines capable of generating 49.5MW of electricity per year.[73] It was completed in January 2008 as a CDM project.

While investment of the scale and ambition displayed by Beijing is difficult to imagine for the vast majority of municipal authorities, national and international drivers, together with partnerships with private-sector companies, are leading to a growing emphasis on renewable and low-carbon energy systems. For example, the US Department of Energy has established a partnership with 25 cities to deliver Solar American Cities. The cities selected will receive a combined US$5 million of funding from the department plus hands-on technical assistance over two years. In Boston, for example, the goal is to achieve 25MW

---

**Box 5.7 Feminist action to gain recognition for women waste pickers in Mumbai, India**

The Parisar Vikas (eco-development) programme was launched in 1998 by the *Stree Mukti Sanghatana* (Women's Liberation Organization), established in 1975, with the cooperation of the Municipal Corporation of Greater Mumbai. The programme aims to address the problems of waste management and of self-employed women engaged in the 'menial' tasks of collecting waste.

The action twins the objectives of improving the social status and economic situation of women waste pickers in Mumbai and recognizing their potential role in achieving the goal of a 'zero waste' Mumbai. In parallel with activities to achieve the active liberation of women waste pickers (such as the organization of training programmes, the arrangement of day-care centres, facilitating their access to health and educational programmes for their children, counselling centres and the development of cultural events which display the reality of these women), *Stree Mukti Sanghatana* is now collaborating with staff at the Baba Nuclear Research Centre to train women in composting, maintenance of biogas plants and fine sorting, and to involve women waste pickers in the operation of a pilot methane gas generation facility.

*Source: Mhapsekar, 2010; see also www.streemuktisanghatana.org, last accessed 18 October 2010*

---

cumulative installed solar capacity by 2015.[74] Although the costs of solar energy in the US are high, in Boston its adoption may be facilitated by the high local energy prices, and the municipality is due to remove some market barriers such as in urban planning charters, zoning regulations, building codes, permitting and inspections, coupled with city-level solar incentives such as solar rebates, financial assistance or tax credits. At the international level, the CDM has the potential to be an important driver for energy-from-waste projects in developing countries, including *Aterro Bandeirantes* and *Aterro San Joao* in São Paulo (Brazil); the Zámbiza landfill methane plant in Quito (Ecuador); the Bordo Poniente landfill biogas capture plant in Mexico City; and – in South Africa – the Bellville South landfill site in Cape Town and the gas-to-energy project in Johannesburg. While such schemes are frequently regarded as 'technical' fixes, there is evidence that they can be used to address broader social issues and may offer significant opportunities for the empowerment of women who work on the lower end of the waste chain (see Box 5.7).

Schemes to generate energy from waste have also proven popular in developed countries where private-sector companies have frequently provided the finances for municipal schemes. In Dallas (US), an interstate 'green' gas sale agreement will allow Dallas Clean Energy LLC to sell bio-methane captured at McCommas Bluff Landfill to Shell Energy North America.[75] The initiative Human Waste to Power the City, announced in June 2009 in Manchester (UK), is a UK£4.3 million two-year demonstration scheme, initiated by National Grid and United Utilities,[76] to convert human waste into bio-methane to power 500 homes.[77] However, despite these initiatives and the increasing interest in waste to energy, research suggests that, beyond small-scale demonstration projects,[78] the development of low-carbon energy systems remains a low priority in cities.[79]

Outside of the energy sector, and beyond the growing interest in generating energy from waste, there is relatively little evidence that municipalities are linking policies for

> The CDM has the potential to be an important driver for energy-from-waste projects in developing countries

> Schemes to generate energy from waste have also proven popular in developed countries

recycling and reducing waste directly to climate change. However, in Nigeria, the Lagos State Waste Management Authority argues that although African cities have comparably less GHG emissions than cities in the developed world, a great portion of these emissions can be attributed to waste management issues. Thus, they expect that their ongoing strategies to improve waste transport planning and the management of landfills, as well as their campaign to reduce the private burning of refuse, will have positive impacts upon the reduction of Lagos GHG emissions.[80] Besides better management, education and awareness initiatives have been demonstrated to be effective in reducing the contribution of landfill sites to GHG emissions, as in the example of Yokohama (Japan) (see Box 5.8). However, such initiatives to reduce the amount of waste sent to landfill may paradoxically weaken the viability of current and future energy-from-waste plants, which rely on a secure stream of waste as a fuel. The potential conflicts between 'technical' and 'behavioural' approaches to reducing GHG emissions from landfill highlight the dilemmas facing urban mitigation efforts where the impacts of policies and measures are uncertain and the benefits and costs of action are divided across a number of different stakeholders and communities.

Initiatives which specifically aim to reduce the carbon intensity of water and sanitation systems at the urban level are also rare. One example is in Mexico City, where the upgrade of network infrastructure will include the upgrade of 2300km of damaged networks and the establishment of 336 separate hydrometric sectors to facilitate the detection and repair of leaks. These actions will require the investment of 2970 million pesos (US$240 million) and may save up to 45,500 tonnes of $CO_2$eq/year.[81] In addition, there is an innovative proposal to generate energy from the flows of water in the network, similar to that under consideration in Durban (South Africa).[82] It is estimated that this measure alone could reduce the city's emissions by 40,700 tonnes $CO_2$eq/year.[83] While the viability of such innovative measures will depend on the particular characteristics of water supply systems, the potential GHG emissions savings that could be achieved through these sorts of maintenance,

*Energy efficiency schemes ... may fail to deliver long-term GHG emissions savings as initial reductions in energy use may be limited by the 'rebound effect'*

*In developing countries ... although transport's share of GHG emissions is low, it is growing much faster than other sectors*

modernization and efficiency measures that are already taking place in many cities may be substantial.

In summary, initiatives in the urban infrastructure domain have focused on energy efficiency schemes, primarily driven by concerns for energy security and financial savings. While such projects are politically and economically attractive, they may fail to deliver long-term GHG emissions savings as initial reductions in energy use may be limited by the 'rebound effect' as demand for energy continues to grow. However, with limited evidence of mitigation initiatives in terms of the development of renewable energy systems – or in the water, sanitation and waste sectors – a key finding from this analysis is that there may be significant hidden potential to mitigate climate change at the urban level through infrastructure networks.

Nevertheless, there remain substantial barriers to the realization of these mitigation gains, not least in terms of the economics and politics of renovating existing infrastructure systems, building new networks, and meeting the basic needs of urban communities, particularly those in informal settlements. Few of these projects address social inclusion issues explicitly, or appear to specifically target low-income groups, disadvantaged areas or slums. In some cases, social inclusion concerns have been at least acknowledged – in anticipation of potential social conflicts generated by these measures – as is the case of the landfill gas to energy project in Johannesburg (South Africa), which will include a public consultation before its completion. In general, however, urban infrastructure projects rely on the assumption that any improvements on current infrastructures will be beneficial for all the inhabitants of the city, an assumption that requires critical scrutiny as climate change tends to deepen the existing inequalities amongst urban populations in terms of access to basic services.

## Transport

The transport sector is a significant contributor to GHG emissions, representing 23 per cent (worldwide) and 30 per cent (OECD) of $CO_2$ emissions from fossil fuel combustion in 2005.[84] In developing countries, especially China, India and other Asian countries, although transport's share of GHG emissions is low, it is growing much faster than other sectors.[85] One of the key reasons for this rising trajectory is the challenge of urban sprawl, discussed above; but the growth in GHG emissions from the transport sector also represents the widespread modal shift that is taking place in cities in developing countries as household incomes and the affordability of motorized individual transport increases and aspirations for such forms of mobility, at both an individual and municipal level, increase. Moreover, urban sprawl may increase the demand for travel in ways that may not be easily met by public transport.[86] For example, 'the transport sector ... [is] the "carbon time bomb" ' in Yogyakarta (Indonesia), as 'the fastest growing fossil fuel consuming sector in the city'[87] in part because 'non-motorized transport modes such as the "*becaks*" (peddycabs) have been banned' due to their perception as being insufficiently 'modern' for municipal aspirations for the city.[88]

However, the growth of private transport is not gender neutral. A Swedish survey from 2007 showed that 75 per cent of all cars were owned by men; moreover, women's cars are generally smaller (thus, generally emit less) than those owned by men.[89] In the UK, 27 per cent more men than women hold driving licences, and women are 38 per cent more likely not to have access to a car, as well as twice as likely to be a non-driver in a household with a car.[90] In the US, men constitute two-thirds of long-distance commuters, while women tend to become dependent on and favourably inclined to using public transport.[91] Perhaps because of their lesser dependence on private transport[92] it has been suggested that women may be more willing than men to accept policies and measures that restrict cars.[93]

At the same time, as the proportion of journeys made in cities by cars and other forms of personalized motorized transport increase, so do the challenges of congestion and air pollution. The synergy between mitigating climate change and these twin issues that are both highly visible in the city and which have popular support has meant that the transport sector is one in which a range of schemes have been developed to reduce GHG emissions (see Table 5.7).

Evidence suggests that there is a contrast between areas where the transport sector features quite prominently in climate change plans and initiatives (such as in Europe and Latin America), and areas where the transport sector has received considerably less attention than other sectors, such as energy infrastructure and the built environment (i.e. North America, Australia and New Zealand).[94] Cities in developing countries show a growing interest in the development of new public transport infrastructure and technical innovation, such as programmes for the introduction of new technologies, fleet replacement with energy-efficient vehicles and fuel switching. Because of the significant investments in infrastructure and technology involved, initiatives to reduce GHG emissions in the transport sector frequently

| Type of scheme | Description |
|---|---|
| New low-carbon transport infrastructure: | The development of new transport infrastructure to encourage low-carbon modes of transportation. |
| Low-carbon infrastructure renewal: | The renewal or upgrading of transport infrastructure to reduce GHG emissions. |
| Fleet replacement: | Replacement of vehicle fleet with energy-efficient or low-carbon vehicles. |
| Fuel switching: | Switch from the use of fossil fuels for powering fleet to alternative low-carbon or renewable fuels. |
| Enhancing energy efficiency: | Measures to enhance the energy efficiency of existing vehicles and their use. |
| Demand-reduction measures: | Measures aimed at reducing the demand for individual motorized transport. |
| Demand-enhancement measures: | Measures aimed at enhancing the demand for alternative forms of travel (e.g. public transport, walking and cycling). |

**Table 5.7**

**Climate change mitigation and transportation**

The growth of private transport is not gender neutral

---

**Box 5.9 Congestion charges: Past, present and future**

Congestion pricing is a system of charging road users a fee for using the road in certain areas at certain times. It has been introduced in a number of large European cities, such as Milan, London, Rome and Stockholm, with the aim of reducing inner-city traffic volumes, reducing air pollution and encouraging the use of more fuel-efficient and environmentally friendly vehicles. Generally, congestion charges apply upon entering a clearly demarcated urban area and are paid on either a daily or per trip basis using a range of methods (online, mobile phone text message, swipe cards, scratch cards or by sensors installed in cars). Sometimes they are adjusted to the time of day, traffic levels or type of vehicle, and usually include some form of exemption for residents, low-emission vehicles, public transport and two-wheeled motorized transport.

The first congestion charge system was introduced in 1975 in Singapore and was combined with car ownership restraints. Initially it was not linked with climate change, but focused on concerns about traffic congestion.

In Rome (Italy), the 'limited traffic zone' was set up in 2001 to improve mobility and limit private vehicle trips in the historic city centre. Around 250,000 vehicles (12 per cent of registered vehicles in Rome) were permitted inside the area, resulting in a 10 per cent decrease in traffic volumes overall, a 20 per cent decrease during the restriction period (06.30am to 18.00pm) and a 6 per cent increase in public transport use.

In Milan (Italy), arguably Europe's third most polluted urban centre, more than half of citizens use private cars and motorcycles, which led the mayor of Milan to introduce 'ecopass' in 2008. This is a pollution-adjusted congestion charge affecting the 8 square kilometre city centre (5 per cent of the city's total area), levied on a sliding scale of engine types (between 07.30am to 19.30pm on weekdays).

In the UK, the London Congestion Charge Zone, one of the largest in the world, was introduced in Central London in 2003 and extended to some parts of West London in 2007. A daily charge of UK£8 allows drivers to enter the 21 square kilometre zone (07.00am to 18.30pm on weekdays). This resulted in traffic volume reductions of 18 per cent at peak times (15 per cent overall); a traffic delay reduction of 39 per cent; increased cycling by 20 per cent; and a 20 per cent increase in taxi and bus use. It should be noted that this success has not been the case across the UK; a similar scheme established in Manchester has not achieved the same results.

In Stockholm (Sweden), congestion fees were implemented on a permanent basis in 2007. These are levied every time a user crosses the cordon area, with the charge varying over the day according to the congestion levels (highest during morning and afternoon peaks, moderate during the middle of the day, and zero during nights and weekends). This scheme has resulted in an overall traffic reduction of 25 per cent; a waiting time reduction of 30 per cent; and a 50 per cent reduction in traffic volume during the evening rush hour.

Overall, evidence of the success of these schemes has been positive and the initial public resistance seems to have waned following their implementation. There have been many implementation problems, especially surrounding the initiation of the schemes. These include resistance from stakeholders and citizens, a lack of alternative infrastructure and problems with payment operations. Some questions about the economical results of the congestion charge in London have also been raised.

*Sources:* Prud'homme and Bocarejo, 2005; Leape, 2006

Table 5.8

Bus rapid transit (BRT)
systems planned or in
operation in different
regions

| Region | Number of cities | Examples | | |
|---|---|---|---|---|
| | | Name | City | Status |
| **Developed countries** | | | | |
| Europe | 21 | Ipswich Rapid | Ipswich, UK | In operation since 2004 |
| North America | 52 | Rapid Ride | Albuquerque, US | In operation since 2004 |
| | | Super Loop | San Diego, US | In operation since 2009 |
| Other | 6 | O-Bahn Busway | Adelaide, Australia | In operation since 1986 |
| | | Northern Busway | Auckland, New Zealand | In operation since 2008 |
| **Developing countries** | | | | |
| Africa | 8 | Lagos BRT | Lagos, Nigeria | In operation since 2008 |
| | | Rea Vaya | Johannesburg, South Africa | In operation since 2010 |
| Asia and the Pacific | 59 | Transjakarta | Jakarta, Indonesia | In operation since 2004 |
| | | Transit Metrobus | Istanbul, Turkey | In operation since 1994 |
| Latin America and the Caribbean | 30 | Trolmerida | Mérida, Venezuela | In operation since 2007 |
| | | Rede Integrada de Transporte | Curitiba, Brazil | In operation since 1980 |

depend on partnerships with private-sector organizations as well as the involvement of national and regional governments. Interventions by grassroots organizations or individuals in the transport sector are normally limited to projects for the promotion of non-motorized transport and demand management initiatives, such as car-sharing schemes.

A recent survey of climate change plans in 30 cities worldwide found that the most common climate change mitigation actions in transport were the development of public transport, the implementation of cleaner technologies, the promotion of non-motorized transport, public awareness campaigns and the implementation of cleaner technologies.[95]

Regulatory measures to manage demand – such as physical restraint (e.g. those implemented in Mexico City and São Paulo, Brazil), parking restraints, establishment of low emissions zones (implemented in Beijing, China, and several European cities) and speed restrictions – appear to be less common, with few examples of economic incentives being used. The examples discussed here also suggest that municipalities have a key role in the provision of infrastructure and the development of new technologies, and that they use a wide range of regulatory tools, including mandatory standards and targets, planning law or planning guidance, performance evaluation and the ban of certain fuels, together with some financial instruments such as subsidies, loans to modernize the fleet and taxes, such as congestion charges (see Box 5.9).

Turning first to issues of transport infrastructure, the mitigation of climate change is one driver behind the development of new mass transportation infrastructure. One of the most common initiatives is the operation of bus rapid transit (BRT) systems, guided bus lines or bus ways to improve the quality and speed of bus services. BRT and similar initiatives – which in many cases may be implemented at a fraction of the cost of an underground metro system – already exist or are planned in cities in all major regions of the world, although not all these initiatives are specifically tied with climate change mitigation objectives (see Table 5.8). The Transmilenio BRT System in Bogotá (Colombia) is often mentioned as a leading example, although it follows the pioneering experience of Curitiba (Brazil).[96] The service, opened in 2000, is administered by

**Initiatives to reduce GHG emissions in the transport sector frequently depend on partnerships with private-sector organizations**

**The mitigation of climate change is one driver behind the development of new mass transportation infrastructure**

Transmilenio S.A., a public company, and operated by private contractors. The system consists of 84km of central bus lines connected with 515km of peripheral lines and 114 passenger-picking stations, and can transport up to 1 million passengers every day. In addition, its 9000 buses are to be replaced with energy-efficient models. However, Transmilenio has been criticized for being overcrowded, too expensive, slow and offering limited access to certain areas of the city. Nevertheless, the experience of Bogotá is often mentioned in other cities as an example of actions to extend or improve existing mass transportation systems. Other public transportation systems such as trams or trains may have received less attention in climate change mitigation plans because of their high costs; but the use of CDM credits may increase the number of these types of projects in developing countries. For example, the Egyptian Ministry of Transport and the National Authority for Tunnels, in cooperation with CDM-Egypt, are planning to build a third line for the Greater Cairo Metro Network between 2010 and 2031.[97] The project will cost €856 million and it is expected to be funded by CDM credits.

A second area in which municipalities have sought to take action is through the development of low-carbon vehicles and fuels. In Germany, Hamburg and Berlin have teamed up in the Clean Energy Partnership,[98] which foresees the development of public fuel cell buses and urban hydrogen filling stations. In Hamburg the aim is to have 10 fuel cell buses in operation by 2010, 500 to 1000 fuel cell vehicles by 2015, together with a public network of filling stations. In Rome (Italy), the urban public agency in charge of local public transport services and the Commune of Rome have been involved in the introduction of over 80 electrically powered buses and 700 methane buses. Stockholm (Sweden) has the largest green fleet in Europe, and is heading for 100 per cent renewable energy in public transport by 2010, with tram and rail being powered by wind and hydroelectricity, and ethanol and biogas fuels used in a large proportion of the city's own fleet, as well as private vehicles (35,000 in total, about 5.3 per cent of vehicles) reducing $CO_2$ emissions by 200,000 tonnes annually. Significantly, cities are also providing arenas for the experimentation and promotion of new technologies, such as in the cases of compressed natural gas use in transport in several cities around the world including Tehran (Iran), Mumbai (India),

Dhaka (Bangladesh) and Bogotá (Colombia),[99] while in Brazil biofuels are promoted in the country's megacities.

A third set of initiatives in the transport sector includes demand-reduction and demand-enhancement measures, led by a wide range of actors and involving different policy instruments, modes of transport and understandings of mobility. For example, a non-profit organization launched by transportation activists, Car Share, has launched City Car Share schemes in several US cities such as San Francisco, Oakland and Berkeley. Public bicycle sharing networks allow people to borrow or rent bicycles so that they can travel around the city without having to own a bicycle, reducing individual purchase and maintenance costs, and storage space requirements. Such programmes are popular in European cities and are used, amongst others, in Barcelona, Spain (*Bicing*); Milan, Italy (*Biciclette Gialle*); Paris, France (*Velib*); Rome, Italy (*Romainbici*); and Stockholm, Sweden (Stockholm City Bikes). A similar programme also exists in Montreal, Canada (Bixi). Municipalities may also impose traffic restrictions, such as congestion charges (see Box 5.9), although this may reduce the access to the city to social groups who cannot afford such tax.

Municipalities can also work with other institutions to reduce demand. For example, the City of Cape Town (South Africa) has a project to develop partnerships with the largest employers within the city to reduce the work-related mobility needs of their employees. However, the introduction of demand management measures is not always straightforward. For example, in Brazil, the Porto Alegre Charter to facilitate and promote pedestrian mobility within the city – which gave new rights to pedestrian and disabled people – had to be substantially modified before it could obtain the approval of local representatives in 2007, although the original proposal of shifting the right of way from motorized vehicles to pedestrians could be maintained.

The dynamism of the transport sector and its interaction with other sectors makes it difficult to anticipate the consequences and future of climate change mitigation measures, particularly when climate change mitigation plans and actions are confronted with the increasing mobility demands of the urban population. Measures to control and reduce demand need to be complemented with alternatives for mass transport and non-motorized transport that often require significant investments in new infrastructure. In many cities, climate change mitigation concerns have been preceded by concerns about urban congestion and air quality, which makes transport a central issue in urban planning and management. Recent transport studies suggest that differential prices of energy sources, based on carbon content, could help to promote better urban transport efficiency.[100] Yet, it is not clear how this could be implemented at the city level. On the other hand, the combination of improved car technologies and traffic management may complement carbon pricing to mitigate climate change while improving the sustainability of current urban transport systems.[101]

## Carbon sequestration

In addition to reducing the amount of GHG emissions that are produced in the city, one means through which urban actors could address the challenge of mitigation is through carbon sequestration. Carbon sequestration involves removing GHG emissions from the atmosphere, either through enhancing natural 'carbon sinks' (e.g. conserving forested areas and enhancing river environments), the development of new carbon sinks (e.g. reforestation or afforestation) or through the capture and storage of GHG being produced within the city. The capture of methane from landfill sites for energy generation[102] is also a form of carbon sequestration. Traditionally, such activities have been peripheral to the main focus of urban mitigation activity. However, new developments in carbon capture and storage technologies, growing interest among national governments in carbon capture and storage, especially in developed countries and the more industrialized developing countries, and the increasing availability of carbon finance through international policy instruments – such as the CDM – are making carbon sequestration schemes more popular at the urban level (see Table 5.9). Regionally, carbon sequestration schemes are more common in developing country cities, often associated with gaining CDM credits or development programmes. However, it should be noted that actions promoting urban tree-planting and restoration, preservation or conservation of carbon sinks may be taken in cities in developed countries for reasons of environmental protection or the preservation of urban green spaces without associating them specifically with climate change mitigation objectives.

Urban carbon sequestration, however, is still in incipient stages. The technology to facilitate carbon capture and storage is still under development, and proposals for its implementation in cities are only now emerging (see Box 5.10). Carbon offset schemes based at the city level are also rare and often reach beyond city limits. In the US, Philadelphia Zoo (in partnership with private actors) has initiated the Footprints scheme to green zoo operations, develop local and international carbon offset projects, and engage with communities in Philadelphia and beyond. The Footprints scheme includes two reforestation projects, one in a former scrub site close to the zoo and another in Sukau, Borneo (Malaysia). Offsetting projects are often led by individuals or NGOs; but sometimes governmental authorities may have a crucial role in mediating the schemes. For example, since 2008 the city of Rio de Janeiro (Brazil) has created its own 'carbon market', which facilitates the participation of private

*In many cities, climate change mitigation concerns have been preceded by concerns about urban congestion and air quality, which makes transport a central issue in urban planning and management*

*New developments in carbon capture and storage technologies ... are making carbon sequestration schemes more popular at the urban level*

**Table 5.9**

**Climate change mitigation and carbon sequestration**

| Type of scheme | Description |
| --- | --- |
| Urban carbon capture and storage: | The development of schemes to capture $CO_2$ emissions from energy generation within the city and place in long-term storage. |
| Urban tree-planting programmes: | Schemes which seek to plant trees to develop the urban 'sink' capacity for $CO_2$. |
| Restoration of carbon sinks: | Schemes which seek to restore areas of natural carbon sinks in the city. |
| Preservation and conservation of carbon sinks: | Schemes which seek to preserve and enhance areas of natural carbon sinks in the city. |
| Carbon offset schemes: | The purchase of carbon sequestration offsets by actors within the city from schemes located either in the city or elsewhere. |

The Rotterdam Climate Initiative combines the city administration, the regional environmental protection agency (DCMR), the Port of Rotterdam, and the businesses in the port. It has set a target of 50 per cent $CO_2$ reduction by 2025 (compared to 1990), two-thirds of which is to be achieved by the use of carbon capture and storage technology. At present, $CO_2$ is piped and sold to horticulturalists to stimulate plant growth. However, through the capture of emissions from two new coal-fired power stations, the process will be scaled up from current volumes (of around 400,000 tonnes per year) to approximately 1 million tonnes per year.

Once carbon capture and storage technology is more fully developed (anticipated by 2020 to 2025), around 20 million tonnes of $CO_2$ per annum will be stored in depleted offshore oil and gas fields. The scheme focuses explicitly on involving stakeholders from the early stages of the projects by presenting a realistic and detailed project timetable and formal consultation procedure for stakeholders, as well as making use of existing infrastructures.

However, carbon capture and storage has been criticized for not providing a long-term solution to greenhouse gas (GHG) reduction due to its high costs and lack of development of technology. Pilot schemes such as in Rotterdam may help to elucidate whether carbon capture and storage can fulfil its low-carbon promise in an urban context.

*Source:* Van Noorden, 2008

**Most carbon sequestration initiatives at the urban level relate to tree-planting schemes and the restoration and preservation of carbon sink**

companies in carbon-offsetting projects by providing them with a methodology for calculating the amount that they need to reforest in order to abate carbon emissions, a helpline, and contacts with potential offset projects.

Most carbon sequestration initiatives at the urban level relate to tree-planting schemes and the restoration and preservation of carbon sinks. Urban tree-planting programmes frequently rely on cooperation between municipalities and citizens. This is the case in several cities in Latin America where the municipality has developed technology transfer and promotion campaigns for urban tree planting. However, the results depend largely on the voluntary and non-monitored intervention of citizens – for example, the Tree Planting Incentives in São Paulo (Brazil), the One House, One Tree programme in Lima (Peru) or the Organic

Since the 1960s, a wide range of actors in both public and private partnerships have been developing Singapore into a 'Garden City'. The effort attempts to increase the aesthetic appeal of the city, providing public open spaces and improving air quality, protecting carbon sinks and reducing the urban heat-island effect. The key strategies to create a Garden City are:

* *tree-planting* on all roads, vacant land and new development sites;
* *providing adequate, attractive and accessible parks*, including 3300ha of parks, as well as larger parks, such as the 185ha East Coast Park along the coastal areas, and smaller parks, such as town parks and precinct gardens; over the next 10 to 15 years, Singapore aims to add another 900ha of park space;
* *linking parks and people* by introducing park connectors such as green corridors for people to stroll, jog and cycle between parks; to date, Singapore has about 100km of park connectors, which is expected to triple to 360km by 2020;
* *retaining natural heritage* in four nature reserves which cover more than 3000ha, or 4.5 per cent of Singapore's land area;
* *building 'gardens in the sky'* by encouraging developers to incorporate green roofing.

*Source:* Singapore Urban Development Authority, 2009

Urban Gardens programme in Caracas (Venezuela). In Johannesburg (South Africa), the Greening Soweto programme was intended to contribute to the preparations for the 2010 FIFA World Cup in addition to its carbon sequestration benefits. The programme started in 2006 with the objective of planting 300,000 trees in Soweto.[103]

The preservation and restoration of carbon sinks is also dependent on government intervention, For example, in Lagos (Nigeria) and surrounding areas, a ban on tree felling has been imposed. So far, more than 3000 trees in the state have been counted and tagged to prevent felling,[104] although is not clear how the ban is enforced. In Bogotá (Colombia), the Botanical Garden – in partnership with local authorities – has started an initiative to improve and regulate urban tree management for the protection and conservation of urban trees, and the creation of a tree registry that may help to preserve individual trees.[105] Carbon sequestration can be combined with city beautification, particularly when a range of measures to create and protect green spaces and facilitate public access are combined, such as in the case of Singapore (see Box 5.11).

In developing countries, CDM mechanisms may help to initiate afforestation and nature conservation programmes with carbon sequestration benefits. In Egypt, for example, the Environmental Affairs Agency, in cooperation with CDM-Egypt, is developing a project (2007 to 2017) for the afforestation of 10km of road (0.5 million trees) around the ring road of greater Cairo. The project, which will cost US$4 million, is expected to contribute to the reduction of GHG emissions (100,000 tonnes of $CO_2$eq per year) and to local sustainable development objectives. However, carbon sequestration programmes need to acknowledge the differential impacts that such programmes may have in different population groups. Further work on the role of gender in urban greening is needed to understand the different services that urban green areas offer to different social groups and their roles in maintaining them.

Despite their current low profile, carbon sequestration projects appear to be gaining ground in at least three ways. *First*, the development of carbon capture and storage technologies may lead to urban pilot projects, although this technology is heavily dependent on economies of scale and carbon storage facilities, with the result that few cities are likely to provide suitable locations. Carbon capture and storage has also been criticized for failing to address the root of the climate change problem in terms of the use of fossil fuels, and any decisions to locate carbon capture and storage plants in urban settings is likely to attract significant opposition. *Second*, the CDM and growing carbon markets may help to finance afforestation and nature conservation programmes in developing countries. It is important to highlight that these programmes may simultaneously provide carbon sequestration functions while also protecting water and soil resources that are crucial for the adaptation of cities to the potential impacts of climate change (see Chapter 6). *Third*, the rapid proliferation of initiatives such as carbon markets or offsetting schemes suggests that these schemes may have more prominence in the future, although often, they may transcend the spatial boundaries of the city.

## Assessing the impact of urban climate change mitigation initiatives

The above discussion suggests that many different initiatives to mitigate climate change are taking place in cities across the world. Despite this, there is still relatively limited information about the individual and collective impact of these measures, especially when they extend beyond municipal buildings and infrastructure systems or involve behavioural change. Municipal networks, such as the CCP campaign, Climate Alliance and C40, have sought to develop indicators of their achievements.[106] The CCP Australia programme, for example, calculated that its 184 members achieved '4.7 million tonnes [$CO_2$eq] abatement – equivalent to over a million cars off the road for one year' as well as 'A$22 million saved by councils and their communities through reduced energy costs'.[107] However, such figures have been limited by their reliance on self-reporting and the lack of a common methodology to enable comparison between different international networks or with cities that lie outside of these networks. One current initiative that may contribute to building a more accurate picture of the impact of urban climate change measures is Project 2°, which aims to provide 'the first global, multi-lingual emissions measurement toolset designed to help cities measure and reduce their greenhouse gas emissions 24 hours a day, seven days week, via the web'.[108] Independent studies explicitly examining the stocks and flows of GHG in cities have also been conducted.[109] Yet, these analyses are focused on understanding historical and future trends rather than any direct assessment of the impact of policies and measures that have been put in place. There is therefore a need for new research which applies these analytical approaches to the assessment of current policy measures. However, it should be noted that the International Standard for Determining Greenhouse Gas Emissions for Cities – jointly launched by the United Nations Environment Programme (UNEP), UN-Habitat and the World Bank at the World Urban Forum in Rio de Janeiro in March 2010 – provides a common method for cities to calculate the amount of GHG emissions produced within their boundaries.[110]

Despite such new methodologies, more general challenges of assessing the impacts of policy interventions, the relatively short time-scales involved, and the fragmented nature of the data available, especially with regard to levels and reductions of GHG emissions across urban communities, will remain. In this context, basic guidance about the potential of different measures may be more useful than measures of emissions reductions. For example, it is clear that systematic efforts to shift from fossil fuel-based energy and transport systems through the use of low-carbon technologies are likely to have a more significant impact upon reducing GHG emissions than small-scale short-term initiatives to improve energy efficiency, which may be compromised by the

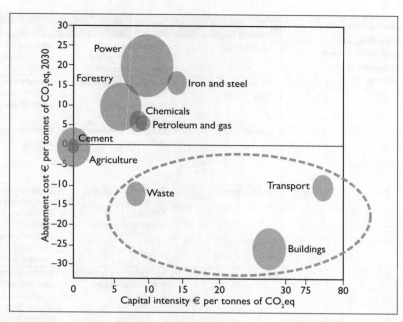

rebound effect once initial financial savings have been achieved. However, in reaching the 'low hanging fruit' – the sectors offering GHG reduction costs that yield long-term returns even without their participation in carbon markets – such schemes may have several additional benefits and act as a means of getting climate change on urban agendas. Figure 5.1 highlights the waste, transport and buildings sectors as the 'low hanging fruits' of urban GHG mitigations.

In short, decisions over which mitigation measures to adopt will be determined by the social, political and economic circumstances in individual cities and guided by the weight given to climate change concerns, rather than by any absolute evaluation of their effectiveness (see Table 5.10). The wide range of actions and the tendency to adopt piecemeal rather than strategic approaches documented in this section point to the multiple drivers and barriers to achieving climate change mitigation in the city. While in developed countries urban actors may be constrained by institutional factors and lack of public support or leadership, in developing countries there is often little incentive for municipalities to mitigate climate change when they cannot address the basic needs of current populations. In the face of these challenges, the following sections elaborate upon the modes of governing that municipalities and other urban actors have adopted to mitigate climate change and the opportunities and constraints that they have encountered.

Based on the discussion of the differences between production and consumption perspectives to the measuring of GHG emissions in Chapter 3, Table 5.11 provides a more specific overview of mitigation activities – from each of these perspectives – that can stop or reduce the current growth in urban GHG emissions.

**Figure 5.1**

**The 'low hanging fruits' of urban GHG mitigation**

*Source: ICLEI, 2010, p9*

Decisions over which mitigation measures to adopt will be determined by the social, political and economic circumstances in individual cities and guided by the weight given to climate change concerns, rather than by any absolute evaluation of their effectiveness

| Type of measure | Examples | Climate change benefits | Additional benefits | Limitations |
|---|---|---|---|---|
| Leadership | • Renewable energy demonstration projects<br>• Education campaigns | • Limited direct impact upon GHG emissions | • Demonstrate commitment to climate action<br>• Encourage action by others | • Impact assessment is difficult<br>• Could be perceived as tokenistic |
| No or low upfront costs | • Energy- and water-efficient behaviour<br>• Waste minimization | • Limited impact upon GHG emissions unless sustained over the long term | • Short-term financial savings<br>• Environmental education | • Difficult to enforce and often involves changing ingrained organizational and cultural practices |
| Cost effective | • Energy- and water-efficient technologies | • Dependent on the scale and timeframe of measures | • Short- to medium-term financial savings<br>• Impacts can be monitored<br>• Address issues of resource poverty and security | • Energy and water savings can be limited by the rebound effect |
| Multiple benefits | • Travel demand-reduction measures<br>• Reforestation and conservation projects | • Dependent on the scale and timeframe of measures<br>• Provides opportunities for working with a wide range of actors and gaining political support for climate change | • Address multiple goals of sustainability and well-being, including air pollution, congestion, urban green space, resource security and meeting basic needs | • Assessment of impacts is difficult<br>• Reliant on the involvement and actions of others<br>• Climate change benefits may be sidelined if they conflict with other objectives |
| Deep cuts | • Low-carbon and renewable energy infrastructure projects | • Large-scale projects may have significant direct impact; small- and medium-sized projects can act as catalysts for change<br>• Provides opportunities for working with a wide range of actors and gaining political support for climate change | • Offer opportunities to update infrastructure networks, provide access to services for poor communities and informal settlements | • High upfront costs and long payback periods<br>• Usually reliant on external sources of funding and partnerships with other public and private actors, which can be fragile |

*Source:* adapted from ICLEI Australia, 2008, p6

| Sector | What can stop or reduce the growth in urban GHG emissions? | |
|---|---|---|
| | **Production perspective** | **Consumption perspective** |
| Energy supply | A shift to less GHG-emitting power generation and distribution; incorporation of electricity-saving devices; an increase in the proportion of electricity generated from renewable energy sources and its integration into the grid; carbon capture and storage. | As in the production perspective, but also a greater focus on less consumption among high-consumption households; a shift to less GHG-intensive consumption. |
| Industry | A shift away from heavy industries and from industry to services; increasing energy efficiency within enterprises; capture of particular GHGs from waste streams. | As in the production perspective but with an extra concern to reduce the GHGs embedded in goods consumed by residents and to discourage consumption with high GHG emissions implications. |
| Forestry and agriculture | N/A (as no emissions are assigned to urban areas). | Encouraging less fossil fuel-intensive production and supply chains for food and forestry products; addressing the very substantial non-$CO_2$ GHG emissions from farming (including livestock); forestry and land-use management practices that contribute to reducing global warming. |
| Transport | Increasing the number of trips made on foot, by bicycle, on public transport; a decrease in the use of private motor vehicles and/or a decrease in their average fuel consumption (including the use of vehicles using alternative fuels); ensuring that urban expansion avoids high levels of private motor vehicle dependence. | As in the production perspective but with a stronger focus on reducing air travel and a concern for lowering the GHG emissions implications of investments in transport infrastructure. |
| Residential/ commercial buildings | Cutting fossil fuel/electricity use, thus cutting GHG emissions from space heating (usually the largest user of fossil fuels in temperate climates) and lighting; much of this is relatively easy and has rapid paybacks. | As in the production perspective but with an added interest in reducing the $CO_2$ emissions embedded in building materials, fixtures and fittings. |
| Waste and wastewater | Reducing volumes of wastes, and waste management that captures GHGs. | As in the production perspective but with a new concern to reduce waste flows that arise from consumption in the city but contribute to GHGs outside its boundaries. |
| Public sector and governance | N/A (as no emissions are acknowledged). | Governance that encourages and supports all the above; also a strong focus on lowering GHG emissions through better management of government-owned buildings and public infrastructure and services; includes a concern for reducing GHG emissions generated in the building of infrastructure and the delivery of services. |

*Notes:* Based on the discussion of GHG emission drivers in Table 3.18.
N/A = not available.

*Source:* based on Satterthwaite et al, 2009b, pp548–549

# URBAN GOVERNANCE FOR CLIMATE CHANGE MITIGATION

As the section above has demonstrated, municipal authorities and other actors have developed a range of strategies and measures for mitigating climate change in different policy sectors. Research suggests that the mechanisms which urban actors use to develop and implement these initiatives can be grouped into distinct 'modes of governing'.[111] First, the section reviews the changing nature of urban governance and the 'modes of governing' that public authorities and private actors use in order to address climate change. Alongside four modes of governing that have been identified within municipal governance – self-governing, provision, regulation and enabling – the growing importance of corporate, donor and civil society actors means that (quasi) private modes of governing – voluntary, private provision and mobilizing – are also becoming important. The section reviews municipal and private modes of governing, in turn, considering the mechanisms and policy instruments involved, comparing their use in the five key policy sectors discussed in the previous section, and their general strengths and limitations.

## Modes of governing climate change mitigation

The term governance can be broadly understood in two different ways. First, in a 'descriptive sense, it refers to the proliferation of institutions, agencies, interests and regulatory systems' involved in managing societies. Second, in a 'normative sense, it refers to an alternative model' for organizing collective affairs, frequently assumed to be based on horizontal coordination between mutually dependent actors where governments may be one among many agencies involved.[112] While there have been many calls to develop 'good governance' in a normative sense, this section focuses on analysing different forms of governance captured in the descriptive definition of the term. Such new forms of governance are thought to have emerged as a result of a 'profound restructuring of the state' evident in:

- 'a relative decline in the role of formal government in the management of social and economic relationships';
- 'the involvement of non-governmental actors in a range of state functions at a variety of spatial levels';
- 'a change from hierarchical forms of government structures to more flexible forms of partnership and networking';
- 'a shift from provision by formal government structures to sharing of responsibilities and service provision between the state and civil society'; and
- 'the devolution and decentralization of formal governmental responsibilities to regional and local governments'.[113]

Understanding the nature, potential and limitations for urban climate change governance involves considering the different ways in which urban governments operate, as well as recognizing the important roles played by a variety of other public and private actors. In this context, research has shown that a small number of distinct '*modes of governing*' are being employed to address climate change in the urban arena.[114] In terms of the modes of governing deployed by municipal authorities, four approaches appear to be important:

1. *self-governing:* the capacity of municipalities to govern their own operations, estate and activities;
2. *provision:* the shaping of practice through the delivery of particular forms of services and resources;
3. *regulation:* the use of traditional forms of authority such as mandates and planning law, and the oversight and implementation of regulation created at other levels of government;[115] and
4. *enabling:* the role of municipalities in facilitating, coordinating and encouraging action through partnership with regional or national governments, private- and voluntary-sector agencies, and through various forms of community engagement.

While municipal modes of governing climate change were dominant during the 1990s, more recently, new modes of urban climate governance are emerging in which private actors (such as foundations, development banks, NGOs and corporations) and public agencies outside the local authorities (donor agencies, international institutions) are initiating schemes and mechanisms to address climate change mitigation activities in the city.[116] Three approaches appear to be gaining ground, which in some ways mirror those being deployed by municipal authorities:

1. *voluntary:* the use of 'soft' forms of regulation to promote action either within an organization or amongst a group of public and private actors, combining features of the self-governing and regulation modes detailed above;
2. *public–private provision* of low-carbon infrastructures and services, either in place of or in parallel to government schemes, including initiatives developed through the auspices of the CDM; and
3. *mobilization*, where private actors seek to engage other organizations in taking action, such as through education campaigns.

Each mode of governing relies on a different combination of policy instruments and mechanisms, and may be more or less effective in mitigating climate change in the urban arena. The following sections review municipal and public–private modes of governing in turn, assessing their use in different policy sectors as well as their strengths and limitations for achieving reductions in GHG emissions.

Four modes of governing ... have been identified within municipal governance – self-governing, provision, regulation and enabling

New modes of urban climate governance are emerging in which private actors ... and public agencies outside the local authorities ... are initiating schemes and mechanisms to address climate change mitigation activities in the city

## Municipal governance

These four approaches of municipal governance – self-governing, provision, regulation and enabling – are not mutually exclusive; rather, municipalities tend to deploy a combination of these modes at any one time. This is indicative of the impact of state restructuring, where – rather than governing in a direct, hierarchical, manner – the task for state authorities is one of 'meta-governance': of articulating and combining different modes of governance.[117] However, research suggests that the self-governing mode remains the dominant approach adopted by municipal authorities in response to climate change. While the self-governing mode has significant limitations in terms of the proportion of urban GHG emissions that can be addressed (see Table 5.12), it offers a visible and often short-term means through which municipal authorities can demonstrate their commitment to climate change. In developed countries, self-governing and enabling modes have been dominant, while initiatives in developing countries are often based on the provision of low-carbon infrastructures and services. While regulation is the least frequently used mode of governing, it is most common in the transport and urban development sectors, reflecting the roles of local authorities in controlling air pollution and land-use planning. The development of

climate change initiatives in the urban infrastructure sector has primarily relied on the provision mode of governing, while the enabling mode dominates in the built environment and carbon sequestration policy sectors.

This analysis suggests that municipal governments are making use of a wide range of policy instruments and mechanisms in seeking to address climate change. Given the cross-cutting nature of climate change as a policy issue, it is perhaps not surprising to find that there is no single 'recipe for success' – with the demands of different policy sectors, as well as different national and local contexts, leading to a 'patchwork' of approaches being adopted. However, the dominance of the self-governing and enabling modes and the limited role played by regulation point to the underlying challenges that municipal governments face in seeking to address climate change. On the one hand, accounting for the impact of regulation, provision and enabling measures – in terms of GHG emissions saved, and the financial and additional benefits accrued – is a complex task. In an era where municipal governments are required to audit their achievements, such measures may be deemed economically and politically unfeasible. At the same time, moving this complex policy issue into concrete actions beyond the areas within which they exercise direct control involves municipalities challenging the deeply ingrained relationship between

**The self-governing mode remains the dominant approach adopted by municipal authorities in response to climate change**

**Table 5.12**

**Municipal modes of governing climate change**

| Mode of governing | Policies and mechanisms | Examples | Advantages | Limitations |
|---|---|---|---|---|
| Self-governing | • Management of local authority estate<br>• Procurement<br>• Leading by example | • Investment in energy-efficient street-lighting<br>• Purchasing renewable energy for municipal buildings<br>• Behavioural change programmes for local authority staff | Self-governing measures are under the direct control of the municipality and can provide quick, verifiable and cost-effective means of reducing GHG emissions. They provide a means for municipalities to demonstrate leadership and commitment to addressing climate change. | Self-governing measures can only address a small proportion of urban GHG emissions. They may be limited to those that can provide a financial return within the (short) time horizons of local governments. |
| Provision | • Operation of municipal infrastructure systems<br>• Green consumer services | • Investment in low-carbon transport systems such as BRT<br>• Household energy surveys and subsidized renovation programmes provided by municipal authority | The provision of low-carbon infrastructure and services has potential for significant reductions in GHG emissions by changing the carbon intensity of utility provision and altering the choices available to households and businesses across the city. The development of new low-carbon infrastructure networks could improve access to basic services and improve livelihoods. | Municipal capacity for providing low-carbon infrastructure and services is hampered by a lack of finances, dependency on the terms and conditions of capital loans, and a limited remit for providing energy, water, waste and transport. In contexts where there is a lack of basic services, developing low-carbon networks is unlikely to be a priority. In addition, the provision of infrastructure and services is only one factor shaping their use and may not lead to an overall reduction in GHG emissions without additional measures. |
| Regulation | • Taxation<br>• Land-use planning<br>• Codes, standards, etc. | • Congestion charging schemes<br>• Requirement for renewable energy technologies in new development<br>• Energy and water efficiency standards for buildings | Regulative measures can provide transparent and effective means for reducing GHG emissions from a variety of policy sectors. They provide a level playing field for the business community. They may also yield additional revenue, which can be invested in additional low-carbon measures. | Regulative measures can be difficult to implement because of concerns about their impact upon businesses or particular sections of the community. Regulations are difficult to apply retrospectively (e.g. to existing buildings) and governments are often reluctant to regulate individual behaviour, meaning that the application of such measures may be confined to a small proportion of total urban GHG emissions. In a context of limited municipal capacity, regulations can be difficult to monitor and enforce. |
| Enabling | • Information and awareness-raising<br>• Incentives<br>• Partnerships | • Education campaigns for walking and cycling<br>• Grants/loans for low-carbon technologies for households/businesses<br>• Development of voluntary GHG emissions reduction schemes for local businesses | Enabling measures can require relatively little financial or political investment. They enable municipal governments to benefit from the resources and capacities of a range of other urban actors in reducing GHG emissions. Through involving a range of different partners, they may increase the democratic mandate for acting on climate change. | Enabling measures are dependent on the goodwill and voluntary actions of businesses and communities who may not be forthcoming. Assessing and verifying the impact of GHG emissions reductions from such measures is often impossible and it may be difficult to evaluate their cost effectiveness. |

*Sources:* Bulkeley and Kern, 2006; Bulkeley et al, 2009; Hammer, 2009; Martinot et al, 2009; ICLEI, 2010

the use of fossil fuels and economic development, and the political and social interests that this sustains.

### ■ Self-governing

Historically, self-governing has been central to municipal efforts to address climate change, particularly in cities in developed countries.[118] In this mode, there are three principal means through which municipal authorities have sought to reduce their own GHG emissions (see Table 5.12). The *first* is through the management of municipal buildings, fleets and services. Local authorities vary considerably in terms of the building stock, vehicles and infrastructure systems that are under their direct control; but in addition to local government buildings, this can include schools, community and health centres, libraries and leisure centres; vehicle fleets for waste collection and road maintenance; as well as energy systems that provide heat and power for municipal buildings or local authority housing. Actions can include technical measures, such as retrofitting buildings with energy efficiency measures – for example, in Yogyakarta (Indonesia), Johannesburg (South Africa) and Mexico City – and demand-reduction programmes for employees (see Box 5.12). In Buenos Aires (Argentina), for example, the local authority expressed concern that, because employees do not bear the cost of energy, energy-saving measures were not being put into place. Thus, employees were given new guidance and training to prevent wasting energy in public buildings. In order to encourage behavioural change, the City of Melbourne (Australia) has implemented a 0.5 per cent performance-related pay increase for staff if they meet targets for improving the environmental performance of the organization.

The *second* is through procurement policies. These can include purchasing renewable energy for the municipality and, in the transport arena, buying alternative low-carbon fuels.

*Third*, local authorities may aim to lead by example, establishing best practice principles, or demonstrate the use of particular technologies or social practices to facilitate their widespread adoption by other local actors. Projects implemented by these means include setting targets for reducing GHG emissions or the use of renewable energy, with a recent survey of 160 cities finding that at least 125 had such targets in place,[119] as well as demonstration projects and promotional campaigns.

Overall, research suggests that within developed and developing countries, the self-governing mode of addressing climate change is prevalent across different urban policy sectors. The 'reasons for embracing [local government], institutional actions are straightforward: they require minimal or no community buy in, creating little political debate; they usually produce direct returns with respect to cost savings; they produce quick, verifiable reductions in emissions'.[120] The ability to demonstrate leadership on climate change mitigation and accrue additional (financial) benefits at relatively little economic, political or social cost has led to a strong emphasis on municipalities undertaking self-governing actions. However, while such measures may provide the initial step towards establishing climate change

---

**Box 5.12 The Green Lighting Programme in Beijing, China**

The Green Lighting Programme was initiated in Beijing in 2004. One of its mandates focuses on replacing normal lights with energy-efficient light bulbs in 2046 primary and middle schools in 18 counties and districts. The result was that it replaced 1,508,889 light bulbs, which saved 14.4MW of electricity valued at 8.21 million RMB (US$1.05 million), and reduced annual $CO_2$ emissions by 14,535 metric tonnes. The project also increased student awareness and knowledge of the concept of saving energy. In 2008, the project was extended to install energy-efficient lighting in 1263 bathrooms inside the 2nd Ring Road, 70 subway stations, 114km of subway tunnels, and in government buildings, hotels, commercial buildings and hospital buildings. The Beijing Development and Reform Commission estimates that 39MW of electricity can be saved each year through the installation of energy-efficient light bulbs.

*Source:* Zhao, 2010

---

policy, the effectiveness of self-governing measures in reducing urban GHG emissions is limited by the extent of the municipal estate and operations. In the majority of cases, municipal GHG emissions constitute a small proportion of the total emissions in a city. In this context, too much emphasis on the self-governing mode may detract attention (and resources) from the broader challenges of reducing GHG emissions across the city.

### ■ Provision

The provision mode for governing climate change involves both the development of low-carbon infrastructure systems and the delivery of 'green' consumer services by municipal governments (see Table 5.12). Historically, municipal authorities have had a strong role in the development of urban infrastructures – energy, water, waste, road and rail networks – and up until the mid 1990s municipalities continued to own their energy generation, water provision, public transport and waste services. In the years following World War II, local governments in the UK and North America began to sell off such assets, and with the rising tide of neo-liberalism in the utilities sector, many such companies in other developed countries were sold during the 1980s and 1990s. As a result, most municipal governments in developed countries have limited capacity and direct responsibilities for delivering low-carbon energy infrastructures, although there are some notable exceptions (see Box 5.13). Rather, these networks are provided by an increasingly diverse set of partnerships and private actors.[121] In developing countries, municipal governments often retain some role in the direct provision of public services and public transport networks alongside new private providers and public–private partnerships, creating the potential for governing climate change mitigation through this mode. However, such networks are limited in their social and spatial coverage, and provide far from universal access to basic services. In contexts where meeting basic needs for energy, sanitation and mobility are pressing, the ability for municipal governments in developing countries to take climate change mitigation into account is limited.

Despite these limitations, municipalities have sought to pursue climate change policy through the provision of infrastructures and services. Municipalities have been involved in the creation of low-carbon communities, such as

*The provision mode for governing climate change involves both the development of low-carbon infrastructure systems and the delivery of 'green' consumer services by municipal governments*

the New Town Development Plan in Seoul (Republic of Korea), which aims to build 277,000 new apartments with district heating, estimated to cost US$2.6 billion.[122] In the built environment, municipal governments have been involved in the provision of energy efficiency measures to existing buildings – for example, Mexico City intends to install 30,000 square metres of green roofs per year until 2012 – as well as providing 'green' services to householders, including energy audits and retrofitting packages undertaken in cities such as Melbourne (Australia) and London (UK). Perhaps most notable have been measures to provide low-carbon mass transport services. In São Paulo (Brazil), the state government planned to invest more than US$7 billion from the Inter-American Development Bank during 2007 to 2010 in order to modernize train lines and provide new bus infra-structure – upgrades which it is thought will reduce emissions by 700,000 tonnes of GHGs that can then be sold in the CDM market.[123]

Seeking to govern climate change through the provision of infrastructure and services has the potential for far-reaching impacts upon urban GHG emissions by changing

> While ... the regulation mode of governing is the least popular approach adopted by municipal governments, it can be very effective in terms of reducing GHG emissions

the carbon intensity of energy, water and waste services, reducing the carbon footprint of the built environment, fostering sustainable forms of urban development and providing low-carbon energy and travel choices for households and businesses. This potential appears to be most significant in cities where municipal governments may retain ownership or control of infrastructure networks and where basic needs have been met. However, such measures also have the potential to be socially progressive, providing the impetus for upgrading social housing and public transport services in deprived urban communities in developed and developing countries (see Box 5.14). In seeking to realize this potential, access to capital investment is likely to be a key barrier, suggesting that donor agencies and development banks may play a central role in making appropriate forms of finance available for the development of low-carbon urban infrastructure networks.

### ■ Regulation

While research suggests that the regulation mode of governing is the least popular approach adopted by municipal governments, it can be very effective in terms of reducing GHG emissions. Three different sets of mechanisms are deployed in this mode. *First*, and least common, is taxation and user fees, which have predominantly been deployed in the transport sector – for example, congestion charging (see Box 5.9) or levies on vehicle pollution.

*Second*, land-use planning, an area where municipal competencies are often strong (at least in developed countries), has been used across different policy sectors to address climate change mitigation. For example, in urban development and design, land-use planning is used to stipulate urban densities and to promote mixed land use in order to reduce the need to travel, and in the built environment sector to mandate particular standards of energy efficiency for new buildings or, as is the case of São Paulo (Brazil) and Barcelona (Spain), to introduce requirements for the compulsory use of solar energy supply in buildings of a certain size. Land-use planning is also being used to foster the development of low-carbon infrastructure. In London (UK), developments over a certain size are required to meet 20 per cent of their projected energy needs through onsite low-carbon or renewable energy generation, measures designed to increase the uptake of decentralized energy tech-nologies.[124]

*Third*, the setting of codes, standards and regulations are most common in the built environment sector, where they are often set by national governments, although examples can also be found at the municipal level, including the ban of certain building products in Vienna (Austria) and Melbourne (Australia); a mandatory energy performance requirement for large office developments in Australia;[125] and mandatory requirements for the use of solar hot water systems for some buildings in Delhi and Bangalore (India).[126] In the transport sector, several municipalities in Europe, such as Paris (France) and Athens (Greece), have experimented with schemes to ban vehicles coming into city centre areas on certain days to reduce congestion and pollution. A further set of indirect measures to reduce GHG emissions includes the implementation of standards for improving the energy efficiency and emission of pollutants from vehicles in cities such as Lima (Peru), Delhi (India) or Bogotá (Colombia).

The regulation mode of governing provides municipal authorities with a set of tried and tested policy instruments through which to address climate change. The directed, transparent and enforceable nature of these instruments means that they can be very effective in achieving reductions of GHG emissions, especially in terms of targeting the use of particular technologies and encouraging behavioural change. However, regulation can be difficult to implement. The characteristics that give it strength – its targeted and enforceable nature – can also attract opposition from those who will be adversely affected by the need to comply with and bear the costs of new standards, plans and taxes. Moreover, local governments may lack the institutional capacity to enforce regulations, particularly in cities in developing countries with limited resources.

### ■ Enabling

Municipalities have also deployed mechanisms to enable other actors to reduce GHG emissions. Research suggests that the enabling mode of governing climate change has been particularly important in developed countries, though it may now also be gaining ground amongst municipal governments in developing countries.[127] Three main approaches have been used by municipalities to facilitate action to reduce GHG emissions within the city. *First*, various forms of information and education campaigns have been implemented. Such initiatives are usually targeted at behavioural change and are therefore most common in the two sectors – built environment and transport – where changes in behaviour can have an impact upon GHG emissions. For example, in Hong Kong (China), the municipality has established a programme to promote energy efficiency in the home through reducing the demand for cooling by keeping indoor environments at 25.5°C.[128] In Durban (South Africa), the municipality has established two energy efficiency clubs with local businesses.[129] Through these clubs, 'participants were introduced to techniques for energy management and auditing, monitoring and targeting, carbon footprint calculations, and making power conservation plans. Members who implemented efficiency measures reported savings of up to R220,000 [South African rand] (US$28,000) for the 1st quarter of 2009, and the concept of "clubs" was generally well received by the industries.'[130] This example is particularly interesting as measures targeted at reducing GHG emissions from large industries are not usually part of urban municipal climate change policy.[131] However, 'the effect of such [public information] campaigns is contested and difficult to measure since they are often part of policy packages'.[132]

*Second*, municipal governments can use incentives of various kinds – including grants, loans and the removal of subsidies or barriers to the adoption of new technologies[133] – to encourage the uptake of low-carbon technologies or to promote behavioural change. Such initiatives can be found in the built environment sector, such as grants for the installation of energy efficiency measures by households, in the urban infrastructure sector, where municipal governments have provided loans and subsidies for the purchase of renewable energy technologies, and in the transport arena, where subsidies for using public transport are common.

*Third*, municipal governments have developed various partnerships with business and civil society organizations to reduce GHG emissions. For example, in Hong Kong (China), the municipal government established a set of guidelines for reporting on and reducing GHG emissions from buildings in 2008, which identified areas for energy efficiency improvement and areas for voluntary action. Since 'its introduction, 37 institutions have signed up as Carbon Audit Green Partners, including private corporations, public hospitals and universities'.[134]

The enabling mode of governing may have significant advantages in terms of its potential impact upon the GHG emissions across the city and its (relatively) low upfront economic and political costs. Seeking to engage a range of communities and businesses in climate change policy can also increase the transparency and legitimacy of urban governance. However, there are also two critical limitations. *First*, such initiatives are restricted to those who are willing to participate. For example, in the Durban energy efficiency clubs, 'not all major players participated fully... Toyota, for example, pulled out after the initial two meetings'.[135] *Second*, the voluntary nature of such initiatives means that they are difficult to monitor and verify, and cannot be 'enforced', but rather depend on the capacity of municipal governments to persuade others to take action:

> ... the effectiveness of urban planning and governance depends not only upon the assumed command-and-control power of a master plan, but upon the persuasive power that can mobilize actions of diverse stakeholders and policy communities to contribute to collective concerns. The likelihood of such enabling power to emerge is higher in the societies where power is more diffused and is transparently exercised... On the contrary, in the societies where power is concentrated, and exercised through corruption and coercion, such consensual processes pose a formidable challenge.[136]

A recent assessment of policy instruments for GHG mitigation in the buildings sector concluded that:

> Although instruments in [the support, information and voluntary action instruments] category might be considered rather 'soft' they can still achieve significant savings and successfully complement other instruments. However, they are usually less effective than regulatory and control measures.[137]

The enabling mode of governing climate change has been particularly important in developed countries

The enabling mode of governing may have significant advantages in terms of its potential impact upon the GHG emissions across the city and its (relatively) low upfront economic and political costs

## Modes of public–private collaboration in urban climate governance

As discussed above,[138] the restructuring of the state has resulted in the increasing involvement of a number of public agencies and private actors in urban climate change governance. In parallel to the approaches developed within municipal authorities, this Global Report identifies three 'modes' of public–private collaboration in urban governance – voluntary, private provision and mobilization – which are being developed in order to address climate change (see Table 5.13). This having been said, it should be noted that, in practice, there is some degree of overlap between these three 'modes'. Importantly, such initiatives do not only seek to reduce GHG emissions from one organization or group of partners, but do so explicitly in the name of one or more city. In this manner, the city has become a key arena within the broader landscape of climate governance.

The evidence reviewed for this chapter suggests that public–private collaboration in climate change governance can be found in both developed and developing countries, and across the urban development, built environment, urban infrastructure, transport and carbon sequestration policy sectors. While limited data on this relatively new phenomenon is available, these approaches appear most likely to be adopted by partnerships or networks than by individual organizations, and to be concentrated on the adoption of voluntary standards for energy and water efficiency, the provision of low-carbon urban developments and infrastructure networks, and the mobilization of behavioural changes to reduce energy and transport use.

Despite their relatively small scale, the emergence of these new forms of urban climate governance may have significant implications for achieving GHG emissions reductions. The involvement of private actors and external public agencies can provide additional sources of expertise and resource, as well as influence over sources of GHG emissions that may otherwise lie outside of the control of municipal authorities. The participation of a range of organizations and communities in addressing climate change can provide a high profile for the issue, easing the path for municipal policies, and potentially offer a means for enhancing the legitimacy and representativeness of local action.

However, partnerships should not be treated as a *panacea*. Coordinated action requires both substantial commitments from the partners and the ability of the organizations to participate effectively (see Box 5.15), and support may suddenly be withdrawn when the partnership fails to meet the objectives of one or some of its members. Partnerships can also be exclusive, serving to promote the interests of one group of actors at the expense of others.[139] This can be especially problematic in developing countries, where empirical evidence suggests that partnerships may lead 'to city government support for projects, programmes, and partnerships with powerful private-sector interests that have very large carbon footprints (in their construction and functioning) and also do little or nothing to address the key needs of low-income urban residents (including addressing the infrastructure deficit)'.[140] Likewise, the implementation of climate change measures by private companies, international networks and external public agencies raises questions about the legitimacy of the decision-making process and how and by whom the benefits, and costs, are borne.[141]

### ■ Voluntary

Voluntary approaches to addressing climate change include those which are based on changes to existing practices

---

*The implementation of climate change measures by private companies, international networks and external public agencies raises questions about the legitimacy of the decision-making process and how and by whom the benefits, and costs, are borne*

**Table 5.13**

**Public–private modes of governing climate change**

| Mode of governing | Policies and mechanisms | Examples | Advantages | Limitations |
|---|---|---|---|---|
| Voluntary | • Changing practices<br>• Demonstration projects<br><br>• Targets and standards | • Voluntary offsetting schemes<br>• Building-integrated photovoltaics<br>• Voluntary energy efficiency standards | Voluntary measures are under the direct control of the organizations involved and can provide quick and cost-effective means of reducing GHG emissions. Adopting voluntary standards or codes of practice can provide a testing ground for future legislative requirements. Voluntary measures are often adopted for reasons of corporate social responsibility and can provide a means for holding private-sector actors accountable for their carbon footprint. | Voluntary measures may be limited to those that can provide a financial return within the (short) time horizons of commercial organizations. Changes in political or economic circumstances can easily derail such initiatives. Undertaking voluntary measures can be a 'stalling' tactic to delay or avoid regulation. Such measures can also lack transparency and accountability with few if any penalties for failing to comply. |
| Private provision | • Urban infrastructure systems<br>• Low-carbon technologies and services | • Investment in waste-to-energy schemes<br>• Energy service companies | The provision of low-carbon infrastructures and services has potential for significant reductions in GHG emissions by changing the carbon intensity of utility provision, altering the choices available to households and businesses across the city. | The provision of low-carbon infrastructure and services may be limited by the terms and conditions attached to investment. In addition, the provision of infrastructure and services is only one factor shaping their use and may not lead to an overall reduction in GHG emissions without additional measures. |
| Mobilization | • Information and awareness-raising<br>• Capacity-building<br>• Incentives | • Energy efficiency advice schemes<br>• Mentoring schemes<br>• Access to subsidized energy efficiency technologies | Mobilizing other private- and public-sector actors to reduce GHG emissions can provide a means of spreading best practice and scaling up demonstration projects. By engaging a range of partners, organizations can limit the costs of acting and reduce any disadvantages of being the 'first mover'. By forming partnerships and networks of like-minded organizations, actors can strengthen their political position and claims to legitimacy. | In order to be effective, mobilization depends on the goodwill and voluntary actions of businesses and communities who may not be forthcoming. Partnerships and networks are reliant on continued interest and investment, which may be difficult to sustain through changes in personnel, politics and economic circumstances. |

within organizations and communities, demonstration projects, and voluntary targets and standards. In the *first* category, for example, are voluntary commitments to change how energy and water are used within buildings, experiments with the use of alternative fuels, and voluntary carbon-offsetting schemes. The *second* category includes, for example, initiatives that seek to demonstrate the potential of energy-efficient buildings, or the economic and social feasibility of low-carbon technologies in the urban infrastructure sector. The *third* category includes schemes which set voluntary benchmarks for achieving GHG emissions reductions, such as those promoted by 'carbon reduction action groups' at the community level.[142]

Community-based climate change initiatives seem to adopt a mixture of these approaches. One such example are 'transition towns', community-based initiatives found in the UK, North America and Australia that seek to reduce GHG emissions and address the challenge of 'peak oil'[143] by encouraging the development of the local economy, local food production, reducing demand for energy and transport, and the use of renewable energy.[144] For example, the Transition Sydney initiative in Sydney (Australia) provides presentations and films for local groups on how to address the challenges of climate change and peak oil, a website for sharing information, and support for community groups seeking to reduce their use of fossil fuels. In Bristol (UK), the Transition initiative offers home energy auditing training and various types of information and support for members seeking to reduce their individual GHG emissions. Another example is the development in Mumbai (India) since 1996 of more than 200 'advanced locality management groups', mainly to organize local waste management programmes, which are now moving into climate change mitigation activities such as the installation of solar water heaters or the development of awareness campaigns in their neighbourhoods.[145]

Such schemes have the potential to offer a progressive and inclusive approach to mitigate climate change, tackling issues of social and environmental justice alongside reducing GHG emissions. However, they are – perhaps necessarily – small in scale and often politically marginal, suggesting that their wider impact upon climate change mitigation may be limited. Their very basis on voluntary action may also be a limitation, with few means to assess the contributions that are being made or for organizations to account for their actions. At the same time, a growing emphasis on voluntary, primarily community-based, responses may serve to shift accountability from actors with responsibilities for the bulk of (urban) GHG emissions to those who have little in the way of power to address either the causes or consequences of climate change.

### ■ Public–private provision

While municipalities can set the frameworks within which new urban development takes place and infrastructure systems are developed, they may have limited jurisdiction over the provision of housing and the development of energy, water, waste and transport services.[146] As a result, partnerships between public and private actors have become a common means through which urban development and infrastructure projects, including those which seek to address climate change, are delivered. In addition, the emergence of the CDM and other carbon markets has led to a range of new partnerships involving municipal governments, urban public utility providers, national governments and carbon 'brokers' in the implementation of low-carbon infrastructure projects, such as energy-from-waste schemes.[147]

A second means through which public–private provision is taking place is through the delivery of low-carbon technologies and services. One example of such an approach has been the establishment of the London Energy Service Company in 2006, a partnership between the London Climate Change Agency and the energy company EDF in order to develop decentralized energy systems.[148] The Clinton Climate Initiative has also sought to develop access to energy service companies amongst its partner cities through developing a 'unique set of contracting terms and conditions, including streamlined procurement, transparency in pricing, and other processes that reduce project cost, development time, and business risk'.[149] While doubts may be expressed about the potential applicability of such projects to a large number of cities, and of the politics of accessing such favourable terms and conditions, it does suggest that alternative business models and financial arrangements can provide a crucial mechanism for achieving reductions of urban GHG emissions.

Given the challenge of urban governance and the privatization of urban utility networks, in most cities municipal authorities have little choice but to work with other actors in the provision of urban infrastructures. As discussed above, partnerships may provide benefits – in terms of resources, knowledge and the pooling of different strengths – but also have significant limitations. In the case of climate change mitigation, these limitations may be exacerbated by the range of actors involved and their diverse interests, ranging from local community groups to international financial organizations and other actors in the carbon market. While it is too early to tell what the impact might be, care needs to be taken that such responses to climate change do not serve to deepen existing urban inequalities.

### ■ Mobilization

A third mode through which public–private urban climate change governance is taking place can be termed mobilization, where partnerships and networks seek to facilitate the reduction of GHG emissions through the provision of advice and information, capacity-building and incentives (see Table 5.13). These approaches can be deployed internally, amongst the members of a partnership or network, or externally, through broader constituencies of business organizations, communities or individuals. Several private organizations, partnerships and networks have sought to mobilize action through providing advice and information. For example, in Beijing (China), Friends of Nature Beijing have led a campaign to maintain indoor temperatures at 26°C and limit the use of air conditioning in order to reduce GHG emissions. In Manchester (UK), a consortium of public and

*Voluntary approaches ... have the potential to offer a progressive and inclusive approach to mitigate climate change, tackling issues of social and environmental justice alongside reducing GHG emissions*

*Partnerships between public and private actors have become a common means through which urban development and infrastructure projects, including those which seek to address climate change, are delivered*

*Given the challenge of urban governance and the privatization of urban utility networks, in most cities municipal authorities have little choice but to work with other actors in the provision of urban infrastructures*

---

**Box 5.15 Manchester Is My Planet: Mobilizing the community?**

In 2005, Manchester Knowledge Capital, a strategic partnership comprised of universities, local authorities, public agencies and leading businesses in the Greater Manchester region (UK) launched Manchester Is My Planet, a programme of initiatives aimed at engaging local communities and individuals in reducing their greenhouse gas (GHG) emissions. Based on pilot studies, and supported by funding (approximately UK£150,000) from the national government, the programme asked people to 'pledge to play my part in reducing Greater Manchester's carbon emissions by 20 per cent before 2010 in order to help the UK meet its international commitment on climate change'. The scheme quickly gathered around 10,000 pledges, resulting in a visit by then Prime Minister Tony Blair and his Cabinet seeking to endorse an example of the successful mobilization of citizens around the climate change agenda. With further funding from the UK's Climate Change Challenge Fund (approximately UK£55,000), a further 8000 pledges were secured by March 2008. However, with reduced funding, work on the programme has been limited to the continued development of its website and the number of pledges currently stands at around 21,000, which according to the programme's organizers suggests an annual saving of 44,600 tonnes of $CO_2$.

This case illustrates the potential for collaboration between public and private actors to mobilize members of the community to act on climate change issues. However, there are a number of limitations to such schemes. First, the request to pledge to reduce GHG emissions has not been accompanied by measures to develop the knowledge and capacity of citizens to take action. Second, undertaking the pledge, as with other voluntary actions, carries no penalties for non-compliance. Third, in the absence of extensive monitoring the impact of such initiatives upon reducing GHG emissions is difficult to determine. Research conducted by Manchester Is My Planet suggests that over 90 per cent of the pledgers took some form of action, while over 70 per cent encouraged others to reduce their energy consumption. However, it is difficult to verify such findings or know whether changes are the result of this particular initiative.

Regardless of the potential impacts upon Manchester's GHG emissions, research suggests that the Manchester Is My Planet initiative has been politically important. First, it helped to establish climate change as an issue on local political agendas, signalling to politicians that members of the public were concerned about the issue. Second, it provided an example of 'best practice' for the national government and for replication by other local authorities and partnerships in the UK. This case therefore suggests that efforts of climate change mitigation taking place 'outside' the state and through the mobilization of individuals and communities can have a direct bearing on the future of urban climate change governance.

*Source: Silver, 2010*

---

> The mobilization mode of governing is becoming an important means through which urban stakeholders and communities are undertaking climate change mitigation

private actors has sought to engage individuals in reducing GHG emissions through a 'pledge' campaign (Box 5.15).

International networks,[150] including C40, ICLEI, the Climate Group[151] and the Clinton Climate Initiative, have developed extensive programmes and tools for providing municipal authorities and private-sector actors with information about current and future levels of GHG emissions and potential strategies to mitigate climate change, including, for example, reducing energy use, and adopting low-carbon forms of urban development and alternative modes of transport. In addition to providing advice and information, these international networks – usually working in partnership with a range of municipal governments, public agencies and private-sector actors – have also developed strategies to build capacity and provide incentives in order to engage urban actors in climate change mitigation (see Box 5.16).

These examples suggest that the mobilization mode of governing is becoming an important means through which urban stakeholders and communities are undertaking climate change mitigation. However, as illustrated by the case of Manchester (UK) (see Box 5.15), the effectiveness of such initiatives in reducing GHG emissions may be limited. In parallel to the enabling mode of governance, mobilization efforts may be hampered by limited participation and its reliance on powers of persuasion. Furthermore, questions can be raised about the mandate of private and public–private partnerships to call on others to act upon climate change, and of the extent to which they can be held to account by those who participate in such initiatives. While mobilization efforts may enable a cross-section of urban stakeholders and communities to respond to climate change mitigation, they may equally serve to promote particular visions of what responding to climate change means at the urban level, failing to account for existing inequalities or challenging the fundamental causes of the problem.

---

**Box 5.16 Climate change mitigation initiatives developed by international city networks**

- *The Climate Group city partnerships* focus on the role of some of the world's biggest cities in demonstrating and delivering the public–private partnerships that, according to them, will build up the low-carbon economy. The initiative includes the partnerships Forward Chicago in the US, and the Mumbai Energy Alliance in India.
- *The C40 Urban Life* programme is a partnership between the C40 and Arup, a consultancy firm, that operates as a co-operative to implement Arup's Sustainable Development Integrated Approach in several cities. The approach will be piloted in Toronto (Canada), and there are plans to extend the programme to five other cities.
- *C40 Carbon Financing* is a capacity-building programme to assist existing and emerging megacities to harness the carbon finance opportunities of the Kyoto Protocol.
- *The Clinton Foundation Building Retrofit Program* focuses on energy efficiency in buildings and has, so far, completed 250 projects in 20 megacities around the world.
- *The Clinton Foundation Transportation Program* focuses both on developing urban transportation systems such as bus rapid transits (BRTs) and advancing carbon-neutral transport technologies such as hybrid cars.

*Source: www.theclimategroup.org; www.c40cities.org; www.clintonfoundation.org, last accessed 18 October 2010*

# OPPORTUNITIES AND CONSTRAINTS

This chapter suggests that significant efforts are taking place to mitigate climate change in urban areas across the world. The level and range of activities being undertaken by cities demonstrate that climate change is an issue firmly on urban policy agendas in both developed and developing countries. What is also clear, however, is that in most cities mitigating climate change remains a marginal issue, and that despite ambitious policy targets the realities of reducing GHG emissions are often more challenging than anticipated.[152] The overall picture is one of policy fragmentation. Islands of best practice can be identified; but comprehensive approaches to addressing climate change remain the exception rather than the rule,[153] and significant gaps between the rhetoric of reducing GHG emissions at the urban level and the realities of putting such policies and schemes into practice can be found.[154] The critical factor shaping urban responses to the challenges of mitigating climate change seems to be governance capacity.[155] In this context, this section reviews the evidence concerning the opportunities and constraints that shape governance capacity according to three broad categories: factors that are institutional, those which are technical or economic, and those which are political in character (see Table 5.14).

## Institutional factors shaping urban governance capacity

Institutional factors which shape urban governance capacity include issues of multilevel governance (municipal competencies and the relationships between different institutions at international, national, regional and local levels); policy implementation and enforcement; and the presence of alternative institutional arrangements, such as international networks and partnerships through which governance capacity can be generated. The first two factors are discussed in the sections below, while the issue of international networks and partnerships is discussed in Chapter 2.

### ■ Multilevel governance

Urban responses to climate change do not take place within a policy or political vacuum. While municipalities are more or less coherent and have varying degrees of autonomy from international policies, and from regional and national governments, the relationship between these arenas of authority is critical in shaping the capacity to govern climate change. The 'multilevel' governance of climate change affects urban responses to climate change in three key ways: by providing the context within which urban responses are framed; by determining the autonomy and competencies – the duties and powers – for municipal authorities to act in response to climate change; and by enabling policy integration within and between local authorities.

*First*, international and national policies have provided the overall framework for municipal responses. National policies have also served as direct drivers for municipal actions. For example, in Sweden approximately 'half of all

| | Examples of opportunities | Examples of constraints |
|---|---|---|
| Institutional | • Proactive national/regional government<br>• Membership of international municipal networks<br>• Formation of partnerships | • Limited formal powers for municipal authority<br>• Absence of policy coordination |
| Technical and economic | • Knowledge of urban GHG emissions<br>• Availability of external funding<br>• Flexible internal finance mechanisms | • Lack of expertise<br>• Lack of financial resources<br>• Suitability of technology |
| Political | • Political champions<br>• Recognition of co-benefits<br>• Political will | • Departure of key personnel<br>• Prioritization of other policy agendas<br>• Conflicts with other critical economic and social issues or sectors |

**Table 5.14**

**Opportunities and constraints for governing climate change mitigation in the city**

municipalities have adopted climate mitigation goals in accordance with the national objective of reduced climate impact as formulated in the Swedish climate strategy',[156] while in China, research suggests that the recent interest in addressing climate change at the local level has not been in response to the issue itself but instead as 'a response to the central government's expectation for these institutions to take action'.[157] However, there are two significant exceptions to this rule which suggest that an enabling national government context is not always necessary for urban responses to climate change. In Australia and the US, the number of cities developing responses to climate change grew exponentially during the late 1990s and early 2000s at a time when both national governments withdrew from the international process of implementing the Kyoto Protocol. However, in both countries, urban responses were organized through international municipal networks, drew heavily on the international policy framework, accessed financial resources from federal government funds, and frequently gained support from cooperative regional-level governments to support the development of urban policy. These examples suggest that an enabling *multilevel* framework is critical in fostering urban capacity even when the political support of the national government is absent.

A *second* critical aspect of multilevel governance concerns 'whether the local authority has broad policy development and implementation powers, or whether these powers are narrowly defined or constrained'[158] in relation to critical policy sectors, such as transport, land-use planning, infrastructure development, building standards and waste. The role of municipalities in these areas is usually defined by central or regional governments and is delegated to local authorities.[159] Municipalities that have specific competencies for the direct provision of waste, transport or energy services, such as is the case in many Northern European countries, can have significant capacity to address climate change that other local authorities lack.[160] However, in general, municipalities have limited powers and responsibilities with respect to energy policy, pricing and supply, the development of urban infrastructure (such as transport systems), the use of economic instruments (such as taxes and charges), as well as energy efficiency standards for buildings and appliances, and more autonomy with regard to land-use planning, education and voluntary programmes.[161] Municipalities can therefore be dependent on the policies and actions of national governments in order to achieve their

Significant efforts are taking place to mitigate climate change in urban areas across the world ... however, ... in most cities mitigating climate change remains a marginal issue

The critical factor shaping urban responses to the challenges of mitigating climate change seems to be governance capacity

There is considerable evidence that municipalities go beyond their direct competencies in undertaking actions to address climate change

the 'cross cutting nature of climate change governance means that environment departments or agencies are frequently not able to implement the policies ... that are required to address the problem'

In many policy areas, municipal authorities, particularly ... in developing countries, are unable or unwilling to enforce regulations and standards

policy goals. For example, in London, the Climate Change Action Plan recognizes the 'difficult truth is that in preparing this action plan we have been unable to present any realistic scenario in which we can achieve the 2025 target ... without major national regulatory and policy change'.[162]

However, actions to reduce GHG emissions may be achieved in cases when municipalities have limited direct or formal powers. First, policy goals can be integrated across different levels of government enabling action. A study of the climate change responses in Helsinki (Finland) shows how energy consumption in the built environment is determined by European Union (EU) regulations, such as the Energy Performance of Buildings, national regulations, municipal regulatory oversight, and voluntary agreements between energy companies and government departments. In this policy area, 'the different levels of governance are working well together ... the city is implementing energy performance policies by implementing the building code and granting energy aid, and also by participating in the voluntary energy conservation agreement scheme',[163] whereas when it comes to the promotion of renewable energy, policy initiatives at the city level remain in contradiction with EU and national policies of increasing renewable energy generation.

A second means through which municipal authorities can overcome limited direct competencies for acting upon climate change is through the development of the limited opportunities that do exist. In Japan:

> ... regional and local governments have the authority to take legislative action when the national government itself has not enacted any specific policies and measures toward climate change, and the national government does not prohibit them from doing so. Using this opening, some governors and mayors have introduced regional and local ordinances which mandate businesses and industries to formulate $CO_2$ reduction plans, introduce emission trading in the regional and local area, or buy renewable energy bonds.[164]

Third, there is considerable evidence that municipalities go beyond their direct competencies in undertaking actions to address climate change. For example, the carbon market created by Rio de Janeiro (Brazil) could have impacts at the national and international levels while creating partnerships between public, private and civil society actors within and beyond the city. The capacity challenges which emerge from limited autonomy and competencies are only partially derived from their relation with national government, but are also dependent on their relation with other partners, and on the ability of local governments to create an 'enabling environment for local civil-society action'.[165] Nonetheless, for many municipalities, a lack of formal powers to address climate change remains a significant barrier and is one reason for the current focus on 'self-governing' across municipalities.[166]

The *third* aspect of multilevel governance that is significant in shaping municipal capacity concerns issues of governance fragmentation at the local level and the internal dynamics of municipalities. At the city-region scale, a key issue concerns the fragmentation of urban governance across multiple authorities. For example, a study of climate responses in Mexico City finds that:

> ... the administrative structure of city's governance differs from its boundaries and carbon-relevant socioeconomic and ecological functioning. Administratively, the city is managed by diverse federal, state and local tiers of government. Yet, the city functions as a complex system; its core area and localities, activities and households are interlinked by economic interchanges and transportation activities, by fluxes of materials and energy.[167]

Research has found that in 'many cities expertise [of climate change] is still concentrated in the environmental department'.[168] This potentially limits municipal capacity for two reasons. First, environmental departments are often marginalized within municipal (and other) authorities and may be in conflict with other parts of the local administration. Second, the 'cross cutting nature of climate change governance means that environment departments or agencies are frequently not able to implement the policies ... that are required to address the problem'.[169] This challenge of horizontal coordination has been exacerbated in many countries in the wake of neo-liberal reforms, which have led to the privatization or contracting-out of what were previously municipal services, increasing the number of actors with whom policy coordination needs to be undertaken. For example, in Johannesburg (South Africa), a process of 'semi-privatization' has occurred within the local authority that 'creates a silo effect where communication between different agencies, utilities and the city administration are fragmented', reducing municipal capacity to address climate change.[170] In this context, 'mainstreaming, coordination, and cooperation across government agencies is vital'.[171] In some cities, this is being achieved through the development of new administrative and institutional structures, such as special units or agencies which coordinate climate change policies. For example, in London (UK), a Climate Change Agency has been established, while in Zurich (Switzerland), an environmental protection unit has been established to supervise climate policy.[172] However, 'where there is a lack of capacity to do this joining up it is clear that the potential of local climate change strategies is curtailed'.[173]

## ■ Policy implementation and enforcement

A second set of institutional factors that shapes urban climate change governance capacity is the ability to implement and enforce policies and measures. In many policy areas, municipal authorities, particularly but not exclusively those in developing countries, are unable or unwilling to enforce regulations and standards. For example, in Nigeria increasing energy efficiency in the built environment and appliance sectors suffers from 'noncompliance resulting from lack of enforcement of the standards ... exposing the Nigerian Energy

Market ... and hence the consumers to all kinds of sub-standard technologies (of course energy inefficient) which may have even been outlawed in their countries of manufacture'.[174] In the Ukraine, research has found a similar situation characterized by a lack of building standards for energy efficiency coupled with the poor enforcement of those that exist.[175] The effectiveness of energy standards may be particularly low in developing countries, given difficulties with enforcement and corruption.[176] However, 'even in developed countries, the estimated savings from energy codes range from 15–16 per cent in the US to 60 per cent in some countries in the EU', suggesting that both the levels at which standards are set and the ways in which they are implemented vary significantly from country to country, in turn affecting the capacity of municipalities to address GHG emissions.[177] The avoidance and corruption of regulations is also a critical challenge. In Indonesia, research found that 'while zoning permit is theoretically supposed to be a tool to control land use, in reality corrupt practices have rendered it ineffective'.[178] However, while at least part of the problem of policy implementation may be laid at the door of corrupt practices or deliberate avoidance, it also stems from the use of inappropriate policy approaches and models. For example, in many developing countries 'the application of imported models of urban planning and government that proved inappropriate to local contexts and possibilities' has served to limit the access of poor communities to land for housing, in turn provoking the emergence of informal settlements and other slums that do not comply with building and planning regulations.[179]

Equally, the challenges of implementation are not confined to municipal authorities. Given the voluntary nature of many of the schemes being developed by the private, civil society and donor communities in cities to address climate change, issues of compliance, monitoring and verification of achievements also affect urban governance capacity. First, significant challenges exist in terms of reliably estimating the GHG emissions reductions attributable to specific schemes, a factor which has so far limited the use of the CDM in urban areas.[180] Second, issues of accountability are also significant. While most schemes rely on self-reporting, there is a growing movement for civil society actors to be involved in processes of verification, such as the development of the Gold Standard for CDM and voluntary carbon markets.[181]

## Technical, material and financial factors shaping urban governance capacity

A second set of factors that provide opportunities and constraints for urban responses to climate change mitigation include issues of technical expertise, the material infrastructures and cultural practices that determine the possibilities for action, as well as the financial resources available.

### ■ Expertise

There are two main ways in which the availability of scientific expertise and knowledge has shaped urban governance capacity for mitigating climate change. *First*, the growing scientific consensus internationally about the nature of the

climate change problem and the need for urgent action has been a motivating factor for many municipalities. As the scientific community has advocated increasingly stringent targets for reducing GHG emissions in order to minimize the risk of exceeding a 2°C warming of the atmosphere, cities have responded with ever more ambitious policy goals. London (UK), for example, has adopted a target of stabilizing 'CO$_2$ emissions in 2025 at 60 per cent below 1990 levels, with steady progress towards this over the next 20 years'.[182] In 2002, the City of Melbourne (Australia) adopted a target of reaching 'zero net emissions' by 2020, an approach that has been adopted by a number of other municipalities in the metropolitan area.[183] Equally, the growing scientific consensus surrounding the issue of climate change has been a significant factor influencing the growing importance of the issue on the agendas of private-sector and civil society organizations, leading to their mobilization and involvement in various initiatives.

*Second*, scientific knowledge has also been significant in the development of local inventories and forecasts of GHG emissions.[184] Such inventories are primarily derived from 'scaling down' regionally and nationally available data, which provides a general overview of the likely pattern of GHG emissions and potential areas of future growth. Some municipalities have sought to derive such inventories from 'the bottom up'. One example is Newcastle (Australia), where community-wide GHG emissions derived from consumption data, and the equivalent GHG emissions from electricity use are updated hourly and reported on the internet, on a billboard in the city and in a weekly television news report.[185] However, most local authorities lack the resources to develop such inventories, while those that have sought to develop a comprehensive picture of GHG emissions across the city have found their efforts constrained by a lack of data, much of which is either not collected on a routine basis or regarded as commercially sensitive by energy providers.[186] While the lack of data and the expertise or resources to gather and assess it is a constraint on the ability of municipalities to measure progress towards policy targets, it is clear that – for the majority of cities – a comprehensive picture of urban GHG emissions may be an impossible goal. It may be better to focus efforts on deriving a general overview of where policy attention should be directed.

Beyond the scientific realm, other sources of expertise are also important. The example of Durban (South Africa) shows the municipality's difficulties in participating in international actions such as the CDM because of a lack of staff training.[187] Once this training is completed, employees may choose to move on to more profitable private-sector jobs to develop the same projects. In addition, local authorities may have little access to recent developments in architectural and engineering professions. For example, in Nigeria, 'lack of information on trends in energy efficient architecture by professionals is a formidable obstacle. This has also encouraged lack of energy conscious building standards and regulations.'[188] Skills shortages, however, are not exclusive to developing countries, although they may be more severe and may affect other aspects of sustainable development not directly connected with climate change.

> The effectiveness of energy standards may be particularly low in developing countries, given difficulties with enforcement and corruption

> The example of Durban (South Africa) shows the municipality's difficulties in participating in international actions such as the CDM because of a lack of staff training

### ■ Urban systems: Infrastructures and cultural practices

Opportunities and constraints facing the urban governance of climate change are also structured by the social and technical networks that constitute cities – a 'seamless web' of material infrastructures and everyday practices that sustain them.[189] The *first* challenge that this raises for the capacity to mitigate climate change is that of urban morphology and design. The opportunities for reducing demand for travel, for example, in a city characterized by urban sprawl or the rapid development of informal settlements at the periphery will be very different from those in a historically compact urban settlement, and urban decision-makers may find their choices heavily constrained by existing infrastructure networks and spatial form.[190] Rather, the comparison of transport systems between Singapore and New York (US) suggests that a range of factors, including fuel pricing and tourism, will be able to shape urban form alongside urban planning.[191] Equally, traditional practices of building design can provide significant barriers to the development and implementation of mitigation measures. In the Ukraine, research finds that:

> ... communal services, predominantly heating, are still very inefficient. Outdated systems in poor condition and high losses due to insufficient maintenance as well as no possibility for heat adjustment are the main reasons for the bad performance ... the efficiency in using energy resources in the building stock in Ukraine is 4–5 times lower than in western countries.[192]

In Nigeria:

> ... most buildings seem to be replicas of buildings in European countries in shape and form despite marked differences in climatic conditions... Window sizes and openings have not responded to physiological comfort thereby necessitating the use of mechanical devices for increased air movement. The choice of windows tends to be in response more too aesthetic needs rather than physiological needs.[193]

Such traditional practices, whether as a result of particular political regimes or the importation of so-called 'modern' design, can have a detrimental effect on the capacity of cities to respond to climate change.

A *second* challenge arises from the nature of the infrastructure systems that supply services, such as energy, water and waste collection, as well as existing building stock. For example, in Helsinki (Finland), the EU target of increasing renewable energy to 20 per cent of energy supply by 2020 is seen as limited by the current district heating network in which biofuels are regarded as the only potential, but still costly and potentially ineffective, option.[194] The introduction of new vehicle fuels, such as in London's (UK) ambitious

plans to create a 'hydrogen economy', are limited by the network of refuelling stations which may encounter local opposition during the planning and development process.[195] In Iran, the programme to substitute fuels by compressed natural gas has been clearly limited by the existence of fuelling stations; while 180 filling stations are planned, the pilot programme in Tehran will include only 2.[196]

### ■ Financial resources

Financial resources are both a driver and a barrier to fostering urban responses to climate change. Municipal authorities lacking the finances to provide even basic services for their constituents are unlikely to invest in climate change mitigation, given the many competing issues on urban agendas. A lack of basic service provision in cities in developing countries, and especially for those living in informal settlements, can reflect 'local governments lacking the resources to meet their responsibilities – and often with very limited capacities to invest (as almost all local revenues go to recurrent expenditures or debt repayment)'.[197]

A lack of finances to invest in basic service provision and in the development of urban infrastructures means that issues of climate change mitigation are far from a priority, and even where there is commitment to act, financial constraints may prevent the implementation and enforcement of policy goals. For example, in Tuzla (Bosnia and Herzegovina), the municipality had to drop proposals to tax the air pollution emissions of the local coal-fired power plant because of the lack of initial resources to administer the programme and the lack of support at the national level.[198] While this is an acute challenge for cities in developing countries, a lack of adequate finance can also act as a barrier to action on climate change mitigation in cities in developed countries. For example, in the UK, local authorities are bound by strict central government controls over their finances, and their ability to provide capital for infrastructure projects and service provision is limited. At the same time, increasing pressure on local government finances has meant that limited funding is available for even small-scale projects.[199] Often, finding a source of finance is not the only problem: allocating the resources in an efficient way is also challenging (see Box 5.17).

Equally, rather than a lack of financial resources, action can be impeded by the financial reporting and distribution mechanisms in place in an organization. For example, in São Paulo (Brazil), research has found that issues of financial resources, surprisingly, are not a key factor shaping the early stages of the development of climate policy and that 'institutional difficulties in reinvesting resources, rather than actual lack of resources, were reported as the main obstacle'.[200] A significant factor shaping the capacity of municipalities and other organizations to respond to climate change at the urban level is therefore the ability to establish novel mechanisms for distributing funding internally to facilitate investment in particular policy measures. This is one area in which political champions or policy entrepreneurs (explored in the section below) have been particularly important in overcoming the 'inflexible budgetary structures'[201] for which municipal authorities are usually renowned.

---

*The opportunities for reducing demand for travel ... in a city characterized by urban sprawl or the rapid development of informal settlements at the periphery will be very different from those in a historically compact urban settlement*

*A lack of finances to invest in basic service provision and in the development of urban infrastructures means that issues of climate change mitigation are far from a priority*

*A significant factor shaping the capacity of municipalities ... to respond to climate change ... is ... the ability to establish novel mechanisms for distributing funding internally to facilitate investment in particular policy measures*

**Box 5.17 Distribution of resources for climate change mitigation in Mexico City**

In 2008 the City of Mexico presented the Climate Change Action Programme, which introduced a number of measures in the fields of energy, transport, water, waste, climate change adaptation and environmental education. Some 60 per cent of the total budget (of some 61 billion pesos) was invested in transport measures and an additional 36 per cent in infrastructure. Only 4 per cent of the budget was invested in measures for the built environment. However, while the measures in the transport and urban infrastructure sectors were expected to reduce carbon emissions by 2.1 million tonnes of $CO_2$eq (47 per cent of projected emission reductions) and 1.9 million tonnes of $CO_2$eq (42 per cent), respectively, built environment measures were projected to reduce the city's carbon emission by 0.5 million tonnes of $CO_2$eq (10 per cent), suggesting that the built environment measures are the most effective in reducing carbon emissions.

The analysis in the figure brings a new dimension into the discussion – namely, the disparity of efficiency of different measures in terms of reducing $CO_2$eq per million of pesos invested. Issues such as the 'rebound effect' that may cancel the energy efficiency gains of built environment programmes (e.g. 'efficient lighting in homes') need to be taken into consideration. Furthermore, the costs and reduction potential of each measure will be different in each city. Overall, the Mexico City approach, which targets a wide range of measures in different sectors, is likely to bring the best results.

*Source:* Ciudad de Mexico, 2008; see also Johnson et al, 2009

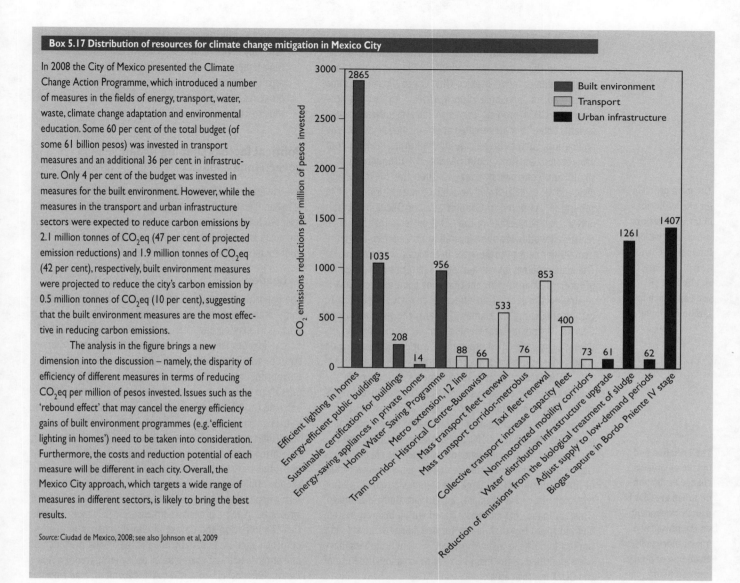

Mechanisms that have been established to leverage internal sources of funding include revolving energy funds (where financial savings from energy efficiency are reinvested in energy conservation or other climate change projects), energy performance contracting and the establishment of energy service companies (where external organizations or companies established by municipalities invest in energy efficiency measures and profit from the financial savings made).[202] In Japan, several local governments are operating local energy service companies that are achieving energy savings of more than 10 per cent.[203]

Access to external sources of funding is also a key factor shaping local capacity to address climate change. Such sources of funding may come from the EU, national governments, through partnership arrangements, or donor organizations. International municipal networks, such as ICLEI's CCP campaign and the C40, have been critical in leveraging funding for municipalities. One recent initiative in which the C40 is involved is the Carbon Finance Capacity Building Programme[204] that 'encourages the use of Carbon Finance to reduce GHG emissions in cities, in particular emerging mega cities' in developing countries.[205] National

governments are also an important source of direct funding for municipal responses to climate change. In The Netherlands, the *Klimaatcovenant* is a multilevel arrangement within which cities are required to complete a performance assessment of their targets, policies and measures and are given funding according to their achievements and population or area for the implementation of climate plans.[206] In the US, many climate change mitigation measures have been associated with philanthropic activities. For example, the development of the Plaza Apartments in San Francisco by the Public Initiatives Development Corporation was supported by grants from a private utility, Pacific Gas, and a partnership between 31 financial and energy multinationals.

One comparatively new source of funding, and one that to date has had little impact upon the development of urban mitigation efforts, is that of carbon finance. As noted in Chapter 2, there are two principal sources of carbon finance: the CDM, and that from emissions trading.[207] In São Paulo (Brazil), the use of the 'methane from the Bandeirantes landfill (one of the largest in the country) for producing electricity' was financed by the CDM, and it has been estimated that this action alone has reduced the city's

**Access to external sources of funding is ... a key factor shaping local capacity to address climate change**

**One ... new source of funding ... is that of carbon finance...: the CDM, and ... emissions trading**

emissions by 11 per cent. The resulting carbon credits were sold, raising significant finances for investments in 'social projects in the area of the landfill'.[208] Similar projects are being implemented in Mexico City, Quito (Ecuador), Lima (Peru) and Johannesburg (South Africa). As this example suggests, CDM projects may also have significant social benefits when they are targeted at low-income sectors of the population, such as is the case in the Kuyasa development project in Cape Town (South Africa),[209] although there is significant controversy over whether the 'development dividend' of the CDM will be realized in this project, or in general.[210] As noted in Chapter 2, as of December 2009, only a small percentage of the CDM projects that had been registered worldwide were located in urban areas, and more than 90 per cent of these were in the solid waste sector.[211] The main problem for city authorities in terms of making use of international funding mechanisms such as the CDM is related to the lack of an effective city-wide approach to carbon financing. In Amman (Jordan), the Amman Green Growth Programme represents an innovation in this field as the first city-wide CDM programme worldwide, focusing on waste, energy, urban transport and urban forestry sectors.[212]

The lack of urban CDM projects reflects both the complexity of the processes involved in the design and verification of projects, the lack of available and consistent data, the problems of ascertaining 'additionality' where a number of factors may shape GHG emissions reductions, and the finances involved, with evidence suggesting that projects that seek to reduce demand for energy generate lower rates of financial return than large-scale energy supply or industrial projects.[213] For example, 'a transport study of Santiago, Chile ... found that costs associated with a bikeway system and improved bus technology for 462 buses achieved very limited benefits as a CDM project'.[214] The following have been identified as the main barriers to expanding the use of the CDM in urban areas:

- *Small individual projects:* typical projects in cities (except for waste management projects in large cities) are small and yield small volumes of emission reductions.
- *Repeated clearances from the same local authority for different projects:* for each project activity, developers need to seek approval, which can be time consuming and cumbersome.
- *Lack of good 'bundling' agents:* due to different budget processes and approval timelines, bundling projects across several cities is a complicated process, which is exacerbated by the fact that very few public agencies have the mandate or capacity to bring together different city governments and mobilize project activities.
- *Lack of strategic planning by the city:* piecemeal assessment of projects, proposed by developers, prevents local authorities from taking a holistic view of their development plans and opportunities to reduce GHG emissions.
- *Lack of opportunity to structurally build the capacity of local authorities to identify GHG mitigation opportunities and to monitor emission reductions:* the lack of strategic

thinking results in continuation of business as usual, generally in the form of breakdown orientation (i.e. replacement only when equipment is broken beyond repair), minimum maintenance (only when reported broken) and least cost-based purchase of equipment (due to budget restraints).[215]

## Political factors shaping urban governance capacity

Political factors that shape the opportunities and constraints of urban climate governance can be considered in terms of issues of leadership, questions of opportunity, the framing of the costs and benefits of acting upon climate change, and underlying structures and processes of political economy.

### ■ Leadership

The opportunities afforded by two different forms of leadership – at individual and organizational levels – have also been critical in shaping governance capacity to address climate change in cities (see Box 5.18). Several studies have demonstrated that individual political champions or policy entrepreneurs have been critical to the development and pursuit of policies and projects at the urban level.[216] One example can be found in London (UK) where former Mayor Ken Livingstone was a key figure in the development of ambitious policy targets and the formation of the C40. Operating both within and outside of the public eye, such individuals are critical in getting climate change on the agenda of municipal and private-sector organizations, countering opposition, forging coalitions, developing policies, and advocating particular goals and measures. Evidence suggests that the initiation and uptake of climate change mitigation on urban agendas is usually dependent on the presence of one or more political champion or policy entrepreneur. A key factor reducing urban governance capacity for addressing climate change may therefore be the lack of committed individuals. However, such individuals are not sufficient for sustaining policy action because of the barriers that they may encounter and the often temporary nature of their role within any one organization.[217] For example, in Durban (South Africa), Mexico City and São Paulo (Brazil), research has found that the effectiveness of individuals and of the coalitions that they form is constrained by the institutional and federal government contexts within which they operate.[218]

At the organizational level, leadership is also an important factor in shaping urban governance capacity. Opportunities to be at the forefront of initiatives amongst a peer group – for example, to be the first municipality to deploy a technology, adopt a certain target or achieve a particular measure – have provided the impetus for action in the urban arena. Such initiatives play on the growing importance of climate change as a means of fostering organizational reputation, both within municipalities and across the corporate sector. International networks seeking to foster urban responses to climate change have, in turn, provided various means for recognizing and rewarding such leadership, such as the Climate Alliance Climate Star award and

CCP Australia's Outstanding Council Initiative award, in turn promoting such forms of leadership. At the same time, being part of a 'leadership group' has also fostered capacity amongst both public- and private-sector actors to address climate change. The C40 is a case in point with its emphasis on being a 'climate leadership group'.[219]

Within cities, public–private partnerships or voluntary agreements amongst private-sector organizations also rely on notions of leadership and innovation. One such example is the recently launched Forward Chicago initiative, orchestrated by the Climate Group and the mayor of Chicago and intended 'to engage Chicago's leading businesses in public–private partnerships to implement selected climate initiatives'.[220] However, this emphasis on the importance of leadership can also constrain municipal capacity to respond to climate change in several important ways. First, it is clearly impossible for every municipality or private-sector actor to 'be the first' to address climate change and the danger arises that an emphasis on innovation will mean that mitigating climate change fails to be adopted as part of mainstream urban policy. Second, leadership groups are, by their very nature, exclusive, with the result that a two-tier approach to addressing climate change in urban areas could be fostered in which the 'best' cities attract resources and political support, leaving 'the rest', where the majority of GHG emissions lie, behind.

## ■ Windows of opportunity

The presence of committed individuals and an institutional framework within which acting on climate change is supported provides a basis upon which windows of opportunity can be used to further climate change policy ambitions. Such opportunities can take the form of specific climate change initiatives, triggering events that create the political and physical space for interventions in the city, or sources of funding or political support that can be diverted for climate change ends.

In terms of climate change initiatives, the participation in international and municipal networks frequently provides windows of opportunity for member municipalities. For Seoul (Republic of Korea), membership of the C40 and the hosting of the 2009 summit of that network provided a basis for galvanizing action in the city. The invitation to Luis Castañeda Lossio, mayor of Lima (Peru), to make a plenary presentation about the climate change initiatives in the city in the 2009 summit propelled the adoption of climate change mitigation measures, including the use of natural gas in city buses and the municipal fleet and the establishment of individual grants for exchanging old cars for gas-fuelled ones.

Research suggests that major urban events, such as sports competitions, can be significant triggers for actions to address climate change, providing both the political profile to what might otherwise be routine infrastructure projects, as well as the finances and motivations to undertake wholesale infrastructure replacement programmes. For example, before the 2010 FIFA World Cup, Cape Town (as well as other major cities in South Africa) was seeking to develop BRT systems[221] with the aim of achieving a 10 per cent

increase in rail transport use and a 10 per cent decrease in private vehicles commuting into city centre between 2005 and 2010. The 2008 Olympic Games in Beijing (China), the 2006 Winter Olympic Games in Turin (Italy) and the 2006 FIFA World Cup in Germany have all been recognized as events triggering significant environmental action (see Box 5.19).

A further means through which windows of opportunity are exploited to address climate change occurs where there is a degree of commitment to action on the issue, and

*Major urban events, such as sports competitions, can be significant triggers for actions to address climate change*

sources of additional funding can be diverted to support policies and measures. For example, in São Paulo (Brazil), the 'need [for] controlling air pollution was a window of opportunity for implementation of climate change related policies'.[222] Similarly, in Rio de Janeiro (Brazil), the federal government's commitment to build 1 million low-cost energy-saving houses in disadvantaged neighbourhoods by 2010 provided the opportunity to experiment with and implement energy-efficient construction materials.[223]

**The bundling of climate change mitigation with other potential social or environmental benefits at the city level may be a potential trigger of climate change action and … may determine the long-term success of the initiatives**

## ■ Issue framing and the realization of co-benefits

The bundling of climate change mitigation with other potential social or environmental benefits at the city level may be a potential trigger of climate change action and a factor that may determine the long-term success of the initiatives. The issues that may influence climate change mitigation actions are varied and depend largely on local conditions.[224]

The examples discussed in this chapter show that a wide range of potential co-benefits may be associated with climate change mitigation. Overall, initiatives in the built environment are often associated with energy savings or with issues of social justice, particularly when actions are associated with developments or improvements targeting low-income population sectors. Energy efficiency programmes are often linked with financial savings. This may be particularly significant for municipalities as 'local governments have come to realize the link between energy saving and climate change. They can claim credit for action on both issues even though they only take action related to energy saving; they are in essence killing two birds with one stone.'[225] Actions related to urban infrastructure may bring direct benefits in terms of improvement in access, affordability and service. In Lagos (Nigeria), climate change mitigation initiatives in the waste sector are linked to improvements in the service and reduction of pollution from waste burning. Actions in the transport sector are associated with reducing congestion and reducing air pollution, through, for example, BRT and congestion charges. Finally, carbon sequestration programmes, particularly those linked with urban tree-planting, are often associated with ideas of city beautification, such as the Greening Soweto proposal in Johannesburg. The combination of social justice and sustainable development concerns may open windows of opportunity for the advancement of climate change mitigation actions.

**Energy efficiency programmes are often linked with financial savings**

**Carbon sequestration programmes … are often associated with … city beautification**

Such strategies may be particularly important in contexts of ambiguous or overtly hostile responses to addressing climate change in cities. However, joining climate change mitigation initiatives with other co-benefits may also have downsides. For example, linking climate change with the local sustainability agenda may mean that climate change actions need to be limited to those about which consensus can be reached, while issues that require a stronger commitment may be dropped. For example, energy efficiency measures can generate consensus between government authorities, industry and civil society about their environmental and economic benefits. On the other hand, measures to control and limit the demand for energy and transport may be discouraged. Similarly, despite its achievements in creating a right of way for pedestrians, the Pedestrian Rights Charter in Porto Alegre (Brazil) was approved only when a number of considerations that restricted individual motorized transport in the city were dropped. Furthermore, the benefits of such initiatives are unlikely to be equally shared, and 'There are many examples of environmental projects in cities that have served only the narrow interests of wealthier groups, or that have included an active anti-poor political agenda.'[226] In this manner, advocating the need to address climate change may further entrench existing inequalities within cities.

**In more affluent urban contexts, efforts to mitigate climate change are often in direct conflict with dominant urban political economies**

## ■ Urban political economies: Conflicting agendas

At the most fundamental level, struggles have emerged over whether cities should or should not be addressing climate change. In many cities, the arguments 'not on my turf' and 'not in my term' are prevalent, particularly in developing countries where resources are limited and other concerns are more pressing.[227] In these cases, 'subnational governments may be overloaded with other local demands, and climate policy may be down on the list of priorities'.[228]

In more affluent urban contexts, efforts to mitigate climate change are often in direct conflict with dominant urban political economies. The very factors that are regarded as driving urban growth – including the availability of cheap land at the urban fringe, short payback periods on capital investment, increased personal mobility, and the growing consumption of energy- and resource-intensive goods and services – are also those which contribute to rising GHG emissions.[229] In this context, initiatives which seek to change patterns of production or to reduce levels of consumption may encounter significant opposition. These issues may be particularly pressing for cities in developing countries, where 'GHG mitigation has a negative connotation because of the perception that this will deny them of their basic right to growth in human services and economic activities; the prospects of "reduced growth" or "no growth" are not feasible'.[230]

Climate change mitigation can contribute to create conditions that favour sustainable development, as discussed above. However, this is not a given, particularly in those cases in which climate change mitigation (and other environmental concerns) have been used by urban elites to attack the interests of the urban poor.[231] In particular, researchers have identified two areas in which mitigation may have serious social consequences for urban populations in developing countries: when it detracts attention from adaptation[232] and when mitigation measures have impacts in particularly disadvantaged sections of the urban population.[233] For example, street-lighting programmes which promote the substitution of standard bulbs by lighting innovations such as light-emitting diodes (LEDs) – such as the one promoted by the Climate Group in Mumbai (India)[234] – may direct investments to affluent areas where lighting infrastructure is already in place, while detracting investments from developing lighting infrastructure in the wide slum areas of the city.

Such tensions between dominant forms of urban growth and climate change mitigation are, however, also discernible in cities in developed countries. In the US, for example, climate change mitigation is likely to be prioritized in those communities which are most likely to be affected by the impacts of climate change, and those with a 'liberal' political constituency.[235] In the UK, climate change initiatives in the transport sector have been undermined by the priority given to economic considerations and the stress on the need to increase the demand for travel.[236] The long-term experience with transport regulation and urban economy in Århus (Denmark), however, suggests that the focus on individual motorized transport is not always the best or the only strategy to improve the local economy.[237] That such alternatives are often overlooked may be due to the ways in which the scope for municipal action on climate change is predetermined by neo-liberal political and economic conditions. For example, research in Portland, Oregon (US), found that climate actions were confined to:

> ... elements of energy consumption that could be influenced in an acceptable way by the municipal government. Energy used in flights to and from Portland International Airport, for instance, was excluded. Also excluded were the significant amounts of energy used in importing and exporting commodities, and the energy actually embodied in commodities.[238]

## COMPARATIVE ANALYSIS

As demonstrated above, cities across the world are undertaking a range of measures to address climate change mitigation. From a handful of pioneering cities during the 1990s, the number of urban municipalities participating in climate change mitigation efforts has expanded significantly over the past two decades. Alongside a growing number of cities in developed countries, the analysis presented in this chapter suggests that climate change mitigation is becoming an increasingly important issue for cities in developing countries as well. Most urban mitigation efforts have been implemented after the adoption of the Kyoto Protocol, with many initiatives, especially in developing countries, dating from the mid 2000s. This reflects the changing international and national climate change policy context in which developing countries with growing contributions to global emissions – including China, India, Brazil, Mexico and South Africa – are becoming involved in mitigation efforts. It is also symptomatic of what has been described as an era of 'governance experimentation' emerging as a result of the fragmentation of authority for governing climate change between public and private actors, and a growing dissatisfaction with the outcomes of national policy processes and international negotiations.[239] Despite the growing profile of climate change as an urban issue, data on the strategies and measures being adopted in cities across the world are limited, especially for cities in developing countries. Equally, where the development of policy and the implementation of measures have been documented, evidence concerning the impacts and effectiveness of climate change mitigation measures is scarce. In this context, detailed comparative analysis of urban climate change mitigation efforts is not possible, though some key trends can be observed.

*First*, the analysis in this chapter suggests that climate change remains a marginal issue for most of the world's cities. Relatively few cities, especially in developing countries, are explicitly seeking to address climate change mitigation, and where this is the case, policy-making is largely confined to the environmental domains of municipal governments and, furthermore, the issue of climate change mitigation is one of concern primarily for urban elites. Although there are growing expectations in developed countries for action on climate change by municipal governments and other urban actors (e.g. in the UK, local authorities are required to prepare climate change mitigation (and adaptation) plans; and in the US, the Mayors Climate Protection Agreement has attracted a significant following), there is limited evidence that this is being approached in a strategic or comprehensive manner.[240] In regions where rapid industrialization and urbanization is taking place (e.g. cities in Latin America and Asia), there is a growing interest in climate change mitigation. This is, for instance, the case for cities such as São Paulo, Porto Alegre and Rio de Janeiro in Brazil; Mexico City; Beijing and Shanghai in China; and Jakarta in Indonesia, where climate change initiatives have proliferated during the last four to five years, not only in a piecemeal fashion, but also in the form of articulated and coordinated climate change action plans. It should also be noted that some developing countries, such as the Philippines,[241] have adopted national frameworks within which municipalities should address climate change mitigation. However, with limited data available, the extent to which such initiatives are taking place in other cities in developing countries is not clear.

Furthermore, the analysis presented in this chapter suggests that governing climate change mitigation is primarily being undertaken by municipal governments, although forms of partnerships and the involvement of private actors is increasingly becoming important. There are relatively few examples of inclusive and participatory approaches to urban climate change mitigation governance. In particular, issues of gender have received minimal attention.[242] Seeking to broaden the basis upon which climate policy is formulated and implemented is a critical challenge for cities. Women's participation in climate change decision-making at the local level may play a specific role in supporting sustainable lifestyles, developing alternative forms of engagement with the environment and challenging traditional patriarchal models of urbanization and planning.

A *second* set of trends indicated by the analysis in this chapter concerns regional differences in terms of what cities are doing and how they are doing it. For example, urban responses to climate change are more common in developed than in developing countries. While international commitments and national policy frameworks have provided important drivers for these cities, the cases of the US and Australia – where significant action has been taken at the urban level despite the withdrawal of both countries from the Kyoto Protocol – highlight the ways in which municipal governments have also pioneered climate policy.[243]

**Climate change remains a marginal issue for most of the world's cities**

**In regions where rapid industrialization and urbanization is taking place ... there is a growing interest in climate change mitigation**

**Some developing countries ... have adopted national frameworks within which municipalities should address climate change mitigation**

**There are relatively few examples of inclusive and participatory approaches to urban climate change mitigation governance**

**Box 5.20 Obstacles to climate change mitigation actions in Durban, South Africa**

For the municipality of Durban, responding to climate change is a major focus of the city's commitment to sustainable development. Durban was one of the first African cities to participate in Local Governments for Sustainability's (ICLEI's) Cities for Climate Protection Campaign (CCP). However, the absence of policy coordination and the existence of competing socio-economic urban policy priorities stand in the way of effectively delivering potential emissions reductions.

Early mitigation projects were landfill gas to electricity (resulting in reductions of 362,000 tonnes of $CO_2$eq per year, or 2 per cent of annual emissions), reduction of energy demand in municipal buildings (reductions of 914 tonnes of GHGs annually), and electricity from micro-turbines integrated within the water piping systems, making use of Durban's uneven topography. While setting municipal climate policy in motion, these initiatives did not result in significant emission reductions. A target of 27.6 per cent reductions by 2020 was proposed in the 2008 Energy Strategy, to be achieved through the use of biofuels in transportation, the creation of residential green energy tariffs, a subsidized residential solar hot water programme, encouraging industrial efficiency, and the encouragement of local energy service companies. Implementing these would demand cross-cutting action across many municipal departments, as well as private partners.

However, projects have been held back by questions regarding who has the resources and the jurisdiction to implement them. For instance, the municipal Department for Environmental Management has the best understanding of the issue. However, it lacks both the resources and the mandate to act upon that knowledge (their remit being primarily biodiversity protection). The entity which is perhaps best positioned to act, the energy provider, is constrained by deeply engrained procedures and relationships (traditionally the intermediary buying electricity from the national grid and selling it on to local customers, they did not see local renewable energy generation within their mandate). And the entity quickest to act, the Department for Water and Sanitation, while effective in making change in its own systems, does not have the desire or reach to coordinate broader changes. Therefore, while substantial opportunities for significant emissions reductions exist, Durban's experience shows that in the absence of integrated planning and streamlining of urban priorities, key barriers may be primarily institutional, not technical.

*Source:* Aylett, 2010

---

*The development and spread of international, national and regional municipal networks has ... provided a key driver for municipal responses in developed countries*

*In developing countries, mitigation initiatives have also often been linked with adaptation responses, taking advantage of the potential synergies between both*

Nonetheless, the development of international and national policy commitments to address climate change in some developing countries – notably, China, India, Brazil, Mexico and South Africa – is also driving a growing policy interest in the issue at the urban level. The development and spread of international, national and regional municipal networks has also provided a key driver for municipal responses in developed countries, and the expansion of these networks to include cities in developing countries is one important reason for their growing participation in climate change mitigation. In developing countries, mitigation initiatives have also often been linked with adaptation responses, taking advantage of the potential synergies between both.[244]

The differences between developing and developed countries, however, are more apparent when examining the measures and mechanisms which have been developed to address climate change mitigation. In developed countries, emphasis has been placed on the energy sector through urban design and development, the built environment and urban infrastructure systems. In developing countries, cities have focused on a more diverse range of urban infrastructure projects, including waste and water systems, as well as issues of carbon sequestration. Those schemes which have been undertaken in the urban development and design sector in cities in developing countries have tended to focus on flagship projects, which are often socially and economically exclusive, in contrast to the involvement of civil society groups and an emphasis on smaller-scale brownfield regeneration projects in developed countries. This may reflect the urban morphologies of these different cities – brownfield sites are likely to be uncommon, particularly in the rapidly industrializing cities in developing countries – as well as the availability of resources for creating 'sustainable' housing.

While projects in the built environment in developed countries have tended to focus on municipal and residential buildings, in developing countries attention has been given to commercial buildings. This reflects the fact that in developed countries, the major challenges in the built environment are related to retrofitting the housing stock, as new developments are gradually incorporating more efficient designs and materials, whereas the contribution of residential dwellings to GHG emissions in developing cities is likely to be minimal for the vast majority of the housing stock. Furthermore, the focus on commercial buildings reflects the growing involvement of private-sector actors in addressing climate change mitigation in developing countries. The evidence presented in this chapter also suggests that initiatives in developed countries are often achieved through processes of self-governing and enabling, while in developing countries, modes of provision, both public and private, have been more significant. Despite these differences, there are relatively few examples of the development and use of alternative energy technologies or of explicit policies to tackle climate change in the transport sector in both developed and developing countries.[245]

However, this broad brush differentiation between developed and developing countries obscures the differences that are emerging within these regions. Urban development and design initiatives in North America, Australia and New Zealand focus on compact city principles and mixed developments to address the historical conditions of suburban development and urban sprawl. However, transport-related initiatives are relatively rare, particularly in terms of limiting and controlling the demand for individual motorized transport and the development of mass transport systems. This contrast with countries in Europe, where there is a growing proliferation of examples to promote demand management and enhancement in transport, while developing or modernizing the public transport infrastructure. Cities in Africa and Latin America and the Caribbean have emphasized actions in urban infrastructure systems, particularly in those cases in which upgrading the infrastructure alone can lead to significant gains, such as is the case in the waste management system in Lagos (Nigeria).[246] In these regions, there is evidence that measures being undertaken in the built environment and urban development and design sectors are seeking to address issues of social equity. However, in Asia, new urban developments are emerging where high-income groups are able to create their own communities – often informed by green values in terms of nature protection and resource conservation, but with less regard for amelioration of social inequalities. Driven through partnerships of private and public agencies, large urban development projects that incorporate climate change mitigation concerns are taking place. However, concerns have been expressed about the impact and effectiveness of such schemes in climate change terms, and also because they may have important environmental justice implications for the social groups who are excluded from these partnerships.

A *third* set of trends relates to the differences in the opportunities and constraints that municipal governments and other actors face in seeking to mitigate climate change. Clearly, the resources available to act upon climate change are significantly different between cities in different regions, as well as between actors within individual cities. For many cities in developed countries a lack of resources is seen as a critical barrier to action, though these challenges are considerably higher for cities in developing countries. Analysis in this chapter also suggests that a lack of expertise, of institutional capacity and of the ability to develop and enforce policy – as well as historic issues of underinvestment in urban infrastructures, informal settlements and persistent poverty – pose significant challenges for cities in developing countries seeking to address climate change mitigation. The example of Durban (see Box 5.20) explains the interaction of multiple obstacles to climate change mitigation in a city in South Africa. In order to address these issues, linking climate change actions with their potential co-benefits appears to be crucial, particularly when these are linked with social and environmental justice objectives to improve the quality of life of the most disadvantaged sectors of the population. Examples such as the Kuyasa housing project in Cape Town (South Africa) and housing projects in Buenos Aires (Argentina) and Rio de Janeiro (Brazil) are encouraging; but their prominence is still relatively low, particularly when compared with the emphasis on the development of exclusive new urban developments. Furthermore, while international policy instruments (such as the CDM), public–private partnerships and international networks may be able to bring a degree of resource and support for climate change activities, there is, to date, mixed evidence of their impact upon fundamental issues of economic deprivation and social inequalities. In developed countries, the impacts of focusing upon co-benefits are less clear cut. While such approaches can generate political support, they could also lead to the watering down of climate change commitments or to a focus only on those initiatives that can yield economic benefits in the relatively short term, detracting attention from more fundamental issues concerning how (and by and for whom) energy is provided, the levels of personal mobility that can be sustained, and the relationship between consumption, growth and climate change.

Despite the significant constraints facing urban climate change mitigation efforts, as the evidence documented in this chapter illustrates, cities are taking important measures to address the issue. The combined effects of institutional structures, financial resources, the social and material make-up of urban infrastructure networks, and political support have created the capacity for significant advances for climate change mitigation. This capacity is not only unevenly distributed regionally and between different countries. Research also suggests that a growing divide may be emerging between cities. Municipal governments and other urban actors with initial capacity in some cities are able to capitalize on opportunities for funding, political influence, access to international organizations and international networks, and partnerships to build on their efforts, while others lack the wherewithal needed to access these resources.[247] Efforts by international networks, private-sector actors and international donor agencies to target a small number of global and megacities as arenas within which to mitigate climate change may exacerbate this divide. As a result, rather than being regionally differentiated, future urban climate change mitigation efforts may be characterized by differences between an elite group of cities with access to substantial resources, those (primarily in developed countries) who may be able to afford to undertake initiatives to pick the 'low hanging fruit', and the vast majority of cities for whom addressing climate change will remain a low priority. Furthermore, the channelling of resources in this manner may also serve to support the interests of urban elites rather than addressing broader issues of sustainable development and well-being. As discussed above, ensuring that climate change mitigation can also address issues of social and environmental justice will necessitate the participation of a broad constituency of actors and, especially in developing countries, a focus on the multiple co-benefits that such initiatives could generate.

# CONCLUDING REMARKS AND LESSONS FOR POLICY

Mitigating climate change is an increasingly pressing urban issue. However, cities have very different starting points in terms of their GHG emissions, related to issues of geography, political economy, infrastructure provision and social practices, and the capacity of governments, private organizations and civil society actors. Historically, cities in developed countries have contributed the vast majority of GHG emissions and bear the major responsibility to act. However, as GHG emissions begin to grow in some developing countries, there is also a need to consider what appropriate and effective urban mitigation efforts might involve, and

*A lack of expertise, of institutional capacity and of the ability to develop and enforce policy … pose significant challenges for cities in developing countries seeking to address climate change mitigation*

*Future urban climate change mitigation efforts may be characterized by differences between an elite group of cities with access to substantial resources …. and the vast majority of cities for whom addressing climate change will remain a low priority*

*Historically, cities in developed countries have contributed the vast majority of GHG emissions and bear the major responsibility to act*

how they might be combined with the more pressing issues of urban adaptation.

This chapter suggests, in line with previous research,[248] that efforts to mitigate climate change in cities face a significant paradox. Those strategies which can be effectively implemented may have the least impact, while those with the potential for the greatest reductions in GHG emissions may be the hardest to achieve. On the one hand, the most commonly implemented and effective strategies are those which focus on reducing GHG emissions from within the municipality (self-governing) and those which aim to improve energy efficiency. As noted in Figure 5.1, the waste, transport and buildings sectors appear to be the 'low hanging fruits' of urban GHG mitigations. It should, however, be kept in mind that the cost efficiency of interventions within these sectors varies considerably (see Box 5.17). This chapter suggests that the complex challenges facing municipal governments – their partial autonomy in critical policy sectors, the splintering of urban infrastructure networks, the difficulties of meeting the basic needs of urban citizens, and the controversial politics that accompany efforts to divert from 'business as usual' – have limited the extent to which urban climate change governance has extended beyond the areas of direct municipal control. At the same time, in contexts of competing aims and conflicting agendas, focusing on energy efficiency has been a means through which urban actors have been able to address multiple agendas, including energy security, financial savings, air pollution and fuel poverty, alongside climate change.

However, there has, to date, been limited assessment of the impact of such measures. While focusing on municipal GHG emissions alone will, in most cases, only account for a small proportion of urban GHG emissions, energy efficiency measures have the potential to achieve significant savings. Examples of individual buildings, new urban developments, the retrofitting of energy-efficient technologies and behavioural programmes documented in this chapter have demonstrated that energy efficiency could provide a crucial component of urban efforts for climate change mitigation. Furthermore, such initiatives have often provided the impetus for the development of comprehensive climate change strategies, as the financial savings and political influence gained within the city drive more ambitious policy goals and the development of additional measures. Nonetheless, such examples remain relatively small scale and isolated. Against a rising trend of energy consumption and GHG emissions, a critical question for future research and the development of policy is therefore the extent to which self-governing and energy efficiency initiatives can lead to widespread and sustained changes in the ways in which energy is used in cities.

On the other hand, measures which may have the greatest impact upon urban GHG emissions, including the provision of low-carbon and renewable energy infrastructure systems, the reduction in demand for personal vehicle travel, as well as enabling and mobilizing actions by communities and stakeholders, have, to date, been less common. While there are some promising signs that such initiatives are taking place – in the development of new urban transit systems in cities in developing countries, projects for urban regeneration, and the growing involvement of a range of private companies and community organizations – these remain the exception rather than the rule. Evidence suggests that such initiatives are most likely to be successful when they demonstrate a range of additional economic, social and environmental benefits, and where they attract the support of key urban actors. While this can be a progressive process, involving communities and stakeholders and addressing issues of social and environmental justice, it can also be one that serves the interests of particular urban elites and leads to a politics of exclusion.

Importantly, the evidence presented in this chapter suggests that the potential for urban climate change mitigation to address issues of social and economic equity is not predetermined by the types of measure or governance mechanisms deployed. For example, projects to generate energy from landfill sites can be undertaken as technical endeavours with little regard for the impacts of such initiatives; but they can also provide new forms of employment, sources of funding for investment in poor communities, and a means of generating secure and affordable energy. Ensuring that mitigating climate change does not come at the expense of addressing issues of inequity and justice is a critical challenge for future policy-making.

**The waste, transport and buildings sectors appear to be the 'low hanging fruits' of urban GHG mitigations**

**Ensuring that mitigating climate change does not come at the expense of addressing issues of inequity and justice is a critical challenge for future policy-making**

# NOTES

1 UNFCCC, 1992, Article 2.
2 See Chapter 2 for details.
3 IEA, 2008.
4 Kern and Bulkeley, 2009.
5 See also Chapter 2.
6 Bulkeley, 2000; Betsill, 2001; Bulkeley and Betsill, 2003; Kousky and Schneider, 2003; Yarnal et al, 2003; Allman et al, 2004; Lindseth, 2004; Davies, 2005; Mackie, 2005; Bulkeley and Kern, 2006.
7 Dhakal, 2004, 2006; Bai, 2007; Holgate, 2007; Romero Lankao, 2007b.
8 Sanchez-Rodriguez et al, 2008.
9 Betsill and Bulkeley, 2007.
10 Satterthwaite, 2008a; Dodman, 2009.
11 Harvey, 1996.
12 Monstadt, 2009, p1927.
13 Alber, 2010; Hemmati, 2008.
14 See Chapter 3.
15 See Chapters 4 and 6.
16 IEA, 2009; UN, 2010.
17 Sassen, 1991.
18 Research undertaken in preparation for this chapter draws on a database of climate change mitigation initiatives taking place in 100 cities. For further information, see www.geography.dur.ac.uk/projects/urbantransitions, last accessed 21 October 2010.
19 See section on 'Urban governance for climate change mitigation'.
20 Kern and Alber, 2008, p4; see also Jollands, 2008.
21 Kern and Alber, 2008, p3.
22 ICLEI, 2006.
23 Bulkeley and Kern, 2006; Betsill and Bulkeley, 2007; Bulkeley et al, 2009.
24 Rutland and Aylett, 2008, p636.
25 Owens, 1992; Banister et al, 1997; Capello et al, 1999; Norman et al, 2006; Lebel et al, 2007. See also discussion on urban form and density in Chapter 3.
26 UN-Habitat, 2009a.
27 UN-Habitat, 2010.
28 UN-Habitat, 2009a.
29 See UN-Habitat, 2009a.
30 City of Cape Town, 2005; City of São Paulo, 2009.
31 For the example of Stockholm (Sweden), see Holden and Norland, 2005.
32 Kolleeny, 2006 (Philadelphia); Energy Planning Knowledge Base, undated (Manchester); A101, 2006 (Moscow); BCIL, 2009 (India).
33 McGray, 2007.
34 Ying, 2009.

35  Moore, 2008.
36  Schifferes, 2007.
37  For example, the backing down from implementing a planned ban on motorized private vehicles on the islands.
38  Pearce, 2009.
39  See also Box 2.7.
40  BSHF Database (available at www.worldhabitatarwards.org) (Boston and San Francisco); WHEDco, 1997 (New York).
41  Alber, 2010.
42  Bulkeley et al, 2009, p43.
43  Akinbami and Lawal, 2009; Bulkeley et al, 2009.
44  Foresight, 2008.
45  Hemmati, 2008; Alber, 2010.
46  Bulkeley et al, 2009, p44.
47  This is discussed in more detail in the section on 'Urban governance for climate change mitigation'.
48  See note 18.
49  US Department of Energy, 2008.
50  Pauzner, 2009.
51  Boardman, 2007. For some interesting examples of how to deal with existing building stock, see also http://thezero-prize.com, last accessed 18 October 2010.
52  For example, innovation in the use of traditional building materials in conventional housing in Bangalore (India) has been led by the pioneering experience and skills development promoted by architect Chitra Viswanath (see www.inika.com/chitra, last accessed 18 October 2010).
53  City of Cape Town, 2005; Ciudad de Mexico, 2008.
54  Bulkeley et al, 2009.
55  Agencianova, 2009. This project follows previous experiences in sustainable building, such as those led by the architect Carlos Levinton from the Special Centre of Production and the University of Buenos Aires to create energy-efficient building materials from recycled products (see Sotello, 2007).
56  Barry, 1943.
57  Zhao and Michaelowa, 2006.
58  Hemmati, 2008.
59  Gender is a critical issue in terms of behavioural patterns relating to climate change. Although commentators have suggested that women may emit less GHGs emissions than men, the evidence is limited by the lack of disaggregation of data about consumption patterns within the household. A study of consumption patterns in single-person households in different European countries supported the hypothesis that women emit less GHG emissions than men (see Alber, 2010); but doubts exist over whether such findings could be

extended over the whole life of individuals or could be applied more generally in different types of households and different countries. On the other hand, women across the world tend to have lower incomes and greater participation in the informal labour market. Even women who are not living in poverty will tend to be less affluent and financially secure than men; hence, they will have more modest consumption associated with lower carbon footprints (Haigh and Vallely, 2010).
60  OECD, 2008.
61  Hemmati, 2008; Haigh and Vallely, 2010.
62  See also Aylett, 2010.
63  Greene et al, 1999.
64  Satterthwaite, 2008b, p11.
65  Hemmati, 2008.
66  UN-Habitat, 2008b.
67  Graham and Marvin, 2001.
68  See Chapter 2.
69  See Bundesministerium für Umwelt, Naturschutz und Reaktorsicherheit (undated).
70  Bulkeley et al. 2009.
71  City of Cape Town, 2005.
72  Greenpeace, 2008; Wu and Zhang, 2008.
73  Greenpeace, 2008.
74  Solar America Cities, 2009.
75  See www.cleanenergy fuels.com/main.html, last accessed 18 October 2010.
76  The scheme was funded by the Department for Environment, Food and Rural Affairs and the Waste Resources Action Programme.
77  Silver, 2010.
78  See also see the sub-section on 'Modes of public–private collaboration in urban climate governance' below.
79  See also Bulkeley et al, 2009; Aylett, 2010.
80  Oresanya, 2009.
81  Ciudad de Mexico, 2008.
82  Aylett, 2010.
83  Ciudad de Mexico, 2008.
84  Short et al, 2008; Bertaud et al, 2009. See also Chapter 3.
85  Karekezi et al, 2003.
86  World Bank, 2009c.
87  Sari, 2007, p129.
88  Sari, 2007, p137.
89  Johnsson-Latham, 2007.
90  Haigh and Vallely, 2010.
91  Johnsson-Latham, 2007.
92  Skutsch, 2002.
93  Alber, 2010.
94  See note 18.
95  Wagner, 2009.
96  See www.transmilenio.gov.co/WebSite/English_Default.aspx, last accessed 18 October 2010.
97  See www.cdm-egypt.org/, last accessed 18 October 2010.
98  The partnership is partly funded by the German federal government's Fuel Cell and Hydrogen Innovation Programme, and the partners

include Daimler, Shell, Total and Vattenfall Europe.
99  Note that this measure is frequently associated with improving the air pollution of the city (e.g. Alam and Rabbani, 2007).
100  Bertaud et al, 2009.
101  Bertaud et al, 2009.
102  See discussion in the sub-section on 'Urban infrastructures' above.
103  Dlamini, 2006. The programme was implemented with financial assistance of the governments of Norway and Denmark and the World Conservation Union (IUCN).
104  Lagos State Government, 2010. Updates about the tree-felling programme can be found at www.lagosstate.gov.ng, last accessed 18 October 2010.
105  Concejo de Bogotá, 2008.
106  See Box 2.7.
107  ICLEI Australia, 2008.
108  This project is a collaboration between the Clinton Climate Initiative, Microsoft Corporation, Autodesk and ICLEI (Clinton Foundation, undated b).
109  See, for example, Bai, 2007; Dhakal, 2009.
110  UNEP et al, 2010. See also Box 2.4.
111  Bulkeley and Kern, 2006.
112  UN-Habitat, 2009a, p73.
113  UN-Habitat, 2009a, p73.
114  Bulkeley and Kern, 2006; Kern and Alber, 2008; Bulkeley et al, 2009.
115  Hammer, 2009.
116  This trend is also reflected in other arenas of climate change governance (see Biermann and Pattberg, 2008; Bulkeley and Newell, 2010).
117  Jessop, 2002, p241; Sørensen and Torfing, 2007, 2009.
118  Bulkeley and Kern, 2006; Gore et al, 2009.
119  Martinot et al, 2009.
120  Gore et al, 2009, p10.
121  See the sub-section on 'Public–private provision' below.
122  Bulkeley et al, 2009.
123  State of São Paulo, 2008.
124  GLA, 2008.
125  Bulkeley and Schroeder, 2008; www.nabers.com.au, last accessed 22 October 2010.
126  Bulkeley et al, 2009.
127  Bulkeley and Kern, 2006; Bulkeley et al, 2009; Gore et al, 2009; Hammer, 2009.
128  Bulkeley et al, 2009.
129  Funded by the Danish International Development Agency.
130  Aylett, 2010.
131  See note 18.
132  UNEP, 2007, p44.
133  In the eastern US, for example, local planning regulations have slowed the establishment of hydrogen refuelling stations.
134  Bulkeley et al, 2009.

135  Aylett, 2010.
136  UN-Habitat, 2009a, p74.
137  UNEP, 2007, p46.
138  See the sub-section on 'Modes of governing climate change mitigation'.
139  UN-Habitat, 2007, p106.
140  Satterthwaite et al, 2009b, p24.
141  Bulkeley and Newell, 2010.
142  See www.carbon rationing.org.uk, last accessed 18 October 2010.
143  Peak oil is the point in time when the maximum rate of global petroleum extraction is reached, after which the rate of production enters terminal decline.
144  See www.transition network.org, last accessed 18 October 2010.
145  Pers comm, 2010.
146  See the sub-section on 'Provision' above.
147  See the sub-section on 'Urban infrastructures' above.
148  The London Energy Service Company is a 'private limited company with shareholdings jointly owned by the London Climate Change Agency Ltd (with a 19 per cent shareholding) and EDF Energy (Projects) Ltd (with an 81 per cent shareholding)' (LCCA, 2007, pp5–6).
149  Clinton Foundation, undated b.
150  See Chapter 2.
151  The Climate Group is an international non-profit organization whose members include national, regional and local governments, as well as private corporations (see www.theclimategroup.org/about-us/, last accessed 18 October 2010).
152  Bulkeley and Betsill, 2003; Romero Lankao, 2007b; Biermann and Pattberg, 2008; Rutland and Aylett, 2008.
153  Kern and Alber, 2008.
154  Betsill and Bulkeley, 2007; Bulkeley et al, 2009; Romero Lankao, 2008.
155  Deangelo and Harvey, 1998; Bulkeley et al, 2009; Hammer, 2009.
156  Granberg and Elander, 2007, p545.
157  Qi et al, 2008, pp397–398.
158  Hammer, 2009.
159  Betsill and Bulkeley, 2007; Bulkeley et al, 2009; Puppim de Oliveira, 2009.
160  Bai, 2007; see also Bulkeley and Kern, 2006; Granberg and Elander, 2007.
161  Collier, 1997; Lebel et al, 2007; Jollands, 2008; Schreurs, 2008; Sugiyama and Takeuchi, 2008; Setzer, 2009.
162  GLA, 2007, p19.
163  Monni and Raes, 2008, p753.
164  Sugiyama and Takeuchi, 2008, p429.
165  Satterthwaite, 2008b, p9.
166  See the sub-section on 'Self-governing' above.
167  Romero Lankao, 2007b, p529.

168  Kern and Alber, 2008.
169  Bulkeley et al, 2009, p23.
170  Holgate, 2007.
171  OECD, 2008, p24; see also Bai, 2007; Crass, 2008; Kern and Alber, 2008.
172  Kern and Alber, 2008, p4.
173  Betsill and Bulkeley, 2007, p450.
174  Akinbami and Lawal, 2009, p12.
175  Schwaiger and Kopets, 2009.
176  UNEP, 2007.
177  Bulkeley et al, 2009, p49.
178  Sari, 2007, p141.
179  Satterthwaite, 2008b, p12.
180  Sippel and Michaelowa, 2009.
181  See www.cdmgold standard.org/, last accessed 18 October 2010.
182  GLA, 2007, p19.
183  Arup, 2008.
184  See Chapter 3.
185  Newcastle City Council, 2008.
186  Allman et al, 2004; Lebel et al, 2007; Sugiyama and Takeuchi, 2008, p432.
187  Aylett, 2010.
188  Akinbami and Lawal, 2009.
189  Akin to previous definitions of large technical systems; see Hughes, 1989.
190  Bertaud et al, 2009, p23.
191  Bertaud et al, 2009.
192  Schwaiger and Kopets, 2009, p3.
193  Akinbami and Lawal, 2009, p10.
194  Monni and Raes, 2008, p749.
195  Hodson and Marvin, 2007.
196  See www.climate-change.ir/en/, last accessed 17 May 2010.
197  Satterthwaite, 2008a, p11.
198  Castán Broto et al, 2007, 2009.

199  Kern and Bulkeley, 2009.
200  Setzer, 2009, p8.
201  Jollands, 2008, p5.
202  Bulkeley and Kern, 2006.
203  Sugiyama and Takeuchi, 2008, p430.
204  The Carbon Finance Capacity Building Programme is an initiative of the World Bank, ECOS, C40, the Swiss Government (SECO) and the Canton of Basel City.
205  See www.lowcarboncities.info/home.html, last accessed 18 October 2010.
206  Kern and Alber, 2008; Jollands, 2008.
207  To date, there is little evidence of the role of the voluntary carbon market in urban climate change governance and this analysis focuses on the CDM.
208  Puppim de Oliveira, 2009, p257. See also Box 3.3.
209  See City of Cape Town, 2005.
210  For an overview of this debate, see Bumpus and Liverman, 2008.
211  World Bank, 2010a.
212  World Bank, 2010c.
213  Sippel and Michaelowa, 2009; Roberts et al, 2009, p13; ICLEI, 2010; Clapp et al, 2010.
214  Roberts et al, 2009, p14.
215  World Bank, 2010a.
216  'Policy entrepreneurs', individuals involved in the innovation of policies or schemes, are defined by 'their willingness to invest their resources – time, energy, reputation, and

sometimes money – in the hope of a future return ... in the form of policies of which they approve, satisfaction from participation, or even personal aggrandizement in the form of job security or career promotion' (Kingdon, 1984, p122). 'Political champions' are individuals who advocate the importance of responding to climate change and who may back particular policies, projects or schemes. See Bulkeley and Betsill, 2003; Bulkeley and Kern, 2006; Qi et al, 2008; Schreurs, 2008.
217  Bulkeley and Kern, 2006, p2253.
218  Romero Lankao, 2007b; Setzer, 2009; Aylett, 2010.
219  See www.c40cities.org/, last accessed 18 October 2010.
220  See www.theclimategroup.org/our-news/news/2009/2/24/the-climate-group-launches-forward-chicago/, last accessed 22 October 2010.
221  City of Cape Town, 2006.
222  Puppim de Oliveira, 2009, p254.
223  Frayssinet, 2009.
224  Bai, 2007; Gore et al, 2009; Betsill and Bulkeley, 2007.
225  Qi et al, 2008, p393.
226  Bartlett et al, 2009, p22; see also McGranahan and Satterthwaite, 2000.
227  Bai, 2007.
228  Puppim de Oliveira, 2009, p25; see also Bai, 2007; Romero

Lankao, 2007b; Jollands, 2008.
229  See Chapter 3.
230  Lasco et al, 2007, p84.
231  Bartlett et al, 2009.
232  See Chapter 6.
233  Bartlett et al, 2009.
234  Urvashi Devidayal, the Climate Group, pers comm, 2010.
235  Zahran et al, 2008.
236  Bulkeley and Betsill, 2003.
237  Flyvbjerg, 2002.
238  Rutland and Aylett 2008, p636.
239  Hoffman, 2011.
240  See the sub-section on 'Municipal policy approaches' above.
241  The Philippines adopted a National Framework Strategy on Climate Change in April 2010 (www.climatechangecommission.gov.ph/link/downloads/nfscc/index.php, last accessed 10 August 2010.
242  Demetriades and Esplen, 2008; Alber, 2010.
243  See the sub-section on 'Multi-level governance' above.
244  See Chapter 6.
245  See note 18.
246  Oresanya, 2009.
247  Granberg and Elander, 2007; Kern and Bulkeley, 2009.
248  For example, Bulkeley and Betsill, 2003; Bulkeley and Kern, 2006; Bai, 2007; Betsill and Bulkeley, 2007; Rutland and Aylett, 2008; Bulkeley et al, 2009; Gore et al, 2009.

# CLIMATE CHANGE ADAPTATION RESPONSES IN URBAN AREAS

The lives and livelihoods of hundreds of millions of people will be affected by what is done (or not done) in urban centres with regard to adapting to climate change over the next decade. Action is urgently needed, both to address current risks and to begin building into urban fabrics and systems resilience to likely future risks. Most urban buildings and infrastructure are long lived; thus, what is designed and built now will have to cope with climate change many decades into the future. As a result, it is generally much easier to make provisions now for likely future climate-related risks – in infrastructure expansion, new buildings and new urban developments – than to have to retrofit buildings, redo infrastructure and readjust settlement layouts in the future.

As noted in Chapter 4, urban centres already concentrate a large proportion of those most at risk from the effects of climate change. This includes a high proportion of urban centres with very large deficits in infrastructure, as well as in the institutional and financial capacity needed to reduce these risks. Urban centres also concentrate the enterprises that generate most of the world's gross domestic product (GDP) and provide livelihoods for around two-thirds of the world's economically active population.[1] In most urban centres, buildings, infrastructure and services will have to cope with an increasing scale and range of climate impacts. Furthermore, as most of the growth in the world's population over the next few decades will occur in the urban centres of developing countries[2] – many (if not most) of which are already unable to provide adequate living conditions for their populations – it is likely that a major proportion of these new urban residents will be living in settlements that do not have the needed resilience to climate change.

Yet, adapting urban areas to climate change is not a new 'standalone' task or responsibility that can be allocated to one single stakeholder. It requires changes in the ways that almost all sectors of government, business and households behave and invest. In addition, much of what is needed to make cities resilient to climate change within the next few decades is no more than 'good development' in the sense of the infrastructure, institutions and services that meet daily needs and reduce disaster risk. As this chapter discusses, however, this is not easily achieved, as particular institutions and funding sources are given responsibilities for 'climate change adaptation' not for 'development that incorporates climate change adaptation'. Many discussions of climate change adaptation start with a discussion of the risks that climate change is bringing or may bring and then consider what needs to be done to address this – without considering how the climate change-related risks fit within other risks. What most urban centres in developing countries need is not a climate change adaptation programme but a development programme – meeting already existing deficits in provision for water, sanitation, drainage, electricity, tenure, healthcare, emergency services, schools, public transport, etc. – within which measures for climate change adaptation are integrated.

The first section of this chapter discusses what is meant by adaptation, adaptive capacity and similar terms, as applied to urban centres. The second section reviews household and community responses to the impacts of climate change and highlights the major challenges to community-based climate change adaptation. This is followed in the third section by a similar review of the responses by city and municipal governments. This review provides the basis for a discussion in the fourth section of the main issues that need to be addressed to develop effective city-based climate change adaptation strategies. The fifth and sixth sections discuss the financing and other key challenges of urban climate change adaptation, respectively. The final section provides some concluding remarks and lessons for policy.

## UNDERSTANDING ADAPTATION

It is important that there is clarity in what is meant by adaptation, adaptive capacity and adaptation deficit. Drawing on the definitions of the Intergovernmental Panel on Climate Change (IPCC),[3] *adaptation* to (human-induced, or 'anthropogenic') climate change is understood to include all actions to reduce the vulnerability of a system (e.g. a city), population group (e.g. a vulnerable population in a city) or an individual or household to the adverse impacts of anticipated climate change. Adaptation to climate variability consists of actions to reduce vulnerability to short-term climate shocks

---

It is ... much easier to make provisions now for likely future climate-related risks ... than to have to retrofit buildings, redo infrastructure and readjust settlement layouts in the future

Adapting urban areas to climate change ... requires changes in the ways that almost all sectors of government, business and households behave and invest

What most urban centres in developing countries need is ... a development programme ... within which measures for climate change adaptation are integrated (whether or not these are influenced by climate change) – for instance, as a city government ensures that the drainage system can cope with monsoon rains. Most of the measures for adapting to climate variability (which will be taking place in most well-governed cities) will also contribute to climate change adaptation (as a co-benefit).

The outcome of successful adaptation is *resilience* – and is a product of governments, enterprises, civil society organizations, households and individuals with strong adaptive capacity.[4] For cities or particular urban neighbourhoods, it indicates a capacity to maintain core functions in the face of hazard threats and impacts, especially for vulnerable populations. It usually requires a capacity to anticipate climate change and plan needed adaptations. The resilience of any population group to climate change interacts with its resilience to other dynamic pressures, including economic change, conflict and violence.

*Adaptive capacity* is the inherent capacity of a system (e.g. a city government), population (e.g. a low-income community in a city) or individual/household to undertake actions that can help to avoid loss and can speed recovery from any impact of climate change. Adaptive capacity is the opposite of *vulnerability*.[5] The risks that have to be reduced by adaptation can be direct, as in larger and/or more frequent floods, or more intense and/or frequent storms or heat waves; or less direct, as climate change negatively affects livelihoods or food supplies (and prices), or access to water needed for domestic consumption or livelihoods. Certain groups may face increased risks or costs from measures taken in response to climate change – including adaptation measures (e.g. measures to protect particular areas of a city from flooding that increase flood risks 'downstream') and mitigation measures (e.g. a greater emphasis on new hydropower schemes that displace large numbers of people from their homes and livelihoods).

Elements of adaptive capacity include knowledge, institutional capacity, and financial and technological resources. Low-income populations in a city will tend to have lower adaptive capacity than high-income populations because of their lower capacity to afford good-quality housing on safe sites. There is also a wide range among city and national governments in their adaptive capacities, relating to the resources available to them, the information base to guide action, the infrastructure in place, and the quality of their institutions and governance systems.

The lack of adaptive capacity to deal with problems caused by climate variability and climate change is strongly related to the scale of what can be termed the *adaptation deficit*: the deficit in infrastructure and service provision and in the institutional and governance system that is meant to be in place to ensure adaptation. Of course, this depends heavily on the competence and capacity of local governments and the quality of the relations between local government and populations at risk within their jurisdiction. In many developing country cities, the main problem is the lack of provision of basic city infrastructure and the lack of capacity to address this. This is one of the central issues with regard to urban climate change adaptation because most discussions on this issue focus on needed adjustments to infrastructure to climate-proof it. However, cities cannot climate-proof infrastructure that is not there. In addition, new sources of funding for climate change adaptation have little value if there is no local capacity to design, implement and maintain the needed adaptation measures, or no interest within local government in working with the populations most at risk (which in many urban contexts, as noted in Chapter 4, are concentrated among low-income households living in informal settlements and slums).

Ultimately, the most important and effective form of adaptation is to stop the process that generates increasing levels of hazard and risk – that is, to slow the growth of, halt and then reduce greenhouse gas (GHG) emissions or other measures to reduce global warming (i.e. mitigation).[6] Failure to mitigate will lead to the failure of adaptation, as climate change risks become increasingly severe. So adaptation and mitigation are not alternative strategies, but complementary ones that need to be pursued together.

It was the failure of the world's governments to reach agreement to reduce GHG emissions during the 1990s that has made the need to greatly increase adaptation capacity so urgent. It is now too late to stop the increase in climate change-related hazards in the short term. Even if the world's governments do reach agreement on the need for rapid reductions in global GHG emissions and actually implement the measures needed to achieve this, the GHG emissions already generated and the time-lags in global systems[7] still mean increasing hazard and risk levels for most urban centres – and, therefore, an increasing need to adapt. Adaptation can reduce the adverse impacts of climate change considerably; but, generally, it cannot remove all adverse impacts – especially if the needed agreements to reduce global emissions have not been achieved. So there are limits to what adaptation can protect. There will also be an increasing number of locations that become permanently beyond adaptation – because the needed measures to protect them are considered too expensive (e.g. particular coastal zones inundated by sea-level rise) or technically unfeasible. Such consequences are often referred to as *residual damage*, and the number of such locations (and populations at risk) is likely to rise without successful mitigation (see Figure 6.4).

As described in more detail later in this chapter, adaptation can be undertaken by different actors – for instance, by individuals, households and commercial enterprises. This may be within government programmes or completely independent of government (in which case it is generally referred to as *autonomous adaptation*). Different levels of government (from national through regional and city-wide to district or ward) and different sectors of government have responsibility for many of the needed adaptations or for providing the regulatory framework – or the carrots or sticks – to encourage other actors to adapt. Adaptations that are planned in anticipation of potential climate change are termed *planned adaptation*. Generally, government agencies have the responsibility to provide information about current and future risks, and provide frameworks that support individual, household, community and private-sector adaptation. However, governments often do not fulfil this role, and community-based and other civil society organizations may

**The outcome of successful adaptation is resilience**

**The lack of adaptive capacity to deal with problems caused by climate variability and climate change is ... related to the scale of ... the adaptation deficit**

**The most important and effective form of adaptation is to stop the process that generates increasing levels of hazard and risk (i.e. mitigation)**

be the initiators and supporters of planned adaptation. As has long been evident in initiatives to improve conditions in informal settlements, a proactive civil society may be required to galvanize government and to demonstrate what can be achieved.[8]

In recognition of the fact that much adaptation to climate variability (and climate change) takes place through the conscious efforts of particular communities, the IPCC has highlighted the importance of what is termed *community-based adaptation*. As discussed in more detail in a later section,[9] community-based adaptation has particular importance where local governments lack adaptive capacity. Yet, it also has importance within effective local government-driven adaptation because of the knowledge and capacity that it can contribute. For urban areas, there is a danger that its relevance will be both overstated and underplayed at the same time. On the one hand, it will be overstated because community-based organization and action cannot provide the city-wide infrastructure and service provision and city-region ecosystem services protection and management that are so central to effective adaptation. On the other hand, however, the importance and effectiveness of community-based adaptation can be underplayed as the policies and practices of governments and international agencies fail to recognize the capacity of community-based organizations to contribute to adaptation or, if they do, they lack the institutional means to support them.[10]

There are also actions and investments that increase rather than reduce risk and vulnerability to the impacts of climate change and these are termed *maladaptation*. Examples of this include the shifting of risk from one social group or place to another; it also includes shifting risk and costs to future generations and/or to ecosystems and ecosystem services. Many investments being made in cities are, in fact, maladaptive rather than adaptive, as they decrease resilience to climate change. Indeed, the very process of 'unmanaged' urban expansion usually brings with it increasing risk as inappropriate sites are developed and as infrastructure provision fails to keep up. Removing maladaptations and the factors that underpin them are often among the first tasks to be addressed before new adaptations.

# HOUSEHOLD AND COMMUNITY RESPONSES TO THE IMPACTS OF CLIMATE CHANGE

National governments are meant to represent the interests of their citizens in international discussions on allocating responsibility for climate change mitigation and in developing international funding sources and institutions and other forms of support for adaptation. Similarly, local (metropolitan, city and municipal) governments are, in principle, responsible for implementing climate change adaptation measures at the local level.

However, risk reduction and resilience to risk also depend on actions taken by households and by community-based organizations. And for a large section of the urban population in developing countries, little can be expected of local and national governments as they currently lack the capacity or willingness to provide the basic infrastructure and services that are central to adaptation.

Where local governments are weak or ineffective, household and community strategies become more important for reducing climate change risks and impacts in urban areas. In such situations, urban residents have long had to cope with a wide range of risks to their lives and livelihoods. Many of the measures that they take to cope with risk are responses to extreme weather, including flooding, extreme temperatures and landslides – although the root cause of the risk is often far more related to the lack of infrastructure or the lack of safer sites that they can afford. In many locations, household and community strategies have developed over years or even decades to prevent loss of life and damage to property. Yet, they have very limited capacities to substitute for government investments in 'hard' infrastructure, which is essential for risk reduction. Since these responses are generally small scale and cannot address the underlying root causes of vulnerability,[11] they have frequently been ignored. However, supporting these local responses should be one aspect of an overall adaptation strategy for urban areas. In doing so, these coping strategies can be enhanced to ensure that the investments made by low-income urban residents contribute to building their resilience.

Studies in informal settlements exemplify the importance of what individuals and households do for themselves – and, for many of these, the importance of family and sometimes of friends and neighbours in providing help. This range of measures taken to help cope with extreme events can be divided into two:

- those that are *preventive* (that remove the hazard or exposure to it); and
- those that are *impact minimizing* or *impact reducing* (better quality defences against the hazard or assets that help recovery).[12]

The discussion below starts by reviewing examples of household and community responses to climate change, and concludes with an assessment of challenges to household- and community-based adaptation.

## Household responses

Individuals and households take measures to reduce risks from extreme weather events such as flooding or extreme temperatures. Likewise, wealth helps individuals or households to buy their way out of risks – for instance, by being able to buy, build or rent homes that can withstand extreme weather in locations that are less at risk from flooding. Higher-income groups can also afford the measures that help them to cope with illness or injury when they are affected (the medical treatment needed, taking time off work) or when their assets are damaged (e.g. through compensation from insurance). Many of these measures also reduce risks for a wide range of hazards; a good-quality secure home with

*Where local governments are weak or ineffective, household and community strategies become more important for reducing climate change risks and impacts in urban areas*

*Supporting ... local responses should be one aspect of an overall adaptation strategy for urban areas*

good infrastructure and services removes or greatly reduces a great range of risks, including most of those related to climate change. Savings schemes can be drawn on to help cope with a wide range of stresses or shocks, including those arising from extreme weather.

Those unable to get or afford these take other measures to reduce the impacts of hazards that they cannot avoid. These can be seen as contributing to adaptation in that they reduce vulnerability to hazards,[13] and many can be considered as strategies in that they include a coherent range of measures that respond to changes in risk levels. A study of this in Indore (India) showed the complex and varied measures by which low-income households living in areas often flooded adapted to flooding.[14] They were prepared to live in homes that flooded regularly because of other advantages that these sites provided – namely, access to low-cost housing, and central city locations close to jobs, to markets for the goods that they made or collected (many earn a living collecting waste), and to health services, schools, electricity and water. Households and enterprises took both temporary and permanent measures to minimize the impacts of flooding – for instance, by raising plinth levels, using flood-resistant building materials, choosing furniture that is less likely to be washed away, and ensuring that shelving and electric wiring are high up the walls, above expected water levels. Many households had suitcases ready, so valuables could be carried to higher ground when floodwaters are rising, and contingency plans for evacuating persons and possessions (e.g. first to move children, older persons and animals to higher ground, then to move electrical goods, then lighter valuables and cooking utensils):

> *When we see very dark clouds up the hills, we expect heavy rains to come. So we get ourselves prepared by transferring our valuable things on our very high beds which are reached by climb-*

*ing ladders. Also, children who sleep on the floor are transferred to the high beds.*[15]

More established residents had also learned how to get compensation from the government for flood damage. None of these measures reduced the flooding; but they certainly reduced the impacts of flooding upon health, assets and livelihoods.

In Lagos (Nigeria), a city with very large deficits in infrastructure and large sections of the population at risk from flooding (see Box 6.1), interviews with the inhabitants of four informal settlements close to the coast showed that they considered flooding as their most serious problem, although flood risks varied by settlement and within each settlement.

A study in Korail (Bangladesh) documented a range of household measures to reduce loss from flooding and high temperatures and facilitate recovery (see Box 6.2). Similarly, a study of flooding problems faced by residents of low-income communities in Accra (Ghana), Kampala (Uganda), Lagos (Nigeria), Maputo (Mozambique) and Nairobi (Kenya)[16] showed a comparable mix of measures to reduce impacts. In Nairobi's informal settlements (where around half the city's population live), responses to flooding included bailing water out of houses, putting children on tables and, if necessary, moving them to nearby unaffected dwellings, digging trenches around houses, constructing temporary dykes or trenches to divert water away from the house, and a range of ways to stop water from coming into homes. Residents also moved to higher ground as floodwaters rose. Similar measures were taken by households in Accra, Lagos and Kampala. In addition, in Kampala, some residents undertook collective work to open up drainage channels. In Lagos, one resident stated the following: 'There has not been assistance from anyone. Neighbours cannot assist because everybody is poor and vulnerable. I am planning to quit this place because it is horrible living here.'[17]

*Low-income households ... were prepared to live in homes that flooded regularly because of other advantages that these sites provided – namely, access to low-cost housing, and central city locations*

---

**Box 6.1 Household and community responses to flooding in informal settlements in Lagos, Nigeria**

The location of Lagos on a narrow low-lying coastal stretch bordering the Atlantic Ocean puts it at risk from sea-level rise and storm surges. However, it is the lack of attention by state and local governments to the needed storm and surface drains and other infrastructure, and also to land-use management, that has created most of the risks from flooding. The city has expanded rapidly and much of the population growth has been housed in informal settlements in marshy areas or near the lagoons. Many new urban developments have taken place on floodplains (as mangroves have been cleared and wetlands filled) or on stilts over the lagoon.

Interviews with inhabitants of four informal settlements close to the coast showed that flooding was the most serious problem that they faced, although flood risks varied by settlement and within each settlement. In one of the communities (Makoko), for instance, residents living next to a channel were more severely affected than other residents. Floodwaters almost always entered homes and floods lasted for up to four days. Over 80 per cent of respondents reported that they had been flooded three or four times during 2008. Most interviewees listed the poor drainage system as the main cause of the floods, with the effects of 'overpopulation' also listed in terms of more household wastes disposed on streets or in drains and the encroachment of drainage channels by buildings.

Almost all respondents highlighted the shortages of potable water after flooding, with 91 per cent mentioning the impacts of flooding upon their health and increased medical expenses. Most also noted how floods deny them job opportunities. There were some community initiatives to clear blocked drainage channels; but most responses were by households as they constructed drains, trenches or walls to try to protect their houses or filling rooms with sand or sawdust. Foodstuffs and other household items were also stored on shelves or cupboards above anticipated flood levels. Three-quarters of respondents received assistance from family and friends after flood events; far fewer received assistance from government or religious organizations.

*Source:* Adelekan, 2010

> **Box 6.2 Household responses to reducing risks from flooding in Korail, Bangladesh**
>
> Korail is one of the largest informal settlements in Dhaka (Bangladesh). It covers 90 acres (36.4ha) and has a population of more than 100,000. When the site was first settled it occupied the high ground; but as the population expanded, houses were built closer to or even over the water of the adjacent lake and reservoir. Despite the risks, this is considered a good location for employment by its residents, as it is near high-end residential and commercial areas. It thus attracts people mostly in service jobs such as cleaners, rickshaw pullers and workers in ready-made garment industries.
>
> Interviews with households living near the water's edge and on higher ground focused on their experience of climate variability, hazards and coping strategies. Those interviewed highlighted how any climate hazard reduces earnings through missed working hours or even days. They took action in response to flooding and water clogging and in response to rainfall that was anticipated (e.g. the regular monsoon rains) and unexpected. Before heavy rainfall, some moved to safer locations. This was not an option for most residents though, as it meant losing assets, disrupting livelihoods and losing the right to stay and live in that location. Most impact-minimizing actions were part of regular practice – for instance, making barriers across door fronts, increasing furniture height (e.g. putting them onto bricks), making higher plinths and arranging higher storage facilities (e.g. placing shelves higher up on the walls). To help cope with very high temperatures, creepers were grown in courtyards to cover roofs and other materials are put on roofs to reduce heat gain; most households used some form of false ceiling or canopy made out of cloth (a popular practice in rural areas, adopted in urban houses).
>
> For houses near or on the water's edge, structures are on stilts, with platforms constructed higher up the stilts. These also have better ventilation than houses inland. Wooden planks for flooring are preferred as they suffer less from water clogging once floods subside after heavy rainfall. Stilts also mean expansion is possible over the lake. During flooding or water clogging, most residents sleep on furniture, use moveable cookers for food preparation (that can be used on shelves or on top of furniture); some shared services with unaffected neighbours. Other measures include making outlets to help get the floodwater out of the house.
>
> Half the households interviewed save regularly with community-savings groups or non-governmental organizations (NGOs), and savings were important for coping with flood impacts. Many households also bought building materials throughout the year so they had these to use in rebuilding, after flooding. Half the households reported that they feel able to ask relatives or friends for help after a disaster.
>
> *Source:* Jabeen et al, 2010

In Dar es Salaam (Tanzania), residents in Tandale (Kinondoni Municipality) take a range of measures to protect themselves and their houses when flooding occurs. These include temporary relocation and placing easily damaged items (such as mattresses) in the ceiling areas of houses. Some households have constructed additional walls around their houses to prevent floodwater from entering.[18]

The above examples show that most household responses are impact-reducing, ad hoc, individual short-term efforts to save lives (e.g. to sleep on high tables or wardrobes and move family members to safer sites), or to protect property (e.g. making barriers to water entry at the door, digging trenches to steer water away from the door, making outlets at the rear of the house so water flows out quickly).

## Community responses

Community-based adaptation is a process that recognizes the importance of local adaptive capacity and the involvement of local residents and their community organizations in facilitating adaptation to climate change.[19] The starting point for community-based adaptation is the individual and collective needs of the residents in a community and their knowledge and capacities. It is based on the premise that local communities have the skills, experience, local knowledge and motivation, and that – through community organizations or networks – they can undertake locally appropriate risk reduction activities that increase resilience to a range of factors, including climate change.[20] It also recognizes (or assumes) a capacity among the residents in any 'community' to work together. The central principles of community-based

adaptation are that it works at the level of the community: it is about communities making choices rather than having them imposed from outside. Advocates of community-based adaptation question the value and effectiveness of top-down adaptation approaches as they see the difficulties of getting these to be pro-poor, locally appropriate and locally accountable.

Community-based adaptation to extreme weather, water constraints or other risks to which climate change contributes is a pragmatic recognition of the limitations or inadequacies of government action on adaptation. It may be the responsibility of government to provide and maintain infrastructure that can deal with extreme events; but for those areas and populations inadequately served by these, community responses can play a significant role in reducing risks or impacts. As such, community-based preparedness is an important part of resilience to extreme weather events whose timing and magnitude are likely to become less predictable as a result of climate change.

To date, community-based adaptation has primarily been practised in rural areas. However, communities in urban areas can also have an important role in determining the most effective responses to help them address the challenges of climate change. For instance, over the last few years, a growing number of studies have examined the responses of low-income households and communities living in informal settlements to extreme weather-related risks, especially floods. In the four informal settlements of Lagos (Nigeria) described above that have to cope with regular flooding (see Box 6.1), there were some community initiatives to clear blocked drainage channels, although most

*The starting point for community-based adaptation is the individual and collective needs of the residents in a community and their knowledge and capacities*

actions were taken by households. The same is true in Korail (Bangladesh), although some households had taken part in initiatives to clean and clear drains (see Box 6.2).

In practice, the development of infrastructure which reduces climate change impacts is often beyond the capabilities of even the best organized and most representative community organizations. For example, developing a drainage system that actually stops or greatly reduces flooding – especially in high-density settlements on high-risk sites with little or no drainage infrastructure and space for new infrastructure – is usually beyond the means of community organizations. This is not to say that it cannot be done; community-directed slum and squatter upgrading has achieved this; but this is where they get appropriate support from government, as in the *Baan Mankong* (Secure Tenure) programme in Thailand.[21] The Orangi Pilot Project Research and Training Institute (in Karachi, Pakistan) has also demonstrated that households in informal settlements can join together to fund and manage the installation of sewers and drains and do so at scale.[22] However, this was facilitated by the fact that most informal settlements in Pakistan's urban areas developed with grid layouts and space for roads and paths (under which the sewers and drains could be installed). In addition, the local government's water and sanitation authority came to support this by providing the trunk sewers and drains into which the neighbourhood initiatives could be integrated. What these and other cases show is how effective risk reduction is possible if household, community and government investments and actions work together in a coordinated manner.

This point is illustrated by discussions with two communities that had experienced serious floods and with emergency managers in the two urban communities of Mansión del Sapo and Maternillo, located in the northeastern municipality of Fajardo in Puerto Rico.[23] These discussions focused on flood hazards, causes and possible solutions. They showed good community knowledge of flood hazards (each community produced a map of the extent of flooding) and its causes. However, the residents' maps differed from those of the emergency managers – especially highlighting the risks for those living close to a drainage channel. They also differed in terms of sources of floodwaters (residents included urban runoff, whereas the emergency managers only considered river overflow).[24] Both communities highlighted solutions that were beyond their own capacities and that set responsibility for addressing the problems with government. Yet, the problem here was both the limitations in what government was likely to do and the limitations in the technical solutions proposed. From a flood risk reduction perspective, it was important to have a stronger community engagement that recognized the need for disaster preparedness because of the limits in what the structural measures that government undertakes or should undertake could achieve. This community engagement should include monitoring local conditions that can cause floods or exacerbate their impacts and acting on this (e.g. drainage channel maintenance) and flood preparation plans (including, where needed, plans for evacuations). Here, resilience to climate change depends not only on technical

measures and structural solutions, but also on household and community capacity to cope better with extreme weather events that are less predictable in their magnitude and timing. This is a point that has relevance for most urban centres and settings.

The constraints on community capacity in the absence of government support are highlighted by a study of 15 disaster-prone slums in El Salvador. Here, too, there was a mix of household and community responses to climate change-related risks. Households recognized that flooding and landslides were the most serious risks to their lives and livelihoods, although earthquakes and windstorms, lack of job opportunities, and water provision and insecurity from violent juvenile crimes were also highlighted. They invested in risk reduction, for instance, by improving their homes, diversifying their livelihoods or having assets that could be sold if a disaster occurred. Many households received remittances from family members working abroad, and these were especially important in providing support for post-disaster recovery. A complex range of issues did, however, limit the effectiveness of community responses. The residents received no support from government agencies. Indeed, most residents viewed local and national governments as unhelpful or even as a hindrance to their efforts.[25] Furthermore, although residents were organized in community-based organizations, none of these were representative of the communities.

Where there are representative community-based organizations, the possibilities of building resilience to climate change are much greater. In many countries, there are now national federations of slum and shack dwellers that have community-based savings groups as their foundation. Although very few of these savings groups have climate change adaptation programmes, almost everything that they do contributes to greater resilience and reduces risks. This often includes many measures taken in response to the extreme weather events that they have long had to cope with. It usually includes measures that make their houses safer – either through support for upgrading (e.g. in Orissa, India, *Mahila Milan* (Women Together) groups developing homes that can withstand cyclones and rainfall) or through acquiring new, safer, more secure land sites upon which to build.

Most of what these federations are doing is building the resilience of low-income households to almost all climate change risks. For instance, a savings account can be drawn on, whatever the shock. Yet, the contribution that these federations make to climate change resilience needs to be appreciated. To give but one example from the 30 or so countries that have national federations of slum/shack dwellers: in Dar es Salaam, the Tanzania Federation of the Urban Poor has been active in building resilience in low-income urban communities through a process of community organization. This began with savings schemes and enumeration exercises (which provide maps and details of all households in informal settlements), and has expanded to include identification and purchasing of land for housing.

The practice of saving regularly has both instrumental benefits (the ability of savers to access funds when neces-

**In practice, the development of infrastructure which reduces climate change impacts is often beyond the capabilities of even the best organized and most representative community organizations**

**Effective risk reduction is possible if household, community and government investments and actions work together in a coordinated manner**

**Almost everything that [community-based savings groups] do contributes to greater resilience and reduces risks**

**Box 6.3 Risk reduction by the Homeless People's Federation of the Philippines**

The Homeless People's Federation of the Philippines is a national network of 161 urban poor community associations, with more than 70,000 individual members. It represents communities and their savings groups from 18 cities and 15 municipalities. The federation and its community associations are engaged in a wide range of initiatives to secure land tenure, to build or improve homes, and to increase economic opportunity. The federation also works with low-income communities residing in areas at high risk from disasters, assisting in reducing risks, or, where needed, in voluntary resettlement; or in community-driven post-disaster reconstruction.

The federation's responses to disaster events provide relevant insights for community-level responses to climate threats. The principles behind, and processes of, disaster risk reduction and climate change adaptation have many similarities. Both address the hazards that will affect particular locations and individuals, and they share an acknowledgement of the importance of addressing root causes of vulnerability.

The federation is engaged in three main activities that build resilience and facilitate adaptation to climate change:

- First, the interventions of the federation have a strong focus on land and shelter. Unsafe housing that cannot resist extreme weather events, located on land that is at risk of a range of climate-related hazards, is often at the core of vulnerability for low-income urban residents. The failure of local and national governments to address this issue is one of the main factors contributing to risk. By working collectively to acquire land and to obtain financing to build more resilient structures, federation members have addressed this aspect of vulnerability.
- Second, collaboration with the state ensures that interventions can take place at a larger scale. An active and well-organized body of citizens and community organizations can provide the impetus for local authorities to support locally based adaptation strategies. In Iloilo, a coastal city that frequently suffers from extreme flooding, the federation has been actively involved in the planning process for a flood control project, and has been able to encourage particular interventions that meet the needs of the group's members.
- Finally, collective savings at the community level act to provide a source of funds that can be used for pre-event preparation and post-event response, as well as for longer-term support of livelihood activities. More importantly, the process of saving builds trust among members of savings groups and enables them to make collective responses to immediate threats and to develop strategies for future actions that strengthen livelihoods and build resilience. Strong local organizations can prevent the sense of dependency that often results after disaster events. In Bikol Province, savings groups helped participants to define and realize their own preferred development response to a devastating mudslide generated by Typhoon Reming in November 2006. According to the federation's regional coordinator, Jocelyn Cantoria, 'the adoption of the savings programme [has shown that the communities] can be

self reliant and not be dependent on government dole-outs ... they have shown that they can collectively contribute to their own development and to that of the municipality as well'.

The Homeless People's Federation has a national programme that includes the organization and mobilization of low-income communities in high-risk areas. For these communities, the federation promotes and supports the scaling-up of community-led processes for identifying and acting on disaster risk that includes secure tenure, adequate housing, basic services, disaster risk management and, when needed, relocation. Activities range from community visits; consultations; preparation of settlement profiles and enumerations; hands-on training; learning exchanges; temporary/transitional housing construction; land acquisition; participatory site and housing design; planning, construction and management; engagements and advocacy and building learning networks among high-risk or disaster-affected communities. A review of lessons learned from the federation's experience highlighted the following:

- Savings groups within the settlements affected helped to provide immediate support for those affected by the disasters.
- Existing community organizations within high-risk settlements can help to provide immediate relief and foster social cohesion with tools to support them taking action to resolve longer-term issues, such as rebuilding or relocation. Representative community organizations are needed to manage difficult issues – such as who gets the temporary accommodation; who gets priority for new housing; and how to design the reblocking that accommodates everyone. In communities lacking such organizations, visiting federation leaders encouraged and supported their formation and capacity to act.
- The visits to the disaster sites by teams of community leaders from the federation and community exchanges that support the survivors' learning on savings management, organizational development, community surveys and house modelling – developing life-size models of houses to see which design and materials produce the best low-cost housing – have proved to be an important stimulus for the development of community organizations.
- Community profiling and surveys helped to mobilize the people who were affected, and also helped them to get organized and to gather data about the residents and the disaster site needed for responses. It also supported them by showing their capabilities to the local government.
- The importance of being able to obtain land on a suitable well-located site, in situations where relocation is necessary, was highlighted.
- The importance of having regional organizations to support each settlement when disasters affect many different settlements was emphasized.
- Supportive local governments and national agencies are important in that they help with much of the above. This is important with respect to getting access to land and/or obtaining land titles, as well as in the form of high-level political support to obtain more rapid response from bureaucracies.

*Sources*: Reyos, 2009; Dodman et al, 2010a

sary) and organizational benefits (the relationships of trust built up within small savings groups that allow their members to work on collective solutions to larger problems). Small-scale loans managed by these savings groups and repaid over short time periods provide much needed capital for livelihood activities, or responses to shocks and stresses. The creation of savings organizations also provides the basis

by which individuals and households can come together to identify and acquire residential land on sites that are less at risk of flooding. Local initiatives have also built resilience through improving the supply of potable water (reconnecting and managing water kiosks); engaging in capacity-building for hygiene promotion; and implementing innovative small-scale solid waste management strategies.

In the examples given above from Orissa and Dar es Salaam, the savings groups and their federations not only organize and act, but also seek partnership with, and support from, government agencies. This is also the case in the Philippines, where there are some interesting and highly relevant examples of community-based responses to extreme weather events that were driven by savings groups formed by low-income groups and the Homeless People's Federation of the Philippines, of which they were members (see Box 6.3). The Philippines is regularly affected by earthquakes, volcanic eruptions, typhoons, storm surges, landslides, floods and droughts. Many low-income urban residents groups live in high-risk sites and have poor-quality housing; they also have little or no protective infrastructure and less resources to call on after disasters. Risk levels have probably increased and are likely to continue increasing because of climate change. The response of the Homeless People's Federation is to get household, community and local governments to work together, as neither of them have the resources and capacities to reduce risks by themselves.

Although communities are taking action to adapt to climate change-related risks, such as floods and high temperatures, they face a number of challenges in this respect. As noted earlier, there are limits to what community-based action can achieve in urban contexts. Much adaptation (and disaster risk reduction) needs the installation and maintenance (and funding) of infrastructure and services that are at a scale and cost beyond the capacity of individuals or communities. However, the limitations in local government capacity – or local government unwillingness to work with those living in informal settlements – mean that what households and communities in informal settlements do are often the only adaptation responses that are actually implemented. Furthermore, it is mostly low-income households and communities who have to rely on community-based actions and community preparedness because they are located in more vulnerable sites, their homes are of poorer quality and they receive less protection from infrastructure or insurance. In this sense, middle- and high-income groups face much lower levels of risk, and usually have much less need for community-based action to remedy deficiencies in infrastructure and services.

There are also difficulties in getting the needed cooperation among community residents for collective responses to climate change risks. This is partly related to the extent to which community organizations comprehensively represent the needs and priorities of those most at risk or most vulnerable. In reality, community organizations are not necessarily accountable to, or fully representative of, all local residents and their needs.[26] In many contexts and societies, women and particular groups within communities (such as racial, ethnic or other minorities) face discrimination from other residents or resident organizations and lack voice. It is not surprising, then, that it is often difficult to get agreement and commitment from all inhabitants of a settlement for community-based actions.

In urban areas in developing countries, an additional challenge relates to the need for community-based adaptation to focus on using, protecting and enhancing the assets available to the urban poor.[27] As such, it includes the use of a range of assets to make livelihoods more resilient so that

| Areas of intervention | Asset-based actions | | |
| --- | --- | --- | --- |
| | Household and neighbourhood | Municipal/city | Regional or national |
| Protection | Household and community-based actions to improve housing and infrastructure. Community-based negotiation for safer sites in locations that serve low-income households. Community-based measures to build disaster-proof assets (e.g. savings) or protect assets (e.g. insurance). | Work with low-income communities to support slum and squatter upgrading informed by hazard mapping and vulnerability analysis. Support increased supply and reduced costs of safe sites for housing. | Government frameworks to support household, neighbourhood and municipal action; risk reduction investments and actions that are needed beyond urban boundaries. |
| Pre-disaster damage limitation | Community-based disaster preparedness and response plans, including ensuring that early warning systems reach everyone, measures to protect houses, safe evacuation sites identified if needed, and provision to help those less able to move quickly. | Early warning systems that reach and serve groups most at risk; preparation of safe sites with services; organization for transport to safe sites; protecting evacuated areas from looting. | National weather systems capable of providing early warning; support for community and municipal actions; upstream flood management. |
| Immediate post-disaster response | Support for immediate household and community responses to reduce risks in affected areas, support the recovery of assets, and develop and implement responses, including cash-based social protection measures; plan and implement repairs. | Encourage and support active engagement of survivors in decisions and responses; draw on resources, skills and social capital of local communities; rapid restoration of infrastructure and services. | Funding and institutional support for community and municipal responses. |
| Rebuilding | Support for households and community organizations to get back to their homes and communities, and plan for rebuilding with greater resilience; support for recovering the household and local economy. | Ensure reconstruction process supports household and community actions, including addressing priorities of women, children and youth; build or rebuild infrastructure and services to more resilient standards. | Funding and institutional support for household, community and municipal action; address deficiencies in regional infrastructure. |

*Source:* adapted from Moser and Satterthwaite, 2008

**Table 6.1**

Examples of asset-based actions at different levels to build resilience to extreme weather

they can cope with a range of challenges, some of which can be predicted and others of which are unforeseen. In this context, community-based adaptation and pro-poor adaptation are intrinsically linked. Pro-poor adaptation raises important questions about the types and aims of responses; who bears any costs; who is involved; and who benefits.[28] It also needs to address the range of reasons why the urban poor are disproportionately vulnerable to climate change, including their greater exposure to hazards, the lack of hazard-reducing infrastructure, the lack of state provision for assistance after extreme events, and the lack of legal and financial protection.[29]

Table 6.1 presents an asset-based framework to support resilience to extreme weather that includes protection (much of it reducing disaster risk), pre-disaster damage limitation, immediate post-disaster response, and rebuilding.[30] An asset-based approach helps to identify the asset vulnerability to climate change of low-income communities, households and individuals, and considers the role of assets in increasing adaptive capacity. Strengthening, protecting and adapting the assets and capabilities of these groups is necessary to reduce urban poverty, while making them better able to cope with gradual climate change and extreme events. However, as illustrated in the table, a number of actions cannot be undertaken by households and communities alone, but need to be addressed at the municipal/city or national level. Such actions are the focus of the next section.

# LOCAL GOVERNMENT RESPONSES TO THE IMPACTS OF CLIMATE CHANGE

As noted in the previous section, the main responsibility for implementing policies to address the impacts of climate change in cities rests with local governments. Yet, many city governments around the world have so far failed to accept and/or act upon this responsibility, with the result that many households and communities have been forced to implement climate change adaptation measures on their own. As the discussion above has shown, however, there are significant limitations as to what community-based adaptation can achieve. A partnership approach – involving households and communities, but also the various levels of government and other partners – is the most effective way to implement climate change adaptation strategies.

In some places, local governments have taken note of the damaging impact of particular storms or heavy rainfall that have highlighted risks that climate change is likely to exacerbate.[31] Elsewhere, the perceived vulnerability of urban economies, populations, assets and infrastructure has encouraged more local government engagement, including some local governments in middle-income countries for whom an adaptation agenda seems more relevant since it addresses local concerns and can include co-benefits with development.[32] These responses have varied from an initial consideration of likely risks and threats to some particular

infrastructural investment and physical interventions, to the development of plans and strategies.

However, and as noted above, the primary responsibility for developing national policies and programmes on climate change adaptation rests with national governments. National governments are also custodians of the interests of urban (and rural) residents in international climate change negotiations, and in the development of international funding sources and institutions, and other forms of support for adaptation. Thus, the first part of this section briefly reviews national frameworks that support climate change adaptation in urban areas. This is followed by a discussion of what is done at the local level with respect to climate change adaptation. It describes how a small, but growing, number of city governments around the world have begun to recognize the threats posed by climate change, first in developed and later in developing countries. It provides examples of cities in countries that are at various stages of climate change impacts assessments, as well as examples of cities that have developed adaptation strategies, before briefly reviewing the links between climate change adaptation and disaster preparedness.

## National frameworks that support adaptation in urban areas

Figure 6.1 outlines the required steps in developing national climate change adaptation policies and programmes. Although the Organisation for Economic Co-operation and Development (OECD) report where this figure originally appeared used the figure to illustrate what was happening at the national government level, the figure fits well in a consideration of what city governments are doing and which of the steps they are taking (and which they are not). As indicated in the figure, adaptation planning and implementation have to be based on an *assessment* of historical and present climate conditions, projections of climate change, as well as current and future implications on vulnerability and impacts. Such assessments are the foundation of adaptation policies, which may be understood as the formulation of *intentions to act*, on the one side, and *adaptation actions*, on the other. The former include identification of adaptation options and discussions of how these fit in with other existing policies. The adaptation actions include the establishment of institutional mechanisms to guide and implement adaptation action; the formulation of new adaptation policies and modification of existing policies to take adaptation into account; and the explicit incorporation of adaptation measures at the project level. Figure 6.1 also illustrates how adaptation actions undertaken now influence the assessment of future climate change impacts.

The report from which Figure 6.1 is drawn also classified OECD countries in three categories with respect to the criteria in Figure 6.1. According to this review (undertaken in 2006), 7 OECD countries were classified as being in early stages of impact assessments; another 27 countries were undertaking advanced impact assessments, but were slow in the development of adaptation responses; while only 5 OECD countries had advanced impact assessments and were

*Community-based adaptation and pro-poor adaptation are intrinsically linked*

*A partnership approach – involving households and communities, but also the various levels of government and other partners – is the most effective way to implement climate change adaptation strategies*

*The primary responsibility for developing national policies and programmes on climate change adaptation rests with national governments*

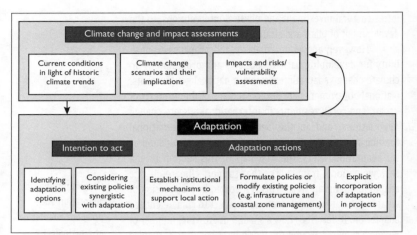

**Figure 6.1**

**The main stages of city-based climate change adaptation**

*Source:* based on Gagnon-Lebrun and Agrawala, 2006, Figure 6

Relatively few national governments are moving towards implementing adaptation initiatives

Funding agencies that support climate change adaptation may judge ... deficits in infrastructure and services as being outside the scope of climate change adaptation

moving towards implementing adaptation. This and other assessments[33] show that relatively few national governments are moving towards implementing adaptation initiatives. A review of what governments in developing countries are doing on adaptation suggested that many are initiating or sponsoring studies of the likely impacts of climate change; but rarely is urban adaptation given much attention.[34] Many countries have developed National Adaptation Programmes of Action (NAPAs)[35] and most recognize the need to strengthen local capacity to plan and act – including changing local building and infrastructure standards and land-use plans. Yet, these NAPAs have rarely engaged the interest of the larger, more powerful national ministries or agencies, or of city or municipal governments. Many give surprisingly little attention to urban areas, given the importance of urban economies to national economic success and, for most countries, to the incomes and livelihoods of much of the population.[36]

It is also difficult to ensure that NAPAs do not become just another policy document that gets little or no action on its recommendations:

> *Countries are already bombarded with international obligations, which place considerable strain on already overloaded institutions with limited capacity, and which may well lead to duplication of effort and reduction in policy coherence.*[37]

It must also be remembered that NAPAs' effectiveness depends on their catalysing and supporting local assessment and action. It has been suggested that what is needed is city-focused City Adaptation Programmes of Action and local-focused Local Adaptation Programmes for Action.[38] As stressed throughout this chapter, risks and vulnerabilities for all aspects of climate change in urban areas are greatly shaped by local contexts and influenced by what local governments do or do not do. Effective adaptation needs to be based on a good understanding of the local context and strong local adaptive capacity. It needs City Adaptation Programmes of Action and, very often, smaller-scale Local Adaptation Programmes for Action that incorporate community-based adaptation – especially for the settlements or areas most at risk. Much more needs to be done in terms of

'mainstreaming' adaptation to climate change within national policy-making processes[39] and putting in place the systems and structures that encourage and support city-driven and locally driven adaptation. Perhaps more to the point, unless adaptation is seen by national and city governments in developing countries to be complementary to development agendas, it will not get considered.

## Local government responses in developing countries

As noted above, there are not many examples of cities in developing countries that have initiated climate change adaptation policies. The bulk of the examples that exist are cities that have started the process outlined in Figure 6.1 by assessing the risks posed by future climate change. Some such examples are outlined below, followed by a discussion of the experiences of cities that have taken this assessment one step further by showing a concrete intention to act through the development of adaptation strategies.

### ■ Assessing climate change risks and the scale of the adaptation deficit

Generally, the first evidence of an interest by city or municipal government in climate change is an interest in assessing the scale and nature of likely risks. Yet, this assessment is not easily done for most developing country cities because of the lack of basic data on environmental hazards and risks (or even of an accurate and detailed map with all settlements on it). It is thus important to note (again) that most climate change-related risks (at least in the next few decades) are an exacerbation of risks already present, which are the result of the inadequacies in local governments' capacities or willingness to manage and govern urban areas. Thus, there is a large deficit in the basic infrastructure and services needed to address not only risks related to extreme weather and water constraints, but also 'everyday' risks. A city where much of the population live in areas that are frequently subjected to floods – because these areas lack storm and surface drains (and often because areas at risk of flooding are among the only areas where low-income groups can buy, build or rent accommodation) – is a city that is more at risk from more frequent or intense rainstorms. The deficits in basic infrastructure and services are not the result of climate change, and funding agencies that support climate change adaptation may judge these (often vast) deficits in infrastructure and services as being outside the scope of climate change adaptation. Box 6.4 gives some examples of the scale and nature of these deficits.

Three examples are provided below of cities for which this first step (i.e. of mapping the tasks at hand) has been taken – namely, Georgetown (Guyana), Bangkok (Thailand) and Dhaka (Bangladesh). These three examples show how climate change risks come to be identified and discussed, and highlight the initial thinking of what measures are needed to address these. Nonetheless, there is still the need to incorporate measures to address these risks into city plans, land-use management, infrastructure investments, service provision, and building and planning codes, and

---

**Box 6.4 The scale of adaptation deficit in selected cities**

*Dar es Salaam.*[a] This is the largest city in Tanzania, with more than 3.3 million inhabitants in 2010, compared to less than 0.2 million in 1960.[b] As a coastal city, it faces climate change-related risks from sea-level rise and coastal erosion, flooding, drought and water scarcity, and the disruption of hydroelectricity generation. These issues are much exacerbated by the mismatch between the growth of the city (and the city economy) and the capacity of the local governments within Dar es Salaam. Some 70 per cent of the population live in informal and/or illegal settlements and most lack adequate provision of basic infrastructure and services, including piped water supplies and provision for sanitation, drainage and solid waste collection. Low-income residents in the city are already coping with a range of climate-related challenges, particularly related to seasonal flooding. Uncollected garbage blocks both natural and artificial drainage channels, which causes flooding after heavy rainfall.

*Dhaka.*[c] Bangladesh is frequently identified as one of the countries most at risk from the effects of climate change. Its large and rapidly growing capital Dhaka is particularly at risk; a population that grew from 0.5 million in 1960 to 14.6 million in 2010[b] has long outstripped the expansion of infrastructure, including flood protection. This is a city already very vulnerable to flooding, especially during the monsoon season – as shown by major floods in 1954, 1955, 1970, 1980, 1987, 1988, 1998 and 2004. The 1988, 1998 and 2004 floods were particularly severe, with very large economic losses. These were mainly caused by the spillover from surrounding rivers. The city has a very large deficit in terms of the proportion of the population living in slum areas, with overcrowded, poor-quality housing that lack piped water, sewers and drains.

*Lagos.*[d] The location of Lagos (Nigeria) on a narrow lowland coastal stretch bordering the Atlantic Ocean puts it at risk from sea-level rise and storm surges; much of the land in and around Lagos is less than 2m above sea level. Yet, it is the lack of attention by state and local governments to the needed storm and surface drains and other infrastructure, and also to land-use management, that has created most of the risks from flooding. The city has expanded rapidly – from less than 0.8 million in 1960 to 10.5 million in 2010[b] – and much of the population growth has been housed in informal settlements in marshy areas or near (or even over) the lagoons. Much of the city lacks the infrastructure needed to limit floods; a high proportion of residents lack not only storm drains, but piped water, sanitation, electricity, all-weather roads and solid waste collection. To this is added the lack of maintenance of storm drains (especially de-silting before the rainy season), the drains and gutters blocked with solid wastes (because of no household solid waste collection service) and the unauthorized buildings that encroach on drains. The expansion of low-income settlements in areas at high risk of flooding (many on stilts) is largely because there are no safer sites available that they can afford.

*Sources:* a draws on Dodman et al, 2010b; b UN, 2010; c draws on Alam and Rabbani, 2007; Ayers and Huq, 2009; Roy, 2009; d draws on Adelekan, 2010; Iwugo et al, 2003; Adeyinka and Taiwo, 2006

---

there is much less evidence of this taking place. The section of the city government that prepares (or commissions) these initial assessments may have little political support within the city government or may be unable to convince the more powerful sectoral agencies within the government to change their plans and investments in response to the risks identified.[40] Inevitably, any forward-looking risk-reducing investment programme that needs serious funding will face competition from other sectors.

In Guyana,[41] the coastal zone that includes Georgetown holds 90 per cent of the country's population and much of the economy. Its highest point is only 1.5m above sea level, with much residential land, including the capital Georgetown, below the sea level at high tide. Large sections of Georgetown's population experience regular floods.[42] Adaptation planning for the densely settled areas around Georgetown has been conducted by an international management consultancy firm, with the intention of identifying and analysing adaptation investment options. Risks were assessed through analysing major climate hazards, identifying the major assets at risk and assessing the vulnerability of these. The main climate hazard facing Guyana, and particularly the densely populated areas near Georgetown, is flooding caused by heavy rains. A variety of scenarios have been developed to estimate the potential for financial losses in the public, agricultural, industrial and commercial, and residential sectors in 2030.

In Georgetown, there is also evidence of the second stage of city-based adaptation – identifying adaptation options and considering existing policies that are synergistic with adaptation (see Figure 6.1). Key adaptation interventions that were identified as being economically attractive included the expansion of early warning infrastructure; the improvement of building codes for new construction; the maintenance of drainage systems; and the upgrading of drainage systems. In each of these cases, there was a cost-benefit ratio of less than 1.0, implying that such measures were economically viable. Several adaptation measures were assessed quantitatively. These include:

- *Infrastructure measures:* repairing and maintaining the sea wall.
- *Health measures:* flood-proofing health clinics, sanitation and water, emergency response system.
- *Financial measures:* cash reserve, contingent capital, strengthening the primary insurance market.

Of these, repairing and maintaining the sea wall, developing an emergency response system and providing contingent capital were seen as generating the most important benefits. Sections of the sea wall are in disrepair and upgrades are needed to protect against coastal flooding; emergency response capabilities currently do not exist; and risk financing can provide money in the case of a crisis event.

> Any forward-looking risk-reducing investment programme that needs serious funding will face competition from other sectors

Additionally, these are relatively low-cost interventions. Thus, some substantial adaptation benefits can be achieved for relatively low costs. This approach has great value in identifying the most cost-effective adaptation responses at the city level, and can help local and national officials to identify the most appropriate interventions. However, it would probably be best used in association with detailed social analysis to ensure that adaptation activities meet human development needs as well as being cost effective from a financial perspective.

The Metropolitan Administration of Bangkok (Thailand) has also begun mapping the climate change-related risks that the city will face; based on this, it is proposing a variety of policy-based, infrastructural and environmental responses (see Table 6.2).[43] Bangkok is vulnerable to a range of climate threats as a result of its location on a low-lying plain affected by subsidence, close to the sea and subjected to regular monsoon rains. A risk management approach is assessing the potential consequences of climate change and identifying appropriate responses. An initial risk assessment – which highlighted flooding, storm surges, drought and risks to the security of the water supply – has been conducted, and these risks will be analysed more extensively to inform adaptation interventions. More overarching adaptation measures will include capacity-building activities, improved communication between scientists and city officials, encouraging the development of climate change risk assessments at the local level, and raising awareness of climate change in homes and communities.

Box 6.4 highlights the climate change-related risks faced by Dhaka (Bangladesh). This is a city with a relatively long history of environmental and climate change awareness, policy and action. It was the first of the least developed countries to complete its National Adaptation Programme of Action (NAPA), and there is a significant effort by the national government to integrate climate change within sectoral plans and policies. The Dhaka Metropolitan Development Plan is intended to meet many climate adaptation needs. For example, a strategic approach to planning could help to enhance response capacity; increased public participation in the planning process could raise public awareness of climate-related threats; and the implementation of sites and services schemes could reduce the vulnerability of the urban poor and enhance their resilience.[44]

At the city level, large-scale flood protection measures are an essential component of an adaptation response. Since 1989, an extensive system of embankments has been constructed, and further investments of this type are currently planned.[45] Canals and drainage systems are currently being renovated, and the banning of polythene bags has helped to reduce the clogging of the city's drainage system.[46]

### ■ Moving from risk assessments to adaptation strategies[47]

Within Africa, South Africa is unusual in having discussions within several city governments on climate change adaptation and thus moving beyond risk assessments to discuss what should be done to address the risks. A number of South African cities have thus developed plans for adapting to climate change. These have been made possible through the strong support of a range of stakeholders, including universi-

> *Within Africa, South Africa is unusual in having discussions within several city governments on climate change adaptation ... moving beyond risk assessments to discuss what should be done to address the risks*

| Climate change impact | Adaptation measures | | |
|---|---|---|---|
| | Community infrastructure and operations | Business and commercial | Residential health and general population |
| General long-term rising temperatures of 3°C–5°C | • Urban design<br>• Tree-planting<br>• Water conservation<br>• Insect and pest controls | • Actions to reduce urban heat island, including building design and green spaces | • Better insulation<br>• Design for efficient cooling<br>• Pest and insect controls<br>• Water conservation |
| Ground and surface water quantity and quality | • Water-use restrictions<br>• Optimize reservoir releases<br>• Expand storage capacity<br>• Greater regulation of surface and groundwater withdrawals | • Water efficiency and conservation programmes<br>• Water pricing<br>• Irrigation practices | • Water efficiency and conservation programmes |
| Sea-level rise (especially in Bang Khuntien District) | • Land-use planning<br>• Construction or improvement of levees and dykes<br>• Creation of water reservoirs | • Coastal protection<br>• Phased retreat<br>• Modifications to operation of port | • Land-use planning<br>• Ecosystem protection |
| Extreme weather-related events (windstorms, prolonged rain, river flooding, drought) | • Emergency preparedness plans<br>• Construction or improvement of levees and dykes<br>• Elevation of buildings<br>• Land-use planning<br>• Increase resilience of electricity network<br>• Improve emergency communications | • Emergency preparedness plans<br>• Flood-proofing of buildings<br>• Elevation of buildings | • Emergency preparedness plans<br>• Flood-proofing of homes<br>• Publicly sponsored flood insurance<br>• Behavioural changes for disaster preparation (e.g. emergency supplies) |
| Increased frequency and intensity of short-duration heavy rains | • Increased size of storm drains, etc.<br>• Increased water-absorbing capacity of urban landscape | • Increase water-absorbing capacity of large paved areas | • Storm sewer protection and maintenance<br>• Landscape design to reduce rapid runoff |
| Increased frequency and intensity of heat waves, droughts and smog episodes | • Use of air conditioners<br>• Heat contingency planning<br>• Reduction of urban traffic<br>• Planting of trees | • Use of air conditioners<br>• Rescheduling protection when necessary | • Use of air conditioners<br>• Public education on behavioural responses |

*Source:* based on Bangkok Metropolitan Administration, 2009, Table 6.1

**Table 6.2**

**Adaptation measures for Bangkok, Thailand**

ties and local authorities. The transition to democracy in 1994 generated new local government structures which included a specific mandate and focus on environmental management, alongside a significantly revised development agenda. This section reviews the experience with developing such climate change adaptation plans in Durban and Cape Town.

Durban has one of the most interesting experiences in developing climate change adaptation plans and strategies because of the innovations that it has demonstrated, and because of the documentation of the internal processes by which it advanced and by which it was constrained.[48] Durban is South Africa's largest port and city on the east coast of Africa, with a population of 2.9 million people in 2010.[49] The local government structure responsible for managing the city is known as eThekwini Municipality. During the 1990s, the municipality had become a leader in the field of local-level environmental management[50] and had also initiated some work on mitigation. The city's planning for adaptation built on these experiences.

Between 2004 and 2006, eThekwini Municipality developed a locally rooted climate change adaptation strategy.[51] This is encapsulated in the Headline Climate Change Adaptation Strategy, which addresses both direct and indirect issues in links between climate change and human health, water and sanitation, coastal zone management, biodiversity, infrastructure and electricity supplies, transportation, food security and agriculture, and disaster risk reduction. Initially, the development of this high-level strategy did not result in any additional innovation or movement from the 'business-as-usual' scenario in terms of municipal functioning and the plans and investments of the larger and more powerful sectors. Climate change risks were seen by other sectors in government as too generic and the risks it outlined too distant; there was also an assumption by many that these were the responsibility of the city's environmental department. There were also other factors drawing attention away from it, such as high existing workloads and urgent development challenges and pressures. The municipality's disaster management unit was an obvious ally – but it lacked capacity and was seen by the municipality as a responsive relief agency, and thus not an influence on infrastructure investments or city planning.

As a result, and in order to engage municipal line functions more effectively in targeted and prioritized climate change adaptation, the adaptation planning process was deepened through the development of more detailed sectoral municipal adaptation plans. At this stage, particular attention was paid to three high-risk sectors (water, health and disaster management) since these form a natural cluster of integrated functions, thereby offering opportunities for cross-sectoral integration and coordination. This sectoral approach has proved to be more successful in facilitating meaningful action, and in time will be rolled out across all relevant municipal sectors. It is through the identification of issues that are relevant to particular sectors within government that their engagement is ensured. Also important for this is that they see climate change adaptation as directly linked to development (and their development and invest-

ment plans). As a staff member from eThekwini Municipality's Environmental Planning and Climate Protection Department has noted:

> *... the more sectoralised approach to adaptation planning now being adopted in Durban has had the effect of encouraging a greater interaction amongst the line functions than occurred during the development of the cross-sectoral [Headline Climate Change Adaptation Strategy]. This can be linked to the clearer definition of tasks and objectives that has emerged from the more detailed understanding of sectoral needs and limitations.[52]*

While climate change has emerged as a significant issue in municipal plans in Durban, and staff and funds have been allocated to climate change issues, the emergence of climate change advocates among local politicians and high-ranking civil servants has been a slower process. However, this is changing, as the mayor and other key officials become more actively engaged in the climate change debate. A process of community-level adaptation planning has now also been facilitated in order to complement and extend the municipal-level interventions. Specific adaptation interventions have included:

- increasing the water-absorbing capacity of the urban landscape;
- improving urban drainage and storm-sewer design;
- increasing natural shoreline stabilization measures;
- utilizing storm water retention/detention ponds and constructed wetlands;
- land-use planning to avoid locating structures in risky areas;
- working with industry to reduce water demand;
- increasing food security;
- using environmental management as the basis for creating 'green jobs'.

The progress in Durban depended on the mobilization of political support for adaptation, and the presence of engaged and motivated stakeholders. However, moving from strategic plans to specific projects will require additional stages of planning and dedicated sources of financing.[53] For this purpose, four institutional markers may be identified for assessing progress in any city towards climate change adaptation:[54]

1 the emergence of an identifiable political/administrative champion(s) for climate change issues;
2 the appearance of climate change as a significant issue in mainstream municipal plans and in stakeholder discussions;
3 the allocation of dedicated resources (human and financial) to climate change issues;
4 incorporating climate change considerations within political and administrative decision-making.

**Climate change risks were seen by other sectors in government as too generic and the risks it outlined too distant; there was also an assumption by many that these were the responsibility of the city's environmental department**

**Moving from strategic plans to specific projects will require additional stages of planning and dedicated sources of financing**

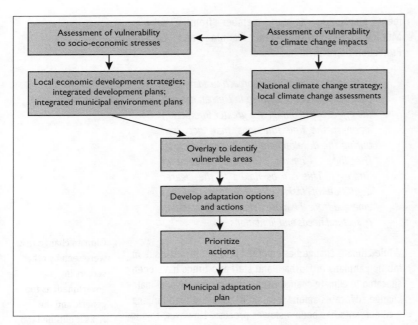

**Process for developing a municipal adaptation plan in Cape Town, South Africa**

*Source:* Mukheibir and Ziervogel, 2007

However, and perhaps obviously, the integration of climate protection considerations within political and administrative decision-making is unlikely to be a smooth process. Anything that affects budget lines and Durban's current desired development path will be contested.[55]

In Cape Town (South Africa), a framework has been proposed for the development of a municipal adaptation plan for climate change (see Figure 6.2) in a document commissioned by the city government. The various steps involved in this process are complemented by two cross-cutting processes: stakeholder engagement is playing an important role in identifying vulnerable sectors and potential initiatives, and also brings politicians and decisions-makers on board; and an assessment of adaptive capacity (the ability of a system to respond to the impacts of climate change). The municipal adaptation plan should be the final output of this process; but a variety of intermediate documents will be produced, including vulnerability maps and assessments to highlight 'hotspots' where social and climate risk interact.

However, in Cape Town, as in all cities, it will be difficult to get the attention of senior politicians and civil servants with regard to climate change adaptation. For instance, in the summary of Cape Town's integrated development plan,[56] no mention is made of climate change adaptation. For Cape Town and many other cities, the first real engagement with climate change adaptation is likely to be through responses to disaster risk. In May 2010, the City of Cape Town's website described the long-range weather forecasts that indicated the possibility of above normal rainfall for the coming winter months and the measures that were being taken to cope with them by various city departments.[57]

**There are many relatively wealthy cities that need major upgrades in their infrastructure that should take account of likely climate change impacts**

## Local government responses in developed countries

Adaptation responses in cities in developed countries are generally much easier to formulate, implement and fund, although not necessarily easier in terms of getting the needed political support. Yet, such cities do not have very large deficits in infrastructure; most or all of their population live in buildings that meet building standards and are served by piped water supplies, sewers, drains and solid waste collection. These cities also have a range of regulations and controls that (when implemented) reduce risks, as well as measures and institutional arrangements that ensure rapid and effective response to disasters, thus limiting their impact when they happen, especially for those who are most at risk.

While the scale of risks and of the populations exposed to them are much smaller and the local capacities to address these much larger, this does not mean that adaptation is necessarily given the priority that it deserves. There are many relatively wealthy cities that need major upgrades in their infrastructure that should take account of likely climate change impacts. In general, most cities in developed countries need to expand their capacity to anticipate and manage extreme weather events. There are also cities that are on sites that are or were relatively safe without climate change, but that now face new levels of risk. For instance, many coastal settlements, whether villages, towns or cities, face increased risks from sea-level rise. Climate change is likely to bring more extreme and frequent heat waves to most regions, with higher risks in large cities or particular 'heat islands' within such cities. Many cities will face constraints on freshwater supplies. However, although adaptation plans for urban centres in developed countries will have many characteristics in common, the particular mix of needed measures will be very specific to each urban centre. For instance, and as discussed below, the measures to adapt to sea-level rise in the adaptation plans of London (UK), Melbourne (Australia) and Rotterdam (The Netherlands) take different forms, and their integration within other measures is specific to each city.

There are also many cities in developed countries where climate change risk, to their governments, seems a distant threat as they are struggling with economic decline. In developed countries, there have been major spatial shifts in where economic growth and new investment concentrate, which have left many cities that were formerly centres of industry and economic success in decline. In such cities, it is difficult to get much attention to climate change adaptation.

As with earlier discussions of cities in developing countries, the first step is to get a sense of what new or increased risks climate change will bring and what impacts these will or may bring. The examples discussed below – from London, Melbourne and Rotterdam – illustrate this first step.[58] The next step after this is the intention to act (see Figure 6.1) – seen in the identification of adaptation options, including all the sector-specific actions needed for this or to support this; this, too, can be seen in these three cities.

The Greater London Authority has developed a climate change adaptation strategy that provides the basis for adaptation actions. As one of the world's wealthiest cities, London has far more abundant financial and technical resources than most other cities.[59] Yet, it faces particular climate risks as a result of its location (on the River Thames), the age of much of the city's infrastructure, and the dense

**Box 6.5 Key risks identified by the climate change adaptation strategy of London, UK**

The climate change adaptation strategy of London identifies responses to three key climate risks: floods, drought and overheating.

The *first* risk, from flooding, is linked to sea-level rise, increased tidal surges up the River Thames (that runs through London), and wetter winters with more frequent and intense heavy rainfall (leading to increases in peak river flows of between 20 and 40 per cent). A series of 'decision pathways' have been developed by the UK Environment Agency (a national governmental body) to respond to this. The Thames Barrier, constructed between 1974 and 1982, is a key part of this strategy – along with 298km of floodwalls, 35 major gates and over 400 minor gates. Although this was not designed with climate change in mind, it is a key part of London's protection against flooding and it has been used far more frequently since 1990. The most recent assessments suggest that in all but the most extreme scenarios, the Thames Barrier will continue to protect London from flooding, although towards the end of the 21st century it may become necessary to use green spaces adjacent to the River Thames to store floodwater.

A city-wide water strategy seeks to reduce the effect of the *second* risk – that is, from water shortages – which are expected to become more frequent as climate change accentuates the seasonality of rainfall. Reducing demand for water will increase the length of time required until drought measures are required – and will also save money and reduce carbon dioxide emissions. The water strategy proposes the following four steps for balancing the supply and demand of water:

1  *Lose less:* reduce the loss of water through better leakage management.
2  *Use less:* improve the efficiency of water use in residential and commercial developments.
3  *Reclaim more:* use reclaimed water for non-potable uses.
4  *Develop new resources:* adopt new resource options that have the least environmental impact.

The *third* risk is from overheating (i.e. when temperatures rise to a point where they affect health and comfort). Overheating also increases demand for energy-intensive cooling (which may lead to power shortages and contribute to increased GHG emissions), a rise in demand for water (increasing pressure on limited water resources) and damage to temperature-sensitive infrastructure. Four courses of action are being used to reduce risks:

1  urban greening to reduce the intensification of temperatures by the urban heat island;
2  designing new and adapting existing buildings and infrastructure to minimize the need for cooling;
3  ensuring that low-carbon energy-efficient measures are used where active cooling is required; and
4  helping urban residents to adapt their behaviour and lifestyles to higher temperatures (a key component of this is ensuring that 'vulnerable' people are identified and provided with suitable social and medical assistance).

*Source: Nickson, 2010*

> The Greater London Authority has … recognized that the provision of ecosystem services … can also generate benefits in responding to climate change

concentration of administrative, commercial and financial activities that are essential to national – and, indeed, global – finance. The adaptation strategy identifies responses to three key climate risks: floods, drought and overheating (see Box 6.5). This strategy relies on the contributions of a range of agencies, operating at the scale of the urban area of London as well as at the national level.

The Greater London Authority has also recognized that the provision of ecosystem services – which may help in the conservation of biodiversity, reduction of pollution or improvement in the aesthetic value of surroundings – can also generate benefits in responding to climate change (see Table 6.3). There are strong co-benefits for adaptation (reduced flood risk and offsetting of urban heat islands), mitigation (reduced energy demand, support biodiversity) and development (reduced noise and air pollution, increased provision for recreation/leisure).

The adaptation strategy of the City of Melbourne (Australia) identifies four main climate risks: reduced rainfall and drought; extreme heat wave; intense rainfall and windstorm; and sea-level rise (see Table 6.4).[60] It also identifies seven urban systems where adaptation actions are needed: water; transport and mobility; buildings and property; social, health and community; business and industry; energy and communications; and emergency services. The risk

management process that was used to analyse these risks included a stage of evaluating risks and deciding whether these are acceptable or not. If the risks are deemed to be unacceptable, then they are treated through a process of adaptation. Throughout, the process is monitored and reviewed, and is linked with communication and consultation. The proposed adaptation measures are intended to reduce the likelihood or consequence of a particular risk or to increase the level of control over it, thereby making it tolerable. These have also been sub-graded to identify whether they fall into the categories of 'control critical', require 'active management', require 'periodic monitoring' or are of 'no major concern'. The risks, key themes and key actions are summarized in Table 6.4.

**Table 6.3**

Ecosystem services provided by green spaces and street trees, London, UK

|  | Green roofs/walls | Street trees | Wetlands | River corridors | Woodlands | Grasslands |
|---|---|---|---|---|---|---|
| Reduce flood risk | ✓✓ | ✓ | ✓✓✓ | ✓✓✓ | ✓✓ | ✓✓ |
| Offset urban heat island | ✓✓ | ✓✓ | ✓✓ | ✓✓ | ✓✓✓ | ✓ |
| Reduce energy demand | ✓✓ | ✓✓ |  |  | ✓ |  |
| Reduce noise/air pollution |  | ✓✓ |  |  | ✓✓ |  |
| Support biodiversity | ✓✓ | ✓ | ✓✓✓ | ✓✓✓ | ✓✓✓ | ✓✓✓ |
| Recreation/leisure | ✓ |  | ✓ | ✓✓ | ✓✓✓ | ✓✓✓ |

*Source: GLA, 2010, Table 7.1*

**Table 6.4**

**Key risks and adaptation strategies in Melbourne, Australia**

| Risk | Key themes | Examples of specific actions |
|------|-----------|------------------------------|
| Drought and reduced rainfall | • Maximize water-use efficiency<br>• Diversify water supply<br>• Maximize water harvesting<br>• Improve waterway and bay health | • Save water through demand management strategies and behavioural change<br>• Structural modifications to treat and/or harvest alternative water supplies<br>• Increase installations of rainwater tanks for toilet flushing<br>• Investigate the use of artificial turf on sports fields |
| Intense rainfall and wind event | • Better drainage and storm water capture<br>• Early public warning system<br>• Integrated emergency services<br>• Better public knowledge and safe behaviour<br>• Minimize debris potential<br>• Increased infrastructure standards | • Drainage improvements at flash flood points on transport system<br>• Continued upgrading of storm water infrastructure<br>• Communications programmes to build capacity for dealing with transport delays in extreme events |
| Heat wave and bushfire | • Cooler surroundings, inside and out, through improved infrastructure<br>• Better public knowledge and safe behaviour<br>• Heat-wave early warning system | • Develop and implement heat wave response plan<br>• Identification and care of high-risk populations<br>• Implement changes to urban form to reduce heat-island effect |
| Sea-level rise | • Future-proof planning for sea-level rise<br>• Better protection for existing low-lying developments<br>• Better flood control through revised drainage planning<br>• Measures to improve resilience to exposed infrastructure | • Modelling of flood risk and infrastructure impacts to sea-level rise<br>• Development of suitable planning guidelines to reflect findings of modelling<br>• More extensive storm water capture and reuse<br>• Alteration of at-risk residential buildings to facilitate entrance and exit during significant floods |

*Source:* City of Melbourne, 2009

Two 'high value' (or cost-effective) adaptation measures have been identified that have the potential to provide benefits across many risks:

1 *storm water harvesting*, which can assist in reducing the impact of flash-flooding events through storing excess storm water while simultaneously storing water for use in times of drought; and

2 *passive cooling*, which can reduce the heat-island effect by reducing temperatures both inside buildings and at street level, therefore reducing overall exposure to the effects of heat waves.

The 1990s brought a shift in the way that disasters and their causes are understood, with much more attention being paid to the links between development and disasters

This concept of 'high value' adaptation can provide a useful tool for adaptation planning, as it indicates the interventions that can have the greatest impact. This is an important consideration, particularly in a context of resource scarcity.

Perhaps not surprisingly, many cities in The Netherlands are considering climate change adaptation measures. The Netherlands has centuries of experience in responding to the challenges faced by being low-lying and coastal. The City of Rotterdam – as a coastal city and one of Europe's largest ports – is particularly aware of these challenges and is aiming to be climate change proof by 2025.[61] The main threat to the city (and the main focus of adaptation measures) is from coastal flooding. Investment in adaptation is necessary to safeguard the health and security of the population, to prevent damage caused by climate change from being unmanageable, to increase the return on investments in the use of public spaces and infrastructure, and to ensure that solutions are innovative and attractive. Responses to climate change in Rotterdam address three key themes:

1 *Knowledge.* Knowledge for climate adaptation is being generated through cooperation with a range of relevant parties, including water and hydraulic engineering institutes, universities, businesses, water boards, housing corporations and developers. New research is being conducted into issues of flooding and heat stress, and knowledge is being exchanged with other port cities, both within and outside The Netherlands.

2 *Action.* This involves the implementation of projects designed to prevent flooding or to reduce its effects. This includes raising dykes, excavating areas to contain extra water, and flood-proofing buildings in areas that are likely to be flooded. In addition, a variety of interventions will be made to ease heat stress in the city – for example, by providing additional shade and cooling.

3 *Marketing.* The City of Rotterdam seeks to be at the forefront of adapting to climate change, and will create a distinct profile for itself as a positive example of a climate-adaptive city in a delta. This is important for relationships with urban residents, major stakeholders (including government agencies and universities) and other cities around the world.

## The links between adaptation and disaster preparedness[62]

The 1990s brought a shift in the way that disasters and their causes are understood, with much more attention being paid to the links between development and disasters.[63] In Latin America, many city governments began to explore this and implement disaster risk reduction measures. This was spurred by the numerous major disasters in the region and supported by decentralization processes and state reforms in many countries.[64] Several countries enacted new legislation that transformed emergency response agencies into national risk reduction systems.[65] Some city governments incorporated disaster risk reduction within development as they changed or adjusted regulatory frameworks, upgraded infrastructure and housing in at-risk informal settlements, and improved urban land-use management with associated zoning and building codes.

This shift by local governments to disaster risk reduction has been driven by different factors. In some countries, it is driven by stronger local democracies (e.g. a shift to elected mayors and city councils) and decentralization (when city governments have a stronger financial base). Sometimes the trigger was a particular disaster event, such as the devastation brought by Hurricane Mitch in Central America (in 1998). Or it was a sequence of events, such as the Popayán earthquake (1983), the Armero mudslide (1985) and other disasters in Colombia. These events encouraged countries, and within these, city and municipal governments, to look more closely at the scale and nature of disaster risk and consider what investments and measures could be put in place to reduce disaster risks. Innovations here include those undertaken by specific local governments, but, as importantly, also those that involve cooperation and coordinated action among groups or associations of local governments. In several countries, there are also national systems to support local authorities and other stakeholders in disaster risk reduction.

These have relevance for climate change adaptation because many are reducing risk levels or exposure to risk for the extreme weather events that climate change is, or is likely, to make more intense, frequent or unpredictable. However, they also have relevance beyond this in that many measures to reduce disaster risk build resilience to a range of hazards. Also, strengthening the capacity to respond rapidly and effectively to disasters and to work with those affected to rebuild their lives, homes and livelihoods will serve all forms of disaster response, whether or not climate change had a role in the disaster.

By 2007 when the IPCC published its Fourth Assessment Report, adaptation to climate change was already taking place in some cities, although these were mostly driven by climate variability. Indeed, societies have a long record of adapting both agriculture and settlements to the impacts of weather and climate through a range of practices that include diversification, water management, disaster risk management and insurance.[66] Yet, climate change poses a new set of risks that may be substantially different from those experienced in the past, and the challenge for adaptation is to ensure that both development needs and the needs imposed by a changing climate (and their link to disaster risk) are met simultaneously.

## TOWARDS EFFECTIVE CITY-BASED CLIMATE CHANGE ADAPTATION STRATEGIES

What can be seen from the examples above are the beginnings of city-based adaptation strategies in *some* cities. These are what might be called the early adapters as well as the early adopters.[67] Getting a more widespread attention by city and municipal governments to climate change adaptation will need clearer and more detailed risk assessments

and a better understanding of how adaptation measures can serve and be integrated within development and disaster risk reduction. It also depends on whether local governments have the knowledge, capacity and willingness to act.

The experiences discussed above indicate that there is an obvious interest in reviewing adaptation responses to potential climate change impacts in different sectors – for instance, in potential damage to infrastructure, to city economies and to public health – and to specific groups that are more vulnerable. There is also an interest in how adaptation responds to the potential social and economic impacts of climate change upon individuals and households, including those relating to displacement and forced migration (and possibly to security). In each of these, there are issues of whose needs are served (and whose are not) by adaptation responses, especially in relation to income level, gender and age. Thus, a whole series of questions might be raised to assess the effectiveness of adaptation policies and practices, including, *inter alia*:

- Do adaptation measures focus on protecting or serving wealthier groups and districts?
- Are those living in informal settlements included and, if so, does this include all informal settlements or only those 'recognized' by the government or those who are more easily accessed?
- Do the particular risks and vulnerabilities women face because of their household, childcare and livelihood responsibilities, or the discrimination they face in getting access to services and finance that can support adaptation, get considered?
- Is the main response in adapting infrastructure to protect what are seen as the most economically important city assets, or to protect city populations with particular attention to those most at risk?

As yet, too few cities have developed coherent adaptation strategies and even fewer have strategies that have begun to have a real influence on public investments and to get needed changes in building and infrastructure standards and land-use management. Most of the literature on climate change adaptation and cities is focusing on what should be done, not on what is being done (because too little is being done). For instance, some city adaptation strategies are justified, in part, by initial figures on the economic assets at risk or by the damage done by extreme weather in the past.[68] In most developed countries and some other countries, revisions to building and infrastructure standards that increase safety margins for likely climate change impacts are being considered. Public health responses to heat waves are being rethought, especially after the limitations revealed by the heat wave in Europe in 2003 – and some cities where heat waves have long been present have strengthened their capacity to reach and serve many of those most at risk. Many local governments have taken measures to manage freshwater resources better because of supply constraints; in many places, these often serve as the first steps for addressing additional water constraints brought by climate change.

By 2007 ... adaptation to climate change was already taking place in some cities, although these were mostly driven by climate variability

Climate change poses a new set of risks ... and the challenge for adaptation is to ensure that both development needs and the needs imposed by a changing climate ... are met simultaneously

Too few cities have developed coherent adaptation strategies and even fewer have strategies that have begun to have a real influence on public investments

| Priority planning area | Preparedness goal | Preparedness actions |
|---|---|---|
| Addressing constraints on freshwater supply | Expand and diversify water supply. | • Develop new groundwater sources.<br>• Construct new surface water reservoirs.<br>• Enhance existing groundwater supplies through aquifer storage and recovery.<br>• Develop advanced wastewater treatment capacity for water reuse. |
| | Reduce demand/improve leak management. | • Increase billing rates for water (possibly with a pricing structure that charges more for high consumption).<br>• Change building codes to require low-flow plumbing fixtures (e.g. shower heads that cut water use).<br>• Provide incentives (e.g. tax breaks, rebates) for switching to more water-efficient processes.<br>• Reduce leakage and unaccounted for water. |
| | Increase drought preparedness. | • Update drought management plans to recognize changing conditions. |
| | Increase public awareness about impacts upon water supplies. | • Provide information on climate change impacts upon water supplies and how residents can reduce water use – for instance, in leaflets sent to water consumers with their bills, newsletters, websites, local newspapers. |
| Storm and floodwater management | Increase capacity to manage storm water. | • Increase capacity of storm water collection systems and ensure their maintenance (which usually includes a need to extend solid waste collection services to all districts).<br>• Modify urban landscaping requirements to reduce storm water runoff.<br>• Preserve ecological buffers (e.g. wetlands). |
| | Reduce property damage from flooding. | • Move or abandon infrastructure in hazardous areas.<br>• Change zoning to discourage or prevent development in flood-hazard areas.<br>• Update building codes to require more flood-resistant structures in floodplains. |
| | Improve early warning systems for storm and flood events. | • Increase the use of climate and weather information in managing risk and events – including the systems that ensure populations at risk get warnings and are able and willing to move temporarily to safe locations when needed.<br>• Update flood maps to reflect changes in risk associated with climate change. |
| Public health | Reduce impacts of extreme heat events. | • Ensure effective early warning systems for extreme heat events with particular attention to reaching those most at risk.<br>• Consider what measures can serve those most at risk with particular attention to those living in heat islands and those most vulnerable to heat stress; can include opening 'cooling' centres during extreme heat events with provision to encourage and support those at risk to move there.<br>• Encourage and promote modifications to the built environment that reduce heat gain, especially the heat-island effect.<br>• Adopt measures within urban centres to reduce urban temperatures, including protection of open space, green space and use of shade trees. |
| | Improve disease surveillance and protection. | • Ensure effective surveillance systems for known diseases and potential diseases moving into the area, and act upon disease prevention and prepare healthcare system to respond.<br>• Increase public education on disease prevention for vector-borne diseases and other diseases that could increase as a result of climate change. |

*Source:* adapted from ICLEI, 2007

There are also issues regarding the social impacts of adaptation measures. For the many cities that need major investments in storm and surface drainage systems, their design and construction have the potential to displace informal settlements – especially those alongside existing drains and rivers – although there are good examples of this being avoided as drainage capacity is increased.[69] Measures to better manage water reservoirs and watersheds might include the displacement of informal settlements – although there are examples showing how this can be avoided.[70] New controls on coastal development to reduce risks from sea-level rise and storms can threaten existing settlements – as they did after the Indian Ocean Tsunami in 2004, although here, too, there are examples of alternative practices that have made coastal settlements more resilient rather than forcing their inhabitants to move.[71]

The first part of this section reviews lessons from the previous section and presents generic lessons for city governments. This is followed by an assessment of adaptation responses in the various economic sectors. The third part takes a closer look at how to build resilience at the local level, while the fourth part reviews the links between

adaptation planning and local governance. The final part of this section presents UN-Habitat's Cities and Climate Change Initiative as an illustration of how international agencies can support climate change adaptation initiatives at the local level.

## Generic lessons for city governments

Table 6.5 provides examples – for city governments – of how climate change adaptation needs to develop preparedness goals and actions for each priority planning area. The table addresses this by focusing on three kinds of impacts that will affect many cities: constraints on freshwater supplies; storm and floodwater management; and impacts upon public health, such as extreme heat and higher risks from diseases spread by certain vectors. The diversity of needed actions also highlights how many different departments of city government need to be involved and to be able to work together.

Drawing on the examples provided in this chapter, it is possible to identify certain key components for developing city adaptation strategies:

- *Build commitment among different stakeholders.* This is an essential first stage. There is a need to get an official recognition by and within cities that climate change impacts need to be considered. This has to include building knowledge and commitment within the different departments of local government, many of whom may see climate change adaptation as drawing resources or attention away from their sectors.[72] Without the commitment of a range of individuals, groups and sectors, it is impossible to address the multiple cross-cutting aspects of adaptation. It is also clear from specific examples of cities developing adaptation strategies that particular individuals had important roles in initiating this – for instance, a mayor or a senior civil servant – although, of course, its success depends on others responding positively.

- *Develop or expand the information base on current conditions.* An important part of this is considering the impact of past extreme weather and other disasters in each city or municipality. This should seek as much detail as possible, ensuring the inclusion of 'small disasters' (disasters that do not get included in international disaster databases), and could draw on the DesInventar methodology developed in Latin America and now widely applied elsewhere, which looks more intensively at disasters in any locality and includes 'small disasters'.[73]

- *Initiate risk/vulnerability assessments for the city.* Such assessments should be built up from community and district assessments (and from global and national projections about climate change impacts). In many cities, this can and should include the kinds of community-driven assessments undertaken by the Philippines Homeless People's Federation that were described earlier (see Box 6.3). It may be seen as laborious and time consuming; but engagements with women and men in settlements and districts that are affected most by extreme weather can produce a more detailed and nuanced understanding of risk and vulnerability – and, thus, a better basis to understand what adaptation is needed – as illustrated by the experiences from the urban communities of Mansión del Sapo and Maternillo in Puerto Rico.[74] Such an assessment should include as much geographic detail as possible. Furthermore, it needs to link hazard maps with details of what is currently located within the hazardous zones – including identifying population groups or settlements most at risk and activities that may pose particular risks (e.g. water treatment plants located in areas at risk from flooding). It is also important that city assessment can draw data from global and national projections about climate change impacts. At present, many such projections are insufficient and imprecise, or at times even contradictory, thus impeding local action. It is, for example, difficult for local governments to plan for appropriate future land use if projections of climate change implications are weak or contradictory.[75]

- *Assess sector-specific vulnerability and responses.* Risks from climate change vary greatly between sectors – and the responsibilities for addressing them vary greatly across the different administrative divisions and departments that make up local and extra-local governments. Adapting to climate change does not only depend on all the key sectors and departments seeing the relevance of actions within their jurisdictions and areas of competence. It is also essential that these departments take appropriate action. However, it is difficult to get all key spending and investing sectors and departments to do this – and the department or division with responsibility for directing attention towards climate change adaptation rarely has more than an advisory role and, moreover, usually has a very limited budget of its own for investment. It needs to convince the departments concerned with public works, public health, housing, solid waste management, schools, etc. to engage with adaptation. The adaptation strategies of Durban (South Africa), London (UK) and Melbourne (Australia)[76] sought to make clear how the main climate threats are linked to specific sectoral responsibilities, and this has made the responsibilities for adaptation much clearer. Agencies responsible for disaster preparedness response have particular importance – although these often need to broaden their focus beyond response to disaster preparedness (and disaster prevention) and all that this implies for their engagement with city- and community-level housing and infrastructure investments. The agencies responsible for disaster response will often see the relevance and importance of their engagement in this; but they too often lack influence and resources, especially in relation to the measures that avoid or prevent disasters.[77] Utility companies, different government departments, and the private sector will all have key roles, too, in addressing specific vulnerabilities.

- *Develop strategic plans for the city as a whole and its surrounds.*[78] Urban authorities should have the key role in developing strategic plans for the city as a whole; but this needs to be done in association with other stakeholders. These strategic plans are necessary to ensure complementarities and coordination between different activities in the urban area. Several of the most effective strategies described above have included strategic adaptation plans. This has been an important part of the process in both Cape Town and Durban (South Africa) – however, because of the commitment from the municipality's Environmental Planning and Climate Protection Department, the plans have moved closer to implementation in Durban. For many major cities, the strategic plans need to encompass the larger region on whose resources and ecosystem services the city depends. This is more easily done when the area under the jurisdiction of the city government includes this larger region; the added complexities politically and institutionally where this is not so are obvious.

- *Support local responses to climate change.* Many of the key adaptations to climate change will require individual and collective action at the community level to build resilience and prevent harmful effects. It is widely accepted that much adaptation will be undertaken

**Risk/vulnerability assessments for the city ... should be built up from community and district assessments (and from global and national projections about climate change impacts)**

**Urban authorities should have the key role in developing strategic plans for the city as a whole; but this needs to be done in association with other stakeholders**

**Table 6.6**

Examples of
specific adaptation
interventions by sector

| Sector | Adaptation option/ strategy | Underlying policy framework | Key constraints to implementation | Key opportunities to implementation |
|---|---|---|---|---|
| Water | Expanded rainwater harvesting; water storage and conservation techniques; water reuse; desalination; water-use and irrigation efficiency. | National water policies and integrated water resources management; water-related hazards management. | Financial and human resources; physical barriers. | Integrated water resources management; synergies with other sectors. |
| Infrastructure and settlements | Relocation; sea walls and storm surge barriers; dune reinforcement; land acquisition and creation of marshlands/wetlands as buffer against sea-level rise and flooding; protection of existing natural barriers. | Standards and regulations that integrate climate change considerations within design; land-use policies; building codes; insurance. | Financial and technological barriers; availability of relocation space. | Integrated policies and management; synergies with sustainable development goals. |
| Human health | Heat–health action plans; emergency medical services; improved climate-sensitive disease surveillance and control; safe water and improved sanitation. | Public health policies that recognize climate risk; strengthened health services; regional and international cooperation. | Limits to human tolerance (vulnerable groups); knowledge limitations; financial capacity. | Upgraded health services; improved quality of life. |
| Tourism | Diversification of tourism attractions and revenues; shifting ski slopes to higher altitudes and glaciers; artificial snow-making. | Integrated planning (e.g. carrying capacity; linkages with other sectors); financial incentives (e.g. subsidies and tax credits). | Appeal/marketing of new attractions; financial and logistical challenges; potential adverse impact upon other sectors (e.g. artificial snow-making may increase energy use). | Revenues from 'new' attractions; involvement of wider group of stakeholders. |
| Transport | Realignment/relocation; design standards and planning for roads, rail and other infrastructure to cope with warming and drainage. | Integrating climate change considerations within national transport policy; investment in research and development for special situations (e.g. permafrost areas). | Financial and technological barriers; availability of less vulnerable routes. | Improved technologies and integration with key sectors (e.g. energy). |
| Energy | Strengthening of overhead transmission and distribution infrastructure; underground cabling for utilities; energy efficiency; use of renewable sources; reduced dependence on single sources of energy; increased efficiency. | National energy policies, regulations, and fiscal and financial incentives to encourage use of alternative sources; incorporating climate change within design standards. | Access to viable alternatives; financial and technological barriers; acceptance of new technologies. | Stimulation of new technologies; use of local resources. |

*Source:* based on Parry et al, 2007b, Table SPM4

What is required ...
is to include
climate-proofing of
infrastructure for
future climate
change

incrementally by individuals and households, and that communities and local organizations also have important roles in this. There are many examples of community-driven 'slum' upgrading that greatly reduced environmental health risks; if served with appropriate information and support, these can include attention to climate change risks (which in the next few decades are mostly increased risk levels from hazards already present). The above examples from the Philippines (see Box 6.3) show that community organizations have the capacity to build resilience and identify appropriate short- and long-term responses to climate events – if they are adequately supported by local authorities. This latter point is an important one; as was noted above, effective climate change strategies require a partnership approach – involving households and communities, but also the various levels of government and other partners, including international organizations.

## Adaptation responses to potential impacts in different economic sectors

It is clear from the discussion above that climate change adaptation action is needed in almost all sectors; Table 6.6, drawn from the IPCC, provides some examples of the kinds of specific adaptation interventions needed by some of the key sectors. Although this table does not highlight this, much of what is listed in the adaptation option/strategy will fall to local government to implement, even if it needs resources and policy and regulation frameworks from higher levels of government.

With regard to infrastructure, most fields of infrastructure management already incorporate measures to cope with climate variability and extreme events – including water, sanitation, transport and energy management. What is required, in addition, is to include climate-proofing of infrastructure for future climate change.[79] Adaptation to climate change will typically involve increases in reserve margins and other kinds of back-up capacity, and attention to system designs that allow adaptation and modifications without major redesigns and that can accommodate more extreme conditions for operations.[80] Infrastructural adaptation can take one of several forms: building retrofitting and strengthening; lifeline infrastructure strengthening; and hazard modification.[81] In Georgetown (Guyana), detailed cost-benefit analyses have been used to assess the most important and cost-effective infrastructural responses to climate change. These have been complemented by a more qualitative approach that seeks to identify costs and benefits from a non-monetary perspective.[82]

Infrastructure can be adapted in a variety of ways, not all of which require complicated technological solutions. Planned adaptation to sea-level rise can involve retreat, accommodation or infrastructural solutions (as is illustrated in Figure 6.3). However, in practical terms, there are strong social, political and economic reasons for protecting land that has already been developed in densely settled urban areas.

There are a growing number of examples of urban areas that have adopted infrastructural solutions to address particular aspects of climate change (although it should be noted that some of the examples provided below, such as that of Venice, are related to natural processes that would require attention even without the added risk brought about by climate change):

- *Responses to flooding.* In Venice (Italy), the *Modulo Sperimentale Elettromeccanico* ('Experimental Electromechanical Module') involves the construction of 79 gates at three lagoon inlets: when waters rise 1.1m above 'normal', air will be injected into these hollow gates, causing them to rise and preventing the city from flooding. In many developing countries, few projects have been implemented; although proposed strategies exist for Nam Dinh Province (Viet Nam), including building reservoirs to retain floodwater, strengthening dyke systems to resist higher flood levels, and constructing emergency spillways along dykes for selective filling of flood retention basins.[83]
- *Water conservation.* Singapore has a Four National Taps Strategy to ensure the future supply of water. The first 'tap' is the supply of water from local catchments based on an integrated system of 15 reservoirs and an extensive drainage system to channel water into these; the second is imported water from Johor (Malaysia); the third is high-grade reclaimed water; and the fourth is desalinated water.[84]
- *Reducing urban temperatures.* 'Cool roofs' and 'porous pavements' are being used in Vancouver (Canada) to reduce the urban heat island. These are covered with light-coloured water sealants that reflect and radiate more heat than dark surfaces, thus reducing the need for mechanical cooling systems.[85]

The World Bank and the Asian Development Bank have been developing their capacity to design and deliver infrastructure that will meet the needs of climate change.[86] Investment in infrastructure can support sustainable socio-economic development, and can also facilitate reconstruction and recovery.

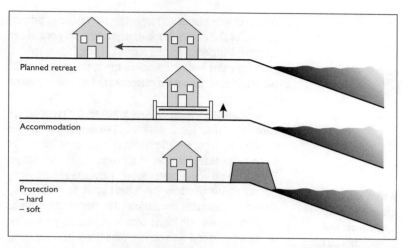

**Figure 6.3**

**Adapting infrastructure to sea-level rise**

*Source:* Parry et al, 2009, p63

However, infrastructural investment is fraught with challenges. Large-scale interventions of this type have frequently failed to take into account the particular social and economic context of the areas in which they are implemented, with negative social consequences, including forced relocations[87] and provision of services in a way that fails to meet the needs of low-income groups.

## Building resilience

The many measures by which low-income households and communities try to cope with extreme weather, and their importance in reducing risks, has been discussed at length already. Many of these measures would fit with the definition of resilience given by the IPCC: the 'ability of a social or ecological system to absorb disturbances while retaining the same basic structure and ways of functioning, the capacity for self-organisation, and the capacity to adapt to stress and change'.[88] Indeed, the many examples of simple pragmatic ways of coping with floods such as having shelves high up walls above anticipated flood levels, and furniture (often raised on bricks) on which the residents can sit or sleep could be included in this definition.

There are important components of resilience beyond 'hard' infrastructure – in part because hard infrastructure will be unable to remove or greatly reduce many risks, especially if the governments of the world do not reach agreement on needed emissions reduction soon.[89] Thus, resilience is also a capacity to live in hazardous, changing and uncertain environments[90] and through assets, social networks and partnerships to have the needed capacity to (in the words of the IPCC) 'absorb disturbances while retaining the same basic structure and ways of functioning'.[91]

Perhaps the building of resilience should be understood as a way of enabling not only coping with added shocks and stresses, but also addressing the myriad challenges that constrain lives and livelihoods. Thus, a key part of building resilience is facilitating poverty reduction and more general improvements to the quality of human lives.[92]

Many interventions being undertaken in urban areas around the world – by local, municipal, national and international stakeholders – contribute to building this resilience through improving housing, infrastructure and services, particularly for the urban poor. Addressing the challenges of

Planned adaptation to sea-level rise can involve retreat, accommodation or infrastructural solutions

The building of resilience should be understood as a way of enabling not only coping with added shocks and stresses, but also addressing the myriad challenges that constrain lives and livelihoods climate change adaptation may not be the explicit or primary purpose of these activities; but, in practice, they provide an essential foundation for the process of adaptation. Indeed, for many cities in developing countries, this is likely to be the single most important component of an overall adaptation strategy.

In addition, many of these urban areas already experience an 'adaptation deficit'. The infrastructure is insufficient to cope with present climatic conditions – let alone those that will arise as a result of climate change. Existing storm drains, water supply networks and transport infrastructure may have been developed decades ago to serve a much smaller population – and before these can be 'adapted' to deal with future climate threats, they must first be upgraded to deal with current conditions. In this respect, it is helpful to consider Stern's definition of adaptation: 'development in a more hostile climate'.[93] Many of the adaptation needs for urban areas in developing countries are based on the need for development that takes a changing climate into account.

A wide range of urban improvement programmes and projects around the world can be seen as contributing to reducing the infrastructure deficit and increasing urban resilience to climate change. Global initiatives to improve urban housing (such as UN-Habitat's World Urban Campaign[94] and its predecessor, the Global Campaign for Secure Tenure) and provide appropriate plans for urban development (such as the City Development Strategies[95] promoted by the Cities Alliance) can form the basis for building urban resilience. However, these large-scale responses require careful analysis to ensure that they remain genuinely pro-poor and meet the needs of the most vulnerable urban residents.[96]

Many communities are already involved in activities that will build the resilience of individuals and households. For many low-income urban residents, savings schemes[97] form the basis for this resilience. The practice of saving regularly with such schemes has both instrumental benefits (the ability to access funds when necessary) and organizational benefits (the relationships of trust built up around small savings groups are central to identifying solutions to larger problems). Small-scale loans repaid over very short time periods provide much needed capital for livelihood activities. They can develop into small-scale loans to help improve or extend housing. Furthermore, organized savings groups have also demonstrated the ability to negotiate for and acquire new land sites that are not vulnerable to climate threats, such as flooding and landslides, upon which to build secure housing and thus provide protection against short- and long-term climatic threats.

Insurance policies for houses, possessions and businesses contribute to resilience where they provide compensation to those whose homes, possessions and businesses have been damaged or destroyed. They could also contribute to building resilience by including financial incentives (such as reduced premiums) for those who have reduced their risks. However, this will not serve those unable to access the formal housing market and/or those who cannot afford insurance. For urban centres in develop-ing countries, this means most of the population and most enterprises. Insurance companies will not offer insurance coverage to cities or to households and businesses on city sites at high risk from climate change because of inappropriate locations or deficits in infrastructure.

## Adaptation planning and local governance

Drawing on the descriptions of household, community-based and local government actions for adaptation in previous sections, this section considers the relative roles for community-based adaptation and for adaptation planning and governance. These tend to operate at different scales (although often with cross-scale linkages) and to involve a distinct balance between individual and collective action, and between behavioural and structural (in terms of both housing and infrastructure) responses. Yet, these frameworks should be viewed as complementary rather than as mutually exclusive. As is evident from a wide range of studies on addressing urban environmental challenges more broadly, there is a need to link structural and behavioural responses. For example, individuals and community groups in cities with limited investment capacities may be best placed to devise the most appropriate sanitation solutions for themselves and their neighbourhood – but these are of little use without larger investments at the scale of the town or city to ensure convenient and easily accessed water supplies for personal hygiene and appropriate provision for the removal of human waste.[98] Conversely, large-scale infrastructural developments to improve drainage and reduce flooding require the knowledge and expertise of engineers; but these interventions will have little value if they do not take into account the needs of those in informal settlements and the social and behavioural norms and expectations of urban residents. Drainage systems also have their capacity to protect cities from flooding much reduced if they are not maintained (which may need community support) and protected from encroachment.

In urban areas of developed countries, citizens take for granted that a range of local structures and organizations provide protection from environmental hazards and help to create resilience to potential disasters. It is assumed that these will also provide for adaptation to climate change. Here, residents do not need to organize themselves to clear drains and collect solid wastes; these are tasks that local authorities do or organize. These urban areas have infrastructure and services that protect them from environmental hazards (through, for instance, the provision of safe water supplies, sewers and drainage) or help them to cope when illness or injury occurs (e.g. through well-managed healthcare and emergency service systems).[99] In urban areas in developing countries, these facilities and services are frequently absent or they serve only a proportion of the population. Local governments lacking capacity and funding, and with large infrastructure and service deficits need the contributions that community-based organizations can bring. There are also the exceptional local governments that have shown how[100] – even with limited resources – effective governance and planning can work towards facilitating urban adaptation.

*A wide range of urban improvement programmes and projects around the world can be seen as contributing to reducing the infrastructure deficit and increasing urban resilience to climate change*

*For many low-income urban residents, savings schemes form the basis for ... resilience*

*Exceptional local governments ... have shown how – even with limited resources – effective governance and planning can work towards facilitating urban adaptation*

Planning for adaptation can take place at a range of scales. As described in the previous section, some urban areas have developed plans for adaptation at both the city and sectoral levels as a key component of their preparation for climate change. The examples of Cape Town and Durban (South Africa) showed how large urban areas can develop municipal-level plans for adaptation that take into account a range of social and environmental challenges.[101] These provide the framework within which local government departments, the private sector, civil society and individuals can prepare and implement their contributions to strategies for adaptation within development or investment plans. There are other examples where city governments have successfully avoided large-scale settlement by low-income populations on dangerous sites that would be at risk from climate change. In Manizales (Colombia), local authorities, universities, NGOs and communities worked together to develop programmes aimed not only at reducing risks, but also at improving the living standards of the poor and at protecting fragile ecological areas. Households were moved off the most dangerous sites, but rehoused nearby, and most of the former housing sites were converted into eco-parks with strong environmental-education components.[102] In Ilo (Peru), long-term engagement by consecutive democratically elected mayors have improved water supply, sanitation, electricity provision, waste collection and public space. Despite the population increasing fivefold between 1960 and 2000, no land invasion or occupation of risk-prone areas by low-income groups looking for housing has taken place, as local authorities have implemented programmes to accommodate this growth in a sustainable way.[103]

There are also the examples of resident groups in cities that organized to influence the future development of their city along more ecologically sustainable paths. These include some where climate change adaptation has been important – as in the city of Tatabánya (Hungary), some 50km from the capital city of Budapest, which offers an example of how community members can be an important driver and resource in climate adaptation (see Box 6.6).[104] Participatory budgeting has become one of the best known and most widely applied forms of citizen engagement in the plans and priorities of city governments,[105] and in some cities, this engagement has included a strong focus on environmental issues.[106]

In London (UK) (see Box 6.5) and Bangkok (Thailand) (see Table 6.2), the approach to climate change adaptation planning has been to identify particular sectors that are 'at risk' and develop plans to address each of these, and with the delegation of responsibility to appropriate agencies. This requires an effective system of oversight and control, and relies on these agents having sufficient financial and technical capacity to make the appropriate investments and interventions. In London, strategies are being developed to address the three key climate risks affecting the city – flooding, drought and overheating. Bangkok will be vulnerable to a similar set of risks, and the Bangkok Metropolitan Administration has proposed adaptation measures to be taken by the community infrastructure sector, the business and commercial sector, and the general population.

---

**Box 6.6 Citizen-driven city adaptation in Tatabánya, Hungary**

Tatabánya is a former mining and industrial town that has approximately 72,000 residents and was known for its high levels of pollution. The residents have formed three groups, each involved in promoting local sustainability:

1   The focus of the 'inhabitants group' is to develop a new vision for the future of the city. They serve in a representative capacity in public decision-making and through their efforts have helped to promote communication between residents and public officials by ensuring that local interests are known.
2   The 'local council of pupils group' is made up of student representatives who engage in a variety of tasks, including participating in local decision-making.
3   The 'local climate group' is comprised of individuals from all walks of life, including students, pensioners, doctors, nurses, teachers, engineers, scientists, public officials, heads of companies and inhabitants. Among their many accomplishments, they have implemented a heat and ultraviolet light alert programme, organized teams to assist in the development of a local climate strategy, initiated a call for tenders on energy-efficient housing, established emissions reduction targets, and implemented educational and information programmes.

What is perhaps most noteworthy of the Tatabánya experience is the commitment of its residents to their city and to addressing both immediate issues and good environmental performance in relation to global systems.

*Sources: Moravcsik and Botos, 2007; Carmin and Zhang, 2009*

---

Urban areas on Mexico's Yucatan coast have been involved in a process of social learning for climate-proofing, based on bringing together a range of stakeholders.[107] This involves a three-stage process of consciousness, institutionalization and implementation:

* *Consciousness* is the process of reflection on established norms and practices with the aim of generating new visions.
* *Institutionalization* is the process of changing stakeholders in urban governance to facilitate new norms and practices.
* *Implementation* is the capacity to enact new practices and activities.

Social learning by civil society is seen as an essential prerequisite to effective adaptation planning. This is particularly so in a context where the government is constrained by a highly competitive and dynamic political culture, with politicians and officials coming in and out of office frequently and seldom building on past knowledge or initiatives.

Urban adaptation planning is therefore intrinsically linked with local governance. A study of ten Asian cities found that preparation for climate change was strongly linked with climate-resilient urban governance.[108] This includes decentralization and autonomy, accountability and transparency, responsiveness and flexibility, participation and inclusion, and experience and support. Urban governance systems that exhibit these characteristics are better able to build resilience through having more effective financial and technical management capacities in 'climate-sensitive' sectors such as waste, water and disaster management. Responsiveness and flexibility are crucial, given the

There are ... examples where city governments have successfully avoided large-scale settlement by low-income populations on dangerous sites that would be at risk from climate change

Urban adaptation planning is ... intrinsically linked with local governance

**Table 6.7**

Adaptation to extreme weather: The role of city/municipal governments

| Role for city/municipal government | Long-term protection | Pre-disaster damage limitation | Immediate post-disaster response | Rebuilding |
|---|---|---|---|---|
| **Built environment** | | | | |
| Building codes | High | | High* | High |
| Land-use regulations and property registration | High | Some | | High |
| Public building construction and maintenance | High | Some | | High |
| Urban planning (including zoning and development controls) | High | | High* | High |
| **Infrastructure** | | | | |
| Piped water, including treatment | High | Some | High | High |
| Sanitation | High | Some | High | High |
| Drainage | High | High** | High | High |
| Roads, bridges, pavements | High | | High | High |
| Electricity | High | Some? | High | High |
| Solid waste disposal facilities | High | Some? | | High |
| Wastewater treatment | High | | | High |
| **Services** | | | | |
| Fire protection | High | Some | High | Some |
| Public order/police/early warning | Medium | High | High | Some |
| Solid waste collection | High | High** | High | High |
| Schools | Medium | Medium | | |
| Healthcare/public health/environmental health/ambulances | Medium | Medium | High | High |
| Public transport | Medium | High | High | High |
| Social welfare (includes provision for childcare and old-age care) | Medium | High | High | High |
| Disaster response (over and above those listed above) | | | High | High |

Notes: * It is important that these do not inhibit rapid responses.
** Clearing/de-silting of drains and ensuring collection of solid wastes has particular importance just prior to extreme rainfall; many cities face serious flooding from extreme rainfall that is expected (e.g. the monsoon rains) and this is often caused or exacerbated by the failure to keep storm and surface drains in good order.

Source: Satterthwaite et al, 2009c

The involvement of the poor and marginalized groups in decision-making ... is key to improving the living conditions of these groups

limited predictability of the consequences of climate change. At the same time, the involvement of the poor and marginalized groups in decision-making, monitoring and evaluation is key to improving the living conditions of these groups. In the context of Mexico, it has been argued that the quality of the governance process is the most important component for enabling climate change adaptation.[109] The need to adapt to climate change and the need to adapt governance systems to be more responsive and effective are therefore closely linked.

Table 6.7 highlights the range of roles that city or municipal governments have in climate change adaptation. It is a reminder of how much adaptation depends on action within many different sectors or parts of local government.[110] This means that adaptation planning needs support not only from public works departments and from development planning and development control, but also from the departments dealing with environmental health, public health, and social and community services (including transport and public space management, and emergency services), as well as those dealing with finance and disaster management.[111] Adaptation to climate change is often taken to mean protection against likely changes (e.g. better drainage systems or coastal defences); but it should also involve three other components listed in Table 6.7: damage limitation measures taken just before an extreme event (that has the potential to cause a disaster), immediate post-extreme event response, and rebuilding. There is also a range of measures that local governments can take that support resilience at the household and community levels. This includes slum and squatter upgrading and schemes that help those with limited incomes to afford to buy, build or

There is ... a range of measures that local governments can take that support resilience at the household and community levels

rent safer, better served accommodation (although to be effective for adaptation, these need to be guided by climate change risk assessments and appropriate responses). It also includes measures to strengthen or support livelihoods and food security for low-income groups. Urban food security depends on households being able to grow or afford food within other needs that have to be purchased.[112] The extent of food insecurity among low-income households in urban areas is given too little consideration,[113] which also means that their vulnerability to the impacts of climate change on agriculture is probably underestimated.

Measures to support resilience include more effective and accessible healthcare services and emergency response services that are prepared for the scale and nature of climate-related (and other) potential disaster risks. It also includes an early warning system that actually reaches all those in need with appropriate information, combined with knowledge of what to do and where to go – and provision to ensure that all can move to identified safe places, when and where needed. It also means a capacity to respond after disasters – as in the measures listed in Table 6.1 for immediate post-disaster response and rebuilding. Within this, there is a clear need for all measures taken to address gender-specific issues of risk management and adaptation, from shelter management to empowerment, and inclusion of women in decision-making at all scales for stronger emphasis on long-term and risk-averse initiatives.

## UN-Habitat's Cities and Climate Change Initiative[114]

The UN-Habitat Cities and Climate Change Initiative

| City | Proposed activities |
|------|---------------------|
| Esmeraldas | • Zoning of riverbanks and preparation of a participatory land-use plan.<br>• Preparation of a risk management plan.<br>• Implementation of an environmental management plan for the Teaone River (including solid waste management and riverside rehabilitation through reforestation). |
| Kampala | • Establishment of national and city climate change network.<br>• Increasing awareness and capacities of Kampala City Council.<br>• Increasing synergies between national and local climate change policies and programmes. |
| Maputo | • Strengthening disaster risk preparedness at the community level.<br>• Localizing the national climate change adaptation plan.<br>• Promoting policy dialogue to strengthen the government response capacity to floods.<br>• Education and public awareness campaigns to create climate change awareness.<br>• Capacity-building with local government and a wider range of partners. |
| Sorsogon | • Development of knowledge products for sharing and cross-fertilization of ideas.<br>• Demonstration of innovative technologies for climate-resilient human settlements, particularly in low-lying urban coastal areas.<br>• Development of the capacity of the city government.<br>• Advocacy, awareness-raising and partnership building on climate change with stakeholders and the general public. |

*Source:* UN-Habitat, 2008a

**Table 6.8**

Proposed and planned activities in the pilot cities of the Cities and Climate Change Initiative

provides an illustration of how international agencies can support local adaptation action. It aims to strengthen the climate change responses of cities and local governments. The initiative is currently being piloted in four cities – Esmeraldas (Ecuador), Kampala (Uganda), Maputo (Mozambique) and Sorsogon City (the Philippines).[115] This initiative brings together local and national governments, academia, NGOs and international organizations to alert cities to the actions that they can take to respond to climate change. Key programme components that are being encouraged for adaptation to climate change include advocacy and policy change, the development and use of toolkits, and knowledge management and dissemination. An important component in this project is the creation of a global network of cities working on adaptation issues, among whom knowledge can be generated and shared.

The four pilot cities in this initiative face a range of challenges related to climate change. Sorsogon City, Maputo and Esmeraldas are all coastal cities affected by frequent flooding and at risk from sea-level rise. In addition, Sorsogon is at risk from tropical cyclones; Esmeraldas has many households living on hillsides and riverbanks; and the protective mangroves around Maputo are disappearing. Kampala is located inland, but is also affected by flooding and the degradation of fragile hill slopes. In all cases, these challenges are compounded by inappropriate management of natural resources and inadequate urban infrastructure.

Various adaptation responses are being planned and implemented in these cities in association with the Cities and Climate Change Initiative (see Table 6.8). Some of these are associated with broader environmental management projects which will simultaneously improve the resilience of communities and the urban area to climate change: the reconstruction of the National Disaster Management Institute in Mozambique will help to improve disaster risk reduction in Maputo and elsewhere in the country; and the flood prevention programme for the Teaone River in Esmeraldas will help to reduce flooding. Other activities involve building networks and capacity: in Kampala, it is proposed to establish a climate change network of various stakeholders addressing climate change, whereas in Maputo it is proposed to support collaboration between the local

government and a range of other partners. Strengthening the capacity of local authorities to address climate change is also a key activity in all four cities – both in terms of awareness of the issues and potential responses to these.

## FINANCING ADAPTATION

In terms of financing for climate change adaptation, there are the two main issues that have to be addressed up front:

1  Will funds will be available to cover the cost of adaptation for urban areas?
2  Is there local capacity in place to use such funds in such a manner that the needed adaptation can take place?

International debates and discussions have tended to focus on the first of these, not the second. Funding for adaptation in developing countries comes (and will come) primarily from two main sources: the dedicated climate change funds available under the United Nations Framework Convention on Climate Change (UNFCCC) (see Box 2.2) and through overseas development assistance. As noted in Chapter 2, the issue of funding has been high on the agenda in international climate negotiations. Ideally, there is wide agreement that international funding for climate change adaptation should be adequate to the task at hand, and should explicitly allocate a fair share of resources to urban settlements. However, in practice, the funds available are, at present, inadequate; furthermore, these funds do not target urban settlements.[116] Moreover, the first consistent approach to identifying adaptation priorities, the NAPAs, generally missed urban priorities. So far, urban priorities also seem to be absent from the funding allocated through the Adaptation Fund.[117]

Adaptation to climate change has become an important priority in the international climate change negotiations during recent years. At the latest meeting of the COP (2010, Cancún, Mexico), Parties reiterated the importance of adaptation and agreed that:

> *adaptation is a challenge faced by all Parties,*
> *and that enhanced action and international*

Funding for adaptation in developing countries comes ... primarily from ... the dedicated climate change funds available under the ...UNFCCC ... and through overseas development assistance

In practice, the funds available are, at present, inadequate; furthermore, these funds do not target urban settlements

*cooperation on adaptation is urgently required to enable and support the implementation of adaptation actions aimed at reducing vulnerability and building resilience in developing country Parties, taking into account the urgent and immediate needs of those developing countries that are particularly vulnerable.*[118]

The Cancún Agreements further reaffirm the commitment made by developed countries to expand the scale of funding available for adaptation during COP-15, including through the US$100 billion which is to be mobilized by 2020 to support action in developing countries. However, the ambiguity on where the increased funding will actually come from remains unresolved. Furthermore, the Cancún Adaptation Framework was established to further enhance action on adaptation.

As noted in Box 2.2, international funding for adaptation through the UNFCCC includes the Special Climate Change Fund, the Least Developed Countries Fund and the Adaptation Fund. The Adaptation Fund was established to finance adaptation projects and programmes in developing countries, with particular attention to those countries that are particularly at risk from the adverse effects of climate change. It is likely to have particular importance because part of its funding comes from a levy on the project activities of the Clean Development Mechanism, and this should give it a considerable and guaranteed source of funding. Thus, unlike the other funds, it is not reliant on negotiating funding from donor agencies. It also has a governance structure in which developing countries have more influence; its independent board has representation from each of the major regions, as well as special seats for the least developed countries and the small island developing states.[119]

A review of financing arrangements for adaptation[120] suggested that there is an opportunity for complementarity between this Adaptation Fund and overseas development assistance. For example, the review suggested that overseas development assistance can help to focus on the drivers of vulnerability that are associated with weak institutional capacity, while the Adaptation Fund supports developing countries' broader climate risk management strategies. It also suggested that the bilateral and multilateral donor agencies can help to build the necessary local and national institutional capacity to receive and make good use of support from the Adaptation Fund. However, this also presupposes a capacity among such agencies to work with civil society and local governments, which is often not present.

This mix of funding might also overcome the contentious issue of the boundary between climate change adaptation and development. Development should certainly include 'adaptation' to all disaster and environmental health risks, including those to which climate change does not contribute or only partially contributes. The large climate change adaptation deficit in most developing countries is also a development deficit. This raises the questions of whether funding for climate change adaptation should include funding for removing this development deficit (which also proves to be an adaptation deficit) or not. In

theory, the governments of developed countries that contribute funding to adaptation will want this to be separated from aid budgets and focus specifically on climate change adaptation. Yet, how can a city adapt to climate change if half of its population live in informal settlements that lack the most rudimentary infrastructure and services? And how can funding for adaptation be managed if there is one funding stream and set of agencies for putting in place needed infrastructure and another for adapting this infrastructure?

Attention should also be paid to the relative costs of mitigation and adaptation. The estimates for the costs of mitigation (achieving the needed reductions in global GHG emissions) appear very high. Many estimates for the costs of adaptation – including those produced by the UNFCCC (see Table 6.9) – are much lower. Based on this, it could be argued that mitigation costs can be reduced by funding for adaptation that allows a less rapid reduction in global GHG emissions. However, if the estimates for the costs of adaptation are far too low and consideration is given to the difficulties in overcoming the lack of adaptive capacity within local governments, it changes the balance. A more realistic assessment of the incapacity and unwillingness of most national, city and municipal governments within developing countries to actually implement needed adaptation measures means that mitigation should receive a much higher priority. In the end, the discussion boils down to the willingness of governments in developed countries (and some industrialized developing countries) to reduce the carbon-intensive consumption patterns of their citizens[121] to benefit others – especially future generations and those who are most at risk and most vulnerable to climate change (most of whom live in developing countries).

This section focuses on the costs of adapting infrastructure to the potential future impacts of climate change. It also includes a discussion of the very large costs involved in remedying the large deficits in infrastructure in urban areas in most developing countries – for instance, the lack of storm and surface drains, paved roads and footpaths, and reliable piped water supplies. Remedying these deficits may not be considered as climate change adaptation; but without remedying these deficits, it is not possible to build resilience to most climate change impacts. Also, if the costs of remedying these infrastructure deficits are considered as part of climate change adaptation, the costs of adaptation increase very considerably.

It is, however, important to note that the discussion below does not include a discussion of many institutional and social adaptation costs. Nor does the discussion touch on the issue of residual damage: the cost incurred in an increasing number of locations that become permanently beyond adaptation – because adaptation is considered too expensive or technically unfeasible. Some such challenges are addressed in the next section.[122]

## The costs of adaptation

The basis for accurate national and global estimates for the costs of adaptation does not exist. The costs of adaptation

---

**The Cancún Agreements ... reaffirm the commitment made by developed countries to expand the scale of funding available for adaptation during COP-15**

**The large climate change adaptation deficit in most developing countries is also a development deficit**

**If the costs of remedying ... infrastructure deficits are considered as part of climate change adaptation, the costs of adaptation increase very considerably**

are so local, so specific to location and to existing levels of housing and infrastructure quality and governance capacity – and there are few examples of locally determined adaptation costs upon which to base national or global estimates. Cost estimates are also greatly influenced by the form that adaptation takes – for instance, what safety margins are built into new infrastructure and what balance is achieved between protection and accommodation.

Most global estimates of the cost of adaptation are based on the costs of climate-related disasters; but these are known to form a very inadequate basis for this. One reason for this is that the cost estimates of climate-related disasters do not include most disasters because they have a very high threshold for a damaging event to be included in their considerations.[123] Where careful local or national reviews of disaster events and their impacts have been carried out, these highlight the very large underestimates, especially with regard to deaths and serious injuries.[124] There is also the problem of assigning costs to disasters based on the value of the properties destroyed – so a disaster that destroys the homes and possessions of hundreds or thousands of households does not appear 'serious' because the monetary value of their homes in informal settlements is low and they had no insurance. It is odd, indeed, to base any estimates for adaptation costs on what insurance companies have had to pay out for extreme weather disasters if almost all those affected by these disasters do not have insurance.

Most estimates for the costs of adaptation that are relevant to urban areas are based on the costs of adapting infrastructure, and thus include roads (of all sizes, from highways to streets and lanes) and bridges, railways, airports, ports, electric power systems, telecommunications, water, sewerage, and drainage/wastewater management systems. The definition of infrastructure is sometimes broadened to include services which make economic and social activities possible – so it would include services such as public transport, healthcare, education and emergency services (which collectively are sometimes termed social infrastructure). A proportion of such infrastructure is outside urban boundaries, although almost all of it is important to the functioning of urban economies. There is also all the ambiguity in what gets included under infrastructure – including housing (sometimes included, sometimes excluded) and the institutions that operate and manage infrastructure.

The UNFCCC secretariat has made estimates for the costs of adapting infrastructure (see Table 6.9); but it does not specify what is included in the term. It is also unclear as to whether housing and the institutions needed to operate and manage infrastructure are included in its estimates.[125] It might be assumed that estimates for the costs of extreme events that draw on records from insurance companies would include housing; but only a very small proportion of households in developing countries have disaster insurance (and thus have their costs included in 'costs' based on insurance claims). The destruction of, or damage to, housing is one of the most common and most serious impacts of many extreme weather events, especially in many developing countries. The damage to, or loss of, housing is usually concentrated among low-income groups and this often also

| Sector | Global costs (US$ billion) | Developed countries (US$ billion) | Developing countries (US$ billion) |
|---|---|---|---|
| Agriculture | 14 | 7 | 7 |
| Water | 11 | 2 | 9 |
| Human health | 5 | Not estimated | 5 |
| Coastal zones | 11 | 7 | 4 |
| Infrastructure | 8–130 | 6–88 | 2–41 |
| Total | 49–171 | 22–105 | 27–66 |

*Note:* All values are in US$ at present day values. The only 'sector' that includes the cost of 'residual damage' in the above estimates comprises the 'coastal zones'.

*Source:* UNFCCC, 2007, cited in Parry et al, 2009

**Table 6.9**

**Annual investment needs by 2030 to cover climate change adaptation costs (estimates)**

includes loss of possessions. Only a very small proportion of the population in developing countries have insurance for this. Assessing the impacts of such events in terms of the value of property damaged or destroyed can be misleading; an event that is devastating to the lives of very large numbers of people (in deaths, injuries and loss of property) may have low economic impacts because of the low value assigned to the housing damaged or destroyed.[126]

For infrastructure, adaptation costs should include the costs of limiting the impacts (as well as preventing them). For many extreme weather events in urban areas with large infrastructure deficits and poor-quality housing, good early warning systems, measures taken just before the extreme event (e.g. reducing the impact of flooding by supporting populations in moving temporarily to high ground or safe sites) and rapid and effective post-event responses (temporary accommodation, restoring access to services, supporting rapid return to settlements damaged and supporting rebuilding) greatly reduce the impacts upon populations and their assets. Yet, these measures might be considered as inadequate or invalid for adaptation funding in that they are not limiting the damage done to infrastructure. The costs of building and maintaining this capacity to reduce the impacts of extreme weather events is not included in figures for infrastructure investments, and these costs are thus not considered in the UNFCCC estimates.

There is also the issue of infrastructural damage that cannot be prevented by adaptation – the so-called 'residual damage' – stemming both from conscious choice (locations/facilities/structures for which full protection is judged to be too costly, or where adaptation is technically not feasible) or from incapacity on the part of those who are at risk and those institutions which have responsibility for reducing this risk (local government, national governments, etc.) (see Figure 6.4). Thus, the UNFCCC estimates for the costs of adapting infrastructure include consideration of a limited part of 'infrastructure' that does not include social infrastructure, disaster-response infrastructure, housing and the institutional infrastructure needed to build, maintain and adapt infrastructure. Thus:

*The UNFCCC estimate of investment needs is probably an under-estimate by a factor of between 2 and 3 for the included sectors. It could be much more if other sectors are considered... For infrastructure it may be several times higher, at the lower end of the cost range.*[127]

The destruction of, or damage to, housing is one of the most common and most serious impacts of many extreme weather events

The UNFCCC estimates for the costs of adapting infrastructure ... does not include social infrastructure, disaster-response infrastructure, housing and the institutional infrastructure needed to build, maintain and adapt infrastructure

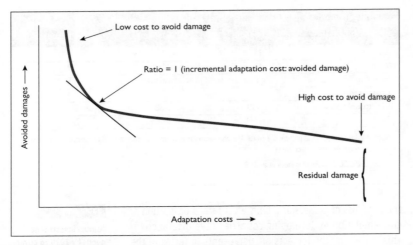

**Figure 6.4**

**Adaptation costs, avoided damages and residual damage**

*Source:* Parry et al, 2009, p12

## The infrastructure deficit

The fact that most developing countries have very large deficiencies in provision for infrastructure has been discussed in detail already. A high proportion of the urban population in Africa and Asia and a significant proportion in Latin America and the Caribbean live in homes and settlements with little or no infrastructure (i.e. no all-weather roads, no drains, no piped water supplies and no provision for electricity). Most urban centres in developing countries have no sewers, including many with several million inhabitants.[128] During the period from 2000 to 2010, the number of slum dwellers in developing countries has increased from 767 million to 828 million, and 'short of drastic action to curb current trend, the slum population worldwide is likely to ... reach a total of 889 million by 2020'.[129] A large proportion of these slums are characterized by inadequate or no provision of basic infrastructure. The lack of provision or inadequacies in the provision of protective infrastructure is perhaps the main reason for the very rapid increase in the number of flood and windstorm 'disasters' since the 1950s.

Reviewing data on disasters also gives some indications of the kinds of impacts that extreme weather events can have upon infrastructure – within the larger costs in terms of death, injury and economic disruption – and loss of livelihood for large numbers of people. Reviewing the 'disasters' registered on an international database[130] between 1996 and 2005 shows not only thousands of people killed and tens of millions affected by floods and windstorms, but also hundreds of billions of dollars worth of damage. For instance, in Asia, floods and windstorms between 1996 and 2005 caused over 70,000 deaths and around US$191 billion worth of economic loss. A large part of these deaths and the economic losses could be attributed to infrastructure deficiencies. UNFCCC notes the following:

> *Evidence for the existence and size of the adaptation deficit can be seen in the mounting losses from extreme weather events such as floods, droughts, tropical cyclones, and other storms. These losses have been mounting at a very rapid rate over the last 50 years. This increase is likely to be mostly due to the expansion of human populations, socio-economic*

**Most of the investment needed for climate change adaptation in terms of infrastructure is required in developed countries, rather than in developing countries**

*activities, real property, and infrastructure of all kinds into zones of high risk. Moreover, much of this property is built at a substandard level and does not conform even to minimal building codes and standards. This widespread failure to build enough weather resistance into existing and expanding human settlements is the main reason for the existence of an adaptation deficit. Real property and socio-economic activities are just not as climate-proof as they could and arguably should be. The evidence suggests strongly that the adaptation deficit continues to increase because losses from extreme events continue to increase. In other words, societies are becoming less well adapted to current climate.*[131]

However, while this recognizes that there is a very large climate change adaptation deficit, much of which is an infrastructure and institutional infrastructure deficit, the UNFCCC report does not consider it appropriate to consider this in estimating adaptation costs for infrastructure.[132]

A review of the basis used by the UNFCCC for estimating the costs of adapting infrastructure[133] suggested that this was based on an incorrect premise – that this can be costed by applying a small increment to existing investment flows into infrastructure that is climate sensitive, with no account taken of the very large infrastructure deficits. This leads to the conclusion that most of the investment needed for climate change adaptation in terms of infrastructure is required in developed countries, rather than in developing countries. It also ends up showing very small sums needed for Africa and other places where there are very low/inadequate investment flows into infrastructure and where many of the countries most at risk from climate change are located. This same review also noted three other assumptions that need to be questioned:[134]

1   The availability of funding from international agencies is the 'solution' for adaptation. In much of Africa and Asia and parts of Latin America and the Caribbean, local governments are weak, ineffective and unaccountable to local populations, so their capacity to design and implement appropriate adaptation strategies that serve those who are most at risk from climate change must be in doubt. This is most obvious in the countries that are often termed 'failed states'; but it is also evident in many other countries. External funding agencies have not proved very effective in addressing this – or even in knowing how to address this.
2   'Adaptation' and 'development' can be kept separate. On the ground, climate change impacts are exacerbating non-climate change impacts and addressing both is inhibited by institutional/governance failures. It is difficult, if not impossible, to separate what proportion of extreme weather damage or water shortages in any locality are caused by climate change. So much of the adaptation deficit for housing and infrastructure is also a development deficit.

3   NAPAs give us an idea of adaptation costs. The focus of most of the NAPAs is a very small part of what these countries will need for adaptation. NAPAs are thus not a good basis for costing adaptation costs.

## The cost of addressing the infrastructure deficit

Detailed cost estimates were undertaken in selected countries to estimate the investments needed to meet the Millennium Development Goals between 2005 and 2015 and these came to US$993 to $1047 per person.[135] Around half of this was for infrastructure (including water and sanitation, energy and roads). Yet, these estimates do not address the elimination of all infrastructure (and other development) deficits. Many of the Millennium Development Goals are only for reducing the problem – for instance, halving the proportion of people without sustainable access to safe drinking water and basic sanitation by 2015. Similarly, the goal for improving the lives of slum dwellers was to reach 100 million slum dwellers by 2020, which would represent only around 13 per cent of the slum population in 2000 (and a much smaller percentage of the likely slum population in 2020). Thus, the total cost to remove the infrastructure deficit is likely to be much higher.

One recent estimate suggested that the cost of removing the housing and infrastructure deficit in developing countries by 2030 would be some US$6.3 trillion – and this would include US$700 billion for expanding housing and infrastructure for expanding urban populations.[136] These estimates are broadly in line with estimates in the 2009 report of the International Strategy for Disaster Response[137] for the investments needed to reduce the deficit in disaster risk avoidance and risk reduction. This suggests that several hundred billion dollars a year are required to address the underlying risk factors for disasters (including those relating to climate change).

However, as was noted in the introduction to this section, the availability of funding is only a part of the solution, as solutions also depend on national and local governments having the competence, capacity and accountability to make the needed investments. It is important to stress that adaptation will require very large capital sums invested in developing countries, but also to recognize that, at the moment, there are no reliable methodologies for estimating these costs accurately. What is more urgent and important is to get serious consideration given to climate change adaptation plans and programmes for particular localities (including cities), and to what resources can be generated for these locally or supported by higher levels of government. Furthermore, there is a need to consider how these plans and programmes can be pro-poor and supportive of general development initiatives. Based on such considerations, it might be possible for the international community to arrive at a more accurate and specific understanding of the international funding mechanisms which are required to support such plans and programmes.

Thus, there is a need for detailed case studies of what adaptation would involve in particular locations and what

component would have to be allocated to infrastructure deficits. The studies described earlier in this chapter are moving in this direction, although most are from cities in developed countries. Such studies need to consider the infrastructure deficit and the needed institutional/governance underpinning for addressing the infrastructure deficit and climate-proofing all new and existing infrastructure and urban developments. From this can come a better idea of the kind of funding needed for adapting infrastructure to climate change risks, and from this, some thoughtful discussion of what these imply for adaptation costs and adaptation funding, in general. It would only take a few such studies of major cities that are particularly at risk from climate change and have large infrastructure deficits to show that the UNFCCC estimates for Africa and for most cities in Asia are far too low. It is also likely that studies of major cities in Latin America at high risk from climate change would also show the UNFCCC estimates for these regions to be far too low.

The UNFCCC notes[138] that even with a growing number of location-based estimates for costs, it will be difficult to extrapolate these to figures for whole regions because of:

- *Very large differences in contexts* (risks and vulnerability), including the scale of the infrastructure deficits and the extent of the local governance failures. In most of the locations with the largest infrastructure deficits and governance failures, much of the data needed to assess such costs are simply not there.
- *Very large differences in costs.* The estimate in London's adaptation plan that it can cost UK£15,000 to make a single dwelling in London cooler in summer could build 15 houses in many Asian and African urban centres.
- *The 'moving target' of urbanization.* United Nations projections suggest that almost all growth in the world's population in the next few decades is expected to be in urban areas in developing countries.[139]
- *Public costs versus private costs.* Many of the costs in adapting cities – particularly in upgrading housing stock – will be borne by private individuals and are even more difficult to account for. Estimates that are based only on the costs of adapting infrastructure are thus certainly not the 'total costs of adaptation'.[140]

# CHALLENGES TO ADAPTATION

Most of the world's urban population and most of its largest cities are now in developing countries. Furthermore, and as noted in the introduction to this chapter, most of the growth in the world's population over the next few decades is likely to occur in the urban centres of developing countries. At the same time, most of the urban centres most at risk from climate change are in developing countries. And it is in urban areas in developing countries that the deficits in infrastructure and services needed to protect populations from climate change are most evident. Yet, most governments and many international agencies still give little or no attention to

*One recent estimate suggested that the cost of removing the housing and infrastructure deficit in developing countries by 2030 would be some US$6.3 trillion*

*The UNFCCC estimates for Africa and for most cities in Asia are far too low*

urban adaptation. Many disaster response agencies are also better equipped to deal with rural disasters than urban disasters.[141]

Perhaps the most pressing challenge for climate change adaptation in urban areas in developing countries is to get it seen and understood as a central dimension of development – and, thus, also a central dimension of economic strength and poverty reduction, including meeting the Millennium Development Goals. If the Millennium Development Goals were met in urban areas, it would certainly increase their resilience to climate change. However, this raises a second challenge of how to get far more effective local action on the ground for development that includes the needed attention to adaptation. A city's economic success may be important for its adaptive capacity – but there are many cities with successful economies where large sections of their population still live in informal settlements that lack the infrastructure and services that reduce climate-related (and many other) risks.

It will also be difficult to balance present and future needs. The adjustment of building and infrastructure standards and designs to address likely increases in extreme weather or water constraints – that may not become evident for 20 or more years in the future – is important as it will be more expensive to rebuild or adjust these in 10 to 20 years. With investment capacity so constrained in most urban centres in developing countries, however, the extra costs of building resilience to future risks will be contested by those who claim that there are more pressing priorities. In this context, it will especially be difficult to get the needed priority to risk reduction for lower-income groups, as wealthier and more powerful interests (residents and businesses) want their risks, vulnerabilities and adaptation needs to be addressed first. City governments that have long ignored the needs and priorities of those living in informal settlements are not likely to become committed to address these deficits.

Effective action on adaptation on the ground also depends on a willingness to act by local governments. The generic lessons that can be drawn from the experiences of cities that have already developed adaptation plans were discussed above.[142] These include building a commitment to act among the different stakeholders, developing the information base on current conditions (and risks), and developing city-wide risk/vulnerability assessments that draw on community and district assessments. City and municipal governments need to consider how to reduce climate-related risks within their plans and investments in infrastructure and land-use management. This usually depends on, or is much enhanced by, civil society organizations, especially those that represent and work with those who are most at risk. An earlier section also discussed the key roles for local governments and for civil society groups of building or supporting the building of resilience to climate-related stresses and shocks.[143] Here, too, there are many co-benefits with development (and poverty reduction).

However, within each country and urban centre, different stakeholders may be working according to very different worldviews of adaptation. This may hamper efforts at creating coherent and holistic adaptation responses that take into consideration the different vulnerabilities highlighted earlier in this Global Report.[144] There is also the growing influence of those who insist that climate change is not happening or that it will bring few costs; a web-based consultation on London's adaptation programme[145] that asked for comments and suggestions produced many remarks to this effect that also showed little or no understanding of climate science.

In addition, little attention is given to urban adaptation by most international agencies, even as they discuss and develop policies on adaptation.[146] Where international funding is available for adaptation, it will be difficult to get the needed attention for addressing the (often very large) deficits in infrastructure and services (the lack of provision for piped water, storm and surface drains, all-weather roads, emergency services, etc.) that arise from governance failures and limitations because these are not seen as climate change related. Getting international support available in a form that allows it to support effective urban adaptation which is integrated within local development (and build local adaptation capacity) is thus a challenge. There are also all the constraints faced by international agencies from their lack of capacity to support local engagement, as they have shifted their support to sector support and basket funding.[147] Channelling funding through recipient governments and supporting their priorities serves development when these governments are competent, representative and accountable; but this is often not the case. This raises issues about the structures and effectiveness of international agencies in supporting needed local action on climate change in the thousands of urban areas where this is required.

Furthermore, official development assistance was not, in the first place, set up to support local governments and civil society groups with regard to adaptation efforts. There is little clarity as yet on how international funding for adaptation (hopefully integrated within development) can work with and serve local governments and civil society groups within each urban centre. The key roles of local government and civil society as designers and implementers of climate change adaptation in urban areas may be better appreciated; but the means by which they can influence climate change negotiations and institutional responses and hold international funders of adaptation to account is not yet clear.

It is important to note that the failure to mitigate sufficiently in developed countries will create ever more adaptation failures, mostly in developing countries, including many countries with insignificant historic and current contributions to climate change. It is also difficult to see any agreement reached on needed mitigation strategies by the governments of developing countries unless the governments in developed countries demonstrate their commitment to mitigation by taking responsibility for their (very high) contribution to global climate change. For (local and national) governments in countries with minimal per capita GHG emissions, it is very difficult to justify to their electorates expenditures on climate change mitigation if they are already unable to provide their populations with basic infrastructure and services.

*Perhaps the most pressing challenge for climate change adaptation in urban areas in developing countries is to get it seen and understood as a central dimension of development*

*City and municipal governments need to consider how to reduce climate-related risks within their plans and investments in infrastructure and land-use management*

*The failure to mitigate sufficiently in developed countries will create ever more adaptation failures, mostly in developing countries, including many countries with insignificant ... contributions to climate change*

Most of the urban populations and places at greatest risk from climate change are not those with large historic or current contributions to GHGs. As noted in Table 1.4, the average African (excluding South Africa) individual emits only 54 per cent of the $CO_2$ emitted by the average Indian, only 16 per cent of that emitted by the average Chinese, and only 4 to 8 per cent of that emitted by the average citizen of the major developed countries. And, in terms of total figures, if $CO_2$ emissions from all African countries (except South Africa) were cut in half, this would only imply a 1.2 per cent reduction in global emissions. In contrast, a similar global emissions reduction could be achieved by the US through a national reduction of $CO_2$ emissions of only 6 per cent. Such issues of *environmental justice* are playing an important part in the increasing focus being given to climate change adaptation in developing countries.

There is also the larger issue for urban adaptation of *population displacement* at a national scale and its influence on migration, including that to urban centres. If cities become the destination of very large flows of rural migrants driven from their homes and livelihoods by, for instance, the damage brought by climate change to agriculture, it will add further to the infrastructure deficit and probably to the scale of settlement on hazardous sites. There are predictions that by 2050, some 200 million people may be forced to leave their homes due to environmental degradation and water shortages caused by climate change.[148] Yet, studies of migration show how population movements are generally rational, well-informed responses by individuals and households to changing circumstances. Thus, they are, in fact, a key part of individual and household adaptation. Land degradation or decreases in rainfall do not inevitably result in migration. Or where they do, most movement is short term, as in response to extreme weather disasters, and short distance, as in migrant responses to drought and land degradation.[149]

Where there are slow-onset impacts from climate change (e.g. rising temperatures and declining rainfall), this can bring negative impacts upon agriculture; but income diversification and short-distance circular migration are likely to be common responses.[150] Where climate change is causing environmental stress for rural livelihoods, it will be one among a number of factors in determining migration. In addition, support for agriculture – including agricultural adaptation initiatives – does not necessarily reduce rural–urban migration. Indeed, successful rural development often supports rapid urban development locally as it generates demand for goods and services from farmers and rural households.[151] Yet, a failure by governments and international agencies to reduce global GHG emissions and to support rural and urban populations to adapt will bring crisis-driven population movements that make those forced to move very vulnerable. Here, migration is no longer planned movement helped by knowledge and contacts in the destination area. The pressures on crisis-driven population movements will also be much increased if developed countries fail to agree on implementing the large reductions in GHG emissions that are needed to avoid dangerous climate change.

So far, there is debate as to whether climate change has yet led to *forced migration* from any location.[152] Yet, there is growing concern about how to address the issue of migrants who are forced to leave their homes due to future climate change. This aspect of 'residual damage', people whose lives and homes cannot be adapted *in situ*, falls outside the scope of most national and international legislation. Under current international law, strictly speaking, those fleeing from environmental pressures are not considered as refugees – this term is reserved for those 'being persecuted for reasons of race, religion, nationality, membership of a particular social group, or political opinion'. Furthermore, the term 'refugee' refers only to people who are 'outside the country of [their] nationality'.[153] In international law, people who are displaced within their own country are referred to as 'internally displaced persons'. Thus:

> *... there is a broad consensus among lawyers considering the issue of climate change migration that current protections at international law do not adequately provide for a number of the categories of person likely to be displaced by climate change.*[154]

There are major consequences of this inadequate protection in human rights law – namely, who will be responsible for assisting this group? If international climate migrants were to be considered refugees, this would imply a responsibility of countries to offer them the same protection as they offer to political refugees. So far, not one country has been willing to accept such a definition.[155] At the same time, the international agency with the primary responsibility for dealing with refugees, and which has been taking on the task of addressing the concerns of internally displaced persons as well – the Office of the High Commissioner for Refugees – is 'already overstretched and ... unable to cope with their current "stock" of refugees'.[156] Thus:

> *Given the nature and magnitude of the problem which climate change displacement presents,* ad hoc *measures based on existing domestic regimes are likely to lead to inconsistency, confusion and conflict.*[157]

There are thus increasing calls for the development of new international legislation to address the concerns of 'climate migrants' – perhaps in the form of an international convention for persons displaced by climate change.[158]

# CONCLUDING REMARKS AND LESSONS FOR POLICY

What needs to be done to support the adaptation of urban areas to climate change has become clearer during the last ten years, in large part because of innovations by civil society groups and local governments, some of which have been described in this chapter. What is much less clear is how to

**If cities become the destination of very large flows of rural migrants driven from their homes ... by ... climate change ..., it will add further to the infrastructure deficit and probably to the scale of settlement on hazardous sites**

**The pressures on crisis-driven population movements will ... [increased] if developed countries fail to agree on implementing ... large reductions in GHG emissions**

**There are ... increasing calls for ... new international legislation to address the concerns of 'climate migrants' – perhaps in the form of an international convention**

translate 'what needs to be done' into 'how to do it', especially in countries and urban areas with weak local governments or local governments unwilling to work with the low-income groups within their jurisdiction.

Clearly, one important way forward is to work with and learn from the innovators – in grassroots organizations, in local governments, in national governments and in international agencies. Another is to encourage the engagement of all key stakeholders in cities (which in the end means almost everyone). This includes far more attention to the needs and capacities of those who are most at risk from climate change. Here, consultations on the ground and risk assessments are not focused on 'climate change', but on all the risks and vulnerabilities that they face – some, most or all of which are likely to be exacerbated by climate change. This can be the basis for risk and vulnerability assessments that inform a 'climate change aware' development agenda. This has to build resilience both to the specific threats identified as certain or likely from climate change and, more generally, to all the stresses and shocks that threaten the well-being and livelihoods of low-income groups. Another important issue here is how to make the adaptation measures provided or financed by the private sector that serve better-off households and businesses extend their range so that they also serve smaller businesses and lower-income households.

Yet, it has to be kept in mind that a focus on community-based adaptation, local assessments or the international transfer of funds only is unlikely to be a successful recipe for climate change adaptation at the city level. Successful adaptation also has to take into account the following major issues:

- Concerted action at the household, community, local government, national government and international levels are required.
- Global and national projections about climate change impacts have to be improved in order to better support measures at the local level. At present, projections are insufficiently precise or, at times, contradictory, which impedes local action.
- The issue of social and environmental justice needs to get appropriate attention, both within cities and countries, but also internationally. As is acknowledged by the UNFCCC, the bulk of funding for climate change

adaptation has to come from those countries that are responsible for global climate change. Also, there is a need to consider who is to pay for the homes and properties lost from the impacts of climate change that cannot be adapted to: the so-called 'residual damage'.
- The emerging international funding for climate change adaptation has to be adequate to the task at hand, and should explicitly allocate a fair share of resources to urban settlements. At present, resources are inadequate and do not target urban settlements.

It is also important that the emerging knowledge about climate change adaptation in urban areas is synthesized and included in the Fifth Assessment Report of the IPCC that will be developed between 2010 and 2014. The work undertaken in preparing this Global Report, as well as other UN-Habitat activities, is already feeding into that process. The Fourth Assessment Report of the IPCC, published in 2007, focused on reviewing and summarizing the evidence for human-induced climate change and presenting the case for the importance of action both on adaptation and mitigation. The Fifth Assessment Report needs to go much further in summarizing and synthesizing what is known about how to achieve adaptation (and mitigation). The initial work for IPCC's Fifth Assessment recognizes the need for more attention to human settlements; in the plans for the Fifth Assessment Report, the Working Group II on 'impacts, adaptation and vulnerability' includes three chapters on 'human settlements, industry and infrastructure', compared to only one in the Fourth Assessment. This includes a chapter on urban areas, another on rural settlements and a third on networked infrastructure that serves all human settlements (including transport, energy and water).[159] There are also measures under way to have closer links between the various working groups on the role of cities and other settlements in considering both adaptation and mitigation; here the interest is in the co-benefits between adaptation and mitigation. It is also being planned that the Fifth Assessment Report should have more detailed coverage on human health, security and livelihoods, and poverty. The ongoing work of the IPCC will thus serve to get the attention of national governments and international agencies to all the measures needed to address climate change adaptation in urban areas discussed in this chapter.

**A focus on community-based adaptation, local assessments or the international transfer of funds only is unlikely to be a successful recipe for climate change adaptation at the city level**

**It is ... important that the emerging knowledge about climate change adaptation in urban areas is synthesized and included in the Fifth Assessment Report of the IPCC**

# NOTES

1   Satterthwaite, 2007.
2   UN, 2010. See also Chapter 1.
3   The definitions used in this section are based on Parry et al (2007b). See also Box 1.1.
4   See Chapter 1 and also UN-Habitat (2007) for an in-depth discussion of resilience with respect to natural and human-made disasters.
5   See also discussion of vulnerability in Chapter 4.
6   See Chapter 5.
7   See Chapter 1.
8   Mitlin, 2008.

9   See section on 'Household and community responses to the impacts of climate change'.
10  Bicknell et al, 2009.
11  See Chapter 4.
12  See Wisner et al, 2004.
13  See Chapter 4 for a more in-depth discussion of vulnerability.
14  Stephens et al, 1996.
15  Mrs Fatu Turay, Kroo Bay Community, Freetown, Sierra Leone. ActionAid International, 2006, p6.
16  Douglas et al, 2008.

17  ActionAid International, 2006, p4.
18  Dodman et al, 2010b.
19  Jones and Rahman, 2007.
20  Mitlin and Dodman, forthcoming.
21  Boonyabancha, 2005, 2009.
22  Hasan, 2006, 2010.
23  This example is drawn from López-Marrero and Tschakert, forthcoming.
24  Similar differences in perception of risks and impacts have been noted in Delhi, India (Diana Reckien, pers comm, 2010).

25  Wamsler, 2007.
26  Mitlin and Dodman, forthcoming; see also Banks, 2008.
27  Sabates-Wheeler et al, 2008.
28  Prowse and Scott, 2008.
29  Dodman and Satterthwaite, 2008.
30  Moser and Satterthwaite, 2008.
31  See Roberts, 2008.
32  Karol and Suarez, 2007; Roberts, 2010a; .
33  Stern, 2006; Satterthwaite et al, 2007a.
34  Satterthwaite et al, 2007a.
35  See Chapter 2.

36   Satterthwaite et al, 2009a.
37   Dalal-Clayton, 2003.
38   Satterthwaite et al, 2007a;
     Satterthwaite et al, 2009a.
39   Huq et al, 2003.
40   The City of Durban has been a
     pioneer within Africa in devel-
     oping a coherent inter-sectoral
     adaptation strategy. See
     Roberts (2008, 2010a) for an
     account of the difficulties in
     getting buy-in from within
     different sectors in govern-
     ment. See also the section
     below on 'Moving from risk
     assessments to adaptation
     strategies'.
41   This section draws on Hintz,
     2009.
42   Pelling, 1997.
43   This section draws on Bangkok
     Metropolitan Administration,
     2009.
44   Roy, 2009.
45   It should be noted that many
     of the actions implemented in
     Bangladesh, particularly during
     the early years, were address-
     ing natural disaster risks, rather
     than climate change adaptation
     *per se*.
46   Alam and Rabbani, 2007.
47   The section on Durban draws
     on Roberts (2008, 2010a,
     2010b); the section on Cape
     Town draws on Mukheibir and
     Ziervogel (2007).
48   Roberts, 2008, 2010a.
49   UN, 2010.
50   Roberts and Diederichs, 2002a,
     2002b.
51   The discussion on Durban
     draws on Roberts, 2008,
     2010a; and Debra Roberts,
     eThekwini Municipality,
     Durban, South Africa, pers
     comm, September 2009.
52   Roberts, 2010a.
53   See www.durban.gov.za/
     durban/services/epcpd/about/
     branches/climate-protection
     -branch.
54   Roberts, 2010b.
55   Roberts, 2010b.
56   City of Cape Town, undated.
57   Departments such as roads
     and storm water, disaster risk
     management and housing (City
     of Cape Town, 2010).
58   In practical terms, this may be
     better understood as the
     second step, as it first needs
     the decision by the city
     government to think about
     climate change adaptation, and
     then to commission the work
     needed to advance such think-
     ing.
59   Nickson, 2010.
60   This draws on City of
     Melbourne, 2009.
61   This draws on City of
     Rotterdam, 2009 and undated.
62   This draws on IFRC, 2010,
     especially Chapter 7.

63   See, for instance, UN, 2009;
     IFRC, 2010.
64   Lungo, 2007.
65   Gavidia, 2006.
66   Adger et al, 2007.
67   Note the literature on the
     early adopters. See, for
     instance, Carmin et al (2009).
68   Satterthwaite et al, 2007a.
69   Boonyabancha, 2005;
     Usavagovitwong and
     Posriprasert, 2006; Some et al,
     2009; see also Hasan, 2006.
70   van Horen, 2001; Torres et al,
     2007.
71   Syukrizal et al, 2009.
72   See Roberts, 2008, 2010a.
73   UN, 2009.
74   See the above section on
     'Community responses'.
75   Osbahr and Roberts, 2007;
     Kehew, 2009.
76   See discussion earlier in this
     chapter.
77   See Roberts, 2010a, for a
     discussion of this in relation to
     Durban, South Africa.
78   For a more elaborate discus-
     sion on urban planning, see
     UN-Habitat, 2009a.
79   See ADB, 2005.
80   Satterthwaite et al, 2007a.
81   Revi, 2008.
82   See Hintz, 2009, and the above
     section on 'Local government
     responses to the impacts of
     climate change'.
83   World Bank, 2008.
84   World Bank, 2008.
85   Bizikova et al, 2008.
86   World Bank, 2008.
87   See the section below on
     'Challenges to adaptation'
88   Parry et al, 2007b, p880. See
     also Box 1.1 and the discussion
     of resilience in UN-Habitat,
     2007.
89   See Chapter 2.
90   See discussion in López-
     Marrero and Tschakert,
     forthcoming.
91   Parry et al, 2007b, p880.
92   Dodman et al, 2009.
93   Stern, 2009.
94   See www.unhabitat.org/
     categories.asp?catid=634, last
     accessed 14 October 2010.
95   See www.citiesalliance.org/
     ca/cds, last accessed 14
     October 2010. It should be
     noted that the City
     Development Strategies
     approach does not yet clearly
     address climate change *per se*.
96   Pieterse, 2008.
97   See discussion in the section
     on 'Household and community
     responses to the impacts of
     climate change' above.
98   Hasan, 2010.
99   Satterthwaite et al, 2007a.
100  Such as Durban (South Africa)
     and Manizales (Colombia).
101  See section on 'Local govern-
     ment responses in developing

countries' above.
102  Velasquez, 1998.
103  Díaz Palacios and Miranda,
     2005.
104  Carmin and Zhang, 2009.
105  See Cabannes, 2004.
106  See, for instance, Menegat,
     2002.
107  Pelling et al, 2008.
108  Tanner et al, 2009.
109  Manuel-Navarrete et al, 2008.
110  Satterthwaite et al, 2009a.
111  See also Roberts, 2010a.
112  Cohen and Garrett, 2010.
113  See Maxwell et al, 1998; Cohen
     and Garrett, 2010; Tolossa,
     2010.
114  This section draws on UN-
     Habitat, 2008a, and the Cities
     and Climate Change Initiative's
     website at www.unhabitat.org/
     content.asp?typeid=19&catid=
     570&cid=6003, last accessed
     14 October 2010.
115  Additional cities are joining the
     initiative in Africa, Asia and
     Latin America.
116  See the section on 'The poten-
     tial of the international climate
     change framework for local
     action' in Chapter 2.
117  It should be noted that the
     Adaptation Fund became
     operational in 2010; see Box
     2.2.
118  UNFCCC, 2010.
119  Ayers, 2009.
120  Ayers, 2009.
121  As can be seen from Table 1.4,
     developed countries and the
     top nine GHG-emitting devel-
     oping countries are
     responsible for 83 per cent of
     all GHG emissions (in 2005)
     and 87 per cent of all $CO_2$
     emission (in 2007).
122  See section on 'Challenges to
     adaptation' below.
123  See UN, 2009; IFRC, 2010.
124  UN, 2009.
125  Dodman and Satterthwaite,
     2009.
126  For more details, see IFRC,
     2010.
127  Parry et al, 2009.
128  Hardoy et al, 2001.
129  UN-Habitat, 2010, p42.
130  Disasters included in the
     Emergency Events Database,
     CRED, Louvain, Belgium
     (www.emdat.be).
131  UNFCCC, 2007, para 371, p90.
132  This having been said, however,
     it is easy to argue that many
     parts of the infrastructure
     deficit are only marginally
     related to the adaptive capacity
     of a community. The adaptive
     capacity of New Orleans
     during Hurricane Katrina in
     2005, for example, would not
     have been much different if, say,
     20 or 80 per cent of the
     population had access to
     sewerage services or piped

water. Yet, in terms of urban
     governance, it is hard for local
     governments, particularly in
     developing countries, to justify
     investments in 'pure' climate
     change adaptation measures if
     a large proportion of the
     population do not have access
     to basic infrastructure and/or
     services.
133  Dodman and Satterthwaite,
     2009.
134  Parry et al, 2009.
135  UN Millennium Project, 2005.
136  Parry et al, 2009.
137  UN, 2009.
138  Dodman and Satterthwaite,
     2009.
139  UN, 2010.
140  Dodman and Satterthwaite,
     2009.
141  Suarez et al, 2008; IFRC, 2010.
142  See the section on 'Generic
     lessons for city governments'
     above.
143  See the section on 'Building
     resilience' above.
144  See Chapter 4.
145  See www.london.gov.uk/
     climatechange, last accessed
     14 October 2010.
146  See, for instance, the lack of
     attention to this in the recent
     *Human Development Report* on
     the topic of climate change
     (UNDP, 2007).
147  Crespin, 2006. 'Basket funding
     is the joint funding by a
     number of donors of a set of
     activities through a common
     account, which keeps the
     basket resources separate
     from all other resources
     intended for the same
     purpose' (Ministry of Foreign
     Affairs of Denmark, 2006, p2).
148  See note 257 in Chapter 4.
149  See Henry et al, 2004; Massey
     et al, 2007; Tacoli, 2009.
150  Tacoli, 2009.
151  See Beauchemin and Bocquier,
     2004; Henry et al, 2004; Massey
     et al, 2007.
152  See, for example, Brown, 2007.
153  Convention relating to the
     Status of Refugees, Article 1,
     http://www2.ohchr.org/english/
     law/pdf/refugees.pdf, last
     accessed 13 October 2010.
154  Hodgkinson et al, undated.
155  Brown, 2007.
156  Brown, 2007, p8.
157  See www.ccdpconvention.com/
     documents/CCDPConvention
     FAQs.pdf, p3, last accessed
     13 October 2010.
158  See www.ccdpconvention.com,
     last accessed 13 October
     2010.
159  See www.ipcc.ch/activities/
     activities.htm for more details,
     last accessed 18 October
     2010.

# CHAPTER 7

# CONCLUSION AND POLICY DIRECTIONS

If concerted action is not undertaken to reduce greenhouse gases (GHGs) and promote more environmentally sustainable and equitable patterns of urban development, there will be a deadly collision between urbanization and climate change. The dangerous course to this collision threatens to have unprecedented negative impacts upon human development, quality of life, economic production, political stability and the health and resilience of the ecosystems upon which human beings depend. However, the coming together of urbanization and climate change will also offer an unprecedented opportunity. Urban areas, with their high concentrations of population, buildings, industries and infrastructure, will face the most severe impacts of climate change. Yet, the same urban areas can become hubs of innovation where alternative options can be designed and tested to promote reductions in GHG emissions (mitigation) and vulnerability to climate change impacts (adaptation).

Significant linkages exist between climate change and development. While climate change is jeopardizing development goals, mitigation and adaptation targets could be greatly threatened by unsustainable pathways of development. Climate change cannot be addressed effectively unless more effective actions are undertaken to reduce emissions, cope with climate changes already under way, and create the conditions to enhance the adaptive capacity of poor countries and population sectors (environmental justice). Mitigation efforts need to focus not only on reducing carbon intensity, or increasing the energy efficiency of infrastructure, buildings, and economic and domestic activities, but also on reducing both the total consumption of fossil fuels and emissions of GHGs through other means. Adaptation strategies cannot be reduced to redesigning buildings and infrastructure, but will also require use of local knowledge, greater participation of key stakeholders, and higher institutional capacity of local governments. In many developing countries, urban centres lack all-weather roads, good-quality homes and other preconditions for successful adaptation (i.e. they suffer from an 'adaptation deficit'). It is therefore necessary to relate adaptation and mitigation responses to development and foster sustainable development with mitigation and adaptation strategies in mind.

Equity is a fundamental dimension of the relationship between climate and development. Because of uneven development patterns and distribution of wealth and infrastructure services at global, national and urban levels, within different sectors, and between different individuals, there is often an inverse relationship between responsibility for climate change and suffering of its consequences. The largest national emitters of GHGs are, by far, the developed countries and a few rapidly industrializing developing countries (see Table 1.4), and this trend can also be seen, generally speaking, in the wealthy sectors within countries and cities around the world. Yet, climate change will deal its heaviest blows on those contributing the least to GHG emissions: poor countries and the poor and vulnerable within their societies.

There are, at present, many actions at different levels designed to respond to the daunting challenge of climate change. Nearly all national governments have signed the United Nations Framework Convention on Climate Change (UNFCCC) and dozens have launched responses at the national level. Numerous provincial/state and local authorities have promoted vigorous, yet varied responses to address climate change, even in the absence of incentives from national governments. Many local authorities are also undertaking a range of mitigation and adaptation measures. Notwithstanding all of these, climate change remains, in practice, a marginal issue for most decision-makers. This Global Report has explored the reasons for this, as well as windows of opportunity that can be used or created to help urban populations and decision-makers reduce their emissions and adapt to climate change in ways that promote sustainable, equitable and resilient paths of urban development.

The purpose of this chapter is to provide an overview of the key findings and messages from all chapters of the Global Report. It will briefly revisit the constraints, challenges and opportunities to mitigation and adaptation actions, and point to some of the linkages among drivers and vulnerabilities. Drawing further from the findings of the previous chapters, this concluding chapter reflects on the multiple linkages, synergies and trade-offs between mitigation, adaptation and urban development. The chapter ends with a set of suggestions on future policy directions, focusing on local, national and international principles and policies for supporting and enhancing urban responses to climate change.

> If concerted action is not undertaken to reduce ... GHGs ... there will be a deadly collision between urbanization and climate change

> There is often an inverse relationship between responsibility for climate change and suffering of its consequences

# KEY FINDINGS AND THEIR IMPLICATIONS

Urbanization and climate change are two human-induced forces that have put humanity at a crossroad of at least two future directions that this Global Report has explored. *First*, there is the plausible future of continuation along a dangerous collision course if national, regional and local governments continue with business as usual. Many of the dysfunctions of the current political, economic and social systems at play could lead inexorably to the very worst outcomes imaginable. For example, it has been difficult for the developed countries, which bear the main responsibility for current GHG emissions,[1] to achieve effective mitigation targets. Notwithstanding decades of development policies, the story of how affluence and poverty affect global climate change is still a tale of two development paths explaining diverse levels of emissions within and across cities. This difference also creates common but differentiated mitigation and adaptation responsibilities (i.e. the wealthy should be most responsible for mitigation and adaptation responses). However, the political reality is that the wealthy also have a greater influence on the political structures at play, making such equitable distribution of responsibility difficult at best. Furthermore, uneven development and inadequate infrastructure and governance structures constrain the ability of populations and local authorities of many urban centres to adapt to existing and future climate change and to other environmental and societal stresses.

A *second* plausible future, and the only option for humanity to avoid the first, is one for which cities have historically proved their talents as sources of innovation, and laboratories for the transition to different and more sustainable (i.e. less carbon intensive and more resilient) pathways of development. The findings of this Global Report, briefly summarized, contribute to making this second option possible.

## Main issues of concern

Trends of urban change in recent decades have a strong bearing upon the present report. Urban population growth has taken place at an unprecedented rate, with a near quintupling of total urban populations between 1950 and 2011. During the same period, the urban population has increased from being less than one third (28.8 per cent in 1950) to more than one half of the global population (50.8 per cent in 2011). The fastest rates of urbanization are currently taking place in developing countries, with the bulk of this growth taking place in smaller urban areas.[2] This, coupled with the increased intensity and frequency of adverse weather events, will have devastating effects precisely where the capacity to deal with the consequences of climate change is weaker, or even lacking. Smaller urban centres in developing countries are often institutionally weak, and unable to promote effective mitigation and adaptation actions. At the same time, a possible advantage also exists as the burgeoning development of these centres may be redirected in ways that reduce their emission levels

to a desired minimum – for example, through the promotion of mono-centric urban structures based on the use of public transportation. Their resilience and ability to cope with climate hazards and other stresses may also be enhanced – for instance, through the development of climate-proof urban infrastructures and effective response systems.

This Global Report aims at contributing to an understanding of the drivers of GHG emissions from urban areas. The purpose of developing this understanding is to help urban policy-makers, enterprises and consumers target effective options for reducing these emissions at the same time that they enhance urban resilience to the impacts of climate change. Last, but certainly not least in importance, the dynamics of urban centres are intimately linked not only to the role of geography in determining a city's need for energy to run heating and air-conditioning systems, or to get access to sources of energy, but also to the role that geography plays in giving cities access to biodiversity, clean water and other ecosystem services at risk from the impacts of climate change. Furthermore, since urban areas have developed over existing ecosystems (or 'ecozones') such as coastal areas, wetlands, drylands, etc., intimately linked to geography and to ecosystem services threatened by changes in the climate system, policies aimed at mitigation and adaptation in these areas should also consider protection or enhancement of natural systems – for example, through tree-planting and coral reef restoration.

Climate change is also interacting with urbanization and, in doing so, increases the magnitude of the developmental and environmental challenges and threats that urban governments are already facing as a result of the pace of current urbanization (each year sees the addition of 67 million new urban dwellers, 91 per cent of whom are added to cities in developing countries).[3] The most recent empirical evidence points unequivocally to the conclusion that the Earth's climate is warming and that this warming has been induced by the massive amounts of GHGs that human beings have pushed into the atmosphere. Human-induced changes in the climate system have been further validated by research that has been published after the release of the Fourth Assessment Report of the Intergovernmental Panel on Climate Change (IPCC) in 2007. According to this assessment, the observed increase in global mean surface temperature since 1990 is 0.33°C. At the same time, changes have been documented in the frequency and severity of storms, precipitation, droughts and other weather extremes of relevance for urban centres.

The main human sources of GHGs are the dramatic rise in energy use, land-use changes and emissions from industrial activities. Increases in GHG emissions have been, to a limited extent, offset by increases in efficiency and/or reductions in the carbon intensity of production and consumption. However, the overall global trend has still been towards large increases in the total amount of anthropogenic (or human-caused) GHG emissions.

Since the onset of the industrial era, urban centres have played a key, though not yet fully understood, role in the unprecedented increases of carbon dioxide ($CO_2$) and methane emissions. Furthermore, emissions are now

---

*Climate change ... increases the magnitude of the developmental and environmental challenges and threats that urban governments are already facing*

*The main human sources of GHGs are the dramatic rise in energy use, land-use changes and emissions from industrial activities*

increasing above the worst scenario established by the IPCC. In this context, humanity is facing two main challenges that urban centres can help address: the *need to adapt*, at least to some amount of continued warming already under way, and the *need to mitigate* (i.e. to achieve development paths that bring about a peaking of emissions by 2015 and a stabilization of GHG concentrations).

While industrialization is certainly responsible for the rapid pace of global climate change, and urbanization is strongly related to industrialization, two questions of primary importance are still being addressed (i.e. the amount of GHG emissions that urban areas are actually responsible for, and the linkages among levels of urbanization, economic development and emissions). Chapter 3 showed that, because of the complexities involved in calculating the urban contribution to GHG emissions and the lack of agreement by researchers on exactly what items to include in the inventories, no precise figures exist of how high a contribution to global warming cities make. Earlier chapters have also illustrated how a dynamic, complex and strong link exists between economic development, urbanization and GHG emissions. However, this relationship is in no way straightforward. Differences in GHG emissions result from the peculiarities and weight of different emitting sectors (such as industries, buildings and transportation). Diverse factors account for the different levels and sources of urban GHG emissions both within and across countries. These include:

- differences in how energy generation, transportation and other emitters operate;
- levels of economic development and affluence as measured by gross domestic product (GDP) per capita;
- technology and technological innovations and acquisition;
- geographic factors;
- demographic structure and dynamics of a city;
- urban functions and a city's economic base;
- urban form (spatial structure) and, related to it, the layout and structure of a city's transportation system;
- city size (i.e. the 'agglomeration' effect);
- climate conditions and natural endowments; and
- market prices and the wider institutional setting of the city and of the broader – national and international – governance structure within which it operates.

The inverse relationship between being the most at fault for the causes of climate change and suffering its most profound consequences springs directly from historical and existing patterns of inequity in development, distribution of wealth, lifestyle and availability of infrastructure services. This inequity exists not only at the global level, where developed countries and a few rapidly industrializing developing countries are the main contributors to total $CO_2$ emissions. It also occurs at the national and local levels, creating differentials in contribution to GHG emissions along several different economic and social lines. These differences can be found within and across cities, between the rich and the poor, the racial or ethnic minority and the majority, the old

and the young, and between men and women. This follows, in general, the differential access to resources, services and political power among and between these groups. As such – even within developing countries – it is the affluent and politically enfranchised enclaves, groups and communities with access to more services and amenities who consume more, travel more and become the highest GHG emitters within their cities, regions and countries. This deeply entrenched inequity lies at the heart of environmental justice issues surrounding climate change mitigation and adaptation actions.

The concentration, within urban centres, of people and their homes, infrastructure, industries and waste within a relatively small area can have two implications for policies aimed at avoiding the negative urban impacts of climate change. On the one hand, urban areas can be dangerous places in which to live and work; their populations can be very vulnerable to extreme weather events or other hazards, with the potential to become disasters. Furthermore, urban settlements can increase the risk of 'concatenated hazards'. Industrialization, inadequate planning and poor design can be key determinants of secondary or technological risks.

On the other hand, the same concentration of people, infrastructure and economic activities in urban centres also means economies of scale or proximity for many of the measures that reduce risks from extreme weather events. Policies on enhancing sustainability and on transition from disaster response to disaster preparedness can help urban settlements to increase their effectiveness at coping with climate hazards.

Not all demographic segments of the urban population are equally affected by the hazards that climate change is predicted to aggravate. The capacity of different urban populations to cope or adapt is influenced not only by age and gender, but also by the context-specific combination of factors such as:

- labour, education, health and the nutrition of the individuals (*human capital*);
- the financial resources available to people (*financial capital*);
- the extent and quality of infrastructure, equipment and services (*physical capital*);
- stocks of environmental productive assets, such as soil, land and atmosphere (*natural capital*);
- the quality and inclusiveness of governance structures and community organizations that provide or manage safety nets and other short- and longer-term responses (*social capital*).

Urban vulnerability to climate change is a dynamic process in many ways: climate change and other stresses – including market integration, governmental policies and environmental change – constantly change, as do the dimensions defining sensitivity and capacity to adapt. Adaptation is also a process of constant adjustments and learning that may evolve in response to different exposures and past experiences. In this context, high adaptive capacity and successful adaptation to one stress (e.g. drought) may result in

**A dynamic, complex and strong link exists between economic development, urbanization and GHG emissions**

**It is the affluent and politically enfranchised ... groups and communities ...who consume more, travel more and become the highest GHG emitters**

exposure to new stresses (such as the urban heat-island effect or water scarcity), some of them provoked by coping responses (such as the use of air conditioning or increased extraction of groundwater). It is therefore important for decision-makers to understand how such complex processes interact and change over time because this understanding can help to inform more successful adaptation and avoid potential negative feedbacks or unintended consequences.

## Cities and the multifaceted nature of climate responses

Representatives from different countries, states and cities are responding at multiple sectoral and governmental levels to the mitigation and adaptation challenges posed by the implications of climate change. These responses go beyond traditional national and state activity, and frequently imply not only multilevel public interventions, but also public–private cooperation and autonomous responses, and self-regulation by individuals and groups. These responses and the issues that they are intended to address are multi-scale in nature because most of the processes involved operate at multiple levels. It is frequently the case that mitigation and adaptation responses do not fit with the issues that they are intended to address. For instance, many of the climatic cause-and-effect relationships are long term and potentially irreversible and, therefore, require preplanning that goes beyond the tenure, administrative power and even the lifetime of most current decision-makers and stakeholders. This makes policy decisions in this area particularly difficult, as uncertainties exist in the understanding of the outcomes and impacts of climate change.

Ideas and policies centred on development, sustainability, climate change and some of their central issues (poverty reduction, disaster management and climate change adaptation) share key characteristics. For instance, in the area of climate change, the notion of development opens the possibility of promoting deep transformations in models of production and lifestyles. The specific nature of these changes has been defined in different ways. The *first* and dominant way is to use new markets to manipulate the inputs and outputs of the existing market system in an attempt to transform them, thus affecting everything beneath the overarching economic system in a cascading or domino-like fashion (such as by using carbon markets to create incentives to curb GHG emissions). The *second* way focuses first on equity and attempts to create transitions based on models of development that include sustainable use of the environment and non-market-driven alternatives to promote human well-being. It is this vision of sustainable and resilient development that has great potential for a movement away from current, unfair and unsustainable patterns of energy use and their dangerous impacts upon the climate system. This alternative model of development would allow urban populations and decision-makers to move towards equity, minimizing human suffering from climate-related disasters and promoting well-being, while creating the conditions for improvements in quality of life for undeveloped areas, including poor urban slum dwellers of

the world. It would create the basis for many alternative development policies and programmes at the international, national, state (or province) and urban levels of governance and in civil society. It would also foster development that can fulfil the twin roles of improving the quality of life of the urban poor, while creating sustainable urban lifestyles that are central to the messages of this report.

Chapter 2 describes the process by which climate change has become part of the international agenda, exploring the main mechanisms, instruments and financing strategies of the UNFCCC and the Kyoto Protocol. The message of climate change, however, only caught public attention with increased scientific knowledge of, and public concern about, global environmental issues that crystallized in the creation of the UNFCCC. This new public awareness was further catalysed by an array of extreme events that are increasingly affecting the world, and the creation of the IPCC. The chapter also identifies the key actors, components and actions of climate governance other than the Climate convention and protocol at the international, regional, national and sub-national levels. The implications of the international climate change milieu for local action at the city level are described and the extent to which actors of this level have benefited from the various funding and support mechanisms currently available is reviewed.

Chapter 2 also outlines some common features that have defined the international climate regime, such as the use of a 'framework' scheme with general formulations that are deliberately ambiguous in order to limit conflicts between the positions of all representatives. The basic principles arrived at are then fleshed out through regular post-agreement meetings of the countries that have adopted the UNFCCC. Particularly in negotiations during key sessions of the Conference of the Parties (COP) to the UNFCCC, little progress is made during most of the negotiation period. Precisely because effective policies to reduce GHG emissions imply deep transformation in energy systems, lifestyles and economic activities, an understandably high contentiousness exists every time the COP discusses how much needs to be mitigated by whom, when and where (burden and timetables of commitments); who will pay for the responses and how (financial assistance and technology transfer); and what institutions and implementation mechanisms need to be in place to ensure participation and compliance.

Conflicts and uncertainties can help to understand, at least partially, the complex and fragmented governance of climate issues. Yet, equally important is to be aware of the fact that rather than being a wholly rational process, policy-making is an incremental undertaking. Climate governance is made up of a patchwork of binding agreements (e.g. the Kyoto Protocol), organizations (such as the UNFCCC secretariat, the IPCC and the United Nations) and networks that are quite different and distinct in their functions and approaches (e.g. rule-setting and information-sharing), their constituencies (private and public), their spatial scope (local, bilateral to global), their focus (e.g. mitigation, adaptation, disaster management and development), and their capacity to steer climate-relevant action. The Climate Convention

*Countries, states and cities are responding at multiple sectoral and governmental levels to the mitigation and adaptation challenges*

also coexists with a set of parallel initiatives and frameworks (e.g. the Hyogo Framework), operating at different sectoral and spatial levels and exerting deep influence on climate issues. For instance, the adaptation and disaster risk management communities share many commonalities, and can learn and benefit from each other's concepts and experience. Yet, outstanding differences also exist, particularly in terms of terminology, actors involved and types of intervention.

A relatively small number of countries, states/provinces and cities have played leading roles in addressing mitigation and – to a lesser extent – adaptation. Some (e.g. London, UK; California, US; King County, Oregon, US; Durban, South Africa) have launched ambitious climate change programmes; have created positive synergies with other tiers of government; and have mobilized the necessary support from the public and private sectors to curb GHG emissions and adapt to climate change. However, even the leaders and frontrunners in climate change action are faced with multiple challenges and difficulties in achieving their mitigation targets (such as the UK). This is true because many proposed actions are voluntary, and policies in many of the existing plans do not appear adequate to address the problem.

Although existing knowledge lags behind the recent explosion in city responses to climate change, it can be said that some urban actors have been able to take advantage of the opportunities offered by the multilevel governance structures briefly described in Chapter 2. More urban authorities than ever currently participate in transnational networks, research-sharing, learning initiatives and advocacy efforts. These urban actors have developed a more aggressive approach, seeking to secure the economic competitiveness of their cities and to input a local voice in international negotiations (such as the World Mayors Council on Climate Change at the COPs) and organizations.

Climate action at the urban level has been shaped by a myriad of factors. These are given by institutional conditions and incentives, such as existing international instruments and financing mechanisms, supra-regional programmes and national regulation systems. This Global Report has provided different examples of this. The emphasis on mitigation strategies and actions by city-relevant local authorities can be partially attributed to the importance of such international mechanisms and programmes as the Clean Development Mechanism (CDM), which were made operational earlier than adaptation funding mechanisms, such as the Adaptation Fund.[4] This mitigation emphasis is also the result of the design, within the European Union, of the European Emissions Trading Scheme – the largest multinational GHG emissions trading scheme in the world – and the leadership of such countries as the UK, Germany and Norway that have been key promoters of climate policies aimed at mitigation. These countries have assembled an array of policies to achieve long-term reductions.

Action on climate change issues – for mitigation or adaptation – is largely a function of knowledge, whether generated by scientific communities or brokered by the media, scientific entrepreneurs or non-governmental organizations (NGOs) at different levels (from the international to the local). It is, hence, necessary for academic institutions, local authorities and key stakeholders to generate the necessary information and create the sense of identity and the buy-in necessary to affect change. Equally important, however, has been the power that different groups have to make their points of view prevail.

Individual and organizational leadership has been another factor shaping climate action and creating windows of opportunity offered by transnational networks. However, administrative structures, party politics, political timetables, inertias and many other institutional constraints need to be overcome, thus requiring a broader-based institutional capacity for climate protection. The absence of this institutional capacity has deterred key mitigation and adaptation efforts. Yet, paradoxically, in some cases (e.g. US actions at the state and urban levels), it has become another source of opportunity for state and local actors to fill a leadership gap.

A fundamental goal of urban actors has been to offer the conditions for business and investment to flourish. This can attract jobs and tax revenue in carbon-relevant sectors (such as renewable energy and production of more efficient appliances). However, it can also create an environmental race for the bottom, as regulations protecting the health and well-being of urban inhabitants are cut in order to promote a business-friendly environment, thus negatively affecting adaptive action.

Creating policies to address climate change is not only about goodwill or institutional capacity, it is also about understanding *the inertia and endurance* characterizing many of the issues that adaptation and mitigation actions are supposed to address. Power plants, refineries and other energy investments have long lifetimes. Similarly, this is also the case with water systems, roads, houses and other components of the built environment at risk from the impacts of climate change. Although increased research, development and actions to reduce emissions are required within the next few years to achieve the target of no more than a 2°C increase in the Earth's average temperature, it will take decades to centuries to move the world's current energy system away from its dependency on fossil fuels, the main source of GHG emissions. Urban form changes at slow rates, cannot be easily shaped by design and takes a very long time to build urban infrastructures.

A key problem outlined in Chapter 2 is that actors and agents of climate change at all levels, including governments, NGOs and civil society, are, most often, preoccupied with immediate and often localized interests and priorities; but these same actors need to move within short timeframes to guarantee long-term and wide-ranging global interests that can seem remote and unpredictable at best. Much action on mitigation and adaptation will need to come from local actors and agents, focusing their work at the local level where all the impacts of climate change will ultimately be felt. Networks of local actors can further energize this movement and may accelerate action at the global level. This work must include education and outreach to build broad-based support for mitigation and adaptation initiatives to increase the adaptive capacity of areas and populations that are most vulnerable to the effects of climate change. It will also

*More urban authorities than ever currently participate in transnational networks, research-sharing, learning initiatives and advocacy efforts*

*Individual and organizational leadership has been ... shaping climate action and creating windows of opportunity*

require a shift in paradigm, from the current focus on international responses to one that is more broad based and inclusive of actions at national and local levels.

## Sources and drivers of cities' GHG emissions

To explore the sources and drivers of urban areas' contribution to climate change is of utmost importance for several reasons. *First*, transportation, energy generation, industrial production and other urban sources are associated with cities and their functioning. Each of these sectors constitutes a universe by itself, not only in terms of the types of GHGs that they generate, or the factors explaining differences in the levels and carbon intensities of their emissions, but also in terms of the mitigation opportunities that they offer, all of which will be briefly discussed here.

*Energy* is by far the most relevant sector for assessing GHG emissions, as the combustion of fossil fuels for electricity generation, heating, cooling, cooking, transportation and industrial production is the major source of GHGs. The energy systems that urban areas rely heavily on are shaped by the quantity of energy used, the energy structure (i.e. the types of energy forms used) and the quality of the energy (e.g. natural gas is less carbon intensive than coal). Variations in emissions by one of the main urban energy sectors, electricity consumption – both between and within urban areas around the world – depend on several factors that policy-makers can address: access to the grid; the type of fuel used to generate electricity; technologies applied; and existence of alternative sources of generation (renewable, nuclear, etc.).

*Transportation* is another key emitter that increases as economies grow, especially in developing countries, and as incomes rise. Emissions by the sector are expected to continue increasing in the coming decades. Particularly in developed countries, urban areas often generate smaller amounts of per capita GHG emissions from ground transportation than rural areas. Density plays a key role in this difference, and is one of the most important factors influencing differences both in the amount of energy used and GHGs emitted across urban areas. This should not lead decision-makers, however, to simply base their actions on a snapshot of *urban form* at a particular moment in time. It should, rather, lead them to address the dynamics of such processes as the extent of automobile use, quality of public transit, land-use planning and governmental policies, all of which determine the impact of urban density upon energy use and emissions by the transport sector. Policies aimed at reducing emissions by the sector need to consider that differences in emissions for a mode of transport (e.g. private vehicles) also depend on several factors: size and types of vehicles, efficiency of engines, maintenance practices, vehicle-trip frequencies and operating speeds, and driving behaviour.

*Commercial and residential buildings* are key sources of direct emissions, indirect emissions and emissions associated with embodied energy (i.e. commercial energy used to make products). These are related to onsite combustion of fuels, public electricity use for street lighting and district heat consumption, and through the materials used for their construction. Decision-makers need to pay attention to such factors that determine emissions from buildings as the need for heating and cooling (determined by climate conditions, but also by cultural preferences and access to monetary resources), the construction of the building, the behaviour of building occupants, the type of fuel used, the size of the space to be heated or cooled, and the orientation of the buildings.

Two other key emitting sectors are *industry and waste*. Because many industrial activities are energy intensive in their operation, their increasing dominance in the economies of such cities as Saldanha Bay in South Africa or Shanghai in China (see Chapter 3) can make up a big part of their emissions. Mitigation policies and strategies need to address the following factors accounting for differences in industrial emissions: location, size and age of the industrial facilities, as well as the carbon intensity of their energy sources. Although waste is a small contributor to global emissions, rates of waste generation have increased during recent years, particularly in rapidly industrializing developing countries that have been experiencing increasing affluence. Waste generation is linked to population, affluence and urbanization; yet – as exemplified by Barcelona (Spain), London (UK) and New York (US) – emissions from waste generation can be greatly reduced by such measures as efficient collection, and technologies for methane capture and storage, as well as for methane to energy conversion.

The *second* reason for highlighting the need for an understanding of the sources and drivers of GHG emissions relates to a twofold purpose of measuring emissions from cities: inventories of emissions provide a basis for comparisons and for inter-urban competition and cooperation; and they constitute a crucial first step in identifying potential solutions. However, cities also rely on inward flows of food, water and consumer goods that result in GHG emissions from areas outside the city.

Notwithstanding the importance of emission inventories, it has been difficult to develop a standardized globally comparable methodology for GHG emissions at the local or municipal level. There are a number of reasons for this:

- It is difficult to attribute to cities emissions by such sectors as aviation and shipping. For example, many of the passengers using major international airports situated in or close to major cities may be from elsewhere in the country, or may only be using these airports for transit purposes.
- The different methodologies used to measure emissions can result in different numbers (i.e. scope issues). For example, Scope 1 inventories only include direct emission sources within the political boundary of a city, while Scope 3 would include all indirect and embodied emissions (such as GHG emissions embedded in food).
- A wide range of boundary definitions are used to define urban areas and cities. It is clear from Chapter 3 that the smaller the scale, the greater the challenges posed by 'boundary problems', which make it increasingly hard to identify which emissions ought or ought not to be allocated to a particular place.

---

*Energy is by far the most relevant sector for assessing GHG emissions*

*It has been difficult to develop a standardized globally comparable methodology for GHG emissions at the local or municipal level*

The above should lead policy-makers to be extremely cautious about statements or numbers on the total contribution of urban areas to GHG emissions – not only because of the lack of an accepted definition of an 'urban area' or 'city', or of globally accepted standards for recording emissions,[5] but also because little clarity exists on the relative allocation of responsibility from 'production-based' or 'consumption-based' approaches. This is illustrated by the fact that vastly different proportions of emissions can be attributed to the manufacturing sector of Chinese or African cities discussed in Chapter 3, which produce goods for consumption in many other locations around the world.

What is clear is that a large measure of the responsibility for the emissions in the producing country should be borne by individuals consuming the produced goods. In contrast with most assessments of the urban contribution to climate change, focused on the emissions that are produced by activities taking place within given territorial boundaries, Chapter 3, therefore, suggests an alternative approach which also considers the emissions associated with the consumption patterns of individuals. This idea acknowledges the fact that many agricultural and manufacturing activities that meet the needs of urban residents take place outside city boundaries, and often in other countries. Furthermore, and most importantly, it leads to the conclusion that unsustainable levels of consumption – as partially determined by the marketing strategies of corporations, but which also drive the processes of production – are crucial to understanding the contribution that urban areas are making to climate change.

Besides patterns of consumption, a variety of overriding factors account for the different contributions of urban areas to GHGs, both within and across countries. The *first* are the various dimensions of geography that can be broadly categorized as climatic situation, altitude and location in relation to sources of energy resources (e.g. hydroelectricity and coal).

The *second* is the demographic composition and dynamics of a society given by changing age structures, and the increasing trend (at least within wealthier groups) towards smaller households. Demographic dynamics relate to GHG emissions in very complex and shifting ways.

Urban form and urban density are the *third* factor, related to a series of social and environmental outcomes. For instance, the extremely low densities of many suburban areas (particularly in North America and Australia) are associated with high levels of household energy consumption and emissions as a result of sprawling buildings and extensive car usage. On the other hand, the extremely high densities of many developing country cities can be related to increased health risks, and high levels of vulnerability to climate change and extreme events. Some of the many factors determining climate risks can be exacerbated by density: coastal location, exposure to the urban heat-island effect, high levels of outdoor and indoor air pollution, and poor sanitation.[6] These same factors, however, can create opportunities for simultaneously improving health and cutting GHG emissions through policies related to transport systems, urban planning, building regulations and household energy supply.

*Finally*, the types of economic activities taking place within an urban centre are another key determinant of GHG emissions, not only because the dominance of industrial activities has a huge influence on patterns of emissions, but also because – as illustrated by many cities in South Africa – extractive activities and energy-intensive manufacturing, particularly if depending on fossil fuels, are obviously associated with higher levels of emissions.

## The multiple urban faces of climate impacts and vulnerabilities

Chapter 4 highlights the potential cumulative and multiplicative effects of loss of lives, damage to infrastructure and of other climate-related impacts. It also addresses the distributional nature of climate change impacts both within and among cities. However, not all of the outcomes of climate change are negative. As illustrated by cities as diverse as Durban (South Africa), Manizales (Colombia), New York (US) and London (UK), the potential also exists for cities to serve as sources of resilience to climate change, taking advantage of opportunities to address multiple developmental problems at once and to lead the world on adaptation efforts.

Chapter 4 also points to current findings on the recent and future trajectory and geographical variations in climate hazards. For instance:

- Although showing significant regional variation, average sea levels have been rising and are expected to continue to rise around the world, thus putting coastal urban areas at risk from property damage, displacement of residents, disruption of transportation and wetlands loss.
- Tropical cyclones and extra-tropical storms have been increasing in intensity since the 1970s.
- Severe precipitation events have become more intense and more frequent, and are predicted to cause a greater incidence of urban flooding.
- As a result of climate change, extreme heat events are predicted to become more intense, longer lasting and more frequent over most land areas.

Urban areas are already facing an array of hazards, with some that are related to climate change and others that are not; but together these hazards may present a complex set of circumstances that will increase impacts. Heat waves coalesce with urban heat islands and air pollution to put urban populations at increased risk from respiratory mortality. When hitting areas subjected to processes of deforestation and erosion, heavy precipitation events result in flooding and landslides, whose primary victims are populations living in slums.

Climate events can also result in different sets of social and environmental impacts upon the economic sectors, infrastructures and population groups of a city. For instance, severe weather events, including heavy precipitation and wind, can decimate the built environment, including homes and places of business. They can also

*Unsustainable levels of consumption ... are crucial to understanding the contribution that urban areas are making to climate change*

*Climate events can ... result in different sets of social and environmental impacts upon the economic sectors, infrastructures and population groups of a city*

disrupt and cause lasting damage to highways, seaports, rivers, bridges and other components of the transportation systems that urban centres depend on. These weather impacts can affect such infrastructures as water supply, sanitation and energy provision. They can also affect the insurance industry and its beneficiaries by increasing the costs of insurance coverage and can negatively affect not only retail and commercial services, but also industrial facilities, especially if they are located in risk-prone areas or depend on climate-sensitive inputs. Furthermore, they can make it difficult for residents of informal settlements to conduct small-scale commerce, petty trading and artisan trades.

When considering how climate change will impact upon urban areas, it is important to distinguish between risk and vulnerability. The same risk factors (such as hurricanes and floods) can be experienced differently by different individuals, demographic groups, cities and countries. Most climate change risks have a high degree of regional and local variation. The level of risk to an urban area from these impacts depends, in part, on how much of the city's population and economic assets are located in high-risk areas (i.e. exposure). In many cases, exposure level will be a function of the location of the city itself. Exposure can also be linked to land-use planning within a city, including continued development in known hazardous zones, and the destruction of natural protective areas.

Similar climate change impacts are not experienced the same way by cities in developing and developed countries. The degree to which urban areas are vulnerable to climate hazards or can suffer negative impacts depends not only on the nature and dynamics of physical hazards, but also on social and environmental factors such as the:

- integrity of infrastructure and urban planning, or lack thereof;
- availability of financial and human resources;
- presence of disease and malnourishment;
- availability of information and level of awareness of risk; and
- extent of dependence on natural resources.

Differences in climate impacts also exist between men and women, the elderly and children, and the wealthy and the poor, both in developed and developing countries. For example, men and women differ in their livelihoods, family roles, behaviours, access to assets and perceptions of risk. Both can be treated differently with respect to planning and relief efforts during and after disasters. Impacts are also frequently more severe for the disabled, and ethnic and other minorities, who are frequently and relatively more disadvantaged in terms of access to assets and adaptation options. The vulnerability of children relates more to their physiological immaturity or to their limited cognitive ability and behavioural experiences, compared to adults. The vulnerability of the elderly can depend upon such factors as poverty (greater in developing countries), social isolation (more common in developed countries) or deteriorating physical health and mobility.

It is also important to note the issue of compounding vulnerabilities – that is, the fact that some population groups fall into more than one such category (e.g. poor old women) and can thus find themselves dramatically constrained in their ability not only to cope with future climate hazards, but even to prepare for and respond to the varied stresses that they already face.

Government intervention can fundamentally improve urban resilience to climate change impacts through targeted adaptive finance, broad institutional strengthening and more attention to the causes of vulnerability. The opposite, however, can also be the case. Maladaptive policies – such as ineffective or completely lacking land-use controls, weak building code standards or ineffective enforcement – have directly resulted in increased vulnerability of urban areas or of households and communities within them to sea-level rise, flooding and coastal storms.

In order to improve resilience to climate impacts, it has been suggested that urban governance should target the most vulnerable populations – namely, the urban poor and individuals living in informal settlements. These two groups have often been ignored in more conventional urban planning and intervention. Policies will need not only to reduce the vulnerabilities of existing populations, but also to address the underlying issues that permit settlement in marginal and vulnerable urban areas in the first place.

## Mitigation responses

Because cities represent concentrations of populations and economic activities with expanding energy demands for heating, cooling, lighting, transportation, industrial processes, water provision, waste disposal and telecommunications, they can be seen as one 'part of the problem' of climate change. Reducing GHG emissions is, therefore, one of the key policy challenges that cities face. Beyond this view of cities as a major culprit in global climate change, however, there is also great promise for these same cities to become 'part of the solution'. Cities may play a major role in mitigation efforts for three reasons. *First*, they have direct or indirect jurisdictional responsibility for some of the key processes that may be involved in the production or reduction of GHGs – processes such as transportation, energy generation and use, land-use planning, and waste collection and disposal. *Second*, by virtue of the fact that cities concentrate populations and businesses, they may also make feasible such potential solutions as mass transit and energy savings in office buildings. *Third*, they may act as a catalyst to other potential levels of action on climate change as municipal governments interact with stakeholders in the private sector and civil society. Over the past two decades, in fact, cities have become a 'crucial arena' where the challenges of climate change are being addressed.

Chapter 5 notes that there are five key sectors where mitigation actions are taking place in urban areas. These sectors are urban form and structure; the built environment; urban infrastructures; transport; and carbon sequestration. In terms of the first of these, *urban sprawl* is an area of concern for both developed and developing countries. This is

---

*Similar climate change impacts are not experienced the same way by cities in developing and developed countries*

*Over the past two decades ... cities have become a 'crucial arena' where the challenges of climate change are being addressed*

so because distances travelled between home and work, education or leisure activities equate to a greater reliance on motorized transportation. Sometimes sprawl has also led to middle-class urban fringe districts where more available land and a release from some of the building constraints of the central city has meant larger house sizes and higher per capita GHG emissions. In other cities, however, sprawl is fuelled by the growth of informal settlements.

In order to address these issues, many strategies have been used to limit urban expansion, reduce travel and increase energy efficiency of the urban form. Some of these take the form of small- and large-scale regeneration projects (urban renewal), and these have taken place primarily in developed countries. Within developing countries there are few mitigation initiatives that make explicit use of urban form; when attempted, they are often constrained by a lack of capacity among local governments to implement them effectively. They also have been criticized for their exclusive nature and potential for exacerbating social inequalities (e.g. the eco-city Dongtan in China).

The design and use of the *built environment* is an area that is critical to urban mitigation efforts, and actions fall into three broad categories. These are economic incentives, regulatory requirements and information programmes. More recently, a growth in voluntary public–private partnerships and a mixing of these approaches has led to an explosion in the range of approaches used, including micro-generation and new building materials. Nonetheless, built environment projects primarily take place within the cities of developed countries, and have sometimes taken the form of efforts to help the urban poor. Some of these projects[7] have been led by grassroots organizations and housing co-operatives, suggesting that innovative forms of social organization are emerging and creating initiatives that address climate change mitigation, while also potentially taking on issues of social and environmental justice. The combination of social and environmental gains may be particularly useful in developing countries and for addressing such issues as fuel poverty.

Many of the *urban infrastructure* initiatives focusing on energy efficiency are primarily driven by such factors as a desire for energy security and financial savings, and – to a lesser extent – by the opportunities offered by international instruments such as the CDM. Both drivers may have helped to keep these projects economically and politically feasible, but, at the same time, may limit their effectiveness in long-term GHG savings, as financial savings have sometimes led to increased usage. Because of this, measures must be taken against the rebound effect, where increased efficiency leads to higher energy consumption. Thus, energy efficiency projects need to be coupled with the development of low-carbon renewable fuel sources and efforts to reduce energy consumption.

As noted above, the *transport* sector is a large contributor to GHG emissions. Growth in GHG emissions also reflects a modal shift, since higher incomes increase the affordability and desirability of private automobiles, and middle- and high-income groups within developing countries move towards the use of private vehicles and away from public transportation. Climate change mitigation schemes in the transport sector may be grouped into seven categories, including low-carbon transport infrastructure; low-carbon infrastructure renewal; fleet replacement; fuel switching; enhancing energy efficiency; demand-reduction measures (for private motorized vehicles); and demand-enhancement measures (for public and other low-carbon modes of transportation, such as human powered).

*Carbon sequestration* involves removing carbon from the atmosphere, either through promoting natural carbon sinks (such as planting trees or protecting forests) or by technological means for carbon capture and storage. While carbon sequestration schemes have mostly been on the periphery of urban mitigation efforts, new technologies for carbon capture and storage and international carbon finance are bringing carbon sequestration to the fore. Most carbon sequestration schemes at the urban level relate to tree-planting schemes, as well as restoration or preservation of carbon sinks. Carbon sequestration projects combine well with city beautification programmes where measures such as creating and protecting green spaces and facilitating public access can be brought together.

Despite the array of mitigation responses by urban centres to date, a piecemeal rather than a strategic approach is very common. Furthermore, notwithstanding the existence of initiatives to measure mitigation outcomes,[8] there is limited information about the individual and collective impact of existing responses, especially when they extend beyond municipal buildings and infrastructure systems or involve behavioural change. The reasons for this include the relatively short time-scales involved and the fragmented nature of the data available, especially with regard to levels and reductions of GHG emissions within and across urban communities and sectors.

Of the four types of municipal governance described in Chapter 5, *self-governing*, the one mostly emphasized by municipal authorities, faces a limitation: frequently, municipal GHG emissions make up a small percentage of the total city emissions. This means that too much attention to the self-governing mode may detract resources from the broader mitigation challenges faced by a city. Seeking to govern climate change through the *provision* of infrastructure and services holds the most potential in cities where municipal governments retain ownership or control of infrastructure networks, and where basic needs have been met. Because of their targeted and enforceable nature, taxation, land-use planning and other *regulation* mechanisms can also be very effective in terms of reducing GHG emissions. Yet, these are also the least popular approaches adopted by municipal governments and, therefore, the most difficult to sustain politically. The *enabling* mode of governing has significant mitigation advantages: it results in (relatively) low upfront economic and political costs, and can also increase the transparency and legitimacy of urban governance. However, enabling initiatives are restricted to those who are willing to participate, and cannot be enforced.

Chapter 5 also explores three modes of public–private governance of climate change action: voluntary, private provision and mobilization. The chapter uses the limited available data on this relatively new phenomenon to draw

**The design and use of the built environment is an area that is critical to urban mitigation efforts**

**Despite the array of mitigation responses by urban centres to date, a piecemeal rather than a strategic approach is very common**

some preliminary conclusions. These approaches tend to be adopted by partnerships or networks, and to focus on the adoption of voluntary standards. They have the potential to offer win–win options (i.e. tackling mitigation through a progressive, inclusive and environmentally fair approach). They are also faced with problems, however: they are small in scale and often politically marginal. They may serve to shift accountability away from actors with much higher responsibilities for the bulk of (urban) GHG emissions. Although partnerships may provide shared resources, knowledge and other benefits, they are often fragile and threatened by the potential of promoting the points of view of a select group of individuals and serving powerful interests at the expense of the disenfranchised.

## Adaptation responses

Because mitigation efforts at all levels have so far not been able to move the global climate system away from its current and dangerous trajectory of change, adaptation actions are urgently needed to address both current and future climate risks in urban areas. What decision-makers at different levels do today to cope and adapt will have an influence on the lives and livelihoods of millions of urban dwellers. Buildings, infrastructures, energy systems and other key components of cities are long lasting. Therefore, what is designed and built now will be fundamental in coping with climate change many decades into the future.

Urban populations have long had to cope with a wide range of risks to their economic activities, lives and livelihoods. In the absence of effective local government actions, these become the most frequent types of response to climate change. Yet, these responses are generally small scale; they do not address the underlying root causes of vulnerability and are therefore best described as coping strategies.

Wealth and access to assets, information or social networks can help individuals to reduce the risk of negative outcomes. Wealth, for instance, allows individuals to buy, build or rent homes that can withstand extreme weather in districts that have not been at risk from flooding. As illustrated by such cases as Dhaka (Bangladesh) and Lagos (Nigeria), populations lacking access to these use other strategies to reduce the risks of negative outcomes. Most of the measures taken to help cope with climate events are impact minimizing or impact reducing, rather than preventive.

Community-based adaptation is important in urban areas, as it helps to address the limitations or inadequacies of governmental intervention (such as in the provision of infrastructure and services); and because it can become an important part in the enhancement of resilience to extreme, and increasingly unpredictable, weather events. Community-based adaptation is based on the premise that local communities have the skills, experience, local knowledge and motivation, and that, through community organizations or networks, they can undertake locally appropriate risk reduction. However, to be effective, community-based adaptation depends on the actual existence of a collective 'community-based' organization in which the needs and priorities of those most at risk or most vulnerable are represented and actually taken care of in effective ways. It also needs to focus on the reasons why the urban poor are disproportionately vulnerable to climate change, such as through their greater exposure to hazards, the lack of hazard-reducing infrastructure, the lack of state provision for assistance after extreme events, and the lack of legal and financial protection.

Equally relevant can be other grassroots organizations. For example, by enhancing the capacity of the urban poor to save regularly, by helping to identify and purchase land for housing, and by promoting other actions of community organization, slum federations (such as in the Philippines and India) are helping to build the resilience of low-income households to many potential shocks.

Yet, community-based adaptation and grassroots organizations are faced with constraints arising from the immense cost, energy and time required to construct, develop and maintain the key determinants of resilience for the inhabitants of many cities in developing countries. These determinants of resilience include infrastructure and services, warning systems, emergency responses, education, etc. In fact, most climate change-related risks exacerbate risks already present, which are the result of inadequacies in local governments' capacities or willingness to manage and govern urban areas. Thus, there is a large deficit in the basic infrastructure and services needed to address not only risks related to extreme weather and water constraints, but also 'everyday' risks.

Cities in developed countries do not have very large infrastructure deficits. Most or all of their inhabitants live in buildings that meet building standards, have access to education and are served by piped water supplies, sewers, drains and solid waste collection. Therefore, their adaptation responses are frequently relatively easier to design, implement and fund. This does not mean that adaptation easily gets the political support that it deserves. Many cities need major upgrades in their infrastructure and should take account of likely climate change impacts. Most cities need to expand their capacity to anticipate and manage extreme weather events. Some cities are located on sites that are at risk from the implications of climate change (such as coastal areas). Finally, key actors in many developed country cities which struggle with economic decline see climate change risks as a distant danger.

Some components of effective adaptation responses can be drawn from the analysis of different case studies presented in Chapter 6. An essential first stage is the recognition among authorities and stakeholders that climate change impacts need to be considered. Then an information base on current conditions (i.e. on impacts of past extreme weather and other disasters) needs to be developed. Involved parties need to build on community and district assessments, as well as projections of future climate change, to develop risk/vulnerability assessments for the city. Strategic plans for the city as a whole and for its different sectors should be developed in association with other stakeholders. Furthermore, measures should be undertaken to

*Adaptation actions are urgently needed to address both current and future climate risks in urban areas*

*Community-based adaptation is important in urban areas, as it helps to address the limitations or inadequacies of governmental intervention*

support those adaptation responses that are already taking place.

Financing for adaptation revolves around two main issues: whether funds will be available to cover the cost of adaptation for urban areas, and whether there is the capacity to use such funds to ensure that the needed adaptation can take place. Financing for adaptation can complement development assistance. While the latter can help focus on the drivers of vulnerability that are associated with weak institutional capacity, the Adaptation Fund (see Box 2.2) can support the broader climate risk management strategies of developing countries. Furthermore, both can help to overcome the contentious issue of the boundary between climate change adaptation and development, if designed with this in mind.

A further related issue refers to the costs of adaptation. Most estimates of the costs of adaptation relevant to urban areas are estimates of the costs of adapting infrastructure, and these are faced with some problems. First is the ambiguity as to what to include under infrastructure (housing, for example, is sometimes included by the IPCC, sometimes excluded). Second is the assumption made that costs can be calculated by applying a small increment to existing investment flows into infrastructure that is climate sensitive, with no account taken of the very large infrastructure deficits. This leads to overestimates of the proportion of investment allocated to developed countries and to underestimates of the sums needed for Africa and other places where there are very low/inadequate investment flows into infrastructure. The third is the belief that the availability of funding from international agencies is the 'solution' for adaptation, forgetting that local governments in many developing countries are often weak, ineffective and unaccountable to local populations. As a result, their capacity to design and implement appropriate adaptation strategies responding to the requirements of those most at risk from climate change must be in doubt. Last, but not least, is the idea that 'adaptation' and 'development' can be kept separate. As noted in earlier chapters, climate change impacts upon the ground are exacerbating non-climate change impacts, and addressing both is inhibited by institutional/governance failures. It is therefore necessary to study carefully what adaptation would involve in particular locations and what component of this is related to the existing infrastructure deficit.

It is also important to keep in mind that it will not be possible to adapt to climate change impacts in a number of locations – because adaptation is considered too expensive or technically unfeasible. Such consequences are often referred to as 'residual damage', and the number of such locations (and populations at risk) is likely to rise without successful mitigation. In addition, the issue of migrants who are forced to leave their homes due to future climate change needs to be addressed. As noted in Chapter 6, people whose lives and homes cannot be adapted *in situ* fall outside the scope of most national and international legislation. Thus, there have been some, though still marginal, calls for the development of new international legislation to address the concerns of 'climate migrants' – perhaps in the form of an international convention for persons displaced by climate change.

# ADDRESSING URBAN GHG EMISSIONS AND VULNERABILITIES: CHALLENGES, CONSTRAINTS AND OPPORTUNITIES

Based on the findings of the previous chapters of this Global Report, this section explores the challenges, constraints and opportunities of efforts to decrease urban GHG emissions and thereby enhance society's resilience to climate change. The global mitigation challenge will be to achieve development paths that will bring down emissions by 2015 and stabilize them by the end of the century at 445 to 490 parts per million $CO_2$ equivalent ($CO_2$eq) by volume.[9] Only in this way can the global average temperature increase be kept below 2°C, which, as recognized in the Copenhagen Accord, is necessary to prevent harmful human interference with the climate system.

Considering an estimated global population of 9 billion by 2050 and an increasing urban share of that population, this means individual carbon footprints around the world will have to be kept at an average of less than 2.2 tonnes per year. Yet, annual per capita emissions in some US cities reach (or even exceed) 20 tonnes of $CO_2$eq.[10] Thus, there is a need to reduce the emissions of many cities and citizens in developed countries (and even in some developing countries) considerably. In order to address this challenge, multilevel and multi-sectoral actions – including many measures at the urban level – will need to achieve:

- reductions in the quantities of fossil fuels used;
- reductions in the carbon content of the fossil fuels used (such as a switch from coal to natural gas); and
- changes in the energy structure (such as increased reliance on renewable energy sources) by switching to other sources of energy, while ensuring that the quality of energy provision is kept.

For example, measures need to be undertaken to ensure that electricity, a key component of urban life, is generated through less carbon-intensive energy sources.[11]

All of these require that the decline in the *carbon intensity of fuels* and the increase in both *energy efficiency* and provision of low carbon-intensity clean fuels are achieved in such a way that the global amount of emissions from fossil fuels is substantially reduced. And as can be concluded from this Global Report, this is not always the case.

Mitigation responses formulated so far have primarily focused on improving energy efficiency or reducing carbon intensity, which does not necessarily translate into a reduction in the total amount of emissions. A focus on energy

*Financing for adaptation can complement development assistance*

*There is a need to reduce the emissions of many cities and citizens in developed countries ... considerably*

efficiency savings in current infrastructure and devices (such as cars) can result in a 'rebound effect' – that is, an increase in consumption (resulting, for instance, from the use of smaller engines but driving longer distances) following financial savings in their operation. Furthermore, such a focus can downplay other more effective options. For example, sizeable renewable energy installations, including wind, solar and hydropower, have received relatively lower emphasis.[12] Thus, regulations and incentives by decision-makers operating at international and national levels need to be set, focusing on a portfolio of energy alternatives (i.e. not only on fossil fuels or biofuels, but on combinations of all fuels that take advantage of and respond to differing circumstances and contexts).

Cities are and can contribute to addressing the mitigation challenges of climate change in several ways:

- as initial seedbeds and niches for entrepreneurial experiments with radically new technologies (by commercial private-sector actors);
- as lively laboratories for experimentation among emerging and future-looking communities (such as Hammarby Sjöstad in Stockholm, Sweden[13]) that share particular perceptions, visions and ideas as to how to move urban communities away from current unsustainable development paths; or
- as communities that build networks and platforms (such as workshops and conferences) to facilitate the exchange of knowledge and experiences, as well as the articulation of best practices.[14]

Depending on their national contexts and histories, urban local authorities have a highly variable level of influence over GHG emissions. They can induce emissions reductions in the energy sector through measures such as retrofitting commercial, domestic and municipal buildings, by switching traffic lights to energy-saving bulbs, etc. Besides having influence over the efficiency of their transportation fleets, they can implement transport planning policies which encourage alternatives to the private car, such as the Transmilenio in Bogotá (Colombia). They can design zoning measures to promote certain patterns of settlement, energy efficiency measures in new buildings, and standards for domestic and commercial buildings, as exemplified by the experiences of London (UK) and Chicago (US). They can implement programmes to reduce GHG emissions in the waste sector, such as through methane capture. Non-governmental actors such as private-sector organizations are now voluntarily involved in actions to decrease energy use in commercial buildings. A similar movement is happening within civil society groups, such as the 'transition towns' movement.[15]

The number of actions currently taking shape show that involved stakeholders do acknowledge the urgency of mitigation, and demonstrate their awareness that action should be taken now to avoid abrupt or irreversible impacts. Mitigation, indeed, is already happening at different levels of government, but not at all levels or with the required effectiveness. As indicated in Chapter 2, there are many challenges faced even by such ambitious endeavours as the

European Emissions Trading Scheme and the mitigation responses of the UK and Germany.[16] Furthermore, although climate change is firmly on the urban policy agendas in both developed and developing countries, it remains a marginal issue in terms of implementation.

Diverse institutional factors explain the challenges, constraints and opportunities of mitigation responses by local authorities. The first relates to the influence of the context-specific interactions between different tiers of government on local authorities' response capacity (multi-level governance). International and national policies provide the enabling – but also constraining – context within which urban responses are framed; determine the autonomy and competencies – the duties and powers – for municipal authorities to act in response to climate change; and enable policy integration within and between local authorities. Another set of institutional factors shaping local authorities' mitigation responses is their institutional ability to implement and enforce policies and measures. In many policy areas, municipal authorities, especially but not exclusively in developing countries, are unable or unwilling to enforce building codes, land-use zoning, fossil fuel standards and other regulations.

Two other factors are critical for the development of mitigation policies – namely, the dynamics of network creation and leadership – the latter both at the individual and institutional levels. Local Governments for Sustainability (ICLEI), organizations compiling and disseminating expert knowledge such as the IPCC, the United Cities and Local Governments and the Urban Leaders Adaptation Initiative, among other international, national and municipal networks of governmental and non-governmental organizations, have also been important in developing municipal capacity.[17] Evidence suggests, however, that these have been more important in developing the capacity of those municipalities that are already leading responses to climate change. Individual political champions and organizations, using climate change as a means and window of opportunity for fostering organizational reputation, have been equally fundamental in shaping action. Yet, if authorities lack the financial and technological resources to execute programmes, the power of leadership and of these networks to affect change can be limited.

Of no lesser relevance as determinants of mitigation responses are the availability of financial resources, of technical expertise, as well as the weight of such structural and enduring factors as the material infrastructure and cultural practices of a city. For instance, the mitigation challenges in the transportation sector will be strongly determined by the urban form of a city, with high-density areas offering compatibility with options to develop metros, trams and other high-efficiency modes of public transportation, while sprawling low-density areas are more compatible with systems of buses and minibuses to cover commuting needs. Options to reduce emissions are constrained by the fact that investments in power plants, industrial facilities and other components of the urban environment have long lifetimes. As for financial resources, given the many competing demands in urban areas, local authorities lacking the money

---

**Mitigation ... is already happening at different levels of government, but not at all levels or with the required effectiveness**

**Determinants of mitigation responses are the availability of financial resources, of technical expertise, as well as ... the material infrastructure and cultural practices of a city**

to provide even basic services for their constituents are unlikely to invest in the mitigation of climate change. Furthermore, the international financial resources available for mitigation (and adaptation) activities under the UNFCCC and the Kyoto Protocol (see Boxes 2.2 and 2.3) are quite simply not sufficient to meet the requirements, particularly of developing countries. As discussed in Chapter 6, this is particularly the case in cities, as very limited resources have so far been made available for initiatives in urban areas.

It is likely that GHG emissions will continue to increase until 2030 even if effective actions are taken now to stabilize emissions around the end of the century, and despite current commitments within the Kyoto Protocol.[18] Thus, adverse impacts of projected climate change and variability are inevitable, and urban centres will be particularly at risk. Regardless of the scale of mitigation undertaken over the next two to three decades, adaptation action will be necessary, which will be another challenging and fundamental dimension of the urgent response to climate change.

The responsibilities of local authorities with regard to the built environment, infrastructure and services that have relevance for adaptation include:

- urban planning and regulatory instruments designed to influence land availability and to authorize and oversee hazardous activities that can produce disasters;
- provision and pricing of various public services, infrastructure and resources; and
- enabling, proactively facilitating and coordinating actions to manage hazards through partnerships with the private sector, the academic sector, non-governmental and grassroots actors (e.g. households and communities) to reduce risk.

Each urban centre may use these areas of responsibility to design adaptation actions. However, the particularities – often determined by the national contexts of these urban centres – will dictate which of these measures will be most effective.

As with mitigation, adaptation is already taking place, at least on a small scale, and the world is witnessing the beginnings of city-based adaptation strategies in some urban centres. As yet, however, too few cities have developed coherent adaptation strategies. Furthermore, in contrast to such sectors as agriculture or forestry, there is relatively much less explicit city-wide attention to urban adaptation. In fact, most of the literature on climate change adaptation in cities is on what should be done, not on what is being done, the main reason being that too little is actually being done. The relatively lower emphasis on adaptation, and particularly on urban adaptation, is partly a result of the existing structure of incentives under the Climate Convention. For instance, funding is available[19] for mitigation activities such as landfill gas capture and for electricity generation, for transportation or carbon capture from reforestation and conservation of forests. However, while very little (only 8.4 per cent of the CDM projects are urban) is readily available for urban mitigation projects, practically nothing is allocated to adaptation efforts at the city level.[20]

Among urban areas, this relatively low interest in adaptation can also be related to the fact – clearly illustrated by Durban (South Africa) – that getting more widespread attention by city and municipal governments to climate change adaptation requires clear and detailed risk assessments (knowledge). It also requires a better understanding of how adaptation measures can serve and be integrated not only within disaster risk reduction, but also within such components of the development agenda as land-use planning, as well as access to water, sanitation and housing. It also depends on diverse institutional factors, in addition to leadership and local authorities' willingness to act. For example, effective adaptation actions can depend on whether authorities have the autonomy, resources and decision-making power to design and implement actions on the built environment, infrastructure and services that have relevance for adaptation; and whether and how adaptation options and challenges are related to such development issues as the need to protect the poor or to provide land and shelter for them (such as Manizales, Colombia, and the Homeless People's Federation of the Philippines).[21]

A fundamental challenge in this context relates not only to whether adaptation is responding effectively to potential climate change impacts in different sectors, but also to social equity issues – that is, whose needs are served (and whose are not) by adaptation responses, especially in relation to income, gender and age. For instance, are the adaptations aimed at protecting the wealthier groups and districts; or those living in informal settlements; or women and their particular risks and vulnerabilities; or the city's most economically important assets; or the city's populations most at risk? Decision-makers can be more effective and legitimate if they include these groups – or at least their genuine representatives – in the process of addressing these questions.[22]

It is not only the city authorities of some early frontrunners that are responding to the adaptation challenges of climate change. Households and communities are already coping with climate-related hazards – for example, by raising plinth levels, saving money and by participating in community initiatives to clear blocked drainage channels to respond to the impacts of flooding (see Chapter 6). However, these cannot be substitutes for serious government investment and action to improve drainage and sanitation, water supplies, roads and other hard infrastructure that is so crucial for risk reduction, or for a city-wide provision of urban services, as well as warning and emergency responses systems.

In the urban areas of many developing countries, household, community and government adaptation responses will need to happen in the context of adaptation (or development) deficits. In many cities of developing countries, at least a percentage of their populations lack water, sanitation, health services, shelter, sound emergency policies and other factors to adapt to the current range of climate variability, let alone to any future climate change impacts. It is impossible to adapt or climate-proof infrastructure, services and emergency responses that do not exist.[23]

**Too few cities have developed coherent adaptation strategies**

**In the urban areas of many developing countries ... adaptation responses will need to happen in the context of adaptation (or development) deficits**

Another key challenge concerns the social impacts of adaptation measures. Actions such as control of urban growth in risk-prone areas and investment in storm and surface drainage systems can increase the vulnerability of some populations. If not carefully designed, they have the potential to displace informal settlements – especially those alongside existing drains and rivers. Furthermore, they can constrain a population's capacity to make their livelihoods; they can shift risks from the populations of one district to the inhabitants of another district; and they can shift risks to future generations.

# ADAPTATION AND MITIGATION: RELATIONSHIPS WITH URBAN DEVELOPMENT AND POLICY

Although a distinction between climate change adaptation and mitigation is deeply set in climate change policy and research, some cities tend to look at the world differently. Early experience with both adaptation and mitigation planning in developed country cities suggests that urban leaders and stakeholders resist focusing on one and not the other, and that they find it difficult to consider either without considering sustainable development goals and development pathways more generally at the same time.[24] In fact, the goal is sustainable development for their cities, and climate change responses are either a means towards that end or impediments to achieving that end. In this context, attention needs to be given not only to the implications of mitigation and adaptation for urban development, but also to the synergies and trade-offs between actions addressing both mitigation and adaptation and other dimensions of policy-making. However, experiences from many cities in developing countries contradict this, as their leaders and stakeholders tend to consider developed countries the culprit of climate change and, thus, responsible for mitigation. Such cities therefore tend to focus on adaptation interventions.

Cities can respond to concerns about climate change impacts in two ways: by contributing to mitigation or by adapting to possible impacts – either of which can promote urban development or impede it.

## Climate change mitigation and urban development

In the coming decades, urban authorities will – in many cases and in many regions – find that the need for global, national, regional and local climate change responses poses significant concerns. The urgency and severity of this challenge cannot be overstated. Recent analyses of potentials for GHG emissions reduction and efficiency improvement, given current global trends and potentials for known technologies, make even relatively moderate goals such as

stabilization at 445 to 490 parts per million of $CO_2$eq (in order to keep average global warming no higher than 2°C) virtually unachievable unless every major technological option realizes the most optimistic hopes for it (including affordable carbon capture and sequestration from coal).[25] In other words, the world seems headed towards climate changes that are even more severe than the sobering descriptions in Chapter 4. Meanwhile, at the Copenhagen COP in December 2009, low-lying island states and other vulnerable developing regions argued that any stabilization level that means average warming above 2°C implies unacceptable levels of impacts and must be avoided. Two apparent crises lie ahead: a crisis of emerging impacts in vulnerable cities as they become ever more urgent and a crisis of global responses to growing pressures for mitigation and adaptation, which are likely to be sources of great controversy and, perhaps, forceful policy developments.

Globally, the mitigation challenge is to reduce GHG emissions from buildings, industry, transportation, energy production and land use, and to reduce or reverse deforestation. As stated earlier, emissions reduction focuses on efficiency improvements in buildings, industry, transportation and energy production, and shifting energy production and use from GHG-emitting fossil sources to alternatives such as renewable energy, nuclear energy and 'decarbonized' energy from fossil sources.[26]

It is important to note that mitigation policies can represent opportunities for cities and their development prospects. As exemplified through the experience with ICLEI's International Local Government GHG Emissions Analysis Protocol, in many cases, actions by cities to reduce their GHG emissions from systems under the jurisdiction of metropolitan governments actually save them money, such as through increases in the efficiency of urban lighting systems or in public transportation systems that reduce costs to the city's budget.[27] Less directly, cities can partner with their own private-sector operators of industrial and transportation systems to seek reductions in GHG emissions from private-sector sources, with city policies (such as taxation) encouraging or discouraging such actions. Even less directly, new energy facilities that help to reduce GHG emissions – such as bioenergy and other renewable energy production facilities – will need to be located somewhere, and cities can seek to be their sites, creating jobs and tax revenues.

But global pressures to push the boundaries of climate change mitigation are likely to be a challenge for urban development as well. Two potential impacts are especially important. *First*, if an urban area's economy depends, even in part, on fossil energy production, it is likely to be adversely affected by any move away from fossil energy. There are many examples of cities whose economies are built, in part, on coal, oil or natural gas production, such as in Nigeria, Angola, China and India.

A *second* impact is that energy costs and prices are likely to increase in most parts of the world as energy systems shift from relatively low-cost fossil energy sources to somewhat more expensive alternative energy systems. Because affordable energy is vitally important as a driver of the development engines of many cities, this could become a

---

*Margin notes:*

Attention needs to be given ... to the synergies and trade-offs between actions addressing both mitigation and adaptation

Global pressures to push the boundaries of climate change mitigation are likely to be a challenge for urban development

challenge for sustainable development – especially in cities whose development paths are likely to be especially energy intensive. In most cases in developing regions, paths for socio-economic and technological development imply *increases* in GHG emissions, not reductions in emissions, including both emissions from the cities themselves and emissions from systems that meet urban needs, such as electric power plants located elsewhere.

However, local authorities often play broader roles, as well, in shaping the development choices of their urban areas, their larger regions and their countries, and these roles have implications for climate change mitigation more broadly than within their own boundaries alone. They are the focus of driving forces for climate change responses, including financial resources, information and communication systems and media, and both technological and policy innovation. Where public decision-making is based on democratic processes, local authorities can dominate 'one-person/one-vote' political contests and thus influence national voting patterns.

There are both synergies and trade-offs between actions addressing the mitigation challenge and other policy dimensions (such as industrial development, energy, health and air pollution).[28] As illustrated by Mexico City, Denver (US) and many Chinese cities, climate change mitigation is more an outcome of efforts driven by economic, security and local environmental concerns, or simply by the need to be at the forefront of initiatives among a peer group of city leaders, rather than being a strategic priority.[29] It is therefore necessary to take advantage of existing synergies between climate protection and other development priorities. For instance, strong synergies exist in the transportation sector between climate change and energy supply and security. Measures replacing oil with domestic biofuels can reduce both emissions and reliance on oil imports (such as in Brazil). A more decentralized electricity system based on new renewable generation may reduce gas imports.

A key question is whether cities have the potential to tap into options such as carbon markets opened by the Kyoto Protocol.[30] For example, could the construction or building materials industries receive financing from the CDM or similar mechanisms for producing cement (or other materials) which incorporates carbon capture and storage? Such carbon credit trading could, potentially, be a way to subsidize the construction of adequate housing for low-income groups in developing countries. This and other options could open a completely new discussion dealing with synergies between GHG emission control and poverty reduction.

Policies addressing other environmental and social problems, such as air pollution or provision of shelter to the poor, can often be adapted at low or no cost in order to reduce GHG emissions and improve the health of the population simultaneously, especially in developing countries. The burning of fossil fuels is linked to both climate change and air pollution. Thus, reductions in the amount of fuel combusted will result in both lower GHG emissions and lower health and environmental impacts from reduced emissions of air pollutants. Aware of these co-benefits, organizations such as the World Health Organization and the US Environmental Protection Agency have applied, at the urban and national levels, environmental assessments of the co-benefits of addressing both air pollution and other issues (such as economic costs and energy). This has helped to introduce policies that address local pollution and GHG emissions together.[31] However, attention needs to be given not only to the synergies, but also to the conflicts between these policy domains. For instance, increases in the energy efficiency of vehicles can result in increased atmospheric emissions, and thus in negative health impacts, if vehicle travel distances increase or drivers switch to vehicles with larger engines (the 'rebound effect').[32]

This means that trade-offs may exist between mitigation and other policy domains. For instance, security arguments may impel countries to increase their dependence on internal reserves of coal rather than relying on natural gas imports.[33] Use of biofuels that are dependent on crops such as corn has been linked to food shortages and cost increases as farmers switch food-producing croplands to more profitable biofuel crop cultivation. This may also be an unintended effect of government subsidies aimed at increasing production of biofuels, but making the growing of food crops less profitable.

## Climate change adaptation and urban development

Adaptation-related issues for urban development across the world include two dimensions: first, the implications of climate change impacts for urban development projects that are likely to call for adaptations; and, second, the relationships between adaptation action to reduce the impacts of climate change, on the one hand, and urban development, on the other.

Climate change poses particularly severe threats for urban development in those areas that are most vulnerable to climate change impacts.[34] For example, many cities are located in coastal areas and river valleys, as well as areas where the economic base is rooted in climate-sensitive sectors, such as agriculture, forestry or tourism, and areas where these regional climate-related activities face increased competition with population and economic growth. If climate change is relatively severe in local contexts rather than moderate, some cities will find that incremental adaptations that protect current activities and ways of life may no longer be adequate.

A current example of what future climate change could mean for urban development can be found in the polar regions of the world, where temperature increases (and emerging sea-level rise) are not only affecting urban infrastructure as the permafrost melts, but are irreversibly destroying polar ecosystems and indigenous ways of life that are closely linked to them.[35] In these and similar cases, adaptations that are 'transformational may be required', such as changes in land uses and movement of investment away from vulnerable areas, or shifts in directions of urban development to different economic sectors or land uses. Climate

There are both synergies and trade-offs between actions addressing the mitigation challenge and other policy dimensions

Some cities will find that incremental adaptations that protect current activities and ways of life may no longer be adequate

change impacts are therefore a critically important challenge for urban development, and if climate change is severe (rather than moderate), the number of cities at risk will be multiplied many times over.

Experience suggests that, given human resources and access to knowledge, urban dwellers often have impressive capabilities to adapt in ways that are good for their development, even with limited financial resources. For example, low-income residents of Lagos (Nigeria), Dhaka (Bangladesh) and Dar es Salaam (Tanzania) already cope with a range of climate-related challenges, especially seasonal flooding.[36] This is particularly evident where effective grassroots organizations are active. This is not to say that decisive action is not needed at all levels; but it is important to note the many actions are already being undertaken by households and communities – frequently in the absence of actions by local government and other stakeholders.

One of the most fundamental challenges in relating climate change adaptation with urban development in many regions, however, is a limited capacity to identify vulnerabilities and adaptation pathways, along with a limited capacity to make adaptation happen. Many small- and medium-sized cities, especially in sub-Saharan Africa, South Asia and Central America, currently show low levels of capacity to adapt to the current range of climate variability, let alone any future climate change impacts. Problems in many such cities include a lack of provision for infrastructure (including all-weather roads, piped water supplies, sewers, drains, electricity, etc.), urban social services (such as health and education) and institutional capacity.

Yet, many cities have shown an ability to adapt to local climate conditions, whether related to climate change or not; and where climate change adaptation is being considered seriously (in urban areas from Bangkok, Thailand, to Melbourne, Australia), in nearly every case adaptation options are being identified that are relatively low cost and have broad constituency support.[37] Some developing country cities have moved beyond option identification to adaptation planning (such as Durban and Cape Town, South Africa).[38] Furthermore, most of the adaptation options offer considerable co-benefits – that is, benefits for urban development and/or environmental stress reduction in the near term, as well as added resilience to impacts of climate change in the longer term, which is often critically important in sustaining attention to adaptation while impacts are gradually emerging.

There are positive examples of cities, such as in Manizales (Colombia) and Ilo (Peru), that are taking steps to promote development and reduce vulnerability at the same time. These cities have implemented actions to prevent rapidly growing low-income populations from settling on dangerous sites. Although these actions have not been driven by climate change concerns, they illustrate how pro-development and pro-poor policies can enhance adaptive capacity. Conflicts and trade-offs between development policies and adaptation are also possible, as in the development of infrastructure whose design and construction have the potential to displace informal settlements.[39]

*Most ... adaptation options offer considerable co-benefits... for urban development and/or environmental stress reduction in the near term*

*Neither mitigation nor adaptation alone can protect the world from the undesirable impacts of climate change*

## Mitigation and adaptation: Seeking synergies rather than conflicts

It is now known that neither mitigation nor adaptation alone can protect the world from the undesirable impacts of climate change. Both must be a part of the global response. Mitigation is essential in order to keep climate change impacts as low as possible; but some impacts can no longer be avoided. This is so because progress is slow on international agreements to implement mitigation, and strategies for GHG emission stabilization in major developing countries are unclear at best. Adaptation is, therefore, essential because some impacts will not be avoided. It is clear that costs will be a constraint for some locations and populations, and adaptation will be limited in its ability to reduce costs from abrupt events, at least in the short run. And, as noted above, some of the impacts are beyond the scope of adaptation: the so-called 'residual damage'. While allocating resources for mitigation and adaptation, it is essential to find solutions for the populations and industries that may be displaced by the impacts of climate change.

Meanwhile, early initiatives with either climate change mitigation or adaptation planning suggest that some urban local authorities and stakeholders are unwilling to discuss mitigation or adaptation options separately, without placing these discussions in the context of where the city and its citizens want to go in the longer run.[40] Cities are one of the most important of all the world's settings for integrating actions to reduce vulnerabilities and mitigation responses as they relate to broader social and economic objectives, such as job creation, improvements in the quality of life, and access to health and water services. The fact that climate change response planning often catalyses these discussions within communities is one of its most important co-benefits.

A major problem is that mitigation and adaptation options often differ in important ways. For instance, they tend to differ as to *when* benefits are realized (mitigation benefits lag in time, while adaptation benefits may be nearer term), *where* benefits are realized (mitigation is global while adaptation benefits are more localized) and *what sectors* are the focus of action (mitigation focused on GHG emitters or sinks, and adaptation focused on activities, infrastructure and population segments sensitive to impacts). Furthermore, it is also important to note that mitigation actions are urgent. If no action is taken within the next ten years or so, the impacts will exponentially increase. This is less the case with adaptation action, which can be phased in time and which will be a continuous process for many decades to come. These differences substantially complicate attempts by urban areas (or by the countries and regions whose policies affect them) to develop integrated climate change response strategies.

Pathways to mitigation and adaptation can be mutually exclusive and competitive alternatives – such as development investments in alternative energies which do not enhance resilience in vulnerable areas versus policies to move development activities away from vulnerable areas;[41] but they may also be complementary and reinforce each other. A simple example of this would be building insulation, which can reduce the need for burning fossil fuels while enabling adaptation to increased temperatures projected with climate change. An important general guideline is that mitigation and adaptation options which offer synergies and complementary pathways should be given special attention. For example, mitigation options that reduce net GHG emissions – such as tree-planting and other biomass sink preservation and/or restoration, along with regional or local renewable energy development – can be complementary parts of an overall mitigation strategy. However, this synergy can be taken one step further with the addition of adaptation benefits. Tree-planting or forest preservation, for instance, can also be an important part of an urban adaptation strategy to prevent heat-island effects, thereby preventing an array of cascading negative effects such as heat-related morbidities, mudslides and coral sedimentation, etc.

In many cases in urban areas, the focus is on invest-ment in major infrastructure that lasts for a number of decades: transportation systems; commercial, residential and government buildings; and industrial development. These investments can profoundly shape both urban mitigation and adaptation not only in the short term, but for as long as half a century or more.

Currently, and with some notable exceptions, most urban initiatives that might be associated with mitigation or adaptation are fragmented, and historically much of the policy attention has been focused mainly on mitigation, with little or no consideration of adaptation. In many cases, the focus is not on climate concerns but on energy security and other development priorities related to economic growth.[42] Even where existing initiatives explicitly address mitigation, they often focus only on one aspect of the whole issue (such as energy efficiency, or even, more narrowly, energy efficiency in metropolitan public-sector functions).[43]

Only a handful of city-wide initiatives – such as in London (UK), Durban (South Africa) and New York (US) – are beginning to grasp the need to address at least some of the complex linkages between mitigation, adaptation and development, and thus have launched mitigation and adapta-tion programmes. For instance, to increase the uptake of decentralized energy technologies in London, developments over a certain size are required to meet 20 per cent of their projected energy needs through onsite low-carbon or renew-able energy generation,[44] thus promoting new economic activities and the creation of green jobs. Furthermore, national and local authorities have already identified adapta-tion responses to three key climate risks – floods, drought and overheating – thus opening alternatives to avoid damage to infrastructure, increased mortality among the aged and other impacts that might constrain the livelihoods of some populations. This means that climate change responses are getting the necessary ingredients to move towards more integrative approaches.[45] However, even these exemplary cases illustrate the challenges of responding to climate change.

# FUTURE POLICY DIRECTIONS

This section explores future policy directions for achieving climate-resilient cities, reflecting on both recent policy developments and more general long-term policy needs. In the face of climate change, policy decisions and actions are not just the responsibility of a city, or of its country or region, or of the international community more broadly – or even of governments alone at any of these levels. Urban development will be shaped by the policies of all levels of government, of private-sector organizations, of non-governmental issue-oriented institutions, of research communities,[46] and of representatives of local communities and civil society organizations. The challenge, and it is an immense one, is to knit together a global response to urban needs and potentials in which a wide variety of partners each contribute what they do best – for instance, combining the resources available at large scales with the innovativeness and knowledge of local realities available at local scales.

From this perspective, this section outlines some principles for policy development at all levels and discusses what policies should be considered at the international, national and local levels and, more briefly, by non-govern-mental partners, to strengthen planning and decision-making in urban areas in response to global climate change.

## Principles for policy development

Several principles are fundamental to an integrated multi-partner approach:

- No single mitigation or adaptation policy is equally well suited to all cities. Reflecting a common saying that 'one size does not fit all', cities are so diverse in terms of the set of societal and environmental drivers of their GHG emissions, their governance structures, their vulnerabil-ities, adaptive capacities and development aims that policy approaches should recognize and be sensitive to the diversity of urban areas worldwide.
- The appropriate approach is not to try to precisely project future climate change and socio-economic conditions, which is burdened by far too many uncer-tainties to encourage decision-making, but to take an *opportunity/risk management* approach in a sustainable development perspective: considering not only emissions but also risks that are present in a range of possible climate and socio-economic futures.
- Policies should emphasize, encourage and reward '*synergies*' and '*co-benefits*' (i.e. what policies can do to *achieve multiple objectives* related to both development and climate change response goals).
- Climate change policies should address both near-term and longer-term issues and needs. Near-term perspec-tives are likely to focus on relatively straightforward 'no regrets' decisions with, *first*, few or no net costs, offer-ing substantial co-benefits for urban development (such as increasing resilience to climate variability; reducing chronic environmental stresses, such as poor drainage;

or addressing critical current needs of especially vulnerable populations who will only get worse with climate change); and, *second*, with broad stakeholder support and representation, not only of the better off, but also of populations more at risk from climate change (the poor, women, children, the elderly, the disabled, ethnic and other minorities, etc.). Longer-term perspectives need to consider risk management for more substantial mitigation pressures and adapting to more significant impacts, focused on contingency planning for a range of possible climate/development futures, monitoring emerging climate and policy conditions, and reassessing risks periodically.

**International public-sector policies related to urban responses to climate change should be supportive and enabling without being directive or constraining**

- Policies need to recognize that institutional roles and potentials differ between scales and sectors of action. In recent history, too often well-intentioned initiatives developed at large scales and implemented top down have discouraged local action by imposing daunting bureaucratic requirements as a condition for access to available resources.[47] Meanwhile, initiatives developed and implemented at small scales and implemented bottom up (such as community-based adaptation) often lack financial and other resources to undertake the necessary investments in urban infrastructure and services, may lack valuable information, and may lead to actions that have adverse consequences for other localities. The challenge is to design new approaches that support multi-scale, multi-sector action, rooted in new kinds of mutual sensitivities, in order to realize the differing and often complementary potentials of a wide range of partners (Chapter 6).[48]

## International policies

International public-sector policies related to urban responses to climate change should be supportive and enabling without being directive or constraining (see also Box 7.1):

---

**Box 7.1 Key principles for urban climate change policy development: The international community**

There are three main areas in which the international community can support and enable more effective urban mitigation and adaptation responses:

1   Financial resources need to be made more directly available to local players – for example, for climate change adaptation in vulnerable cities, investment in a portfolio of alternative energy options, and investment resources for mitigation partnerships between local governments and local private-sector systems.
2   Bureaucratic burdens on local access to international support should be eased. The international community can help to create direct communication and accountability channels between local actors and international donors through intermediary organizations that can help disperse resources and monitor performance.
3   Information of climate change science and options for mitigation and adaptation responses should be more widely available. The Intergovernmental Panel on Climate Change (IPCC), the United Nations and other international organizations need to widen the spectrum of available knowledge on observed and future climate change impacts upon urban centres; on mitigation and adaptation alternatives available for urban responses; and on the costs, benefits, potentials and limits of these options.

---

- *Resources.* The international community has access to vitally important financial resources that can be provided to support many vulnerable cities that need additional resources to respond to climate change. For example, international policies should include much more significant financial support for climate change adaptation in vulnerable cities, for investment in a portfolio of alternative energy options, and to support mitigation partnerships between local governments and local private-sector actors. In particular, it is essential that action is taken to facilitate the use of the Adaptation Fund and the CDM for initiatives in urban areas.
- *Information and options.* The IPCC is already helping cities and influencing development pathways by providing information about climate change science and response options, alerting local leaders (and the people to whom they listen) to emerging issues and resolving some disputes about scientific facts. International policies should continue these roles, with increased attention both to widening the spectrum of mitigation and adaptation alternatives available for urban responses, and to improving information about the costs, benefits, potentials and limits of the options. A similar role has been played by the Clinton Climate Initiative and ICLEI (see Box 2.7), which have been prominent in the exchange of ideas, best practices and experiences, at least for urban areas that are already at the forefront of climate change responses.
- *Reduced bureaucratic burdens.* International policies should do a better job of balancing legitimate concerns about accountability (such as establishing 'additionality' through detailed quantitative analysis) with a need to make access to support much easier, simpler and less bound up in expensive analysis. Options might include a wider use of third-party intermediary ('boundary') organizations to disperse resources and monitor performance. Likewise, streamlined approaches for approving investments in certain types of projects that, time and again, have been shown to yield climate change-related benefits should be developed and approved (e.g. through the CDM). Likewise, in order to ease access to carbon finance for cities in developing countries, the CDM Executive Board should approve the new city-wide programme of activities methodology that was recently submitted for their consideration.[49]

## National policies

As illustrated by some countries – such as the UK, Germany, Norway, Brazil and the Republic of Korea – the mitigation responses of national governments can go beyond the endorsement of international climate conventions and accords. Driven by reasons as diverse as energy security and an actual concern for the implications of climate change, they may engage in the design and implementation of national mitigation strategies and adaptation planning. However, from the perspective of urban areas, national governments generally assist development by determining sets of policy and market conditions and rules that shape decentralized activities – incentives, limits and expectations for the future – and by helping to coordinate responses that involve a wide range of individual partners. They also play essential roles in looking beyond current conditions and priorities, both for countries as a whole and cities in particular, towards longer-term changes in conditions and the possible need for changes in rules that define development pathways and risk management 'playing fields' (see also Box 7.2):

- *Enabling framework.* National (and local) governments should facilitate the climate change mitigation and adaptation interventions of all stakeholders. The example of the Philippines (see Box 6.3) illustrates how governments, through the provision of an enabling framework, can enhance the effectiveness of interventions of other actors, particularly people living in poverty.
- *Incentives.* Some countries already offer incentives for climate change mitigation actions, while many countries, in effect, discourage mitigation and adaptation actions through policies enacted with other issues in mind – or during earlier periods before climate change became a reality. Countries can promote urban area initiatives related to climate change mitigation or adaptation by removing 'maladaptations' that are counterproductive and by providing incentives such as favourable tax treatment, eligibility for federal funding support and high-visibility public recognition.
- *Coordination.* As cities, sectors, regions and other parties act to support mitigation and adaptation, these dispersed activities need coordination in order to ensure that they are mutually reinforcing rather than causing problems in other contexts. For example, a decision to convert a natural forest area to a bioenergy plantation can contribute to mitigation by reducing the need for fossil fuels, but it can threaten biodiversity protection. A decision by one city to protect coastal areas with barriers can have impacts upon wetland ecologies that are important to the economic base of other cities inland. Countries should make it standard policy to ensure information-sharing about localized plans and to provide mechanisms to resolve conflicts as they arise.
- *Risk-sharing.* Countries can contribute to mitigation and adaptation by cities in two ways related to risk-sharing. On the adaptation side, a frequent concern is with climate change threats that have high probabilities of occurring at a national level but low probabilities for any single city, such as extreme weather events. Here, countries can work together with private, non-governmental (such as slum federations) and public providers of insurance and reinsurance to offer protection to each city without requiring each to make a sizeable investment in order to reduce risks from a particular kind of low-probability threat. On the mitigation side, some possible actions involve technologies that are so innovative that their economic value has not

| **Box 7.2 Key principles for urban climate change policy development: National governments** |

National governments should primarily use the following mechanisms to enable mitigation and adaptation actions at the local level:

- Engage in the design and implementation of national mitigation strategies and adaptation planning.
- Offer tax rebates, tax exemptions and other incentives for investments in alternative energy sources, energy-efficient appliances, climate-proof infrastructures, houses and appliances, among other climate change mitigation and adaptation actions.
- Encourage appropriate climate responses. For example, redesign policies enacted with other issues in mind or during periods prior to climate change, such as policies that use the definition of a 100-year floodplain, which can result in maladaptations.
- Enhance coordination and streamlining between sectoral and administrative entities. For instance, make sure that decisions by one city to protect coastal areas with barriers do not have impacts upon basins that are suppliers of fresh water, or wetland ecologies that are important to the economic base of that city or other cities inland.
- Develop partnerships with non-governmental actors to share risks (see also Box 7.4). For example, national governments can work with private insurance providers to offer protection to each city without requiring each to make a sizeable investment in order to reduce risks from a particular kind of low-probability threat.
- Anticipate and plan for the possibility of much more substantial climate change impacts and adaptation needs in the longer term than those that are currently anticipated in the next decades.

yet been fully established. Here, countries can encourage innovation through such policies as offering partial loan guarantees in case the technology does not perform as well as hoped.

- *Assistance where transformational adaptations are required.* Countries should help their cities in looking ahead to the possibilities of much more substantial climate change impacts and adaptation needs in the longer term than those that are currently anticipated in the next decades. An example might be a city located in a vulnerable coastal area subject to threats from more severe storms and sea-level rise over the next half century, where in the longer run, moving some populations and economic activities away from the most vulnerable areas might need to be considered. As indicated earlier in this chapter, policies should support contingency planning, monitoring of emerging conditions and the development of response alternatives.

## City policies

Urban areas are the main loci of action, rooted in local development aspirations and preferences, local knowledge of needs and options, local awareness of realities that shape choices, and local potentials for innovation. One of the major challenges for policies in most urban areas, however, is to broaden the discourse about policy directions beyond conventional structures of political power and government action, and to engage their communities much more inclusively (see Chapter 5). With this challenge in mind, urban policies should (see also Box 7.3):

**As cities, sectors, regions and other parties act to support mitigation and adaptation, these dispersed activities need coordination**

Urban policy-makers should begin from an awareness of local development aspirations and preferences, local knowledge of needs and options, local realities that shape choices, and local potentials for innovation. Urban authorities should:

• Develop a vision of where they want their future development to go and find ways of relating climate change responses to urban development aspirations.
• Expand the scope of community participation and action by representatives of the private sector, neighbourhoods (especially the poor) and grassroots groups, as well as opinion leaders of all kinds in order to ensure that a broad-based collection of perspectives is gathered.
• Using an inclusive participatory process (as referred to above), cities should conduct vulnerability assessments to identify common and differentiated risks to their urban development plans and their different demographic sectors, and decide on objectives and ways to reduce those risks.
• Pay particular attention to the importance of adding climate-sensitive features to major infrastructure, especially when they are being designed, as the cost of adding these features will almost always be smaller before the infrastructure is built than they would after it is in place.

In order to achieve more effective policies, local governments need to expand the scope, accountability and effectiveness of participation and engagement of NGOs, such as community and grassroots groups, the academic sector, the private sector and opinion leaders. This will serve multiple purposes:

• It will become a source of innovative options, as well as both scientific and locally relevant knowledge.
• It will allow participants to understand and mediate the diverse perspectives and interests at play.
• It will provide a broad-based support for decisions and promote knowledge on the causes of emissions and vulnerabilities, as well as mitigation and adaptation options thus achieved.

Partnerships with the private sector and NGOs are of special relevance in this context. For example:

• Resources from international, national and local private organizations can be mobilized to invest in the development of new technologies, housing projects and climate-proof infrastructures, and to aid in the development of climate change risk assessments.
• The widespread involvement of NGOs in climate arenas as diverse as climate awareness and education and disaster relief should be welcomed rather than making attempts to keep them outside of these structures and interactions. The inputs and perspectives of these organizations can be harnessed to help develop a more integrated urban development planning.

Broad-based oversight organizations, such as advisory boards, representing the interests of all actors should be created to help avoid the danger that private or sectarian interests may distort local action – for instance, by investing in technologies, infrastructures and housing that only benefit a minority, or by hijacking the benefits of grassroots funding. This is especially of concern in urban areas within countries that have experienced strong centralized control in the hands of local elites and state agents; but the principle of broad-based oversight can and should be practised everywhere.

• *Develop a vision of the future.* A city is not in a position to evaluate how climate change responses relate to its urban development unless it has a vision of where it wants that development to go. This requires not only the development of possible scenarios of future economic, demographic and land-use futures, along with resource requirements, but also richer 'narratives' of a set of futures that help to explain why they are desirable from the city's perspective.
• *Expand the scope of community participation and action.* In connection with developing its vision, a city needs to become a community of communities – reaching beyond formal governmental structures to the private sector, neighbourhoods and grassroots groups, as well as opinion leaders of all kinds in order to ensure a broad-based collection of perspectives is gathered. This is crucial for ensuring knowledge, innovation and broad-based support for a city's response strategy (see Chapter 6).
• *Conduct participatory risk assessments and turn the assessments into action plans.* Using inclusive participatory processes, in which both women and men, as well as all socio-economic and age groups are represented, supplemented by scientific knowledge, cities should assess risks to their urban development plans and objectives, identify ways to reduce those risks through actions in the near term that offer development co-benefits, develop a plan of action to take high-priority actions, and consider longer-term risks that may require larger-scale planning and strategy development (see Chapter 6).
• *Pay particular attention to the importance of investment in major infrastructure.* Major infrastructure casts long shadows through time for both mitigation and adaptation. Particularly important is investment in small- and medium-sized urban centres, including in large residential and commercial developments, government structures, industrial structures, transportation systems, energy facilities, and other facilities such as water supply and waste disposal systems. The time to consider mitigation and adaptation is when these types of infrastructure are being designed, when the cost of climate-sensitive features is almost always smaller than after the infrastructure is in place. An example of a policy option in vulnerable coastal or riverine cities would be a building code for new infrastructure developments that requires them to be able to withstand significant future flooding.

## Policies of other partners in a global policy response

Governments do not, in isolation, determine appropriate responses to climate change in development contexts. The private sector and NGOs are critically important partners. Other organizations may be important in some urban areas as well, such as community and/or faith-based organizations (see also Box 7.4):

- *The private sector.* Positive connections between climate change responses and urban development will only become mainstreamed when they become part of normal day-to-day decision-making in local markets and local economic institutions. Ranging from activities of large multinational corporations to local informal industries, the private sector must be included in urban, national and global policy-making on climate-related issues. For localities, this starts with including the private sector in discussions of urban needs and alternatives; encouraging private-sector organizations to conduct their own climate change risk assessments; looking for roles that they can play better than the public sector (such as stockpiling and providing emergency supplies); and encouraging innovative thinking about how private-sector business strategies can find opportunities in helping cities strengthen their commitments to climate change mitigation and adaptation.

- *Non-governmental organizations.* NGOs range from international environmental groups that provide information, technical assistance and policy advocacy; philanthropic foundations that take the lead in developing urban climate change response initiatives when governments and the private sector prove unwilling to move ahead quickly enough; and local community organizations, formal and informal, that play major roles in emergency response situations in cities – and are stepping forward to represent the interests and concerns of especially vulnerable populations in many cities. Here, the policy challenge is to incorporate these roles within integrated urban development planning rather than hold them outside these structures and interactions.

## CONCLUDING REMARKS

In summary, policy directions for linking climate change responses with urban development offer abundant opportunities; but they call for new philosophies about how to think about the future and how to connect different roles of different levels of government and different parts of the urban community. In many cases, this implies changes in how urban areas operate – fostering closer coordination between local governments and local economic institutions, and building new connections between central power structures and parts of the population who have often been kept outside of the circle of consultation and discourse.

The difficulties involved in changing deeply set patterns of interaction and decision-making in urban areas should not be underestimated. Because it is so difficult, successful experiences need to be identified, described and widely publicized as models for others. However, where this challenge is met, it is likely not only to increase opportunities and reduce threats to urban development in profoundly important ways, but to make the urban area a more effective socio-political entity, in general – a better city in how it works day to day and how it solves a myriad of problems as they emerge – far beyond climate change connections alone.

It is in this sense that climate change responses can be catalysts for socially inclusive, economically productive and environmentally friendly urban development, helping to pioneer new patterns of stakeholder communication and participation.[50]

> Climate change responses can be catalysts for socially inclusive, economically productive and environmentally friendly urban development

## NOTES

1   I.e. the Annex 1 countries of the UNFCCC; see Chapter 2.
2   UN, 2010.
3   UN, 2010.
4   The Adaptation Fund only became operational in 2010. See also Boxes 2.2 and 2.3.
5   However, and as noted earlier, an International Standard for Determining Greenhouse Gas Emissions for Cities was launched by UNEP, UN-Habitat and the World Bank at the World Urban Forum in Rio de Janeiro, Brazil, in March 2010.
6   It should, however, be noted that the provision of modern sanitation facilities becomes less expensive with densification.
7   Such as housing co-operatives in Tel Aviv, Israel (see Chapter 5).
8   Such as the Project 2° (see Chapter 5).
9   See Chapter 1.

10   Such as Denver and Washington, DC (see Chapter 3).
11   See Chapter 5.
12   See Chapter 5.
13   See Box 5.4.
14   In line with the activities and recommendations of the C40 and ICLEI (see Chapters 2 and 5).
15   See Chapter 5.
16   See Chapters 2 and 5.
17   See Chapter 2.
18   Sims et al, 2007.
19   Through the CDM (see Box 2.3) and through such programmes as the United Nations Collaborative Programme on Reducing Emissions from Deforestation and Forest Degradation in Developing Countries (UN-REDD) (see Table 2.2).
20   See Chapter 2.
21   See Chapter 6.
22   See Chapters 4 and 6.

23   See Chapter 6.
24   See NRC, 2010.
25   NRC, 2009, 2010; Greene et al, 2010.
26   'Decarbonized' as a result of carbon capture and storage initiatives (see Chapter 5).
27   See Chapter 3.
28   Barker et al, 2007.
29   See Chapter 5.
30   See Chapter 2.
31   Barker et al, 2007.
32   See Chapter 5.
33   Barker et al, 2007.
34   See Chapters 4 and 6.
35   ACIA, 2004.
36   See Chapter 6.
37   This having been said, it is important to note that some climate change adaptation interventions can be very costly and/or contentious.
38   See Chapter 6.
39   See Chapter 6.
40   NRC, 2010.
41   Wilbanks and Sathaye, 2007.

42   Such as in Beijing (China) (see Chapter 5).
43   See Chapter 5.
44   See Chapter 5.
45   NRC, 2010.
46   See, for example, Rosenzweig et al, 2011.
47   See Chapter 5.
48   Wilbanks, 2007.
49   The proposal was submitted by the World Bank in July 2010. Under present rules, the CDM Executive Board cannot approve programmes of activities that use multiple methodologies. By their very nature, city-wide programmes draw on a range of methodologies that support GHG mitigation technologies; but as such they cannot be considered for approval through the CDM – unless the guidelines of the CDM Executive Board are revised.
50   Wilbanks, 2003.

# STATISTICAL ANNEX

## GENERAL DISCLAIMER

The designations employed and presentation of the data do not imply the expression of any opinion whatsoever on the part of the Secretariat of the United Nations concerning the legal status of any country, city or area or of its authorities, or concerning the delimitation of its frontiers or boundaries.

# TECHNICAL NOTES

The Statistical Annex comprises 16 tables covering such broad statistical categories as demography, housing, economic and social indicators. The Annex is divided into three sections presenting data at the regional, country and city levels. Tables A.1 to A.4 present regional-level data grouped by selected criteria of economic and development achievements, as well as geographic distribution. Tables B.1 to B.8 contain country-level data and Tables C.1 to C.3 are devoted to city-level data. Data have been compiled from various international sources, from national statistical offices and from the United Nations.

## EXPLANATION OF SYMBOLS

The following symbols have been used in presenting data throughout the Statistical Annex:

category not applicable    ..
data not available         ...
magnitude zero             –

## COUNTRY GROUPINGS AND STATISTICAL AGGREGATES

### World major groupings

**More developed regions:** All countries and areas of Europe and Northern America, as well as Australia, Japan and New Zealand.

**Less developed regions:** All countries and areas of Africa, Latin America, Asia (excluding Japan) and Oceania (excluding Australia and New Zealand).

**Least developed countries:** Afghanistan, Angola, Bangladesh, Benin, Bhutan, Burkina Faso, Burundi, Cambodia, Central African Republic, Chad, Comoros, Democratic Republic of the Congo, Djibouti, Equatorial Guinea, Eritrea, Ethiopia, Gambia, Guinea, Guinea-Bissau, Haiti, Kiribati, Lao People's Democratic Republic, Lesotho, Liberia, Madagascar, Malawi, Maldives, Mali, Mauritania, Mozambique, Myanmar, Nepal, Niger, Rwanda, Samoa, São

Tomé and Príncipe, Senegal, Sierra Leone, Solomon Islands, Somalia, Sudan, Timor-Leste, Togo, Tuvalu, Uganda, United Republic of Tanzania, Vanuatu, Yemen, Zambia.

**Small Island Developing States:**[1] American Samoa, Anguilla, Antigua and Barbuda, Aruba, Bahamas, Bahrain, Barbados, Belize, British Virgin Islands, Cape Verde, Comoros, Cook Islands, Cuba, Dominica, Dominican Republic, Fiji, French Polynesia, Grenada, Guam, Guinea-Bissau, Guyana, Haiti, Jamaica, Kiribati, Maldives, Marshall Islands, Mauritius, Micronesia (Federated States of), Montserrat, Nauru, Netherlands Antilles, New Caledonia, Niue, Northern Mariana Islands, Palau, Papua New Guinea, Puerto Rico, Saint Kitts and Nevis, Saint Lucia, Saint Vincent and the Grenadines, Samoa, São Tomé and Príncipe, Seychelles, Solomon Islands, Suriname, Timor-Leste, Tonga, Trinidad and Tobago, Tuvalu, United States Virgin Islands, Vanuatu.

**Sub-Saharan Africa:** Angola, Benin, Botswana, Burkina Faso, Burundi, Cameroon, Cape Verde, Central African Republic, Chad, Comoros, Congo, Côte d'Ivoire, Democratic Republic of the Congo, Djibouti, Egypt, Equatorial Guinea, Eritrea, Ethiopia, Gabon, Gambia, Ghana, Guinea, Guinea-Bissau, Kenya, Lesotho, Liberia, Madagascar, Malawi, Mali, Mauritania, Mauritius, Mayotte, Morocco, Mozambique, Namibia, Niger, Nigeria, Réunion, Rwanda, Saint Helena, São Tomé and Príncipe, Senegal, Seychelles, Sierra Leone, Somalia, South Africa, Sudan, Swaziland, Togo, Uganda, United Republic of Tanzania, Zambia, Zimbabwe.

### Countries in the Human Development Index aggregates[2]

**Very high human development:** Andorra, Australia, Austria, Bahrain, Barbados, Belgium, Brunei Darussalam, Canada, Cyprus, Czech Republic, Denmark, Estonia, Finland, France, Germany, Greece, Hong Kong SAR of China, Hungary, Iceland, Ireland, Israel, Italy, Japan, Liechtenstein, Luxembourg, Malta, Netherlands, New Zealand, Norway, Poland, Portugal, Qatar, Republic of Korea, Singapore, Slovakia, Slovenia, Spain, Sweden, Switzerland, United Arab Emirates, United Kingdom, United States of America.

**High human development:** Albania, Algeria, Argentina, Armenia, Azerbaijan, Bahamas, Belarus, Belize, Bosnia and Herzegovina, Brazil, Bulgaria, Chile, Colombia, Costa Rica, Croatia, Ecuador, Georgia, Iran (Islamic Republic of), Jamaica, Jordan, Kazakhstan, Kuwait, Latvia, Libyan Arab Jamahiriya, Lithuania, Malaysia, Mauritius, Mexico, Montenegro, Panama, Peru, Romania, Russian Federation, Saudi Arabia, Serbia, The former Yugoslav Republic of Macedonia, Tonga, Trinidad and Tobago, Tunisia, Turkey, Ukraine, Uruguay, Venezuela (Bolivarian Republic of).

**Medium human development:** Bolivia, Botswana, Cambodia, Cape Verde, China, Congo, Dominican Republic, Egypt, El Salvador, Equatorial Guinea, Fiji, Gabon, Guatemala, Guyana, Honduras, India, Indonesia, Kyrgyzstan, Lao People's Democratic Republic, Maldives, Micronesia (Federated States of), Moldova, Mongolia, Morocco, Namibia, Nicaragua, Pakistan, Paraguay, Philippines, São Tomé and Príncipe, Solomon Islands, South Africa, Sri Lanka, Suriname, Swaziland, Syrian Arab Republic, Tajikistan, Thailand, Timor-Leste, Turkmenistan, Uzbekistan, Viet Nam.

**Low human development:** Afghanistan, Angola, Bangladesh, Benin, Burkina Faso, Burundi, Cameroon, Central African Republic, Chad, Comoros, Côte d'Ivoire, Democratic Republic of the Congo, Djibouti, Ethiopia, Gambia, Ghana, Guinea, Guinea-Bissau, Haiti, Kenya, Lesotho, Liberia, Madagascar, Malawi, Mali, Mauritania, Mozambique, Myanmar, Nepal, Niger, Nigeria, Papua New Guinea, Rwanda, Senegal, Sierra Leone, Sudan, Togo, Uganda, United Republic of Tanzania, Yemen, Zambia, Zimbabwe.

## Countries in the income aggregates[3]

The World Bank classifies all member economies and all other economies with populations of more than 30,000. In the *World Development Report 2011*, economies are divided among income groups according to 2009 GNI per capita, calculated using the World Bank Atlas method. The groups are:

**High income:** Andorra, Aruba, Australia, Austria, Bahamas, Bahrain, Barbados, Belgium, Bermuda, Brunei Darussalam, Canada, Cayman Islands, Channel Islands, Croatia, Cyprus, Czech Republic, Denmark, Equatorial Guinea, Estonia, Faeroe Islands, Finland, France, French Polynesia, Germany, Gibraltar, Greece, Greenland, Guam, Hong Kong SAR of China, Hungary, Iceland, Ireland, Isle of Man, Israel, Italy, Japan, Kuwait, Latvia, Liechtenstein, Luxembourg, Macao SAR of China, Malta, Monaco, Netherlands Antilles, Netherlands, New Caledonia, New Zealand, Northern Mariana Islands, Norway, Oman, Poland, Portugal, Puerto Rico, Qatar, Republic of Korea, San Marino, Saudi Arabia, Singapore, Slovakia, Slovenia, Spain, Sweden, Switzerland, Trinidad and Tobago, Turks and Caicos Islands, United Arab Emirates, United Kingdom, United States of America, United States Virgin Islands.

**Upper-middle income:** Albania, Algeria, American Samoa, Antigua and Barbuda, Argentina, Azerbaijan, Belarus, Bosnia and Herzegovina, Botswana, Brazil, Bulgaria, Chile, Colombia, Costa Rica, Cuba, Dominica, Dominican Republic, Fiji, Gabon, Grenada, Iran (Islamic Republic of), Jamaica, Kazakhstan, Lebanon, Libyan Arab Jamahiriya, Lithuania, Malaysia, Mauritius, Mayotte, Mexico, Montenegro, Namibia, Palau, Panama, Peru, Romania, Russian Federation, Serbia, Seychelles, South Africa, Saint Kitts and Nevis, Saint Lucia, Saint Vincent and the Grenadines, Suriname, The former Yugoslav Republic of Macedonia, Turkey, Uruguay, Venezuela (Bolivarian Republic of).

**Lower-middle income:** Angola, Armenia, Belize, Bhutan, Bolivia, Cameroon, Cape Verde, China, Congo, Côte d'Ivoire, Djibouti, Ecuador, Egypt, El Salvador, Georgia, Guatemala, Guyana, Honduras, India, Indonesia, Iraq, Jordan, Kiribati, Lesotho, Maldives, Marshall Islands, Micronesia (Federated States of), Moldova, Mongolia, Morocco, Nicaragua, Nigeria, Occupied Palestinian Territory, Pakistan, Papua New Guinea, Paraguay, Philippines, Samoa, São Tomé and Príncipe, Senegal, Sri Lanka, Sudan, Swaziland, Syrian Arab Republic, Thailand, Timor-Leste, Tonga, Tunisia, Turkmenistan, Tuvalu, Ukraine, Uzbekistan, Vanuatu, Viet Nam, Yemen.

**Low income:** Afghanistan, Bangladesh, Benin, Burkina Faso, Burundi, Cambodia, Central African Republic, Chad, Comoros, Democratic People's Republic of Korea, Democratic Republic of the Congo, Eritrea, Ethiopia, Gambia, Ghana, Guinea, Guinea-Bissau, Haiti, Kenya, Kyrgyzstan, Lao People's Democratic Republic, Liberia, Madagascar, Malawi, Mali, Mauritania, Mozambique, Myanmar, Nepal, Niger, Rwanda, Sierra Leone, Solomon Islands, Somalia, Tajikistan, Togo, Uganda, United Republic of Tanzania, Zambia, Zimbabwe.

## Sub-regional aggregates

### ■ Africa

**Eastern Africa:** Burundi, Comoros, Djibouti, Eritrea, Ethiopia, Kenya, Madagascar, Malawi, Mauritius, Mayotte, Mozambique, Réunion, Rwanda, Seychelles, Somalia, Uganda, United Republic of Tanzania, Zambia, Zimbabwe.
**Middle Africa:** Angola, Cameroon, Central African Republic, Chad, Congo, Democratic Republic of the Congo, Equatorial Guinea, Gabon, São Tomé and Príncipe.
**Northern Africa:** Algeria, Egypt, Libyan Arab Jamahiriya, Morocco, Sudan, Tunisia, Western Sahara.
**Southern Africa:** Botswana, Lesotho, Namibia, South Africa, Swaziland.
**Western Africa:** Benin, Burkina Faso, Cape Verde, Côte d'Ivoire, Gambia, Ghana, Guinea, Guinea-Bissau, Liberia, Mali, Mauritania, Niger, Nigeria, Saint Helena, Senegal, Sierra Leone, Togo.

### ■ Asia

**Eastern Asia:** China, Hong Kong SAR of China, Macao SAR of China, Democratic People's Republic of Korea, Japan, Mongolia, Republic of Korea.

**South-Central Asia:** Afghanistan, Bangladesh, Bhutan, India, Iran (Islamic Republic of), Kazakhstan, Kyrgyzstan, Maldives, Nepal, Pakistan, Sri Lanka, Tajikistan, Turkmenistan, Uzbekistan.

**South-Eastern Asia:** Brunei Darussalam, Cambodia, Indonesia, Lao People's Democratic Republic, Malaysia, Myanmar, Philippines, Singapore, Thailand, Timor-Leste, Viet Nam.

**Western Asia:** Armenia, Azerbaijan, Bahrain, Cyprus, Georgia, Iraq, Israel, Jordan, Kuwait, Lebanon, Occupied Palestinian Territory, Oman, Qatar, Saudi Arabia, Syrian Arab Republic, Turkey, United Arab Emirates, Yemen.

#### ■ Europe

**Eastern Europe:** Belarus, Bulgaria, Czech Republic, Hungary, Moldova, Poland, Romania, Russian Federation, Slovakia, Ukraine.

**Northern Europe:** Channel Islands, Denmark, Estonia, Faeroe Islands, Finland, Iceland, Ireland, Isle of Man, Latvia, Lithuania, Norway, Sweden, United Kingdom.

**Southern Europe:** Albania, Andorra, Bosnia and Herzegovina, Croatia, Gibraltar, Greece, Holy See, Italy, Malta, Montenegro, Portugal, San Marino, Serbia, Slovenia, Spain, The former Yugoslav Republic of Macedonia.

**Western Europe:** Austria, Belgium, France, Germany, Liechtenstein, Luxembourg, Monaco, Netherlands, Switzerland.

#### ■ Latin America and the Caribbean

**Caribbean:** Anguilla, Antigua and Barbuda, Aruba, Bahamas, Barbados, British Virgin Islands, Cayman Islands, Cuba, Dominica, Dominican Republic, Grenada, Guadeloupe, Haiti, Jamaica, Martinique, Montserrat, Netherlands Antilles, Puerto Rico, Saint Kitts and Nevis, Saint Lucia, Saint Vincent and the Grenadines, Trinidad and Tobago, Turks and Caicos Islands, United States Virgin Islands.

**Central America:** Belize, Costa Rica, El Salvador, Guatemala, Honduras, Mexico, Nicaragua, Panama.

**South America:** Argentina, Bolivia, Brazil, Chile, Colombia, Ecuador, Falkland Islands (Malvinas), French Guiana, Guyana, Paraguay, Peru, Suriname, Uruguay, Venezuela (Bolivarian Republic of).

#### ■ Northern America

Bermuda, Canada, Greenland, Saint-Pierre-et-Miquelon, United States of America.

#### ■ Oceania

**Australia/New Zealand:** Australia, New Zealand.

**Melanesia:** Fiji, New Caledonia, Papua New Guinea, Solomon Islands, Vanuatu.

**Micronesia:** Guam, Kiribati, Marshall Islands, Micronesia (Federated States of), Nauru, Northern Mariana Islands, Palau.

**Polynesia:** American Samoa, Cook Islands, French Polynesia, Niue, Pitcairn, Samoa, Tokelau, Tonga, Tuvalu, Wallis and Futuna Islands.

## NOMENCLATURE AND ORDER OF PRESENTATION

Tables A.1 to A.4 contain regional data, grouped in income, human development and geographic aggregates. Tables B.1 to B.8 and C.1 to C.3 contain country and city-level data, respectively. In these tables, the countries or areas are listed in English alphabetical order within the macro-regions of Africa, Asia, Europe, Latin America, Northern America and Oceania. Countries or area names are presented in the form commonly used within the United Nations Secretariat for statistical purposes. Due to space limitations, the short name is used – for example, the United Kingdom of Great Britain and Northern Ireland is referred to as 'United Kingdom'.

## DEFINITION OF TERMS

**Access to electricity:** percentage of households which, within their housing unit are connected to electricity.

**Access to piped water:** percentage of households which, for source of drinking water are connected to piped water within their housing unit, tap placed in the yard or plot outside the house.

**Access to sewerage:** percentage of households which, within their housing unit are connected to sewerage.

**Access to telephone:** percentage of households which, within their housing unit are connected to telephone.

**Access to mobile:** percentage of households with mobile phone.

**Gini index:** the extent to which the distribution of income (or, in some cases, consumption expenditure) or assets (such as land) among individuals or households within an economy deviates from a perfectly equal distribution. A Lorenz curve plots the cumulative percentages of total income received against the cumulative number of recipients, starting with the poorest individual or household. The Gini index measures the area between the Lorenz curve and a hypothetical line of absolute equality, expressed as a percentage of the maximum area under the line. Thus, a Gini index of 0 represents perfect equality, while an index of 1 implies absolute inequality.

**Greenhouse gas emissions, carbon dioxide:** emissions from the burning of fossil fuels and the manufacture of cement and include carbon dioxide produced during the consumption of solid, liquid, and gas fuels and gas flaring.

**Greenhouse gas emissions, methane:** emissions from human activities such as agriculture and from industrial methane production.

**Greenhouse gas emissions, nitrous oxide:** emissions from agricultural biomass burning, industrial activities, and livestock management.

**Greenhouse gas emissions, other:** the by-product emissions of hydrofluorocarbons (by-product emissions of fluoroform from chlorodifluoromethane manufacture and use of hydrofluorocarbons), perfluorocarbons (by-product emissions of tetrafluoromethane and hexafluoroethane from primary aluminium production and use of fluorocarbons, in particular for semiconductor manufacturing), and sulphur hexafluoride (various sources, the largest being the use and manufacture of gas insulated switchgear used in electricity distribution networks).

**Greenhouse gas emissions, percentage change:** (calculated by UN-Habitat) refers to the average annual percentage growth rate of metric tonnes of emissions produced during the indicated period for each country, major regions and global totals.

**Gross national income:** the sum of value added by all resident producers plus any product taxes (less subsidies) not included in the valuation of output plus net receipts of primary income (compensation of employees and property income) from abroad. Data are in current US dollars converted using the World Bank Atlas method.

**Gross national income per capita:** gross national income (GNI) divided by mid-year population. GNI per capita in US dollars is converted using the World Bank Atlas method.

**Gross national income PPP:** gross national income converted to international dollars using purchasing power parity (PPP) rates. An international dollar has the same purchasing power over GNI as a US dollar has in the United States of America.

**Household:** the concept of household is based on the arrangements made by persons, individually or in groups, for providing themselves with food or other essentials for living. A household may be either:

1. A one-person household: a person who makes provision for his or her own food or other essentials for living without combining with any other person to form a part of a multi-person household.

2. A multi-person household: a group of two or more persons living together who make common provision for food or other essentials for living. The persons in the group may pool their incomes and may, to a greater or lesser extent, have a common budget; they may be related or unrelated persons or constitute a combination of persons both related and unrelated. This concept of household is known as the 'housekeeping' concept. It does not assume that the number of households and housing units is equal. Although the concept of housing unit implies that it is a space occupied by one household, it may also be occupied by more than one household or by a part of a household (e.g. two nuclear households that share one housing unit for economic reasons or one household in a polygamous society routinely occupying two or more housing units).

**Household connection to improved drinking water:** percentage of households which, within their housing unit, are connected to any of the following types of water supply for drinking: piped water, public tap, borehole or pump, protected well, protected spring or rainwater.

**Improved drinking water coverage:** percentage of people using improved drinking water sources or delivery points. Improved drinking water technologies are more likely to provide safe drinking water than those characterized as unimproved. Improved drinking water sources: piped water into dwelling, plot or yard; public tap/standpipe; tube well /borehole; protected dug well; protected spring; rainwater collection. Unimproved drinking water sources: unprotected dug well; unprotected spring; cart with small tank/drum; bottled water;[4] tanker-truck; surface water (river, dam, lake, pond, stream, canal, irrigation channels).

**Improved sanitation coverage:** percentage of people using improved sanitation facilities. Improved sanitation facilities are more likely to prevent human contact with human excreta than unimproved facilities.

**International poverty line:** based on nationally representative primary household surveys conducted by national statistical offices or by private agencies under the supervision of government or international agencies and obtained from government statistical offices and World Bank country departments. **Population below US$1.25 a day and population below US$2 a day:** percentages of the population living on less than $1.25 a day and $2 a day at 2005 international prices. As a result of revisions in PPP exchange rates, poverty rates for individual countries cannot be compared with poverty rates reported in earlier editions

**Level of urbanization:** percentage of the population residing in places classified as urban. Urban and rural settlements are defined in the national context and vary among countries (the definitions of urban are generally national definitions incorporated within the latest census).

**Motor vehicles:** include cars, buses and freight vehicles but not two-wheelers.

**National population below national poverty line:** percentage of the country's population living below the national poverty line. National estimates are based on population weighted subgroup estimates from household surveys.

**Persons in housing units:** number of persons resident in housing units.

**Population, rural:** mid-year estimates and projections (medium variant) of the population residing in human settlements classified as rural (see also 'Population, urban' below).

**Population, total:** mid-year population estimates and projections for the world, region, countries or areas. The Population Division of the United Nations Department of Economic and Social Affairs updates, every two years, population estimates and projections by incorporating new data, new estimates and new analyses of data on population, fertility, mortality and international migration. Data from new population censuses and/or demographic surveys are used to verify and update old estimates of population or demographic indicators, or to make new ones and to check the validity of the assumptions made in the projections. Population rate of change (calculated by UN-Habitat) refers to the average annual percentage change of population during the indicated period for each country, major regions and global totals. The formula used throughout the Annex is as follows: $r = [(1/t) \times \ln(A2/A1)] \times 100$, where 'A1' is a value at any given year; 'A2' is a value at any given year later than the year of 'A1'; 't' is the year interval between 'A1' and 'A2'; and 'ln' is the natural logarithm function.

**Population, urban:** mid-year population of areas defined as urban in each country and reported to the United Nations. Estimates of the world's urban population would change significantly if China, India, and a few other populous nations were to change their definition of urban centres. According to China's State Statistical Bureau, by the end of 1996 urban residents accounted for about 43 per cent of China's population, while in 1994 only 20 per cent of the population was considered urban. In addition to the continuous migration of people from rural to urban areas, one of the main reasons for this shift was the rapid growth in the hundreds of towns reclassified as cities in recent years. Because the estimates in the table are based on national definitions of what constitutes a city or metropolitan area, cross-country comparisons should be made with caution.

**Population density:** mid-year population divided by land area in square kilometres.

**Railways:** length of railway route available for train service, irrespective of the number of parallel tracks. Passengers carried by railway are the number of passengers transported by rail multiplied by kilometres travelled. Goods hauled by railway are the volume of goods transported by railway, measured in metric tons multiplied by kilometres travelled.

**Roads:** motorways, highways, main or national roads, and secondary or regional roads. A motorway is a road specially designed and built for motor vehicles that separates the traffic flowing in opposite directions. **Total road network:** includes motorways, highways and main or national roads, secondary or regional roads, and all other roads in a country. **Paved roads:** roads surfaced with crushed stone (macadam) and hydrocarbon binder or bitumized agents, with concrete or with cobblestones, as a percentage of all of the country's roads measured in length. Goods hauled by road are the volume of goods transported by road vehicles, measured in millions of metric tons multiplied by kilometres travelled.

**Survey year:** the year in which the underlying data were collected.

**Urban poverty rate:** percentage of the urban population living below the national urban poverty line.

**Urban slum dwellers:** individuals residing in housing with one or more of the following conditions: inadequate drinking water; inadequate sanitation; poor structural quality/durability of housing; overcrowding; and insecurity of tenure.

**Urban agglomerations and capital cities:** the term 'urban agglomeration' refers to the population contained within the contours of a contiguous territory inhabited at urban density levels without regard to administrative boundaries. It usually incorporates the population in a city or town plus that in the suburban areas lying outside of but being adjacent to the city boundaries. Whenever possible, data classified according to the concept of urban agglomeration are used. However, some countries do not produce data according to the concept of urban agglomeration but use instead that of metropolitan area or city proper. If possible, such data are adjusted to conform to the concept of urban agglomeration. When sufficient information is not available to permit such an adjustment, data based on the concept of city proper or metropolitan area are used. The sources listed online indicate whether data were adjusted to conform to the urban agglomeration concept or whether a different concept was used. Table C.1 contains revised estimates and projections for all urban agglomerations comprising 750,000 or more inhabitants.

## SOURCES OF DATA

The Statistical Tables have been compiled from the following UN-Habitat databases:

United Nations Human Settlements Programme (UN-Habitat), Global Urban Indicators Database 2010
United Nations Human Settlements Programme (UN-Habitat), Urban Info 2010

In addition, various statistical publications from the United Nations and other international organizations have been used. These include:

United Nations Development Programme (2010) *Human Development Report 2010*, New York, http://hdr.undp.org/en/reports/global/hdr2010/
United Nations, Department of Economic and Social Affairs, Population Division (2009), *World Population Prospects: The 2008 Revision*, New York
United Nations, Department of Economic and Social Affairs, Population Division (2010) *World Urbanization Prospects: The 2009 Revision*, United Nations, New York
World Bank (2005) *World Development Indicators 2005*, World Bank, Washington, DC
World Bank (2006) *World Development Report 2006*, World Bank, Washington, DC
World Bank (2010) *World Development Indicators 2010*, World Bank, Washington, DC
World Bank (2010) *World Development Indicators Online database*, http://data.worldbank.org/indicator
World Bank (2010) *World Development Report 2011*, World Bank, Washington, DC
World Health Organization (WHO) and United Nations Children's Fund (UNICEF) Joint Monitoring Programme for Water Supply and Sanitation (JMP) (2010) *Progress on Sanitation and Drinking-Water 2010 Update*, WHO and UNICEF, Geneva and New York, www.who.int/water_sanitation_health/publications/9789241563956/en/index.html

# NOTES

1.  As classified by United Nations Department of Economic and Social Affairs (UNDESA); see http://www.sidsnet.org/sids_list.html for detail.

2.  As classified by the United Nations Development Programme (UNDP); see *Human Development Report 2010* for detail. The following countries and territories were not classified: American Samoa, Anguilla, Antigua and Barbuda, Aruba, Bermuda, Bhutan, British Virgin Islands, Cayman Islands, Channel Islands, Cook Islands, Cuba, Democratic People's Republic of Korea, Dominica, Eritrea, Faeroe Islands, Falkland Islands (Malvinas), French Guiana, French Polynesia, Gibraltar, Greenland, Grenada, Guadeloupe, Guam, Holy See, Iraq, Isle of Man, Kiribati, Lebanon, Macao SAR of China, Marshall Islands, Martinique, Mayotte, Monaco, Montserrat, Nauru, Netherlands Antilles, New Caledonia, Niue, Northern Mariana Islands, Occupied Palestinian Territory, Oman, Palau, Pitcairn, Puerto Rico, Réunion, Saint Helena, Saint Kitts and Nevis, Saint Lucia, Saint Vincent and the Grenadines, Saint-Pierre-et-Miquelon, Samoa, San Marino, Seychelles, Somalia, Tokelau, Turks and Caicos Islands, Tuvalu, United States Virgin Islands, Vanuatu, Wallis and Futuna Islands, and Western Sahara.

3.  As classified by the World Bank; see *World Development Report 2011* for detail. The following countries and territories were not classified: Anguilla, British Virgin Islands, Cook Islands, Falkland Islands (Malvinas), French Guiana, Guadeloupe, Holy See, Martinique, Montserrat, Nauru, Niue, Pitcairn, Réunion, Saint Helena, Saint-Pierre-et-Miquelon, Tokelau, Wallis and Futuna Islands, and Western Sahara.

4.  Bottled water is considered improved only when the household uses water from an improved source for cooking and personal hygiene.

# DATA TABLES

## REGIONAL AGGREGATES
## TABLE A.1

### Total Population Size, Rate of Change and Population Density

| | Estimates and projections ('000) | | | | Rate of change (%) | | | Population density (people/km²) | |
|---|---|---|---|---|---|---|---|---|---|
| | 2000 | 2010 | 2020 | 2030 | 2000–2010 | 2010–2020 | 2020–2030 | 2000 | 2030 |
| **WORLD** | 6,115,367 | 6,908,688 | 7,674,833 | 8,308,895 | 1.22 | 1.05 | 0.79 | 45 | 61 |
| **World Major Aggregates** | | | | | | | | | |
| More Developed Regions | 1,194,967 | 1,237,228 | 1,268,343 | 1,281,628 | 0.35 | 0.25 | 0.10 | 22 | 24 |
| Less Developed Regions | 4,920,400 | 5,671,460 | 6,406,489 | 7,027,267 | 1.42 | 1.22 | 0.92 | 59 | 85 |
| Least Developed Countries | 677,368 | 855,209 | 1,060,067 | 1,272,279 | 2.33 | 2.15 | 1.82 | 33 | 61 |
| Other Less Developed Countries | 4,243,033 | 4,816,251 | 5,346,422 | 5,754,988 | 1.27 | 1.04 | 0.74 | 68 | 93 |
| Less Developed Regions, excluding China | 3,646,339 | 4,309,696 | 4,967,045 | 5,556,003 | 1.67 | 1.42 | 1.12 | 50 | 76 |
| Small Island Developing States | 52,809 | 59,642 | 66,205 | 72,097 | 1.22 | 1.04 | 0.85 | 42 | 58 |
| Sub-Saharan Africa | 674,842 | 863,314 | 1,081,114 | 1,307,831 | 2.46 | 2.25 | 1.90 | 28 | 54 |
| **Human Development Index Aggregates** | | | | | | | | | |
| Very High Human Development | 993,772 | 1,055,971 | 1,101,353 | 1,129,453 | 0.61 | 0.42 | 0.25 | 31 | 35 |
| High Human Development | 970,891 | 1,052,377 | 1,124,577 | 1,175,057 | 0.81 | 0.66 | 0.44 | 20 | 24 |
| Medium Human Development | 3,190,507 | 3,597,308 | 3,967,424 | 4,239,713 | 1.20 | 0.98 | 0.66 | 117 | 155 |
| Low Human Development | 871,324 | 1,099,018 | 1,360,204 | 1,626,493 | 2.32 | 2.13 | 1.79 | 38 | 71 |
| **Income Aggregates** | | | | | | | | | |
| High Income | 1,036,187 | 1,106,127 | 1,158,870 | 1,193,450 | 0.65 | 0.47 | 0.29 | 27 | 32 |
| Middle Income | 4,387,508 | 4,939,256 | 5,452,303 | 5,845,635 | 1.18 | 0.99 | 0.70 | 55 | 73 |
| Upper-Middle Income | 931,193 | 1,015,174 | 1,085,244 | 1,134,305 | 0.86 | 0.67 | 0.44 | 19 | 23 |
| Lower-Middle Income | 3,456,315 | 3,924,082 | 4,367,059 | 4,711,330 | 1.27 | 1.07 | 0.76 | 109 | 148 |
| Low Income | 691,678 | 863,301 | 1,063,654 | 1,269,812 | 2.22 | 2.09 | 1.77 | 39 | 71 |
| **Geographic Aggregates** | | | | | | | | | |
| **Africa** | 819,462 | 1,033,043 | 1,276,369 | 1,524,187 | 2.32 | 2.12 | 1.77 | 27 | 50 |
| Eastern Africa | 252,710 | 327,186 | 420,200 | 518,064 | 2.58 | 2.50 | 2.09 | 40 | 81 |
| Middle Africa | 98,060 | 128,909 | 164,284 | 201,602 | 2.74 | 2.42 | 2.05 | 15 | 30 |
| Northern Africa | 179,525 | 212,921 | 247,564 | 277,351 | 1.71 | 1.51 | 1.14 | 21 | 33 |
| Southern Africa | 51,387 | 57,968 | 61,134 | 64,037 | 1.21 | 0.53 | 0.46 | 19 | 24 |
| Western Africa | 237,781 | 306,058 | 383,187 | 463,133 | 2.52 | 2.25 | 1.89 | 39 | 75 |
| **Asia** | 3,698,296 | 4,166,741 | 4,596,256 | 4,916,701 | 1.19 | 0.98 | 0.67 | 116 | 154 |
| Eastern Asia | 1,472,444 | 1,563,951 | 1,640,388 | 1,666,372 | 0.60 | 0.48 | 0.16 | 125 | 142 |
| South-Central Asia | 1,518,322 | 1,780,473 | 2,028,786 | 2,231,846 | 1.59 | 1.31 | 0.95 | 141 | 207 |
| South-Eastern Asia | 517,193 | 589,615 | 653,541 | 706,492 | 1.31 | 1.03 | 0.78 | 115 | 157 |
| Western Asia | 190,336 | 232,702 | 273,541 | 311,991 | 2.01 | 1.62 | 1.32 | 39 | 65 |
| **Europe** | 726,568 | 732,759 | 732,952 | 723,373 | 0.08 | 0.00 | -0.13 | 32 | 31 |
| Eastern Europe | 304,088 | 291,485 | 281,511 | 268,320 | -0.42 | -0.35 | -0.48 | 16 | 14 |
| Northern Europe | 94,359 | 98,909 | 103,400 | 107,221 | 0.47 | 0.44 | 0.36 | 52 | 59 |
| Southern Europe | 145,119 | 153,778 | 157,455 | 157,228 | 0.58 | 0.24 | -0.01 | 110 | 119 |
| Western Europe | 183,001 | 188,587 | 190,585 | 190,605 | 0.30 | 0.11 | 0.00 | 165 | 172 |
| **Latin America and the Caribbean** | 521,228 | 588,649 | 645,543 | 689,859 | 1.22 | 0.92 | 0.66 | 25 | 34 |
| Caribbean | 38,650 | 42,312 | 45,470 | 47,922 | 0.91 | 0.72 | 0.53 | 165 | 205 |
| Central America | 135,171 | 153,115 | 169,861 | 183,885 | 1.25 | 1.04 | 0.79 | 55 | 74 |
| South America | 347,407 | 393,221 | 430,212 | 458,052 | 1.24 | 0.90 | 0.63 | 19 | 26 |
| **Northern America** | 318,654 | 351,659 | 383,384 | 410,204 | 0.99 | 0.86 | 0.68 | 15 | 19 |
| **Oceania** | 31,160 | 35,838 | 40,329 | 44,572 | 1.40 | 1.18 | 1.00 | 4 | 5 |
| Australia/New Zealand | 23,039 | 25,815 | 28,344 | 30,627 | 1.14 | 0.93 | 0.77 | 3 | 4 |
| Melanesia | 7,010 | 8,778 | 10,613 | 12,452 | 2.25 | 1.90 | 1.60 | 13 | 23 |
| Micronesia | 497 | 573 | 646 | 713 | 1.43 | 1.19 | 1.00 | 160 | 230 |
| Polynesia | 614 | 672 | 727 | 779 | 0.90 | 0.78 | 0.69 | 73 | 93 |

Sources: United Nations Department of Economic and Social Affairs, Population Division (2009) *World Urbanization Prospects: The 2008 Revision*, United Nations, New York; United Nations Department of Economic and Social Affairs, Population Division (2010) World Urbanization Prospects: The 2009 Revision, United Nations, New York. Figures in regional, income or development aggregates are calculated on the basis of country/area level data from Table B.1.

Note: Lists of countries/areas in aggregates are presented in the Technical Notes.

# TABLE A.2

## Urban and Rural Population Size and Rate of Change

| | Urban population | | | | | | | Rural population | | | | | | |
|---|---|---|---|---|---|---|---|---|---|---|---|---|---|---|
| | Estimates and projections ('000) | | | | Rate of change (%) | | | Estimates and projections ('000) | | | | Rate of change (%) | | |
| | 2000 | 2010 | 2020 | 2030 | 2000–2010 | 2010–2020 | 2020–2030 | 2000 | 2010 | 2020 | 2030 | 2000–2010 | 2010–2020 | 2020–2030 |
| **WORLD** | 2,837,431 | 3,486,326 | 4,176,234 | 4,899,858 | 2.06 | 1.81 | 1.60 | 3,277,937 | 3,422,362 | 3,498,599 | 3,409,038 | 0.43 | 0.22 | -0.26 |
| **World Major Aggregates** | | | | | | | | | | | | | | |
| More Developed Regions | 869,233 | 929,851 | 988,130 | 1,036,550 | 0.67 | 0.61 | 0.48 | 325,734 | 307,377 | 280,214 | 245,078 | -0.58 | -0.93 | -1.34 |
| Less Developed Regions | 1,968,198 | 2,556,475 | 3,188,104 | 3,863,308 | 2.62 | 2.21 | 1.92 | 2,952,203 | 3,114,985 | 3,218,385 | 3,163,960 | 0.54 | 0.33 | -0.17 |
| Least Developed Countries | 167,181 | 249,442 | 366,150 | 519,537 | 4.00 | 3.84 | 3.50 | 510,186 | 605,767 | 693,917 | 752,742 | 1.72 | 1.36 | 0.81 |
| Other Less Developed Countries | 1,801,016 | 2,307,033 | 2,821,954 | 3,343,771 | 2.48 | 2.01 | 1.70 | 2,442,016 | 2,509,218 | 2,524,468 | 2,411,218 | 0.27 | 0.06 | -0.46 |
| Less Developed Regions, excluding China | 1,508,061 | 1,913,018 | 2,393,054 | 2,949,063 | 2.38 | 2.24 | 2.09 | 2,138,278 | 2,396,678 | 2,573,992 | 2,606,941 | 1.14 | 0.71 | 0.13 |
| Small Island Developing States | 27,682 | 33,269 | 39,014 | 44,839 | 1.84 | 1.59 | 1.39 | 25,118 | 26,374 | 27,197 | 27,260 | 0.49 | 0.31 | 0.02 |
| Sub-Saharan Africa | 220,606 | 321,400 | 456,580 | 626,683 | 3.76 | 3.51 | 3.17 | 454,236 | 541,914 | 624,534 | 681,148 | 1.76 | 1.42 | 0.87 |
| **Human Development Index Aggregates** | | | | | | | | | | | | | | |
| Very High Human Development | 743,983 | 818,351 | 882,751 | 936,113 | 0.95 | 0.76 | 0.59 | 249,794 | 237,620 | 218,604 | 193,339 | -0.50 | -0.83 | -1.23 |
| High Human Development | 698,381 | 797,979 | 890,670 | 965,152 | 1.33 | 1.10 | 0.80 | 272,506 | 254,397 | 233,904 | 209,900 | -0.69 | -0.84 | -1.08 |
| Medium Human Development | 1,090,497 | 1,436,933 | 1,794,395 | 2,170,060 | 2.76 | 2.22 | 1.90 | 2,100,009 | 2,160,379 | 2,173,029 | 2,069,654 | 0.28 | 0.06 | -0.49 |
| Low Human Development | 247,830 | 366,677 | 529,624 | 734,793 | 3.92 | 3.68 | 3.27 | 623,497 | 732,343 | 830,587 | 891,698 | 1.61 | 1.26 | 0.71 |
| **Income Aggregates** | | | | | | | | | | | | | | |
| High Income | 774,391 | 855,606 | 926,830 | 986,780 | 1.00 | 0.80 | 0.63 | 259,682 | 247,945 | 229,084 | 203,462 | -0.46 | -0.79 | -1.19 |
| Middle Income | 1,887,460 | 2,378,715 | 2,887,033 | 3,403,810 | 2.31 | 1.94 | 1.65 | 2,500,044 | 2,560,546 | 2,565,274 | 2,441,817 | 0.24 | 0.02 | -0.49 |
| Upper-Middle Income | 666,588 | 766,942 | 856,953 | 929,145 | 1.40 | 1.11 | 0.81 | 264,603 | 248,234 | 228,293 | 205,152 | -0.64 | -0.84 | -1.07 |
| Lower-Middle Income | 1,220,872 | 1,611,773 | 2,030,080 | 2,474,665 | 2.78 | 2.31 | 1.98 | 2,235,441 | 2,312,312 | 2,336,981 | 2,236,665 | 0.34 | 0.11 | -0.44 |
| Low Income | 173,725 | 249,727 | 359,728 | 506,362 | 3.63 | 3.65 | 3.42 | 517,957 | 613,576 | 703,930 | 763,447 | 1.69 | 1.37 | 0.81 |
| **Geographic Aggregates** | | | | | | | | | | | | | | |
| **Africa** | 294,602 | 412,990 | 569,117 | 761,293 | 3.38 | 3.21 | 2.91 | 524,861 | 620,053 | 707,253 | 762,895 | 1.67 | 1.32 | 0.76 |
| Eastern Africa | 52,641 | 77,194 | 116,130 | 172,766 | 3.83 | 4.08 | 3.97 | 200,069 | 249,992 | 304,070 | 345,298 | 2.23 | 1.96 | 1.27 |
| Middle Africa | 36,486 | 55,592 | 81,493 | 112,727 | 4.21 | 3.82 | 3.24 | 61,574 | 73,318 | 82,791 | 88,875 | 1.75 | 1.22 | 0.71 |
| Northern Africa | 85,656 | 108,912 | 137,341 | 167,876 | 2.40 | 2.32 | 2.01 | 93,868 | 104,009 | 110,224 | 109,475 | 1.03 | 0.58 | -0.07 |
| Southern Africa | 27,657 | 34,021 | 38,809 | 43,741 | 2.07 | 1.32 | 1.20 | 23,730 | 23,947 | 22,325 | 20,295 | 0.09 | -0.70 | -0.95 |
| Western Africa | 92,162 | 137,271 | 195,344 | 264,182 | 3.98 | 3.53 | 3.02 | 145,620 | 168,787 | 187,843 | 198,951 | 1.48 | 1.07 | 0.57 |
| **Asia** | 1,360,900 | 1,757,314 | 2,168,798 | 2,598,358 | 2.56 | 2.10 | 1.81 | 2,337,395 | 2,409,427 | 2,427,458 | 2,318,343 | 0.30 | 0.07 | -0.46 |
| Eastern Asia | 594,676 | 784,688 | 940,684 | 1,061,980 | 2.77 | 1.81 | 1.21 | 877,768 | 779,263 | 699,704 | 604,392 | -1.19 | -1.08 | -1.46 |
| South-Central Asia | 447,425 | 571,112 | 733,039 | 936,279 | 2.44 | 2.50 | 2.45 | 1,070,897 | 1,209,360 | 1,295,746 | 1,295,567 | 1.22 | 0.69 | 0.00 |
| South-Eastern Asia | 197,360 | 246,701 | 305,412 | 373,411 | 2.23 | 2.13 | 2.01 | 319,833 | 342,914 | 348,130 | 333,081 | 0.70 | 0.15 | -0.44 |
| Western Asia | 121,438 | 154,813 | 189,664 | 226,688 | 2.43 | 2.03 | 1.78 | 68,897 | 77,889 | 83,877 | 85,303 | 1.23 | 0.74 | 0.17 |
| **Europe** | 514,422 | 533,295 | 552,486 | 567,403 | 0.36 | 0.35 | 0.27 | 212,146 | 199,464 | 180,465 | 155,970 | -0.62 | -1.00 | -1.46 |
| Eastern Europe | 207,409 | 200,938 | 199,963 | 198,744 | -0.32 | -0.05 | -0.06 | 96,679 | 90,546 | 81,548 | 69,575 | -0.66 | -1.05 | -1.59 |
| Northern Europe | 73,502 | 78,217 | 83,704 | 89,282 | 0.62 | 0.68 | 0.65 | 20,857 | 20,691 | 19,695 | 17,939 | -0.08 | -0.49 | -0.93 |
| Southern Europe | 95,015 | 104,209 | 111,664 | 117,473 | 0.92 | 0.69 | 0.51 | 50,104 | 49,569 | 45,791 | 39,755 | -0.11 | -0.79 | -1.41 |
| Western Europe | 138,495 | 149,931 | 157,155 | 161,904 | 0.79 | 0.47 | 0.30 | 44,506 | 38,656 | 33,430 | 28,701 | -1.41 | -1.45 | -1.53 |
| **Latin America and the Caribbean** | 393,420 | 468,757 | 533,147 | 585,490 | 1.75 | 1.29 | 0.94 | 127,807 | 119,892 | 112,395 | 104,369 | -0.64 | -0.65 | -0.74 |
| Caribbean | 23,708 | 28,278 | 32,510 | 36,143 | 1.76 | 1.39 | 1.06 | 14,941 | 14,034 | 12,960 | 11,779 | -0.63 | -0.80 | -0.96 |
| Central America | 92,948 | 110,251 | 127,463 | 143,535 | 1.71 | 1.45 | 1.19 | 42,222 | 42,865 | 42,398 | 40,350 | 0.15 | -0.11 | -0.50 |
| South America | 276,764 | 330,228 | 373,175 | 405,812 | 1.77 | 1.22 | 0.84 | 70,643 | 62,993 | 57,037 | 52,240 | -1.15 | -0.99 | -0.88 |
| **Northern America** | 252,154 | 288,803 | 324,279 | 355,499 | 1.36 | 1.16 | 0.92 | 66,500 | 62,856 | 59,105 | 54,705 | -0.56 | -0.62 | -0.77 |
| **Oceania** | 21,932 | 25,167 | 28,406 | 31,816 | 1.38 | 1.21 | 1.13 | 9,227 | 10,671 | 11,924 | 12,756 | 1.45 | 1.11 | 0.67 |
| Australia/New Zealand | 20,024 | 22,878 | 25,516 | 27,948 | 1.33 | 1.09 | 0.91 | 3,015 | 2,937 | 2,827 | 2,679 | -0.26 | -0.38 | -0.54 |
| Melanesia | 1,329 | 1,614 | 2,110 | 2,964 | 1.94 | 2.68 | 3.40 | 5,680 | 7,164 | 8,503 | 9,488 | 2.32 | 1.71 | 1.10 |
| Micronesia | 326 | 390 | 454 | 523 | 1.80 | 1.52 | 1.40 | 171 | 183 | 191 | 191 | 0.69 | 0.45 | -0.04 |
| Polynesia | 253 | 285 | 325 | 380 | 1.20 | 1.31 | 1.58 | 361 | 387 | 402 | 399 | 0.69 | 0.38 | -0.08 |

*Sources:* United Nations Department of Economic and Social Affairs, Population Division (2010) *World Urbanization Prospects: The 2009 Revision*, United Nations, New York. Figures in regional, income or development aggregates are calculated on the basis of country/area level data from Table B.2.

*Note:* Lists of countries/areas in aggregates are presented in the Technical Notes.

# TABLE A.3

## Urbanization

| | Level of urbanization | | | | | | |
|---|---|---|---|---|---|---|---|
| | Estimates and projections (%) | | | | Rate of change (%) | | |
| | **2000** | **2010** | **2020** | **2030** | **2000–2010** | **2010–2020** | **2020–2030** |
| **WORLD** | 46.4 | 50.5 | 54.4 | 59.0 | 0.84 | 0.75 | 0.80 |
| **World Major Aggregates** | | | | | | | |
| More Developed Regions | 72.7 | 75.2 | 77.9 | 80.9 | 0.33 | 0.36 | 0.37 |
| Less Developed Regions | 40.0 | 45.1 | 49.8 | 55.0 | 1.19 | 0.99 | 1.00 |
| Least Developed Countries | 24.7 | 29.2 | 34.5 | 40.8 | 1.67 | 1.69 | 1.67 |
| Other Less Developed Countries | 42.4 | 47.9 | 52.8 | 58.1 | 1.21 | 0.97 | 0.96 |
| Less Developed Regions, excluding China | 41.4 | 44.4 | 48.2 | 53.1 | 0.71 | 0.82 | 0.97 |
| Small Island Developing States | 52.4 | 55.8 | 58.9 | 62.2 | 0.62 | 0.55 | 0.54 |
| Sub-Saharan Africa | 32.7 | 37.2 | 42.2 | 47.9 | 1.30 | 1.26 | 1.26 |
| **Human Development Index Aggregates** | | | | | | | |
| Very High Human Development | 74.9 | 77.5 | 80.2 | 82.9 | 0.35 | 0.34 | 0.33 |
| High Human Development | 71.9 | 75.8 | 79.2 | 82.1 | 0.53 | 0.44 | 0.36 |
| Medium Human Development | 34.2 | 39.9 | 45.2 | 51.2 | 1.56 | 1.24 | 1.24 |
| Low Human Development | 28.4 | 33.4 | 38.9 | 45.2 | 1.60 | 1.54 | 1.49 |
| **Income Aggregates** | | | | | | | |
| High Income | 74.9 | 77.5 | 80.2 | 82.9 | 0.35 | 0.34 | 0.33 |
| Middle Income | 43.0 | 48.2 | 53.0 | 58.2 | 1.13 | 0.95 | 0.95 |
| Upper-Middle Income | 71.6 | 75.5 | 79.0 | 81.9 | 0.54 | 0.44 | 0.37 |
| Lower-Middle Income | 35.3 | 41.1 | 46.5 | 52.5 | 1.51 | 1.24 | 1.22 |
| Low Income | 25.1 | 28.9 | 33.8 | 39.9 | 1.41 | 1.56 | 1.65 |
| **Geographic Aggregates** | | | | | | | |
| **Africa** | 36.0 | 40.0 | 44.6 | 49.9 | 1.06 | 1.09 | 1.13 |
| Eastern Africa | 20.8 | 23.6 | 27.6 | 33.3 | 1.25 | 1.58 | 1.88 |
| Middle Africa | 37.2 | 43.1 | 49.6 | 55.9 | 1.48 | 1.40 | 1.20 |
| Northern Africa | 47.7 | 51.2 | 55.5 | 60.5 | 0.70 | 0.81 | 0.87 |
| Southern Africa | 53.8 | 58.7 | 63.5 | 68.3 | 0.87 | 0.78 | 0.73 |
| Western Africa | 38.8 | 44.9 | 51.0 | 57.0 | 1.46 | 1.28 | 1.12 |
| **Asia** | 36.8 | 42.2 | 47.2 | 52.8 | 1.36 | 1.12 | 1.13 |
| Eastern Asia | 40.4 | 50.2 | 57.3 | 63.7 | 2.17 | 1.34 | 1.06 |
| South-Central Asia | 29.5 | 32.1 | 36.1 | 42.0 | 0.85 | 1.19 | 1.49 |
| South-Eastern Asia | 38.2 | 41.8 | 46.7 | 52.9 | 0.92 | 1.11 | 1.23 |
| Western Asia | 63.8 | 66.5 | 69.3 | 72.7 | 0.42 | 0.41 | 0.47 |
| **Europe** | 70.8 | 72.8 | 75.4 | 78.4 | 0.28 | 0.35 | 0.40 |
| Eastern Europe | 68.2 | 68.9 | 71.0 | 74.1 | 0.11 | 0.30 | 0.42 |
| Northern Europe | 77.9 | 79.1 | 81.0 | 83.3 | 0.15 | 0.23 | 0.28 |
| Southern Europe | 65.5 | 67.8 | 70.9 | 74.7 | 0.34 | 0.45 | 0.52 |
| Western Europe | 75.7 | 79.5 | 82.5 | 84.9 | 0.49 | 0.37 | 0.30 |
| **Latin America and the Caribbean** | 75.5 | 79.6 | 82.6 | 84.9 | 0.54 | 0.36 | 0.27 |
| Caribbean | 61.3 | 66.8 | 71.5 | 75.4 | 0.86 | 0.67 | 0.53 |
| Central America | 68.8 | 72.0 | 75.0 | 78.1 | 0.46 | 0.41 | 0.39 |
| South America | 79.7 | 84.0 | 86.7 | 88.6 | 0.53 | 0.32 | 0.21 |
| **Northern America** | 79.1 | 82.1 | 84.6 | 86.7 | 0.37 | 0.29 | 0.24 |
| **Oceania** | 70.4 | 70.2 | 70.4 | 71.4 | -0.02 | 0.03 | 0.13 |
| Australia/New Zealand | 86.9 | 88.6 | 90.0 | 91.3 | 0.19 | 0.16 | 0.14 |
| Melanesia | 19.0 | 18.4 | 19.9 | 23.8 | -0.31 | 0.78 | 1.80 |
| Micronesia | 65.6 | 68.1 | 70.4 | 73.3 | 0.37 | 0.33 | 0.41 |
| Polynesia | 41.2 | 42.4 | 44.7 | 48.8 | 0.30 | 0.52 | 0.88 |

*Sources:* United Nations Department of Economic and Social Affairs, Population Division (2010) *World Urbanization Prospects: The 2009 Revision*, United Nations, New York. Figures in regional, income or development aggregates are calculated on the basis of country/area level data from Table B.3.

*Note:* Lists of countries/areas in aggregates are presented in the Technical Notes.

# TABLE A.4

## Urban Agglomerations

| | Number of urban agglomerations Estimates and projections | | | Distribution of urban population by size of agglomerations (%) | | | Population estimates and projections ('000) | | |
|---|---|---|---|---|---|---|---|---|---|
| | 2000 | 2010 | 2020 | 2000 | 2010 | 2020 | 2000 | 2010 | 2020 |
| **WORLD** | | | | | | | | | |
| 10 million or more | 16 | 21 | 28 | 8.2 | 9.3 | 10.4 | 231,624 | 324,190 | 436,308 |
| 5 to 10 million | 28 | 33 | 43 | 6.9 | 6.7 | 7.0 | 195,644 | 233,827 | 290,456 |
| 1 to 5 million | 305 | 388 | 467 | 20.6 | 22.1 | 22.0 | 584,050 | 772,084 | 917,985 |
| 500,000 to 1 million | 402 | 516 | 608 | 9.6 | 10.2 | 10.2 | 273,483 | 355,619 | 425,329 |
| Fewer than 500,000 | ... | ... | ... | 54.7 | 51.6 | 50.4 | 1,552,631 | 1,800,607 | 2,106,156 |
| **World Major Aggregates** | | | | | | | | | |
| More Developed Regions | | | | | | | | | |
| 10 million or more | 5 | 6 | 6 | 9.8 | 10.9 | 10.5 | 85,279 | 101,228 | 103,834 |
| 5 to 10 million | 5 | 7 | 9 | 4.2 | 4.9 | 5.9 | 36,472 | 45,595 | 58,692 |
| 1 to 5 million | 98 | 102 | 104 | 22.5 | 22.0 | 21.2 | 195,393 | 204,587 | 209,392 |
| 500,000 to 1 million | 117 | 126 | 132 | 9.1 | 9.1 | 9.1 | 78,818 | 84,750 | 89,863 |
| Fewer than 500,000 | ... | ... | ... | 54.4 | 53.1 | 53.3 | 473,271 | 493,691 | 526,350 |
| Less Developed Regions | | | | | | | | | |
| 10 million or more | 11 | 15 | 22 | 7.4 | 8.7 | 10.4 | 146,345 | 222,962 | 332,474 |
| 5 to 10 million | 23 | 26 | 34 | 8.1 | 7.4 | 7.3 | 159,172 | 188,232 | 231,764 |
| 1 to 5 million | 207 | 286 | 363 | 19.7 | 22.2 | 22.2 | 388,657 | 567,497 | 708,593 |
| 500,000 to 1 million | 285 | 390 | 476 | 9.9 | 10.6 | 10.5 | 194,664 | 270,868 | 335,466 |
| Fewer than 500,000 | ... | ... | ... | 54.8 | 51.1 | 49.6 | 1,079,360 | 1,306,916 | 1,579,806 |
| Least Developed Countries | | | | | | | | | |
| 10 million or more | 1 | 1 | 2 | 6.2 | 5.9 | 8.6 | 10,285 | 14,648 | 31,509 |
| 5 to 10 million | 1 | 2 | 6 | 3.4 | 5.6 | 10.0 | 5,611 | 13,926 | 36,755 |
| 1 to 5 million | 20 | 27 | 38 | 21.9 | 23.2 | 19.7 | 36,567 | 57,905 | 72,152 |
| 500,000 to 1 million | 19 | 28 | 34 | 7.7 | 8.1 | 6.4 | 12,915 | 20,269 | 23,438 |
| Fewer than 500,000 | ... | ... | ... | 60.9 | 57.2 | 55.2 | 101,803 | 142,693 | 202,296 |
| Other Less Developed Countries | | | | | | | | | |
| 10 million or more | 10 | 14 | 20 | 7.6 | 9.0 | 10.7 | 136,060 | 208,314 | 300,965 |
| 5 to 10 million | 22 | 24 | 28 | 8.5 | 7.6 | 6.9 | 153,561 | 174,306 | 195,009 |
| 1 to 5 million | 187 | 259 | 325 | 19.5 | 22.1 | 22.6 | 352,090 | 509,591 | 636,441 |
| 500,000 to 1 million | 266 | 362 | 442 | 10.1 | 10.9 | 11.1 | 181,749 | 250,599 | 312,028 |
| Fewer than 500,000 | ... | ... | ... | 54.3 | 50.5 | 48.8 | 977,555 | 1,164,222 | 1,377,510 |
| Less Developed Regions, excluding China | | | | | | | | | |
| 10 million or more | 10 | 13 | 17 | 8.8 | 10.1 | 11.2 | 133,121 | 194,002 | 267,576 |
| 5 to 10 million | 16 | 18 | 25 | 7.3 | 6.7 | 7.1 | 110,003 | 127,795 | 170,661 |
| 1 to 5 million | 156 | 207 | 250 | 19.3 | 21.3 | 20.7 | 291,625 | 408,322 | 496,275 |
| 500,000 to 1 million | 189 | 242 | 293 | 8.7 | 8.8 | 8.6 | 130,840 | 168,013 | 206,053 |
| Fewer than 500,000 | ... | ... | ... | 55.9 | 53.1 | 52.3 | 842,473 | 1,014,887 | 1,252,489 |
| Sub-Saharan Africa | | | | | | | | | |
| 10 million or more | — | 1 | 2 | — | 3.3 | 5.9 | — | 10,578 | 26,949 |
| 5 to 10 million | 2 | 2 | 5 | 5.8 | 4.3 | 6.6 | 12,844 | 13,926 | 29,931 |
| 1 to 5 million | 28 | 40 | 55 | 23.4 | 26.1 | 24.8 | 51,706 | 83,765 | 113,394 |
| 500,000 to 1 million | 32 | 51 | 58 | 10.3 | 10.9 | 9.1 | 22,795 | 34,940 | 41,453 |
| Fewer than 500,000 | ... | ... | ... | 60.4 | 55.4 | 53.6 | 133,262 | 178,191 | 244,853 |
| **Geographic Aggregates** | | | | | | | | | |
| **Africa** | | | | | | | | | |
| Eastern Africa | | | | | | | | | |
| 10 million or more | — | — | 2 | — | — | 8.9 | — | — | 10,296 |
| 5 to 10 million | — | — | — | — | — | — | — | — | — |
| 1 to 5 million | 9 | 10 | 14 | 26.5 | 26.6 | 23.0 | 13,929 | 20,519 | 26,686 |
| 500,000 to 1 million | 4 | 10 | 11 | 4.5 | 9.5 | 6.8 | 2,393 | 7,324 | 7,895 |
| Fewer than 500,000 | ... | ... | ... | 69.0 | 63.9 | 61.4 | 36,318 | 49,351 | 71,254 |
| Middle Africa | | | | | | | | | |
| 10 million or more | — | — | 1 | — | — | 15.7 | — | — | 12,788 |
| 5 to 10 million | 1 | 1 | 1 | 15.4 | 15.7 | 8.7 | 5,611 | 8,754 | 7,080 |
| 1 to 5 million | 3 | 7 | 9 | 14.3 | 25.3 | 20.5 | 5,216 | 14,087 | 16,712 |
| 500,000 to 1 million | 9 | 9 | 12 | 17.4 | 11.4 | 10.7 | 6,334 | 6,321 | 8,679 |
| Fewer than 500,000 | ... | ... | ... | 53.0 | 47.5 | 44.5 | 19,326 | 26,430 | 36,234 |
| Northern Africa | | | | | | | | | |
| 10 million or more | 1 | 1 | 1 | 11.9 | 10.1 | 9.1 | 10,170 | 11,001 | 12,540 |
| 5 to 10 million | — | 1 | 2 | — | 4.7 | 8.9 | — | 5,172 | 12,206 |
| 1 to 5 million | 6 | 6 | 7 | 17.9 | 13.3 | 10.2 | 15,369 | 14,446 | 14,064 |
| 500,000 to 1 million | 8 | 15 | 20 | 6.2 | 9.1 | 10.5 | 5,319 | 9,961 | 14,410 |
| Fewer than 500,000 | ... | ... | ... | 64.0 | 62.7 | 61.2 | 54,798 | 68,331 | 84,121 |
| Southern Africa | | | | | | | | | |
| 10 million or more | — | — | — | — | — | — | — | — | — |
| 5 to 10 million | — | — | — | — | — | — | — | — | — |
| 1 to 5 million | 5 | 7 | 7 | 40.6 | 49.4 | 47.3 | 11,227 | 16,795 | 18,337 |
| 500,000 to 1 million | 2 | 1 | 2 | 6.7 | 1.8 | 3.1 | 1,855 | 615 | 1,211 |
| Fewer than 500,000 | ... | ... | ... | 52.7 | 48.8 | 49.6 | 14,575 | 16,611 | 19,261 |
| Western Africa | | | | | | | | | |
| 10 million or more | — | 1 | 1 | — | 7.7 | 7.2 | — | 10,578 | 14,162 |
| 5 to 10 million | 1 | — | 1 | 7.8 | 0.0 | 2.8 | 7,233 | — | 5,550 |
| 1 to 5 million | 10 | 16 | 24 | 18.9 | 23.6 | 25.9 | 17,384 | 32,364 | 50,598 |
| 500,000 to 1 million | 17 | 26 | 27 | 13.3 | 12.9 | 9.9 | 12,213 | 17,749 | 19,309 |
| Fewer than 500,000 | ... | ... | ... | 60.0 | 55.8 | 54.1 | 55,332 | 76,580 | 105,725 |

# TABLE A.4
*continued*

| | Number of urban agglomerations Estimates and projections | | | Distribution of urban population by size of agglomerations (%) | | | Population estimates and projections ('000) | | |
|---|---|---|---|---|---|---|---|---|---|
| | 2000 | 2010 | 2020 | 2000 | 2010 | 2020 | 2000 | 2010 | 2020 |
| **Asia** | | | | | | | | | |
| Eastern Asia | | | | | | | | | |
| 10 million or more | 3 | 4 | 7 | 9.9 | 9.8 | 12.1 | 58,839 | 76,966 | 113,354 |
| 5 to 10 million | 8 | 9 | 10 | 9.9 | 8.9 | 7.5 | 59,086 | 70,210 | 70,870 |
| 1 to 5 million | 64 | 93 | 129 | 21.3 | 24.3 | 26.2 | 126,523 | 190,704 | 246,437 |
| 500,000 to 1 million | 112 | 163 | 198 | 12.5 | 14.5 | 14.9 | 74,165 | 113,597 | 139,749 |
| Fewer than 500,000 | ... | ... | ... | 46.4 | 42.5 | 39.4 | 276,063 | 333,211 | 370,274 |
| South-Central Asia | | | | | | | | | |
| 10 million or more | 5 | 5 | 5 | 14.6 | 15.0 | 14.2 | 65,180 | 85,523 | 103,854 |
| 5 to 10 million | 5 | 7 | 10 | 6.6 | 8.2 | 10.0 | 29,694 | 46,607 | 73,161 |
| 1 to 5 million | 41 | 57 | 69 | 15.2 | 17.4 | 16.8 | 68,047 | 99,505 | 123,437 |
| 500,000 to 1 million | 49 | 68 | 88 | 7.8 | 8.1 | 8.2 | 34,884 | 46,266 | 60,441 |
| Fewer than 500,000 | ... | ... | ... | 55.8 | 51.3 | 50.8 | 249,620 | 293,211 | 372,146 |
| South-Eastern Asia | | | | | | | | | |
| 10 million or more | — | 1 | 2 | — | 4.7 | 7.8 | — | 11,628 | 23,943 |
| 5 to 10 million | 3 | 3 | 4 | 12.5 | 9.1 | 8.7 | 24,680 | 22,354 | 26,644 |
| 1 to 5 million | 14 | 17 | 21 | 14.9 | 13.7 | 12.0 | 29,437 | 33,688 | 36,772 |
| 500,000 to 1 million | 15 | 20 | 28 | 4.9 | 5.6 | 6.0 | 9,727 | 13,785 | 18,235 |
| Fewer than 500,000 | ... | ... | ... | 67.7 | 67.0 | 65.4 | 133,516 | 165,246 | 199,818 |
| Western Asia | | | | | | | | | |
| 10 million or more | — | 1 | 1 | — | 6.8 | 6.2 | — | 10,525 | 11,689 |
| 5 to 10 million | 2 | 1 | 2 | 11.5 | 3.8 | 6.9 | 13,944 | 5,891 | 13,131 |
| 1 to 5 million | 18 | 24 | 30 | 26.7 | 31.4 | 31.6 | 32,391 | 48,593 | 59,883 |
| 500,000 to 1 million | 18 | 24 | 31 | 10.1 | 10.9 | 11.3 | 12,231 | 16,878 | 21,353 |
| Fewer than 500,000 | ... | ... | ... | 51.8 | 47.1 | 44.1 | 62,873 | 72,927 | 83,609 |
| **Europe** | | | | | | | | | |
| Eastern Europe | | | | | | | | | |
| 10 million or more | 1 | 1 | 1 | 4.8 | 5.3 | 5.3 | 10,005 | 10,550 | 10,662 |
| 5 to 10 million | — | — | — | — | — | — | — | — | — |
| 1 to 5 million | 23 | 20 | 19 | 16.4 | 15.4 | 15.1 | 34,034 | 30,975 | 30,190 |
| 500,000 to 1 million | 29 | 34 | 34 | 8.9 | 11.2 | 11.5 | 18,556 | 22,459 | 22,993 |
| Fewer than 500,000 | ... | ... | ... | 69.8 | 68.2 | 68.1 | 144,814 | 136,954 | 136,119 |
| Northern Europe | | | | | | | | | |
| 10 million or more | — | — | — | — | — | — | — | — | — |
| 5 to 10 million | 1 | 1 | 1 | 11.2 | 11.0 | 10.5 | 8,225 | 8,631 | 8,753 |
| 1 to 5 million | 7 | 8 | 8 | 14.3 | 15.3 | 15.0 | 10,501 | 11,958 | 12,520 |
| 500,000 to 1 million | 9 | 9 | 11 | 9.0 | 8.1 | 9.1 | 6,618 | 6,315 | 7,608 |
| Fewer than 500,000 | ... | ... | ... | 65.5 | 65.6 | 65.5 | 48,159 | 51,313 | 54,823 |
| Southern Europe | | | | | | | | | |
| 10 million or more | — | — | — | — | — | — | — | — | — |
| 5 to 10 million | 1 | 2 | 2 | 5.3 | 10.5 | 10.6 | 5,014 | 10,935 | 11,823 |
| 1 to 5 million | 9 | 8 | 8 | 24.3 | 18.1 | 17.2 | 23,083 | 18,823 | 19,209 |
| 500,000 to 1 million | 18 | 18 | 19 | 12.5 | 11.6 | 11.6 | 11,834 | 12,090 | 12,998 |
| Fewer than 500,000 | ... | ... | ... | 58.0 | 59.8 | 60.6 | 55,083 | 62,361 | 67,634 |
| Western Europe | | | | | | | | | |
| 10 million or more | — | 1 | 1 | — | 7.0 | 6.9 | — | 10,485 | 10,880 |
| 5 to 10 million | 1 | — | — | 7.0 | 0.0 | 0.0 | 9,739 | — | — |
| 1 to 5 million | 10 | 12 | 13 | 11.1 | 12.3 | 12.8 | 15,434 | 18,374 | 20,072 |
| 500,000 to 1 million | 19 | 20 | 20 | 9.2 | 8.5 | 8.1 | 12,710 | 12,780 | 12,729 |
| Fewer than 500,000 | ... | ... | ... | 72.6 | 72.2 | 72.2 | 100,612 | 108,291 | 113,474 |
| **Latin America and the Caribbean** | | | | | | | | | |
| Caribbean | | | | | | | | | |
| 10 million or more | — | — | — | — | — | — | — | — | — |
| 5 to 10 million | — | — | — | — | — | — | — | — | — |
| 1 to 5 million | 4 | 4 | 4 | 33.5 | 32.5 | 31.6 | 7,930 | 9,196 | 10,278 |
| 500,000 to 1 million | 1 | 3 | 3 | 2.4 | 6.3 | 6.0 | 580 | 1,772 | 1,946 |
| Fewer than 500,000 | ... | ... | ... | 64.1 | 61.2 | 62.4 | 15,197 | 17,310 | 20,286 |
| Central America | | | | | | | | | |
| 10 million or more | 1 | 1 | 1 | 19.4 | 17.7 | 16.1 | 18,022 | 19,460 | 20,476 |
| 5 to 10 million | — | — | — | — | — | — | — | — | — |
| 1 to 5 million | 11 | 16 | 19 | 19.9 | 25.1 | 27.0 | 18,461 | 27,655 | 34,476 |
| 500,000 to 1 million | 22 | 24 | 25 | 16.4 | 15.4 | 14.4 | 15,262 | 17,016 | 18,349 |
| Fewer than 500,000 | ... | ... | ... | 44.3 | 41.8 | 42.5 | 41,204 | 46,120 | 54,162 |
| South America | | | | | | | | | |
| 10 million or more | 3 | 3 | 5 | 14.4 | 13.7 | 18.3 | 39,749 | 45,287 | 68,124 |
| 5 to 10 million | 3 | 4 | 2 | 6.8 | 8.9 | 3.4 | 18,925 | 29,244 | 12,828 |
| 1 to 5 million | 28 | 35 | 36 | 20.6 | 22.7 | 23.1 | 57,124 | 74,976 | 86,069 |
| 500,000 to 1 million | 31 | 30 | 34 | 7.7 | 6.5 | 6.9 | 21,421 | 21,328 | 25,635 |
| Fewer than 500,000 | ... | ... | ... | 50.4 | 48.3 | 48.4 | 139,545 | 159,394 | 180,519 |
| Northern America | | | | | | | | | |
| 10 million or more | 2 | 2 | 2 | 11.8 | 11.1 | 10.4 | 29,659 | 32,187 | 33,837 |
| 5 to 10 million | 2 | 4 | 6 | 5.4 | 9.0 | 11.8 | 13,494 | 26,029 | 38,116 |
| 1 to 5 million | 37 | 42 | 44 | 33.8 | 32.9 | 29.8 | 85,310 | 95,001 | 96,533 |
| 500,000 to 1 million | 39 | 40 | 43 | 10.9 | 9.8 | 9.4 | 27,380 | 28,282 | 30,579 |
| Fewer than 500,000 | ... | ... | ... | 38.2 | 37.2 | 38.6 | 96,311 | 107,304 | 125,214 |

# TABLE A.4

## *continued*

| | Number of urban agglomerations Estimates and projections | | | Distribution of urban population by size of agglomerations (%) | | | Population estimates and projections ('000) | | |
|---|---|---|---|---|---|---|---|---|---|
| | 2000 | 2010 | 2020 | 2000 | 2010 | 2020 | 2000 | 2010 | 2020 |
| **Oceania** | | | | | | | | | |
| 10 million or more | — | — | — | — | — | — | — | — | — |
| 5 to 10 million | — | — | — | — | — | — | — | — | — |
| 1 to 5 million | 6 | 6 | 6 | 57.7 | 57.3 | 55.3 | 12,652 | 14,423 | 15,711 |
| 500,000 to 1 million | — | 2 | 2 | — | 4.3 | 4.3 | — | 1,082 | 1,210 |
| Fewer than 500,000 | ... | ... | ... | 42.3 | 38.4 | 40.4 | 9,280 | 9,663 | 11,484 |
| Australia/New Zealand | | | | | | | | | |
| 10 million or more | — | — | — | — | — | — | — | — | — |
| 5 to 10 million | — | — | — | — | — | — | — | — | — |
| 1 to 5 million | 6 | 6 | 6 | 63.2 | 63.0 | 61.6 | 12,652 | 14,423 | 15,711 |
| 500,000 to 1 million | — | 2 | 2 | — | 4.7 | 4.7 | — | 1,082 | 1,210 |
| Fewer than 500,000 | ... | ... | ... | 36.8 | 32.2 | 33.7 | 7,372 | 7,374 | 8,595 |
| Melanesia | | | | | | | | | |
| 10 million or more | — | — | — | — | — | — | — | — | — |
| 5 to 10 million | — | — | — | — | — | — | — | — | — |
| 1 to 5 million | — | — | — | — | — | — | — | — | — |
| 500,000 to 1 million | — | — | — | — | — | — | — | — | — |
| Fewer than 500,000 | ... | ... | ... | 100.0 | 100.0 | 100.0 | 1,329 | 1,614 | 2,110 |
| Micronesia | | | | | | | | | |
| 10 million or more | — | — | — | — | — | — | — | — | — |
| 5 to 10 million | — | — | — | — | — | — | — | — | — |
| 1 to 5 million | — | — | — | — | — | — | — | — | — |
| 500,000 to 1 million | — | — | — | — | — | — | — | — | — |
| Fewer than 500,000 | ... | ... | ... | 100.0 | 100.0 | 100.0 | 326 | 390 | 454 |
| Polynesia | | | | | | | | | |
| 10 million or more | — | — | — | — | — | — | — | — | — |
| 5 to 10 million | — | — | — | — | — | — | — | — | — |
| 1 to 5 million | — | — | — | — | — | — | — | — | — |
| 500,000 to 1 million | — | — | — | — | — | — | — | — | — |
| Fewer than 500,000 | ... | ... | ... | 100.0 | 100.0 | 100.0 | 253 | 285 | 325 |

*Source:* United Nations Department of Economic and Social Affairs, Population Division (2010) *World Urbanization Prospects: The 2009 Revision*, United Nations, New York. The figures in regional aggregates are not consistent with city data in table C.1.

*Note:* Lists of countries/areas in aggregates are presented in the Technical Notes.

# COUNTRY LEVEL DATA
# TABLE B.1

## Total Population Size, Rate of Change and Population Density

| | Estimates and projections ('000) | | | | Rate of change (%) | | | Population density (people/km²) | |
|---|---|---|---|---|---|---|---|---|---|
| | 2000 | 2010 | 2020 | 2030 | 2000–2010 | 2010–2020 | 2020–2030 | 2000 | 2030 |
| **AFRICA** | | | | | | | | | |
| Algeria | 30,506 | 35,423 | 40,630 | 44,726 | 1.49 | 1.37 | 0.96 | 13 | 19 |
| Angola | 14,280 | 18,993 | 24,507 | 30,416 | 2.85 | 2.55 | 2.16 | 11 | 24 |
| Benin | 6,659 | 9,212 | 12,177 | 15,399 | 3.25 | 2.79 | 2.35 | 59 | 137 |
| Botswana | 1,723 | 1,978 | 2,227 | 2,434 | 1.38 | 1.19 | 0.89 | 3 | 4 |
| Burkina Faso | 11,676 | 16,287 | 21,871 | 27,940 | 3.33 | 2.95 | 2.45 | 43 | 102 |
| Burundi | 6,473 | 8,519 | 10,318 | 11,936 | 2.75 | 1.92 | 1.46 | 233 | 429 |
| Cameroon | 15,865 | 19,958 | 24,349 | 28,602 | 2.30 | 1.99 | 1.61 | 33 | 60 |
| Cape Verde | 439 | 513 | 584 | 645 | 1.56 | 1.30 | 0.99 | 109 | 160 |
| Central African Republic | 3,746 | 4,506 | 5,340 | 6,150 | 1.85 | 1.70 | 1.41 | 6 | 10 |
| Chad | 8,402 | 11,506 | 14,897 | 19,018 | 3.14 | 2.58 | 2.44 | 7 | 15 |
| Comoros | 552 | 691 | 838 | 975 | 2.25 | 1.93 | 1.51 | 297 | 524 |
| Congo | 3,036 | 3,759 | 4,699 | 5,479 | 2.14 | 2.23 | 1.54 | 9 | 16 |
| Côte d'Ivoire | 17,281 | 21,571 | 26,954 | 32,551 | 2.22 | 2.23 | 1.89 | 54 | 101 |
| Democratic Republic of the Congo | 50,829 | 67,827 | 87,640 | 108,594 | 2.88 | 2.56 | 2.14 | 22 | 46 |
| Djibouti | 730 | 879 | 1,027 | 1,192 | 1.86 | 1.56 | 1.49 | 31 | 51 |
| Egypt | 70,174 | 84,474 | 98,638 | 110,907 | 1.85 | 1.55 | 1.17 | 70 | 111 |
| Equatorial Guinea | 529 | 693 | 875 | 1,067 | 2.70 | 2.33 | 1.98 | 19 | 38 |
| Eritrea | 3,657 | 5,224 | 6,719 | 8,086 | 3.57 | 2.52 | 1.85 | 31 | 69 |
| Ethiopia | 65,515 | 84,976 | 107,964 | 131,561 | 2.60 | 2.39 | 1.98 | 59 | 119 |
| Gabon | 1,233 | 1,501 | 1,779 | 2,044 | 1.97 | 1.70 | 1.39 | 5 | 8 |
| Gambia | 1,302 | 1,751 | 2,227 | 2,736 | 2.96 | 2.40 | 2.06 | 115 | 242 |
| Ghana | 19,529 | 24,333 | 29,567 | 34,884 | 2.20 | 1.95 | 1.65 | 82 | 146 |
| Guinea | 8,384 | 10,324 | 13,467 | 16,897 | 2.08 | 2.66 | 2.27 | 34 | 69 |
| Guinea-Bissau | 1,304 | 1,647 | 2,065 | 2,536 | 2.34 | 2.26 | 2.05 | 36 | 70 |
| Kenya | 31,441 | 40,863 | 52,034 | 63,199 | 2.62 | 2.42 | 1.94 | 54 | 109 |
| Lesotho | 1,889 | 2,084 | 2,244 | 2,359 | 0.98 | 0.74 | 0.50 | 62 | 78 |
| Liberia | 2,824 | 4,102 | 5,253 | 6,470 | 3.73 | 2.47 | 2.08 | 25 | 58 |
| Libyan Arab Jamahiriya | 5,346 | 6,546 | 7,699 | 8,519 | 2.03 | 1.62 | 1.01 | 3 | 5 |
| Madagascar | 15,275 | 20,146 | 25,687 | 31,528 | 2.77 | 2.43 | 2.05 | 26 | 54 |
| Malawi | 11,831 | 15,692 | 20,537 | 25,897 | 2.82 | 2.69 | 2.32 | 100 | 219 |
| Mali | 10,523 | 13,323 | 16,767 | 20,467 | 2.36 | 2.30 | 1.99 | 8 | 17 |
| Mauritania | 2,604 | 3,366 | 4,091 | 4,791 | 2.57 | 1.95 | 1.58 | 3 | 5 |
| Mauritius[1] | 1,195 | 1,297 | 1,372 | 1,420 | 0.82 | 0.56 | 0.34 | 586 | 696 |
| Mayotte | 149 | 199 | 250 | 302 | 2.89 | 2.28 | 1.89 | 397 | 808 |
| Morocco | 28,827 | 32,381 | 36,200 | 39,259 | 1.16 | 1.11 | 0.81 | 65 | 88 |
| Mozambique | 18,249 | 23,406 | 28,545 | 33,894 | 2.49 | 1.98 | 1.72 | 23 | 42 |
| Namibia | 1,824 | 2,212 | 2,614 | 2,993 | 1.93 | 1.67 | 1.35 | 2 | 4 |
| Niger | 11,031 | 15,891 | 22,947 | 32,563 | 3.65 | 3.67 | 3.50 | 9 | 26 |
| Nigeria | 124,842 | 158,259 | 193,252 | 226,651 | 2.37 | 2.00 | 1.59 | 135 | 245 |
| Réunion | 724 | 837 | 931 | 1,009 | 1.45 | 1.06 | 0.80 | 288 | 402 |
| Rwanda | 7,958 | 10,277 | 13,233 | 16,104 | 2.56 | 2.53 | 1.96 | 302 | 611 |
| Saint Helena[2] | 5 | 4 | 4 | 5 | -2.23 | 0.00 | 2.23 | 42 | 38 |
| São Tomé and Príncipe | 140 | 165 | 197 | 234 | 1.64 | 1.77 | 1.72 | 145 | 242 |
| Senegal | 9,902 | 12,861 | 16,197 | 19,541 | 2.61 | 2.31 | 1.88 | 50 | 99 |
| Seychelles | 81 | 85 | 89 | 93 | 0.48 | 0.46 | 0.44 | 178 | 205 |
| Sierra Leone | 4,228 | 5,836 | 7,318 | 8,943 | 3.22 | 2.26 | 2.01 | 59 | 125 |
| Somalia | 7,394 | 9,359 | 12,246 | 15,744 | 2.36 | 2.69 | 2.51 | 12 | 25 |
| South Africa | 44,872 | 50,492 | 52,671 | 54,726 | 1.18 | 0.42 | 0.38 | 37 | 45 |
| Sudan | 34,904 | 43,192 | 52,309 | 60,995 | 2.13 | 1.92 | 1.54 | 14 | 24 |
| Swaziland | 1,080 | 1,202 | 1,376 | 1,524 | 1.07 | 1.35 | 1.02 | 62 | 88 |
| Togo | 5,247 | 6,780 | 8,445 | 10,115 | 2.56 | 2.20 | 1.80 | 92 | 178 |
| Tunisia | 9,452 | 10,374 | 11,366 | 12,127 | 0.93 | 0.91 | 0.65 | 58 | 74 |
| Uganda | 24,433 | 33,796 | 46,319 | 60,819 | 3.24 | 3.15 | 2.72 | 101 | 252 |
| United Republic of Tanzania | 34,131 | 45,040 | 59,603 | 75,498 | 2.77 | 2.80 | 2.36 | 36 | 80 |
| Western Sahara | 315 | 530 | 723 | 819 | 5.20 | 3.11 | 1.25 | 1 | 3 |
| Zambia | 10,467 | 13,257 | 16,916 | 20,889 | 2.36 | 2.44 | 2.11 | 14 | 28 |
| Zimbabwe | 12,455 | 12,644 | 15,571 | 17,917 | 0.15 | 2.08 | 1.40 | 32 | 46 |
| **ASIA** | | | | | | | | | |
| Afghanistan | 20,536 | 29,117 | 39,585 | 50,649 | 3.49 | 3.07 | 2.46 | 31 | 78 |
| Armenia | 3,076 | 3,090 | 3,175 | 3,170 | 0.05 | 0.27 | -0.02 | 103 | 106 |
| Azerbaijan | 8,121 | 8,934 | 9,838 | 10,323 | 0.95 | 0.96 | 0.48 | 94 | 119 |
| Bahrain | 650 | 807 | 953 | 1,085 | 2.16 | 1.66 | 1.30 | 937 | 1,564 |
| Bangladesh | 140,767 | 164,425 | 185,552 | 203,214 | 1.55 | 1.21 | 0.91 | 978 | 1,411 |
| Bhutan | 561 | 708 | 820 | 902 | 2.33 | 1.47 | 0.95 | 12 | 19 |
| Brunei Darussalam | 333 | 407 | 478 | 547 | 2.01 | 1.61 | 1.35 | 58 | 95 |
| Cambodia | 12,760 | 15,053 | 17,707 | 20,100 | 1.65 | 1.62 | 1.27 | 70 | 111 |
| China[3] | 1,266,954 | 1,354,146 | 1,431,155 | 1,462,468 | 0.67 | 0.55 | 0.22 | 132 | 152 |
| China, Hong Kong SAR[4] | 6,667 | 7,069 | 7,701 | 8,185 | 0.59 | 0.86 | 0.61 | 6,066 | 7,448 |
| China, Macao SAR[5] | 441 | 548 | 588 | 611 | 2.17 | 0.70 | 0.38 | 16,958 | 23,507 |
| Cyprus | 787 | 880 | 970 | 1,053 | 1.12 | 0.97 | 0.82 | 85 | 114 |
| Democratic People's Republic of Korea | 22,859 | 23,991 | 24,802 | 25,301 | 0.48 | 0.33 | 0.20 | 190 | 210 |
| Georgia | 4,745 | 4,219 | 3,982 | 3,779 | -1.17 | -0.58 | -0.52 | 68 | 54 |
| India | 1,042,590 | 1,214,464 | 1,367,225 | 1,484,598 | 1.53 | 1.18 | 0.82 | 317 | 452 |
| Indonesia | 205,280 | 232,517 | 254,218 | 271,485 | 1.25 | 0.89 | 0.66 | 108 | 143 |
| Iran (Islamic Republic of) | 66,903 | 75,078 | 83,740 | 89,936 | 1.15 | 1.09 | 0.71 | 41 | 55 |
| Iraq | 24,652 | 31,467 | 40,228 | 48,909 | 2.44 | 2.46 | 1.95 | 56 | 112 |

# TABLE B.1

**continued**

| | Estimates and projections ('000) | | | | Rate of change (%) | | | Population density (people/km²) | |
|---|---|---|---|---|---|---|---|---|---|
| | 2000 | 2010 | 2020 | 2030 | 2000–2010 | 2010–2020 | 2020–2030 | 2000 | 2030 |
| Israel | 6,084 | 7,285 | 8,307 | 9,219 | 1.80 | 1.31 | 1.04 | 275 | 416 |
| Japan | 126,706 | 126,995 | 123,664 | 117,424 | 0.02 | -0.27 | -0.52 | 335 | 311 |
| Jordan | 4,853 | 6,472 | 7,519 | 8,616 | 2.88 | 1.50 | 1.36 | 54 | 96 |
| Kazakhstan | 14,957 | 15,753 | 16,726 | 17,244 | 0.52 | 0.60 | 0.30 | 5 | 6 |
| Kuwait | 2,228 | 3,051 | 3,690 | 4,273 | 3.14 | 1.90 | 1.47 | 125 | 240 |
| Kyrgyzstan | 4,955 | 5,550 | 6,159 | 6,543 | 1.13 | 1.04 | 0.60 | 25 | 33 |
| Lao People's Democratic Republic | 5,403 | 6,436 | 7,651 | 8,854 | 1.75 | 1.73 | 1.46 | 23 | 37 |
| Lebanon | 3,772 | 4,255 | 4,587 | 4,858 | 1.20 | 0.75 | 0.57 | 363 | 467 |
| Malaysia | 23,274 | 27,914 | 32,017 | 35,275 | 1.82 | 1.37 | 0.97 | 71 | 107 |
| Maldives | 272 | 314 | 362 | 403 | 1.44 | 1.42 | 1.07 | 914 | 1,352 |
| Mongolia | 2,389 | 2,701 | 3,002 | 3,236 | 1.23 | 1.06 | 0.75 | 2 | 2 |
| Myanmar | 46,610 | 50,496 | 55,497 | 59,353 | 0.80 | 0.94 | 0.67 | 69 | 88 |
| Nepal | 24,432 | 29,853 | 35,269 | 40,646 | 2.00 | 1.67 | 1.42 | 166 | 276 |
| Occupied Palestinian Territory | 3,149 | 4,409 | 5,806 | 7,320 | 3.37 | 2.75 | 2.32 | 523 | 1,216 |
| Oman | 2,402 | 2,905 | 3,495 | 4,048 | 1.90 | 1.85 | 1.47 | 8 | 13 |
| Pakistan | 148,132 | 184,753 | 226,187 | 265,690 | 2.21 | 2.02 | 1.61 | 186 | 334 |
| Philippines | 77,689 | 93,617 | 109,683 | 124,384 | 1.86 | 1.58 | 1.26 | 259 | 415 |
| Qatar | 617 | 1,508 | 1,740 | 1,951 | 8.94 | 1.43 | 1.14 | 56 | 177 |
| Republic of Korea | 46,429 | 48,501 | 49,475 | 49,146 | 0.44 | 0.20 | -0.07 | 466 | 494 |
| Saudi Arabia | 20,808 | 26,246 | 31,608 | 36,545 | 2.32 | 1.86 | 1.45 | 10 | 17 |
| Singapore | 4,018 | 4,837 | 5,219 | 5,460 | 1.86 | 0.76 | 0.45 | 5,883 | 7,994 |
| Sri Lanka | 18,767 | 20,410 | 21,713 | 22,194 | 0.84 | 0.62 | 0.22 | 286 | 338 |
| Syrian Arab Republic | 16,511 | 22,505 | 26,475 | 30,560 | 3.10 | 1.62 | 1.43 | 89 | 165 |
| Tajikistan | 6,173 | 7,075 | 8,446 | 9,618 | 1.36 | 1.77 | 1.30 | 43 | 67 |
| Thailand | 62,347 | 68,139 | 71,443 | 73,462 | 0.89 | 0.47 | 0.28 | 122 | 143 |
| Timor-Leste | 815 | 1,171 | 1,618 | 2,125 | 3.62 | 3.23 | 2.73 | 55 | 143 |
| Turkey | 66,460 | 75,705 | 83,873 | 90,375 | 1.30 | 1.02 | 0.75 | 85 | 115 |
| Turkmenistan | 4,502 | 5,177 | 5,816 | 6,276 | 1.40 | 1.16 | 0.76 | 9 | 13 |
| United Arab Emirates | 3,238 | 4,707 | 5,660 | 6,555 | 3.74 | 1.84 | 1.47 | 39 | 78 |
| Uzbekistan | 24,776 | 27,794 | 31,185 | 33,933 | 1.15 | 1.15 | 0.84 | 55 | 76 |
| Viet Nam | 78,663 | 89,029 | 98,011 | 105,447 | 1.24 | 0.96 | 0.73 | 237 | 318 |
| Yemen | 18,182 | 24,256 | 31,635 | 39,350 | 2.88 | 2.66 | 2.18 | 34 | 75 |
| **EUROPE** | | | | | | | | | |
| Albania | 3,068 | 3,169 | 3,338 | 3,416 | 0.32 | 0.52 | 0.23 | 107 | 119 |
| Andorra | 66 | 87 | 100 | 113 | 2.76 | 1.39 | 1.22 | 142 | 242 |
| Austria | 8,005 | 8,387 | 8,539 | 8,637 | 0.47 | 0.18 | 0.11 | 95 | 103 |
| Belarus | 10,054 | 9,588 | 9,112 | 8,564 | -0.47 | -0.51 | -0.62 | 48 | 41 |
| Belgium | 10,193 | 10,698 | 11,048 | 11,303 | 0.48 | 0.32 | 0.23 | 334 | 370 |
| Bosnia and Herzegovina | 3,694 | 3,760 | 3,677 | 3,520 | 0.18 | -0.22 | -0.44 | 72 | 69 |
| Bulgaria | 8,006 | 7,497 | 7,017 | 6,469 | -0.66 | -0.66 | -0.81 | 72 | 58 |
| Channel Islands[6] | 147 | 150 | 151 | 151 | 0.20 | 0.07 | 0.00 | 752 | 776 |
| Croatia | 4,505 | 4,410 | 4,318 | 4,180 | -0.21 | -0.21 | -0.32 | 80 | 74 |
| Czech Republic | 10,224 | 10,411 | 10,568 | 10,520 | 0.18 | 0.15 | -0.05 | 130 | 133 |
| Denmark | 5,335 | 5,481 | 5,557 | 5,616 | 0.27 | 0.14 | 0.11 | 124 | 130 |
| Estonia | 1,370 | 1,339 | 1,333 | 1,301 | -0.23 | -0.04 | -0.24 | 30 | 29 |
| Faeroe Islands | 46 | 50 | 53 | 56 | 0.83 | 0.58 | 0.55 | 33 | 40 |
| Finland[7] | 5,173 | 5,346 | 5,496 | 5,544 | 0.33 | 0.28 | 0.09 | 15 | 16 |
| France | 59,128 | 62,637 | 64,931 | 66,474 | 0.58 | 0.36 | 0.23 | 107 | 121 |
| Germany | 82,075 | 82,057 | 80,422 | 77,854 | 0.00 | -0.20 | -0.32 | 230 | 218 |
| Gibraltar | 29 | 31 | 32 | 31 | 0.67 | 0.32 | -0.32 | 4,818 | 5,240 |
| Greece | 10,942 | 11,183 | 11,284 | 11,234 | 0.22 | 0.09 | -0.04 | 83 | 85 |
| Holy See[8] | 1 | 1 | 1 | 1 | 0.00 | 0.00 | 0.00 | 1,789 | 1,739 |
| Hungary | 10,215 | 9,973 | 9,766 | 9,509 | -0.24 | -0.21 | -0.27 | 110 | 102 |
| Iceland | 281 | 329 | 370 | 392 | 1.58 | 1.17 | 0.58 | 3 | 4 |
| Ireland | 3,804 | 4,589 | 5,145 | 5,573 | 1.88 | 1.14 | 0.80 | 54 | 79 |
| Isle of Man | 77 | 80 | 81 | 80 | 0.38 | 0.12 | -0.12 | 134 | 140 |
| Italy | 57,116 | 60,098 | 60,408 | 59,549 | 0.51 | 0.05 | -0.14 | 190 | 198 |
| Latvia | 2,374 | 2,240 | 2,153 | 2,049 | -0.58 | -0.40 | -0.50 | 37 | 32 |
| Liechtenstein | 33 | 36 | 39 | 42 | 0.87 | 0.80 | 0.74 | 205 | 259 |
| Lithuania | 3,501 | 3,255 | 3,058 | 2,909 | -0.73 | -0.62 | -0.50 | 54 | 45 |
| Luxembourg | 437 | 492 | 550 | 615 | 1.19 | 1.11 | 1.12 | 169 | 238 |
| Malta | 389 | 410 | 422 | 427 | 0.53 | 0.29 | 0.12 | 1,231 | 1,351 |
| Moldova | 4,100 | 3,576 | 3,378 | 3,182 | -1.37 | -0.57 | -0.60 | 121 | 94 |
| Monaco | 32 | 33 | 34 | 35 | 0.31 | 0.30 | 0.29 | 21,478 | 23,738 |
| Montenegro | 661 | 626 | 631 | 634 | -0.54 | 0.08 | 0.05 | 48 | 46 |
| Netherlands | 15,915 | 16,653 | 17,143 | 17,498 | 0.45 | 0.29 | 0.20 | 383 | 421 |
| Norway[9] | 4,484 | 4,855 | 5,200 | 5,518 | 0.79 | 0.69 | 0.59 | 12 | 14 |
| Poland | 38,433 | 38,038 | 37,497 | 36,187 | -0.10 | -0.14 | -0.36 | 119 | 112 |
| Portugal | 10,226 | 10,732 | 10,767 | 10,620 | 0.48 | 0.03 | -0.14 | 111 | 115 |
| Romania | 22,138 | 21,190 | 20,380 | 19,489 | -0.44 | -0.39 | -0.45 | 93 | 82 |
| Russian Federation | 146,670 | 140,367 | 135,406 | 128,864 | -0.44 | -0.36 | -0.50 | 9 | 8 |
| San Marino | 27 | 32 | 33 | 33 | 1.70 | 0.31 | 0.00 | 442 | 548 |
| Serbia | 10,134 | 9,856 | 9,783 | 9,644 | -0.28 | -0.07 | -0.14 | 115 | 109 |
| Slovakia | 5,379 | 5,412 | 5,442 | 5,348 | 0.06 | 0.06 | -0.17 | 110 | 109 |
| Slovenia | 1,985 | 2,025 | 2,053 | 2,037 | 0.20 | 0.14 | -0.08 | 98 | 101 |
| Spain | 40,264 | 45,317 | 48,564 | 49,772 | 1.18 | 0.69 | 0.25 | 80 | 98 |
| Sweden | 8,860 | 9,293 | 9,713 | 10,076 | 0.48 | 0.44 | 0.37 | 20 | 22 |
| Switzerland | 7,184 | 7,595 | 7,879 | 8,148 | 0.56 | 0.37 | 0.34 | 174 | 197 |
| TFYR Macedonia[10] | 2,012 | 2,043 | 2,046 | 2,016 | 0.15 | 0.01 | -0.15 | 78 | 78 |

# TABLE B.1

## continued

| | Estimates and projections ('000) | | | | Rate of change (%) | | | Population density (people/km²) | |
|---|---|---|---|---|---|---|---|---|---|
| | 2000 | 2010 | 2020 | 2030 | 2000–2010 | 2010–2020 | 2020–2030 | 2000 | 2030 |
| Ukraine | 48,870 | 45,433 | 42,945 | 40,188 | -0.73 | -0.56 | -0.66 | 81 | 67 |
| United Kingdom | 58,907 | 61,899 | 65,090 | 67,956 | 0.50 | 0.50 | 0.43 | 243 | 280 |
| **LATIN AMERICA AND THE CARIBBEAN** | | | | | | | | | |
| Anguilla | 11 | 15 | 18 | 19 | 3.10 | 1.82 | 0.54 | 123 | 207 |
| Antigua and Barbuda | 77 | 89 | 97 | 105 | 1.45 | 0.86 | 0.79 | 175 | 237 |
| Argentina | 36,939 | 40,666 | 44,304 | 47,255 | 0.96 | 0.86 | 0.64 | 13 | 17 |
| Aruba | 91 | 107 | 111 | 112 | 1.62 | 0.37 | 0.09 | 504 | 625 |
| Bahamas | 305 | 346 | 384 | 418 | 1.26 | 1.04 | 0.85 | 22 | 30 |
| Barbados | 252 | 257 | 262 | 260 | 0.20 | 0.19 | -0.08 | 585 | 606 |
| Belize | 252 | 313 | 375 | 430 | 2.17 | 1.81 | 1.37 | 11 | 19 |
| Bolivia | 8,317 | 10,031 | 11,638 | 13,034 | 1.87 | 1.49 | 1.13 | 8 | 12 |
| Brazil | 174,174 | 195,423 | 209,051 | 217,146 | 1.15 | 0.67 | 0.38 | 20 | 26 |
| British Virgin Islands | 21 | 23 | 25 | 27 | 0.91 | 0.83 | 0.77 | 136 | 178 |
| Cayman Islands | 40 | 57 | 61 | 65 | 3.54 | 0.68 | 0.64 | 153 | 246 |
| Chile | 15,419 | 17,135 | 18,639 | 19,779 | 1.06 | 0.84 | 0.59 | 20 | 26 |
| Colombia | 39,773 | 46,300 | 52,278 | 57,264 | 1.52 | 1.21 | 0.91 | 35 | 50 |
| Costa Rica | 3,931 | 4,640 | 5,250 | 5,762 | 1.66 | 1.24 | 0.93 | 77 | 113 |
| Cuba | 11,087 | 11,204 | 11,193 | 11,019 | 0.10 | -0.01 | -0.16 | 100 | 99 |
| Dominica | 68 | 67 | 67 | 69 | -0.15 | 0.00 | 0.29 | 91 | 91 |
| Dominican Republic | 8,830 | 10,225 | 11,451 | 12,431 | 1.47 | 1.13 | 0.82 | 182 | 256 |
| Ecuador | 12,310 | 13,775 | 15,376 | 16,679 | 1.12 | 1.10 | 0.81 | 43 | 59 |
| El Salvador | 5,945 | 6,194 | 6,618 | 7,177 | 0.41 | 0.66 | 0.81 | 283 | 341 |
| Falkland Islands (Malvinas) | 3 | 3 | 3 | 3 | 0.00 | 0.00 | 0.00 | 0 | 0 |
| French Guiana | 165 | 231 | 292 | 354 | 3.36 | 2.34 | 1.93 | 2 | 4 |
| Grenada | 101 | 104 | 108 | 108 | 0.29 | 0.38 | 0.00 | 294 | 315 |
| Guadeloupe | 429 | 467 | 484 | 492 | 0.85 | 0.36 | 0.16 | 252 | 288 |
| Guatemala | 11,231 | 14,377 | 18,091 | 21,692 | 2.47 | 2.30 | 1.82 | 103 | 199 |
| Guyana | 756 | 761 | 745 | 714 | 0.07 | -0.21 | -0.43 | 4 | 3 |
| Haiti | 8,648 | 10,188 | 11,722 | 13,196 | 1.64 | 1.40 | 1.18 | 312 | 476 |
| Honduras | 6,230 | 7,616 | 9,136 | 10,492 | 2.01 | 1.82 | 1.38 | 56 | 94 |
| Jamaica | 2,568 | 2,730 | 2,834 | 2,873 | 0.61 | 0.37 | 0.14 | 234 | 261 |
| Martinique | 385 | 406 | 415 | 418 | 0.53 | 0.22 | 0.07 | 349 | 379 |
| Mexico | 99,531 | 110,645 | 119,682 | 126,457 | 1.06 | 0.79 | 0.55 | 51 | 65 |
| Montserrat | 5 | 6 | 6 | 7 | 1.82 | 0.00 | 1.54 | 49 | 66 |
| Netherlands Antilles | 181 | 201 | 210 | 209 | 1.05 | 0.44 | -0.05 | 226 | 262 |
| Nicaragua | 5,101 | 5,822 | 6,682 | 7,387 | 1.32 | 1.38 | 1.00 | 39 | 57 |
| Panama | 2,951 | 3,508 | 4,027 | 4,488 | 1.73 | 1.38 | 1.08 | 39 | 59 |
| Paraguay | 5,350 | 6,460 | 7,533 | 8,483 | 1.89 | 1.54 | 1.19 | 13 | 21 |
| Peru | 26,004 | 29,496 | 32,881 | 36,006 | 1.26 | 1.09 | 0.91 | 20 | 28 |
| Puerto Rico | 3,819 | 3,998 | 4,135 | 4,195 | 0.46 | 0.34 | 0.14 | 430 | 473 |
| Saint Kitts and Nevis | 46 | 52 | 59 | 64 | 1.23 | 1.26 | 0.81 | 176 | 244 |
| Saint Lucia | 157 | 174 | 190 | 204 | 1.03 | 0.88 | 0.71 | 292 | 378 |
| Saint Vincent and the Grenadines | 108 | 109 | 110 | 113 | 0.09 | 0.09 | 0.27 | 278 | 290 |
| Suriname | 467 | 524 | 568 | 602 | 1.15 | 0.81 | 0.58 | 3 | 4 |
| Trinidad and Tobago | 1,295 | 1,344 | 1,384 | 1,382 | 0.37 | 0.29 | -0.01 | 252 | 269 |
| Turks and Caicos Islands | 19 | 33 | 36 | 39 | 5.52 | 0.87 | 0.80 | 44 | 90 |
| United States Virgin Islands | 109 | 109 | 106 | 99 | 0.00 | -0.28 | -0.68 | 313 | 284 |
| Uruguay | 3,321 | 3,372 | 3,493 | 3,588 | 0.15 | 0.35 | 0.27 | 19 | 21 |
| Venezuela (Bolivarian Republic of) | 24,408 | 29,044 | 33,412 | 37,145 | 1.74 | 1.40 | 1.06 | 27 | 41 |
| **NORTHERN AMERICA** | | | | | | | | | |
| Bermuda | 63 | 65 | 66 | 66 | 0.31 | 0.15 | 0.00 | 1,186 | 1,243 |
| Canada | 30,687 | 33,890 | 37,101 | 40,096 | 0.99 | 0.91 | 0.78 | 3 | 4 |
| Greenland | 56 | 57 | 57 | 55 | 0.18 | 0.00 | -0.36 | 0 | 0 |
| Saint-Pierre-et-Miquelon | 6 | 6 | 6 | 6 | 0.00 | 0.00 | 0.00 | 26 | 25 |
| United States of America | 287,842 | 317,641 | 346,153 | 369,981 | 0.99 | 0.86 | 0.67 | 30 | 38 |
| **OCEANIA** | | | | | | | | | |
| American Samoa | 58 | 69 | 80 | 91 | 1.74 | 1.48 | 1.29 | 290 | 460 |
| Australia[11] | 19,171 | 21,512 | 23,675 | 25,656 | 1.15 | 0.96 | 0.80 | 2 | 3 |
| Cook Islands | 18 | 20 | 21 | 22 | 1.05 | 0.49 | 0.47 | 74 | 93 |
| Fiji | 802 | 854 | 888 | 918 | 0.63 | 0.39 | 0.33 | 44 | 50 |
| French Polynesia | 236 | 272 | 304 | 329 | 1.42 | 1.11 | 0.79 | 59 | 82 |
| Guam | 155 | 180 | 201 | 220 | 1.50 | 1.10 | 0.90 | 283 | 401 |
| Kiribati | 84 | 100 | 115 | 131 | 1.74 | 1.40 | 1.30 | 116 | 180 |
| Marshall Islands | 52 | 63 | 75 | 83 | 1.92 | 1.74 | 1.01 | 288 | 457 |
| Micronesia (Federated States of) | 107 | 111 | 118 | 125 | 0.37 | 0.61 | 0.58 | 153 | 178 |
| Nauru | 10 | 10 | 11 | 11 | 0.00 | 0.95 | 0.00 | 478 | 527 |
| New Caledonia | 215 | 254 | 288 | 318 | 1.67 | 1.26 | 0.99 | 12 | 17 |
| New Zealand | 3,868 | 4,303 | 4,669 | 4,972 | 1.07 | 0.82 | 0.63 | 14 | 18 |
| Niue | 2 | 1 | 1 | 1 | -6.93 | 0.00 | 0.00 | 7 | 4 |
| Northern Mariana Islands | 69 | 88 | 104 | 119 | 2.43 | 1.67 | 1.35 | 149 | 256 |
| Palau | 19 | 21 | 22 | 25 | 1.00 | 0.47 | 1.28 | 42 | 53 |
| Papua New Guinea | 5,388 | 6,888 | 8,468 | 10,058 | 2.46 | 2.07 | 1.72 | 12 | 22 |
| Pitcairn | 0 | 0 | 0 | 0 | 0.00 | 0.00 | 0.00 | 12 | 11 |
| Samoa | 177 | 179 | 184 | 191 | 0.11 | 0.28 | 0.37 | 62 | 68 |
| Solomon Islands | 416 | 536 | 662 | 788 | 2.53 | 2.11 | 1.74 | 14 | 27 |
| Tokelau | 2 | 1 | 1 | 1 | -6.93 | 0.00 | 0.00 | 128 | 101 |
| Tonga | 99 | 104 | 108 | 115 | 0.49 | 0.38 | 0.63 | 152 | 178 |

# TABLE B.1

## *continued*

| | Estimates and projections ('000) | | | | Rate of change (%) | | | Population density (people/km²) | |
|---|---|---|---|---|---|---|---|---|---|
| | 2000 | 2010 | 2020 | 2030 | 2000–2010 | 2010–2020 | 2020–2030 | 2000 | 2030 |
| Tuvalu | 10 | 10 | 10 | 11 | 0.00 | 0.00 | 0.95 | 367 | 419 |
| Vanuatu | 190 | 246 | 307 | 369 | 2.58 | 2.22 | 1.84 | 16 | 30 |
| Wallis and Futuna Islands | 15 | 15 | 17 | 17 | 0.00 | 1.25 | 0.00 | 73 | 85 |

*Sources:* United Nations Department of Economic and Social Affairs, Population Division (2010) *World Urbanization Prospects: The 2009 Revision,* United Nations, New York; United Nations Department of Economic and Social Affairs, Population Division (2009) *World Population Prospects: The 2008 Revision,* United Nations, New York.

Notes:
(1) Including Agalega, Rodrigues, and Saint Brandon.
(2) Including Ascension, and Tristan da Cunha.
(3) For statistical purposes, the data for China do not include Hong Kong and Macao, Special Administrative Regions (SAR) of China.
(4) As of 1 July 1997, Hong Kong became a Special Administrative Region (SAR) of China.
(5) As of 20 December 1999, Macao became a Special Administrative Region (SAR) of China.
(6) Refers to Guernsey and Jersey.
(7) Including Åland Islands.
(8) Refers to the Vatican City State.
(9) Including Svalbard and Jan Mayen Islands.
(10) The former Yugoslav Republic of Macedonia.
(11) Including Christmas Island, Cocos (Keeling) Islands, and Norfolk Island.

# TABLE B.2

## Urban and Rural Population Size and Rate of Change

| | Urban population | | | | | | | Rural population | | | | | | |
|---|---|---|---|---|---|---|---|---|---|---|---|---|---|---|
| | Estimates and projections ('000) | | | | Rate of change (%) | | | Estimates and projections ('000) | | | | Rate of change (%) | | |
| | 2000 | 2010 | 2020 | 2030 | 2000–2010 | 2010–2020 | 2020–2030 | 2000 | 2010 | 2020 | 2030 | 2000–2010 | 2010–2020 | 2020–2030 |
| **AFRICA** | | | | | | | | | | | | | | |
| Algeria | 18,246 | 23,555 | 29,194 | 34,097 | 2.55 | 2.15 | 1.55 | 12,260 | 11,868 | 11,436 | 10,630 | -0.32 | -0.37 | -0.73 |
| Angola | 6,995 | 11,112 | 16,184 | 21,784 | 4.63 | 3.76 | 2.97 | 7,284 | 7,881 | 8,323 | 8,631 | 0.79 | 0.55 | 0.36 |
| Benin | 2,553 | 3,873 | 5,751 | 8,275 | 4.17 | 3.95 | 3.64 | 4,107 | 5,339 | 6,426 | 7,124 | 2.62 | 1.85 | 1.03 |
| Botswana | 917 | 1,209 | 1,506 | 1,769 | 2.76 | 2.20 | 1.61 | 806 | 769 | 722 | 665 | -0.47 | -0.63 | -0.82 |
| Burkina Faso | 2,083 | 4,184 | 7,523 | 11,958 | 6.97 | 5.87 | 4.63 | 9,593 | 12,103 | 14,348 | 15,982 | 2.32 | 1.70 | 1.08 |
| Burundi | 536 | 937 | 1,524 | 2,362 | 5.59 | 4.86 | 4.38 | 5,937 | 7,582 | 8,794 | 9,574 | 2.45 | 1.48 | 0.85 |
| Cameroon | 7,910 | 11,655 | 15,941 | 20,304 | 3.88 | 3.13 | 2.42 | 7,955 | 8,303 | 8,408 | 8,298 | 0.43 | 0.13 | -0.13 |
| Cape Verde | 235 | 313 | 394 | 468 | 2.87 | 2.30 | 1.72 | 204 | 199 | 190 | 177 | -0.25 | -0.46 | -0.71 |
| Central African Republic | 1,410 | 1,755 | 2,268 | 2,978 | 2.19 | 2.56 | 2.72 | 2,336 | 2,751 | 3,072 | 3,171 | 1.64 | 1.10 | 0.32 |
| Chad | 1,964 | 3,179 | 5,054 | 7,843 | 4.82 | 4.64 | 4.39 | 6,438 | 8,328 | 9,843 | 11,174 | 2.57 | 1.67 | 1.27 |
| Comoros | 155 | 195 | 259 | 356 | 2.30 | 2.84 | 3.18 | 397 | 496 | 580 | 619 | 2.23 | 1.56 | 0.65 |
| Congo | 1,770 | 2,335 | 3,118 | 3,883 | 2.77 | 2.89 | 2.19 | 1,265 | 1,424 | 1,582 | 1,596 | 1.18 | 1.05 | 0.09 |
| Côte d'Ivoire | 7,524 | 10,906 | 15,574 | 20,873 | 3.71 | 3.56 | 2.93 | 9,757 | 10,664 | 11,380 | 11,678 | 0.89 | 0.65 | 0.26 |
| Democratic Republic of the Congo | 15,168 | 23,887 | 36,834 | 53,382 | 4.54 | 4.33 | 3.71 | 35,662 | 43,940 | 50,806 | 55,212 | 2.09 | 1.45 | 0.83 |
| Djibouti | 555 | 670 | 798 | 956 | 1.88 | 1.75 | 1.81 | 175 | 209 | 230 | 237 | 1.78 | 0.96 | 0.30 |
| Egypt | 30,032 | 36,664 | 45,301 | 56,477 | 2.00 | 2.12 | 2.21 | 40,142 | 47,810 | 53,336 | 54,430 | 1.75 | 1.09 | 0.20 |
| Equatorial Guinea | 205 | 275 | 379 | 527 | 2.94 | 3.21 | 3.30 | 324 | 418 | 496 | 540 | 2.55 | 1.71 | 0.85 |
| Eritrea | 650 | 1,127 | 1,845 | 2,780 | 5.50 | 4.93 | 4.10 | 3,007 | 4,097 | 4,874 | 5,305 | 3.09 | 1.74 | 0.85 |
| Ethiopia | 9,762 | 14,158 | 20,800 | 31,383 | 3.72 | 3.85 | 4.11 | 55,753 | 70,818 | 87,165 | 100,178 | 2.39 | 2.08 | 1.39 |
| Gabon | 989 | 1,292 | 1,579 | 1,853 | 2.67 | 2.01 | 1.60 | 245 | 210 | 200 | 192 | -1.54 | -0.49 | -0.41 |
| Gambia | 639 | 1,018 | 1,449 | 1,943 | 4.66 | 3.53 | 2.93 | 663 | 733 | 779 | 793 | 1.00 | 0.61 | 0.18 |
| Ghana | 8,584 | 12,524 | 17,274 | 22,565 | 3.78 | 3.22 | 2.67 | 10,945 | 11,808 | 12,293 | 12,319 | 0.76 | 0.40 | 0.02 |
| Guinea | 2,603 | 3,651 | 5,580 | 8,219 | 3.38 | 4.24 | 3.87 | 5,781 | 6,673 | 7,887 | 8,678 | 1.43 | 1.67 | 0.96 |
| Guinea-Bissau | 387 | 494 | 678 | 979 | 2.44 | 3.17 | 3.67 | 917 | 1,153 | 1,387 | 1,557 | 2.29 | 1.85 | 1.16 |
| Kenya | 6,204 | 9,064 | 13,826 | 20,884 | 3.79 | 4.22 | 4.12 | 25,237 | 31,799 | 38,208 | 42,315 | 2.31 | 1.84 | 1.02 |
| Lesotho | 377 | 560 | 775 | 999 | 3.96 | 3.25 | 2.54 | 1,511 | 1,524 | 1,469 | 1,360 | 0.09 | -0.37 | -0.77 |
| Liberia | 1,252 | 1,961 | 2,739 | 3,725 | 4.49 | 3.34 | 3.07 | 1,572 | 2,141 | 2,514 | 2,745 | 3.09 | 1.61 | 0.88 |
| Libyan Arab Jamahiriya | 4,083 | 5,098 | 6,181 | 7,060 | 2.22 | 1.93 | 1.33 | 1,263 | 1,447 | 1,517 | 1,459 | 1.36 | 0.47 | -0.39 |
| Madagascar | 4,143 | 6,082 | 8,953 | 13,048 | 3.84 | 3.87 | 3.77 | 11,132 | 14,064 | 16,734 | 18,480 | 2.34 | 1.74 | 0.99 |
| Malawi | 1,796 | 3,102 | 5,240 | 8,395 | 5.46 | 5.24 | 4.71 | 10,036 | 12,590 | 15,297 | 17,502 | 2.27 | 1.95 | 1.35 |
| Mali | 2,982 | 4,777 | 7,325 | 10,491 | 4.71 | 4.27 | 3.59 | 7,541 | 8,546 | 9,442 | 9,976 | 1.25 | 1.00 | 0.55 |
| Mauritania | 1,041 | 1,395 | 1,859 | 2,478 | 2.93 | 2.87 | 2.87 | 1,563 | 1,971 | 2,232 | 2,313 | 2.32 | 1.24 | 0.36 |
| Mauritius[1] | 510 | 542 | 595 | 681 | 0.61 | 0.93 | 1.35 | 685 | 754 | 777 | 738 | 0.96 | 0.30 | -0.51 |
| Mayotte | 71 | 100 | 129 | 168 | 3.42 | 2.55 | 2.64 | 78 | 99 | 121 | 134 | 2.38 | 2.01 | 1.02 |
| Morocco | 15,375 | 18,859 | 23,158 | 27,157 | 2.04 | 2.05 | 1.59 | 13,452 | 13,523 | 13,042 | 12,102 | 0.05 | -0.36 | -0.75 |
| Mozambique | 5,601 | 8,996 | 13,208 | 18,199 | 4.74 | 3.84 | 3.21 | 12,649 | 14,410 | 15,338 | 15,695 | 1.30 | 0.62 | 0.23 |
| Namibia | 590 | 840 | 1,161 | 1,541 | 3.53 | 3.24 | 2.83 | 1,234 | 1,372 | 1,453 | 1,452 | 1.06 | 0.57 | -0.01 |
| Niger | 1,785 | 2,719 | 4,417 | 7,641 | 4.21 | 4.85 | 5.48 | 9,246 | 13,173 | 18,529 | 24,922 | 3.54 | 3.41 | 2.96 |
| Nigeria | 53,078 | 78,818 | 109,859 | 144,116 | 3.95 | 3.32 | 2.71 | 71,765 | 79,441 | 83,394 | 82,534 | 1.02 | 0.49 | -0.10 |
| Réunion | 650 | 787 | 891 | 972 | 1.91 | 1.24 | 0.87 | 73 | 50 | 40 | 37 | -3.78 | -2.23 | -0.78 |
| Rwanda | 1,096 | 1,938 | 2,993 | 4,550 | 5.70 | 4.35 | 4.19 | 6,862 | 8,340 | 10,241 | 11,554 | 1.95 | 2.05 | 1.21 |
| Saint Helena[2] | 2 | 2 | 2 | 2 | 0.00 | 0.00 | 0.00 | 3 | 3 | 3 | 2 | 0.00 | 0.00 | -4.05 |
| São Tomé and Príncipe | 75 | 103 | 136 | 173 | 3.17 | 2.78 | 2.41 | 65 | 62 | 61 | 61 | -0.47 | -0.16 | 0.00 |
| Senegal | 3,995 | 5,450 | 7,524 | 10,269 | 3.11 | 3.22 | 3.11 | 5,907 | 7,410 | 8,673 | 9,273 | 2.27 | 1.57 | 0.67 |
| Seychelles | 41 | 47 | 54 | 62 | 1.37 | 1.39 | 1.38 | 40 | 38 | 35 | 31 | -0.51 | -0.82 | -1.21 |
| Sierra Leone | 1,501 | 2,241 | 3,134 | 4,384 | 4.01 | 3.35 | 3.36 | 2,727 | 3,595 | 4,184 | 4,559 | 2.76 | 1.52 | 0.86 |
| Somalia | 2,458 | 3,505 | 5,268 | 7,851 | 3.55 | 4.07 | 3.99 | 4,936 | 5,854 | 6,978 | 7,893 | 1.71 | 1.76 | 1.23 |
| South Africa | 25,528 | 31,155 | 35,060 | 39,032 | 1.99 | 1.18 | 1.07 | 19,344 | 19,338 | 17,611 | 15,694 | 0.00 | -0.94 | -1.15 |
| Sudan | 11,661 | 17,322 | 24,804 | 33,267 | 3.96 | 3.59 | 2.94 | 23,243 | 25,871 | 27,505 | 27,728 | 1.07 | 0.61 | 0.08 |
| Swaziland | 244 | 257 | 307 | 400 | 0.52 | 1.78 | 2.65 | 835 | 945 | 1,069 | 1,125 | 1.24 | 1.23 | 0.51 |
| Togo | 1,917 | 2,945 | 4,261 | 5,795 | 4.29 | 3.69 | 3.07 | 3,331 | 3,835 | 4,183 | 4,319 | 1.41 | 0.87 | 0.32 |
| Tunisia | 5,996 | 6,980 | 8,096 | 9,115 | 1.52 | 1.48 | 1.19 | 3,456 | 3,394 | 3,270 | 3,012 | -0.18 | -0.37 | -0.82 |
| Uganda | 2,952 | 4,493 | 7,381 | 12,503 | 4.20 | 4.96 | 5.27 | 21,481 | 29,303 | 38,939 | 48,315 | 3.11 | 2.84 | 2.16 |
| United Republic of Tanzania | 7,614 | 11,883 | 18,945 | 29,190 | 4.45 | 4.66 | 4.32 | 26,517 | 33,157 | 40,658 | 46,308 | 2.23 | 2.04 | 1.30 |
| Western Sahara | 264 | 434 | 606 | 704 | 4.97 | 3.34 | 1.50 | 51 | 96 | 116 | 115 | 6.33 | 1.89 | -0.09 |
| Zambia | 3,643 | 4,733 | 6,584 | 9,340 | 2.62 | 3.30 | 3.50 | 6,824 | 8,524 | 10,332 | 11,549 | 2.22 | 1.92 | 1.11 |
| Zimbabwe | 4,205 | 4,837 | 6,839 | 9,086 | 1.40 | 3.46 | 2.84 | 8,251 | 7,807 | 8,732 | 8,832 | -0.55 | 1.12 | 0.11 |
| **ASIA** | | | | | | | | | | | | | | |
| Afghanistan | 4,148 | 6,581 | 10,450 | 16,296 | 4.62 | 4.62 | 4.44 | 16,388 | 22,537 | 29,134 | 34,353 | 3.19 | 2.57 | 1.65 |
| Armenia | 1,989 | 1,984 | 2,087 | 2,186 | -0.03 | 0.51 | 0.46 | 1,087 | 1,107 | 1,088 | 983 | 0.18 | -0.17 | -1.01 |
| Azerbaijan | 4,158 | 4,639 | 5,332 | 6,044 | 1.09 | 1.39 | 1.25 | 3,964 | 4,294 | 4,506 | 4,279 | 0.80 | 0.48 | -0.52 |
| Bahrain | 574 | 715 | 852 | 984 | 2.20 | 1.75 | 1.44 | 76 | 92 | 101 | 102 | 1.91 | 0.93 | 0.10 |
| Bangladesh | 33,208 | 46,149 | 62,886 | 83,408 | 3.29 | 3.09 | 2.82 | 107,559 | 118,276 | 122,667 | 119,807 | 0.95 | 0.36 | -0.24 |
| Bhutan | 143 | 246 | 348 | 451 | 5.42 | 3.47 | 2.59 | 419 | 463 | 472 | 451 | 1.00 | 0.19 | -0.46 |
| Brunei Darussalam | 237 | 308 | 379 | 450 | 2.62 | 2.07 | 1.72 | 96 | 99 | 99 | 97 | 0.31 | 0.00 | -0.20 |
| Cambodia | 2,157 | 3,027 | 4,214 | 5,870 | 3.39 | 3.31 | 3.31 | 10,603 | 12,026 | 13,493 | 14,230 | 1.26 | 1.15 | 0.53 |
| China[3] | 453,029 | 635,839 | 786,761 | 905,494 | 3.39 | 2.13 | 1.41 | 813,925 | 718,307 | 644,394 | 557,019 | -1.25 | -1.09 | -1.46 |
| China, Hong Kong SAR[4] | 6,667 | 7,069 | 7,701 | 8,185 | 0.59 | 0.86 | 0.61 | — | — | — | — | | | |
| China, Macao SAR[5] | 441 | 548 | 588 | 611 | 2.17 | 0.70 | 0.38 | — | — | — | — | | | |
| Cyprus | 540 | 619 | 705 | 797 | 1.37 | 1.30 | 1.23 | 247 | 261 | 265 | 256 | 0.55 | 0.15 | -0.35 |
| Democratic People's Republic of Korea | 13,581 | 14,446 | 15,413 | 16,633 | 0.62 | 0.65 | 0.76 | 9,278 | 9,545 | 9,389 | 8,668 | 0.28 | -0.16 | -0.80 |
| Georgia | 2,498 | 2,225 | 2,177 | 2,218 | -1.16 | -0.22 | 0.19 | 2,247 | 1,994 | 1,806 | 1,561 | -1.19 | -0.99 | -1.46 |
| India | 288,430 | 364,459 | 463,328 | 590,091 | 2.34 | 2.40 | 2.42 | 754,160 | 850,005 | 903,896 | 894,507 | 1.20 | 0.61 | -0.10 |
| Indonesia | 86,219 | 102,960 | 122,257 | 145,776 | 1.77 | 1.72 | 1.76 | 119,061 | 129,557 | 131,961 | 125,709 | 0.84 | 0.18 | -0.49 |
| Iran (Islamic Republic of) | 42,952 | 53,120 | 63,596 | 71,767 | 2.12 | 1.80 | 1.21 | 23,951 | 21,958 | 20,145 | 18,169 | -0.87 | -0.86 | -1.03 |

# TABLE B.2

**continued**

| | Urban population | | | | | | | Rural population | | | | | | |
|---|---|---|---|---|---|---|---|---|---|---|---|---|---|---|
| | Estimates and projections ('000) | | | | Rate of change (%) | | | Estimates and projections ('000) | | | | Rate of change (%) | | |
| | 2000 | 2010 | 2020 | 2030 | 2000–2010 | 2010–2020 | 2020–2030 | 2000 | 2010 | 2020 | 2030 | 2000–2010 | 2010–2020 | 2020–2030 |
| Iraq | 16,722 | 20,822 | 26,772 | 33,930 | 2.19 | 2.51 | 2.37 | 7,931 | 10,644 | 13,455 | 14,979 | 2.94 | 2.34 | 1.07 |
| Israel | 5,563 | 6,692 | 7,673 | 8,583 | 1.85 | 1.37 | 1.12 | 521 | 594 | 634 | 636 | 1.31 | 0.65 | 0.03 |
| Japan | 82,633 | 84,875 | 85,848 | 85,700 | 0.27 | 0.11 | -0.02 | 44,073 | 42,120 | 37,817 | 31,724 | -0.45 | -1.08 | -1.76 |
| Jordan | 3,798 | 5,083 | 5,998 | 7,063 | 2.91 | 1.66 | 1.63 | 1,055 | 1,390 | 1,520 | 1,554 | 2.76 | 0.89 | 0.22 |
| Kazakhstan | 8,417 | 9,217 | 10,417 | 11,525 | 0.91 | 1.22 | 1.01 | 6,539 | 6,537 | 6,309 | 5,718 | 0.00 | -0.36 | -0.98 |
| Kuwait | 2,188 | 3,001 | 3,637 | 4,218 | 3.16 | 1.92 | 1.48 | 40 | 49 | 53 | 55 | 2.03 | 0.78 | 0.37 |
| Kyrgyzstan | 1,744 | 1,918 | 2,202 | 2,625 | 0.95 | 1.38 | 1.76 | 3,211 | 3,633 | 3,957 | 3,918 | 1.23 | 0.85 | -0.10 |
| Lao People's Democratic Republic | 1,187 | 2,136 | 3,381 | 4,699 | 5.88 | 4.59 | 3.29 | 4,216 | 4,300 | 4,269 | 4,155 | 0.20 | -0.07 | -0.27 |
| Lebanon | 3,244 | 3,712 | 4,065 | 4,374 | 1.35 | 0.91 | 0.73 | 528 | 543 | 522 | 484 | 0.28 | -0.39 | -0.76 |
| Malaysia | 14,424 | 20,146 | 25,128 | 28,999 | 3.34 | 2.21 | 1.43 | 8,849 | 7,768 | 6,889 | 6,277 | -1.30 | -1.20 | -0.93 |
| Maldives | 75 | 126 | 186 | 242 | 5.19 | 3.89 | 2.63 | 197 | 188 | 175 | 161 | -0.47 | -0.72 | -0.83 |
| Mongolia | 1,358 | 1,675 | 2,010 | 2,316 | 2.10 | 1.82 | 1.42 | 1,031 | 1,026 | 992 | 920 | -0.05 | -0.34 | -0.75 |
| Myanmar | 12,956 | 16,990 | 22,570 | 28,545 | 2.71 | 2.84 | 2.35 | 33,654 | 33,505 | 32,927 | 30,808 | -0.04 | -0.17 | -0.67 |
| Nepal | 3,281 | 5,559 | 8,739 | 12,902 | 5.27 | 4.52 | 3.90 | 21,150 | 24,294 | 26,529 | 27,744 | 1.39 | 0.88 | 0.45 |
| Occupied Palestinian Territory | 2,267 | 3,269 | 4,447 | 5,810 | 3.66 | 3.08 | 2.67 | 883 | 1,140 | 1,359 | 1,510 | 2.55 | 1.76 | 1.05 |
| Oman | 1,719 | 2,122 | 2,645 | 3,184 | 2.11 | 2.20 | 1.85 | 683 | 783 | 850 | 864 | 1.37 | 0.82 | 0.16 |
| Pakistan | 49,088 | 66,318 | 90,199 | 121,218 | 3.01 | 3.08 | 2.96 | 99,045 | 118,435 | 135,987 | 144,472 | 1.79 | 1.38 | 0.61 |
| Philippines | 37,283 | 45,781 | 57,657 | 72,555 | 2.05 | 2.31 | 2.30 | 40,406 | 47,836 | 52,026 | 51,829 | 1.69 | 0.84 | -0.04 |
| Qatar | 586 | 1,445 | 1,679 | 1,891 | 9.03 | 1.50 | 1.19 | 31 | 63 | 62 | 60 | 7.09 | -0.16 | -0.33 |
| Republic of Korea | 36,967 | 40,235 | 42,362 | 43,086 | 0.85 | 0.52 | 0.17 | 9,462 | 8,265 | 7,113 | 6,060 | -1.35 | -1.50 | -1.60 |
| Saudi Arabia | 16,615 | 21,541 | 26,617 | 31,516 | 2.60 | 2.12 | 1.69 | 4,193 | 4,705 | 4,991 | 5,030 | 1.15 | 0.59 | 0.08 |
| Singapore | 4,018 | 4,837 | 5,219 | 5,460 | 1.86 | 0.76 | 0.45 | — | — | — | — | — | — | — |
| Sri Lanka | 2,971 | 2,921 | 3,360 | 4,339 | -0.17 | 1.40 | 2.56 | 15,796 | 17,489 | 18,353 | 17,855 | 1.02 | 0.48 | -0.28 |
| Syrian Arab Republic | 8,577 | 12,545 | 15,948 | 19,976 | 3.80 | 2.40 | 2.25 | 7,934 | 9,961 | 10,527 | 10,584 | 2.28 | 0.55 | 0.05 |
| Tajikistan | 1,635 | 1,862 | 2,364 | 3,121 | 1.30 | 2.39 | 2.78 | 4,538 | 5,213 | 6,083 | 6,497 | 1.39 | 1.54 | 0.66 |
| Thailand | 19,417 | 23,142 | 27,800 | 33,624 | 1.76 | 1.83 | 1.90 | 42,930 | 44,997 | 43,643 | 39,838 | 0.47 | -0.31 | -0.91 |
| Timor-Leste | 198 | 329 | 538 | 848 | 5.08 | 4.92 | 4.55 | 617 | 842 | 1,080 | 1,277 | 3.11 | 2.49 | 1.68 |
| Turkey | 43,027 | 52,728 | 62,033 | 70,247 | 2.03 | 1.63 | 1.24 | 23,433 | 22,977 | 21,840 | 20,128 | -0.20 | -0.51 | -0.82 |
| Turkmenistan | 2,062 | 2,562 | 3,175 | 3,793 | 2.17 | 2.15 | 1.78 | 2,440 | 2,614 | 2,642 | 2,483 | 0.69 | 0.11 | -0.62 |
| United Arab Emirates | 2,599 | 3,956 | 4,915 | 5,821 | 4.20 | 2.17 | 1.69 | 639 | 751 | 745 | 735 | 1.62 | -0.08 | -0.14 |
| Uzbekistan | 9,273 | 10,075 | 11,789 | 14,500 | 0.83 | 1.57 | 2.07 | 15,502 | 17,720 | 19,396 | 19,433 | 1.34 | 0.90 | 0.02 |
| Viet Nam | 19,263 | 27,046 | 36,269 | 46,585 | 3.39 | 2.93 | 2.50 | 59,400 | 61,983 | 61,743 | 58,862 | 0.43 | -0.04 | -0.48 |
| Yemen | 4,776 | 7,714 | 12,082 | 17,844 | 4.79 | 4.49 | 3.90 | 13,406 | 16,542 | 19,553 | 21,506 | 2.10 | 1.67 | 0.95 |
| **EUROPE** | | | | | | | | | | | | | | |
| Albania | 1,280 | 1,645 | 2,027 | 2,301 | 2.51 | 2.09 | 1.27 | 1,787 | 1,524 | 1,311 | 1,115 | -1.59 | -1.51 | -1.62 |
| Andorra | 61 | 76 | 85 | 96 | 2.20 | 1.12 | 1.22 | 5 | 10 | 15 | 17 | 6.93 | 4.05 | 1.25 |
| Austria | 5,267 | 5,666 | 6,003 | 6,372 | 0.73 | 0.58 | 0.60 | 2,738 | 2,722 | 2,537 | 2,265 | -0.06 | -0.70 | -1.13 |
| Belarus | 7,030 | 7,162 | 7,219 | 7,070 | 0.19 | 0.08 | -0.21 | 3,023 | 2,426 | 1,894 | 1,494 | -2.20 | -2.48 | -2.37 |
| Belgium | 9,899 | 10,421 | 10,792 | 11,070 | 0.51 | 0.35 | 0.25 | 294 | 277 | 256 | 233 | -0.60 | -0.79 | -0.94 |
| Bosnia and Herzegovina | 1,597 | 1,828 | 2,028 | 2,170 | 1.35 | 1.04 | 0.68 | 2,097 | 1,932 | 1,648 | 1,349 | -0.82 | -1.59 | -2.00 |
| Bulgaria | 5,516 | 5,357 | 5,215 | 5,012 | -0.29 | -0.27 | -0.40 | 2,490 | 2,140 | 1,802 | 1,456 | -1.51 | -1.72 | -2.13 |
| Channel Islands[6] | 45 | 47 | 52 | 59 | 0.43 | 1.01 | 1.26 | 102 | 103 | 100 | 92 | 0.10 | -0.30 | -0.83 |
| Croatia | 2,504 | 2,546 | 2,657 | 2,781 | 0.17 | 0.43 | 0.46 | 2,001 | 1,864 | 1,661 | 1,399 | -0.71 | -1.15 | -1.72 |
| Czech Republic | 7,565 | 7,656 | 7,929 | 8,202 | 0.12 | 0.35 | 0.34 | 2,660 | 2,755 | 2,639 | 2,318 | 0.35 | -0.43 | -1.30 |
| Denmark | 4,540 | 4,761 | 4,923 | 5,058 | 0.48 | 0.33 | 0.27 | 795 | 720 | 634 | 558 | -0.99 | -1.27 | -1.28 |
| Estonia | 951 | 931 | 942 | 955 | -0.21 | 0.12 | 0.14 | 419 | 409 | 390 | 347 | -0.24 | -0.48 | -1.17 |
| Faeroe Islands | 17 | 20 | 23 | 26 | 1.63 | 1.40 | 1.23 | 29 | 30 | 31 | 30 | 0.34 | 0.33 | -0.33 |
| Finland[7] | 4,252 | 4,549 | 4,805 | 4,947 | 0.68 | 0.55 | 0.29 | 922 | 797 | 691 | 597 | -1.46 | -1.43 | -1.46 |
| France | 45,466 | 53,398 | 58,267 | 61,043 | 1.61 | 0.87 | 0.47 | 13,662 | 9,238 | 6,664 | 5,431 | -3.91 | -3.27 | -2.05 |
| Germany | 59,970 | 60,598 | 60,827 | 60,993 | 0.10 | 0.04 | 0.03 | 22,105 | 21,458 | 19,595 | 16,862 | -0.30 | -0.91 | -1.50 |
| Gibraltar | 29 | 31 | 32 | 31 | 0.67 | 0.32 | -0.32 | — | — | — | — | | | |
| Greece | 6,537 | 6,868 | 7,307 | 7,785 | 0.49 | 0.62 | 0.63 | 4,406 | 4,315 | 3,977 | 3,449 | -0.21 | -0.82 | -1.42 |
| Holy See[8] | 1 | 1 | 1 | 1 | 0.00 | 0.00 | 0.00 | | | | | | | |
| Hungary | 6,596 | 6,791 | 7,011 | 7,180 | 0.29 | 0.32 | 0.24 | 3,619 | 3,182 | 2,755 | 2,329 | -1.29 | -1.44 | -1.68 |
| Iceland | 260 | 308 | 349 | 372 | 1.69 | 1.25 | 0.64 | 21 | 22 | 21 | 20 | 0.47 | -0.47 | -0.49 |
| Ireland | 2,250 | 2,842 | 3,370 | 3,889 | 2.34 | 1.70 | 1.43 | 1,554 | 1,747 | 1,775 | 1,684 | 1.17 | 0.16 | -0.53 |
| Isle of Man | 40 | 41 | 41 | 43 | 0.25 | 0.00 | 0.48 | 37 | 40 | 39 | 37 | 0.78 | -0.25 | -0.53 |
| Italy | 38,395 | 41,083 | 42,840 | 44,395 | 0.68 | 0.42 | 0.36 | 18,721 | 19,015 | 17,569 | 15,154 | 0.16 | -0.79 | -1.48 |
| Latvia | 1,616 | 1,517 | 1,471 | 1,453 | -0.63 | -0.31 | -0.12 | 758 | 723 | 681 | 596 | -0.47 | -0.60 | -1.33 |
| Liechtenstein | 5 | 5 | 6 | 7 | 0.00 | 1.82 | 1.54 | 28 | 31 | 33 | 34 | 1.02 | 0.63 | 0.30 |
| Lithuania | 2,345 | 2,181 | 2,096 | 2,080 | -0.73 | -0.40 | -0.08 | 1,156 | 1,075 | 962 | 828 | -0.73 | -1.11 | -1.50 |
| Luxembourg | 366 | 419 | 480 | 547 | 1.35 | 1.36 | 1.31 | 71 | 73 | 70 | 67 | 0.28 | -0.42 | -0.44 |
| Malta | 359 | 388 | 405 | 413 | 0.78 | 0.43 | 0.20 | 30 | 22 | 17 | 14 | -3.10 | -2.58 | -1.94 |
| Moldova | 1,828 | 1,679 | 1,833 | 1,938 | -0.85 | 0.88 | 0.56 | 2,272 | 1,897 | 1,546 | 1,244 | -1.80 | -2.05 | -2.17 |
| Monaco | 32 | 33 | 34 | 35 | 0.31 | 0.30 | 0.29 | — | — | — | — | | | |
| Montenegro | 387 | 384 | 394 | 417 | -0.08 | 0.26 | 0.57 | 274 | 241 | 237 | 217 | -1.28 | -0.17 | -0.88 |
| Netherlands | 12,222 | 13,799 | 14,824 | 15,501 | 1.21 | 0.72 | 0.45 | 3,692 | 2,854 | 2,319 | 1,997 | -2.57 | -2.08 | -1.49 |
| Norway[9] | 3,411 | 3,856 | 4,297 | 4,700 | 1.23 | 1.08 | 0.90 | 1,073 | 1,000 | 903 | 818 | -0.70 | -1.02 | -0.99 |
| Poland | 23,719 | 23,187 | 23,135 | 23,481 | -0.23 | -0.02 | 0.15 | 14,714 | 14,851 | 14,362 | 12,705 | 0.09 | -0.33 | -1.23 |
| Portugal | 5,563 | 6,515 | 7,148 | 7,585 | 1.58 | 0.93 | 0.59 | 4,663 | 4,218 | 3,619 | 3,034 | -1.00 | -1.53 | -1.76 |
| Romania | 11,734 | 12,177 | 12,839 | 13,296 | 0.37 | 0.53 | 0.35 | 10,404 | 9,013 | 7,541 | 6,192 | -1.44 | -1.78 | -1.97 |
| Russian Federation | 107,582 | 102,702 | 100,892 | 99,153 | -0.46 | -0.18 | -0.17 | 39,088 | 37,665 | 34,513 | 29,711 | -0.37 | -0.87 | -1.50 |
| San Marino | 25 | 30 | 31 | 32 | 1.82 | 0.33 | 0.32 | 2 | 2 | 2 | 2 | 0.00 | 0.00 | 0.00 |
| Serbia | 5,369 | 5,525 | 5,871 | 6,252 | 0.29 | 0.61 | 0.63 | 4,765 | 4,331 | 3,911 | 3,392 | -0.95 | -1.02 | -1.42 |
| Slovakia | 3,025 | 2,975 | 3,031 | 3,168 | -0.17 | 0.19 | 0.44 | 2,354 | 2,437 | 2,411 | 2,179 | 0.35 | -0.11 | -1.01 |
| Slovenia | 1,008 | 1,002 | 1,035 | 1,110 | -0.06 | 0.32 | 0.70 | 978 | 1,022 | 1,018 | 927 | 0.44 | -0.04 | -0.94 |

# TABLE B.2

## *continued*

| | Urban population | | | | | | | Rural population | | | | | | |
|---|---|---|---|---|---|---|---|---|---|---|---|---|---|---|
| | Estimates and projections ('000) | | | | Rate of change (%) | | | Estimates and projections ('000) | | | | Rate of change (%) | | |
| | 2000 | 2010 | 2020 | 2030 | 2000–2010 | 2010–2020 | 2020–2030 | 2000 | 2010 | 2020 | 2030 | 2000–2010 | 2010–2020 | 2020–2030 |
| Spain | 30,707 | 35,073 | 38,542 | 40,774 | 1.33 | 0.94 | 0.56 | 9,558 | 10,243 | 10,021 | 8,998 | 0.69 | -0.22 | -1.08 |
| Sweden | 7,445 | 7,870 | 8,333 | 8,799 | 0.56 | 0.57 | 0.54 | 1,415 | 1,424 | 1,380 | 1,277 | 0.06 | -0.31 | -0.78 |
| Switzerland | 5,268 | 5,591 | 5,922 | 6,336 | 0.60 | 0.58 | 0.68 | 1,917 | 2,003 | 1,957 | 1,812 | 0.44 | -0.23 | -0.77 |
| TFYR Macedonia[10] | 1,194 | 1,212 | 1,260 | 1,331 | 0.15 | 0.39 | 0.55 | 818 | 831 | 785 | 685 | 0.16 | -0.57 | -1.36 |
| Ukraine | 32,814 | 31,252 | 30,860 | 30,243 | -0.49 | -0.13 | -0.20 | 16,056 | 14,181 | 12,085 | 9,946 | -1.24 | -1.60 | -1.95 |
| United Kingdom | 46,331 | 49,295 | 53,001 | 56,901 | 0.62 | 0.72 | 0.71 | 12,576 | 12,604 | 12,089 | 11,055 | 0.02 | -0.42 | -0.89 |
| **LATIN AMERICA AND THE CARIBBEAN** | | | | | | | | | | | | | | |
| Anguilla | 11 | 15 | 18 | 19 | 3.10 | 1.82 | 0.54 | — | — | — | — | — | — | — |
| Antigua and Barbuda | 25 | 27 | 32 | 40 | 0.77 | 1.70 | 2.23 | 52 | 62 | 66 | 65 | 1.76 | 0.63 | -0.15 |
| Argentina | 33,291 | 37,572 | 41,554 | 44,726 | 1.21 | 1.01 | 0.74 | 3,648 | 3,093 | 2,750 | 2,529 | -1.65 | -1.18 | -0.84 |
| Aruba | 42 | 50 | 54 | 59 | 1.74 | 0.77 | 0.89 | 48 | 57 | 57 | 53 | 1.72 | 0.00 | -0.73 |
| Bahamas | 250 | 291 | 331 | 367 | 1.52 | 1.29 | 1.03 | 55 | 55 | 54 | 51 | 0.00 | -0.18 | -0.57 |
| Barbados | 97 | 114 | 134 | 151 | 1.61 | 1.62 | 1.19 | 155 | 142 | 128 | 110 | -0.88 | -1.04 | -1.52 |
| Belize | 120 | 164 | 213 | 268 | 3.12 | 2.61 | 2.30 | 131 | 149 | 161 | 162 | 1.29 | 0.77 | 0.06 |
| Bolivia | 5,143 | 6,675 | 8,265 | 9,799 | 2.61 | 2.14 | 1.70 | 3,174 | 3,356 | 3,373 | 3,235 | 0.56 | 0.05 | -0.42 |
| Brazil | 141,416 | 169,098 | 187,104 | 197,874 | 1.79 | 1.01 | 0.56 | 32,759 | 26,326 | 21,947 | 19,272 | -2.19 | -1.82 | -1.30 |
| British Virgin Islands | 8 | 10 | 11 | 14 | 2.23 | 0.95 | 2.41 | 12 | 14 | 14 | 13 | 1.54 | 0.00 | -0.74 |
| Cayman Islands | 40 | 57 | 61 | 65 | 3.54 | 0.68 | 0.64 | — | — | — | — | — | — | — |
| Chile | 13,252 | 15,251 | 16,958 | 18,247 | 1.40 | 1.06 | 0.73 | 2,167 | 1,884 | 1,681 | 1,532 | -1.40 | -1.14 | -0.93 |
| Colombia | 28,666 | 34,758 | 40,800 | 46,357 | 1.93 | 1.60 | 1.28 | 11,107 | 11,542 | 11,478 | 10,907 | 0.38 | -0.06 | -0.51 |
| Costa Rica | 2,321 | 2,989 | 3,643 | 4,259 | 2.53 | 1.98 | 1.56 | 1,610 | 1,651 | 1,607 | 1,503 | 0.25 | -0.27 | -0.67 |
| Cuba | 8,382 | 8,429 | 8,462 | 8,550 | 0.06 | 0.04 | 0.10 | 2,705 | 2,776 | 2,732 | 2,469 | 0.26 | -0.16 | -1.01 |
| Dominica | 46 | 45 | 47 | 50 | -0.22 | 0.43 | 0.62 | 22 | 22 | 21 | 18 | 0.00 | -0.47 | -1.54 |
| Dominican Republic | 5,452 | 7,074 | 8,560 | 9,793 | 2.60 | 1.91 | 1.35 | 3,378 | 3,151 | 2,890 | 2,638 | -0.70 | -0.86 | -0.91 |
| Ecuador | 7,423 | 9,222 | 11,152 | 12,813 | 2.17 | 1.90 | 1.39 | 4,887 | 4,553 | 4,223 | 3,866 | -0.71 | -0.75 | -0.88 |
| El Salvador | 3,503 | 3,983 | 4,583 | 5,287 | 1.28 | 1.40 | 1.43 | 2,443 | 2,211 | 2,035 | 1,890 | -1.00 | -0.83 | -0.74 |
| Falkland Islands (Malvinas) | 2 | 2 | 2 | 3 | 0.00 | 0.00 | 4.05 | 1 | 1 | 1 | 1 | 0.00 | 0.00 | 0.00 |
| French Guiana | 124 | 177 | 229 | 288 | 3.56 | 2.58 | 2.29 | 41 | 55 | 62 | 66 | 2.94 | 1.20 | 0.63 |
| Grenada | 37 | 41 | 48 | 55 | 1.03 | 1.58 | 1.36 | 65 | 63 | 60 | 53 | -0.31 | -0.49 | -1.24 |
| Guadeloupe | 422 | 460 | 476 | 485 | 0.86 | 0.34 | 0.19 | 7 | 7 | 7 | 7 | 0.00 | 0.00 | 0.00 |
| Guatemala | 5,068 | 7,111 | 9,893 | 13,153 | 3.39 | 3.30 | 2.85 | 6,163 | 7,266 | 8,198 | 8,539 | 1.65 | 1.21 | 0.41 |
| Guyana | 217 | 218 | 233 | 265 | 0.05 | 0.67 | 1.29 | 539 | 544 | 512 | 449 | 0.09 | -0.61 | -1.31 |
| Haiti | 3,079 | 5,307 | 7,546 | 9,450 | 5.44 | 3.52 | 2.25 | 5,569 | 4,881 | 4,177 | 3,746 | -1.32 | -1.56 | -1.09 |
| Honduras | 2,832 | 3,930 | 5,263 | 6,656 | 3.28 | 2.92 | 2.35 | 3,398 | 3,686 | 3,874 | 3,835 | 0.81 | 0.50 | -0.10 |
| Jamaica | 1,330 | 1,420 | 1,521 | 1,660 | 0.65 | 0.69 | 0.87 | 1,237 | 1,310 | 1,313 | 1,213 | 0.57 | 0.02 | -0.79 |
| Martinique | 345 | 362 | 370 | 376 | 0.48 | 0.22 | 0.16 | 40 | 44 | 45 | 42 | 0.95 | 0.22 | -0.69 |
| Mexico | 74,372 | 86,113 | 96,558 | 105,300 | 1.47 | 1.14 | 0.87 | 25,159 | 24,532 | 23,125 | 21,157 | -0.25 | -0.59 | -0.89 |
| Montserrat | 1 | 1 | 1 | 1 | 0.00 | 0.00 | 0.00 | 4 | 5 | 5 | 5 | 2.23 | 0.00 | 0.00 |
| Netherlands Antilles | 163 | 187 | 199 | 200 | 1.37 | 0.62 | 0.05 | 18 | 14 | 11 | 9 | -2.51 | -2.41 | -2.01 |
| Nicaragua | 2,792 | 3,337 | 4,077 | 4,860 | 1.78 | 2.00 | 1.76 | 2,309 | 2,485 | 2,605 | 2,527 | 0.73 | 0.47 | -0.30 |
| Panama | 1,941 | 2,624 | 3,233 | 3,751 | 3.01 | 2.09 | 1.49 | 1,010 | 884 | 794 | 736 | -1.33 | -1.07 | -0.76 |
| Paraguay | 2,960 | 3,972 | 5,051 | 6,102 | 2.94 | 2.40 | 1.89 | 2,390 | 2,487 | 2,482 | 2,380 | 0.40 | -0.02 | -0.42 |
| Peru | 18,994 | 22,688 | 26,389 | 29,902 | 1.78 | 1.51 | 1.25 | 7,010 | 6,808 | 6,492 | 6,103 | -0.29 | -0.48 | -0.62 |
| Puerto Rico | 3,614 | 3,949 | 4,112 | 4,178 | 0.89 | 0.40 | 0.16 | 204 | 49 | 23 | 18 | -14.26 | -7.56 | -2.45 |
| Saint Kitts and Nevis | 15 | 17 | 21 | 26 | 1.25 | 2.11 | 2.14 | 31 | 35 | 38 | 37 | 1.21 | 0.82 | -0.27 |
| Saint Lucia | 44 | 49 | 58 | 74 | 1.08 | 1.69 | 2.44 | 113 | 125 | 132 | 130 | 1.01 | 0.54 | -0.15 |
| Saint Vincent and the Grenadines | 49 | 54 | 60 | 68 | 0.97 | 1.05 | 1.25 | 59 | 55 | 50 | 44 | -0.70 | -0.95 | -1.28 |
| Suriname | 303 | 364 | 418 | 466 | 1.83 | 1.38 | 1.09 | 164 | 161 | 150 | 137 | -0.18 | -0.71 | -0.91 |
| Trinidad and Tobago | 140 | 186 | 250 | 328 | 2.84 | 2.96 | 2.72 | 1,155 | 1,157 | 1,133 | 1,054 | 0.02 | -0.21 | -0.72 |
| Turks and Caicos Islands | 16 | 31 | 35 | 38 | 6.61 | 1.21 | 0.82 | 3 | 2 | 1 | 1 | -4.05 | -6.93 | 0.00 |
| United States Virgin Islands | 101 | 104 | 102 | 96 | 0.29 | -0.19 | -0.61 | 8 | 5 | 4 | 3 | -4.70 | -2.23 | -2.88 |
| Uruguay | 3,033 | 3,119 | 3,264 | 3,382 | 0.28 | 0.45 | 0.36 | 288 | 254 | 229 | 206 | -1.26 | -1.04 | -1.06 |
| Venezuela (Bolivarian Republic of) | 21,940 | 27,113 | 31,755 | 35,588 | 2.12 | 1.58 | 1.14 | 2,468 | 1,931 | 1,658 | 1,556 | -2.45 | -1.52 | -0.63 |
| **NORTHERN AMERICA** | | | | | | | | | | | | | | |
| Bermuda | 63 | 65 | 66 | 66 | 0.31 | 0.15 | 0.00 | — | — | — | — | — | — | — |
| Canada | 24,389 | 27,309 | 30,426 | 33,680 | 1.13 | 1.08 | 1.02 | 6,298 | 6,581 | 6,675 | 6,416 | 0.44 | 0.14 | -0.40 |
| Greenland | 46 | 48 | 49 | 49 | 0.43 | 0.21 | 0.00 | 10 | 9 | 8 | 6 | -1.05 | -1.18 | -2.88 |
| Saint-Pierre-et-Miquelon | 6 | 5 | 6 | 6 | -1.82 | 1.82 | 0.00 | 1 | 1 | 0 | 0 | 0.00 | -1.00 | 0.00 |
| United States of America | 227,651 | 261,375 | 293,732 | 321,698 | 1.38 | 1.17 | 0.91 | 60,191 | 56,266 | 52,421 | 48,283 | -0.67 | -0.71 | -0.82 |
| **OCEANIA** | | | | | | | | | | | | | | |
| American Samoa | 51 | 64 | 76 | 87 | 2.27 | 1.72 | 1.35 | 6 | 5 | 4 | 4 | -1.82 | -2.23 | 0.00 |
| Australia[11] | 16,710 | 19,169 | 21,459 | 23,566 | 1.37 | 1.13 | 0.94 | 2,461 | 2,343 | 2,216 | 2,089 | -0.49 | -0.56 | -0.59 |
| Cook Islands | 11 | 15 | 17 | 19 | 3.10 | 1.25 | 1.11 | 6 | 5 | 4 | 3 | -1.82 | -2.23 | -2.88 |
| Fiji | 384 | 443 | 501 | 566 | 1.43 | 1.23 | 1.22 | 418 | 411 | 387 | 352 | -0.17 | -0.60 | -0.95 |
| French Polynesia | 124 | 140 | 160 | 186 | 1.21 | 1.34 | 1.51 | 112 | 132 | 144 | 143 | 1.64 | 0.87 | -0.07 |
| Guam | 144 | 168 | 188 | 208 | 1.54 | 1.12 | 1.01 | 11 | 12 | 13 | 13 | 0.87 | 0.80 | 0.00 |
| Kiribati | 36 | 44 | 54 | 67 | 2.01 | 2.05 | 2.16 | 48 | 56 | 62 | 63 | 1.54 | 1.02 | 0.16 |
| Marshall Islands | 36 | 45 | 56 | 65 | 2.23 | 2.19 | 1.49 | 16 | 18 | 18 | 18 | 1.18 | 0.00 | 0.00 |
| Micronesia (Federated States of) | 24 | 25 | 29 | 38 | 0.41 | 1.48 | 2.70 | 83 | 86 | 88 | 87 | 0.36 | 0.23 | -0.11 |
| Nauru | 10 | 10 | 11 | 11 | 0.00 | 0.95 | 0.00 | — | — | — | — | — | — | — |
| New Caledonia | 127 | 146 | 169 | 200 | 1.39 | 1.46 | 1.68 | 88 | 108 | 120 | 119 | 2.05 | 1.05 | -0.08 |
| New Zealand | 3,314 | 3,710 | 4,058 | 4,382 | 1.13 | 0.90 | 0.77 | 554 | 594 | 611 | 590 | 0.70 | 0.28 | -0.35 |
| Niue | 1 | 1 | 1 | 1 | 0.00 | 0.00 | 0.00 | 1 | 1 | 1 | 1 | 0.00 | 0.00 | 0.00 |
| Northern Mariana Islands | 62 | 81 | 96 | 111 | 2.67 | 1.70 | 1.45 | 7 | 8 | 8 | 8 | 1.34 | 0.00 | 0.00 |
| Palau | 13 | 17 | 20 | 23 | 2.68 | 1.63 | 1.40 | 6 | 3 | 2 | 2 | -6.93 | -4.05 | 0.00 |
| Papua New Guinea | 711 | 863 | 1,194 | 1,828 | 1.94 | 3.25 | 4.26 | 4,676 | 6,026 | 7,275 | 8,230 | 2.54 | 1.88 | 1.23 |

# TABLE B.2

**continued**

| | Urban population | | | | | | | Rural population | | | | | | |
|---|---|---|---|---|---|---|---|---|---|---|---|---|---|---|
| | Estimates and projections ('000) | | | | Rate of change (%) | | | Estimates and projections ('000) | | | | Rate of change (%) | | |
| | 2000 | 2010 | 2020 | 2030 | 2000–2010 | 2010–2020 | 2020–2030 | 2000 | 2010 | 2020 | 2030 | 2000–2010 | 2010–2020 | 2020–2030 |
| Pitcairn | — | — | — | — | — | — | — | 138 | 143 | 146 | 146 | 0.36 | 0.21 | 0.00 |
| Samoa | 39 | 36 | 38 | 46 | -0.80 | 0.54 | 1.91 | 350 | 436 | 510 | 558 | 2.20 | 1.57 | 0.90 |
| Solomon Islands | 65 | 99 | 152 | 230 | 4.21 | 4.29 | 4.14 | 2 | 1 | 1 | 1 | -6.93 | 0.00 | 0.00 |
| Tokelau | — | — | — | — | — | — | — | 76 | 80 | 81 | 80 | 0.51 | 0.12 | -0.12 |
| Tonga | 23 | 24 | 28 | 35 | 0.43 | 1.54 | 2.23 | 5 | 5 | 5 | 4 | 0.00 | 0.00 | -2.23 |
| Tuvalu | 4 | 5 | 6 | 7 | 2.23 | 1.82 | 1.54 | 149 | 183 | 212 | 229 | 2.06 | 1.47 | 0.77 |
| Vanuatu | 41 | 63 | 95 | 140 | 4.30 | 4.11 | 3.88 | 15 | 15 | 17 | 17 | 0.00 | 1.25 | 0.00 |
| Wallis and Futuna Islands | — | — | — | — | — | — | — | | | | | | | |

*Sources:* United Nations Department of Economic and Social Affairs, Population Division (2010) *World Urbanization Prospects: The 2009 Revision*, United Nations, New York.

*Notes:*
(1) Including Agalega, Rodrigues, and Saint Brandon.
(2) Including Ascension, and Tristan da Cunha.
(3) For statistical purposes, the data for China do not include Hong Kong and Macao, Special Administrative Regions (SAR) of China.
(4) As of 1 July 1997, Hong Kong became a Special Administrative Region (SAR) of China.
(5) As of 20 December 1999, Macao became a Special Administrative Region (SAR) of China.
(6) Refers to Guernsey and Jersey.
(7) Including Åland Islands.
(8) Refers to the Vatican City State.
(9) Including Svalbard and Jan Mayen Islands.
(10) The former Yugoslav Republic of Macedonia.
(11) Including Christmas Island, Cocos (Keeling) Islands, and Norfolk Island.

# TABLE B.3

## Urbanization and Urban Slum Dwellers

| | Level of urbanization | | | | | | | Urban slum dwellers | | | | | | | |
|---|---|---|---|---|---|---|---|---|---|---|---|---|---|---|---|
| | Estimates and projections (%) | | | | Rate of change (%) | | | Estimate ('000) | | | | | Rate of change (%) | | |
| | 2000 | 2010 | 2020 | 2030 | 2000–2010 | 2010–2020 | 2020–2030 | 1990 | 1995 | 2000 | 2005 | 2007 | 1990–1995 | 1995–2000 | 2000–2005 |
| **AFRICA** | | | | | | | | | | | | | | | |
| Algeria | 59.8 | 66.5 | 71.9 | 76.2 | 1.06 | 0.78 | 0.58 | ... | ... | ... | ... | ... | ... | ... | ... |
| Angola | 49.0 | 58.5 | 66.0 | 71.6 | 1.77 | 1.21 | 0.81 | ... | ... | ... | 86.5 | ... | ... | ... | ... |
| Benin | 38.3 | 42.0 | 47.2 | 53.7 | 0.92 | 1.17 | 1.29 | 79.3 | 76.8 | 74.3 | 71.8 | 70.8 | -0.64 | -0.66 | -0.68 |
| Botswana | 53.2 | 61.1 | 67.6 | 72.7 | 1.38 | 1.01 | 0.73 | ... | ... | ... | ... | ... | ... | ... | ... |
| Burkina Faso | 17.8 | 25.7 | 34.4 | 42.8 | 3.67 | 2.92 | 2.18 | 78.8 | 72.4 | 65.9 | 59.5 | 59.5 | -1.71 | -1.87 | -2.06 |
| Burundi | 8.3 | 11.0 | 14.8 | 19.8 | 2.82 | 2.97 | 2.91 | ... | ... | ... | 64.3 | ... | ... | ... | ... |
| Cameroon | 49.9 | 58.4 | 65.5 | 71.0 | 1.57 | 1.15 | 0.81 | 50.8 | 49.6 | 48.4 | 47.4 | 46.6 | -0.49 | -0.51 | -0.40 |
| Cape Verde | 53.4 | 61.1 | 67.4 | 72.5 | 1.35 | 0.98 | 0.73 | ... | ... | ... | ... | ... | ... | ... | ... |
| Central African Republic | 37.6 | 38.9 | 42.5 | 48.4 | 0.34 | 0.89 | 1.30 | 87.5 | 89.7 | 91.9 | 94.1 | 95.0 | 0.50 | 0.49 | 0.48 |
| Chad | 23.4 | 27.6 | 33.9 | 41.2 | 1.65 | 2.06 | 1.95 | 98.9 | 96.4 | 93.9 | 91.3 | 90.3 | -0.52 | -0.53 | -0.55 |
| Comoros | 28.1 | 28.2 | 30.8 | 36.5 | 0.04 | 0.88 | 1.70 | 65.4 | 65.4 | 65.4 | 68.9 | 68.9 | 0.00 | 0.00 | 1.05 |
| Congo | 58.3 | 62.1 | 66.3 | 70.9 | 0.63 | 0.65 | 0.67 | ... | ... | ... | 53.4 | ... | ... | ... | ... |
| Côte d'Ivoire | 43.5 | 50.6 | 57.8 | 64.1 | 1.51 | 1.33 | 1.03 | 53.4 | 54.3 | 55.3 | 56.2 | 56.6 | 0.35 | 0.35 | 0.34 |
| Democratic Republic of the Congo | 29.8 | 35.2 | 42.0 | 49.2 | 1.67 | 1.77 | 1.58 | ... | ... | ... | 76.4 | ... | ... | ... | ... |
| Djibouti | 76.0 | 76.2 | 77.6 | 80.2 | 0.03 | 0.18 | 0.33 | ... | ... | ... | ... | ... | ... | ... | ... |
| Egypt | 42.8 | 43.4 | 45.9 | 50.9 | 0.14 | 0.56 | 1.03 | 50.2 | 39.2 | 28.1 | 17.1 | 17.1 | -4.96 | -6.62 | -9.95 |
| Equatorial Guinea | 38.8 | 39.7 | 43.3 | 49.4 | 0.23 | 0.87 | 1.32 | ... | ... | ... | 66.3 | ... | ... | ... | ... |
| Eritrea | 17.8 | 21.6 | 27.5 | 34.4 | 1.93 | 2.41 | 2.24 | ... | ... | ... | ... | ... | ... | ... | ... |
| Ethiopia | 14.9 | 16.7 | 19.3 | 23.9 | 1.14 | 1.45 | 2.14 | 95.5 | 95.5 | 88.6 | 81.8 | 79.1 | 0.00 | -1.48 | -1.60 |
| Gabon | 80.1 | 86.0 | 88.8 | 90.6 | 0.71 | 0.32 | 0.20 | ... | ... | ... | 38.7 | ... | ... | ... | ... |
| Gambia | 49.1 | 58.1 | 65.0 | 71.0 | 1.68 | 1.12 | 0.88 | ... | ... | ... | 45.4 | ... | ... | ... | ... |
| Ghana | 44.0 | 51.5 | 58.4 | 64.7 | 1.57 | 1.26 | 1.02 | 65.5 | 58.8 | 52.1 | 45.4 | 42.8 | -2.15 | -2.41 | -2.74 |
| Guinea | 31.0 | 35.4 | 41.4 | 48.6 | 1.33 | 1.57 | 1.60 | 80.4 | 68.8 | 57.3 | 45.7 | 45.7 | -3.11 | -3.68 | -4.51 |
| Guinea-Bissau | 29.7 | 30.0 | 32.8 | 38.6 | 0.10 | 0.89 | 1.63 | ... | ... | ... | 83.1 | ... | ... | ... | ... |
| Kenya | 19.7 | 22.2 | 26.6 | 33.0 | 1.19 | 1.81 | 2.16 | 54.9 | 54.8 | 54.8 | 54.8 | 54.8 | -0.01 | -0.01 | -0.01 |
| Lesotho | 20.0 | 26.9 | 34.5 | 42.4 | 2.96 | 2.49 | 2.06 | 35.1 | 35.1 | 35.1 | 35.1 | 35.1 | 0.00 | 0.00 | 0.00 |
| Liberia | 44.3 | 47.8 | 52.1 | 57.6 | 0.76 | 0.86 | 1.00 | ... | ... | ... | ... | ... | ... | ... | ... |
| Libyan Arab Jamahiriya | 76.4 | 77.9 | 80.3 | 82.9 | 0.19 | 0.30 | 0.32 | ... | ... | ... | ... | ... | ... | ... | ... |
| Madagascar | 27.1 | 30.2 | 34.9 | 41.4 | 1.08 | 1.45 | 1.71 | 93.0 | 88.6 | 84.1 | 80.6 | 78.0 | -0.97 | -1.02 | -0.86 |
| Malawi | 15.2 | 19.8 | 25.5 | 32.4 | 2.64 | 2.53 | 2.39 | 66.4 | 66.4 | 66.4 | 66.4 | 67.7 | 0.00 | 0.00 | 0.00 |
| Mali | 28.3 | 35.9 | 43.7 | 51.3 | 2.38 | 1.97 | 1.60 | 94.2 | 84.8 | 75.4 | 65.9 | 65.9 | -2.11 | -2.36 | -2.67 |
| Mauritania | 40.0 | 41.4 | 45.4 | 51.7 | 0.34 | 0.92 | 1.30 | ... | ... | ... | ... | ... | ... | ... | ... |
| Mauritius[1] | 42.7 | 41.8 | 43.4 | 48.0 | -0.21 | 0.38 | 1.01 | ... | ... | ... | ... | ... | ... | ... | ... |
| Mayotte | 47.7 | 50.1 | 51.6 | 55.7 | 0.48 | 0.30 | 0.77 | ... | ... | ... | ... | ... | ... | ... | ... |
| Morocco | 53.3 | 58.2 | 64.0 | 69.2 | 0.88 | 0.95 | 0.78 | 37.4 | 35.2 | 24.2 | 13.1 | 13.1 | -1.21 | -7.54 | -12.23 |
| Mozambique | 30.7 | 38.4 | 46.3 | 53.7 | 2.24 | 1.87 | 1.48 | 75.6 | 76.9 | 78.2 | 79.5 | 80.0 | 0.34 | 0.33 | 0.33 |
| Namibia | 32.4 | 38.0 | 44.4 | 51.5 | 1.59 | 1.56 | 1.48 | 34.4 | 34.1 | 33.9 | 33.9 | 33.6 | -0.13 | -0.14 | -0.01 |
| Niger | 16.2 | 17.1 | 19.3 | 23.5 | 0.54 | 1.21 | 1.97 | 83.6 | 83.1 | 82.6 | 82.1 | 81.9 | -0.12 | -0.12 | -0.12 |
| Nigeria | 42.5 | 49.8 | 56.8 | 63.6 | 1.59 | 1.32 | 1.13 | 77.3 | 73.5 | 69.6 | 65.8 | 64.2 | -1.02 | -1.08 | -1.14 |
| Réunion | 89.9 | 94.0 | 95.7 | 96.3 | 0.45 | 0.18 | 0.06 | ... | ... | ... | ... | ... | ... | ... | ... |
| Rwanda | 13.8 | 18.9 | 22.6 | 28.3 | 3.14 | 1.79 | 2.25 | 96.0 | 87.9 | 79.7 | 71.6 | 68.3 | -1.77 | -1.95 | -2.16 |
| Saint Helena[2] | 39.7 | 39.7 | 41.7 | 46.4 | 0.00 | 0.49 | 1.07 | ... | ... | ... | ... | ... | ... | ... | ... |
| São Tomé and Príncipe | 53.4 | 62.2 | 69.0 | 74.0 | 1.53 | 1.04 | 0.70 | ... | ... | ... | ... | ... | ... | ... | ... |
| Senegal | 40.3 | 42.4 | 46.5 | 52.5 | 0.51 | 0.92 | 1.21 | 70.6 | 59.8 | 48.9 | 38.1 | 38.1 | -3.33 | -4.00 | -5.01 |
| Seychelles | 51.0 | 55.3 | 61.1 | 66.6 | 0.81 | 1.00 | 0.86 | ... | ... | ... | ... | ... | ... | ... | ... |
| Sierra Leone | 35.5 | 38.4 | 42.8 | 49.0 | 0.79 | 1.08 | 1.35 | ... | ... | ... | 97.0 | ... | ... | ... | ... |
| Somalia | 33.2 | 37.4 | 43.0 | 49.9 | 1.19 | 1.40 | 1.49 | ... | ... | ... | 73.5 | ... | ... | ... | ... |
| South Africa | 56.9 | 61.7 | 66.6 | 71.3 | 0.81 | 0.76 | 0.68 | 46.2 | 39.7 | 33.2 | 28.7 | 28.7 | -3.03 | -3.58 | -2.91 |
| Sudan | 33.4 | 40.1 | 47.4 | 54.5 | 1.83 | 1.67 | 1.40 | ... | ... | ... | 94.2 | ... | ... | ... | ... |
| Swaziland | 22.6 | 21.4 | 22.3 | 26.2 | -0.55 | 0.41 | 1.61 | ... | ... | ... | ... | ... | ... | ... | ... |
| Togo | 36.5 | 43.4 | 50.5 | 57.3 | 1.73 | 1.52 | 1.26 | ... | ... | ... | 62.1 | ... | ... | ... | ... |
| Tunisia | 63.4 | 67.3 | 71.2 | 75.2 | 0.60 | 0.56 | 0.55 | ... | ... | ... | ... | ... | ... | ... | ... |
| Uganda | 12.1 | 13.3 | 15.9 | 20.6 | 0.95 | 1.79 | 2.59 | 75.0 | 75.0 | 75.0 | 66.7 | 63.4 | 0.00 | 0.00 | -2.35 |
| United Republic of Tanzania | 22.3 | 26.4 | 31.8 | 38.7 | 1.69 | 1.86 | 1.96 | 77.4 | 73.7 | 70.1 | 66.4 | 65.0 | -0.97 | -1.01 | -1.07 |
| Western Sahara | 83.9 | 81.8 | 83.9 | 85.9 | -0.25 | 0.25 | 0.24 | ... | ... | ... | ... | ... | ... | ... | ... |
| Zambia | 34.8 | 35.7 | 38.9 | 44.7 | 0.26 | 0.86 | 1.39 | 57.0 | 57.1 | 57.2 | 57.2 | 57.3 | 0.02 | 0.02 | 0.01 |
| Zimbabwe | 33.8 | 38.3 | 43.9 | 50.7 | 1.25 | 1.36 | 1.44 | 4.0 | 3.7 | 3.3 | 17.9 | 17.9 | -1.56 | -2.11 | 33.64 |
| **ASIA** | | | | | | | | | | | | | | | |
| Afghanistan | 20.2 | 22.6 | 26.4 | 32.2 | 1.12 | 1.55 | 1.99 | ... | ... | ... | ... | ... | ... | ... | ... |
| Armenia | 64.7 | 64.2 | 65.7 | 69.0 | -0.08 | 0.23 | 0.49 | ... | ... | ... | ... | ... | ... | ... | ... |
| Azerbaijan | 51.2 | 51.9 | 54.2 | 58.7 | 0.14 | 0.43 | 0.78 | ... | ... | ... | ... | ... | ... | ... | ... |
| Bahrain | 88.4 | 88.6 | 89.4 | 90.6 | 0.02 | 0.09 | 0.13 | ... | ... | ... | ... | ... | ... | ... | ... |
| Bangladesh | 23.6 | 28.1 | 33.9 | 41.0 | 1.75 | 1.88 | 1.90 | 87.3 | 84.7 | 77.8 | 70.8 | 70.8 | -0.60 | -1.71 | -1.87 |
| Bhutan | 25.4 | 34.7 | 42.4 | 50.0 | 3.12 | 2.00 | 1.65 | ... | ... | ... | ... | ... | ... | ... | ... |
| Brunei Darussalam | 71.1 | 75.7 | 79.3 | 82.3 | 0.63 | 0.46 | 0.37 | ... | ... | ... | ... | ... | ... | ... | ... |
| Cambodia | 16.9 | 20.1 | 23.8 | 29.2 | 1.73 | 1.69 | 2.04 | ... | ... | ... | 78.9 | ... | ... | ... | ... |
| China[3] | 35.8 | 47.0 | 55.0 | 61.9 | 2.72 | 1.57 | 1.18 | 43.6 | 40.5 | 37.3 | 32.9 | 31.0 | -1.49 | -1.61 | -2.52 |
| China, Hong Kong SAR[4] | 100.0 | 100.0 | 100.0 | 100.0 | 0.00 | 0.00 | 0.00 | ... | ... | ... | ... | ... | ... | ... | ... |
| China, Macao SAR[5] | 100.0 | 100.0 | 100.0 | 100.0 | 0.00 | 0.00 | 0.00 | ... | ... | ... | ... | ... | ... | ... | ... |
| Cyprus | 68.6 | 70.3 | 72.7 | 75.7 | 0.24 | 0.34 | 0.40 | ... | ... | ... | ... | ... | ... | ... | ... |
| Democratic People's Republic of Korea | 59.4 | 60.2 | 62.1 | 65.7 | 0.13 | 0.31 | 0.56 | ... | ... | ... | ... | ... | ... | ... | ... |
| Georgia | 52.6 | 52.7 | 54.7 | 58.7 | 0.02 | 0.37 | 0.71 | ... | ... | ... | ... | ... | ... | ... | ... |
| India | 27.7 | 30.0 | 33.9 | 39.7 | 0.80 | 1.22 | 1.58 | 54.9 | 48.2 | 41.5 | 34.8 | 32.1 | -2.61 | -3.00 | -3.53 |
| Indonesia | 42.0 | 44.3 | 48.1 | 53.7 | 0.53 | 0.82 | 1.10 | 50.8 | 42.6 | 34.4 | 26.3 | 23.0 | -3.51 | -4.26 | -5.42 |
| Iran (Islamic Republic of) | 64.2 | 70.8 | 75.9 | 79.8 | 0.98 | 0.70 | 0.50 | 16.9 | 16.9 | 16.9 | 52.8 | 52.8 | 0.00 | 0.00 | 22.77 |

# TABLE B.3

## continued

| | Level of urbanization | | | | | | | Urban slum dwellers | | | | | | | |
|---|---|---|---|---|---|---|---|---|---|---|---|---|---|---|---|
| | Estimates and projections (%) | | | | Rate of change (%) | | | Estimate ('000) | | | | | Rate of change (%) | | |
| | 2000 | 2010 | 2020 | 2030 | 2000–2010 | 2010–2020 | 2020–2030 | 1990 | 1995 | 2000 | 2005 | 2007 | 1990–1995 | 1995–2000 | 2000–2005 |
| Iraq | 67.8 | 66.2 | 66.6 | 69.4 | -0.24 | 0.06 | 0.41 | ... | ... | ... | ... | ... | ... | ... | ... |
| Israel | 91.4 | 91.9 | 92.4 | 93.1 | 0.05 | 0.05 | 0.08 | ... | ... | ... | ... | ... | ... | ... | ... |
| Japan | 65.2 | 66.8 | 69.4 | 73.0 | 0.24 | 0.38 | 0.51 | ... | ... | ... | ... | ... | ... | ... | ... |
| Jordan | 78.3 | 78.5 | 79.8 | 82.0 | 0.03 | 0.16 | 0.27 | ... | ... | ... | 15.8 | ... | ... | ... | ... |
| Kazakhstan | 56.3 | 58.5 | 62.3 | 66.8 | 0.38 | 0.63 | 0.70 | ... | ... | ... | ... | ... | ... | ... | ... |
| Kuwait | 98.2 | 98.4 | 98.6 | 98.7 | 0.02 | 0.02 | 0.01 | ... | ... | ... | ... | ... | ... | ... | ... |
| Kyrgyzstan | 35.2 | 34.5 | 35.7 | 40.1 | -0.20 | 0.34 | 1.16 | ... | ... | ... | ... | ... | ... | ... | ... |
| Lao People's Democratic Republic | 22.0 | 33.2 | 44.2 | 53.1 | 4.12 | 2.86 | 1.83 | ... | ... | ... | 79.3 | ... | ... | ... | ... |
| Lebanon | 86.0 | 87.2 | 88.6 | 90.0 | 0.14 | 0.16 | 0.16 | ... | ... | ... | 53.1 | ... | ... | ... | ... |
| Malaysia | 62.0 | 72.2 | 78.5 | 82.2 | 1.52 | 0.84 | 0.46 | ... | ... | ... | ... | ... | ... | ... | ... |
| Maldives | 27.7 | 40.1 | 51.5 | 60.1 | 3.70 | 2.50 | 1.54 | ... | ... | ... | ... | ... | ... | ... | ... |
| Mongolia | 56.9 | 62.0 | 67.0 | 71.6 | 0.86 | 0.78 | 0.66 | 68.5 | 66.7 | 64.9 | 57.9 | 57.9 | -0.53 | -0.55 | -2.28 |
| Myanmar | 27.8 | 33.6 | 40.7 | 48.1 | 1.89 | 1.92 | 1.67 | ... | ... | ... | 45.6 | ... | ... | ... | ... |
| Nepal | 13.4 | 18.6 | 24.8 | 31.7 | 3.28 | 2.88 | 2.45 | 70.6 | 67.3 | 64.0 | 60.7 | 59.4 | -0.96 | -1.00 | -1.06 |
| Occupied Palestinian Territory | 72.0 | 74.1 | 76.6 | 79.4 | 0.29 | 0.33 | 0.36 | ... | ... | ... | ... | ... | ... | ... | ... |
| Oman | 71.6 | 73.0 | 75.7 | 78.7 | 0.19 | 0.36 | 0.39 | ... | ... | ... | ... | ... | ... | ... | ... |
| Pakistan | 33.1 | 35.9 | 39.9 | 45.6 | 0.81 | 1.06 | 1.34 | 51.0 | 49.8 | 48.7 | 47.5 | 47.0 | -0.46 | -0.47 | -0.49 |
| Philippines | 48.0 | 48.9 | 52.6 | 58.3 | 0.19 | 0.73 | 1.03 | 54.3 | 50.8 | 47.2 | 43.7 | 42.3 | -1.35 | -1.45 | -1.56 |
| Qatar | 94.9 | 95.8 | 96.5 | 96.9 | 0.09 | 0.07 | 0.04 | ... | ... | ... | ... | ... | ... | ... | ... |
| Republic of Korea | 79.6 | 83.0 | 85.6 | 87.7 | 0.42 | 0.31 | 0.24 | ... | ... | ... | ... | ... | ... | ... | ... |
| Saudi Arabia | 79.8 | 82.1 | 84.2 | 86.2 | 0.28 | 0.25 | 0.23 | ... | ... | ... | 18.0 | ... | ... | ... | ... |
| Singapore | 100.0 | 100.0 | 100.0 | 100.0 | 0.00 | 0.00 | 0.00 | ... | ... | ... | ... | ... | ... | ... | ... |
| Sri Lanka | 15.8 | 14.3 | 15.5 | 19.6 | -1.00 | 0.81 | 2.35 | ... | ... | ... | ... | ... | ... | ... | ... |
| Syrian Arab Republic | 51.9 | 55.7 | 60.2 | 65.4 | 0.71 | 0.78 | 0.83 | ... | ... | ... | 10.5 | ... | ... | ... | ... |
| Tajikistan | 26.5 | 26.3 | 28.0 | 32.5 | -0.08 | 0.63 | 1.49 | ... | ... | ... | ... | ... | ... | ... | ... |
| Thailand | 31.1 | 34.0 | 38.9 | 45.8 | 0.89 | 1.35 | 1.63 | ... | ... | ... | 26.0 | ... | ... | ... | ... |
| Timor-Leste | 24.3 | 28.1 | 33.2 | 39.9 | 1.45 | 1.67 | 1.84 | ... | ... | ... | ... | ... | ... | ... | ... |
| Turkey | 64.7 | 69.6 | 74.0 | 77.7 | 0.73 | 0.61 | 0.49 | 23.4 | 20.7 | 17.9 | 15.5 | 14.1 | -2.49 | -2.84 | -2.91 |
| Turkmenistan | 45.8 | 49.5 | 54.6 | 60.4 | 0.78 | 0.98 | 1.01 | ... | ... | ... | ... | ... | ... | ... | ... |
| United Arab Emirates | 80.3 | 84.1 | 86.8 | 88.8 | 0.46 | 0.32 | 0.23 | ... | ... | ... | ... | ... | ... | ... | ... |
| Uzbekistan | 37.4 | 36.2 | 37.8 | 42.7 | -0.33 | 0.43 | 1.22 | ... | ... | ... | ... | ... | ... | ... | ... |
| Viet Nam | 24.5 | 30.4 | 37.0 | 44.2 | 2.16 | 1.96 | 1.78 | 60.5 | 54.6 | 48.8 | 41.3 | 38.3 | -2.04 | -2.27 | -3.33 |
| Yemen | 26.3 | 31.8 | 38.2 | 45.3 | 1.90 | 1.83 | 1.70 | ... | ... | ... | 67.2 | ... | ... | ... | ... |
| **EUROPE** | | | | | | | | | | | | | | | |
| Albania | 41.7 | 51.9 | 60.7 | 67.4 | 2.19 | 1.57 | 1.05 | ... | ... | ... | ... | ... | ... | ... | ... |
| Andorra | 92.4 | 88.0 | 84.9 | 85.1 | -0.49 | -0.36 | 0.02 | ... | ... | ... | ... | ... | ... | ... | ... |
| Austria | 65.8 | 67.6 | 70.3 | 73.8 | 0.27 | 0.39 | 0.49 | ... | ... | ... | ... | ... | ... | ... | ... |
| Belarus | 69.9 | 74.7 | 79.2 | 82.6 | 0.66 | 0.58 | 0.42 | ... | ... | ... | ... | ... | ... | ... | ... |
| Belgium | 97.1 | 97.4 | 97.7 | 97.9 | 0.03 | 0.03 | 0.02 | ... | ... | ... | ... | ... | ... | ... | ... |
| Bosnia and Herzegovina | 43.2 | 48.6 | 55.2 | 61.7 | 1.18 | 1.27 | 1.11 | ... | ... | ... | ... | ... | ... | ... | ... |
| Bulgaria | 68.9 | 71.5 | 74.3 | 77.5 | 0.37 | 0.38 | 0.42 | ... | ... | ... | ... | ... | ... | ... | ... |
| Channel Islands[6] | 30.5 | 31.4 | 34.2 | 39.1 | 0.29 | 0.85 | 1.34 | ... | ... | ... | ... | ... | ... | ... | ... |
| Croatia | 55.6 | 57.7 | 61.5 | 66.5 | 0.37 | 0.64 | 0.78 | ... | ... | ... | ... | ... | ... | ... | ... |
| Czech Republic | 74.0 | 73.5 | 75.0 | 78.0 | -0.07 | 0.20 | 0.39 | ... | ... | ... | ... | ... | ... | ... | ... |
| Denmark | 85.1 | 86.9 | 88.6 | 90.1 | 0.21 | 0.19 | 0.17 | ... | ... | ... | ... | ... | ... | ... | ... |
| Estonia | 69.4 | 69.5 | 70.7 | 73.4 | 0.01 | 0.17 | 0.37 | ... | ... | ... | ... | ... | ... | ... | ... |
| Faeroe Islands | 36.3 | 40.3 | 42.2 | 46.6 | 1.05 | 0.46 | 0.99 | ... | ... | ... | ... | ... | ... | ... | ... |
| Finland[7] | 82.2 | 85.1 | 87.4 | 89.2 | 0.35 | 0.27 | 0.20 | ... | ... | ... | ... | ... | ... | ... | ... |
| France | 76.9 | 85.3 | 89.7 | 91.8 | 1.04 | 0.50 | 0.23 | ... | ... | ... | ... | ... | ... | ... | ... |
| Germany | 73.1 | 73.8 | 75.6 | 78.3 | 0.10 | 0.24 | 0.35 | ... | ... | ... | ... | ... | ... | ... | ... |
| Gibraltar | 100.0 | 100.0 | 100.0 | 100.0 | 0.00 | 0.00 | 0.00 | ... | ... | ... | ... | ... | ... | ... | ... |
| Greece | 59.7 | 61.4 | 64.8 | 69.3 | 0.28 | 0.54 | 0.67 | ... | ... | ... | ... | ... | ... | ... | ... |
| Holy See[8] | 100.0 | 100.0 | 100.0 | 100.0 | 0.00 | 0.00 | 0.00 | ... | ... | ... | ... | ... | ... | ... | ... |
| Hungary | 64.6 | 68.1 | 71.8 | 75.5 | 0.53 | 0.53 | 0.50 | ... | ... | ... | ... | ... | ... | ... | ... |
| Iceland | 92.4 | 93.4 | 94.3 | 95.0 | 0.11 | 0.10 | 0.07 | ... | ... | ... | ... | ... | ... | ... | ... |
| Ireland | 59.1 | 61.9 | 65.5 | 69.8 | 0.46 | 0.57 | 0.64 | ... | ... | ... | ... | ... | ... | ... | ... |
| Isle of Man | 51.8 | 50.6 | 51.2 | 53.9 | -0.23 | 0.12 | 0.51 | ... | ... | ... | ... | ... | ... | ... | ... |
| Italy | 67.2 | 68.4 | 70.9 | 74.6 | 0.18 | 0.36 | 0.51 | ... | ... | ... | ... | ... | ... | ... | ... |
| Latvia | 68.1 | 67.7 | 68.4 | 70.9 | -0.06 | 0.10 | 0.36 | ... | ... | ... | ... | ... | ... | ... | ... |
| Liechtenstein | 15.1 | 14.3 | 15.0 | 18.0 | -0.54 | 0.48 | 1.82 | ... | ... | ... | ... | ... | ... | ... | ... |
| Lithuania | 67.0 | 67.0 | 68.5 | 71.5 | 0.00 | 0.22 | 0.43 | ... | ... | ... | ... | ... | ... | ... | ... |
| Luxembourg | 83.8 | 85.2 | 87.4 | 89.1 | 0.17 | 0.25 | 0.19 | ... | ... | ... | ... | ... | ... | ... | ... |
| Malta | 92.4 | 94.7 | 96.0 | 96.6 | 0.25 | 0.14 | 0.06 | ... | ... | ... | ... | ... | ... | ... | ... |
| Moldova | 44.6 | 47.0 | 54.2 | 60.9 | 0.52 | 1.43 | 1.17 | ... | ... | ... | ... | ... | ... | ... | ... |
| Monaco | 100.0 | 100.0 | 100.0 | 100.0 | 0.00 | 0.00 | 0.00 | ... | ... | ... | ... | ... | ... | ... | ... |
| Montenegro | 58.5 | 61.5 | 62.4 | 65.7 | 0.50 | 0.15 | 0.52 | ... | ... | ... | ... | ... | ... | ... | ... |
| Netherlands | 76.8 | 82.9 | 86.5 | 88.6 | 0.76 | 0.43 | 0.24 | ... | ... | ... | ... | ... | ... | ... | ... |
| Norway[9] | 76.1 | 79.4 | 82.6 | 85.2 | 0.42 | 0.40 | 0.31 | ... | ... | ... | ... | ... | ... | ... | ... |
| Poland | 61.7 | 61.0 | 61.7 | 64.9 | -0.11 | 0.11 | 0.51 | ... | ... | ... | ... | ... | ... | ... | ... |
| Portugal | 54.4 | 60.7 | 66.4 | 71.4 | 1.10 | 0.90 | 0.73 | ... | ... | ... | ... | ... | ... | ... | ... |
| Romania | 53.0 | 57.5 | 63.0 | 68.2 | 0.81 | 0.91 | 0.79 | ... | ... | ... | ... | ... | ... | ... | ... |
| Russian Federation | 73.4 | 73.2 | 74.5 | 76.9 | -0.03 | 0.18 | 0.32 | ... | ... | ... | ... | ... | ... | ... | ... |
| San Marino | 93.4 | 94.1 | 94.4 | 94.9 | 0.07 | 0.03 | 0.05 | ... | ... | ... | ... | ... | ... | ... | ... |
| Serbia | 53.0 | 56.1 | 60.0 | 64.8 | 0.57 | 0.67 | 0.77 | ... | ... | ... | ... | ... | ... | ... | ... |
| Slovakia | 56.2 | 55.0 | 55.7 | 59.2 | -0.22 | 0.13 | 0.61 | ... | ... | ... | ... | ... | ... | ... | ... |
| Slovenia | 50.8 | 49.5 | 50.4 | 54.5 | -0.26 | 0.18 | 0.78 | ... | ... | ... | ... | ... | ... | ... | ... |

# TABLE B.3

*continued*

| | Level of urbanization | | | | | | | Urban slum dwellers | | | | | | | |
|---|---|---|---|---|---|---|---|---|---|---|---|---|---|---|---|
| | Estimates and projections (%) | | | | Rate of change (%) | | | Estimate ('000) | | | | | Rate of change (%) | | |
| | 2000 | 2010 | 2020 | 2030 | 2000–2010 | 2010–2020 | 2020–2030 | 1990 | 1995 | 2000 | 2005 | 2007 | 1990–1995 | 1995–2000 | 2000–2005 |
| Spain | 76.3 | 77.4 | 79.4 | 81.9 | 0.14 | 0.26 | 0.31 | ... | ... | ... | ... | ... | ... | ... | ... |
| Sweden | 84.0 | 84.7 | 85.8 | 87.3 | 0.08 | 0.13 | 0.17 | ... | ... | ... | ... | ... | ... | ... | ... |
| Switzerland | 73.3 | 73.6 | 75.2 | 77.8 | 0.04 | 0.22 | 0.34 | ... | ... | ... | ... | ... | ... | ... | ... |
| TFYR Macedonia[10] | 59.4 | 59.3 | 61.6 | 66.0 | -0.02 | 0.38 | 0.69 | ... | ... | ... | ... | ... | ... | ... | ... |
| Ukraine | 67.1 | 68.8 | 71.9 | 75.3 | 0.25 | 0.44 | 0.46 | ... | ... | ... | ... | ... | ... | ... | ... |
| United Kingdom | 78.7 | 79.6 | 81.4 | 83.7 | 0.11 | 0.22 | ^28 | ... | ... | ... | ... | ... | ... | ... | ... |
| **LATIN AMERICA AND THE CARIBBEAN** | | | | | | | | | | | | | | | |
| Anguilla | 100.0 | 100.0 | 100.0 | 100.0 | 0.00 | 0.00 | 0.00 | ... | ... | ... | ... | ... | ... | ... | ... |
| Antigua and Barbuda | 32.1 | 30.3 | 32.5 | 38.4 | -0.58 | 0.70 | 1.67 | ... | ... | ... | ... | ... | ... | ... | ... |
| Argentina | 90.1 | 92.4 | 93.8 | 94.6 | 0.25 | 0.15 | 0.08 | 30.5 | 31.7 | 32.9 | 26.2 | 23.5 | 0.77 | 0.74 | -4.55 |
| Aruba | 46.7 | 46.9 | 48.8 | 52.5 | 0.04 | 0.40 | 0.73 | ... | ... | ... | ... | ... | ... | ... | ... |
| Bahamas | 82.0 | 84.1 | 86.1 | 87.9 | 0.25 | 0.24 | 0.21 | ... | ... | ... | ... | ... | ... | ... | ... |
| Barbados | 38.3 | 44.5 | 51.1 | 57.9 | 1.50 | 1.38 | 1.25 | ... | ... | ... | ... | ... | ... | ... | ... |
| Belize | 47.8 | 52.2 | 56.9 | 62.3 | 0.88 | 0.86 | 0.91 | ... | ... | ... | 47.3 | ... | ... | ... | ... |
| Bolivia | 61.8 | 66.5 | 71.0 | 75.2 | 0.73 | 0.65 | 0.57 | 62.2 | 58.2 | 54.3 | 50.4 | 48.8 | -1.30 | -1.40 | -1.50 |
| Brazil | 81.2 | 86.5 | 89.5 | 91.1 | 0.63 | 0.34 | 0.18 | 36.7 | 34.1 | 31.5 | 29.0 | 28.0 | -1.45 | -1.56 | -1.69 |
| British Virgin Islands | 39.4 | 41.0 | 45.2 | 51.6 | 0.40 | 0.98 | 1.32 | ... | ... | ... | ... | ... | ... | ... | ... |
| Cayman Islands | 100.0 | 100.0 | 100.0 | 100.0 | 0.00 | 0.00 | 0.00 | ... | ... | ... | ... | ... | ... | ... | ... |
| Chile | 85.9 | 89.0 | 91.0 | 92.3 | 0.35 | 0.22 | 0.14 | ... | ... | ... | 9.0 | ... | ... | ... | ... |
| Colombia | 72.1 | 75.1 | 78.0 | 81.0 | 0.41 | 0.38 | 0.38 | 31.2 | 26.8 | 22.3 | 17.9 | 16.1 | -3.07 | -3.62 | -4.43 |
| Costa Rica | 59.0 | 64.4 | 69.4 | 73.9 | 0.88 | 0.75 | 0.63 | ... | ... | ... | 10.9 | ... | ... | ... | ... |
| Cuba | 75.6 | 75.2 | 75.6 | 77.6 | -0.05 | 0.05 | 0.26 | ... | ... | ... | ... | ... | ... | ... | ... |
| Dominica | 67.2 | 67.2 | 69.4 | 73.1 | 0.00 | 0.32 | 0.52 | ... | ... | ... | ... | ... | ... | ... | ... |
| Dominican Republic | 61.7 | 69.2 | 74.8 | 78.8 | 1.15 | 0.78 | 0.52 | 27.9 | 24.4 | 21.0 | 17.6 | 16.2 | -2.63 | -3.03 | -3.57 |
| Ecuador | 60.3 | 66.9 | 72.5 | 76.8 | 1.04 | 0.80 | 0.58 | ... | ... | ... | 21.5 | ... | ... | ... | ... |
| El Salvador | 58.9 | 64.3 | 69.3 | 73.7 | 0.88 | 0.75 | 0.62 | ... | ... | ... | 28.9 | ... | ... | ... | ... |
| Falkland Islands (Malvinas) | 67.6 | 73.6 | 78.2 | 81.6 | 0.85 | 0.61 | 0.43 | ... | ... | ... | ... | ... | ... | ... | ... |
| French Guiana | 75.1 | 76.4 | 78.6 | 81.4 | 0.17 | 0.28 | 0.35 | ... | ... | ... | 10.5 | ... | ... | ... | ... |
| Grenada | 35.9 | 39.3 | 44.5 | 51.2 | 0.90 | 1.24 | 1.40 | ... | ... | ... | 6.0 | ... | ... | ... | ... |
| Guadeloupe | 98.4 | 98.4 | 98.5 | 98.6 | 0.00 | 0.01 | 0.01 | ... | ... | ... | 5.4 | ... | ... | ... | ... |
| Guatemala | 45.1 | 49.5 | 54.7 | 60.6 | 0.93 | 1.00 | 1.02 | 58.6 | 53.3 | 48.1 | 42.9 | 40.8 | -1.87 | -2.06 | -2.30 |
| Guyana | 28.7 | 28.6 | 31.3 | 37.2 | -0.03 | 0.90 | 1.73 | ... | ... | ... | 33.7 | ... | ... | ... | ... |
| Haiti | 35.6 | 52.1 | 64.4 | 71.6 | 3.81 | 2.12 | 1.06 | 93.4 | 93.4 | 93.4 | 70.1 | 70.1 | 0.00 | 0.00 | -5.74 |
| Honduras | 45.5 | 51.6 | 57.6 | 63.4 | 1.26 | 1.10 | 0.96 | ... | ... | ... | 34.9 | ... | ... | ... | ... |
| Jamaica | 51.8 | 52.0 | 53.7 | 57.8 | 0.04 | 0.32 | 0.74 | ... | ... | ... | 60.5 | ... | ... | ... | ... |
| Martinique | 89.7 | 89.0 | 89.1 | 90.0 | -0.08 | 0.01 | 0.10 | ... | ... | ... | ... | ... | ... | ... | ... |
| Mexico | 74.7 | 77.8 | 80.7 | 83.3 | 0.41 | 0.37 | 0.32 | 23.1 | 21.5 | 19.9 | 14.4 | 14.4 | -1.44 | -1.55 | -6.47 |
| Montserrat | 11.0 | 14.3 | 16.9 | 21.6 | 2.62 | 1.67 | 2.45 | ... | ... | ... | ... | ... | ... | ... | ... |
| Netherlands Antilles | 90.2 | 93.2 | 94.7 | 95.5 | 0.33 | 0.16 | 0.08 | ... | ... | ... | ... | ... | ... | ... | ... |
| Nicaragua | 54.7 | 57.3 | 61.0 | 65.8 | 0.46 | 0.63 | 0.76 | 89.1 | 74.5 | 60.0 | 45.5 | 45.5 | -3.56 | -4.34 | -5.55 |
| Panama | 65.8 | 74.8 | 80.3 | 83.6 | 1.28 | 0.71 | 0.40 | ... | ... | ... | 23.0 | ... | ... | ... | ... |
| Paraguay | 55.3 | 61.5 | 67.1 | 71.9 | 1.06 | 0.87 | 0.69 | ... | ... | ... | 17.6 | ... | ... | ... | ... |
| Peru | 73.0 | 76.9 | 80.3 | 83.0 | 0.52 | 0.43 | 0.33 | 66.4 | 56.3 | 46.2 | 36.1 | 36.1 | -3.30 | -3.96 | -4.94 |
| Puerto Rico | 94.6 | 98.8 | 99.5 | 99.6 | 0.43 | 0.07 | 0.01 | ... | ... | ... | ... | ... | ... | ... | ... |
| Saint Kitts and Nevis | 32.8 | 32.4 | 35.4 | 41.6 | -0.12 | 0.89 | 1.61 | ... | ... | ... | ... | ... | ... | ... | ... |
| Saint Lucia | 28.0 | 28.0 | 30.6 | 36.1 | 0.00 | 0.89 | 1.65 | ... | ... | ... | 11.9 | ... | ... | ... | ... |
| Saint Vincent and the Grenadines | 45.2 | 49.3 | 54.6 | 60.7 | 0.87 | 1.02 | 1.06 | ... | ... | ... | ... | ... | ... | ... | ... |
| Suriname | 64.9 | 69.4 | 73.5 | 77.3 | 0.67 | 0.57 | 0.50 | ... | ... | ... | 3.9 | ... | ... | ... | ... |
| Trinidad and Tobago | 10.8 | 13.9 | 18.1 | 23.7 | 2.52 | 2.64 | 2.70 | ... | ... | ... | 24.7 | ... | ... | ... | ... |
| Turks and Caicos Islands | 84.6 | 93.3 | 96.5 | 97.4 | 0.98 | 0.34 | 0.09 | ... | ... | ... | ... | ... | ... | ... | ... |
| United States Virgin Islands | 92.6 | 95.3 | 96.5 | 97.0 | 0.29 | 0.13 | 0.05 | ... | ... | ... | ... | ... | ... | ... | ... |
| Uruguay | 91.3 | 92.5 | 93.4 | 94.3 | 0.13 | 0.10 | 0.10 | ... | ... | ... | ... | ... | ... | ... | ... |
| Venezuela (Bolivarian Republic of) | 89.9 | 93.4 | 95.0 | 95.8 | 0.38 | 0.17 | 0.08 | ... | ... | ... | 32.0 | ... | ... | ... | ... |
| **NORTHERN AMERICA** | | | | | | | | | | | | | | | |
| Bermuda | 100.0 | 100.0 | 100.0 | 100.0 | 0.00 | 0.00 | 0.00 | ... | ... | ... | ... | ... | ... | ... | ... |
| Canada | 79.5 | 80.6 | 82.0 | 84.0 | 0.14 | 0.17 | 0.24 | ... | ... | ... | ... | ... | ... | ... | ... |
| Greenland | 81.6 | 84.2 | 86.5 | 88.4 | 0.31 | 0.27 | 0.22 | ... | ... | ... | ... | ... | ... | ... | ... |
| Saint-Pierre-et-Miquelon | 89.1 | 90.6 | 91.8 | 92.8 | 0.17 | 0.13 | 0.11 | ... | ... | ... | ... | ... | ... | ... | ... |
| United States of America | 79.1 | 82.3 | 84.9 | 87.0 | 0.40 | 0.31 | 0.24 | ... | ... | ... | ... | ... | ... | ... | ... |
| **OCEANIA** | | | | | | | | | | | | | | | |
| American Samoa | 88.8 | 93.0 | 94.8 | 95.6 | 0.46 | 0.19 | 0.08 | ... | ... | ... | ... | ... | ... | ... | ... |
| Australia[11] | 87.2 | 89.1 | 90.6 | 91.9 | 0.22 | 0.17 | 0.14 | ... | ... | ... | ... | ... | ... | ... | ... |
| Cook Islands | 65.2 | 75.3 | 81.4 | 84.9 | 1.44 | 0.78 | 0.42 | ... | ... | ... | ... | ... | ... | ... | ... |
| Fiji | 47.9 | 51.9 | 56.4 | 61.7 | 0.80 | 0.83 | 0.90 | ... | ... | ... | ... | ... | ... | ... | ... |
| French Polynesia | 52.4 | 51.4 | 52.7 | 56.6 | -0.19 | 0.25 | 0.71 | ... | ... | ... | ... | ... | ... | ... | ... |
| Guam | 93.1 | 93.2 | 93.5 | 94.2 | 0.01 | 0.03 | 0.07 | ... | ... | ... | ... | ... | ... | ... | ... |
| Kiribati | 43.0 | 43.9 | 46.5 | 51.7 | 0.21 | 0.58 | 1.06 | ... | ... | ... | ... | ... | ... | ... | ... |
| Marshall Islands | 68.4 | 71.8 | 75.3 | 78.8 | 0.49 | 0.48 | 0.45 | ... | ... | ... | ... | ... | ... | ... | ... |
| Micronesia (Federated States of) | 22.3 | 22.7 | 25.1 | 30.3 | 0.18 | 1.01 | 1.88 | ... | ... | ... | ... | ... | ... | ... | ... |
| Nauru | 100.0 | 100.0 | 100.0 | 100.0 | 0.00 | 0.00 | 0.00 | ... | ... | ... | ... | ... | ... | ... | ... |
| New Caledonia | 59.2 | 57.4 | 58.5 | 62.7 | -0.31 | 0.19 | 0.69 | ... | ... | ... | ... | ... | ... | ... | ... |
| New Zealand | 85.7 | 86.2 | 86.9 | 88.1 | 0.06 | 0.08 | 0.14 | ... | ... | ... | ... | ... | ... | ... | ... |
| Niue | 33.1 | 37.5 | 43.0 | 49.4 | 1.25 | 1.37 | 1.39 | ... | ... | ... | ... | ... | ... | ... | ... |
| Northern Mariana Islands | 90.2 | 91.3 | 92.4 | 93.3 | 0.12 | 0.12 | 0.10 | ... | ... | ... | ... | ... | ... | ... | ... |
| Palau | 70.0 | 83.4 | 89.6 | 92.0 | 1.75 | 0.72 | 0.26 | ... | ... | ... | ... | ... | ... | ... | ... |
| Papua New Guinea | 13.2 | 12.5 | 14.1 | 18.2 | -0.54 | 1.20 | 2.55 | ... | ... | ... | ... | ... | ... | ... | ... |

## TABLE B.3

*continued*

| | Level of urbanization | | | | | | | Urban slum dwellers | | | | | | | |
|---|---|---|---|---|---|---|---|---|---|---|---|---|---|---|---|
| | Estimates and projections (%) | | | | Rate of change (%) | | | Estimate ('000) | | | | | Rate of change (%) | | |
| | 2000 | 2010 | 2020 | 2030 | 2000–2010 | 2010–2020 | 2020–2030 | 1990 | 1995 | 2000 | 2005 | 2007 | 1990–1995 | 1995–2000 | 2000–2005 |
| Pitcairn | 0.0 | 0.0 | 0.0 | 0.0 | 0.00 | 0.00 | 0.00 | ... | ... | ... | ... | ... | ... | ... | ... |
| Samoa | 22.0 | 20.2 | 20.5 | 24.0 | -0.85 | 0.15 | 1.58 | ... | ... | ... | ... | ... | ... | ... | ... |
| Solomon Islands | 15.7 | 18.6 | 23.0 | 29.2 | 1.70 | 2.12 | 2.39 | ... | ... | ... | ... | ... | ... | ... | ... |
| Tokelau | 0.0 | 0.0 | 0.0 | 0.0 | 0.00 | 0.00 | 0.00 | ... | ... | ... | ... | ... | ... | ... | ... |
| Tonga | 23.0 | 23.4 | 25.6 | 30.4 | 0.17 | 0.90 | 1.72 | ... | ... | ... | ... | ... | ... | ... | ... |
| Tuvalu | 46.0 | 50.4 | 55.6 | 61.5 | 0.91 | 0.98 | 1.01 | ... | ... | ... | ... | ... | ... | ... | ... |
| Vanuatu | 21.7 | 25.6 | 31.0 | 38.0 | 1.65 | 1.91 | 2.04 | ... | ... | ... | ... | ... | ... | ... | ... |
| Wallis and Futuna Islands | 0.0 | 0.0 | 0.0 | 0.0 | 0.00 | 0.00 | 0.00 | ... | ... | ... | ... | ... | ... | ... | ... |

*Sources:* United Nations Department of Economic and Social Affairs, Population Division (2010) *World Urbanization Prospects: The 2009 Revision*, United Nations, New York; United Nations Human Settlements Programme (UN-Habitat), Urban Info 2010.

*Notes:*
(1) Including Agalega, Rodrigues, and Saint Brandon.
(2) Including Ascension, and Tristan da Cunha.
(3) For statistical purposes, the data for China do not include Hong Kong and Macao, Special Administrative Regions (SAR) of China.
(4) As of 1 July 1997, Hong Kong became a Special Administrative Region (SAR) of China.
(5) As of 20 December 1999, Macao became a Special Administrative Region (SAR) of China.
(6) Refers to Guernsey and Jersey.
(7) Including Åland Islands.
(8) Refers to the Vatican City State.
(9) Including Svalbard and Jan Mayen Islands.
(10) The former Yugoslav Republic of Macedonia.
(11) Including Christmas Island, Cocos (Keeling) Islands, and Norfolk Island.

# TABLE B.4

## Access to Drinking Water and Sanitation

| | Improved drinking water coverage | | | | | | Household connection to improved drinking water | | | | | | Improved sanitation coverage | | | | | |
|---|---|---|---|---|---|---|---|---|---|---|---|---|---|---|---|---|---|---|
| | Total (%) | | Urban (%) | | Rural (%) | | Total (%) | | Urban (%) | | Rural (%) | | Total (%) | | Urban (%) | | Rural (%) | |
| | 1990 | 2008 | 1990 | 2008 | 1990 | 2008 | 1990 | 2008 | 1990 | 2008 | 1990 | 2008 | 1990 | 2008 | 1990 | 2008 | 1990 | 2008 |
| **AFRICA** | | | | | | | | | | | | | | | | | | |
| Algeria | 94 | 83 | 100 | 85 | 88 | 79 | 68 | 72 | 87 | 80 | 48 | 56 | 88 | 95 | 99 | 98 | 77 | 88 |
| Angola | 36 | 50 | 30 | 60 | 40 | 38 | 0 | 20 | 1 | 34 | 0 | 1 | 25 | 57 | 58 | 86 | 6 | 18 |
| Benin | 56 | 75 | 72 | 84 | 47 | 69 | 7 | 12 | 19 | 26 | 0 | 2 | 5 | 12 | 14 | 24 | 1 | 4 |
| Botswana | 93 | 95 | 100 | 99 | 88 | 90 | 24 | 62 | 39 | 80 | 13 | 35 | 36 | 60 | 58 | 74 | 20 | 39 |
| Burkina Faso | 41 | 76 | 73 | 95 | 36 | 72 | 2 | 4 | 12 | 21 | 0 | 0 | 6 | 11 | 28 | 33 | 2 | 6 |
| Burundi | 70 | 72 | 97 | 83 | 68 | 71 | 3 | 6 | 32 | 47 | 1 | 1 | 44 | 46 | 41 | 49 | 44 | 46 |
| Cameroon | 50 | 74 | 77 | 92 | 31 | 51 | 11 | 15 | 25 | 25 | 2 | 3 | 47 | 47 | 65 | 56 | 35 | 35 |
| Cape Verde | ... | 84 | ... | 85 | ... | 82 | ... | 38 | ... | 46 | ... | 27 | ... | 54 | ... | 65 | ... | 38 |
| Central African Republic | 58 | 67 | 78 | 92 | 47 | 51 | 3 | 2 | 8 | 6 | 0 | 0 | 11 | 34 | 21 | 43 | 5 | 28 |
| Chad | 38 | 50 | 48 | 67 | 36 | 44 | 2 | 5 | 10 | 17 | 0 | 1 | 6 | 9 | 20 | 23 | 2 | 4 |
| Comoros[1] | 87 | 95 | 98 | 91 | 83 | 97 | 16 | 30 | 31 | 53 | 10 | 21 | 17 | 36 | 34 | 50 | 11 | 30 |
| Congo | ... | 71 | ... | 95 | ... | 34 | ... | 28 | ... | 43 | ... | 3 | ... | 30 | ... | 31 | ... | 29 |
| Côte d'Ivoire | 76 | 80 | 90 | 93 | 67 | 68 | 22 | 40 | 49 | 67 | 5 | 14 | 20 | 23 | 38 | 36 | 8 | 11 |
| Democratic Republic of the Congo | 45 | 46 | 90 | 80 | 27 | 28 | 14 | 9 | 51 | 23 | 0 | 2 | 9 | 23 | 23 | 23 | 4 | 23 |
| Djibouti | 77 | 92 | 80 | 98 | 69 | 52 | 57 | 72 | 69 | 82 | 19 | 3 | 66 | 56 | 73 | 63 | 45 | 10 |
| Egypt | 90 | 99 | 96 | 100 | 86 | 98 | 61 | 92 | 90 | 99 | 39 | 87 | 72 | 94 | 91 | 97 | 57 | 92 |
| Equatorial Guinea | ... | ... | ... | ... | ... | ... | 4 | ... | 12 | ... | 0 | 0 | ... | ... | ... | ... | ... | ... |
| Eritrea | 43 | 61 | 62 | 74 | 39 | 57 | 6 | 9 | 40 | 42 | 0 | 0 | ... | 14 | 58 | 52 | 0 | 4 |
| Ethiopia | 17 | 38 | 77 | 98 | 8 | 26 | 1 | 7 | 10 | 40 | 0 | 0 | 4 | 12 | 21 | 29 | 1 | 8 |
| Gabon | ... | 87 | ... | 95 | ... | 41 | ... | 43 | ... | 49 | ... | 10 | ... | 33 | ... | 33 | ... | 30 |
| Gambia | 77 | 92 | 85 | 96 | 67 | 86 | 9 | 33 | 24 | 55 | 0 | 5 | ... | 67 | ... | 68 | ... | 65 |
| Ghana | 54 | 82 | 84 | 90 | 37 | 74 | 16 | 17 | 41 | 30 | 2 | 3 | 7 | 13 | 11 | 18 | 4 | 7 |
| Guinea | 52 | 71 | 87 | 89 | 38 | 61 | 6 | 10 | 21 | 26 | 0 | 1 | 9 | 19 | 18 | 34 | 6 | 11 |
| Guinea-Bissau | ... | 61 | ... | 83 | 37 | 51 | 2 | 9 | 6 | 27 | 0 | 1 | ... | 21 | ... | 49 | ... | 9 |
| Kenya | 43 | 59 | 91 | 83 | 32 | 52 | 19 | 19 | 57 | 44 | 10 | 12 | 26 | 31 | 24 | 27 | 27 | 32 |
| Lesotho | 61 | 85 | 88 | 97 | 57 | 81 | 4 | 19 | 19 | 59 | 1 | 5 | 32 | 29 | 29 | 40 | 32 | 25 |
| Liberia | 58 | 68 | 86 | 79 | 34 | 51 | 11 | 2 | 21 | 3 | 3 | 0 | 11 | 17 | 21 | 25 | 3 | 4 |
| Libyan Arab Jamahiriya | 54 | ... | 54 | ... | 55 | ... | ... | ... | ... | ... | ... | ... | 97 | 97 | 97 | 97 | 96 | 96 |
| Madagascar | 31 | 41 | 78 | 71 | 16 | 29 | 6 | 7 | 25 | 14 | 0 | 4 | 8 | 11 | 14 | 15 | 6 | 10 |
| Malawi | 40 | 80 | 90 | 95 | 33 | 77 | 7 | 7 | 45 | 26 | 2 | 2 | 42 | 56 | 50 | 51 | 41 | 57 |
| Mali | 29 | 56 | 54 | 81 | 22 | 44 | 4 | 12 | 17 | 34 | 0 | 1 | 26 | 36 | 36 | 45 | 23 | 32 |
| Mauritania | 30 | 49 | 36 | 52 | 26 | 47 | 6 | 22 | 15 | 34 | 0 | 14 | 16 | 26 | 29 | 50 | 8 | 9 |
| Mauritius | 99 | 99 | 100 | 100 | 99 | 99 | 99 | 99 | 100 | 100 | 99 | 99 | 91 | 91 | 93 | 93 | 90 | 90 |
| Mayotte[2] | ... | ... | ... | ... | ... | ... | ... | ... | ... | ... | ... | ... | ... | ... | ... | ... | ... | ... |
| Morocco | 74 | 81 | 94 | 98 | 55 | 60 | 38 | 58 | 74 | 88 | 5 | 19 | 53 | 69 | 81 | 83 | 27 | 52 |
| Mozambique | 36 | 47 | 73 | 77 | 26 | 29 | 5 | 8 | 22 | 20 | 1 | 1 | 11 | 17 | 36 | 38 | 4 | 4 |
| Namibia | 64 | 92 | 99 | 99 | 51 | 88 | 33 | 44 | 82 | 72 | 14 | 27 | 25 | 33 | 66 | 60 | 9 | 17 |
| Niger | 35 | 48 | 57 | 96 | 31 | 39 | 3 | 7 | 21 | 37 | 0 | 1 | 5 | 9 | 19 | 34 | 2 | 4 |
| Nigeria | 47 | 58 | 79 | 75 | 30 | 42 | 14 | 6 | 32 | 11 | 4 | 2 | 37 | 32 | 39 | 36 | 36 | 28 |
| Réunion | ... | ... | ... | ... | ... | ... | ... | ... | ... | ... | ... | ... | ... | ... | ... | ... | ... | ... |
| Rwanda | 68 | 65 | 96 | 77 | 66 | 62 | 2 | 4 | 32 | 15 | 0 | 1 | 23 | 54 | 35 | 50 | 22 | 55 |
| Saint Helena | ... | ... | ... | ... | ... | ... | ... | ... | ... | ... | ... | ... | ... | ... | ... | ... | ... | ... |
| São Tomé and Príncipe | ... | 89 | ... | 89 | ... | 88 | ... | 26 | ... | 32 | ... | 18 | ... | 26 | ... | 30 | ... | 19 |
| Senegal | 61 | 69 | 88 | 92 | 43 | 52 | 19 | 38 | 45 | 74 | 3 | 12 | 38 | 51 | 62 | 69 | 22 | 38 |
| Seychelles | ... | ... | ... | 100 | ... | ... | ... | ... | ... | 100 | ... | ... | ... | ... | ... | 97 | ... | ... |
| Sierra Leone | ... | 49 | ... | 86 | ... | 26 | ... | 6 | ... | 15 | 1 | 1 | ... | 13 | ... | 24 | ... | 6 |
| Somalia | ... | 30 | ... | 67 | ... | 9 | ... | 19 | ... | 51 | 0 | 0 | ... | 23 | ... | 52 | ... | 6 |
| South Africa | 83 | 91 | 98 | 99 | 66 | 78 | 56 | 67 | 85 | 89 | 25 | 32 | 69 | 77 | 80 | 84 | 58 | 65 |
| Sudan | 65 | 57 | 85 | 64 | 58 | 52 | 34 | 28 | 76 | 47 | 19 | 14 | 34 | 34 | 63 | 55 | 23 | 18 |
| Swaziland | ... | 69 | ... | 92 | ... | 61 | ... | 32 | ... | 67 | ... | 21 | ... | 55 | ... | 61 | ... | 53 |
| Togo | 49 | 60 | 79 | 87 | 36 | 41 | 4 | 6 | 14 | 12 | 0 | 1 | 13 | 12 | 25 | 24 | 8 | 3 |
| Tunisia | 81 | 94 | 95 | 99 | 62 | 84 | 61 | 76 | 89 | 94 | 22 | 39 | 74 | 85 | 95 | 96 | 44 | 64 |
| Uganda | 43 | 67 | 78 | 91 | 39 | 64 | 1 | 3 | 9 | 19 | 0 | 1 | 39 | 48 | 35 | 38 | 40 | 49 |
| United Republic of Tanzania | 55 | 54 | 94 | 80 | 46 | 45 | 7 | 8 | 34 | 23 | 1 | 3 | 24 | 24 | 27 | 32 | 23 | 21 |
| Western Sahara | ... | ... | ... | ... | ... | ... | ... | ... | ... | ... | ... | ... | ... | ... | ... | ... | ... | ... |
| Zambia | 49 | 60 | 89 | 87 | 23 | 46 | 20 | 14 | 49 | 37 | 1 | 1 | 46 | 49 | 62 | 59 | 36 | 43 |
| Zimbabwe | 78 | 82 | 99 | 99 | 70 | 72 | 32 | 36 | 94 | 88 | 7 | 5 | 43 | 44 | 58 | 56 | 37 | 37 |
| **ASIA** | | | | | | | | | | | | | | | | | | |
| Afghanistan | ... | 48 | ... | 78 | ... | 39 | ... | 4 | ... | 16 | ... | 0 | ... | 37 | ... | 60 | ... | 30 |
| Armenia | ... | 96 | 99 | 98 | ... | 93 | 84 | 87 | 96 | 97 | 59 | 70 | ... | 90 | 95 | 95 | ... | 80 |
| Azerbaijan | 70 | 80 | 88 | 88 | 49 | 71 | 44 | 50 | 67 | 78 | 17 | 20 | ... | 45 | ... | 51 | ... | 39 |
| Bahrain | ... | ... | 100 | 100 | ... | ... | ... | ... | 100 | 100 | ... | ... | ... | ... | 100 | 100 | ... | ... |
| Bangladesh | 78 | 80 | 88 | 85 | 76 | 78 | 6 | 6 | 28 | 24 | 0 | 0 | 39 | 53 | 59 | 56 | 34 | 52 |
| Bhutan | ... | 92 | ... | 99 | ... | 88 | ... | 57 | ... | 81 | ... | 45 | ... | 65 | ... | 87 | ... | 54 |
| Brunei Darussalam | ... | ... | ... | ... | ... | ... | ... | ... | ... | ... | ... | ... | ... | ... | ... | ... | ... | ... |
| Cambodia | 35 | 61 | 52 | 81 | 33 | 56 | 2 | 16 | 17 | 55 | 0 | 5 | 9 | 29 | 38 | 67 | 5 | 18 |
| China | 67 | 89 | 97 | 98 | 56 | 82 | 54 | 83 | 86 | 96 | 42 | 73 | 41 | 55 | 48 | 58 | 38 | 52 |
| China, Hong Kong SAR | ... | ... | ... | ... | ... | ... | ... | ... | ... | ... | ... | ... | ... | ... | ... | ... | ... | ... |
| China, Macao SAR | ... | ... | ... | ... | ... | ... | ... | ... | ... | ... | ... | ... | ... | ... | ... | ... | ... | ... |
| Cyprus | 100 | 100 | 100 | 100 | 100 | 100 | 100 | 100 | 100 | 100 | 100 | 100 | 100 | 100 | 100 | 100 | 100 | 100 |
| Democratic People's Republic of Korea | 100 | 100 | 100 | 100 | 100 | 100 | ... | ... | ... | ... | ... | ... | ... | ... | ... | ... | ... | ... |
| Georgia | 81 | 98 | 94 | 100 | 66 | 96 | 53 | 73 | 81 | 92 | 19 | 51 | 96 | 95 | 97 | 96 | 95 | 93 |
| India | 72 | 88 | 90 | 96 | 66 | 84 | 19 | 22 | 52 | 48 | 8 | 11 | 18 | 31 | 49 | 54 | 7 | 21 |
| Indonesia | 71 | 80 | 92 | 89 | 62 | 71 | 9 | 23 | 24 | 37 | 2 | 8 | 33 | 52 | 58 | 67 | 22 | 36 |
| Iran (Islamic Republic of) | 91 | ... | 98 | 98 | 84 | ... | 84 | ... | 96 | 96 | 69 | ... | 83 | ... | 86 | ... | 78 | ... |
| Iraq | 81 | 79 | 97 | 91 | 44 | 55 | ... | 76 | ... | 90 | ... | 49 | ... | 73 | ... | 76 | ... | 66 |

# TABLE B.4

## continued

| | Improved drinking water coverage | | | | | | Household connection to improved drinking water | | | | | | Improved sanitation coverage | | | | | |
|---|---|---|---|---|---|---|---|---|---|---|---|---|---|---|---|---|---|---|
| | Total (%) | | Urban (%) | | Rural (%) | | Total (%) | | Urban (%) | | Rural (%) | | Total (%) | | Urban (%) | | Rural (%) | |
| | 1990 | 2008 | 1990 | 2008 | 1990 | 2008 | 1990 | 2008 | 1990 | 2008 | 1990 | 2008 | 1990 | 2008 | 1990 | 2008 | 1990 | 2008 |
| Israel | 100 | 100 | 100 | 100 | 100 | 100 | 100 | 100 | 100 | 100 | 98 | 98 | 100 | 100 | 100 | 100 | 100 | 100 |
| Japan | 100 | 100 | 100 | 100 | 100 | 100 | 93 | 98 | 97 | 99 | 86 | 95 | 100 | 100 | 100 | 100 | 100 | 100 |
| Jordan | 97 | 96 | 99 | 98 | 91 | 91 | 95 | 91 | 98 | 94 | 87 | 79 | ... | 98 | 98 | 98 | ... | 97 |
| Kazakhstan | 96 | 95 | 99 | 99 | 92 | 90 | 63 | 58 | 91 | 82 | 28 | 24 | 96 | 97 | 96 | 97 | 97 | 98 |
| Kuwait | 99 | 99 | 99 | 99 | 99 | 99 | ... | ... | ... | ... | ... | ... | 100 | 100 | 100 | 100 | 100 | 100 |
| Kyrgyzstan | ... | 90 | 98 | 99 | ... | 85 | 44 | 54 | 75 | 89 | 25 | 34 | ... | 93 | 94 | 94 | ... | 93 |
| Lao People's Democratic Republic | ... | 57 | ... | 72 | ... | 51 | ... | 20 | ... | 55 | ... | 4 | ... | 53 | ... | 86 | ... | 38 |
| Lebanon | 100 | 100 | 100 | 100 | 100 | 100 | ... | ... | 100 | 100 | ... | ... | ... | ... | 100 | 100 | ... | ... |
| Malaysia | 88 | 100 | 94 | 100 | 82 | 99 | 72 | 97 | 86 | 99 | 59 | 91 | 84 | 96 | 88 | 96 | 81 | 95 |
| Maldives | 90 | 91 | 100 | 99 | 87 | 86 | 12 | 37 | 47 | 95 | 0 | 2 | 69 | 98 | 100 | 100 | 58 | 96 |
| Mongolia | 58 | 76 | 81 | 97 | 27 | 49 | 30 | 19 | 52 | 32 | 0 | 2 | ... | 50 | ... | 64 | ... | 32 |
| Myanmar | 57 | 71 | 87 | 75 | 47 | 69 | 5 | 6 | 19 | 15 | 1 | 2 | ... | 81 | ... | 86 | ... | 79 |
| Nepal | 76 | 88 | 96 | 93 | 74 | 87 | 8 | 17 | 43 | 52 | 5 | 10 | 11 | 31 | 41 | 51 | 8 | 27 |
| Occupied Palestinian Territory | ... | 91 | 100 | 91 | ... | 91 | ... | 78 | ... | 84 | ... | 64 | ... | 89 | ... | 91 | ... | 84 |
| Oman | 80 | 88 | 84 | 92 | 72 | 77 | 21 | 54 | 29 | 68 | 6 | 18 | 85 | ... | 97 | 97 | 61 | ... |
| Pakistan | 86 | 90 | 96 | 95 | 81 | 87 | 24 | 33 | 57 | 55 | 9 | 20 | 28 | 45 | 73 | 72 | 8 | 29 |
| Philippines | 84 | 91 | 93 | 93 | 76 | 87 | 24 | 48 | 40 | 60 | 8 | 25 | 58 | 76 | 70 | 80 | 46 | 69 |
| Qatar | 100 | 100 | 100 | 100 | 100 | 100 | ... | ... | ... | ... | ... | ... | 100 | 100 | 100 | 100 | 100 | 100 |
| Republic of Korea | ... | 98 | 97 | 100 | ... | 88 | ... | 93 | 96 | 99 | ... | 64 | 100 | 100 | 100 | 100 | 100 | 100 |
| Saudi Arabia | 89 | ... | 97 | 97 | 63 | ... | 88 | ... | 97 | 97 | 60 | ... | ... | ... | 100 | 100 | ... | ... |
| Singapore | 100 | 100 | 100 | 100 | .. | .. | 100 | 100 | 100 | 100 | .. | .. | 99 | 100 | 99 | 100 | .. | .. |
| Sri Lanka | 67 | 90 | 91 | 98 | 62 | 88 | 11 | 28 | 37 | 65 | 6 | 22 | 70 | 91 | 85 | 88 | 67 | 92 |
| Syrian Arab Republic | 85 | 89 | 96 | 94 | 75 | 84 | 72 | 83 | 93 | 93 | 51 | 71 | 83 | 96 | 94 | 96 | 72 | 95 |
| Tajikistan | ... | 70 | ... | 94 | ... | 61 | ... | 40 | ... | 83 | ... | 25 | ... | 94 | 93 | 95 | ... | 94 |
| Thailand | 91 | 98 | 97 | 99 | 89 | 98 | 33 | 54 | 78 | 85 | 14 | 39 | 80 | 96 | 93 | 95 | 74 | 96 |
| Timor-Leste | ... | 69 | ... | 86 | ... | 63 | ... | 16 | ... | 28 | ... | 1 | ... | 50 | ... | 76 | ... | 40 |
| Turkey | 85 | 99 | 94 | 100 | 73 | 96 | 76 | 96 | 91 | 98 | 54 | 92 | 84 | 90 | 96 | 97 | 66 | 75 |
| Turkmenistan | ... | ... | 97 | 97 | ... | ... | ... | ... | ... | ... | ... | ... | 98 | 98 | 99 | 99 | 97 | 97 |
| United Arab Emirates | 100 | 100 | 100 | 100 | 100 | 100 | ... | 78 | ... | 80 | ... | 70 | 97 | 97 | 98 | 98 | 95 | 95 |
| Uzbekistan | 90 | 87 | 97 | 98 | 85 | 81 | 57 | 48 | 86 | 85 | 37 | 26 | 84 | 100 | 95 | 100 | 76 | 100 |
| Viet Nam | 58 | 94 | 88 | 99 | 51 | 92 | 9 | 22 | 45 | 56 | 0 | 9 | 35 | 75 | 61 | 94 | 29 | 67 |
| Yemen | ... | 62 | ... | 72 | ... | 57 | ... | 28 | ... | 54 | ... | 17 | 18 | 52 | 64 | 94 | 6 | 33 |
| **EUROPE** | | | | | | | | | | | | | | | | | | |
| Albania | ... | 97 | 100 | 96 | ... | 98 | ... | 86 | 98 | 91 | ... | 82 | ... | 98 | ... | 98 | ... | 98 |
| Andorra | 100 | 100 | 100 | 100 | 100 | 100 | ... | ... | 100 | 100 | ... | ... | 100 | 100 | 100 | 100 | 100 | 100 |
| Austria | 100 | 100 | 100 | 100 | 100 | 100 | 100 | 100 | 100 | 100 | 100 | 100 | 100 | 100 | 100 | 100 | 100 | 100 |
| Belarus | 100 | 100 | 100 | 100 | 99 | 99 | ... | 89 | ... | 95 | ... | 72 | ... | 93 | ... | 91 | ... | 97 |
| Belgium | 100 | 100 | 100 | 100 | 100 | 100 | 100 | 100 | 100 | 100 | 96 | 100 | 100 | 100 | 100 | 100 | 100 | 100 |
| Bosnia and Herzegovina | ... | 99 | ... | 100 | ... | 98 | ... | 82 | ... | 94 | ... | 71 | ... | 95 | ... | 99 | ... | 92 |
| Bulgaria | 100 | 100 | 100 | 100 | 99 | 100 | 88 | ... | 96 | 96 | 72 | ... | 99 | 100 | 100 | 100 | 98 | 100 |
| Channel Islands | ... | ... | ... | ... | ... | ... | ... | ... | ... | ... | ... | ... | ... | ... | ... | ... | ... | ... |
| Croatia | ... | 99 | ... | 100 | ... | 97 | ... | 88 | ... | 96 | ... | 77 | ... | 99 | ... | 99 | ... | 98 |
| Czech Republic | 100 | 100 | 100 | 100 | 100 | 100 | ... | 95 | 97 | 97 | ... | 91 | 100 | 98 | 100 | 99 | 98 | 97 |
| Denmark | 100 | 100 | 100 | 100 | 100 | 100 | 100 | ... | 100 | ... | 100 | 100 | 100 | 100 | 100 | 100 | 100 | 100 |
| Estonia | 98 | 98 | 99 | 99 | 97 | 97 | 80 | 90 | 92 | 97 | 51 | 75 | ... | 95 | ... | 96 | ... | 94 |
| Faeroe Islands | ... | ... | ... | ... | ... | ... | ... | ... | ... | ... | ... | ... | ... | ... | ... | ... | ... | ... |
| Finland | 100 | 100 | 100 | 100 | 100 | 100 | 92 | ... | 96 | 100 | 85 | ... | 100 | 100 | 100 | 100 | 100 | 100 |
| France | 100 | 100 | 100 | 100 | 100 | 100 | 99 | 100 | 100 | 100 | 95 | 100 | 100 | 100 | 100 | 100 | 100 | 100 |
| Germany | 100 | 100 | 100 | 100 | 100 | 100 | 99 | 99 | 100 | 100 | 97 | 97 | 100 | 100 | 100 | 100 | 100 | 100 |
| Gibraltar | ... | ... | ... | ... | ... | ... | ... | ... | ... | ... | ... | ... | ... | ... | ... | ... | ... | ... |
| Greece | 96 | 100 | 99 | 100 | 92 | 99 | 92 | 100 | 99 | 100 | 82 | 99 | 97 | 98 | 100 | 99 | 92 | 97 |
| Holy See | ... | ... | ... | ... | ... | ... | ... | ... | ... | ... | ... | ... | ... | ... | ... | ... | ... | ... |
| Hungary | 96 | 100 | 98 | 100 | 91 | 100 | 86 | 94 | 94 | 95 | 72 | 93 | 100 | 100 | 100 | 100 | 100 | 100 |
| Iceland | 100 | 100 | 100 | 100 | 100 | 100 | 100 | 100 | 100 | 100 | 100 | 100 | 100 | 100 | 100 | 100 | 100 | 100 |
| Ireland | 100 | 100 | 100 | 100 | 100 | 100 | 100 | 100 | 100 | 100 | 99 | 99 | 99 | 99 | 100 | 100 | 98 | 98 |
| Isle of Man | ... | ... | ... | ... | ... | ... | ... | ... | ... | ... | ... | ... | ... | ... | ... | ... | ... | ... |
| Italy | 100 | 100 | 100 | 100 | 100 | 100 | 99 | 100 | 100 | 100 | 96 | 100 | ... | ... | ... | ... | ... | ... |
| Latvia | 99 | 99 | 100 | 100 | 96 | 96 | ... | 82 | ... | 93 | ... | 59 | ... | 78 | ... | 82 | ... | 71 |
| Liechtenstein | ... | ... | ... | ... | ... | ... | ... | ... | ... | ... | ... | ... | ... | ... | ... | ... | ... | ... |
| Lithuania | ... | ... | ... | ... | ... | ... | 76 | ... | 89 | ... | 49 | ... | ... | ... | ... | ... | ... | ... |
| Luxembourg | 100 | 100 | 100 | 100 | 100 | 100 | 100 | 100 | 100 | 100 | 98 | 98 | 100 | 100 | 100 | 100 | 100 | 100 |
| Malta | 100 | 100 | 100 | 100 | 98 | 100 | 100 | 100 | 100 | 100 | 98 | 100 | 100 | 100 | 100 | 100 | 100 | 100 |
| Moldova | ... | 90 | ... | 96 | ... | 85 | ... | 40 | ... | 79 | ... | 13 | ... | 79 | ... | 85 | ... | 74 |
| Monaco | 100 | 100 | 100 | 100 | .. | .. | 100 | 100 | 100 | 100 | .. | .. | 100 | 100 | 100 | 100 | .. | .. |
| Montenegro | ... | 98 | ... | 100 | ... | 96 | ... | 85 | ... | 98 | ... | 66 | ... | 92 | ... | 96 | ... | 86 |
| Netherlands | 100 | 100 | 100 | 100 | 100 | 100 | 98 | 100 | 100 | 100 | 95 | 100 | 100 | 100 | 100 | 100 | 100 | 100 |
| Norway | 100 | 100 | 100 | 100 | 100 | 100 | 100 | 100 | 100 | 100 | 100 | 100 | 100 | 100 | 100 | 100 | 100 | 100 |
| Poland | 100 | 100 | 100 | 100 | 100 | 100 | 88 | 98 | 97 | 99 | 73 | 96 | ... | 90 | 96 | 96 | ... | 80 |
| Portugal | 96 | 99 | 98 | 99 | 94 | 100 | 87 | 99 | 95 | 99 | 80 | 100 | 92 | 100 | 97 | 100 | 87 | 100 |
| Romania | ... | ... | ... | ... | ... | ... | 47 | 61 | 85 | 91 | 3 | 26 | 71 | 72 | 88 | 88 | 52 | 54 |
| Russian Federation | 93 | 96 | 98 | 98 | 81 | 89 | 76 | 78 | 87 | 92 | 45 | 40 | 87 | 87 | 93 | 93 | 70 | 70 |
| San Marino | ... | ... | ... | ... | ... | ... | ... | ... | ... | ... | ... | ... | ... | ... | ... | ... | ... | ... |
| Serbia | ... | 99 | ... | 99 | ... | 98 | ... | 81 | ... | 97 | ... | 63 | ... | 92 | ... | 96 | ... | 88 |
| Slovakia | ... | 100 | ... | 100 | ... | 100 | 95 | 94 | 100 | 94 | 89 | 94 | 100 | 100 | 100 | 100 | 100 | 99 |
| Slovenia | 100 | 99 | 100 | 100 | 99 | 99 | 100 | 99 | 100 | 100 | 99 | 99 | 100 | 100 | 100 | 100 | 100 | 100 |
| Spain | 100 | 100 | 100 | 100 | 100 | 100 | 99 | 99 | 99 | 99 | 100 | 100 | 100 | 100 | 100 | 100 | 100 | 100 |
| Sweden | 100 | 100 | 100 | 100 | 100 | 100 | 100 | 100 | 100 | 100 | 100 | 100 | 100 | 100 | 100 | 100 | 100 | 100 |

# TABLE B.4

## continued

| | Improved drinking water coverage | | | | | | Household connection to improved drinking water | | | | | | Improved sanitation coverage | | | | | |
|---|---|---|---|---|---|---|---|---|---|---|---|---|---|---|---|---|---|---|
| | Total (%) | | Urban (%) | | Rural (%) | | Total (%) | | Urban (%) | | Rural (%) | | Total (%) | | Urban (%) | | Rural (%) | |
| | 1990 | 2008 | 1990 | 2008 | 1990 | 2008 | 1990 | 2008 | 1990 | 2008 | 1990 | 2008 | 1990 | 2008 | 1990 | 2008 | 1990 | 2008 |
| Switzerland | 100 | 100 | 100 | 100 | 100 | 100 | 100 | 100 | 100 | 100 | 99 | 99 | 100 | 100 | 100 | 100 | 100 | 100 |
| TFYR Macedonia[3] | ... | 100 | ... | 100 | ... | 99 | ... | 92 | ... | 96 | ... | 84 | ... | 89 | ... | 92 | ... | 82 |
| Ukraine | ... | 98 | 99 | 98 | ... | 97 | ... | 67 | 93 | 87 | ... | 25 | 95 | 95 | 97 | 97 | 91 | 90 |
| United Kingdom | 100 | 100 | 100 | 100 | 100 | 100 | 100 | 100 | 100 | 100 | 98 | 98 | 100 | 100 | 100 | 100 | 100 | 100 |
| **LATIN AMERICA AND THE CARIBBEAN** | | | | | | | | | | | | | | | | | | |
| Anguilla | ... | ... | ... | ... | .. | .. | ... | ... | .. | .. | .. | .. | 99 | 99 | 99 | 99 | .. | .. |
| Antigua and Barbuda | ... | ... | 95 | 95 | ... | ... | ... | ... | ... | ... | ... | ... | ... | ... | 98 | 98 | ... | ... |
| Argentina | 94 | 97 | 97 | 98 | 72 | 80 | 69 | 80 | 76 | 83 | 22 | 45 | 90 | 90 | 93 | 91 | 73 | 77 |
| Aruba | 100 | 100 | 100 | 100 | 100 | 100 | 100 | 100 | 100 | 100 | 100 | 100 | ... | ... | ... | ... | ... | ... |
| Bahamas | ... | ... | 98 | 98 | ... | ... | ... | ... | ... | ... | ... | ... | 100 | 100 | 100 | 100 | 100 | 100 |
| Barbados | 100 | 100 | 100 | 100 | 100 | 100 | ... | ... | 98 | 100 | ... | ... | 100 | 100 | 100 | 100 | 100 | 100 |
| Belize | 75 | 99 | 85 | 99 | 63 | 100 | 47 | 74 | 77 | 87 | 20 | 61 | 74 | 90 | 73 | 93 | 75 | 86 |
| Bolivia | 70 | 86 | 92 | 96 | 42 | 67 | 50 | 77 | 78 | 93 | 14 | 47 | 19 | 25 | 29 | 34 | 6 | 9 |
| Brazil | 88 | 97 | 96 | 99 | 65 | 84 | 78 | 91 | 92 | 96 | 35 | 62 | 69 | 80 | 81 | 87 | 35 | 37 |
| British Virgin Islands | 98 | 98 | 98 | 98 | 98 | 98 | 97 | 97 | 97 | 97 | 97 | 97 | 100 | 100 | 100 | 100 | 100 | 100 |
| Cayman Islands | ... | 95 | ... | 95 | .. | .. | 37 | 92 | 37 | 92 | .. | .. | 96 | 96 | 96 | 96 | .. | .. |
| Chile | 90 | 96 | 99 | 99 | 48 | 75 | 84 | 93 | 97 | 99 | 22 | 47 | 84 | 96 | 91 | 98 | 48 | 83 |
| Colombia | 88 | 92 | 98 | 99 | 68 | 73 | 86 | 84 | 98 | 94 | 59 | 56 | 68 | 74 | 80 | 81 | 43 | 55 |
| Costa Rica | 93 | 97 | 99 | 100 | 86 | 91 | 82 | 96 | 92 | 100 | 71 | 89 | 93 | 95 | 94 | 95 | 91 | 96 |
| Cuba | 82 | 94 | 93 | 96 | 53 | 89 | 64 | 75 | 77 | 82 | 30 | 54 | 80 | 91 | 86 | 94 | 64 | 81 |
| Dominica | ... | ... | ... | ... | ... | ... | ... | ... | ... | ... | ... | ... | ... | ... | ... | ... | ... | ... |
| Dominican Republic | 88 | 86 | 98 | 87 | 76 | 84 | 73 | 72 | 94 | 80 | 46 | 54 | 73 | 83 | 83 | 87 | 61 | 74 |
| Ecuador | 72 | 94 | 81 | 97 | 62 | 88 | 47 | 88 | 66 | 96 | 24 | 74 | 69 | 92 | 86 | 96 | 48 | 84 |
| El Salvador | 74 | 87 | 90 | 94 | 58 | 76 | 43 | 65 | 72 | 80 | 14 | 42 | 75 | 87 | 88 | 89 | 62 | 83 |
| Falkland Islands (Malvinas) | ... | ... | ... | ... | ... | ... | ... | ... | ... | ... | ... | ... | ... | ... | ... | ... | ... | ... |
| French Guiana | ... | ... | ... | ... | ... | ... | ... | ... | ... | ... | ... | ... | ... | ... | ... | ... | ... | ... |
| Grenada | ... | ... | 97 | 97 | ... | ... | ... | ... | ... | ... | ... | ... | 97 | 97 | 96 | 96 | 97 | 97 |
| Guadeloupe | ... | ... | 98 | 98 | ... | ... | ... | ... | 98 | 98 | ... | ... | ... | ... | ... | 95 | ... | ... |
| Guatemala | 82 | 94 | 91 | 98 | 75 | 90 | 49 | 81 | 68 | 95 | 35 | 68 | 65 | 81 | 84 | 89 | 51 | 73 |
| Guyana | ... | 94 | ... | 98 | ... | 93 | ... | 67 | ... | 76 | ... | 63 | ... | 81 | ... | 85 | ... | 80 |
| Haiti | 47 | 63 | 62 | 71 | 41 | 55 | 9 | 12 | 27 | 21 | 2 | 4 | 26 | 17 | 44 | 24 | 19 | 10 |
| Honduras | 72 | 86 | 91 | 95 | 59 | 77 | 58 | 83 | 82 | 94 | 42 | 72 | 44 | 71 | 68 | 80 | 28 | 62 |
| Jamaica | 93 | 94 | 98 | 98 | 88 | 89 | 61 | 70 | 89 | 91 | 33 | 47 | 83 | 83 | 82 | 82 | 83 | 84 |
| Martinique | ... | ... | 100 | 100 | ... | ... | ... | ... | 99 | 99 | ... | ... | ... | ... | ... | 95 | ... | ... |
| Mexico | 85 | 94 | 94 | 96 | 64 | 87 | 77 | 87 | 88 | 92 | 50 | 72 | 66 | 85 | 80 | 90 | 30 | 68 |
| Montserrat | 100 | 100 | 100 | 100 | 100 | 100 | 12 | 15 | 98 | 98 | 0 | 0 | 96 | 96 | 96 | 96 | 96 | 96 |
| Netherlands Antilles | ... | ... | ... | ... | ... | ... | ... | ... | ... | ... | ... | ... | ... | ... | ... | ... | ... | ... |
| Nicaragua | 74 | 85 | 92 | 98 | 54 | 68 | 52 | 62 | 83 | 88 | 18 | 27 | 43 | 52 | 59 | 63 | 26 | 37 |
| Panama | 84 | 93 | 99 | 97 | 66 | 83 | 80 | 89 | 97 | 93 | 60 | 79 | 58 | 69 | 73 | 75 | 40 | 51 |
| Paraguay | 52 | 86 | 81 | 99 | 25 | 66 | 29 | 65 | 59 | 85 | 0 | 35 | 37 | 70 | 61 | 90 | 15 | 40 |
| Peru | 75 | 82 | 88 | 90 | 45 | 61 | 55 | 70 | 73 | 84 | 15 | 35 | 54 | 68 | 71 | 81 | 16 | 36 |
| Puerto Rico | ... | ... | ... | ... | ... | ... | ... | ... | ... | ... | ... | ... | ... | ... | ... | ... | ... | ... |
| Saint Kitts and Nevis | 99 | 99 | 99 | 99 | 99 | 99 | ... | ... | ... | ... | ... | ... | 96 | 96 | 96 | 96 | 96 | 96 |
| Saint Lucia | 98 | 98 | 98 | 98 | 98 | 98 | ... | ... | ... | ... | ... | ... | ... | ... | ... | ... | ... | ... |
| Saint Vincent and the Grenadines | ... | ... | ... | ... | ... | ... | ... | ... | ... | ... | ... | ... | ... | ... | ... | ... | ... | ... |
| Suriname | ... | 93 | 99 | 97 | ... | 81 | ... | 70 | 94 | 78 | ... | 45 | ... | 84 | 90 | 90 | ... | 66 |
| Trinidad and Tobago | 88 | 94 | 92 | 98 | 88 | 93 | 69 | 76 | 81 | 88 | 68 | 74 | 93 | 92 | 93 | 92 | 93 | 92 |
| Turks and Caicos Islands | 100 | 100 | 100 | 100 | 100 | 100 | ... | ... | ... | ... | ... | ... | ... | ... | 98 | 98 | ... | ... |
| United States Virgin Islands | ... | ... | ... | ... | ... | ... | ... | ... | ... | ... | ... | ... | ... | ... | ... | ... | ... | ... |
| Uruguay | 96 | 100 | 98 | 100 | 79 | 100 | 89 | 98 | 94 | 98 | 50 | 92 | 94 | 100 | 95 | 100 | 83 | 99 |
| Venezuela (Bolivarian Republic of) | 90 | ... | 93 | ... | 71 | ... | 80 | ... | 87 | ... | 44 | ... | 82 | ... | 89 | ... | 45 | ... |
| **NORTHERN AMERICA** | | | | | | | | | | | | | | | | | | |
| Bermuda | ... | ... | ... | ... | ... | ... | ... | ... | ... | ... | ... | ... | ... | ... | ... | ... | ... | ... |
| Canada | 100 | 100 | 100 | 100 | 99 | 99 | ... | ... | 100 | 100 | ... | ... | 100 | 100 | 100 | 100 | 99 | 99 |
| Greenland | ... | ... | ... | ... | ... | ... | ... | ... | ... | ... | ... | ... | ... | ... | ... | ... | ... | ... |
| Saint-Pierre-et-Miquelon | ... | ... | ... | ... | ... | ... | ... | ... | ... | ... | ... | ... | ... | ... | ... | ... | 96 | 96 |
| United States of America | 99 | 99 | 100 | 100 | 94 | 94 | 84 | 88 | 97 | 97 | 46 | 46 | 100 | 100 | 100 | 100 | 99 | 99 |
| **OCEANIA** | | | | | | | | | | | | | | | | | | |
| American Samoa | ... | ... | ... | ... | ... | ... | ... | ... | ... | ... | ... | ... | ... | ... | ... | ... | ... | ... |
| Australia | 100 | 100 | 100 | 100 | 100 | 100 | ... | ... | ... | ... | ... | ... | 100 | 100 | 100 | 100 | 100 | 100 |
| Cook Islands | 94 | ... | 99 | 98 | 87 | ... | ... | ... | ... | ... | ... | ... | 96 | 100 | 100 | 100 | 91 | 100 |
| Fiji | ... | ... | 92 | ... | ... | ... | ... | ... | ... | ... | ... | ... | ... | ... | 92 | ... | ... | ... |
| French Polynesia | 100 | 100 | 100 | 100 | 100 | 100 | 98 | 98 | 99 | 99 | 96 | 96 | 98 | 98 | 99 | 99 | 97 | 97 |
| Guam | 100 | 100 | 100 | 100 | 100 | 100 | ... | ... | ... | ... | ... | ... | 99 | 99 | 99 | 99 | 98 | 98 |
| Kiribati | 48 | ... | 76 | ... | 33 | ... | 25 | ... | 46 | ... | 13 | ... | 26 | ... | 36 | ... | 21 | ... |
| Marshall Islands | 95 | 94 | 94 | 92 | 97 | 99 | ... | 1 | ... | 1 | ... | 0 | 64 | 73 | 77 | 83 | 41 | 53 |
| Micronesia (Federated States of) | 89 | ... | 93 | 95 | 87 | ... | ... | ... | ... | ... | ... | ... | 29 | ... | 55 | ... | 20 | ... |
| Nauru | ... | 90 | ... | 90 | .. | .. | ... | ... | ... | ... | .. | .. | ... | 50 | ... | 50 | .. | .. |
| New Caledonia | ... | ... | ... | ... | ... | ... | ... | ... | ... | ... | ... | ... | ... | ... | ... | ... | ... | ... |
| New Zealand | 100 | 100 | 100 | 100 | 100 | 100 | 100 | 100 | 100 | 100 | 100 | 100 | ... | ... | ... | ... | 88 | ... |
| Niue | 100 | 100 | 100 | 100 | 100 | 100 | ... | ... | ... | ... | ... | ... | 100 | 100 | 100 | 100 | 100 | 100 |
| Northern Mariana Islands | 98 | 98 | 98 | 98 | 100 | 97 | ... | ... | ... | ... | ... | ... | 84 | 85 | ... | ... | 78 | 96 |
| Palau | 81 | ... | 73 | ... | 98 | ... | ... | ... | ... | ... | ... | ... | 69 | ... | 76 | 96 | 54 | ... |
| Papua New Guinea | 41 | 40 | 89 | 87 | 32 | 33 | 13 | 10 | 61 | 57 | 4 | 3 | 47 | 45 | 78 | 71 | 42 | 41 |
| Pitcairn | ... | ... | ... | ... | ... | ... | ... | ... | ... | ... | ... | ... | ... | ... | ... | ... | ... | ... |
| Samoa | 91 | ... | 99 | ... | 89 | ... | ... | ... | ... | ... | ... | ... | 98 | 100 | 100 | 100 | 98 | 100 |
| Solomon Islands | ... | ... | ... | ... | ... | ... | ... | ... | 76 | ... | ... | ... | ... | ... | 98 | 98 | ... | ... |

# TABLE B.4

## *continued*

| | Improved drinking water coverage | | | | | | Household connection to improved drinking water | | | | | | Improved sanitation coverage | | | | | |
|---|---|---|---|---|---|---|---|---|---|---|---|---|---|---|---|---|---|---|
| | Total (%) | | Urban (%) | | Rural (%) | | Total (%) | | Urban (%) | | Rural (%) | | Total (%) | | Urban (%) | | Rural (%) | |
| | 1990 | 2008 | 1990 | 2008 | 1990 | 2008 | 1990 | 2008 | 1990 | 2008 | 1990 | 2008 | 1990 | 2008 | 1990 | 2008 | 1990 | 2008 |
| Tokelau | 90 | 97 | .. | .. | 90 | 97 | ... | ... | .. | .. | ... | ... | 41 | 93 | .. | .. | 41 | 93 |
| Tonga | ... | 100 | ... | 100 | ... | 100 | ... | ... | ... | ... | ... | ... | 96 | 96 | 98 | 98 | 96 | 96 |
| Tuvalu | 90 | 97 | 92 | 98 | 89 | 97 | ... | 97 | ... | 97 | ... | 97 | 80 | 84 | 86 | 88 | 76 | 81 |
| Vanuatu | 57 | 83 | 91 | 96 | 49 | 79 | 37 | 44 | 79 | 79 | 27 | 33 | ... | 52 | ... | 66 | ... | 48 |
| Wallis and Futuna Islands | 100 | 100 | .. | .. | 100 | 100 | 80 | 81 | .. | .. | 80 | 81 | 96 | 96 | .. | .. | 96 | 96 |

*Source:* World Health Organization (WHO) and United Nations Children's Fund (UNICEF) Joint Monitoring Programme for Water Supply and Sanitation (JMP) (2010) *Progress on Sanitation and Drinking-Water 2010 Update*, WHO and UNICEF, Geneva.

*Notes:*
(1) This includes the island of Mayotte.
(2) Data for Mayotte is included in the data on Comoros.
(3) The former Yugoslav Republic of Macedonia.

# TABLE B.5

## Poverty and Inequality

| | Gross national income PPP $/capita | | Inequality | | | | National poverty line | | | | International poverty line | | |
|---|---|---|---|---|---|---|---|---|---|---|---|---|---|
| | | | Income / consumption | | Land | | | Population | | | | Population | |
| | 2000 | 2008[1] | Survey year[2] | Gini index | Survey year | Gini index | Survey year | Rural % | Urban % | National % | Survey year[3] | Below US$1.25/day | Below US$2/day |
| **AFRICA** | | | | | | | | | | | | | |
| Algeria | 5,120 | 7,890 | 1995 | 0.35 | | ... | 1995 | 30.3 | 14.7 | 22.6 | 1995 | 6.8 | 23.6 |
| Angola | 1,850 | 4,830 | 2000 | 0.59 | | ... | | ... | ... | ... | 2000 | 54.3 | 70.2 |
| Benin | 1,120 | 1,470 | 2003 | 0.37 | | ... | 2003 | 46.0 | 29.0 | 39.0 | 2003 | 47.3 | 75.3 |
| Botswana | 8,340 | 13,310 | 1993-94 | 0.61 | | ... | | ... | ... | ... | 1993-94 | 31.2 | 49.4 |
| Burkina Faso | 810 | 1,160 | 2003 | 0.40 | 1993 | 0.42 | 2003 | 52.4 | 19.2 | 46.4 | 2003 | 56.5 | 81.2 |
| Burundi | 310 | 380 | 2006 | 0.33 | | ... | | ... | ... | ... | 2006 | 81.3 | 93.4 |
| Cameroon | 1,520 | 2,170 | 2001 | 0.45 | | ... | 2007[11] | 55.0 | 12.2 | 39.9 | 2001 | 32.8 | 57.7 |
| Cape Verde | 1,970 | 3,090 | 2001 | 0.50 | | ... | | ... | ... | ... | 2001 | 20.6 | 40.2 |
| Central African Republic | 660 | 730 | 2003 | 0.44 | | ... | | ... | ... | ... | 2003 | 62.4 | 81.9 |
| Chad | 640 | 1,070 | 2002-03 | 0.40 | | ... | | ... | ... | ... | 2002-03 | 61.9 | 83.3 |
| Comoros | 970 | 1,170 | 2004 | 0.64 | | ... | | ... | ... | ... | 2004 | 46.1 | 65.0 |
| Congo | 2,020 | 2,810 | 2005 | 0.47 | | ... | | ... | ... | ... | 2005 | 54.1 | 74.4 |
| Côte d'Ivoire | 1,430 | 1,580 | 2002 | 0.48 | | ... | | ... | ... | ... | 2002 | 23.3 | 46.8 |
| Democratic Republic of the Congo | 200 | 280 | | ... | | ... | | ... | ... | ... | 2005-06 | 59.2 | 79.5 |
| Djibouti | 1,600 | 2,320 | 2002 | 0.40 | | ... | | ... | ... | ... | 2002 | 18.8 | 41.2 |
| Egypt | 3,570 | 5,470 | 2004-05 | 0.32 | 1990 | 0.65 | 1999-2000 | | | 16.7 | 2004-05 | <2 | 18.4 |
| Equatorial Guinea | 5,330 | 21,720 | | ... | | ... | | ... | ... | ... | | ... | ... |
| Eritrea | 610 | 640[5] | | ... | | ... | | ... | ... | ... | | ... | ... |
| Ethiopia | 460 | 870 | 2005 | 0.30 | 2001 | 0.47 | 1999-2000 | 45.0 | 37.0 | 44.2 | 2005 | 39.0 | 77.5 |
| Gabon | 9,940 | 12,400 | 2005 | 0.42 | | ... | | ... | ... | ... | 2005 | 4.8 | 19.6 |
| Gambia | 920 | 1,280 | 2003 | 0.47 | | ... | 2003 | 63.0 | 57.0 | 61.3 | 2003 | 34.3 | 56.7 |
| Ghana | 900 | 1,320 | 2006 | 0.43 | | ... | 2005-06 | 39.2 | 10.8 | 28.5 | 2006 | 30.0 | 53.6 |
| Guinea | 760 | 970 | 2003 | 0.43 | | ... | 1994 | | | 40.0 | 2003 | 70.1 | 87.2 |
| Guinea-Bissau | 480 | 520 | 2002 | 0.36 | 1988 | 0.62 | | ... | ... | ... | 2002 | 48.8 | 77.9 |
| Kenya | 1,120 | 1,560 | 2005-06 | 0.48 | | ... | 2005/06 | 49.7 | 34.4 | 46.6 | 2005-06 | 19.7 | 39.9 |
| Lesotho | 1,320 | 1,970 | 2002-03 | 0.53 | 1989-90 | 0.49 | 2002/03[11] | 60.5 | 41.5 | 56.3 | 2002-03 | 43.4 | 62.2 |
| Liberia | 290 | 310 | 2007 | 0.53 | | ... | | ... | ... | ... | 2007 | 83.7 | 94.8 |
| Libyan Arab Jamahiriya | ... | 16,270 | | ... | | ... | | ... | ... | ... | | ... | ... |
| Madagascar | 790 | 1,050 | 2005 | 0.47 | | ... | 2005[11] | 53.5 | 52.0 | 68.7 | 2005 | 67.8 | 89.6 |
| Malawi | 600 | 810 | 2004-05 | 0.39 | 1993 | 0.52 | 2004-05 | 55.9 | 25.4 | 52.4 | 2004-05[14] | 73.9 | 90.4 |
| Mali | 710 | 1,100 | 2006 | 0.39 | | ... | | ... | ... | ... | 2006 | 51.4 | 77.1 |
| Mauritania | 1,410 | ... | 2000 | 0.39 | | ... | 2000 | 61.2 | 25.4 | 46.3 | 2000 | 21.2 | 44.1 |
| Mauritius | 8,040 | 12,580 | | ... | | ... | | ... | ... | ... | | ... | ... |
| Mayotte | ... | ... | | ... | | ... | | ... | ... | ... | | ... | ... |
| Morocco | 2,510 | 4,190 | 2007 | 0.41 | 1996 | 0.62 | 1998-99 | 27.2 | 12.0 | 19.0 | 2007 | 2.5 | 14.0 |
| Mozambique | 420 | 770 | 2002-03 | 0.47 | | ... | 2002-03 | 54.1 | 51.6 | 55.2 | 2002-03 | 74.7 | 90.0 |
| Namibia | 4,160 | 6,250 | 1993[9] | 0.74 | 1997 | 0.36 | | ... | ... | ... | 1993[15] | 49.1 | 62.2 |
| Niger | 500 | 680 | 2005 | 0.44 | | ... | 1989-93 | 66.0 | 52.0 | 63.0 | 2005 | 65.9 | 85.6 |
| Nigeria | 1,130 | 1,980 | 2003-04 | 0.43 | | ... | 1992-93 | 36.4 | 30.4 | 34.1 | 2003-04 | 64.4 | 83.9 |
| Réunion | ... | ... | | ... | | ... | | ... | ... | ... | | ... | ... |
| Rwanda | 580 | 1,110 | 2000 | 0.47 | | ... | 2005-06[11] | 62.5 | ... | 56.9 | 2000 | 76.6 | 90.3 |
| Saint Helena | ... | ... | | ... | | ... | | ... | ... | ... | | ... | ... |
| São Tomé and Príncipe | ... | 1,790 | 2000-01 | 0.51 | | ... | | ... | ... | ... | 2000-01 | 28.4 | 56.6 |
| Senegal | 1,270 | 1,780 | 2005 | 0.39 | 1998 | 0.50 | 1992 | 40.4 | 23.7 | 33.4 | 2005 | 33.5 | 60.3 |
| Seychelles | 15,310 | 19,650 | 2006-07 | 0.02 | | ... | | ... | ... | ... | 2006-07 | <2 | <2 |
| Sierra Leone | 360 | 770 | 2003 | 0.43 | | ... | 2003-04 | 79.0 | 56.4 | 70.2 | 2003 | 53.4 | 76.1 |
| Somalia | ... | ... | | ... | | ... | | ... | ... | ... | | ... | ... |
| South Africa | 6,470 | 9,790 | 2000 | 0.58 | | ... | 2008[11] | ... | ... | 22.0 | 2000 | 26.2 | 42.9 |
| Sudan | 1,070 | 1,920 | | ... | | ... | | ... | ... | ... | | ... | ... |
| Swaziland | 3,650 | 5,000 | 2000-01 | 0.51 | | ... | | ... | ... | ... | 2000-01 | 62.9 | 81.0 |
| Togo | 690 | 830 | 2006 | 0.34 | | ... | | ... | ... | ... | 2006 | 38.7 | 69.3 |
| Tunisia | 4,590 | 7,460 | 2000 | 0.41 | 1993 | 0.70 | 1995 | 13.9 | 3.6 | 7.6 | 2000 | 2.6 | 12.8 |
| Uganda | 680 | 1,140 | 2005 | 0.43 | 1991 | 0.59 | 2005-06[11] | 34.2 | 13.7 | 31.1 | 2005 | 51.5 | 75.6 |
| United Republic of Tanzania | 770 | 1,260[6] | 2000-01 | 0.35 | | ... | 2000-01 | 38.7 | 29.5 | 35.7 | 2000-01 | 88.5 | 96.6 |
| Western Sahara | ... | ...[7] | | ... | | ... | | ... | ... | ... | | ... | ... |
| Zambia | 840 | 1,230 | 2004-05 | 0.51 | | ... | 2004 | 78.0 | 53.0 | 68.0 | 2004-05 | 64.3 | 81.5 |
| Zimbabwe | 210 | ... | 1995 | 0.50 | | ... | 1995-96 | 48.0 | 7.9 | 34.9 | | ... | ... |
| **ASIA** | | | | | | | | | | | | | |
| Afghanistan | ... | 1,100[5] | | ... | | ... | 2007 | 45.0 | 27.0 | 42.0 | | ... | ... |
| Armenia | 2,090 | 6,310 | 2007 | 0.30 | | ... | 2001 | 48.7 | 51.9 | 50.9 | 2007 | 3.7 | 21.0 |
| Azerbaijan | 2,060 | 7,770 | 2005 | 0.17 | | ... | 2001 | 42.0 | 55.0 | 49.6 | 2005 | <2 | <2 |
| Bahrain | 20,030 | 33,430 | | ... | | ... | | ... | ... | ... | | ... | ... |
| Bangladesh | 820 | 1,450 | 2005 | 0.31 | 1996 | 0.62 | 2005 | 43.8 | 28.4 | 40.0 | 2005 | 49.6 | 81.3[17] |
| Bhutan | 2,330 | 4,820 | 2003 | 0.47 | | ... | | ... | ... | ... | 2003 | 26.2 | 49.5 |
| Brunei Darussalam | 42,050 | ... | | ... | | ... | | ... | ... | ... | | ... | ... |
| Cambodia | 860 | 1,870 | 2007 | 0.44 | | ... | 2007 | 34.7 | .. | 30.1 | 2007 | 25.8 | 57.8 |
| China | 2,330 | 6,010 | 2005[9] | 0.42 | | ... | 2006[11] | 2.5 | ... | ... | 2005 | 15.9 | 36.3[18] |
| China, Hong Kong SAR | 26,520 | 44,000 | 1996[9] | 0.43 | | ... | | ... | ... | ... | | ... | ... |
| China, Macao SAR | 20,250 | ... | | ... | | ... | | ... | ... | ... | | ... | ... |
| Cyprus | 18,710 | 24,980 | | ... | | ... | | ... | ... | ... | | ... | ... |
| Democratic People's Republic of Korea | ... | ... | | ... | | ... | | ... | ... | ... | | ... | ... |
| Georgia | 2,140 | 4,920 | 2005 | 0.41 | | ... | 2003 | 52.7 | 56.2 | 54.5 | 2005 | 13.4 | 30.4 |
| India | 1,500 | 2,930 | 2004-05 | 0.37 | | ... | 1999-2000 | 30.2 | 24.7 | 28.6 | 2004-05 | 41.6 | 75.6[18] |
| Indonesia | 2,200 | 3,600 | 2007 | 0.38 | 1993 | 0.46 | 2004 | 20.1 | 12.1 | 16.7 | 2007 | 29.4 | 60.0 |

# TABLE B.5

**continued**

| | Gross national income PPP $/capita | | Inequality | | | | National poverty line | | | | International poverty line | | |
|---|---|---|---|---|---|---|---|---|---|---|---|---|---|
| | | | Income / consumption | | Land | | | Population | | | | Population | |
| | 2000 | 2008[1] | Survey year[2] | Gini index | Survey year | Gini index | Survey year | Rural % | Urban % | National % | Survey year[3] | Below US$1.25/day | Below US$2/day |
| Iran (Islamic Republic of) | 6,790 | ... | 2005 | 0.38 | | ... | | ... | ... | ... | 2005 | <2 | 8.0 |
| Iraq | ... | ... | | ... | | ... | | ... | ... | ... | | ... | ... |
| Israel | 21,480 | 27,450 | 2001[9] | 0.39 | | ... | | ... | ... | ... | | ... | ... |
| Japan | 25,950 | 35,190 | 1993[9] | 0.25 | 1995 | 0.59 | | ... | ... | ... | | ... | ... |
| Jordan | 3,240 | 5,720 | 2006 | 0.38 | 1997 | 0.78 | 2002 | 18.7 | 12.9 | 14.2 | 2006 | <2 | 3.5 |
| Kazakhstan | 4,450 | 9,720 | 2007 | 0.31 | | ... | 2002 | ... | ... | 15.4 | 2007 | <2 | <2 |
| Kuwait | 35,410 | ... | | ... | | ... | | ... | ... | ... | | ... | ... |
| Kyrgyzstan | 1,250 | 2,150 | 2007 | 0.34 | | ... | 2005 | 50.8 | 29.8 | 43.1 | 2007 | 3.4 | 27.5 |
| Lao People's Democratic Republic | 1,130 | 2,050 | 2002-03 | 0.33 | 1999 | 0.39 | 2002-03 | ... | ... | 33.5 | 2002-03 | 44.0 | 76.8[17] |
| Lebanon | 7,710 | 11,750 | | ... | | ... | | ... | ... | ... | | ... | ... |
| Malaysia | 8,350 | 13,740 | 2004[9] | 0.38 | ... | 1989 | ... | ... | 15.5 | 2004 | 15 | <2 | 7.8 |
| Maldives | 2,920 | 5,290 | 2004 | 0.37 | | ... | | ... | ... | ... | | ... | ... |
| Mongolia | 1,790 | 3,470 | 2007-08 | 0.37 | | ... | 2002 | 43.4 | 30.3 | 36.1 | 2007-08 | 2.2 | 13.6 |
| Myanmar | ... | ... | | ... | | ... | | ... | ... | ... | | ... | ... |
| Nepal | 800 | 1,120 | 2003-04 | 0.47 | 1992 | 0.45 | 2003-04 | 34.6 | 9.6 | 30.9 | 2003-04 | 55.1 | 77.6 |
| Occupied Palestinian Territory | ... | ... | | ... | | ... | | ... | ... | ... | | ... | ... |
| Oman | 15,100 | ... | | ... | | ... | | ... | ... | ... | | ... | ... |
| Pakistan | 1,690 | 2,590 | 2004-05 | 0.31 | 1990 | 0.57 | 1998-99 | 35.9 | 24.2 | 32.6 | 2004-05 | 22.6 | 60.3 |
| Philippines | 2,430 | 3,900 | 2006 | 0.44 | 1991 | 0.55 | 1997 | 36.9 | 11.9 | 25.1 | 2006 | 22.6 | 45.0 |
| Qatar | ... | ... | 2006-07 | 0.41 | | ... | | ... | ... | ... | | ... | ... |
| Republic of Korea | 17,130 | 27,840 | 1998[9] | 0.32 | 1990 | 0.34 | | ... | ... | ... | | ... | ... |
| Saudi Arabia | 17,500 | 24,500 | | ... | | ... | | ... | ... | ... | | ... | ... |
| Singapore | 32,870 | 47,970 | 1998[9] | 0.43 | | ... | | ... | ... | ... | | ... | ... |
| Sri Lanka | 2,660 | 4,460 | 2002 | 0.41 | | ... | 2002 | 7.9 | 24.7 | 22.7 | 2002 | 14.0 | 39.7 |
| Syrian Arab Republic | 3,150 | 4,490 | | ... | | ... | | ... | ... | ... | | ... | ... |
| Tajikistan | 800 | 1,870 | 2004 | 0.34 | | ... | 2007 | 55.0 | 49.4 | 53.5 | 2004 | 21.5 | 50.8 |
| Thailand | 4,850 | 7,770 | 2004 | 0.43 | 1993 | 0.47 | 1998 | ... | ... | 13.6 | 2004 | <2 | 11.5 |
| Timor-Leste | 790 | 4,690[5] | 2007 | 0.32 | | ... | | ... | ... | ... | 2007 | 37.2 | 72.8 |
| Turkey | 8,730 | 13,420 | 2006 | 0.41 | 1991 | 0.61 | 2002 | 34.5 | 22.0 | 27.0 | 2006 | 2.6 | 8.2 |
| Turkmenistan | 1,930 | 6,130 | 1998 | 0.41 | | ... | | ... | ... | ... | 1998 | 24.8 | 49.6 |
| United Arab Emirates | 41,610 | ... | | ... | | ... | | ... | ... | ... | | ... | ... |
| Uzbekistan | 1,420 | 2,660[5] | 2003 | 0.37 | | ... | 2003 | 29.8 | 22.6 | 27.2 | | ... | ... |
| Viet Nam | 1,390 | 2,700 | 2006 | 0.38 | 1994 | 0.53 | 2002 | 35.6 | 6.6 | 28.9 | 2006 | 21.5 | 48.4 |
| Yemen | 1,710 | 2,220 | 2005 | 0.38 | | ... | 1998 | 45.0 | 30.8 | 41.8 | 2005 | 17.5 | 46.6 |
| **EUROPE** | | | | | | | | | | | | | |
| Albania | 4,100 | 7,520 | 2005 | 0.33 | 1998 | 0.84 | 2005 | 24.2 | 11.2 | 18.5 | 2005 | <2 | 7.8 |
| Andorra | ... | ... | | ... | | ... | | ... | ... | ... | | ... | ... |
| Austria | 28,290 | 37,360 | 2000[9] | 0.29 | 1999-2000 | 0.59 | | ... | ... | ... | | ... | ... |
| Belarus | 5,120 | 12,120 | 2007 | 0.29 | | ... | 2004 | ... | ... | 17.4 | 2007 | <2 | <2 |
| Belgium | 28,180 | 35,380 | 2000[9] | 0.33 | 1999-2000 | 0.56 | | ... | ... | ... | | ... | ... |
| Bosnia and Herzegovina | 4,620 | 8,360 | 2007 | 0.36 | | ... | | ... | ... | ... | 2007 | <2 | <2 |
| Bulgaria | 6,180 | 11,370 | 2003 | 0.29 | | ... | 2001 | ... | ... | 12.8 | 2003 | <2 | <2 |
| Channel Islands | ... | ... | | ... | | ... | | ... | ... | ... | | ... | ... |
| Croatia | 10,910 | 17,050 | 2005 | 0.29 | | ... | 2004 | ... | ... | 11.1 | 2005 | <2 | <2 |
| Czech Republic | 14,650 | 22,890 | 1996[9] | 0.26 | 2000 | 0.92 | | ... | ... | ... | 1996[15] | <2 | <2 |
| Denmark | 28,220 | 37,530 | 1997[9] | 0.25 | 1999-2000 | 0.51 | | ... | ... | ... | | ... | ... |
| Estonia | 9,530 | 19,320 | 2004 | 0.36 | 2001 | 0.79 | | ... | ... | ... | 2004 | <2 | <2 |
| Faeroe Islands | ... | ... | | ... | | ... | | ... | ... | ... | | ... | ... |
| Finland | 25,490 | 35,940 | 2000[9] | 0.27 | 1999-2000 | 0.27 | | ... | ... | ... | | ... | ... |
| France | 25,680 | 33,280 | 1995[9] | 0.33 | 1999-2000 | 0.58 | | ... | ... | ... | | ... | ... |
| Germany | 25,700 | 35,950 | 2000[9] | 0.28 | 1999-2000 | 0.63 | | ... | ... | ... | | ... | ... |
| Gibraltar | ... | ... | | ... | | ... | | ... | ... | ... | | ... | ... |
| Greece | 18,460 | 28,300 | 2000[9] | 0.34 | 1999-2000 | 0.58 | | ... | ... | ... | | ... | ... |
| Holy See | ... | ... | | ... | | ... | | ... | ... | ... | | ... | ... |
| Hungary | 11,740 | 18,210 | 2004 | 0.30 | | ... | 1997 | ... | ... | 17.3 | 2004 | <2 | <2 |
| Iceland | 28,060 | 25,300 | | ... | | ... | | ... | ... | ... | | ... | ... |
| Ireland | 24,690 | 35,710 | 2000[9] | 0.34 | | ... | | ... | ... | ... | | ... | ... |
| Isle of Man | ... | ... | | ... | | ... | | ... | ... | ... | | ... | ... |
| Italy | 25,400 | 30,800 | 2000[9] | 0.36 | 1999-2000 | 0.73 | | ... | ... | ... | | ... | ... |
| Latvia | 8,260 | 16,010 | 2007 | 0.36 | 2001 | 0.58 | 2004 | 12.7 | ... | 5.9 | 2007 | <2 | <2 |
| Liechtenstein | ... | ... | | ... | | ... | | ... | ... | ... | | ... | ... |
| Lithuania | 8,720 | 17,170 | 2004 | 0.36 | | ... | | ... | ... | ... | 2004 | <2 | <2 |
| Luxembourg | 46,750 | 52,770 | | ... | 1999-2000 | 0.48 | | ... | ... | ... | | ... | ... |
| Malta | 18,380 | ... | | ... | | ... | | ... | ... | ... | | ... | ... |
| Moldova | 1,490 | 3,270[8] | 2007 | 0.37 | | ... | 2002 | 67.2 | 42.6 | 48.5 | 2007 | 2.4 | 11.5 |
| Monaco | ... | ... | 2007-08 | 0.37 | | ... | | ... | ... | ... | | ... | ... |
| Montenegro | 5,940 | 13,420 | 2007 | 0.37 | | ... | | ... | ... | ... | 2007 | <2 | <2 |
| Netherlands | 30,040 | 40,620 | 1999[9] | 0.31 | 1999-2000 | 0.57 | | ... | ... | ... | | ... | ... |
| Norway | 35,640 | 59,250 | 2000[9] | 0.26 | 1999 | 0.18 | | ... | ... | ... | | ... | ... |
| Poland | 10,470 | 16,710 | 2005 | 0.35 | 2002 | 0.69 | 2001 | ... | ... | 14.8 | 2005 | <2 | <2 |
| Portugal | 16,670 | 22,330 | 1997[9] | 0.39 | 1999-2000 | 0.74 | | ... | ... | ... | | ... | ... |
| Romania | 5,780 | 13,380 | 2007 | 0.32 | | ... | 2002 | ... | ... | 28.9 | 2007 | <2 | 4.1 |
| Russian Federation | 7,420 | 15,460 | 2007 | 0.44 | | ... | 2002 | ... | ... | 19.6 | 2007 | <2 | <2 |
| San Marino | ... | ... | | ... | | ... | | ... | ... | ... | | ... | ... |
| Serbia | 5,630 | 10,380 | 2008 | 0.28 | | ... | | ... | ... | ... | 2008 | <2 | <2 |

# TABLE B.5

## continued

| | Gross national income PPP $/capita | | Inequality | | | | National poverty line | | | | International poverty line | | |
|---|---|---|---|---|---|---|---|---|---|---|---|---|---|
| | | | Income / consumption | | Land | | | Population | | | | Population | |
| | 2000 | 2008[1] | Survey year[2] | Gini index | Survey year | Gini index | Survey year | Rural % | Urban % | National % | Survey year[3] | Below US$1.25/day | Below US$2/day |
| Slovakia | 10,810 | 21,460 | 1996[9] | 0.26 | | ... | | ... | ... | ... | 1996[15] | <2 | <2 |
| Slovenia | 17,490 | 27,160 | 2004 | 0.31 | 1991 | 0.62 | | ... | ... | ... | 2004 | <2 | <2 |
| Spain | 21,140 | 30,830 | 2000[9] | 0.35 | 1999-2000 | 0.77 | | ... | ... | ... | | ... | ... |
| Sweden | 27,530 | 37,780 | 2000[9] | 0.25 | 1999-2000 | 0.32 | | ... | ... | ... | | ... | ... |
| Switzerland | 34,060 | 39,210 | 2000[9] | 0.34 | 1999 | 0.50 | | ... | ... | ... | | ... | ... |
| TFYR Macedonia[4] | 6,030 | 9,250 | 2006 | 0.43 | | ... | 2003 | 22.3 | ... | 21.7 | 2006 | <2 | 5.3 |
| Ukraine | 3,170 | 7,210 | 2008 | 0.28 | | ... | 2003 | 28.4 | ... | 19.5 | 2008 | <2 | <2 |
| United Kingdom | 26,020 | 36,240 | 1999[9] | 0.36 | 1999-2000 | 0.66 | | ... | ... | ... | | ... | ... |
| **LATIN AMERICA AND THE CARIBBEAN** | | | | | | | | | | | | | |
| Anguilla | ... | ... | | ... | | ... | | ... | ... | ... | | ... | ... |
| Antigua and Barbuda | 11,420 | 19,660 | | ... | | ... | | ... | ... | ... | | ... | ... |
| Argentina | 8,850 | 14,000 | 2006[9,10] | 0.49 | 1988 | 0.83 | 2001[11] | ... | 35.9 | ... | 2004[10,15] | 4.5 | 11.3 |
| Aruba | ... | ... | | ... | | ... | | ... | ... | ... | | ... | ... |
| Bahamas | ... | ... | | ... | | ... | | ... | ... | ... | | ... | ... |
| Barbados | ... | ... | | ... | | ... | | ... | ... | ... | | ... | ... |
| Belize | 4,630 | 5,940[5] | 1995 | 0.60 | | ... | | ... | ... | ... | | ... | ... |
| Bolivia | 2,930 | 4,140 | 2007 | 0.57 | | ... | 2007 | 63.9 | 23.7 | 37.7 | 2007[16] | 11.9 | 21.9 |
| Brazil | 6,810 | 10,080 | 2007[9] | 0.55 | 1996 | 0.85 | 2002-03 | 41.0 | 17.5 | 21.5 | 2007[15] | 5.2 | 12.7 |
| British Virgin Islands | ... | ... | | ... | | ... | | ... | ... | ... | | ... | ... |
| Cayman Islands | ... | ... | | ... | | ... | | ... | ... | ... | | ... | ... |
| Chile | 8,910 | 13,250 | 2006[9] | 0.52 | | ... | 2006[11] | ... | ... | 13.7 | 2006[15] | <2 | 2.4 |
| Colombia | 5,550 | 8,430 | 2006[9] | 0.59 | 2001 | 0.80 | 2006 | 62.1 | 39.1 | 45.1 | 2006[15] | 16.0 | 27.9 |
| Costa Rica | 6,610 | 10,960 | 2007[9] | 0.49 | | ... | 2004 | 28.3 | 20.8 | 23.9 | 2007[15] | <2 | 4.3 |
| Cuba | ... | ... | | ... | | ... | | ... | ... | ... | | ... | ... |
| Dominica | 5,300 | 8,300 | | ... | | ... | | ... | ... | ... | | ... | ... |
| Dominican Republic | 4,760 | 7,800[5] | 2007[9] | 0.48 | | ... | 2007[11] | 54.1 | 45.4 | 48.5 | 2007[15] | 4.4 | 12.3 |
| Ecuador | 4,430 | 7,780 | 2007[9] | 0.54 | | ... | 2006[11] | 61.5 | 24.9 | 38.3 | 2007[15] | 4.7 | 12.8 |
| El Salvador | 4,500 | 6,630[5] | 2007[9] | 0.47 | | ... | 2006[12] | 36.0 | 27.8 | 30.7 | 2007[15] | 6.4 | 13.2 |
| Falkland Islands (Malvinas) | ... | ... | | ... | | ... | | ... | ... | ... | | ... | ... |
| French Guiana | ... | ... | | ... | | ... | | ... | ... | ... | | ... | ... |
| Grenada | 5,910 | 8,430[5] | | ... | | ... | | ... | ... | ... | | ... | ... |
| Guadeloupe | ... | ... | | ... | | ... | | ... | ... | ... | | ... | ... |
| Guatemala | 3,460 | 4,690[5] | 2006[9] | 0.54 | | ... | 2006 | 72.0 | 28.0 | 51.0 | 2006[15] | 11.7 | 24.3 |
| Guyana | 1,980 | 3,030 | 1998[9] | 0.43 | | ... | 1995 | 66.0 | ... | ... | 1998[15] | 7.7 | 16.8 |
| Haiti | ... | ... | 2001[9] | 0.60 | | ... | 1995 | 66.0 | ... | ... | 2001[15] | 54.9 | 72.1 |
| Honduras | 2,490 | 3,830[5] | 2006[9] | 0.55 | 1993 | 0.66 | 2004 | 70.4 | 29.5 | 50.7 | 2006[15] | 18.2 | 29.7 |
| Jamaica | 5,560 | 7,370 | 2004 | 0.46 | | ... | 2000 | 25.1 | 12.8 | 18.7 | 2004 | <2 | 5.8 |
| Martinique | ... | ... | | ... | | ... | | ... | ... | ... | | ... | ... |
| Mexico | 8,960 | 14,340 | 2008[9] | 0.52 | | ... | 2004 | 56.9 | 41.0 | 47.0 | 2008[15] | 4.0 | 8.2 |
| Montserrat | ... | ... | | ... | | ... | | ... | ... | ... | | ... | ... |
| Netherlands Antilles | ... | ... | | ... | | ... | | ... | ... | ... | | ... | ... |
| Nicaragua | 1,780 | 2,620[5] | 2005[9] | 0.52 | 2001 | 0.72 | 2001 | 64.3 | 28.7 | 45.8 | 2005[15] | 15.8 | 31.8 |
| Panama | 6,830 | 12,630 | 2006[9] | 0.55 | 2001 | 0.52 | 2003 | ... | ... | 36.8 | 2006[15] | 9.5 | 17.8 |
| Paraguay | 3,360 | 4,660 | 2007[9] | 0.53 | 1991 | 0.93 | 1990[13] | 28.5 | 19.7 | 20.5 | 2007[15] | 6.5 | 14.2 |
| Peru | 4,750 | 7,950 | 2007[9] | 0.51 | 1994 | 0.86 | 2004 | 72.5 | 40.3 | 51.6 | 2007[15] | 7.7 | 17.8 |
| Puerto Rico | ... | ... | | ... | | ... | | ... | ... | ... | | ... | ... |
| Saint Kitts and Nevis | 9,720 | 15,490 | | ... | | ... | | ... | ... | ... | | ... | ... |
| Saint Lucia | 6,860 | 9,020[5] | 1995[9] | 0.43 | | ... | | ... | ... | ... | 1995[15] | 20.9 | 40.6 |
| Saint Vincent and the Grenadines | 5,010 | 8,570 | | ... | | ... | | ... | ... | ... | | ... | ... |
| Suriname | 4,400 | 6,680[5] | 1999[9] | 0.53 | | ... | | ... | ... | ... | 1999[15] | 15.5 | 27.2 |
| Trinidad and Tobago | 10,790 | 24,240 | 1992[9] | 0.40 | | ... | 1992 | 20.0 | 24.0 | 21.0 | 1992[15] | 4.2 | 13.5 |
| Turks and Caicos Islands | ... | ... | | ... | | ... | | ... | ... | ... | | ... | ... |
| United States Virgin Islands | ... | ... | | ... | | ... | | ... | ... | ... | | ... | ... |
| Uruguay | 8,170 | 12,550 | 2007[9] | 0.47 | 2000 | 0.79 | 1998 | ... | 24.7 | ... | 2007[15] | <2 | 4.3 |
| Venezuela (Bolivarian Republic of) | 8,360 | 12,850 | 2006[9] | 0.43 | 1996-97 | 0.88 | 1997-99 | ... | ... | 52.0 | 2006[15] | 3.5 | 10.2 |
| **NORTHERN AMERICA** | | | | | | | | | | | | | |
| Bermuda | ... | ... | | ... | | ... | | ... | ... | ... | | ... | ... |
| Canada | 27,670 | 38,710 | 2000[9] | 0.33 | 1991 | 0.64 | | ... | ... | ... | | ... | ... |
| Greenland | ... | ... | | ... | | ... | | ... | ... | ... | | ... | ... |
| Saint-Pierre-et-Miquelon | ... | ... | | ... | | ... | | ... | ... | ... | | ... | ... |
| United States of America | 35,190 | 46,790 | 2000[9] | 0.41 | 1997 | 0.76 | | ... | ... | ... | | ... | ... |
| **OCEANIA** | | | | | | | | | | | | | |
| American Samoa | ... | ... | | ... | | ... | | ... | ... | ... | | ... | ... |
| Australia | 26,690 | 37,250 | 1994[9] | 0.35 | | ... | | ... | ... | ... | | ... | ... |
| Cook Islands | ... | ... | | ... | | ... | | ... | ... | ... | | ... | ... |
| Fiji | 3,500 | 4,320 | | ... | | ... | | ... | ... | ... | | ... | ... |
| French Polynesia | ... | ... | | ... | | ... | | ... | ... | ... | | ... | ... |
| Guam | ... | ... | | ... | | ... | | ... | ... | ... | | ... | ... |
| Kiribati | 3,100 | 3,620 | | ... | | ... | | ... | ... | ... | | ... | ... |
| Marshall Islands | ... | ... | | ... | | ... | | ... | ... | ... | | ... | ... |
| Micronesia (Federated States of) | 2,830 | 3,270[5] | 2000 | 0.01 | | ... | | ... | ... | ... | | ... | ... |
| Nauru | ... | ... | | ... | | ... | | ... | ... | ... | | ... | ... |
| New Caledonia | ... | ... | | ... | | ... | | ... | ... | ... | | ... | ... |

# TABLE B.5
## continued

| | Gross national income PPP $/capita | | Inequality | | | | National poverty line | | | | International poverty line | | |
|---|---|---|---|---|---|---|---|---|---|---|---|---|---|
| | | | Income / consumption | | Land | | | Population | | | | Population | |
| | 2000 | 2008[1] | Survey year[2] | Gini index | Survey year | Gini index | Survey year | Rural % | Urban % | National % | Survey year[3] | Below US$1.25/day | Below US$2/day |
| New Zealand | 19,450 | 25,200 | 1997[9] | 0.36 | | ... | | ... | ... | ... | | ... | ... |
| Niue | ... | ... | | ... | | ... | | ... | ... | ... | | ... | ... |
| Northern Mariana Islands | ... | ... | | ... | | ... | | ... | ... | ... | | ... | ... |
| Palau | ... | ... | | ... | | ... | | ... | ... | ... | | ... | ... |
| Papua New Guinea | 1,620 | 2,030[5] | 1996 | 0.51 | | ... | 1996 | 41.3 | 16.1 | 37.5 | 1996 | 35.8 | 57.4 |
| Pitcairn | ... | ... | | ... | | ... | | ... | ... | ... | | ... | ... |
| Samoa | 2,810 | 4,410[5] | | ... | | ... | | ... | ... | ... | | ... | ... |
| Solomon Islands | 1,970 | 2,230 | | ... | | ... | | ... | ... | ... | | ... | ... |
| Tokelau | ... | ... | | ... | | ... | | ... | ... | ... | | ... | ... |
| Tonga | 2,960 | 3,980[5] | | ... | | ... | | ... | ... | ... | | ... | ... |
| Tuvalu | ... | ... | | ... | | ... | | ... | ... | ... | | ... | ... |
| Vanuatu | 2,930 | ... | | ... | | ... | | ... | ... | ... | | ... | ... |
| Wallis and Futuna Islands | ... | ... | | ... | | ... | | ... | ... | ... | | ... | ... |

*Sources:* World Bank (2010) *World Development Indicators 2010*, World Bank, Washington, DC; World Bank (2006) *World Development Report 2006*, World Bank, Washington, DC.

*Notes:*
(1) Data are extrapolated from the 2005 International Comparison Program benchmark estimates, unless otherwise specified.
(2) Data refers to expenditure shares by percentiles of population, ranked by per capita expenditure, unless otherwise specfied.
(3) Data are expenditure based unless otherwise specified.
(4) The former Yugoslav Republic of Macedonia.
(5) Estimate is based on regression.
(6) Data covers mainland Tanzania only.
(7) Data for Western Sahara is included in the data on Morocco.
(8) Data excludes Transnistria.
(9) Data refers to income shares by percentiles of population, ranked by per capita income.
(10) Data includes urban areas only.
(11) Data are from national sources.
(12) Data refers to share of households rather than share of population.
(13) Data covers Asuncion metropolitan area only.
(14) Due to change in survey design, the most recent survey is not strictly comparable with the previous one.
(15) Data is income based.
(16) In purchasing power parity (PPP) dollars imputed using regression.
(17) Refers to data adjusted by spatial consumer price index information.
(18) Data covers weighted average of urban and rural estimates.

# TABLE B.6

## Transport Infrastructure

| | Roads | | | | Motor vehicles | | Railways | | |
|---|---|---|---|---|---|---|---|---|---|
| | Total (km) | Paved (%) | Passengers (m-p-km) | Goods hauled (m-t-km) | Number per 1000 population | | Route (km) | Passengers (m-p-km) | Goods hauled (m-t-km) |
| | 2000–2007[1] | 2000–2007[1] | 2000–2007[1] | 2000–2007[1] | 1990 | 2007 | 2000–2008[1] | 2000–2008[1] | 2000–2008[1] |
| **AFRICA** | | | | | | | | | |
| Algeria | 108,302 | 70.2 | ... | ... | 55 | 91 | 3,572 | 937 | 1,562 |
| Angola | 51,429 | 10.4 | 166,045 | 4,709 | 19 | 40 | ... | ... | ... |
| Benin | 19,000 | 9.5 | ... | ... | 3 | 21 | 758 | ... | 36 |
| Botswana | 25,798 | 33.2 | ... | ... | 18 | 113 | 888 | 94 | 674 |
| Burkina Faso | 92,495 | 4.2 | ... | ... | 4 | 11 | 622 | ... | ... |
| Burundi | 12,322 | 10.4 | ... | ... | ... | 6 | ... | ... | ... |
| Cameroon | 51,346 | 8.4 | ... | ... | 10 | ... | 977 | 379 | 978 |
| Cape Verde | ... | ... | ... | ... | ... | ... | ... | ... | ... |
| Central African Republic | 24,307 | ... | ... | ... | 1 | 0 | ... | ... | ... |
| Chad | 40,000 | 0.8 | ... | ... | 2 | 6 | ... | ... | ... |
| Comoros | ... | ... | ... | ... | ... | ... | ... | ... | ... |
| Congo | 17,289 | 5.0 | ... | ... | 18 | 26 | 795 | 211 | 234 |
| Côte d'Ivoire | 80,000 | 8.1 | ... | ... | 24 | ... | 639 | ... | 675 |
| Democratic Republic of the Congo | 153,497 | 1.8 | ... | ... | ... | 5 | 4,007 | 95 | 352 |
| Djibouti | ... | ... | ... | ... | ... | ... | ... | ... | ... |
| Egypt | 92,370 | 81.0 | ... | ... | 29 | ... | 5,063 | 40,830 | 4,188 |
| Equatorial Guinea | ... | ... | ... | ... | ... | ... | ... | ... | ... |
| Eritrea | 4,010 | 21.8 | ... | ... | 1 | 11 | ... | ... | ... |
| Ethiopia | 42,429 | 12.8 | 219,113 | 2,456 | 1 | 3 | ... | ... | ... |
| Gabon | 9,170 | 10.2 | ... | ... | 32 | ... | 810 | 99 | 2,502 |
| Gambia | 3,742 | 19.3 | 16 | ... | 13 | 7 | ... | ... | ... |
| Ghana | 57,614 | 14.9 | ... | ... | 8 | 33 | 953 | 85 | 181 |
| Guinea | 44,348 | 9.8 | ... | ... | 4 | ... | ... | ... | ... |
| Guinea-Bissau | 3,455 | 27.9 | ... | ... | 7 | 33 | ... | ... | ... |
| Kenya | 63,265 | 14.1 | ... | 22 | 12 | 21 | 1,917 | 250 | 1,399 |
| Lesotho | 5,940 | 18.3 | ... | ... | 11 | ... | ... | ... | ... |
| Liberia | 10,600 | 6.2 | ... | ... | 14 | 3 | ... | ... | ... |
| Libyan Arab Jamahiriya | 83,200 | 57.2 | ... | ... | ... | 291 | ... | ... | ... |
| Madagascar | 49,827 | 11.6 | ... | ... | 6 | ... | 854 | 10 | 1 |
| Malawi | 15,451 | 45.0 | ... | ... | 4 | 9 | 797 | 44 | 33 |
| Mali | 18,709 | 18.0 | ... | ... | 3 | 9 | ... | ... | ... |
| Mauritania | 11,066 | 26.8 | ... | ... | 10 | ... | 728 | 47 | 7,622 |
| Mauritius | 2,028 | 98.0 | ... | ... | 59 | 150 | ... | ... | ... |
| Mayotte | ... | ... | ... | ... | ... | ... | ... | ... | ... |
| Morocco | 57,799 | 62.0 | ... | 1,212 | 37 | 71 | 1,989 | 3,836 | 4,959 |
| Mozambique | 30,400 | 18.7 | ... | ... | 4 | 10 | 3,116 | 114 | 695 |
| Namibia | 42,237 | 12.8 | 47 | 591 | 71 | 109 | ... | ... | ... |
| Niger | 18,951 | 20.7 | ... | ... | 6 | 5 | ... | ... | ... |
| Nigeria | 193,200 | 15.0 | ... | ... | 30 | 31 | 3,528 | 174 | 77 |
| Réunion | ... | ... | ... | ... | ... | ... | ... | ... | ... |
| Rwanda | 14,008 | 19.0 | ... | ... | 2 | 4 | ... | ... | ... |
| Saint Helena | ... | ... | ... | ... | ... | ... | ... | ... | ... |
| São Tomé and Príncipe | ... | ... | ... | ... | ... | ... | ... | ... | ... |
| Senegal | 13,576 | 29.3 | ... | ... | 11 | 20 | ... | 129 | 384 |
| Seychelles | ... | ... | ... | ... | ... | ... | ... | ... | ... |
| Sierra Leone | 11,300 | 8.0 | ... | ... | 10 | 5 | ... | ... | ... |
| Somalia | 22,100 | 11.8 | ... | ... | 2 | ... | ... | ... | ... |
| South Africa | 362,099 | 17.3 | ... | 434 | 139 | 159 | 24,487 | 13,865 | 106,014 |
| Sudan | 11,900 | 36.3 | ... | ... | 9 | 28 | 4,578 | 34 | 766 |
| Swaziland | 3,594 | 30.0 | ... | ... | 66 | 89 | 300 | 0 | 2 |
| Togo | 7,520 | 31.6 | ... | ... | 24 | 2 | ... | ... | ... |
| Tunisia | 19,232 | 65.8 | ... | 16,611 | 48 | 103 | 2,218 | 1,487 | 2,197 |
| Uganda | 70,746 | 23.0 | ... | ... | 2 | 7 | ... | ... | ... |
| United Republic of Tanzania | 78,891 | 8.6 | ... | ... | 5 | 12 | 2,600[2] | 475[2] | 728[2] |
| Western Sahara | ... | ... | ... | ... | ... | ... | ... | ... | ... |
| Zambia | 66,781 | 22.0 | ... | ... | 14 | 18 | ... | ... | ... |
| Zimbabwe | 97,267 | 19.0 | ... | ... | 32 | 106 | 2,583 | ... | 1,580 |
| **ASIA** | | | | | | | | | |
| Afghanistan | 42,150 | 29.3 | ... | ... | ... | 23 | ... | ... | ... |
| Armenia | 7,515 | 89.8 | 2,693 | 434 | 5 | 105 | 845 | 27 | 354 |
| Azerbaijan | 59,141 | 49.4 | 11,786 | 8,222 | 52 | 61 | 2,099 | 1,047 | 10,021 |
| Bahrain | ... | ... | ... | ... | ... | ... | ... | ... | ... |
| Bangladesh | 239,226 | 9.5 | ... | ... | 1 | 2 | 2,835 | 5,609 | 870 |
| Bhutan | ... | ... | ... | ... | ... | ... | ... | ... | ... |
| Brunei Darussalam | ... | ... | ... | ... | ... | ... | ... | ... | ... |
| Cambodia | 38,257 | 6.3 | 201 | 3 | 1 | ... | ... | ... | ... |
| China | 3,583,715 | 70.7 | 1,150,677 | 975,420 | 5 | 32 | 60,809 | 772,834 | 2,511,804 |
| China, Hong Kong SAR | 2,009 | 100.0 | ... | ... | 66 | 72 | ... | ... | ... |
| China, Macao SAR | ... | ... | ... | ... | ... | ... | ... | ... | ... |
| Cyprus | ... | ... | ... | ... | ... | ... | ... | ... | ... |
| Democratic People's Republic of Korea | 25,554 | 2.8 | ... | ... | ... | ... | ... | ... | ... |
| Georgia | 20,329 | 38.6 | 5,269 | 586 | 107 | 116 | 1,513 | 774 | 6,928 |
| India | 3,316,452 | 47.4 | ... | ... | 4 | 12 | 63,327 | 769,956 | 521,371 |
| Indonesia | 391,009 | 55.4 | ... | ... | 16 | 76 | 3,370 | 14,344 | 4,390 |
| Iran (Islamic Republic of) | 172,927 | 72.8 | ... | ... | 34 | 16 | 7,335 | 13,900 | 21,829 |
| Iraq | 45,550 | 84.3 | ... | ... | 14 | ... | 2,032 | 61 | 640 |

# TABLE B.6

## continued

| | Roads | | | | Motor vehicles | | Railways | | |
|---|---|---|---|---|---|---|---|---|---|
| | Total (km) | Paved (%) | Passengers (m-p-km) | Goods hauled (m-t-km) | Number per 1000 population | | Route (km) | Passengers (m-p-km) | Goods hauled (m-t-km) |
| | 2000–2007 | 2000–2007 | 2000–2007 | 2000–2007 | 1990 | 2007 | 2000–2008 | 2000–2008 | 2000–2008 |
| Israel | 17,870 | 100.0 | ... | ... | 210 | 305 | 1,005 | 1,968 | 1,055 |
| Japan | 1,196,999 | 79.3 | 947,562 | 327,632 | 469 | 595 | 20,048 | 255,865 | 23,032 |
| Jordan | 7,768 | 100.0 | ... | ... | 60 | 137 | 251 | ... | 789 |
| Kazakhstan | 93,123 | 90.3 | 103,381 | 53,816 | 76 | 170 | 14,205 | 14,450 | 214,907 |
| Kuwait | 5,749 | 85.0 | ... | ... | ... | 502 | ... | ... | ... |
| Kyrgyzstan | 34,000 | 91.1 | 6,468 | 819 | 44 | 59 | 417 | 60 | 849 |
| Lao People's Democratic Republic | 29,811 | 13.4 | ... | ... | 9 | 21 | ... | ... | ... |
| Lebanon | 6,970 | ... | ... | ... | 321 | ... | ... | ... | ... |
| Malaysia | 93,109 | 79.8 | ... | ... | 124 | 272 | 1,665 | 2,268 | 1,350 |
| Maldives | ... | ... | ... | ... | ... | ... | ... | ... | ... |
| Mongolia | 49,250 | 3.5 | 557 | 242 | 21 | 61 | 1,810 | 1,400 | 8,261 |
| Myanmar | 27,000 | 11.9 | ... | ... | 2 | 7 | ... | 4,163 | 885 |
| Nepal | 17,280 | 56.9 | ... | ... | ... | 5 | ... | ... | ... |
| Occupied Palestinian Territory | 5,147 | 100 | ... | ... | ... | 16 | ... | ... | ... |
| Oman | 48,874 | 41.3 | ... | ... | 130 | 225 | ... | ... | ... |
| Pakistan | 260,420 | 65.4 | 263,788 | 129,249 | 6 | 11 | 7,791 | 24,731 | 6,187 |
| Philippines | 200,037 | 9.9 | ... | ... | 10 | 32 | 479 | 83 | ... |
| Qatar | 7,790 | 90.0 | ... | ... | ... | 724 | ... | ... | ... |
| Republic of Korea | 102,061 | 77.6 | 97,854 | 12,545 | 79 | 338 | 3,381 | 32,025 | 11,566 |
| Saudi Arabia | 221,372 | 21.5 | ... | ... | 165 | ... | 2,758 | 337 | 1,748 |
| Singapore | 3,297 | 100.0 | ... | ... | 130 | 149 | ... | ... | ... |
| Sri Lanka | 97,286 | 81.0 | 21,067 | ... | 21 | 58 | 1,463 | 4,767 | 135 |
| Syrian Arab Republic | 40,032 | 100.0 | 589 | ... | 26 | 52 | 2,139 | 1,120 | 2,370 |
| Tajikistan | 27,767 | ... | 150 | 14,572 | 3 | 38 | 616 | 53 | 1,274 |
| Thailand | 180,053 | 98.5 | ... | ... | 46 | ... | 4,429 | 8,037 | 3,161 |
| Timor-Leste | ... | ... | ... | ... | ... | ... | ... | ... | ... |
| Turkey | 426,951 | ... | 209,115 | 177,399 | 50 | 131 | 8,699 | 5,097 | 10,104 |
| Turkmenistan | 24,000 | 81.2 | ... | ... | ... | 106 | 3,181 | 1,570 | 10,973 |
| United Arab Emirates | 4,030 | 100.0 | ... | ... | 121 | 313 | ... | ... | ... |
| Uzbekistan | 81,600 | 87.3 | ... | 1,200 | ... | ... | 4,230 | 2,264 | 21,594 |
| Viet Nam | 160,089 | 47.6 | 49,372 | 20,537 | ... | 13 | 3,147 | 4,659 | 3,910 |
| Yemen | 71,300 | 8.7 | ... | ... | 34 | 35 | ... | ... | ... |
| **EUROPE** | | | | | | | | | |
| Albania | 18,000 | 39.0 | 197 | 2,200 | 11 | 102 | 423 | 51 | 53 |
| Andorra | ... | ... | ... | ... | ... | ... | ... | ... | ... |
| Austria | 107,206 | 100.0 | 69,000 | 26,411 | 421 | 556 | 5,755 | 10,275 | 18,710 |
| Belarus | 94,797 | 88.6 | 9,353 | 15,779 | 61 | 282 | 5,491 | 8,188 | 47,933 |
| Belgium | 153,070 | 78.2 | 130,868 | 51,572 | 423 | 539 | 3,513 | 10,403 | 7,882 |
| Bosnia and Herzegovina | 21,846 | 52.3 | ... | 300 | 114 | 170 | 1,016 | 78 | 1,237 |
| Bulgaria | 40,231 | 98.4 | 13,688 | 11,843 | 163 | 295 | 4,159 | 2,335 | 4,673 |
| Channel Islands | ... | ... | ... | ... | ... | ... | ... | ... | ... |
| Croatia | 29,038 | 89.1 | 3,277 | 10,175 | ... | 377 | 2,722 | 1,810 | 3,312 |
| Czech Republic | 128,511 | 100.0 | 90,055 | 46,600 | 246 | 470 | 9,487 | 6,759 | 15,961 |
| Denmark | 72,412 | 100.0 | 70,635 | 11,495 | 368 | 466 | 2,133 | 5,843 | ... |
| Estonia | 58,034 | 28.8 | 3,190 | 7,641 | 211 | 444 | 816 | 274 | 5,683 |
| Faeroe Islands | ... | ... | ... | ... | ... | ... | ... | ... | ... |
| Finland | 78,889 | 65.4 | 71,300 | 26,400 | 441 | 559 | 5,919 | 4,052 | 10,777 |
| France | 951,125 | 100.0 | 775,000 | 313,000 | 494 | 600 | 29,901 | 88,283 | 41,530 |
| Germany | 644,471 | 100.0 | 966,692 | 461,900 | 405 | 623 | 33,862 | 76,997 | 91,178 |
| Gibraltar | ... | ... | ... | ... | ... | ... | ... | ... | ... |
| Greece | 117,533 | 91.8 | ... | 18,360 | 248 | ...4 | 2,552 | 2,003 | 786 |
| Holy See | ... | ... | ... | ... | ... | ... | ... | ... | ... |
| Hungary | 195,719 | 37.7 | 11,784 | 30,495 | 212 | 384 | 7,942 | 5,927 | 7,786 |
| Iceland | ... | ... | ... | ... | ... | ... | ... | ... | ... |
| Ireland | 96,602 | 100.0 | ... | 15,900 | 270 | 537 | 1,919 | 1,976 | 103 |
| Isle of Man | ... | ... | ... | ... | ... | ... | ... | ... | ... |
| Italy | 487,700 | 100.0 | 97,560 | 192,700 | 529 | 677 | 16,862 | 46,998 | 19,918 |
| Latvia | 69,687 | 100.0 | 2,664 | 2,729 | 135 | 459 | 2,263 | 951 | 17,704 |
| Liechtenstein | ... | ... | ... | ... | ... | ... | ... | ... | ... |
| Lithuania | 80,715 | 28.6 | 42,739 | 18,134 | 160 | 479 | 1,765 | 398 | 14,748 |
| Luxembourg | ... | ... | ... | ... | ... | ... | ... | ... | ... |
| Malta | ... | ... | ... | ... | ... | ... | ... | ... | ... |
| Moldova | 12,755 | 85.7 | 1,640 | 1,577 | 53 | 120 | 1,156 | 485 | 3,092 |
| Monaco | ... | ... | ... | ... | ... | ... | ... | ... | ... |
| Montenegro | ... | ... | ... | ... | ... | ... | ... | ... | ... |
| Netherlands | 126,100 | 90.0 | ... | 77,100 | 405 | 503 | 2,896 | 15,313 | ... |
| Norway | 92,920 | 79.6 | 60,597 | 14,966 | 458 | 572 | 4,114 | 2,705 | ... |
| Poland | 258,910 | 90.3 | 27,359 | 136,490 | 168 | 451 | 19,627 | 17,958 | 39,200 |
| Portugal | 82,900 | 86.0 | ... | 45,032 | 222 | 507 | 2,842 | 3,814 | 2,550 |
| Romania | 198,817 | 30.2 | 7,985 | 51,531 | 72 | 180 | 10,784 | 6,880 | 12,861 |
| Russian Federation | 933,000 | 80.9 | 78,000 | 199,000 | 87 | 245 | 84,158 | 175,800 | 2,400,000 |
| San Marino | ... | ... | ... | ... | ... | ... | ... | ... | ... |
| Serbia | 39,184 | 62.7 | 3,865 | 452 | 137 | 244 | 4,058 | 749 | 4,214 |
| Slovakia | 43,761 | 87.0 | 7,816 | 22,114 | 194 | 282 | 3,592 | 2,279 | 9,004 |
| Slovenia | 38,708 | 100.0 | 817 | 12,112 | 306 | 547 | 1,228 | 834 | 3,520 |
| Spain | 666,292 | 99.0 | 397,117 | 132,868 | 360 | 601 | 15,046 | 23,344 | 10,224 |
| Sweden | 427,045 | 31.7 | 109,300 | 40,123 | 464 | 523 | 9,830 | 7,156 | 11,500 |

# TABLE B.6

## *continued*

| | Roads | | | | Motor vehicles | | Railways | | |
|---|---|---|---|---|---|---|---|---|---|
| | Total (km) | Paved (%) | Passengers (m-p-km) | Goods hauled (m-t-km) | Number per 1000 population | | Route (km) | Passengers (m-p-km) | Goods hauled (m-t-km) |
| | 2000–2007 | 2000–2007 | 2000–2007 | 2000–2007 | 1990 | 2007 | 2000–2008 | 2000–2008 | 2000–2008 |
| Switzerland | 71,354 | 100.0 | 94,250 | 16,337 | 491 | 569 | 3,499 | 18,367 | 16,227 |
| TFYR Macedonia[3] | 13,840 | ... | 1,027 | 8,299 | 132 | 136 | 699 | 148 | 743 |
| Ukraine | 169,422 | 97.8 | 55,446 | 26,625 | 63 | 140 | 21,676 | 53,056 | 257,006 |
| United Kingdom | 420,009 | 100.0 | 736,000 | 166,728 | 400 | 527 | 16,321 | 51,759 | 12,512 |
| **LATIN AMERICA AND THE CARIBBEAN** | | | | | | | | | |
| Anguilla | ... | ... | ... | ... | ... | ... | ... | ... | ... |
| Antigua and Barbuda | ... | ... | ... | ... | ... | ... | ... | ... | ... |
| Argentina | 231,374 | 30.0 | ... | ... | 181 | 314 | 35,753 | ... | 12,871 |
| Aruba | ... | ... | ... | ... | ... | ... | ... | ... | ... |
| Bahamas | ... | ... | ... | ... | ... | ... | ... | ... | ... |
| Barbados | ... | ... | ... | ... | ... | ... | ... | ... | ... |
| Belize | ... | ... | ... | ... | ... | ... | ... | ... | ... |
| Bolivia | 62,479 | 7.0 | | | 41 | 68 | 2,866 | 313 | 1,060 |
| Brazil | 1,751,868 | 5.5 | | | 88 | 198 | 29,817 | ... | 267,700 |
| British Virgin Islands | ... | ... | | | ... | ... | ... | ... | ... |
| Cayman Islands | ... | ... | | | ... | ... | ... | ... | ... |
| Chile | 79,814 | 20.2 | ... | | 81 | 164 | 5,898 | 759 | 4,296 |
| Colombia | 164,278 | ... | 157 | 39,726 | 39 | 66 | 1,663 | ... | 9,049 |
| Costa Rica | 36,654 | 25.5 | 27 | ... | 87 | 152 | ... | ... | ... |
| Cuba | 60,856 | 49.0 | 5,266 | 2,133 | 37 | 38 | 5,076 | 1,285 | 1,351 |
| Dominica | ... | ... | ... | ... | ... | ... | ... | ... | ... |
| Dominican Republic | 12,600 | 49.4 | ... | ... | 75 | 123 | ... | ... | ... |
| Ecuador | 43,670 | 14.8 | 11,819 | 5,453 | 35 | 63 | ... | ... | ... |
| El Salvador | 10,029 | 19.8 | ... | ... | 33 | 84 | ... | ... | ... |
| Falkland Islands (Malvinas) | ... | ... | ... | ... | ... | ... | ... | ... | ... |
| French Guiana | ... | ... | ... | ... | ... | ... | ... | ... | ... |
| Grenada | ... | ... | ... | ... | ... | ... | ... | ... | ... |
| Guadeloupe | ... | ... | ... | ... | ... | ... | ... | ... | ... |
| Guatemala | 14,095 | 34.5 | ... | ... | 21 | 117 | ... | ... | ... |
| Guyana | ... | ... | ... | ... | ... | ... | ... | ... | ... |
| Haiti | 4,160 | 24.3 | ... | ... | 8 | ... | ... | ... | ... |
| Honduras | 13,600 | 20.4 | ... | ... | 22 | 97 | ... | ... | ... |
| Jamaica | 22,121 | 73.3 | ... | ... | 52 | 188 | ... | ... | ... |
| Martinique | ... | ... | ... | ... | ... | ... | ... | ... | ... |
| Mexico | 360,075 | 38.2 | 449,917 | 209,392 | 119 | 244 | 26,677 | 84 | 71,136 |
| Montserrat | ... | ... | ... | ... | ... | ... | ... | ... | ... |
| Netherlands Antilles | ... | ... | ... | ... | ... | ... | ... | ... | ... |
| Nicaragua | 18,669 | 11.4 | ... | ... | 19 | 48 | ... | ... | ... |
| Panama | 11,643 | 34.6 | ... | ... | 75 | 188 | ... | ... | ... |
| Paraguay | 29,500 | 50.8 | ... | ... | 27 | 82 | ... | ... | ... |
| Peru | 78,986 | 13.9 | ... | ... | ... | 52 | 2,020 | 55 | 627 |
| Puerto Rico | 25,645 | 95.0 | ... | 10 | 295 | 642 | ... | ... | ... |
| Saint Kitts and Nevis | ... | ... | ... | ... | ... | ... | ... | ... | ... |
| Saint Lucia | ... | ... | ... | ... | ... | ... | ... | ... | ... |
| Saint Vincent and the Grenadines | ... | ... | ... | ... | ... | ... | ... | ... | ... |
| Suriname | ... | ... | ... | ... | ... | ... | ... | ... | ... |
| Trinidad and Tobago | 8,320 | 51.1 | ... | ... | 117 | 351 | ... | ... | ... |
| Turks and Caicos Islands | ... | ... | ... | ... | ... | ... | ... | ... | ... |
| United States Virgin Islands | ... | ... | ... | ... | ... | ... | ... | ... | ... |
| Uruguay | 77,732 | 10.0 | 2,032 | ... | 138 | 176 | 2,993 | 15 | 284 |
| Venezuela (Bolivarian Republic of) | 96,155 | 33.6 | ... | ... | 93 | 147 | 336 | | 81 |
| **NORTHERN AMERICA** | | | | | | | | | |
| Bermuda | ... | ... | ... | ... | ... | ... | ... | ... | ... |
| Canada | 1,409,000 | 39.9 | 493,814 | 184,774 | 605 | 597 | 57,216 | 3,056 | 358,154 |
| Greenland | ... | ... | ... | ... | ... | ... | ... | ... | ... |
| Saint-Pierre-et-Miquelon | ... | ... | ... | ... | ... | ... | ... | ... | ... |
| United States of America | 6,544,257 | 65.3 | 7,940,003 | 1,889,923 | 758 | 814[5] | 227,058 | 9,935 | 2,788,230[6] |
| **OCEANIA** | | | | | | | | | |
| American Samoa | ... | ... | ... | ... | ... | ... | ... | ... | ... |
| Australia | 815,074 | ... | 301,550 | 173,000 | 530 | 653 | 9,661 | 1,526 | 61,019 |
| Cook Islands | ... | ... | ... | ... | ... | ... | ... | ... | ... |
| Fiji | ... | ... | ... | ... | ... | ... | ... | ... | ... |
| French Polynesia | ... | ... | ... | ... | ... | ... | ... | ... | ... |
| Guam | ... | ... | ... | ... | ... | ... | ... | ... | ... |
| Kiribati | ... | ... | ... | ... | ... | ... | ... | ... | ... |
| Marshall Islands | ... | ... | ... | ... | ... | ... | ... | ... | ... |
| Micronesia (Federated States of) | ... | ... | ... | ... | ... | ... | ... | ... | ... |
| Nauru | ... | ... | ... | ... | ... | ... | ... | ... | ... |
| New Caledonia | ... | ... | ... | ... | ... | ... | ... | ... | ... |
| New Zealand | 93,748 | 65.4 | ... | ... | 524 | 729 | ... | ... | ... |
| Niue | ... | ... | ... | ... | ... | ... | ... | ... | ... |
| Northern Mariana Islands | ... | ... | ... | ... | ... | ... | ... | ... | ... |
| Palau | ... | ... | ... | ... | ... | ... | ... | ... | ... |
| Papua New Guinea | 19,600 | 3.5 | ... | ... | 27 | 9 | ... | ... | ... |
| Pitcairn | ... | ... | ... | ... | ... | ... | ... | ... | ... |
| Samoa | ... | ... | ... | ... | ... | ... | ... | ... | ... |

# TABLE B.6

*continued*

| | Roads | | | | Motor vehicles | | Railways | | |
|---|---|---|---|---|---|---|---|---|---|
| | Total (km) | Paved (%) | Passengers (m-p-km) | Goods hauled (m-t-km) | Number per 1000 population | | Route (km) | Passengers (m-p-km) | Goods hauled (m-t-km) |
| | 2000–2007[1] | 2000–2007[1] | 2000–2007[1] | 2000–2007[1] | 1990 | 2007 | 2000–2008[1] | 2000–2008[1] | 2000–2008[1] |
| Solomon Islands | ... | ... | ... | ... | ... | ... | ... | ... | ... |
| Tokelau | ... | ... | ... | ... | ... | ... | ... | ... | ... |
| Tonga | ... | ... | ... | ... | ... | ... | ... | ... | ... |
| Tuvalu | ... | ... | ... | ... | ... | ... | ... | ... | ... |
| Vanuatu | ... | ... | ... | ... | ... | ... | ... | ... | ... |
| Wallis and Futuna Islands | ... | ... | ... | ... | ... | ... | ... | ... | ... |

*Sources:* World Bank (2005) *World Development Indicators 2005*, World Bank, Washington, DC.; World Bank (2010) *World Development Indicators 2010*, World Bank, Washington, DC.

*Notes:*
(1) Data are for the latest year available in the period shown.
(2) Includes Tazara railway.
(3) The former Yugoslav Republic of Macedonia.
(4) The number of passenger cars per 1,000 people is 429. Passenger cars as a subset of motor vehicles, are road motor vehicles other than two-wheelers, intended for the carriage of passengers and designed to seat no more than nine peole (including the driver).
(5) Data are from the US Federal Highway Adminstration.
(6) Refers to class 1 railways only.

# TABLE B.7

## Greenhouse Gas Emissions and Rate of Change

| | Carbon dioxide emissions | | Methane emissions | | | Nitrous oxide emissions | | | Other greenhouse gas emissions | | Total greenhouse gas emissions | |
|---|---|---|---|---|---|---|---|---|---|---|---|---|
| | '000 tonnes | % change | '000 tonnes of CO$_2$ equivalent | From non-agricultural sources % | % change | '000 tonnes of CO$_2$ equivalent | From non-agricultural sources % | % change | '000 tonnes of CO$_2$ equivalent | % change | '000 tonnes of CO$_2$ equivalent | % change |
| | 2005 | 1990–2005 | 2005 | 2005 | 1990–2005 | 2005 | 2005 | 1990–2005 | 2005 | 1990–2005 | 2005 | 1990–2005 |
| **AFRICA** | | | | | | | | | | | | |
| Algeria | 138,078 | 5.0 | 24,310 | 84.7 | 2.1 | 10,330 | 10.9 | 1.2 | 110 | -3.5 | 172,828 | 4.2 |
| Angola | 9,849 | 8.2 | 37,020 | 60.9 | 11.4 | 28,350 | 64.1 | 30.3 | 0 | ... | 75,219 | ... |
| Benin | 2,565 | 17.3 | 4,840 | 52.5 | 5.2 | 4,660 | 32.0 | 8.0 | 0 | ... | 12,065 | ... |
| Botswana | 4,521 | 7.2 | 4,480 | 28.1 | 223.1 | 2,460 | 3.7 | ... | 0 | ... | 11,461 | ... |
| Burkina Faso | 788 | 2.3 | ... | ... | ... | ... | ... | ... | ... | ... | ... | ... |
| Burundi | 169 | -3.0 | ... | ... | ... | ... | ... | ... | ... | ... | ... | ... |
| Cameroon | 3,715 | 7.6 | 15,110 | 44.0 | 2.9 | 14,540 | 15.0 | 5.0 | 890 | 0.7 | 34,255 | 4.0 |
| Cape Verde | 297 | 15.8 | ... | ... | ... | ... | ... | ... | ... | ... | ... | ... |
| Central African Republic | 234 | 1.2 | ... | ... | ... | ... | ... | ... | ... | ... | ... | ... |
| Chad | 392 | 11.2 | ... | ... | ... | ... | ... | ... | ... | ... | ... | ... |
| Comoros | 88 | 1.0 | ... | ... | ... | ... | ... | ... | ... | ... | ... | ... |
| Congo | 1,605 | 2.3 | 50,320 | 73.7 | 5.4 | 38,680 | 76.8 | 6.6 | 0 | ... | 90,605 | ... |
| Côte d'Ivoire | 8,160 | 2.7 | 15,320 | 79.4 | 12.2 | 12,350 | 75.0 | 26.8 | 0 | ... | 35,830 | ... |
| Democratic Republic of the Congo | 2,143 | -3.2 | 5,750 | 88.2 | 7.7 | 2,250 | 84.4 | 11.6 | 0 | ... | 10,143 | ... |
| Djibouti | 473 | 1.2 | ... | ... | ... | ... | ... | ... | ... | ... | ... | ... |
| Egypt | 173,355 | 8.6 | 32,960 | 55.8 | 2.8 | 27,810 | 14.4 | 4.3 | 1,820 | -1.3 | 235,945 | 6.6 |
| Equatorial Guinea | 4,338 | 232.5 | ... | ... | ... | ... | ... | ... | ... | ... | ... | ... |
| Eritrea | 751 | ... | 2,410 | 22.4 | 1.0 | 2,350 | 0.9 | 5.0 | 0 | ... | 5,511 | ... |
| Ethiopia | 5,485 | 5.5 | 47,740 | 22.8 | 1.5 | 63,130 | 1.4 | 1.6 | 0 | ... | 116,355 | ... |
| Gabon | 1,869 | -4.6 | 2,040 | 95.6 | -2.3 | 420 | 42.9 | -5.2 | 0 | ... | 4,329 | ... |
| Gambia | 319 | 4.5 | ... | ... | ... | ... | ... | ... | ... | ... | ... | ... |
| Ghana | 7,467 | 6.0 | 8,630 | 50.4 | 4.2 | 10,520 | 11.4 | 8.8 | 170 | -0.7 | 26,787 | 6.1 |
| Guinea | 1,359 | 1.9 | ... | ... | ... | ... | ... | ... | ... | ... | ... | ... |
| Guinea-Bissau | 271 | 0.5 | ... | ... | ... | ... | ... | ... | ... | ... | ... | ... |
| Kenya | 10,944 | 5.9 | 20,310 | 35.0 | 0.3 | 19,060 | 3.6 | -0.8 | 0 | ... | 50,314 | ... |
| Lesotho | ... | ... | ... | ... | ... | ... | ... | ... | ... | ... | ... | ... |
| Liberia | 736 | 3.5 | ... | ... | ... | ... | ... | ... | ... | ... | ... | ... |
| Libyan Arab Jamahiriya | 54,854 | 2.4 | 8,540 | 91.1 | -0.2 | 2,050 | 8.3 | -1.9 | 290 | 12.7 | 65,734 | 1.8 |
| Madagascar | 2,796 | 12.2 | ... | ... | ... | ... | ... | ... | ... | ... | ... | ... |
| Malawi | 1,048 | 4.8 | ... | ... | ... | ... | ... | ... | ... | ... | ... | ... |
| Mali | 568 | 2.3 | ... | ... | ... | ... | ... | ... | ... | ... | ... | ... |
| Mauritania | 1,649 | -2.5 | ... | ... | ... | ... | ... | ... | ... | ... | ... | ... |
| Mauritius | 3,408 | 8.9 | ... | ... | ... | ... | ... | ... | ... | ... | ... | ... |
| Mayotte | ... | ... | ... | ... | ... | ... | ... | ... | ... | ... | ... | ... |
| Morocco | 47,496 | 6.8 | 13,240 | 58.4 | 3.1 | 15,510 | 24.8 | 0.5 | 0 | ... | 76,246 | ... |
| Mozambique | 1,854 | 5.7 | 11,680 | 35.7 | 1.6 | 9,930 | 0.3 | 15.8 | 0 | ... | 23,464 | ... |
| Namibia | 2,722 | 2470.0 | 4,260 | 10.1 | -0.1 | 4,620 | 0.9 | 0.6 | 0 | ... | 11,602 | ... |
| Niger | 927 | -0.8 | ... | ... | ... | ... | ... | ... | ... | ... | ... | ... |
| Nigeria | 113,786 | 10.1 | 78,290 | 66.3 | 2.1 | 39,030 | 12.9 | 2.6 | 80 | -2.2 | 231,186 | 4.9 |
| Réunion | ... | ... | ... | ... | ... | ... | ... | ... | ... | ... | ... | ... |
| Rwanda | 766 | 0.8 | ... | ... | ... | ... | ... | ... | ... | ... | ... | ... |
| Saint Helena | ... | ... | ... | ... | ... | ... | ... | ... | ... | ... | ... | ... |
| São Tomé and Príncipe | 103 | 3.7 | ... | ... | ... | ... | ... | ... | ... | ... | ... | ... |
| Senegal | 5,573 | 5.0 | 6,340 | 24.1 | 0.9 | 10,250 | 1.0 | 4.3 | 10 | ... | 22,173 | ... |
| Seychelles | 696 | 34.2 | ... | ... | ... | ... | ... | ... | ... | ... | ... | ... |
| Sierra Leone | 1,004 | 10.6 | ... | ... | ... | ... | ... | ... | ... | ... | ... | ... |
| Somalia | 253 | 85.3 | ... | ... | ... | ... | ... | ... | ... | ... | ... | ... |
| South Africa | 408,792 | 1.5 | 59,200 | 76.2 | 0.9 | 29,250 | 17.3 | 0.7 | 2,600 | 5.3 | 499,842 | 1.4 |
| Sudan | 10,992 | 6.5 | 67,310 | 26.7 | 4.6 | 59,750 | 3.8 | 3.4 | 0 | ... | 138,052 | ... |
| Swaziland | 1,019 | 9.3 | ... | ... | ... | ... | ... | ... | ... | ... | ... | ... |
| Togo | 1,337 | 4.9 | 2,840 | 51.4 | 3.9 | 5,470 | 11.2 | 11.7 | 0 | ... | 9,647 | ... |
| Tunisia | 22,783 | 4.8 | 4,390 | 65.8 | 1.2 | 7,230 | 5.8 | 4.6 | 30 | ... | 34,433 | ... |
| Uganda | 2,338 | 12.4 | ... | ... | ... | ... | ... | ... | ... | ... | ... | ... |
| United Republic of Tanzania | 5,082 | 7.6 | 39,460 | 36.5 | 3.1 | 31,690 | 15.7 | 2.4 | 0 | ... | 76,232 | ... |
| Western Sahara | ... | ... | ... | ... | ... | ... | ... | ... | ... | ... | ... | ... |
| Zambia | 2,363 | -0.2 | 16,770 | 31.4 | 4.7 | 11,410 | 34.9 | 9.2 | 0 | ... | 30,543 | ... |
| Zimbabwe | 11,542 | -2.0 | 10,400 | 39.6 | -0.3 | 10,160 | 2.9 | 0.9 | 20 | ... | 32,122 | ... |
| **ASIA** | | | | | | | | | | | | |
| Afghanistan | 700 | -4.9 | ... | ... | ... | ... | ... | ... | ... | ... | ... | ... |
| Armenia | 4,346 | ... | 2,300 | 49.1 | -1.7 | 450 | 6.7 | -3.4 | 10 | ... | 7,106 | ... |
| Azerbaijan | 35,259 | ... | 11,550 | 54.6 | -1.4 | 4,040 | 6.4 | 0.0 | 50 | -4.8 | 50,899 | ... |
| Bahrain | 19,668 | 4.4 | 1,970 | 99.5 | 1.6 | 60 | 66.7 | 1.3 | 190 | -6.0 | 21,888 | 2.8 |
| Bangladesh | 40,080 | 10.6 | 92,530 | 30.8 | 0.9 | 37,100 | 8.1 | 4.4 | 0 | ... | 169,710 | ... |
| Bhutan | 392 | 13.7 | ... | ... | ... | ... | ... | ... | ... | ... | ... | ... |
| Brunei Darussalam | 5,903 | -0.5 | 2,060 | 99.0 | 1.7 | 370 | 97.3 | 28.6 | 0 | ... | 8,333 | ... |
| Cambodia | 3,719 | 48.3 | 14,890 | 28.5 | ... | 3,820 | 25.9 | 0.0 | 0 | ... | 22,429 | ... |
| China | 5,621,470 | 8.9 | 995,760 | 50.0 | 0.7 | 566,680 | 7.3 | 1.6 | 119,720 | 85.7 | 7,303,630 | 6.2 |
| China, Hong Kong SAR | 41,062 | 3.2 | 1,090 | 99.1 | -0.5 | 200 | 95.0 | -0.3 | 330 | ... | 42,682 | ... |
| China, Macao SAR | 2,308 | 8.2 | ... | ... | ... | ... | ... | ... | ... | ... | ... | ... |
| Cyprus | 7,497 | 4.1 | 330 | 48.5 | 1.5 | 640 | 9.4 | 1.2 | 0 | ... | 8,467 | ... |
| Democratic People's Republic of Korea | 83,411 | -4.4 | 10,650 | 63.6 | 0.6 | 23,160 | 2.5 | 10.1 | 860 | 12.4 | 118,081 | -3.7 |
| Georgia | 4,796 | ... | 4,330 | 48.3 | -1.7 | 3,390 | 50.7 | 0.0 | 10 | ... | 12,526 | ... |
| India | 1,422,808 | 7.1 | 712,330 | 35.2 | 0.9 | 300,680 | 7.0 | 2.2 | 9,510 | 1.2 | 2,445,328 | 3.9 |
| Indonesia | 330,537 | 8.0 | 224,330 | 58.8 | 1.6 | 69,910 | 27.4 | 1.1 | 900 | -2.3 | 625,677 | 4.0 |
| Iran (Islamic Republic of) | 435,719 | 6.1 | 95,060 | 78.2 | 4.9 | 66,140 | 2.4 | 2.4 | 1,560 | -1.8 | 598,479 | 5.3 |

# TABLE B.7

## continued

| | Carbon dioxide emissions | | Methane emissions | | | Nitrous oxide emissions | | | Other greenhouse gas emissions | | Total greenhouse gas emissions | |
|---|---|---|---|---|---|---|---|---|---|---|---|---|
| | '000 tonnes | % change | '000 tonnes of $CO_2$ equivalent | From non-agricultural sources % | % change | '000 tonnes of $CO_2$ equivalent | From non-agricultural sources % | % change | '000 tonnes of $CO_2$ equivalent | % change | '000 tonnes of $CO_2$ equivalent | % change |
| | 2005 | 1990–2005 | 2005 | 2005 | 1990–2005 | 2005 | 2005 | 1990–2005 | 2005 | 1990–2005 | 2005 | 1990–2005 |
| Iraq | 88,566 | 4.6 | 10,980 | 85.3 | -0.1 | 3,990 | 7.0 | -2.6 | 470 | 1.4 | 104,006 | 3.2 |
| Israel | 63,618 | 6.0 | 1,170 | 63.2 | 1.1 | 1,820 | 16.5 | -0.3 | 1,140 | 2.4 | 67,748 | 5.5 |
| Japan | 1,299,243 | 0.7 | 53,480 | 86.6 | -0.5 | 23,590 | 50.7 | -1.7 | 70,570 | 11.0 | 1,446,883 | 0.8 |
| Jordan | 21,317 | 7.0 | 1,610 | 75.8 | 3.3 | 1,240 | 6.5 | 0.5 | 10 | ... | 24,177 | ... |
| Kazakhstan | 177,110 | ... | 28,270 | 62.1 | -3.3 | 5,530 | 9.8 | -5.1 | 0 | ... | 210,910 | ... |
| Kuwait | 89,805 | 8.0 | 11,200 | 98.5 | 4.3 | 540 | 18.5 | 7.7 | 390 | 3.7 | 101,935 | 7.5 |
| Kyrgyzstan | 5,566 | ... | 3,520 | 27.8 | -1.7 | 3,260 | 1.2 | -1.5 | 60 | ... | 12,406 | ... |
| Lao People's Democratic Republic | 1,407 | 33.3 | ... | ... | ... | ... | ... | ... | ... | ... | ... | ... |
| Lebanon | 17,481 | 6.2 | 980 | 81.6 | 2.3 | 1,020 | 6.9 | 2.5 | 0 | ... | 19,481 | ... |
| Malaysia | 183,171 | 14.9 | 25,510 | 77.7 | 1.3 | 9,920 | 35.7 | -1.0 | 530 | -3.0 | 219,131 | 9.5 |
| Maldives | 678 | 22.7 | ... | ... | ... | ... | ... | ... | ... | ... | ... | ... |
| Mongolia | 8,805 | -0.8 | 4,840 | 16.1 | -2.3 | 22,850 | 0.4 | 8.6 | 0 | ... | 36,495 | ... |
| Myanmar | 10,464 | 9.7 | 60,840 | 30.0 | 3.4 | 25,900 | 33.2 | 5.3 | 10 | ... | 97,214 | ... |
| Nepal | 3,166 | 26.6 | 36,040 | 19.5 | 0.4 | 7,100 | 11.5 | 1.6 | 0 | ... | 46,306 | ... |
| Occupied Palestinian Territory | 2,752 | ... | ... | ... | ... | ... | ... | ... | ... | ... | ... | ... |
| Oman | 31,444 | 13.6 | 4,260 | 87.1 | 7.4 | 1,140 | 3.5 | 2.1 | 0 | ... | 36,844 | ... |
| Pakistan | 133,960 | 6.4 | 110,300 | 33.7 | 2.2 | 80,040 | 3.6 | 3.0 | 620 | -0.8 | 324,920 | 3.8 |
| Philippines | 76,369 | 4.8 | 44,860 | 33.3 | 1.0 | 18,940 | 4.4 | 0.4 | 350 | 16.7 | 140,519 | 2.6 |
| Qatar | 46,676 | 19.8 | 5,190 | 98.5 | 8.8 | 280 | 14.3 | 3.7 | 0 | ... | 52,146 | ... |
| Republic of Korea | 473,836 | 6.4 | 31,280 | 68.9 | 0.9 | 22,020 | 63.9 | 8.8 | 8,700 | 4.1 | 535,836 | 5.9 |
| Saudi Arabia | 366,766 | 4.7 | 63,500 | 98.1 | 4.0 | 7,720 | 7.9 | -0.4 | 1,530 | -2.2 | 439,516 | 4.4 |
| Singapore | 59,514 | 1.8 | 1,260 | 95.2 | 4.7 | 7,970 | 99.2 | 288.5 | 1,300 | 15.0 | 70,044 | 3.0 |
| Sri Lanka | 11,582 | 13.8 | 10,280 | 38.2 | 0.0 | 3,130 | 10.9 | 2.0 | 0 | ... | 24,992 | ... |
| Syrian Arab Republic | 66,549 | 5.2 | 7,960 | 65.3 | 2.5 | 9,430 | 5.1 | 1.3 | 0 | ... | 83,939 | ... |
| Tajikistan | 5,800 | ... | 3,270 | 31.5 | -0.8 | 1,590 | 0.6 | -3.3 | 120 | 3.3 | 10,780 | ... |
| Thailand | 270,894 | 12.2 | 78,840 | 23.9 | 1.0 | 27,990 | 12.1 | 2.1 | 940 | -2.7 | 378,664 | 6.8 |
| Timor-Leste | 176 | ... | ... | ... | ... | ... | ... | ... | ... | ... | ... | ... |
| Turkey | 248,295 | 4.6 | 23,140 | 40.5 | -1.0 | 47,950 | 12.0 | 0.6 | 1,480 | -3.2 | 320,865 | 3.0 |
| Turkmenistan | 41,726 | ... | 23,060 | 84.8 | -2.0 | 3,200 | 21.3 | -1.5 | 250 | ... | 68,236 | ... |
| United Arab Emirates | 135,594 | 9.8 | 34,250 | 98.3 | 5.3 | 2,730 | 9.5 | 12.9 | 480 | 7.9 | 173,054 | 8.7 |
| Uzbekistan | 112,481 | ... | 51,480 | 76.8 | 1.6 | 14,660 | 1.7 | 0.2 | 760 | ... | 179,381 | ... |
| Viet Nam | 101,764 | 25.0 | 75,080 | 33.2 | 2.8 | 37,470 | 5.1 | 11.3 | 10 | ... | 214,324 | ... |
| Yemen | 20,159 | ... | 9,040 | 72.3 | 6.4 | 7,080 | 1.1 | 2.6 | 10 | ... | 36,289 | ... |
| **EUROPE** | | | | | | | | | | | | |
| Albania | 4,532 | -2.6 | 2,170 | 30.0 | -0.2 | 1,390 | 2.9 | -2.7 | 50 | ... | 8,142 | ... |
| Andorra | ... | ... | ... | ... | ... | ... | ... | ... | ... | ... | ... | ... |
| Austria | 72,767 | 1.3 | 7,210 | 49.9 | -0.8 | 4,620 | 14.7 | -1.3 | 3,310 | 11.6 | 87,907 | 1.1 |
| Belarus | 64,289 | ... | 16,620 | 61.2 | -0.9 | 10,360 | 34.4 | -2.1 | 440 | ... | 91,709 | ... |
| Belgium | 109,312 | 0.1 | 7,610 | 40.3 | -1.7 | 9,650 | 34.6 | -0.9 | 9,380 | 474.4 | 135,952 | 0.4 |
| Bosnia and Herzegovina | 25,593 | ... | 2,850 | 67.4 | 2.8 | 1,020 | 17.6 | -0.7 | 850 | 5.7 | 30,313 | ... |
| Bulgaria | 46,958 | -2.6 | 6,140 | 67.3 | -2.4 | 5,880 | 35.5 | -3.7 | 650 | ... | 59,628 | ... |
| Channel Islands | ... | ... | ... | ... | ... | ... | ... | ... | ... | ... | ... | ... |
| Croatia | 23,600 | ... | 3,690 | 70.2 | -0.4 | 3,590 | 36.2 | 0.4 | 720 | 0.5 | 31,600 | ... |
| Czech Republic | 114,636 | ... | 14,930 | 82.8 | -2.2 | 6,570 | 25.0 | -2.6 | 3,530 | 1170.0 | 139,666 | ... |
| Denmark | 46,793 | -0.5 | 4,920 | 32.3 | -0.9 | 7,380 | 21.4 | -1.7 | 1,460 | 30.8 | 60,553 | -0.6 |
| Estonia | 18,203 | ... | 1,230 | 65.0 | -3.5 | 610 | 16.4 | -4.2 | 60 | ... | 20,103 | ... |
| Faeroe Islands | 678 | 0.6 | ... | ... | ... | ... | ... | ... | ... | ... | ... | ... |
| Finland | 54,755 | 0.5 | 5,470 | 69.7 | -1.7 | 5,330 | 40.5 | -0.8 | 1,030 | 24.5 | 66,585 | 0.2 |
| France | 394,360 | -0.1 | 43,520 | 28.9 | -1.6 | 78,090 | 22.7 | -0.8 | 27,010 | 10.1 | 542,980 | -0.1 |
| Germany | 803,065 | ... | 58,100 | 60.8 | -3.1 | 69,470 | 25.8 | -0.7 | 41,980 | 18.3 | 972,615 | ... |
| Gibraltar | ... | ... | ... | ... | ... | ... | ... | ... | ... | ... | ... | ... |
| Greece | 98,847 | 2.4 | 7,410 | 60.9 | 1.1 | 13,090 | 8.7 | 0.0 | 1,620 | 7.0 | 120,967 | 2.0 |
| Holy See | ... | ... | ... | ... | ... | ... | ... | ... | ... | ... | ... | ... |
| Hungary | 58,778 | -0.3 | 11,050 | 81.7 | -1.5 | 8,760 | 24.0 | -1.8 | 1,540 | 6.8 | 80,128 | -0.7 |
| Iceland | 2,184 | 0.4 | 330 | 45.5 | -0.4 | 650 | 40.0 | 3.7 | 80 | -6.0 | 3,244 | -0.7 |
| Ireland | 44,001 | 2.8 | 3,660 | 68.0 | -4.6 | 12,320 | 7.4 | -0.3 | 2,050 | 117.6 | 62,031 | 0.8 |
| Isle of Man | ... | ... | ... | ... | ... | ... | ... | ... | ... | ... | ... | ... |
| Italy | 469,798 | 0.7 | 36,670 | 62.3 | -0.9 | 37,200 | 29.5 | 0.3 | 27,710 | 32.1 | 571,378 | 0.8 |
| Latvia | 7,057 | ... | 2,290 | 70.7 | -3.1 | 1,390 | 11.5 | -3.2 | 110 | ... | 10,847 | ... |
| Liechtenstein | ... | ... | ... | ... | ... | ... | ... | ... | ... | ... | ... | ... |
| Lithuania | 13,989 | ... | 3,650 | 61.9 | -3.5 | 2,860 | 9.8 | -2.1 | 150 | ... | 20,649 | ... |
| Luxembourg | 11,318 | 1.0 | 180 | 100.0 | -1.7 | 80 | 100.0 | 4.0 | 50 | ... | 11,628 | ... |
| Malta | 2,587 | 1.0 | 100 | 60.0 | 0.7 | 50 | 20.0 | 0.0 | 0 | ... | 2,737 | ... |
| Moldova | 8,138 | ... | 2,590 | 69.1 | -3.1 | 970 | 5.2 | -4.7 | 360 | ... | 12,058 | ... |
| Monaco | ... | ... | ... | ... | ... | ... | ... | ... | ... | ... | ... | ... |
| Montenegro | ... | ... | ... | ... | ... | ... | ... | ... | ... | ... | ... | ... |
| Netherlands | 174,890 | 0.3 | 15,180 | 50.8 | -1.4 | 16,800 | 48.5 | -0.9 | 5,300 | -0.7 | 212,170 | 0.0 |
| Norway | 60,811 | 6.3 | 12,080 | 85.7 | 3.9 | 4,680 | 47.0 | -0.8 | 1,770 | -4.3 | 79,341 | 4.1 |
| Poland | 303,346 | -0.8 | 60,060 | 81.6 | -2.2 | 26,110 | 27.5 | -1.2 | 1,270 | 11.7 | 390,786 | -1.1 |
| Portugal | 65,413 | 3.2 | 7,140 | 47.1 | -0.3 | 7,000 | 19.3 | 0.1 | 1,050 | 47.2 | 80,603 | 2.5 |
| Romania | 91,791 | -2.8 | 23,260 | 69.9 | -3.0 | 11,790 | 30.4 | -3.5 | 2,220 | 3.2 | 129,061 | -2.9 |
| Russian Federation | 1,514,412 | ... | 501,380 | 92.1 | -1.4 | 42,650 | 23.8 | -4.5 | 56,600 | 12.8 | 2,115,042 | ... |
| San Marino | ... | ... | ... | ... | ... | ... | ... | ... | ... | ... | ... | ... |
| Serbia | ... | ... | ... | ... | ... | ... | ... | ... | ... | ... | ... | ... |
| Slovakia | 37,666 | ... | 5,290 | 80.5 | -1.9 | 2,760 | 42.0 | -2.7 | 710 | 466.7 | 46,426 | ... |

# TABLE B.7

## *continued*

| | Carbon dioxide emissions | | Methane emissions | | | Nitrous oxide emissions | | | Other greenhouse gas emissions | | Total greenhouse gas emissions | |
|---|---|---|---|---|---|---|---|---|---|---|---|---|
| | '000 tonnes | % change | '000 tonnes of $CO_2$ equivalent | From non-agricultural sources % | % change | '000 tonnes of $CO_2$ equivalent | From non-agricultural sources % | % change | '000 tonnes of $CO_2$ equivalent | % change | '000 tonnes of $CO_2$ equivalent | % change |
| | 2005 | 1990–2005 | 2005 | 2005 | 1990–2005 | 2005 | 2005 | 1990–2005 | 2005 | 1990–2005 | 2005 | 1990–2005 |
| Slovenia | 14,916 | ... | 1,630 | 52.1 | -0.4 | 1,100 | 11.8 | 0.2 | 210 | -4.3 | 17,856 | ... |
| Spain | 356,196 | 3.7 | 38,010 | 55.9 | 1.3 | 48,520 | 14.3 | 2.5 | 15,050 | 15.9 | 457,776 | 3.5 |
| Sweden | 51,454 | 0.0 | 6,460 | 58.5 | -1.1 | 6,070 | 23.2 | -0.3 | 1,620 | 4.2 | 65,604 | -0.1 |
| Switzerland | 41,323 | -0.3 | 4,150 | 32.0 | -0.9 | 2,840 | 21.8 | -0.7 | 3,310 | 22.4 | 51,623 | 0.0 |
| TFYR Macedonia[1] | 11,230 | ... | ... | ... | ... | ... | ... | ... | ... | ... | ... | ... |
| Ukraine | 326,997 | ... | 75,640 | 84.3 | -3.2 | 23,270 | 45.8 | -4.4 | 1,390 | 147.8 | 427,297 | ... |
| United Kingdom | 553,238 | -0.2 | 39,400 | 49.3 | -2.8 | 65,480 | 47.8 | -0.3 | 14,030 | 9.2 | 672,148 | -0.4 |
| **LATIN AMERICA AND THE CARIBBEAN** | | | | | | | | | | | | |
| Anguilla | ... | ... | ... | ... | ... | ... | ... | ... | ... | ... | ... | ... |
| Antigua and Barbuda | 410 | 2.4 | ... | ... | ... | ... | ... | ... | ... | ... | ... | ... |
| Argentina | 158,823 | 2.7 | 94,340 | 36.1 | 1.0 | 83,410 | 2.3 | 1.9 | 930 | -3.4 | 337,503 | 1.9 |
| Aruba | 2,308 | 1.7 | ... | ... | ... | ... | ... | ... | ... | ... | ... | ... |
| Bahamas | 2,107 | 0.5 | ... | ... | ... | ... | ... | ... | ... | ... | ... | ... |
| Barbados | 1,315 | 1.5 | ... | ... | ... | ... | ... | ... | ... | ... | ... | ... |
| Belize | 817 | 10.8 | ... | ... | ... | ... | ... | ... | ... | ... | ... | ... |
| Bolivia | 9,559 | 4.9 | 27,120 | 65.5 | 5.0 | 28,300 | 56.7 | 6.5 | 0 | ... | 64,979 | ... |
| Brazil | 349,696 | 4.5 | 421,820 | 32.9 | 3.2 | 300,300 | 25.6 | 2.1 | 7,760 | 3.1 | 1,079,576 | 3.2 |
| British Virgin Islands | ... | ... | ... | ... | ... | ... | ... | ... | ... | ... | ... | ... |
| Cayman Islands | 502 | 6.6 | ... | ... | ... | ... | ... | ... | ... | ... | ... | ... |
| Chile | 59,397 | 4.5 | 19,560 | 70.1 | 2.5 | 12,590 | 11.3 | 3.6 | 10 | ... | 91,557 | ... |
| Colombia | 59,130 | 0.2 | 61,690 | 44.9 | 1.7 | 24,530 | 22.0 | 1.1 | 330 | 4.9 | 145,680 | 0.9 |
| Costa Rica | 7,273 | 9.8 | 2,450 | 42.0 | -2.3 | 2,850 | 1.1 | -1.1 | 0 | ... | 12,573 | ... |
| Cuba | 24,853 | -1.7 | 9,490 | 37.6 | -0.3 | 8,330 | 12.6 | -2.6 | 110 | ... | 42,783 | ... |
| Dominica | 114 | 6.3 | ... | ... | ... | ... | ... | ... | ... | ... | ... | ... |
| Dominican Republic | 19,877 | 7.2 | 5,960 | 37.9 | 0.9 | 2,850 | 3.9 | -2.1 | 0 | ... | 28,687 | ... |
| Ecuador | 30,646 | 5.5 | 12,890 | 42.6 | 0.4 | 8,500 | 2.4 | -0.3 | 0 | ... | 52,036 | ... |
| El Salvador | 6,287 | 9.4 | 3,200 | 51.9 | 1.1 | 2,250 | 4.9 | 0.7 | 0 | ... | 11,737 | ... |
| Falkland Islands (Malvinas) | ... | ... | ... | ... | ... | ... | ... | ... | ... | ... | ... | ... |
| French Guiana | ... | ... | ... | ... | ... | ... | ... | ... | ... | ... | ... | ... |
| Grenada | 234 | 6.3 | ... | ... | ... | ... | ... | ... | ... | ... | ... | ... |
| Guadeloupe | ... | ... | ... | ... | ... | ... | ... | ... | ... | ... | ... | ... |
| Guatemala | 11,860 | 8.9 | 8,990 | 57.3 | 3.5 | 7,980 | 29.2 | 4.5 | 0 | ... | 28,830 | ... |
| Guyana | 1,491 | 2.1 | ... | ... | ... | ... | ... | ... | ... | ... | ... | ... |
| Haiti | 1,766 | 5.2 | 3,740 | 38.8 | 2.0 | 4,290 | 1.6 | 4.9 | 0 | ... | 9,796 | ... |
| Honduras | 7,779 | 13.4 | 5,380 | 28.1 | 0.5 | 3,860 | 2.1 | 0.6 | 0 | ... | 17,019 | ... |
| Jamaica | 10,157 | 1.8 | 1,160 | 52.6 | -0.3 | 1,020 | 3.9 | -1.1 | 0 | ... | 12,337 | ... |
| Martinique | ... | ... | ... | ... | ... | ... | ... | ... | ... | ... | ... | ... |
| Mexico | 429,065 | 0.8 | 120,100 | 60.4 | 1.7 | 75,500 | 9.9 | 0.5 | 3,160 | 4.2 | 627,825 | 0.9 |
| Montserrat | ... | ... | ... | ... | ... | ... | ... | ... | ... | ... | ... | ... |
| Netherlands Antilles | 3,752 | -2.6 | 110 | 90.9 | 1.5 | 60 | 66.7 | 6.7 | 0 | ... | 3,922 | ... |
| Nicaragua | 4,151 | 3.8 | 6,350 | 19.8 | 2.4 | 3,210 | 3.1 | -1.0 | 0 | ... | 13,711 | ... |
| Panama | 5,976 | 6.1 | 3,040 | 27.6 | 0.2 | 2,070 | 4.3 | -1.2 | 0 | ... | 11,086 | ... |
| Paraguay | 3,829 | 4.6 | 17,750 | 29.1 | 3.5 | 12,870 | 18.2 | 1.9 | 0 | ... | 34,449 | ... |
| Peru | 37,135 | 5.0 | 21,510 | 51.9 | 1.6 | 18,720 | 10.6 | 2.1 | 80 | ... | 77,445 | ... |
| Puerto Rico | ... | ... | ... | ... | ... | ... | ... | ... | ... | ... | ... | ... |
| Saint Kitts and Nevis | 136 | 7.0 | ... | ... | ... | ... | ... | ... | ... | ... | ... | ... |
| Saint Lucia | 374 | 8.4 | ... | ... | ... | ... | ... | ... | ... | ... | ... | ... |
| Saint Vincent and the Grenadines | 194 | 9.4 | ... | ... | ... | ... | ... | ... | ... | ... | ... | ... |
| Suriname | 2,378 | 2.1 | ... | ... | ... | ... | ... | ... | ... | ... | ... | ... |
| Trinidad and Tobago | 30,931 | 5.5 | 3,820 | 99.0 | 3.5 | 360 | 8.3 | 0.4 | 0 | ... | 35,111 | ... |
| Turks and Caicos Islands | ... | ... | ... | ... | ... | ... | ... | ... | ... | ... | ... | ... |
| United States Virgin Islands | ... | ... | ... | ... | ... | ... | ... | ... | ... | ... | ... | ... |
| Uruguay | 5,987 | 3.3 | 17,700 | 9.7 | 1.7 | 15,630 | 0.4 | 0.2 | 20 | ... | 39,337 | ... |
| Venezuela (Bolivarian Republic of) | 152,419 | 1.7 | 65,730 | 66.4 | 3.9 | 26,460 | 22.2 | 1.5 | 2,300 | 4.9 | 246,909 | 2.2 |
| **NORTHERN AMERICA** | | | | | | | | | | | | |
| Bermuda | 564 | -0.4 | ... | ... | ... | ... | ... | ... | ... | ... | ... | ... |
| Canada | 559,376 | 1.6 | 103,830 | 77.8 | 1.7 | 51,390 | 13.3 | 0.1 | 11,010 | -0.9 | 725,606 | 1.4 |
| Greenland | 557 | 0.0 | ... | ... | ... | ... | ... | ... | ... | ... | ... | ... |
| Saint-Pierre-et-Miquelon | ... | ... | ... | ... | ... | ... | ... | ... | ... | ... | ... | ... |
| United States of America | 5,837,067 | 1.3 | 810,280 | 81.6 | -0.4 | 456,210 | 25.3 | 0.7 | 108,420 | 1.3 | 7,211,977 | 1.1 |
| **OCEANIA** | | | | | | | | | | | | |
| American Samoa | ... | ... | ... | ... | ... | ... | ... | ... | ... | ... | ... | ... |
| Australia | 365,524 | 1.7 | 116,840 | 38.5 | 0.8 | 114,500 | 5.1 | 0.5 | 4,580 | 5.0 | 601,444 | 1.3 |
| Cook Islands | ... | ... | ... | ... | ... | ... | ... | ... | ... | ... | ... | ... |
| Fiji | 1,663 | 6.9 | ... | ... | ... | ... | ... | ... | ... | ... | ... | ... |
| French Polynesia | 854 | 2.4 | ... | ... | ... | ... | ... | ... | ... | ... | ... | ... |
| Guam | ... | ... | ... | ... | ... | ... | ... | ... | ... | ... | ... | ... |
| Kiribati | 26 | 1.1 | ... | ... | ... | ... | ... | ... | ... | ... | ... | ... |
| Marshall Islands | 84 | 5.1 | ... | ... | ... | ... | ... | ... | ... | ... | ... | ... |
| Micronesia (Federated States of) | ... | ... | ... | ... | ... | ... | ... | ... | ... | ... | ... | ... |
| Nauru | ... | ... | ... | ... | ... | ... | ... | ... | ... | ... | ... | ... |
| New Caledonia | 2,799 | 4.8 | ... | ... | ... | ... | ... | ... | ... | ... | ... | ... |
| New Zealand | 30,081 | 2.2 | 27,490 | 17.7 | 0.0 | 27,960 | 0.6 | -1.2 | 820 | 7.0 | 86,351 | 0.2 |
| Niue | ... | ... | ... | ... | ... | ... | ... | ... | ... | ... | ... | ... |
| Northern Mariana Islands | ... | ... | ... | ... | ... | ... | ... | ... | ... | ... | ... | ... |

# TABLE B.7
## *continued*

| | Carbon dioxide emissions | | Methane emissions | | | Nitrous oxide emissions | | | Other greenhouse gas emissions | | Total greenhouse gas emissions | |
|---|---|---|---|---|---|---|---|---|---|---|---|---|
| | '000 tonnes | % change | '000 tonnes of $CO_2$ equivalent | From non-agricultural sources % | % change | '000 tonnes of $CO_2$ equivalent | From non-agricultural sources % | % change | '000 tonnes of $CO_2$ equivalent | % change | '000 tonnes of $CO_2$ equivalent | % change |
| | 2005 | 1990–2005 | 2005 | 2005 | 1990–2005 | 2005 | 2005 | 1990–2005 | 2005 | 1990–2005 | 2005 | 1990–2005 |
| Palau | 117 | ... | ... | ... | ... | ... | ... | ... | ... | ... | ... | ... |
| Papua New Guinea | 4,609 | 7.7 | ... | ... | ... | ... | ... | ... | ... | ... | ... | ... |
| Pitcairn | ... | ... | ... | ... | ... | ... | ... | ... | ... | ... | ... | ... |
| Samoa | 158 | 1.8 | ... | ... | ... | ... | ... | ... | ... | ... | ... | ... |
| Solomon Islands | 180 | 0.8 | ... | ... | ... | ... | ... | ... | ... | ... | ... | ... |
| Tokelau | ... | ... | ... | ... | ... | ... | ... | ... | ... | ... | ... | ... |
| Tonga | 132 | 4.8 | ... | ... | ... | ... | ... | ... | ... | ... | ... | ... |
| Tuvalu | ... | ... | ... | ... | ... | ... | ... | ... | ... | ... | ... | ... |
| Vanuatu | 88 | 1.8 | ... | ... | ... | ... | ... | ... | ... | ... | ... | ... |
| Wallis and Futuna Islands | ... | ... | ... | ... | ... | ... | ... | ... | ... | ... | ... | ... |

*Source:* World Bank (2010). Data retrieved 17 June 2010 from World Development Indicators Online (WDI) database.

*Notes:*
(1) The former Yugoslav Republic of Macedonia.

# TABLE B.8

## Greenhouse Gas Emissions per Capita and as Proportion of World Total

| | Greenhouse gas emissions per capita metric tonnes of $CO_2$ equivalent | | | | | Greenhouse gas emissions as percetage of world total | | | | |
|---|---|---|---|---|---|---|---|---|---|---|
| | Carbon dioxide 2005 | Methane 2005 | Nitrous oxide 2005 | Other 2005 | Total 2005 | Carbon dioxide[1] 2005 | Methane[2] 2005 | Nitrous oxide[3] 2005 | Other[4] 2005 | Total[5] 2005 |
| **AFRICA** | | | | | | | | | | |
| Algeria | 4.20 | 0.74 | 0.31 | 0.00 | 5.25 | 0.50 | 0.37 | 0.27 | 0.02 | 0.45 |
| Angola | 0.59 | 2.23 | 1.71 | 0.00 | 4.53 | 0.04 | 0.56 | 0.75 | 0.00 | 0.19 |
| Benin | 0.33 | 0.62 | 0.59 | 0.00 | 1.54 | 0.01 | 0.07 | 0.12 | 0.00 | 0.03 |
| Botswana | 2.46 | 2.44 | 1.34 | 0.00 | 6.24 | 0.02 | 0.07 | 0.06 | 0.00 | 0.03 |
| Burkina Faso | 0.06 | ... | ... | ... | ... | 0.00 | ... | ... | ... | ... |
| Burundi | 0.02 | ... | ... | ... | ... | 0.00 | ... | ... | ... | ... |
| Cameroon | 0.21 | 0.85 | 0.82 | 0.05 | 1.93 | 0.01 | 0.23 | 0.38 | 0.15 | 0.09 |
| Cape Verde | 0.62 | ... | ... | ... | ... | 0.00 | ... | ... | ... | ... |
| Central African Republic | 0.06 | ... | ... | ... | ... | 0.00 | ... | ... | ... | ... |
| Chad | 0.04 | ... | ... | ... | ... | 0.00 | ... | ... | ... | ... |
| Comoros | 0.15 | ... | ... | ... | ... | 0.00 | ... | ... | ... | ... |
| Congo | 0.04 | 0.10 | 0.04 | 0.00 | 0.18 | 0.01 | 0.09 | 0.06 | 0.00 | 0.03 |
| Côte d'Ivoire | 0.42 | 0.80 | 0.64 | 0.00 | 1.86 | 0.03 | 0.23 | 0.33 | 0.00 | 0.09 |
| Democratic Republic of the Congo | 0.47 | 14.73 | 11.32 | 0.00 | 26.52 | 0.01 | 0.76 | 1.02 | 0.00 | 0.23 |
| Djibouti | 0.59 | ... | ... | ... | ... | 0.00 | ... | ... | ... | ... |
| Egypt | 2.25 | 0.43 | 0.36 | 0.02 | 3.06 | 0.63 | 0.50 | 0.73 | 0.30 | 0.61 |
| Equatorial Guinea | 7.13 | ... | ... | ... | ... | 0.02 | ... | ... | ... | ... |
| Eritrea | 0.17 | 0.54 | 0.53 | 0.00 | 1.24 | ... | 0.04 | 0.06 | 0.00 | 0.01 |
| Ethiopia | 0.07 | 0.64 | 0.85 | 0.00 | 1.56 | 0.02 | 0.72 | 1.67 | 0.00 | 0.30 |
| Gabon | 1.37 | 1.49 | 0.31 | 0.00 | 3.17 | 0.01 | 0.03 | 0.01 | 0.00 | 0.01 |
| Gambia | 0.21 | ... | ... | ... | ... | 0.00 | ... | ... | ... | ... |
| Ghana | 0.34 | 0.39 | 0.48 | 0.01 | 1.22 | 0.03 | 0.13 | 0.28 | 0.03 | 0.07 |
| Guinea | 0.15 | ... | ... | ... | ... | 0.00 | ... | ... | ... | ... |
| Guinea-Bissau | 0.18 | ... | ... | ... | ... | 0.00 | ... | ... | ... | ... |
| Kenya | 0.31 | 0.57 | 0.53 | 0.00 | 1.41 | 0.04 | 0.31 | 0.50 | 0.00 | 0.13 |
| Lesotho | ... | ... | ... | ... | ... | ... | ... | ... | ... | ... |
| Liberia | 0.22 | ... | ... | ... | ... | 0.00 | ... | ... | ... | ... |
| Libyan Arab Jamahiriya | 9.26 | 1.44 | 0.35 | 0.05 | 11.10 | 0.20 | 0.13 | 0.05 | 0.05 | 0.17 |
| Madagascar | 0.16 | ... | ... | ... | ... | 0.01 | ... | ... | ... | ... |
| Malawi | 0.08 | ... | ... | ... | ... | 0.00 | ... | ... | ... | ... |
| Mali | 0.05 | ... | ... | ... | ... | 0.00 | ... | ... | ... | ... |
| Mauritania | 0.55 | ... | ... | ... | ... | 0.01 | ... | ... | ... | ... |
| Mauritius | 2.74 | ... | ... | ... | ... | 0.01 | ... | ... | ... | ... |
| Mayotte | ... | ... | ... | ... | ... | ... | ... | ... | ... | ... |
| Morocco | 1.56 | 0.43 | 0.51 | 0.00 | 2.50 | 0.17 | 0.20 | 0.41 | 0.00 | 0.20 |
| Mozambique | 0.09 | 0.56 | 0.48 | 0.00 | 1.13 | 0.01 | 0.18 | 0.26 | 0.00 | 0.06 |
| Namibia | 1.35 | 2.12 | 2.30 | 0.00 | 5.77 | 0.01 | 0.06 | 0.12 | 0.00 | 0.03 |
| Niger | 0.07 | ... | ... | ... | ... | 0.00 | ... | ... | ... | ... |
| Nigeria | 0.81 | 0.56 | 0.28 | 0.00 | 1.65 | 0.41 | 1.19 | 1.03 | 0.01 | 0.60 |
| Réunion | ... | ... | ... | ... | ... | ... | ... | ... | ... | ... |
| Rwanda | 0.09 | ... | ... | ... | ... | 0.00 | ... | ... | ... | ... |
| Saint Helena | ... | ... | ... | ... | ... | ... | ... | ... | ... | ... |
| São Tomé and Príncipe | 0.67 | ... | ... | ... | ... | 0.00 | ... | ... | ... | ... |
| Senegal | 0.49 | 0.56 | 0.91 | 0.00 | 1.96 | 0.02 | 0.10 | 0.27 | 0.00 | 0.06 |
| Seychelles | 8.40 | ... | ... | ... | ... | 0.00 | ... | ... | ... | ... |
| Sierra Leone | 0.20 | ... | ... | ... | ... | 0.00 | ... | ... | ... | ... |
| Somalia | 0.03 | ... | ... | ... | ... | 0.00 | ... | ... | ... | ... |
| South Africa | 8.72 | 1.26 | 0.62 | 0.06 | 10.66 | 1.48 | 0.90 | 0.77 | 0.44 | 1.29 |
| Sudan | 0.28 | 1.74 | 1.54 | 0.00 | 3.56 | 0.04 | 1.02 | 1.58 | 0.00 | 0.36 |
| Swaziland | 0.91 | ... | ... | ... | ... | 0.00 | ... | ... | ... | ... |
| Togo | 0.22 | 0.47 | 0.91 | 0.00 | 1.60 | 0.00 | 0.04 | 0.14 | 0.00 | 0.02 |
| Tunisia | 2.27 | 0.44 | 0.72 | 0.00 | 3.43 | 0.08 | 0.07 | 0.19 | 0.01 | 0.09 |
| Uganda | 0.08 | ... | ... | ... | ... | 0.01 | ... | ... | ... | ... |
| United Republic of Tanzania | 0.13 | 1.01 | 0.81 | 0.00 | 1.95 | 0.02 | 0.60 | 0.84 | 0.00 | 0.20 |
| Western Sahara | ... | ... | ... | ... | ... | ... | ... | ... | ... | ... |
| Zambia | 0.20 | 1.43 | 0.97 | 0.00 | 2.60 | 0.01 | 0.25 | 0.30 | 0.00 | 0.08 |
| Zimbabwe | 0.93 | 0.83 | 0.81 | 0.00 | 2.57 | 0.04 | 0.16 | 0.27 | 0.00 | 0.08 |
| **ASIA** | | | | | | | | | | |
| Afghanistan | 0.03 | ... | ... | ... | ... | 0.00 | ... | ... | ... | ... |
| Armenia | 1.42 | 0.75 | 0.15 | 0.00 | 2.32 | 0.02 | 0.03 | 0.01 | 0.00 | 0.02 |
| Azerbaijan | 4.20 | 1.38 | 0.48 | 0.01 | 6.07 | 0.13 | 0.17 | 0.11 | 0.01 | 0.13 |
| Bahrain | 27.03 | 2.71 | 0.08 | 0.26 | 30.08 | 0.07 | 0.03 | 0.00 | 0.03 | 0.06 |
| Bangladesh | 0.26 | 0.60 | 0.24 | 0.00 | 1.10 | 0.14 | 1.40 | 0.98 | 0.00 | 0.44 |
| Bhutan | 0.60 | ... | ... | ... | ... | 0.00 | ... | ... | ... | ... |
| Brunei Darussalam | 15.95 | 5.57 | 1.00 | 0.00 | 22.52 | 0.02 | 0.03 | 0.01 | 0.00 | 0.02 |
| Cambodia | 0.27 | 1.07 | 0.28 | 0.00 | 1.62 | 0.01 | 0.23 | 0.10 | 0.00 | 0.06 |
| China | 4.31 | 0.76 | 0.43 | 0.09 | 5.59 | 20.32 | 15.08 | 14.97 | 20.05 | 18.89 |
| China, Hong Kong SAR | 6.03 | 0.16 | 0.03 | 0.05 | 6.27 | 0.15 | 0.02 | 0.01 | 0.06 | 0.11 |
| China, Macao SAR | 4.73 | ... | ... | ... | ... | 0.01 | ... | ... | ... | ... |
| Cyprus | 8.97 | 0.39 | 0.77 | 0.00 | 10.13 | 0.03 | 0.00 | 0.02 | 0.00 | 0.02 |
| Democratic People's Republic of Korea | 3.55 | 0.45 | 0.98 | 0.04 | 5.02 | 0.30 | 0.16 | 0.61 | 0.14 | 0.31 |
| Georgia | 1.07 | 0.97 | 0.76 | 0.00 | 2.80 | 0.02 | 0.07 | 0.09 | 0.00 | 0.03 |
| India | 1.30 | 0.65 | 0.27 | 0.01 | 2.23 | 5.14 | 10.79 | 7.94 | 1.59 | 6.33 |
| Indonesia | 1.51 | 1.02 | 0.32 | 0.00 | 2.85 | 1.19 | 3.40 | 1.85 | 0.15 | 1.62 |
| Iran (Islamic Republic of) | 6.31 | 1.38 | 0.96 | 0.02 | 8.67 | 1.57 | 1.44 | 1.75 | 0.26 | 1.55 |
| Iraq | 3.11 | 0.39 | 0.14 | 0.02 | 3.66 | 0.32 | 0.17 | 0.11 | 0.08 | 0.27 |
| Israel | 9.18 | 0.17 | 0.26 | 0.16 | 9.77 | 0.23 | 0.02 | 0.05 | 0.19 | 0.18 |

# TABLE B.8

## *continued*

| | Greenhouse gas emissions per capita metric tonnes of CO$_2$ equivalent | | | | | Greenhouse gas emissions as percetage of world total | | | | |
|---|---|---|---|---|---|---|---|---|---|---|
| | Carbon dioxide 2005 | Methane 2005 | Nitrous oxide 2005 | Other 2005 | Total 2005 | Carbon dioxide[1] 2005 | Methane[2] 2005 | Nitrous oxide[3] 2005 | Other[4] 2005 | Total[5] 2005 |
| Japan | 10.17 | 0.42 | 0.18 | 0.55 | 11.32 | 4.70 | 0.81 | 0.62 | 11.82 | 3.74 |
| Jordan | 3.94 | 0.30 | 0.23 | 0.00 | 4.47 | 0.08 | 0.02 | 0.03 | 0.00 | 0.06 |
| Kazakhstan | 11.69 | 1.87 | 0.37 | 0.00 | 13.93 | 0.64 | 0.43 | 0.15 | 0.00 | 0.55 |
| Kuwait | 35.42 | 4.42 | 0.21 | 0.15 | 40.20 | 0.32 | 0.17 | 0.01 | 0.07 | 0.26 |
| Kyrgyzstan | 1.08 | 0.68 | 0.63 | 0.01 | 2.40 | 0.02 | 0.05 | 0.09 | 0.01 | 0.03 |
| Lao People's Democratic Republic | 0.24 | ... | ... | ... | ... | 0.01 | ... | ... | ... | ... |
| Lebanon | 4.28 | 0.24 | 0.25 | 0.00 | 4.77 | 0.06 | 0.01 | 0.03 | 0.00 | 0.05 |
| Malaysia | 7.15 | 1.00 | 0.39 | 0.02 | 8.56 | 0.66 | 0.39 | 0.26 | 0.09 | 0.57 |
| Maldives | 2.32 | ... | ... | ... | ... | 0.00 | ... | ... | ... | ... |
| Mongolia | 3.45 | 1.90 | 8.96 | 0.00 | 14.31 | 0.03 | 0.07 | 0.60 | 0.00 | 0.09 |
| Myanmar | 0.22 | 1.26 | 0.54 | 0.00 | 2.02 | 0.04 | 0.92 | 0.68 | 0.00 | 0.25 |
| Nepal | 0.12 | 1.32 | 0.26 | 0.00 | 1.70 | 0.01 | 0.55 | 0.19 | 0.00 | 0.12 |
| Occupied Palestinian Territory | 0.77 | ... | ... | ... | ... | 0.01 | ... | ... | ... | ... |
| Oman | 12.01 | 1.63 | 0.44 | 0.00 | 14.08 | 0.11 | 0.06 | 0.03 | 0.00 | 0.10 |
| Pakistan | 0.86 | 0.71 | 0.51 | 0.00 | 2.08 | 0.48 | 1.67 | 2.11 | 0.10 | 0.84 |
| Philippines | 0.89 | 0.52 | 0.22 | 0.00 | 1.63 | 0.28 | 0.68 | 0.50 | 0.06 | 0.36 |
| Qatar | 52.72 | 5.86 | 0.32 | 0.00 | 58.90 | 0.17 | 0.08 | 0.01 | 0.00 | 0.13 |
| Republic of Korea | 9.84 | 0.65 | 0.46 | 0.18 | 11.13 | 1.71 | 0.47 | 0.58 | 1.46 | 1.39 |
| Saudi Arabia | 15.86 | 2.75 | 0.33 | 0.07 | 19.01 | 1.33 | 0.96 | 0.20 | 0.26 | 1.14 |
| Singapore | 13.95 | 0.30 | 1.87 | 0.30 | 16.42 | 0.22 | 0.02 | 0.21 | 0.22 | 0.18 |
| Sri Lanka | 0.59 | 0.52 | 0.16 | 0.00 | 1.27 | 0.04 | 0.16 | 0.08 | 0.00 | 0.06 |
| Syrian Arab Republic | 3.48 | 0.42 | 0.49 | 0.00 | 4.39 | 0.24 | 0.12 | 0.25 | 0.00 | 0.22 |
| Tajikistan | 0.89 | 0.50 | 0.24 | 0.02 | 1.65 | 0.02 | 0.05 | 0.04 | 0.02 | 0.03 |
| Thailand | 4.11 | 1.20 | 0.42 | 0.01 | 5.74 | 0.98 | 1.19 | 0.74 | 0.16 | 0.98 |
| Timor-Leste | 0.18 | ... | ... | ... | ... | 0.00 | ... | ... | ... | ... |
| Turkey | 3.49 | 0.33 | 0.67 | 0.02 | 4.51 | 0.90 | 0.35 | 1.27 | 0.25 | 0.83 |
| Turkmenistan | 8.62 | 4.76 | 0.66 | 0.05 | 14.09 | 0.15 | 0.35 | 0.08 | 0.04 | 0.18 |
| United Arab Emirates | 33.16 | 8.38 | 0.67 | 0.12 | 42.33 | 0.49 | 0.52 | 0.07 | 0.08 | 0.45 |
| Uzbekistan | 4.30 | 1.97 | 0.56 | 0.03 | 6.86 | 0.41 | 0.78 | 0.39 | 0.13 | 0.46 |
| Viet Nam | 1.22 | 0.90 | 0.45 | 0.00 | 2.57 | 0.37 | 1.14 | 0.99 | 0.00 | 0.55 |
| Yemen | 0.96 | 0.43 | 0.34 | 0.00 | 1.73 | 0.07 | 0.14 | 0.19 | 0.00 | 0.09 |
| **EUROPE** | | | | | | | | | | |
| Albania | 1.46 | 0.70 | 0.45 | 0.02 | 2.63 | 0.02 | 0.03 | 0.04 | 0.01 | 0.02 |
| Andorra | ... | ... | ... | ... | ... | ... | ... | ... | ... | ... |
| Austria | 8.84 | 0.88 | 0.56 | 0.40 | 10.68 | 0.26 | 0.11 | 0.12 | 0.55 | 0.23 |
| Belarus | 6.58 | 1.70 | 1.06 | 0.05 | 9.39 | 0.23 | 0.25 | 0.27 | 0.07 | 0.24 |
| Belgium | 10.43 | 0.73 | 0.92 | 0.90 | 12.98 | 0.40 | 0.12 | 0.25 | 1.57 | 0.35 |
| Bosnia and Herzegovina | 6.77 | 0.75 | 0.27 | 0.22 | 8.01 | 0.09 | 0.04 | 0.03 | 0.14 | 0.08 |
| Bulgaria | 6.07 | 0.79 | 0.76 | 0.08 | 7.70 | 0.17 | 0.09 | 0.16 | 0.11 | 0.15 |
| Channel Islands | ... | ... | ... | ... | ... | ... | ... | ... | ... | ... |
| Croatia | 5.31 | 0.83 | 0.81 | 0.16 | 7.11 | 0.09 | 0.06 | 0.09 | 0.12 | 0.08 |
| Czech Republic | 11.20 | 1.46 | 0.64 | 0.34 | 13.64 | 0.41 | 0.23 | 0.17 | 0.59 | 0.36 |
| Denmark | 8.64 | 0.91 | 1.36 | 0.27 | 11.18 | 0.17 | 0.07 | 0.19 | 0.24 | 0.16 |
| Estonia | 13.52 | 0.91 | 0.45 | 0.04 | 14.92 | 0.07 | 0.02 | 0.02 | 0.01 | 0.05 |
| Faeroe Islands | 14.05 | ... | ... | ... | ... | 0.00 | ... | ... | ... | ... |
| Finland | 10.44 | 1.04 | 1.02 | 0.20 | 12.70 | 0.20 | 0.08 | 0.14 | 0.17 | 0.17 |
| France | 6.48 | 0.71 | 1.28 | 0.44 | 8.91 | 1.43 | 0.66 | 2.06 | 4.52 | 1.40 |
| Germany | 9.74 | 0.70 | 0.84 | 0.51 | 11.79 | 2.90 | 0.88 | 1.83 | 7.03 | 2.52 |
| Gibraltar | ... | ... | ... | ... | ... | ... | ... | ... | ... | ... |
| Greece | 8.90 | 0.67 | 1.18 | 0.15 | 10.90 | 0.36 | 0.11 | 0.35 | 0.27 | 0.31 |
| Holy See | ... | ... | ... | ... | ... | ... | ... | ... | ... | ... |
| Hungary | 5.83 | 1.10 | 0.87 | 0.15 | 7.95 | 0.21 | 0.17 | 0.23 | 0.26 | 0.21 |
| Iceland | 7.36 | 1.11 | 2.19 | 0.27 | 10.93 | 0.01 | 0.00 | 0.02 | 0.01 | 0.01 |
| Ireland | 10.58 | 0.88 | 2.96 | 0.49 | 14.91 | 0.16 | 0.06 | 0.33 | 0.34 | 0.16 |
| Isle of Man | ... | ... | ... | ... | ... | ... | ... | ... | ... | ... |
| Italy | 8.02 | 0.63 | 0.63 | 0.47 | 9.75 | 1.70 | 0.56 | 0.98 | 4.64 | 1.48 |
| Latvia | 3.07 | 1.00 | 0.60 | 0.05 | 4.72 | 0.03 | 0.03 | 0.04 | 0.02 | 0.03 |
| Liechtenstein | ... | ... | ... | ... | ... | ... | ... | ... | ... | ... |
| Lithuania | 4.10 | 1.07 | 0.84 | 0.04 | 6.05 | 0.05 | 0.06 | 0.08 | 0.03 | 0.05 |
| Luxembourg | 24.33 | 0.39 | 0.17 | 0.11 | 25.00 | 0.04 | 0.00 | 0.00 | 0.01 | 0.03 |
| Malta | 6.41 | 0.25 | 0.12 | 0.00 | 6.78 | 0.01 | 0.00 | 0.00 | 0.00 | 0.01 |
| Moldova | 2.16 | 0.69 | 0.26 | 0.10 | 3.21 | 0.03 | 0.04 | 0.03 | 0.06 | 0.03 |
| Monaco | ... | ... | ... | ... | ... | ... | ... | ... | ... | ... |
| Montenegro | ... | ... | ... | ... | ... | ... | ... | ... | ... | ... |
| Netherlands | 10.72 | 0.93 | 1.03 | 0.32 | 13.00 | 0.63 | 0.23 | 0.44 | 0.89 | 0.55 |
| Norway | 13.15 | 2.61 | 1.01 | 0.38 | 17.15 | 0.22 | 0.18 | 0.12 | 0.30 | 0.21 |
| Poland | 7.95 | 1.57 | 0.68 | 0.03 | 10.23 | 1.10 | 0.91 | 0.69 | 0.21 | 1.01 |
| Portugal | 6.20 | 0.68 | 0.66 | 0.10 | 7.64 | 0.24 | 0.11 | 0.18 | 0.18 | 0.21 |
| Romania | 4.24 | 1.08 | 0.54 | 0.10 | 5.96 | 0.33 | 0.35 | 0.31 | 0.37 | 0.33 |
| Russian Federation | 10.58 | 3.50 | 0.30 | 0.40 | 14.78 | 5.47 | 7.59 | 1.13 | 9.48 | 5.47 |
| San Marino | ... | ... | ... | ... | ... | ... | ... | ... | ... | ... |
| Serbia | ... | ... | ... | ... | ... | ... | ... | ... | ... | ... |
| Slovakia | 6.99 | 0.98 | 0.51 | 0.13 | 8.61 | 0.14 | 0.08 | 0.07 | 0.12 | 0.12 |
| Slovenia | 7.46 | 0.81 | 0.55 | 0.10 | 8.92 | 0.05 | 0.02 | 0.03 | 0.04 | 0.05 |
| Spain | 8.21 | 0.88 | 1.12 | 0.35 | 10.56 | 1.29 | 0.58 | 1.28 | 2.52 | 1.18 |
| Sweden | 5.70 | 0.72 | 0.67 | 0.18 | 7.27 | 0.19 | 0.10 | 0.16 | 0.27 | 0.17 |
| Switzerland | 5.56 | 0.56 | 0.38 | 0.45 | 6.95 | 0.15 | 0.06 | 0.08 | 0.55 | 0.13 |
| TFYR Macedonia[6] | 5.52 | ... | ... | ... | ... | 0.04 | ... | ... | ... | ... |

# TABLE B.8

## continued

| | Greenhouse gas emissions per capita metric tonnes of $CO_2$ equivalent | | | | | Greenhouse gas emissions as percetage of world total | | | | |
|---|---|---|---|---|---|---|---|---|---|---|
| | Carbon dioxide 2005 | Methane 2005 | Nitrous oxide 2005 | Other 2005 | Total 2005 | Carbon dioxide[1] 2005 | Methane[2] 2005 | Nitrous oxide[3] 2005 | Other[4] 2005 | Total[5] 2005 |
| Ukraine | 6.94 | 1.61 | 0.49 | 0.03 | 9.07 | 1.18 | 1.15 | 0.61 | 0.23 | 1.11 |
| United Kingdom | 9.19 | 0.65 | 1.09 | 0.23 | 11.16 | 2.00 | 0.60 | 1.73 | 2.35 | 1.74 |
| **LATIN AMERICA AND THE CARIBBEAN** | | | | | | | | | | |
| Anguilla | ... | ... | ... | ... | ... | ... | ... | ... | ... | ... |
| Antigua and Barbuda | 4.91 | ... | ... | ... | ... | 0.00 | ... | ... | ... | ... |
| Argentina | 4.10 | 2.44 | 2.15 | 0.02 | 8.71 | 0.57 | 1.43 | 2.20 | 0.16 | 0.87 |
| Aruba | 22.84 | ... | ... | ... | ... | 0.01 | ... | ... | ... | ... |
| Bahamas | 6.47 | ... | ... | ... | ... | 0.01 | ... | ... | ... | ... |
| Barbados | 5.19 | ... | ... | ... | ... | 0.00 | ... | ... | ... | ... |
| Belize | 2.80 | ... | ... | ... | ... | 0.00 | ... | ... | ... | ... |
| Bolivia | 1.04 | 2.95 | 3.08 | 0.00 | 7.07 | 0.03 | 0.41 | 0.75 | 0.00 | 0.17 |
| Brazil | 1.88 | 2.27 | 1.61 | 0.04 | 5.80 | 1.26 | 6.39 | 7.93 | 1.30 | 2.79 |
| British Virgin Islands | ... | ... | ... | ... | ... | ... | ... | ... | ... | ... |
| Cayman Islands | 11.31 | ... | ... | ... | ... | 0.00 | ... | ... | ... | ... |
| Chile | 3.64 | 1.20 | 0.77 | 0.00 | 5.61 | 0.21 | 0.30 | 0.33 | 0.00 | 0.24 |
| Colombia | 1.37 | 1.43 | 0.57 | 0.01 | 3.38 | 0.21 | 0.93 | 0.65 | 0.06 | 0.38 |
| Costa Rica | 1.68 | 0.57 | 0.66 | 0.00 | 2.91 | 0.03 | 0.04 | 0.08 | 0.00 | 0.03 |
| Cuba | 2.22 | 0.85 | 0.74 | 0.01 | 3.82 | 0.09 | 0.14 | 0.22 | 0.02 | 0.11 |
| Dominica | 1.58 | ... | ... | ... | ... | 0.00 | ... | ... | ... | ... |
| Dominican Republic | 2.08 | 0.63 | 0.30 | 0.00 | 3.01 | 0.07 | 0.09 | 0.08 | 0.00 | 0.07 |
| Ecuador | 2.35 | 0.99 | 0.65 | 0.00 | 3.99 | 0.11 | 0.20 | 0.22 | 0.00 | 0.13 |
| El Salvador | 1.04 | 0.53 | 0.37 | 0.00 | 1.94 | 0.02 | 0.05 | 0.06 | 0.00 | 0.03 |
| Falkland Islands (Malvinas) | ... | ... | ... | ... | ... | ... | ... | ... | ... | ... |
| French Guiana | ... | ... | ... | ... | ... | ... | ... | ... | ... | ... |
| Grenada | 2.28 | ... | ... | ... | ... | 0.00 | ... | ... | ... | ... |
| Guadeloupe | ... | ... | ... | ... | ... | ... | ... | ... | ... | ... |
| Guatemala | 0.93 | 0.71 | 0.63 | 0.00 | 2.27 | 0.04 | 0.14 | 0.21 | 0.00 | 0.07 |
| Guyana | 1.95 | ... | ... | ... | ... | 0.01 | ... | ... | ... | ... |
| Haiti | 0.19 | 0.40 | 0.46 | 0.00 | 1.05 | 0.01 | 0.06 | 0.11 | 0.00 | 0.03 |
| Honduras | 1.13 | 0.78 | 0.56 | 0.00 | 2.47 | 0.03 | 0.08 | 0.10 | 0.00 | 0.04 |
| Jamaica | 3.83 | 0.44 | 0.38 | 0.00 | 4.65 | 0.04 | 0.02 | 0.03 | 0.00 | 0.03 |
| Martinique | ... | ... | ... | ... | ... | ... | ... | ... | ... | ... |
| Mexico | 4.16 | 1.17 | 0.73 | 0.03 | 6.09 | 1.55 | 1.82 | 1.99 | 0.53 | 1.62 |
| Montserrat | ... | ... | ... | ... | ... | ... | ... | ... | ... | ... |
| Netherlands Antilles | 20.12 | 0.59 | 0.32 | 0.00 | 21.03 | 0.01 | 0.00 | 0.00 | 0.00 | 0.01 |
| Nicaragua | 0.76 | 1.16 | 0.59 | 0.00 | 2.51 | 0.02 | 0.10 | 0.08 | 0.00 | 0.04 |
| Panama | 1.85 | 0.94 | 0.64 | 0.00 | 3.43 | 0.02 | 0.05 | 0.05 | 0.00 | 0.03 |
| Paraguay | 0.65 | 3.01 | 2.18 | 0.00 | 5.84 | 0.01 | 0.27 | 0.34 | 0.00 | 0.09 |
| Peru | 1.33 | 0.77 | 0.67 | 0.00 | 2.77 | 0.13 | 0.33 | 0.49 | 0.01 | 0.20 |
| Puerto Rico | ... | ... | ... | ... | ... | ... | ... | ... | ... | ... |
| Saint Kitts and Nevis | 2.83 | ... | ... | ... | ... | 0.00 | ... | ... | ... | ... |
| Saint Lucia | 2.27 | ... | ... | ... | ... | 0.00 | ... | ... | ... | ... |
| Saint Vincent and the Grenadines | 1.78 | ... | ... | ... | ... | 0.00 | ... | ... | ... | ... |
| Suriname | 4.76 | ... | ... | ... | ... | 0.01 | ... | ... | ... | ... |
| Trinidad and Tobago | 23.46 | 2.90 | 0.27 | 0.00 | 26.63 | 0.11 | 0.06 | 0.01 | 0.00 | 0.09 |
| Turks and Caicos Islands | ... | ... | ... | ... | ... | ... | ... | ... | ... | ... |
| United States Virgin Islands | ... | ... | ... | ... | ... | ... | ... | ... | ... | ... |
| Uruguay | 1.81 | 5.35 | 4.73 | 0.01 | 11.90 | 0.02 | 0.27 | 0.41 | 0.00 | 0.10 |
| Venezuela (Bolivarian Republic of) | 5.73 | 2.47 | 1.00 | 0.09 | 9.29 | 0.55 | 1.00 | 0.70 | 0.39 | 0.64 |
| **NORTHERN AMERICA** | | | | | | | | | | |
| Bermuda | 8.87 | ... | ... | ... | ... | 0.00 | ... | ... | ... | ... |
| Canada | 17.31 | 3.21 | 1.59 | 0.34 | 22.45 | 2.02 | 1.57 | 1.36 | 1.84 | 1.88 |
| Greenland | 9.78 | ... | ... | ... | ... | 0.00 | ... | ... | ... | ... |
| Saint-Pierre-et-Miquelon | ... | ... | ... | ... | ... | ... | ... | ... | ... | ... |
| United States of America | 19.75 | 2.74 | 1.54 | 0.37 | 24.40 | 21.10 | 12.27 | 12.05 | 18.16 | 18.66 |
| **OCEANIA** | | | | | | | | | | |
| American Samoa | ... | ... | ... | ... | ... | ... | ... | ... | ... | ... |
| Australia | 17.92 | 5.73 | 5.61 | 0.22 | 29.48 | 1.32 | 1.77 | 3.02 | 0.77 | 1.56 |
| Cook Islands | ... | ... | ... | ... | ... | ... | ... | ... | ... | ... |
| Fiji | 2.01 | ... | ... | ... | ... | 0.01 | ... | ... | ... | ... |
| French Polynesia | 3.34 | ... | ... | ... | ... | 0.00 | ... | ... | ... | ... |
| Guam | ... | ... | ... | ... | ... | ... | ... | ... | ... | ... |
| Kiribati | 0.28 | ... | ... | ... | ... | 0.00 | ... | ... | ... | ... |
| Marshall Islands | 1.51 | ... | ... | ... | ... | 0.00 | ... | ... | ... | ... |
| Micronesia (Federated States of) | ... | ... | ... | ... | ... | ... | ... | ... | ... | ... |
| Nauru | ... | ... | ... | ... | ... | ... | ... | ... | ... | ... |
| New Caledonia | 11.94 | ... | ... | ... | ... | 0.01 | ... | ... | ... | ... |
| New Zealand | 7.28 | 6.65 | 6.76 | 0.20 | 20.89 | 0.11 | 0.42 | 0.74 | 0.14 | 0.22 |
| Niue | ... | ... | ... | ... | ... | ... | ... | ... | ... | ... |
| Northern Mariana Islands | ... | ... | ... | ... | ... | ... | ... | ... | ... | ... |
| Palau | 5.87 | ... | ... | ... | ... | 0.00 | ... | ... | ... | ... |
| Papua New Guinea | 0.75 | ... | ... | ... | ... | 0.02 | ... | ... | ... | ... |
| Pitcairn | ... | ... | ... | ... | ... | ... | ... | ... | ... | ... |
| Samoa | 0.88 | ... | ... | ... | ... | 0.00 | ... | ... | ... | ... |
| Solomon Islands | 0.38 | ... | ... | ... | ... | 0.00 | ... | ... | ... | ... |
| Tokelau | ... | ... | ... | ... | ... | ... | ... | ... | ... | ... |
| Tonga | 1.30 | ... | ... | ... | ... | 0.00 | ... | ... | ... | ... |

# TABLE B.8

## *continued*

| | Greenhouse gas emissions per capita metric tonnes of $CO_2$ equivalent | | | | | Greenhouse gas emissions as percetage of world total | | | | |
|---|---|---|---|---|---|---|---|---|---|---|
| | Carbon dioxide 2005 | Methane 2005 | Nitrous oxide 2005 | Other 2005 | Total 2005 | Carbon dioxide[1] 2005 | Methane[2] 2005 | Nitrous oxide[3] 2005 | Other[4] 2005 | Total[5] 2005 |
| Tuvalu | ... | ... | ... | ... | ... | ... | ... | ... | ... | ... |
| Vanuatu | 0.41 | ... | ... | ... | ... | 0.00 | ... | ... | ... | ... |
| Wallis and Futuna Islands | ... | ... | ... | ... | ... | ... | ... | ... | ... | ... |

*Source:* World Bank (2010). Data retrieved 17 June 2010 from World Development Indicators Online (WDI) database.

*Notes:*
(1) Percentages are based on a total of 27,668,659 tonnes allocated to countries, exluding 1,537,085 tonnes emitted globally and not accounted for in national inventories.
(2) Percentages are based on a total of 6,603,040 tonnes allocated to countries, exluding 4,450 tonnes emitted globally and not accounted for in national inventories.
(3) Percentages are based on a total of 3,786,400 tonnes allocated to countries, exluding 1,400 tonnes emitted globally and not accounted for in national inventories.
(4) Percentages are based on a total of 597,090 tonnes allocated to countries, exluding 4,800 tonnes emitted globally and not accounted for in national inventories.
(5) Percentages are based on a total of 38,655,189 tonnes allocated to countries, exluding 1,547,735 tonnes emitted globally and not accounted for in national inventories.
(6) The former Yugoslav Republic of Macedonia.

# CITY LEVEL DATA
## TABLE C.1

### Urban Agglomerations with 750,000 Inhabitants or More: Population Size and Rate of Change

| | | Estimates and projections ('000) | | | Annual rate of change (%) | | Share in national urban population (%) | | |
|---|---|---|---|---|---|---|---|---|---|
| | | 2000 | 2010 | 2020 | 2000–2010 | 2010–2020 | 2000 | 2010 | 2020 |
| **AFRICA** | | | | | | | | | |
| Algeria | El Djazaïr (Algiers) | 2,254 | 2,800 | 3,371 | 2.17 | 1.86 | 12.4 | 11.9 | 11.5 |
| Algeria | Wahran (Oran) | 705 | 770 | 902 | 0.88 | 1.58 | 3.9 | 3.3 | 3.1 |
| Angola | Huambo | 578 | 1,034 | 1,551 | 5.82 | 4.05 | 8.3 | 9.3 | 9.6 |
| Angola | Luanda | 2,591 | 4,772 | 7,080 | 6.11 | 3.95 | 37.0 | 42.9 | 43.7 |
| Benin | Cotonou | 642 | 844 | 1,217 | 2.74 | 3.66 | 25.1 | 21.8 | 21.2 |
| Burkina Faso | Ouagadougou | 921 | 1,908 | 3,457 | 7.28 | 5.94 | 44.2 | 45.6 | 46.0 |
| Cameroon | Douala | 1,432 | 2,125 | 2,815 | 3.95 | 2.81 | 18.1 | 18.2 | 17.7 |
| Cameroon | Yaoundé | 1,192 | 1,801 | 2,392 | 4.13 | 2.84 | 15.1 | 15.5 | 15.0 |
| Chad | N'Djaména | 647 | 829 | 1,170 | 2.48 | 3.45 | 32.9 | 26.1 | 23.1 |
| Congo | Brazzaville | 986 | 1,323 | 1,703 | 2.94 | 2.52 | 55.7 | 56.7 | 54.6 |
| Côte d'Ivoire | Abidjan | 3,032 | 4,125 | 5,550 | 3.08 | 2.97 | 40.3 | 37.8 | 35.6 |
| Côte d'Ivoire | Yamoussoukro | 348 | 885 | 1,559 | 9.34 | 5.66 | 4.6 | 8.1 | 10.0 |
| Democratic Republic of the Congo | Kananga | 552 | 878 | 1,324 | 4.64 | 4.11 | 3.6 | 3.7 | 3.6 |
| Democratic Republic of the Congo | Kinshasa | 5,611 | 8,754 | 12,788 | 4.45 | 3.79 | 37.0 | 36.6 | 34.7 |
| Democratic Republic of the Congo | Kisangani | 535 | 812 | 1,221 | 4.17 | 4.07 | 3.5 | 3.4 | 3.3 |
| Democratic Republic of the Congo | Lubumbashi | 995 | 1,543 | 2,304 | 4.39 | 4.01 | 6.6 | 6.5 | 6.3 |
| Democratic Republic of the Congo | Mbuji-Mayi | 924 | 1,488 | 2,232 | 4.76 | 4.05 | 6.1 | 6.2 | 6.1 |
| Egypt | Al-Iskandariyah (Alexandria) | 3,592 | 4,387 | 5,201 | 2.00 | 1.70 | 12.0 | 12.0 | 11.5 |
| Egypt | Al-Qahirah (Cairo) | 10,170 | 11,001 | 12,540 | 0.79 | 1.31 | 33.9 | 30.0 | 27.7 |
| Ethiopia | Addis Ababa | 2,376 | 2,930 | 3,981 | 2.10 | 3.07 | 24.3 | 20.7 | 19.1 |
| Ghana | Accra | 1,674 | 2,342 | 3,110 | 3.36 | 2.84 | 19.5 | 18.7 | 18.0 |
| Ghana | Kumasi | 1,187 | 1,834 | 2,448 | 4.35 | 2.89 | 13.8 | 14.6 | 14.2 |
| Guinea | Conakry | 1,219 | 1,653 | 2,427 | 3.05 | 3.84 | 46.8 | 45.3 | 43.5 |
| Kenya | Mombasa | 687 | 1,003 | 1,479 | 3.78 | 3.88 | 11.1 | 11.1 | 10.7 |
| Kenya | Nairobi | 2,230 | 3,523 | 5,192 | 4.57 | 3.88 | 35.9 | 38.9 | 37.6 |
| Liberia | Monrovia | 836 | 827 | 807 | -0.11 | -0.24 | 66.8 | 42.2 | 29.5 |
| Libyan Arab Jamahiriya | Tarabulus (Tripoli) | 1,022 | 1,108 | 1,286 | 0.81 | 1.49 | 25.0 | 21.7 | 20.8 |
| Madagascar | Antananarivo | 1,361 | 1,879 | 2,658 | 3.23 | 3.47 | 32.9 | 30.9 | 29.7 |
| Mali | Bamako | 1,110 | 1,699 | 2,514 | 4.26 | 3.92 | 37.2 | 35.6 | 34.3 |
| Morocco | Agadir | 609 | 783 | 948 | 2.51 | 1.91 | 4.0 | 4.2 | 4.1 |
| Morocco | Dar-el-Beida (Casablanca) | 3,043 | 3,284 | 3,816 | 0.76 | 1.50 | 19.8 | 17.4 | 16.5 |
| Morocco | Fès | 870 | 1,065 | 1,277 | 2.02 | 1.82 | 5.7 | 5.6 | 5.5 |
| Morocco | Marrakech | 755 | 928 | 1,114 | 2.06 | 1.83 | 4.9 | 4.9 | 4.8 |
| Morocco | Rabat | 1,507 | 1,802 | 2,139 | 1.79 | 1.71 | 9.8 | 9.6 | 9.2 |
| Morocco | Tanger | 591 | 788 | 958 | 2.86 | 1.96 | 3.8 | 4.2 | 4.1 |
| Mozambique | Maputo | 1,096 | 1,655 | 2,350 | 4.12 | 3.51 | 19.6 | 18.4 | 17.8 |
| Mozambique | Matola | 504 | 793 | 1,139 | 4.54 | 3.62 | 9.0 | 8.8 | 8.6 |
| Niger | Niamey | 680 | 1,048 | 1,643 | 4.33 | 4.50 | 38.1 | 38.5 | 37.2 |
| Nigeria | Aba | 614 | 785 | 1,058 | 2.46 | 2.98 | 1.2 | 1.0 | 1.0 |
| Nigeria | Abuja | 832 | 1,995 | 2,977 | 8.75 | 4.00 | 1.6 | 2.5 | 2.7 |
| Nigeria | Benin City | 975 | 1,302 | 1,758 | 2.89 | 3.00 | 1.8 | 1.7 | 1.6 |
| Nigeria | Ibadan | 2,236 | 2,837 | 3,760 | 2.38 | 2.82 | 4.2 | 3.6 | 3.4 |
| Nigeria | Ilorin | 653 | 835 | 1,125 | 2.46 | 2.98 | 1.2 | 1.1 | 1.0 |
| Nigeria | Jos | 627 | 802 | 1,081 | 2.47 | 2.98 | 1.2 | 1.0 | 1.0 |
| Nigeria | Kaduna | 1,220 | 1,561 | 2,087 | 2.46 | 2.90 | 2.3 | 2.0 | 1.9 |
| Nigeria | Kano | 2,658 | 3,395 | 4,495 | 2.45 | 2.81 | 5.0 | 4.3 | 4.1 |
| Nigeria | Lagos | 7,233 | 10,578 | 14,162 | 3.80 | 2.92 | 13.6 | 13.4 | 12.9 |
| Nigeria | Maiduguri | 758 | 970 | 1,303 | 2.47 | 2.95 | 1.4 | 1.2 | 1.2 |
| Nigeria | Ogbomosho | 798 | 1,032 | 1,389 | 2.57 | 2.97 | 1.5 | 1.3 | 1.3 |
| Nigeria | Port Harcourt | 863 | 1,104 | 1,482 | 2.46 | 2.94 | 1.6 | 1.4 | 1.3 |
| Nigeria | Zaria | 752 | 963 | 1,295 | 2.47 | 2.96 | 1.4 | 1.2 | 1.2 |
| Rwanda | Kigali | 497 | 939 | 1,392 | 6.36 | 3.94 | 45.3 | 48.5 | 46.5 |
| Senegal | Dakar | 2,029 | 2,863 | 3,796 | 3.44 | 2.82 | 50.8 | 52.5 | 50.5 |
| Sierra Leone | Freetown | 688 | 901 | 1,219 | 2.70 | 3.02 | 45.8 | 40.2 | 38.9 |
| Somalia | Muqdisho (Mogadishu) | 1,201 | 1,500 | 2,156 | 2.22 | 3.63 | 48.9 | 42.8 | 40.9 |
| South Africa | Cape Town | 2,715 | 3,405 | 3,701 | 2.26 | 0.83 | 10.6 | 10.9 | 10.6 |
| South Africa | Durban | 2,370 | 2,879 | 3,133 | 1.95 | 0.85 | 9.3 | 9.2 | 8.9 |
| South Africa | Ekurhuleni (East Rand) | 2,326 | 3,202 | 3,497 | 3.20 | 0.88 | 9.1 | 10.3 | 10.0 |
| South Africa | Johannesburg | 2,732 | 3,670 | 3,996 | 2.95 | 0.85 | 10.7 | 11.8 | 11.4 |
| South Africa | Port Elizabeth | 958 | 1,068 | 1,173 | 1.09 | 0.94 | 3.8 | 3.4 | 3.3 |
| South Africa | Pretoria | 1,084 | 1,429 | 1,575 | 2.76 | 0.97 | 4.2 | 4.6 | 4.5 |
| South Africa | Vereeniging | 897 | 1,143 | 1,262 | 2.42 | 0.99 | 3.5 | 3.7 | 3.6 |
| Sudan | Al-Khartum (Khartoum) | 3,949 | 5,172 | 7,005 | 2.70 | 3.03 | 33.9 | 29.9 | 28.2 |
| Togo | Lomé | 1,020 | 1,667 | 2,398 | 4.91 | 3.64 | 53.2 | 56.6 | 56.3 |
| Uganda | Kampala | 1,097 | 1,598 | 2,504 | 3.76 | 4.49 | 37.2 | 35.6 | 33.9 |
| United Republic of Tanzania | Dar es Salaam | 2,116 | 3,349 | 5,103 | 4.59 | 4.21 | 27.8 | 28.2 | 26.9 |
| Zambia | Lusaka | 1,073 | 1,451 | 1,941 | 3.02 | 2.91 | 29.5 | 30.7 | 29.5 |
| Zimbabwe | Harare | 1,379 | 1,632 | 2,170 | 1.68 | 2.85 | 32.8 | 33.7 | 31.7 |
| **ASIA** | | | | | | | | | |
| Afghanistan | Kabul | 1,963 | 3,731 | 5,665 | 6.42 | 4.18 | 47.3 | 56.7 | 54.2 |
| Armenia | Yerevan | 1,111 | 1,112 | 1,132 | 0.01 | 0.18 | 55.9 | 56.0 | 54.2 |
| Azerbaijan | Baku | 1,806 | 1,972 | 2,190 | 0.88 | 1.05 | 43.4 | 42.5 | 41.1 |
| Bangladesh | Chittagong | 3,308 | 4,962 | 6,447 | 4.05 | 2.62 | 10.0 | 10.8 | 10.3 |
| Bangladesh | Dhaka | 10,285 | 14,648 | 18,721 | 3.54 | 2.45 | 31.0 | 31.7 | 29.8 |
| Bangladesh | Khulna | 1,285 | 1,682 | 2,211 | 2.69 | 2.73 | 3.9 | 3.6 | 3.5 |
| Bangladesh | Rajshahi | 678 | 878 | 1,164 | 2.58 | 2.82 | 2.0 | 1.9 | 1.9 |
| Cambodia | Phnum Pénh (Phnom Penh) | 1,160 | 1,562 | 2,093 | 2.98 | 2.93 | 53.8 | 51.6 | 49.7 |

# TABLE C.1

**continued**

| | | Estimates and projections ('000) | | | Annual rate of change (%) | | Share in national urban population (%) | | |
|---|---|---|---|---|---|---|---|---|---|
| | | 2000 | 2010 | 2020 | 2000–2010 | 2010–2020 | 2000 | 2010 | 2020 |
| China | Anshan, Liaoning | 1,384 | 1,663 | 1,990 | 1.84 | 1.80 | 0.3 | 0.3 | 0.3 |
| China | Anyang | 753 | 1,130 | 1,326 | 4.06 | 1.60 | 0.2 | 0.2 | 0.2 |
| China | Baoding | 884 | 1,213 | 1,524 | 3.16 | 2.28 | 0.2 | 0.2 | 0.2 |
| China | Baotou | 1,406 | 1,932 | 2,243 | 3.18 | 1.49 | 0.3 | 0.3 | 0.3 |
| China | Beijing | 9,757 | 12,385 | 14,296 | 2.39 | 1.43 | 2.2 | 1.9 | 1.8 |
| China | Bengbu | 687 | 914 | 1,142 | 2.85 | 2.23 | 0.2 | 0.1 | 0.1 |
| China | Benxi | 857 | 969 | 1,136 | 1.23 | 1.59 | 0.2 | 0.2 | 0.1 |
| China | Changchun | 2,730 | 3,597 | 4,409 | 2.76 | 2.04 | 0.6 | 0.6 | 0.6 |
| China | Changde | 735 | 849 | 994 | 1.44 | 1.58 | 0.2 | 0.1 | 0.1 |
| China | Changsha, Hunan | 2,077 | 2,415 | 2,885 | 1.51 | 1.78 | 0.5 | 0.4 | 0.4 |
| China | Changzhou, Jiangsu | 1,068 | 2,062 | 2,466 | 6.58 | 1.79 | 0.2 | 0.3 | 0.3 |
| China | Chengdu | 3,919 | 4,961 | 5,886 | 2.36 | 1.71 | 0.9 | 0.8 | 0.7 |
| China | Chifeng | 677 | 842 | 1,020 | 2.18 | 1.92 | 0.1 | 0.1 | 0.1 |
| China | Chongqing | 6,039 | 9,401 | 10,514 | 4.43 | 1.12 | 1.3 | 1.5 | 1.3 |
| China | Cixi | 650 | 781 | 928 | 1.83 | 1.72 | 0.1 | 0.1 | 0.1 |
| China | Dalian | 2,833 | 3,306 | 3,896 | 1.54 | 1.64 | 0.6 | 0.5 | 0.5 |
| China | Dandong | 679 | 795 | 947 | 1.58 | 1.75 | 0.1 | 0.1 | 0.1 |
| China | Daqing | 1,082 | 1,546 | 1,981 | 3.57 | 2.48 | 0.2 | 0.2 | 0.3 |
| China | Datong, Shanxi | 1,049 | 1,251 | 1,500 | 1.76 | 1.82 | 0.2 | 0.2 | 0.2 |
| China | Dongguan, Guangdong | 3,631 | 5,347 | 6,483 | 3.87 | 1.93 | 0.8 | 0.8 | 0.8 |
| China | Foshan | 754 | 4,969 | 5,903 | 18.86 | 1.72 | 0.2 | 0.8 | 0.8 |
| China | Fushun, Liaoning | 1,358 | 1,378 | 1,544 | 0.15 | 1.14 | 0.3 | 0.2 | 0.2 |
| China | Fuxin | 667 | 821 | 999 | 2.08 | 1.96 | 0.1 | 0.1 | 0.1 |
| China | Fuyang | 695 | 874 | 1,045 | 2.29 | 1.79 | 0.2 | 0.1 | 0.1 |
| China | Fuzhou, Fujian | 1,978 | 2,787 | 3,509 | 3.43 | 2.30 | 0.4 | 0.4 | 0.4 |
| China | Guangzhou, Guangdong | 7,330 | 8,884 | 10,409 | 1.92 | 1.58 | 1.6 | 1.4 | 1.3 |
| China | Guilin | 757 | 991 | 1,231 | 2.69 | 2.17 | 0.2 | 0.2 | 0.2 |
| China | Guiyang | 1,860 | 2,154 | 2,519 | 1.47 | 1.57 | 0.4 | 0.3 | 0.3 |
| China | Haerbin | 3,419 | 4,251 | 4,800 | 2.18 | 1.21 | 0.8 | 0.7 | 0.6 |
| China | Handan | 811 | 1,249 | 1,652 | 4.32 | 2.80 | 0.2 | 0.2 | 0.2 |
| China | Hangzhou | 2,411 | 3,860 | 4,470 | 4.71 | 1.47 | 0.5 | 0.6 | 0.6 |
| China | Hefei | 1,532 | 2,404 | 2,850 | 4.51 | 1.70 | 0.3 | 0.4 | 0.4 |
| China | Hengyang | 793 | 1,099 | 1,393 | 3.26 | 2.37 | 0.2 | 0.2 | 0.2 |
| China | Huizhou | 551 | 1,384 | 1,713 | 9.22 | 2.13 | 0.1 | 0.2 | 0.2 |
| China | Huai'an | 818 | 998 | 1,195 | 1.99 | 1.80 | 0.2 | 0.2 | 0.2 |
| China | Huaibei | 617 | 962 | 1,275 | 4.44 | 2.82 | 0.1 | 0.2 | 0.2 |
| China | Huainan | 1,049 | 1,396 | 1,738 | 2.86 | 2.19 | 0.2 | 0.2 | 0.2 |
| China | Hohhot | 1,005 | 1,589 | 2,118 | 4.58 | 2.87 | 0.2 | 0.2 | 0.3 |
| China | Huludao | 529 | 795 | 1,045 | 4.08 | 2.74 | 0.1 | 0.1 | 0.1 |
| China | Jiamusi | 619 | 817 | 1,020 | 2.78 | 2.22 | 0.1 | 0.1 | 0.1 |
| China | Jiangmen | 519 | 1,103 | 1,355 | 7.55 | 2.06 | 0.1 | 0.2 | 0.2 |
| China | Jiaozuo | 631 | 900 | 1,155 | 3.55 | 2.49 | 0.1 | 0.1 | 0.1 |
| China | Jieyang | 608 | 855 | 1,081 | 3.41 | 2.35 | 0.1 | 0.1 | 0.1 |
| China | Jilin | 1,435 | 1,888 | 2,338 | 2.74 | 2.14 | 0.3 | 0.3 | 0.3 |
| China | Jinan, Shandong | 2,592 | 3,237 | 3,813 | 2.22 | 1.64 | 0.6 | 0.5 | 0.5 |
| China | Jingzhou | 761 | 1,039 | 1,302 | 3.12 | 2.25 | 0.2 | 0.2 | 0.2 |
| China | Jining, Shandong | 856 | 1,077 | 1,304 | 2.30 | 1.91 | 0.2 | 0.2 | 0.2 |
| China | Jinjiang | 456 | 858 | 1,216 | 6.31 | 3.49 | 0.1 | 0.1 | 0.2 |
| China | Jinzhou | 770 | 857 | 998 | 1.07 | 1.52 | 0.2 | 0.1 | 0.1 |
| China | Jixi, Heilongjiang | 823 | 1,042 | 1,278 | 2.36 | 2.04 | 0.2 | 0.2 | 0.2 |
| China | Kaohsiung | 1,488 | 1,611 | 1,850 | 0.79 | 1.38 | 0.3 | 0.3 | 0.2 |
| China | Kunming | 2,561 | 3,116 | 3,691 | 1.96 | 1.69 | 0.6 | 0.5 | 0.5 |
| China | Lanzhou | 1,890 | 2,285 | 2,724 | 1.90 | 1.76 | 0.4 | 0.4 | 0.3 |
| China | Lianyungang | 567 | 878 | 1,105 | 4.37 | 2.30 | 0.1 | 0.1 | 0.1 |
| China | Linyi, Shandong | 1,932 | 2,177 | 2,594 | 1.19 | 1.75 | 0.4 | 0.3 | 0.3 |
| China | Liuzhou | 1,027 | 1,352 | 1,675 | 2.75 | 2.14 | 0.2 | 0.2 | 0.2 |
| China | Lufeng | 556 | 889 | 1,192 | 4.69 | 2.94 | 0.1 | 0.1 | 0.2 |
| China | Luoyang | 1,213 | 1,539 | 1,875 | 2.38 | 1.97 | 0.3 | 0.2 | 0.2 |
| China | Luzhou | 649 | 850 | 1,049 | 2.69 | 2.10 | 0.1 | 0.1 | 0.1 |
| China | Maoming | 617 | 803 | 983 | 2.63 | 2.03 | 0.1 | 0.1 | 0.1 |
| China | Mianyang, Sichuan | 758 | 1,006 | 1,244 | 2.83 | 2.12 | 0.2 | 0.2 | 0.2 |
| China | Mudanjiang | 665 | 783 | 933 | 1.63 | 1.75 | 0.1 | 0.1 | 0.1 |
| China | Nanchang | 1,648 | 2,701 | 3,236 | 4.94 | 1.81 | 0.4 | 0.4 | 0.4 |
| China | Nanchong | 606 | 808 | 1,006 | 2.88 | 2.19 | 0.1 | 0.1 | 0.1 |
| China | Nanjing, Jiangsu | 3,472 | 4,519 | 5,524 | 2.64 | 2.01 | 0.8 | 0.7 | 0.7 |
| China | Nanning | 1,445 | 2,096 | 2,508 | 3.72 | 1.79 | 0.3 | 0.3 | 0.3 |
| China | Nantong | 607 | 1,423 | 1,734 | 8.52 | 1.98 | 0.1 | 0.2 | 0.2 |
| China | Nanyang, Henan | 672 | 867 | 1,060 | 2.55 | 2.01 | 0.1 | 0.1 | 0.1 |
| China | Neijiang | 685 | 883 | 1,088 | 2.54 | 2.09 | 0.2 | 0.1 | 0.1 |
| China | Ningbo | 1,303 | 2,217 | 2,782 | 5.31 | 2.27 | 0.3 | 0.3 | 0.4 |
| China | Panjin | 593 | 813 | 1,028 | 3.16 | 2.35 | 0.1 | 0.1 | 0.1 |
| China | Pingdingshan, Henan | 852 | 1,024 | 1,222 | 1.84 | 1.77 | 0.2 | 0.2 | 0.2 |
| China | Puning | 603 | 911 | 1,172 | 4.13 | 2.52 | 0.1 | 0.1 | 0.1 |
| China | Putian | 439 | 1,085 | 1,241 | 9.05 | 1.34 | 0.1 | 0.2 | 0.2 |
| China | Qingdao | 2,659 | 3,323 | 3,923 | 2.23 | 1.66 | 0.6 | 0.5 | 0.5 |
| China | Qinhuangdao | 702 | 893 | 1,088 | 2.41 | 1.98 | 0.2 | 0.1 | 0.1 |
| China | Qiqihaer | 1,331 | 1,588 | 1,894 | 1.77 | 1.76 | 0.3 | 0.2 | 0.2 |
| China | Quanzhou | 728 | 1,068 | 1,367 | 3.83 | 2.47 | 0.2 | 0.2 | 0.2 |
| China | Rizhao | 613 | 816 | 1,014 | 2.87 | 2.17 | 0.1 | 0.1 | 0.1 |
| China | Shanghai | 13,224 | 16,575 | 19,094 | 2.26 | 1.41 | 2.9 | 2.6 | 2.4 |
| China | Shantou | 1,247 | 3,502 | 3,983 | 10.33 | 1.29 | 0.3 | 0.6 | 0.5 |

# TABLE C.1

**continued**

| | | Estimates and projections ('000) | | | Annual rate of change (%) | | Share in national urban population (%) | | |
|---|---|---|---|---|---|---|---|---|---|
| | | 2000 | 2010 | 2020 | 2000–2010 | 2010–2020 | 2000 | 2010 | 2020 |
| China | Shaoguan | 517 | 845 | 995 | 4.91 | 1.63 | 0.1 | 0.1 | 0.1 |
| China | Shaoxing | 608 | 853 | 1,077 | 3.39 | 2.33 | 0.1 | 0.1 | 0.1 |
| China | Shenyang | 4,562 | 5,166 | 6,108 | 1.24 | 1.68 | 1.0 | 0.8 | 0.8 |
| China | Shenzhen | 6,069 | 9,005 | 10,585 | 3.95 | 1.62 | 1.3 | 1.4 | 1.3 |
| China | Shijiazhuang | 1,914 | 2,487 | 3,044 | 2.62 | 2.02 | 0.4 | 0.4 | 0.4 |
| China | Suzhou, Jiangsu | 1,316 | 2,398 | 2,842 | 6.00 | 1.70 | 0.3 | 0.4 | 0.4 |
| China | Taian, Shandong | 910 | 1,239 | 1,548 | 3.09 | 2.23 | 0.2 | 0.2 | 0.2 |
| China | Taichung | 978 | 1,251 | 1,538 | 2.46 | 2.07 | 0.2 | 0.2 | 0.2 |
| China | Tainan | 723 | 777 | 895 | 0.72 | 1.41 | 0.2 | 0.1 | 0.1 |
| China | Taipei | 2,630 | 2,633 | 2,921 | 0.01 | 1.04 | 0.6 | 0.4 | 0.4 |
| China | Taiyuan, Shanxi | 2,503 | 3,154 | 3,812 | 2.31 | 1.89 | 0.6 | 0.5 | 0.5 |
| China | Taizhou, Jiangsu | 535 | 795 | 1,028 | 3.95 | 2.57 | 0.1 | 0.1 | 0.1 |
| China | Taizhou, Zhejiang | 1,190 | 1,338 | 1,566 | 1.17 | 1.57 | 0.3 | 0.2 | 0.2 |
| China | Tangshan, Hebei | 1,390 | 1,870 | 2,335 | 2.97 | 2.22 | 0.3 | 0.3 | 0.3 |
| China | Tianjin | 6,670 | 7,884 | 9,216 | 1.67 | 1.56 | 1.5 | 1.2 | 1.2 |
| China | Ürümqi (Wulumqi) | 1,705 | 2,398 | 3,040 | 3.41 | 2.37 | 0.4 | 0.4 | 0.4 |
| China | Weifang | 1,235 | 1,698 | 2,131 | 3.18 | 2.27 | 0.3 | 0.3 | 0.3 |
| China | Wenzhou | 1,565 | 2,659 | 3,436 | 5.30 | 2.56 | 0.3 | 0.4 | 0.4 |
| China | Wuhan | 6,638 | 7,681 | 8,868 | 1.46 | 1.44 | 1.5 | 1.2 | 1.1 |
| China | Wuhu, Anhui | 634 | 908 | 1,169 | 3.59 | 2.53 | 0.1 | 0.1 | 0.1 |
| China | Wuxi, Jiangsu | 1,409 | 2,682 | 3,206 | 6.44 | 1.78 | 0.3 | 0.4 | 0.4 |
| China | Xiamen | 1,416 | 2,207 | 2,926 | 4.44 | 2.82 | 0.3 | 0.3 | 0.4 |
| China | Xi'an, Shaanxi | 3,690 | 4,747 | 5,414 | 2.52 | 1.31 | 0.8 | 0.7 | 0.7 |
| China | Xiangfan, Hubei | 847 | 1,399 | 1,674 | 5.02 | 1.79 | 0.2 | 0.2 | 0.2 |
| China | Xiangtan, Hunan | 698 | 926 | 1,155 | 2.83 | 2.21 | 0.2 | 0.1 | 0.1 |
| China | Xianyang, Shaanxi | 790 | 1,019 | 1,247 | 2.55 | 2.02 | 0.2 | 0.2 | 0.2 |
| China | Xining | 844 | 1,261 | 1,649 | 4.02 | 2.68 | 0.2 | 0.2 | 0.2 |
| China | Xinxiang | 762 | 1,016 | 1,267 | 2.88 | 2.21 | 0.2 | 0.2 | 0.2 |
| China | Xuzhou | 1,367 | 2,142 | 2,833 | 4.49 | 2.80 | 0.3 | 0.3 | 0.4 |
| China | Yancheng, Jiangsu | 671 | 1,289 | 1,622 | 6.53 | 2.30 | 0.1 | 0.2 | 0.2 |
| China | Yangzhou | 702 | 1,080 | 1,430 | 4.32 | 2.80 | 0.2 | 0.2 | 0.2 |
| China | Yantai | 1,218 | 1,526 | 1,836 | 2.25 | 1.85 | 0.3 | 0.2 | 0.2 |
| China | Yichang | 692 | 959 | 1,132 | 3.26 | 1.66 | 0.2 | 0.2 | 0.1 |
| China | Yichun, Heilongjiang | 815 | 779 | 856 | -0.45 | 0.94 | 0.2 | 0.1 | 0.1 |
| China | Yinchuan | 571 | 911 | 1,225 | 4.67 | 2.96 | 0.1 | 0.1 | 0.2 |
| China | Yingkou | 624 | 848 | 1,072 | 3.07 | 2.34 | 0.1 | 0.1 | 0.1 |
| China | Yiyang, Hunan | 678 | 820 | 974 | 1.90 | 1.72 | 0.1 | 0.1 | 0.1 |
| China | Yueyang | 881 | 1,096 | 1,317 | 2.18 | 1.84 | 0.2 | 0.2 | 0.2 |
| China | Zaozhuang | 853 | 1,175 | 1,473 | 3.20 | 2.26 | 0.2 | 0.2 | 0.2 |
| China | Zhangjiakou | 797 | 1,043 | 1,294 | 2.69 | 2.16 | 0.2 | 0.2 | 0.2 |
| China | Zhanjiang | 818 | 996 | 1,198 | 1.97 | 1.85 | 0.2 | 0.2 | 0.2 |
| China | Zhengzhou | 2,438 | 2,966 | 3,519 | 1.96 | 1.71 | 0.5 | 0.5 | 0.4 |
| China | Zhenjiang, Jiangsu | 679 | 1,007 | 1,308 | 3.94 | 2.62 | 0.1 | 0.2 | 0.2 |
| China | Zhongshan | 1,376 | 2,211 | 2,927 | 4.75 | 2.81 | 0.3 | 0.3 | 0.4 |
| China | Zhuhai | 799 | 1,252 | 1,420 | 4.49 | 1.26 | 0.2 | 0.2 | 0.2 |
| China | Zhuzhou | 819 | 1,025 | 1,244 | 2.24 | 1.94 | 0.2 | 0.2 | 0.2 |
| China | Zibo | 1,874 | 2,456 | 3,004 | 2.70 | 2.01 | 0.4 | 0.4 | 0.4 |
| China | Zigong | 592 | 918 | 1,067 | 4.39 | 1.50 | 0.1 | 0.1 | 0.1 |
| China | Zunyi | 541 | 843 | 1,118 | 4.44 | 2.82 | 0.1 | 0.1 | 0.1 |
| China, Hong Kong SAR | Hong Kong | 6,667 | 7,069 | 7,701 | 0.59 | 0.86 | 100.0 | 100.0 | 100.0 |
| Democratic People's Rep. of Korea | P'yongyang | 2,777 | 2,833 | 2,894 | 0.20 | 0.21 | 20.4 | 19.6 | 18.8 |
| Georgia | Tbilisi | 1,100 | 1,120 | 1,138 | 0.18 | 0.16 | 44.0 | 50.3 | 52.3 |
| India | Agra | 1,293 | 1,703 | 2,089 | 2.75 | 2.04 | 0.4 | 0.5 | 0.5 |
| India | Ahmadabad | 4,427 | 5,717 | 6,892 | 2.56 | 1.87 | 1.5 | 1.6 | 1.5 |
| India | Aligarh | 653 | 863 | 1,068 | 2.79 | 2.13 | 0.2 | 0.2 | 0.2 |
| India | Allahabad | 1,035 | 1,277 | 1,570 | 2.10 | 2.07 | 0.4 | 0.4 | 0.3 |
| India | Amritsar | 990 | 1,297 | 1,597 | 2.70 | 2.08 | 0.3 | 0.4 | 0.3 |
| India | Asansol | 1,065 | 1,423 | 1,751 | 2.90 | 2.07 | 0.4 | 0.4 | 0.4 |
| India | Aurangabad | 868 | 1,198 | 1,478 | 3.22 | 2.10 | 0.3 | 0.3 | 0.3 |
| India | Bangalore | 5,567 | 7,218 | 8,674 | 2.60 | 1.84 | 1.9 | 2.0 | 1.9 |
| India | Bareilly | 722 | 868 | 1,072 | 1.84 | 2.11 | 0.3 | 0.2 | 0.2 |
| India | Bhiwandi | 603 | 859 | 1,066 | 3.54 | 2.16 | 0.2 | 0.2 | 0.2 |
| India | Bhopal | 1,426 | 1,843 | 2,257 | 2.57 | 2.03 | 0.5 | 0.5 | 0.5 |
| India | Bhubaneswar | 637 | 912 | 1,131 | 3.59 | 2.15 | 0.2 | 0.3 | 0.2 |
| India | Kolkata (Calcutta) | 13,058 | 15,552 | 18,449 | 1.75 | 1.71 | 4.5 | 4.3 | 4.0 |
| India | Chandigarh | 791 | 1,049 | 1,296 | 2.82 | 2.11 | 0.3 | 0.3 | 0.3 |
| India | Jammu | 588 | 857 | 1,064 | 3.77 | 2.16 | 0.2 | 0.2 | 0.2 |
| India | Chennai (Madras) | 6,353 | 7,547 | 9,043 | 1.72 | 1.81 | 2.2 | 2.1 | 2.0 |
| India | Coimbatore | 1,420 | 1,807 | 2,212 | 2.41 | 2.02 | 0.5 | 0.5 | 0.5 |
| India | Delhi | 15,730 | 22,157 | 26,272 | 3.43 | 1.70 | 5.5 | 6.1 | 5.7 |
| India | Dhanbad | 1,046 | 1,328 | 1,633 | 2.39 | 2.07 | 0.4 | 0.4 | 0.4 |
| India | Durg-Bhilainagar | 905 | 1,172 | 1,445 | 2.59 | 2.09 | 0.3 | 0.3 | 0.3 |
| India | Guwahati (Gauhati) | 797 | 1,053 | 1,300 | 2.79 | 2.11 | 0.3 | 0.3 | 0.3 |
| India | Gwalior | 855 | 1,039 | 1,280 | 1.95 | 2.06 | 0.3 | 0.3 | 0.3 |
| India | Hubli-Dharwad | 776 | 946 | 1,168 | 1.98 | 2.11 | 0.3 | 0.3 | 0.3 |
| India | Hyderabad | 5,445 | 6,751 | 8,110 | 2.15 | 1.83 | 1.9 | 1.9 | 1.8 |
| India | Indore | 1,597 | 2,173 | 2,659 | 3.08 | 2.02 | 0.6 | 0.6 | 0.6 |
| India | Jabalpur | 1,100 | 1,367 | 1,679 | 2.17 | 2.06 | 0.4 | 0.4 | 0.4 |
| India | Jaipur | 2,259 | 3,131 | 3,813 | 3.26 | 1.97 | 0.8 | 0.9 | 0.8 |
| India | Jalandhar | 694 | 917 | 1,134 | 2.79 | 2.12 | 0.2 | 0.3 | 0.2 |
| India | Jamshedpur | 1,081 | 1,387 | 1,705 | 2.49 | 2.06 | 0.4 | 0.4 | 0.4 |

# TABLE C.I

**continued**

| | | Estimates and projections ('000) | | | Annual rate of change (%) | | Share in national urban population (%) | | |
|---|---|---|---|---|---|---|---|---|---|
| | | 2000 | 2010 | 2020 | 2000–2010 | 2010–2020 | 2000 | 2010 | 2020 |
| India | Jodhpur | 842 | 1,061 | 1,308 | 2.31 | 2.10 | 0.3 | 0.3 | 0.3 |
| India | Kanpur | 2,641 | 3,364 | 4,084 | 2.42 | 1.94 | 0.9 | 0.9 | 0.9 |
| India | Kochi (Cochin) | 1,340 | 1,610 | 1,971 | 1.83 | 2.03 | 0.5 | 0.4 | 0.4 |
| India | Kota | 692 | 884 | 1,093 | 2.45 | 2.12 | 0.2 | 0.2 | 0.2 |
| India | Kozhikode (Calicut) | 875 | 1,007 | 1,240 | 1.41 | 2.08 | 0.3 | 0.3 | 0.3 |
| India | Lucknow | 2,221 | 2,873 | 3,497 | 2.57 | 1.97 | 0.8 | 0.8 | 0.8 |
| India | Ludhiana | 1,368 | 1,760 | 2,156 | 2.52 | 2.03 | 0.5 | 0.5 | 0.5 |
| India | Madurai | 1,187 | 1,365 | 1,674 | 1.40 | 2.04 | 0.4 | 0.4 | 0.4 |
| India | Meerut | 1,143 | 1,494 | 1,836 | 2.68 | 2.06 | 0.4 | 0.4 | 0.4 |
| India | Moradabad | 626 | 845 | 1,048 | 3.00 | 2.15 | 0.2 | 0.2 | 0.2 |
| India | Mumbai (Bombay) | 16,086 | 20,041 | 23,719 | 2.20 | 1.68 | 5.6 | 5.5 | 5.1 |
| India | Mysore | 776 | 942 | 1,163 | 1.94 | 2.11 | 0.3 | 0.3 | 0.3 |
| India | Nagpur | 2,089 | 2,607 | 3,175 | 2.22 | 1.97 | 0.7 | 0.7 | 0.7 |
| India | Nashik | 1,117 | 1,588 | 1,954 | 3.52 | 2.07 | 0.4 | 0.4 | 0.4 |
| India | Patna | 1,658 | 2,321 | 2,839 | 3.36 | 2.01 | 0.6 | 0.6 | 0.6 |
| India | Pune (Poona) | 3,655 | 5,002 | 6,050 | 3.14 | 1.90 | 1.3 | 1.4 | 1.3 |
| India | Raipur | 680 | 943 | 1,167 | 3.27 | 2.13 | 0.2 | 0.3 | 0.3 |
| India | Rajkot | 974 | 1,357 | 1,672 | 3.32 | 2.09 | 0.3 | 0.4 | 0.4 |
| India | Ranchi | 844 | 1,119 | 1,380 | 2.82 | 2.10 | 0.3 | 0.3 | 0.3 |
| India | Salem | 736 | 932 | 1,152 | 2.36 | 2.12 | 0.3 | 0.3 | 0.2 |
| India | Solapur | 853 | 1,133 | 1,398 | 2.84 | 2.10 | 0.3 | 0.3 | 0.3 |
| India | Srinagar | 954 | 1,216 | 1,497 | 2.43 | 2.08 | 0.3 | 0.3 | 0.3 |
| India | Surat | 2,699 | 4,168 | 5,071 | 4.35 | 1.96 | 0.9 | 1.1 | 1.1 |
| India | Thiruvananthapuram | 885 | 1,006 | 1,239 | 1.28 | 2.08 | 0.3 | 0.3 | 0.3 |
| India | Tiruchirappalli | 837 | 1,010 | 1,245 | 1.88 | 2.09 | 0.3 | 0.3 | 0.3 |
| India | Vadodara | 1,465 | 1,872 | 2,292 | 2.45 | 2.02 | 0.5 | 0.5 | 0.5 |
| India | Varanasi (Benares) | 1,199 | 1,432 | 1,756 | 1.78 | 2.04 | 0.4 | 0.4 | 0.4 |
| India | Vijayawada | 999 | 1,207 | 1,484 | 1.89 | 2.07 | 0.3 | 0.3 | 0.3 |
| India | Visakhapatnam | 1,309 | 1,625 | 1,992 | 2.16 | 2.04 | 0.5 | 0.4 | 0.4 |
| Indonesia | Bandar Lampung | 743 | 799 | 903 | 0.73 | 1.22 | 0.9 | 0.8 | 0.7 |
| Indonesia | Bandung | 2,138 | 2,412 | 2,739 | 1.21 | 1.27 | 2.5 | 2.3 | 2.2 |
| Indonesia | Bogor | 751 | 1,044 | 1,251 | 3.29 | 1.81 | 0.9 | 1.0 | 1.0 |
| Indonesia | Jakarta | 8,390 | 9,210 | 10,256 | 0.93 | 1.08 | 9.7 | 8.9 | 8.4 |
| Indonesia | Malang | 757 | 786 | 891 | 0.38 | 1.25 | 0.9 | 0.8 | 0.7 |
| Indonesia | Medan | 1,912 | 2,131 | 2,419 | 1.08 | 1.27 | 2.2 | 2.1 | 2.0 |
| Indonesia | Palembang | 1,459 | 1,244 | 1,356 | -1.59 | 0.86 | 1.7 | 1.2 | 1.1 |
| Indonesia | Semarang | 1,427 | 1,296 | 1,424 | -0.96 | 0.94 | 1.7 | 1.3 | 1.2 |
| Indonesia | Surabaya | 2,611 | 2,509 | 2,738 | -0.40 | 0.87 | 3.0 | 2.4 | 2.2 |
| Indonesia | Pekan Baru | 588 | 891 | 1,128 | 4.16 | 2.36 | 0.7 | 0.9 | 0.9 |
| Indonesia | Ujung Pandang | 1,031 | 1,294 | 1,512 | 2.27 | 1.56 | 1.2 | 1.3 | 1.2 |
| Iran (Islamic Republic of) | Ahvaz | 868 | 1,060 | 1,249 | 2.00 | 1.64 | 2.0 | 2.0 | 2.0 |
| Iran (Islamic Republic of) | Esfahan | 1,382 | 1,742 | 2,056 | 2.32 | 1.66 | 3.2 | 3.3 | 3.2 |
| Iran (Islamic Republic of) | Karaj | 1,087 | 1,584 | 1,937 | 3.77 | 2.01 | 2.5 | 3.0 | 3.0 |
| Iran (Islamic Republic of) | Kermanshah | 729 | 837 | 974 | 1.38 | 1.52 | 1.7 | 1.6 | 1.5 |
| Iran (Islamic Republic of) | Mashhad | 2,073 | 2,652 | 3,128 | 2.46 | 1.65 | 4.8 | 5.0 | 4.9 |
| Iran (Islamic Republic of) | Qom | 843 | 1,042 | 1,232 | 2.12 | 1.67 | 2.0 | 2.0 | 1.9 |
| Iran (Islamic Republic of) | Shiraz | 1,115 | 1,299 | 1,510 | 1.53 | 1.51 | 2.6 | 2.4 | 2.4 |
| Iran (Islamic Republic of) | Tabriz | 1,264 | 1,483 | 1,724 | 1.60 | 1.51 | 2.9 | 2.8 | 2.7 |
| Iran (Islamic Republic of) | Tehran | 6,880 | 7,241 | 8,059 | 0.51 | 1.07 | 16.0 | 13.6 | 12.7 |
| Iraq | Al-Basrah (Basra) | 759 | 923 | 1,139 | 1.96 | 2.10 | 4.5 | 4.4 | 4.3 |
| Iraq | Al-Mawsil (Mosul) | 1,056 | 1,447 | 1,885 | 3.15 | 2.64 | 6.3 | 6.9 | 7.0 |
| Iraq | Baghdad | 5,200 | 5,891 | 7,321 | 1.25 | 2.17 | 31.1 | 28.3 | 27.3 |
| Iraq | Irbil (Erbil) | 757 | 1,009 | 1,301 | 2.87 | 2.54 | 4.5 | 4.8 | 4.9 |
| Iraq | Sulaimaniya | 580 | 836 | 1,121 | 3.66 | 2.93 | 3.5 | 4.0 | 4.2 |
| Israel | Hefa (Haifa) | 888 | 1,036 | 1,144 | 1.54 | 0.99 | 16.0 | 15.5 | 14.9 |
| Israel | Jerusalem | 651 | 782 | 901 | 1.83 | 1.42 | 11.7 | 11.7 | 11.7 |
| Israel | Tel Aviv-Yafo (Tel Aviv-Jaffa) | 2,752 | 3,272 | 3,689 | 1.73 | 1.20 | 49.5 | 48.9 | 48.1 |
| Japan | Fukuoka-Kitakyushu | 2,716 | 2,816 | 2,834 | 0.36 | 0.06 | 3.3 | 3.3 | 3.3 |
| Japan | Hiroshima | 2,044 | 2,081 | 2,088 | 0.18 | 0.03 | 2.5 | 2.5 | 2.4 |
| Japan | Kyoto | 1,806 | 1,804 | 1,804 | -0.01 | 0.00 | 2.2 | 2.1 | 2.1 |
| Japan | Nagoya | 3,122 | 3,267 | 3,295 | 0.45 | 0.09 | 3.8 | 3.8 | 3.8 |
| Japan | Osaka-Kobe | 11,165 | 11,337 | 11,368 | 0.15 | 0.03 | 13.5 | 13.4 | 13.2 |
| Japan | Sapporo | 2,508 | 2,687 | 2,721 | 0.69 | 0.13 | 3.0 | 3.2 | 3.2 |
| Japan | Sendai | 2,184 | 2,376 | 2,413 | 0.84 | 0.15 | 2.6 | 2.8 | 2.8 |
| Japan | Tokyo | 34,450 | 36,669 | 37,088 | 0.62 | 0.11 | 41.7 | 43.2 | 43.2 |
| Jordan | Amman | 1,007 | 1,105 | 1,272 | 0.93 | 1.41 | 26.5 | 21.7 | 21.2 |
| Kazakhstan | Almaty | 1,159 | 1,383 | 1,554 | 1.77 | 1.17 | 13.8 | 15.0 | 14.9 |
| Kuwait | Al Kuwayt (Kuwait City) | 1,499 | 2,305 | 2,790 | 4.30 | 1.91 | 68.5 | 76.8 | 76.7 |
| Kyrgyzstan | Bishkek | 770 | 864 | 967 | 1.15 | 1.13 | 44.2 | 45.0 | 43.9 |
| Lebanon | Bayrut (Beirut) | 1,487 | 1,937 | 2,090 | 2.64 | 0.76 | 45.8 | 52.2 | 51.4 |
| Malaysia | Johore Bharu | 630 | 999 | 1,295 | 4.61 | 2.60 | 4.4 | 5.0 | 5.2 |
| Malaysia | Klang | 631 | 1,128 | 1,503 | 5.81 | 2.87 | 4.4 | 5.6 | 6.0 |
| Malaysia | Kuala Lumpur | 1,306 | 1,519 | 1,820 | 1.51 | 1.81 | 9.1 | 7.5 | 7.2 |
| Mongolia | Ulaanbaatar | 764 | 966 | 1,129 | 2.35 | 1.56 | 56.3 | 57.7 | 56.2 |
| Myanmar | Mandalay | 810 | 1,034 | 1,331 | 2.44 | 2.52 | 6.3 | 6.1 | 5.9 |
| Myanmar | Nay Pyi Taw | — | 1,024 | 1,344 | .. | 2.72 | — | 6.0 | 6.0 |
| Myanmar | Yangon | 3,553 | 4,350 | 5,456 | 2.02 | 2.27 | 27.4 | 25.6 | 24.2 |
| Nepal | Kathmandu | 644 | 1,037 | 1,589 | 4.76 | 4.27 | 19.6 | 18.7 | 18.2 |
| Pakistan | Faisalabad | 2,140 | 2,849 | 3,704 | 2.86 | 2.62 | 4.4 | 4.3 | 4.1 |
| Pakistan | Gujranwala | 1,224 | 1,652 | 2,165 | 3.00 | 2.70 | 2.5 | 2.5 | 2.4 |
| Pakistan | Hyderabad | 1,222 | 1,590 | 2,084 | 2.63 | 2.71 | 2.5 | 2.4 | 2.3 |

# TABLE C.1

*continued*

| | | Estimates and projections ('000) | | | Annual rate of change (%) | | Share in national urban population (%) | | |
|---|---|---|---|---|---|---|---|---|---|
| | | 2000 | 2010 | 2020 | 2000–2010 | 2010–2020 | 2000 | 2010 | 2020 |
| Pakistan | Islamabad | 595 | 856 | 1,132 | 3.64 | 2.79 | 1.2 | 1.3 | 1.3 |
| Pakistan | Karachi | 10,021 | 13,125 | 16,693 | 2.70 | 2.40 | 20.4 | 19.8 | 18.5 |
| Pakistan | Lahore | 5,449 | 7,132 | 9,150 | 2.69 | 2.49 | 11.1 | 10.8 | 10.1 |
| Pakistan | Multan | 1,263 | 1,659 | 2,174 | 2.73 | 2.70 | 2.6 | 2.5 | 2.4 |
| Pakistan | Peshawar | 1,066 | 1,422 | 1,868 | 2.88 | 2.73 | 2.2 | 2.1 | 2.1 |
| Pakistan | Quetta | 614 | 841 | 1,113 | 3.15 | 2.80 | 1.3 | 1.3 | 1.2 |
| Pakistan | Rawalpindi | 1,520 | 2,026 | 2,646 | 2.87 | 2.67 | 3.1 | 3.1 | 2.9 |
| Philippines | Cebu | 721 | 860 | 1,046 | 1.76 | 1.96 | 1.9 | 1.9 | 1.8 |
| Philippines | Davao | 1,152 | 1,519 | 1,881 | 2.77 | 2.14 | 3.1 | 3.3 | 3.3 |
| Philippines | Manila | 9,958 | 11,628 | 13,687 | 1.55 | 1.63 | 26.7 | 25.4 | 23.7 |
| Philippines | Zamboanga | 605 | 854 | 1,082 | 3.45 | 2.37 | 1.6 | 1.9 | 1.9 |
| Republic of Korea | Goyang | 744 | 961 | 1,025 | 2.56 | 0.64 | 2.0 | 2.4 | 2.4 |
| Republic of Korea | Bucheon | 763 | 909 | 960 | 1.75 | 0.55 | 2.1 | 2.3 | 2.3 |
| Republic of Korea | Incheon | 2,464 | 2,583 | 2,630 | 0.47 | 0.18 | 6.7 | 6.4 | 6.2 |
| Republic of Korea | Gwangju | 1,346 | 1,476 | 1,524 | 0.92 | 0.32 | 3.6 | 3.7 | 3.6 |
| Republic of Korea | Busan | 3,673 | 3,425 | 3,409 | -0.70 | -0.05 | 9.9 | 8.5 | 8.0 |
| Republic of Korea | Seongnam | 911 | 955 | 983 | 0.47 | 0.29 | 2.5 | 2.4 | 2.3 |
| Republic of Korea | Seoul | 9,917 | 9,773 | 9,767 | -0.15 | -0.01 | 26.8 | 24.3 | 23.1 |
| Republic of Korea | Suweon | 932 | 1,132 | 1,193 | 1.94 | 0.52 | 2.5 | 2.8 | 2.8 |
| Republic of Korea | Daegu | 2,478 | 2,458 | 2,481 | -0.08 | 0.09 | 6.7 | 6.1 | 5.9 |
| Republic of Korea | Daejon | 1,362 | 1,509 | 1,562 | 1.02 | 0.35 | 3.7 | 3.8 | 3.7 |
| Republic of Korea | Ulsan | 1,011 | 1,081 | 1,116 | 0.67 | 0.32 | 2.7 | 2.7 | 2.6 |
| Saudi Arabia | Al-Madinah (Medina) | 795 | 1,104 | 1,351 | 3.28 | 2.02 | 4.8 | 5.1 | 5.1 |
| Saudi Arabia | Ar-Riyadh (Riyadh) | 3,567 | 4,848 | 5,809 | 3.07 | 1.81 | 21.5 | 22.5 | 21.8 |
| Saudi Arabia | Ad-Dammam | 639 | 902 | 1,109 | 3.45 | 2.07 | 3.8 | 4.2 | 4.2 |
| Saudi Arabia | Jiddah | 2,509 | 3,234 | 3,868 | 2.54 | 1.79 | 15.1 | 15.0 | 14.5 |
| Saudi Arabia | Makkah (Mecca) | 1,168 | 1,484 | 1,789 | 2.39 | 1.87 | 7.0 | 6.9 | 6.7 |
| Singapore | Singapore | 4,018 | 4,837 | 5,219 | 1.86 | 0.76 | 100.0 | 100.0 | 100.0 |
| Syrian Arab Republic | Dimashq (Damascus) | 2,063 | 2,597 | 3,213 | 2.30 | 2.13 | 24.1 | 20.7 | 20.1 |
| Syrian Arab Republic | Halab (Aleppo) | 2,204 | 3,087 | 3,864 | 3.37 | 2.25 | 25.7 | 24.6 | 24.2 |
| Syrian Arab Republic | Hamah | 495 | 897 | 1,180 | 5.96 | 2.74 | 5.8 | 7.2 | 7.4 |
| Syrian Arab Republic | Hims (Homs) | 856 | 1,328 | 1,702 | 4.39 | 2.48 | 10.0 | 10.6 | 10.7 |
| Thailand | Krung Thep (Bangkok) | 6,332 | 6,976 | 7,902 | 0.97 | 1.25 | 32.6 | 30.1 | 28.4 |
| Turkey | Adana | 1,123 | 1,361 | 1,556 | 1.92 | 1.34 | 2.6 | 2.6 | 2.5 |
| Turkey | Ankara | 3,179 | 3,906 | 4,401 | 2.06 | 1.19 | 7.4 | 7.4 | 7.1 |
| Turkey | Antalya | 595 | 838 | 969 | 3.42 | 1.45 | 1.4 | 1.6 | 1.6 |
| Turkey | Bursa | 1,180 | 1,588 | 1,816 | 2.97 | 1.34 | 2.7 | 3.0 | 2.9 |
| Turkey | Gaziantep | 844 | 1,109 | 1,274 | 2.73 | 1.39 | 2.0 | 2.1 | 2.1 |
| Turkey | Istanbul | 8,744 | 10,525 | 11,689 | 1.85 | 1.05 | 20.3 | 20.0 | 18.8 |
| Turkey | Izmir | 2,216 | 2,723 | 3,083 | 2.06 | 1.24 | 5.2 | 5.2 | 5.0 |
| Turkey | Konya | 734 | 978 | 1,125 | 2.87 | 1.40 | 1.7 | 1.9 | 1.8 |
| United Arab Emirates | Dubayy (Dubai) | 906 | 1,567 | 1,934 | 5.48 | 2.10 | 34.9 | 39.6 | 39.3 |
| Uzbekistan | Tashkent | 2,135 | 2,210 | 2,420 | 0.35 | 0.91 | 23.0 | 21.9 | 20.5 |
| Viet Nam | Da Nang - CP | 570 | 838 | 1,146 | 3.85 | 3.13 | 3.0 | 3.1 | 3.2 |
| Viet Nam | Hai Phòng | 1,704 | 1,970 | 2,432 | 1.45 | 2.11 | 8.8 | 7.3 | 6.7 |
| Viet Nam | Hà Noi | 1,631 | 2,814 | 4,056 | 5.45 | 3.66 | 8.5 | 10.4 | 11.2 |
| Viet Nam | Thành Pho Ho Chí Minh (Ho Chi Minh City) | 4,336 | 6,167 | 8,067 | 3.52 | 2.69 | 22.5 | 22.8 | 22.2 |
| Yemen | Sana'a' | 1,365 | 2,342 | 3,585 | 5.40 | 4.26 | 28.6 | 30.4 | 29.7 |
| **EUROPE** | | | | | | | | | |
| Austria | Wien (Vienna) | 1,549 | 1,706 | 1,779 | 0.97 | 0.42 | 29.4 | 30.1 | 29.6 |
| Belarus | Minsk | 1,700 | 1,852 | 1,917 | 0.86 | 0.34 | 24.2 | 25.9 | 26.6 |
| Belgium | Antwerpen | 925 | 965 | 984 | 0.42 | 0.19 | 9.3 | 9.3 | 9.1 |
| Belgium | Bruxelles-Brussel | 1,776 | 1,904 | 1,948 | 0.70 | 0.23 | 17.9 | 18.3 | 18.1 |
| Bulgaria | Sofia | 1,128 | 1,196 | 1,215 | 0.59 | 0.16 | 20.4 | 22.3 | 23.3 |
| Czech Republic | Praha (Prague) | 1,172 | 1,162 | 1,168 | -0.09 | 0.05 | 15.5 | 15.2 | 14.7 |
| Denmark | København (Copenhagen) | 1,077 | 1,186 | 1,238 | 0.96 | 0.43 | 23.7 | 24.9 | 25.1 |
| Finland | Helsinki | 1,019 | 1,117 | 1,170 | 0.92 | 0.46 | 24.0 | 24.6 | 24.3 |
| France | Bordeaux | 763 | 838 | 899 | 0.94 | 0.70 | 1.7 | 1.6 | 1.5 |
| France | Lille | 1,004 | 1,033 | 1,092 | 0.28 | 0.56 | 2.2 | 1.9 | 1.9 |
| France | Lyon | 1,362 | 1,468 | 1,559 | 0.75 | 0.60 | 3.0 | 2.7 | 2.7 |
| France | Marseille-Aix-en-Provence | 1,363 | 1,469 | 1,560 | 0.75 | 0.60 | 3.0 | 2.8 | 2.7 |
| France | Nice-Cannes | 899 | 977 | 1,045 | 0.83 | 0.67 | 2.0 | 1.8 | 1.8 |
| France | Paris | 9,739 | 10,485 | 10,880 | 0.74 | 0.37 | 21.4 | 19.6 | 18.7 |
| France | Toulouse | 778 | 912 | 989 | 1.59 | 0.81 | 1.7 | 1.7 | 1.7 |
| Germany | Berlin | 3,384 | 3,450 | 3,498 | 0.19 | 0.14 | 5.6 | 5.7 | 5.8 |
| Germany | Hamburg | 1,710 | 1,786 | 1,825 | 0.43 | 0.22 | 2.9 | 2.9 | 3.0 |
| Germany | Köln (Cologne) | 963 | 1,001 | 1,018 | 0.39 | 0.17 | 1.6 | 1.7 | 1.7 |
| Germany | München (Munich) | 1,202 | 1,349 | 1,412 | 1.15 | 0.46 | 2.0 | 2.2 | 2.3 |
| Greece | Athínai (Athens) | 3,179 | 3,257 | 3,312 | 0.24 | 0.17 | 48.6 | 47.4 | 45.3 |
| Greece | Thessaloniki | 797 | 837 | 868 | 0.49 | 0.36 | 12.2 | 12.2 | 11.9 |
| Hungary | Budapest | 1,787 | 1,706 | 1,711 | -0.46 | 0.03 | 27.1 | 25.1 | 24.4 |
| Ireland | Dublin | 989 | 1,099 | 1,261 | 1.05 | 1.38 | 44.0 | 38.7 | 37.4 |
| Italy | Milano (Milan) | 2,985 | 2,967 | 2,981 | -0.06 | 0.05 | 7.8 | 7.2 | 7.0 |
| Italy | Napoli (Naples) | 2,232 | 2,276 | 2,293 | 0.20 | 0.07 | 5.8 | 5.5 | 5.4 |
| Italy | Palermo | 855 | 875 | 891 | 0.23 | 0.18 | 2.2 | 2.1 | 2.1 |
| Italy | Roma (Rome) | 3,385 | 3,362 | 3,376 | -0.07 | 0.04 | 8.8 | 8.2 | 7.9 |
| Italy | Torino (Turin) | 1,694 | 1,665 | 1,679 | -0.17 | 0.08 | 4.4 | 4.1 | 3.9 |
| Netherlands | Amsterdam | 1,005 | 1,049 | 1,097 | 0.43 | 0.45 | 8.2 | 7.6 | 7.4 |
| Netherlands | Rotterdam | 991 | 1,010 | 1,044 | 0.19 | 0.33 | 8.1 | 7.3 | 7.0 |
| Norway | Oslo | 774 | 888 | 985 | 1.37 | 1.04 | 22.7 | 23.0 | 22.9 |

# TABLE C.I

*continued*

| | | Estimates and projections ('000) | | | Annual rate of change (%) | | Share in national urban population (%) | | |
|---|---|---|---|---|---|---|---|---|---|
| | | 2000 | 2010 | 2020 | 2000–2010 | 2010–2020 | 2000 | 2010 | 2020 |
| Poland | Kraków (Cracow) | 756 | 756 | 756 | 0.00 | 0.00 | 3.2 | 3.3 | 3.3 |
| Poland | Warszawa (Warsaw) | 1,666 | 1,712 | 1,722 | 0.27 | 0.06 | 7.0 | 7.4 | 7.4 |
| Portugal | Lisboa (Lisbon) | 2,672 | 2,824 | 2,973 | 0.55 | 0.51 | 48.0 | 43.3 | 41.6 |
| Portugal | Porto | 1,254 | 1,355 | 1,448 | 0.77 | 0.66 | 22.5 | 20.8 | 20.3 |
| Romania | Bucuresti (Bucharest) | 1,949 | 1,934 | 1,959 | -0.08 | 0.13 | 16.6 | 15.9 | 15.3 |
| Russian Federation | Chelyabinsk | 1,082 | 1,094 | 1,095 | 0.11 | 0.01 | 1.0 | 1.1 | 1.1 |
| Russian Federation | Yekaterinburg | 1,303 | 1,344 | 1,376 | 0.31 | 0.24 | 1.2 | 1.3 | 1.4 |
| Russian Federation | Kazan | 1,096 | 1,140 | 1,164 | 0.39 | 0.21 | 1.0 | 1.1 | 1.2 |
| Russian Federation | Krasnoyarsk | 911 | 961 | 998 | 0.53 | 0.38 | 0.8 | 0.9 | 1.0 |
| Russian Federation | Moskva (Moscow) | 10,005 | 10,550 | 10,662 | 0.53 | 0.11 | 9.3 | 10.3 | 10.6 |
| Russian Federation | Nizhniy Novgorod | 1,331 | 1,267 | 1,253 | -0.49 | -0.11 | 1.2 | 1.2 | 1.2 |
| Russian Federation | Novosibirsk | 1,426 | 1,397 | 1,398 | -0.21 | 0.01 | 1.3 | 1.4 | 1.4 |
| Russian Federation | Omsk | 1,136 | 1,124 | 1,112 | -0.11 | -0.11 | 1.1 | 1.1 | 1.1 |
| Russian Federation | Perm | 1,014 | 982 | 972 | -0.32 | -0.10 | 0.9 | 1.0 | 1.0 |
| Russian Federation | Rostov-na-Donu (Rostov-on-Don) | 1,061 | 1,046 | 1,038 | -0.14 | -0.08 | 1.0 | 1.0 | 1.0 |
| Russian Federation | Samara | 1,173 | 1,131 | 1,119 | -0.36 | -0.11 | 1.1 | 1.1 | 1.1 |
| Russian Federation | Sankt Peterburg (Saint Petersburg) | 4,719 | 4,575 | 4,557 | -0.31 | -0.04 | 4.4 | 4.5 | 4.5 |
| Russian Federation | Saratov | 878 | 822 | 798 | -0.66 | -0.30 | 0.8 | 0.8 | 0.8 |
| Russian Federation | Ufa | 1,049 | 1,023 | 1,016 | -0.25 | -0.07 | 1.0 | 1.0 | 1.0 |
| Russian Federation | Volgograd | 1,010 | 977 | 965 | -0.33 | -0.12 | 0.9 | 1.0 | 1.0 |
| Russian Federation | Voronezh | 854 | 842 | 838 | -0.14 | -0.05 | 0.8 | 0.8 | 0.8 |
| Serbia | Beograd (Belgrade) | 1,122 | 1,117 | 1,149 | -0.04 | 0.28 | 20.9 | 20.2 | 19.6 |
| Spain | Barcelona | 4,560 | 5,083 | 5,443 | 1.09 | 0.68 | 14.9 | 14.5 | 14.1 |
| Spain | Madrid | 5,014 | 5,851 | 6,379 | 1.54 | 0.86 | 16.3 | 16.7 | 16.6 |
| Spain | Valencia | 795 | 814 | 857 | 0.24 | 0.51 | 2.6 | 2.3 | 2.2 |
| Sweden | Stockholm | 1,206 | 1,285 | 1,327 | 0.63 | 0.32 | 16.2 | 16.3 | 15.9 |
| Switzerland | Zürich (Zurich) | 1,078 | 1,150 | 1,196 | 0.65 | 0.39 | 20.5 | 20.6 | 20.2 |
| Ukraine | Dnipropetrovs'k | 1,077 | 1,004 | 967 | -0.70 | -0.38 | 3.3 | 3.2 | 3.1 |
| Ukraine | Donets'k | 1,026 | 966 | 941 | -0.60 | -0.26 | 3.1 | 3.1 | 3.0 |
| Ukraine | Kharkiv | 1,484 | 1,453 | 1,444 | -0.21 | -0.06 | 4.5 | 4.6 | 4.7 |
| Ukraine | Kyiv (Kiev) | 2,606 | 2,805 | 2,914 | 0.74 | 0.38 | 7.9 | 9.0 | 9.4 |
| Ukraine | Odesa | 1,037 | 1,009 | 1,011 | -0.27 | 0.02 | 3.2 | 3.2 | 3.3 |
| Ukraine | Zaporizhzhya | 822 | 775 | 758 | -0.59 | -0.22 | 2.5 | 2.5 | 2.5 |
| United Kingdom | Birmingham | 2,285 | 2,302 | 2,375 | 0.07 | 0.31 | 4.9 | 4.7 | 4.5 |
| United Kingdom | Glasgow | 1,171 | 1,170 | 1,218 | -0.01 | 0.40 | 2.5 | 2.4 | 2.3 |
| United Kingdom | Liverpool | 818 | 819 | 857 | 0.01 | 0.45 | 1.8 | 1.7 | 1.6 |
| United Kingdom | London | 8,225 | 8,631 | 8,753 | 0.48 | 0.14 | 17.8 | 17.5 | 16.5 |
| United Kingdom | Manchester | 2,248 | 2,253 | 2,325 | 0.02 | 0.31 | 4.9 | 4.6 | 4.4 |
| United Kingdom | Newcastle upon Tyne | 880 | 891 | 932 | 0.12 | 0.45 | 1.9 | 1.8 | 1.8 |
| United Kingdom | West Yorkshire | 1,495 | 1,547 | 1,606 | 0.34 | 0.37 | 3.2 | 3.1 | 3.0 |
| **LATIN AMERICA AND THE CARIBBEAN** | | | | | | | | | |
| Argentina | Buenos Aires | 11,847 | 13,074 | 13,606 | 0.99 | 0.40 | 35.6 | 34.8 | 32.7 |
| Argentina | Córdoba | 1,348 | 1,493 | 1,601 | 1.02 | 0.70 | 4.0 | 4.0 | 3.9 |
| Argentina | Mendoza | 838 | 917 | 990 | 0.90 | 0.77 | 2.5 | 2.4 | 2.4 |
| Argentina | Rosario | 1,152 | 1,231 | 1,322 | 0.66 | 0.71 | 3.5 | 3.3 | 3.2 |
| Argentina | San Miguel de Tucumán | 722 | 831 | 899 | 1.41 | 0.79 | 2.2 | 2.2 | 2.2 |
| Bolivia | La Paz | 1,390 | 1,673 | 2,005 | 1.85 | 1.81 | 27.0 | 25.1 | 24.3 |
| Bolivia | Santa Cruz | 1,054 | 1,649 | 2,103 | 4.48 | 2.43 | 20.5 | 24.7 | 25.4 |
| Brazil | Aracaju | 606 | 782 | 883 | 2.55 | 1.21 | 0.4 | 0.5 | 0.5 |
| Brazil | Baixada Santista[1] | 1,468 | 1,819 | 2,014 | 2.14 | 1.02 | 1.0 | 1.1 | 1.1 |
| Brazil | Belém | 1,748 | 2,191 | 2,427 | 2.26 | 1.02 | 1.2 | 1.3 | 1.3 |
| Brazil | Belo Horizonte | 4,659 | 5,852 | 6,420 | 2.28 | 0.93 | 3.3 | 3.5 | 3.4 |
| Brazil | Brasília | 2,746 | 3,905 | 4,433 | 3.52 | 1.27 | 1.9 | 2.3 | 2.4 |
| Brazil | Campinas | 2,264 | 2,818 | 3,109 | 2.19 | 0.98 | 1.6 | 1.7 | 1.7 |
| Brazil | Cuiabá | 686 | 772 | 843 | 1.18 | 0.88 | 0.5 | 0.5 | 0.5 |
| Brazil | Curitiba | 2,494 | 3,462 | 3,913 | 3.28 | 1.22 | 1.8 | 2.0 | 2.1 |
| Brazil | Florianópolis | 734 | 1,049 | 1,210 | 3.57 | 1.43 | 0.5 | 0.6 | 0.6 |
| Brazil | Fortaleza | 2,875 | 3,719 | 4,130 | 2.57 | 1.05 | 2.0 | 2.2 | 2.2 |
| Brazil | Goiânia | 1,635 | 2,146 | 2,405 | 2.72 | 1.14 | 1.2 | 1.3 | 1.3 |
| Brazil | Grande São Luís | 1,066 | 1,283 | 1,415 | 1.85 | 0.98 | 0.8 | 0.8 | 0.8 |
| Brazil | Grande Vitória | 1,398 | 1,848 | 2,078 | 2.79 | 1.17 | 1.0 | 1.1 | 1.1 |
| Brazil | João Pessoa | 827 | 1,015 | 1,129 | 2.05 | 1.06 | 0.6 | 0.6 | 0.6 |
| Brazil | Londrina | 613 | 814 | 925 | 2.84 | 1.28 | 0.4 | 0.5 | 0.5 |
| Brazil | Maceió | 952 | 1,192 | 1,329 | 2.25 | 1.09 | 0.7 | 0.7 | 0.7 |
| Brazil | Manaus | 1,392 | 1,775 | 1,979 | 2.43 | 1.09 | 1.0 | 1.0 | 1.1 |
| Brazil | Natal | 910 | 1,316 | 1,519 | 3.69 | 1.43 | 0.6 | 0.8 | 0.8 |
| Brazil | Norte/Nordeste Catarinense[2] | 815 | 1,069 | 1,207 | 2.71 | 1.21 | 0.6 | 0.6 | 0.6 |
| Brazil | Pôrto Alegre | 3,505 | 4,092 | 4,428 | 1.55 | 0.79 | 2.5 | 2.4 | 2.4 |
| Brazil | Recife | 3,230 | 3,871 | 4,219 | 1.81 | 0.86 | 2.3 | 2.3 | 2.3 |
| Brazil | Rio de Janeiro | 10,803 | 11,950 | 12,617 | 1.01 | 0.54 | 7.6 | 7.1 | 6.7 |
| Brazil | Salvador | 2,968 | 3,918 | 4,370 | 2.78 | 1.09 | 2.1 | 2.3 | 2.3 |
| Brazil | São Paulo | 17,099 | 20,262 | 21,628 | 1.70 | 0.65 | 12.1 | 12.0 | 11.6 |
| Brazil | Teresina | 789 | 900 | 984 | 1.32 | 0.89 | 0.6 | 0.5 | 0.5 |
| Chile | Santiago | 5,275 | 5,952 | 6,408 | 1.21 | 0.74 | 39.8 | 39.0 | 37.8 |
| Chile | Valparaíso | 803 | 873 | 946 | 0.84 | 0.80 | 6.1 | 5.7 | 5.6 |
| Colombia | Barranquilla | 1,531 | 1,867 | 2,145 | 1.98 | 1.39 | 5.3 | 5.4 | 5.3 |
| Colombia | Bucaramanga | 855 | 1,092 | 1,303 | 2.45 | 1.77 | 3.0 | 3.1 | 3.2 |
| Colombia | Cali | 1,950 | 2,401 | 2,800 | 2.08 | 1.54 | 6.8 | 6.9 | 6.9 |
| Colombia | Cartagena | 737 | 962 | 1,158 | 2.66 | 1.85 | 2.6 | 2.8 | 2.8 |
| Colombia | Medellín | 632 | 774 | 910 | 2.03 | 1.62 | 2.2 | 2.2 | 2.2 |

# TABLE C.1

*continued*

| | | Estimates and projections ('000) | | | Annual rate of change (%) | | Share in national urban population (%) | | |
|---|---|---|---|---|---|---|---|---|---|
| | | 2000 | 2010 | 2020 | 2000–2010 | 2010–2020 | 2000 | 2010 | 2020 |
| Colombia | Bogotá | 6,356 | 8,500 | 10,129 | 2.91 | 1.75 | 22.2 | 24.5 | 24.8 |
| Costa Rica | San José | 1,032 | 1,461 | 1,799 | 3.48 | 2.08 | 44.5 | 48.9 | 49.4 |
| Cuba | La Habana (Havana) | 2,187 | 2,130 | 2,095 | -0.26 | -0.17 | 26.1 | 25.3 | 24.8 |
| Dominican Republic | Santo Domingo | 1,813 | 2,180 | 2,552 | 1.84 | 1.58 | 33.3 | 30.8 | 29.8 |
| Ecuador | Guayaquil | 2,077 | 2,690 | 3,153 | 2.59 | 1.59 | 28.0 | 29.2 | 28.3 |
| Ecuador | Quito | 1,357 | 1,846 | 2,188 | 3.08 | 1.70 | 18.3 | 20.0 | 19.6 |
| El Salvador | San Salvador | 1,248 | 1,565 | 1,789 | 2.26 | 1.34 | 35.6 | 39.3 | 39.0 |
| Guatemala | Ciudad de Guatemala (Guatemala City) | 908 | 1,104 | 1,481 | 1.95 | 2.94 | 17.9 | 15.5 | 15.0 |
| Haiti | Port-au-Prince | 1,693 | 2,143 | 2,868 | 2.36 | 2.91 | 55.0 | 40.4 | 38.0 |
| Honduras | Tegucigalpa | 793 | 1,028 | 1,339 | 2.60 | 2.64 | 28.0 | 26.2 | 25.4 |
| Mexico | Aguascalientes | 734 | 926 | 1,039 | 2.32 | 1.15 | 1.0 | 1.1 | 1.1 |
| Mexico | Chihuahua | 683 | 840 | 939 | 2.07 | 1.11 | 0.9 | 1.0 | 1.0 |
| Mexico | Ciudad de México (Mexico City) | 18,022 | 19,460 | 20,476 | 0.77 | 0.51 | 24.2 | 22.6 | 21.2 |
| Mexico | Ciudad Juárez | 1,225 | 1,394 | 1,528 | 1.29 | 0.92 | 1.6 | 1.6 | 1.6 |
| Mexico | Culiacán | 749 | 836 | 918 | 1.10 | 0.94 | 1.0 | 1.0 | 1.0 |
| Mexico | Guadalajara | 3,703 | 4,402 | 4,796 | 1.73 | 0.86 | 5.0 | 5.1 | 5.0 |
| Mexico | Hermosillo | 616 | 781 | 878 | 2.38 | 1.18 | 0.8 | 0.9 | 0.9 |
| Mexico | León de los Aldamas | 1,290 | 1,571 | 1,739 | 1.97 | 1.02 | 1.7 | 1.8 | 1.8 |
| Mexico | Mérida | 848 | 1,015 | 1,127 | 1.80 | 1.05 | 1.1 | 1.2 | 1.2 |
| Mexico | Mexicali | 770 | 934 | 1,040 | 1.93 | 1.07 | 1.0 | 1.1 | 1.1 |
| Mexico | Monterrey | 3,266 | 3,896 | 4,253 | 1.76 | 0.88 | 4.4 | 4.5 | 4.4 |
| Mexico | Puebla | 1,907 | 2,315 | 2,551 | 1.94 | 0.97 | 2.6 | 2.7 | 2.6 |
| Mexico | Querétaro | 795 | 1,031 | 1,160 | 2.60 | 1.18 | 1.1 | 1.2 | 1.2 |
| Mexico | Saltillo | 643 | 801 | 897 | 2.20 | 1.13 | 0.9 | 0.9 | 0.9 |
| Mexico | San Luis Potosí | 858 | 1,049 | 1,168 | 2.01 | 1.07 | 1.2 | 1.2 | 1.2 |
| Mexico | Tampico | 659 | 761 | 842 | 1.43 | 1.01 | 0.9 | 0.9 | 0.9 |
| Mexico | Tijuana | 1,287 | 1,664 | 1,861 | 2.57 | 1.12 | 1.7 | 1.9 | 1.9 |
| Mexico | Toluca de Lerdo | 1,417 | 1,582 | 1,725 | 1.10 | 0.87 | 1.9 | 1.8 | 1.8 |
| Mexico | Torreón | 1,014 | 1,199 | 1,325 | 1.68 | 1.00 | 1.4 | 1.4 | 1.4 |
| Nicaragua | Managua | 887 | 944 | 1,103 | 0.62 | 1.56 | 31.8 | 28.3 | 27.1 |
| Panama | Ciudad de Panamá (Panama City) | 1,072 | 1,378 | 1,652 | 2.51 | 1.81 | 55.2 | 52.5 | 51.1 |
| Paraguay | Asunción | 1,507 | 2,030 | 2,505 | 2.98 | 2.10 | 50.9 | 51.1 | 49.6 |
| Peru | Arequipa | 678 | 789 | 903 | 1.52 | 1.35 | 3.6 | 3.5 | 3.4 |
| Peru | Lima | 7,294 | 8,941 | 10,145 | 2.04 | 1.26 | 38.4 | 39.4 | 38.4 |
| Puerto Rico | San Juan | 2,237 | 2,743 | 2,763 | 2.04 | 0.07 | 61.9 | 69.5 | 67.2 |
| Uruguay | Montevideo | 1,605 | 1,635 | 1,653 | 0.19 | 0.11 | 52.9 | 52.4 | 50.6 |
| Venezuela (Bolivarian Republic of) | Barquisimeto | 946 | 1,180 | 1,350 | 2.21 | 1.35 | 4.3 | 4.4 | 4.3 |
| Venezuela (Bolivarian Republic of) | Caracas | 2,864 | 3,090 | 3,467 | 0.76 | 1.15 | 13.1 | 11.4 | 10.9 |
| Venezuela (Bolivarian Republic of) | Maracaibo | 1,724 | 2,192 | 2,488 | 2.40 | 1.27 | 7.9 | 8.1 | 7.8 |
| Venezuela (Bolivarian Republic of) | Maracay | 898 | 1,057 | 1,208 | 1.63 | 1.34 | 4.1 | 3.9 | 3.8 |
| Venezuela (Bolivarian Republic of) | Valencia | 1,392 | 1,770 | 2,014 | 2.40 | 1.29 | 6.3 | 6.5 | 6.3 |
| **NORTHERN AMERICA** | | | | | | | | | |
| Canada | Calgary | 953 | 1,182 | 1,315 | 2.15 | 1.07 | 3.9 | 4.3 | 4.3 |
| Canada | Edmonton | 924 | 1,113 | 1,227 | 1.86 | 0.98 | 3.8 | 4.1 | 4.0 |
| Canada | Montréal | 3,471 | 3,783 | 4,048 | 0.86 | 0.68 | 14.2 | 13.9 | 13.3 |
| Canada | Ottawa-Gatineau | 1,079 | 1,182 | 1,285 | 0.91 | 0.84 | 4.4 | 4.3 | 4.2 |
| Canada | Toronto | 4,607 | 5,449 | 5,875 | 1.68 | 0.75 | 18.9 | 20.0 | 19.3 |
| Canada | Vancouver | 1,959 | 2,220 | 2,400 | 1.25 | 0.78 | 8.0 | 8.1 | 7.9 |
| United States of America | Atlanta | 3,542 | 4,691 | 5,036 | 2.81 | 0.71 | 1.6 | 1.8 | 1.7 |
| United States of America | Austin | 913 | 1,215 | 1,329 | 2.86 | 0.90 | 0.4 | 0.5 | 0.5 |
| United States of America | Baltimore | 2,083 | 2,320 | 2,508 | 1.08 | 0.78 | 0.9 | 0.9 | 0.9 |
| United States of America | Boston | 4,049 | 4,593 | 4,920 | 1.26 | 0.69 | 1.8 | 1.8 | 1.7 |
| United States of America | Bridgeport-Stamford | 894 | 1,055 | 1,154 | 1.66 | 0.90 | 0.4 | 0.4 | 0.4 |
| United States of America | Buffalo | 977 | 1,045 | 1,142 | 0.67 | 0.89 | 0.4 | 0.4 | 0.4 |
| United States of America | Charlotte | 769 | 1,043 | 1,144 | 3.05 | 0.92 | 0.3 | 0.4 | 0.4 |
| United States of America | Chicago | 8,333 | 9,204 | 9,758 | 0.99 | 0.58 | 3.7 | 3.5 | 3.3 |
| United States of America | Cincinnati | 1,508 | 1,686 | 1,831 | 1.12 | 0.83 | 0.7 | 0.6 | 0.6 |
| United States of America | Cleveland | 1,789 | 1,942 | 2,104 | 0.82 | 0.80 | 0.8 | 0.7 | 0.7 |
| United States of America | Columbus, Ohio | 1,138 | 1,313 | 1,432 | 1.43 | 0.87 | 0.5 | 0.5 | 0.5 |
| United States of America | Dallas-Fort Worth | 4,172 | 4,951 | 5,301 | 1.71 | 0.68 | 1.8 | 1.9 | 1.8 |
| United States of America | Dayton | 706 | 800 | 878 | 1.25 | 0.93 | 0.3 | 0.3 | 0.3 |
| United States of America | Denver-Aurora | 1,998 | 2,394 | 2,590 | 1.81 | 0.79 | 0.9 | 0.9 | 0.9 |
| United States of America | Detroit | 3,909 | 4,200 | 4,500 | 0.72 | 0.69 | 1.7 | 1.6 | 1.5 |
| United States of America | El Paso | 678 | 779 | 856 | 1.39 | 0.94 | 0.3 | 0.3 | 0.3 |
| United States of America | Hartford | 853 | 942 | 1,031 | 0.99 | 0.90 | 0.4 | 0.4 | 0.4 |
| United States of America | Honolulu | 720 | 812 | 891 | 1.20 | 0.93 | 0.3 | 0.3 | 0.3 |
| United States of America | Houston | 3,849 | 4,605 | 4,937 | 1.79 | 0.70 | 1.7 | 1.8 | 1.7 |
| United States of America | Indianapolis | 1,228 | 1,490 | 1,623 | 1.93 | 0.86 | 0.5 | 0.6 | 0.6 |
| United States of America | Jacksonville, Florida | 886 | 1,022 | 1,119 | 1.43 | 0.91 | 0.4 | 0.4 | 0.4 |
| United States of America | Kansas City | 1,365 | 1,513 | 1,645 | 1.03 | 0.84 | 0.6 | 0.6 | 0.6 |
| United States of America | Las Vegas | 1,335 | 1,916 | 2,086 | 3.61 | 0.85 | 0.6 | 0.7 | 0.7 |
| United States of America | Los Angeles-Long Beach-Santa Ana | 11,814 | 12,762 | 13,463 | 0.77 | 0.53 | 5.2 | 4.9 | 4.6 |
| United States of America | Louisville | 866 | 979 | 1,071 | 1.23 | 0.90 | 0.4 | 0.4 | 0.4 |
| United States of America | McAllen | 532 | 789 | 870 | 3.94 | 0.98 | 0.2 | 0.3 | 0.3 |
| United States of America | Memphis | 976 | 1,117 | 1,221 | 1.35 | 0.89 | 0.4 | 0.4 | 0.4 |
| United States of America | Miami | 4,946 | 5,750 | 6,142 | 1.51 | 0.66 | 2.2 | 2.2 | 2.1 |
| United States of America | Milwaukee | 1,311 | 1,428 | 1,554 | 0.85 | 0.85 | 0.6 | 0.5 | 0.5 |
| United States of America | Minneapolis-St. Paul | 2,397 | 2,693 | 2,905 | 1.16 | 0.76 | 1.1 | 1.0 | 1.0 |
| United States of America | Nashville-Davidson | 755 | 911 | 999 | 1.88 | 0.92 | 0.3 | 0.3 | 0.3 |
| United States of America | New Orleans | 1,009 | 858 | 984 | -1.62 | 1.37 | 0.4 | 0.3 | 0.3 |
| United States of America | New York-Newark | 17,846 | 19,425 | 20,374 | 0.85 | 0.48 | 7.8 | 7.4 | 6.9 |

# TABLE C.1
**continued**

| | | Estimates and projections ('000) | | | Annual rate of change (%) | | Share in national urban population (%) | | |
|---|---|---|---|---|---|---|---|---|---|
| | | **2000** | **2010** | **2020** | **2000–2010** | **2010–2020** | **2000** | **2010** | **2020** |
| United States of America | Oklahoma City | 748 | 812 | 891 | 0.82 | 0.93 | 0.3 | 0.3 | 0.3 |
| United States of America | Orlando | 1,165 | 1,400 | 1,526 | 1.84 | 0.86 | 0.5 | 0.5 | 0.5 |
| United States of America | Philadelphia | 5,160 | 5,626 | 6,004 | 0.86 | 0.65 | 2.3 | 2.2 | 2.0 |
| United States of America | Phoenix-Mesa | 2,934 | 3,684 | 3,965 | 2.28 | 0.74 | 1.3 | 1.4 | 1.3 |
| United States of America | Pittsburgh | 1,755 | 1,887 | 2,045 | 0.73 | 0.80 | 0.8 | 0.7 | 0.7 |
| United States of America | Portland | 1,595 | 1,944 | 2,110 | 1.98 | 0.82 | 0.7 | 0.7 | 0.7 |
| United States of America | Providence | 1,178 | 1,317 | 1,435 | 1.12 | 0.86 | 0.5 | 0.5 | 0.5 |
| United States of America | Raleigh | 549 | 769 | 848 | 3.37 | 0.97 | 0.2 | 0.3 | 0.3 |
| United States of America | Richmond | 822 | 944 | 1,034 | 1.38 | 0.91 | 0.4 | 0.4 | 0.4 |
| United States of America | Riverside-San Bernardino | 1,516 | 1,807 | 1,962 | 1.76 | 0.82 | 0.7 | 0.7 | 0.7 |
| United States of America | Rochester | 696 | 780 | 857 | 1.14 | 0.94 | 0.3 | 0.3 | 0.3 |
| United States of America | Sacramento | 1,402 | 1,660 | 1,805 | 1.69 | 0.84 | 0.6 | 0.6 | 0.6 |
| United States of America | Salt Lake City | 890 | 997 | 1,091 | 1.14 | 0.90 | 0.4 | 0.4 | 0.4 |
| United States of America | San Antonio | 1,333 | 1,521 | 1,655 | 1.32 | 0.84 | 0.6 | 0.6 | 0.6 |
| United States of America | San Diego | 2,683 | 2,999 | 3,231 | 1.11 | 0.75 | 1.2 | 1.1 | 1.1 |
| United States of America | San Francisco-Oakland | 3,236 | 3,541 | 3,804 | 0.90 | 0.72 | 1.4 | 1.4 | 1.3 |
| United States of America | San Jose | 1,543 | 1,718 | 1,865 | 1.07 | 0.82 | 0.7 | 0.7 | 0.6 |
| United States of America | Seattle | 2,727 | 3,171 | 3,415 | 1.51 | 0.74 | 1.2 | 1.2 | 1.2 |
| United States of America | St. Louis | 2,081 | 2,259 | 2,442 | 0.82 | 0.78 | 0.9 | 0.9 | 0.8 |
| United States of America | Tampa-St Petersburg | 2,072 | 2,387 | 2,581 | 1.42 | 0.78 | 0.9 | 0.9 | 0.9 |
| United States of America | Tucson | 724 | 853 | 936 | 1.64 | 0.93 | 0.3 | 0.3 | 0.3 |
| United States of America | Virginia Beach | 1,397 | 1,534 | 1,668 | 0.94 | 0.84 | 0.6 | 0.6 | 0.6 |
| United States of America | Washington, DC | 3,949 | 4,460 | 4,779 | 1.22 | 0.69 | 1.7 | 1.7 | 1.6 |
| **OCEANIA** | | | | | | | | | |
| Australia | Adelaide | 1,102 | 1,168 | 1,263 | 0.58 | 0.78 | 6.6 | 6.1 | 5.9 |
| Australia | Brisbane | 1,603 | 1,970 | 2,178 | 2.06 | 1.00 | 9.6 | 10.3 | 10.1 |
| Australia | Melbourne | 3,433 | 3,853 | 4,152 | 1.15 | 0.75 | 20.5 | 20.1 | 19.3 |
| Australia | Perth | 1,373 | 1,599 | 1,753 | 1.52 | 0.92 | 8.2 | 8.3 | 8.2 |
| Australia | Sydney | 4,078 | 4,429 | 4,733 | 0.83 | 0.66 | 24.4 | 23.1 | 22.1 |
| New Zealand | Auckland | 1,063 | 1,404 | 1,631 | 2.79 | 1.50 | 32.1 | 37.9 | 40.2 |

*Source:* United Nations Department of Economic and Social Affairs, Population Division (2010) *World Urbanization Prospects: The 2009 Revision*, United Nations, New York.

*Notes:*
(1) Including Santos.
(2) Including Jointville.

# TABLE C.2

## Population of Capital Cities (2009)

| | | City population ('000) | City population as a percentage of | |
| --- | --- | --- | --- | --- |
| | | | Urban population (%) | Total population (%) |
| **AFRICA** | | | | |
| Algeria | El Djazaïr (Algiers) | 2,740 | 11.9 | 7.9 |
| Angola | Luanda | 4,511 | 42.3 | 24.4 |
| Benin[1] | Cotonou | 815 | 21.9 | 9.1 |
| Botswana | Gaborone | 196 | 16.6 | 10.0 |
| Burkina Faso | Ouagadougou | 1,777 | 45.4 | 11.3 |
| Burundi | Bujumbura | 455 | 51.3 | 5.5 |
| Cameroon | Yaoundé | 1,739 | 15.5 | 8.9 |
| Cape Verde | Praia | 125 | 41.0 | 24.8 |
| Central African Republic | Bangui | 702 | 41.0 | 15.9 |
| Chad | N'Djaména | 808 | 26.6 | 7.2 |
| Comoros | Moroni | 49 | 25.6 | 7.2 |
| Congo | Brazzaville | 1,292 | 56.9 | 35.1 |
| Côte d'Ivoire[2] | Abidjan | 4,009 | 38.2 | 19.0 |
| Côte d'Ivoire[2] | Yamoussoukro | 808 | 7.7 | 3.8 |
| Democratic Republic of the Congo | Kinshasa | 8,401 | 36.8 | 12.7 |
| Djibouti | Djibouti | 567 | 86.1 | 65.6 |
| Egypt | Al-Qahirah (Cairo) | 10,903 | 30.3 | 13.1 |
| Equatorial Guinea | Malabo | 128 | 47.8 | 18.9 |
| Eritrea | Asmara | 649 | 60.6 | 12.8 |
| Ethiopia | Addis Ababa | 2,863 | 21.0 | 3.5 |
| Gabon | Libreville | 619 | 49.0 | 42.0 |
| Gambia | Banjul | 436 | 44.6 | 25.6 |
| Ghana | Accra | 2,269 | 18.8 | 9.5 |
| Guinea | Conakry | 1,597 | 45.5 | 15.9 |
| Guinea-Bissau | Bissau | 302 | 62.7 | 18.7 |
| Kenya | Nairobi | 3,375 | 38.8 | 8.5 |
| Lesotho | Maseru | 220 | 40.7 | 10.6 |
| Liberia | Monrovia | 882 | 47.0 | 22.3 |
| Libyan Arab Jamahiriya | Tarabulus (Tripoli) | 1,095 | 22.0 | 17.1 |
| Madagascar | Antananarivo | 1,816 | 31.0 | 9.3 |
| Malawi | Lilongwe | 821 | 27.9 | 5.4 |
| Mali | Bamako | 1,628 | 35.7 | 12.5 |
| Mauritania | Nouakchott | 709 | 52.3 | 21.5 |
| Mauritius | Port Louis | 149 | 27.7 | 11.6 |
| Mayotte | Mamoudzou | 6 | 6.2 | 3.1 |
| Morocco | Rabat | 1,770 | 9.6 | 5.5 |
| Mozambique | Maputo | 1,589 | 18.4 | 6.9 |
| Namibia | Windhoek | 342 | 42.1 | 15.7 |
| Niger | Niamey | 1,004 | 38.7 | 6.6 |
| Nigeria | Abuja | 1,857 | 2.4 | 1.2 |
| Réunion | Saint-Denis | 141 | 18.2 | 17.0 |
| Rwanda | Kigali | 909 | 49.0 | 9.1 |
| Saint Helena | Jamestown | 1 | 39.5 | 15.7 |
| São Tomé and Príncipe | São Tomé | 60 | 59.9 | 36.8 |
| Senegal | Dakar | 2,777 | 52.6 | 22.2 |
| Seychelles | Victoria | 26 | 56.3 | 30.9 |
| Sierra Leone | Freetown | 875 | 40.4 | 15.4 |
| Somalia | Muqdisho (Mogadishu) | 1,353 | 40.1 | 14.8 |
| South Africa[3] | Bloemfontein | 436 | 1.4 | 0.9 |
| South Africa[3] | Cape Town | 3,353 | 10.9 | 6.7 |
| South Africa[3] | Pretoria | 1,404 | 4.6 | 2.8 |
| Sudan | Al-Khartum (Khartoum) | 5,021 | 30.2 | 11.9 |
| Swaziland[4] | Lobamba | … | … | … |
| Swaziland[4] | Mbabane | 74 | 29.1 | 6.2 |
| Togo | Lomé | 1,593 | 56.3 | 24.1 |
| Tunisia | Tunis | 759 | 11.0 | 7.4 |
| Uganda | Kampala | 1,535 | 35.8 | 4.7 |
| United Republic of Tanzania | Dodoma | 200 | 1.8 | 0.5 |
| Western Sahara | El Aaiún | 213 | 50.9 | 41.5 |
| Zambia | Lusaka | 1,413 | 30.7 | 10.9 |
| Zimbabwe | Harare | 1,606 | 34.0 | 12.8 |
| **ASIA** | | | | |
| Afghanistan | Kabul | 3,573 | 56.9 | 12.7 |
| Armenia | Yerevan | 1,110 | 56.1 | 36.0 |
| Azerbaijan | Baku | 1,950 | 42.6 | 22.1 |
| Bahrain | Al-Manamah (Manama) | 163 | 23.3 | 20.6 |
| Bangladesh | Dhaka | 14,251 | 31.9 | 8.8 |
| Bhutan | Thimphu | 89 | 37.8 | 12.8 |
| Brunei Darussalam | Bandar Seri Begawan | 22 | 7.4 | 5.6 |
| Cambodia | Phnum Pénh (Phnom Penh) | 1,519 | 51.8 | 10.3 |
| China | Beijing | 12,214 | 2.0 | 0.9 |
| China, Hong Kong SAR[5] | Hong Kong | 7,022 | 100.0 | 100.0 |
| China, Macao SAR[6] | Macao | 538 | 100.0 | 100.0 |
| Cyprus | Lefkosia (Nicosia) | 240 | 39.3 | 27.5 |
| Democratic People's Republic of Korea | P'yongyang | 2,828 | 19.7 | 11.8 |
| Georgia | Tbilisi | 1,115 | 49.7 | 26.2 |
| India[7] | Delhi | 21,720 | 6.1 | 1.8 |
| Indonesia | Jakarta | 9,121 | 9.0 | 4.0 |

# TABLE C.2

## *continued*

| | | City population | City population as a percentage of | |
|---|---|---|---|---|
| | | | Urban population | Total population |
| | | ('000) | (%) | (%) |
| Iran (Islamic Republic of) | Tehran | 7,190 | 13.8 | 9.7 |
| Iraq | Baghdad | 5,751 | 28.2 | 18.7 |
| Israel | Jerusalem | 768 | 11.7 | 10.7 |
| Japan | Tokyo | 36,507 | 43.1 | 28.7 |
| Jordan | Amman | 1,088 | 22.0 | 17.2 |
| Kazakhstan | Astana | 650 | 7.1 | 4.2 |
| Kuwait | Al Kuwayt (Kuwait City) | 2,230 | 75.9 | 74.7 |
| Kyrgyzstan | Bishkek | 854 | 45.0 | 15.6 |
| Lao People's Democratic Republic | Vientiane | 799 | 39.5 | 12.6 |
| Lebanon | Bayrut (Beirut) | 1,909 | 51.9 | 45.2 |
| Malaysia[8] | Kuala Lumpur | 1,494 | 7.6 | 5.4 |
| Maldives | Male | 120 | 100.0 | 38.9 |
| Mongolia | Ulaanbaatar | 949 | 57.7 | 35.5 |
| Myanmar | Nay Pyi Taw | 992 | 6.0 | 2.0 |
| Nepal | Kathmandu | 990 | 18.7 | 3.4 |
| Occupied Palestinian Territory | Ramallah | 69 | 2.2 | 1.6 |
| Oman | Masqat (Muscat) | 634 | 30.6 | 22.3 |
| Pakistan | Islamabad | 832 | 1.3 | 0.5 |
| Philippines | Manila | 11,449 | 25.6 | 12.4 |
| Qatar | Ad-Dawhah (Doha) | 427 | 31.6 | 30.3 |
| Republic of Korea | Seoul | 9,778 | 24.5 | 20.2 |
| Saudi Arabia | Ar-Riyadh (Riyadh) | 4,725 | 22.4 | 18.4 |
| Singapore | Singapore | 4,737 | 100.0 | 100.0 |
| Sri Lanka[9] | Colombo | 681 | 23.5 | 3.4 |
| Sri Lanka[9] | Sri Jayewardenepura Kotte | 123 | 4.2 | 0.6 |
| Syrian Arab Republic | Dimashq (Damascus) | 2,527 | 20.8 | 11.5 |
| Tajikistan | Dushanbe | 704 | 38.5 | 10.1 |
| Thailand | Krung Thep (Bangkok) | 6,902 | 30.3 | 10.2 |
| Timor-Leste | Dili | 166 | 53.0 | 14.7 |
| Turkey | Ankara | 3,846 | 7.4 | 5.1 |
| Turkmenistan | Ashgabat | 637 | 25.4 | 12.5 |
| United Arab Emirates | Abu Zaby (Abu Dhabi) | 666 | 17.3 | 14.5 |
| Uzbekistan | Tashkent | 2,201 | 22.1 | 8.0 |
| Viet Nam | Hà Noi | 2,668 | 10.2 | 3.0 |
| Yemen | Sana'a' | 2,229 | 30.3 | 9.5 |
| **EUROPE** | | | | |
| Albania | Tiranë (Tirana) | 433 | 26.9 | 13.7 |
| Andorra | Andorra la Vella | 25 | 32.9 | 29.1 |
| Austria | Wien (Vienna) | 1,693 | 30.1 | 20.2 |
| Belarus | Minsk | 1,837 | 25.7 | 19.1 |
| Belgium | Bruxelles-Brussel | 1,892 | 18.2 | 17.8 |
| Bosnia and Herzegovina | Sarajevo | 392 | 21.7 | 10.4 |
| Bulgaria | Sofia | 1,192 | 22.2 | 15.8 |
| Channel Islands[10] | St. Helier and St. Peter Port | 30 | 63.4 | 19.8 |
| Croatia | Zagreb | 685 | 27.0 | 15.5 |
| Czech Republic | Praha (Prague) | 1,162 | 15.2 | 11.2 |
| Denmark | København (Copenhagen) | 1,174 | 24.8 | 21.5 |
| Estonia | Tallinn | 399 | 42.9 | 29.8 |
| Faeroe Islands | Tórshavn | 20 | 100.0 | 40.3 |
| Finland | Helsinki | 1,107 | 24.5 | 20.8 |
| France | Paris | 10,410 | 19.7 | 16.7 |
| Germany | Berlin | 3,438 | 5.7 | 4.2 |
| Gibraltar | Gibraltar | 31 | 100.0 | 100.0 |
| Greece | Athínai (Athens) | 3,252 | 47.6 | 29.1 |
| Holy See | Vatican City | 1 | 100.0 | 100.0 |
| Hungary | Budapest | 1,705 | 25.2 | 17.1 |
| Iceland | Reykjavík | 198 | 65.8 | 61.4 |
| Ireland | Dublin | 1,084 | 39.0 | 24.0 |
| Isle of Man | Douglas | 26 | 63.9 | 32.4 |
| Italy | Roma (Rome) | 3,357 | 8.2 | 5.6 |
| Latvia | Riga | 711 | 46.6 | 31.6 |
| Liechtenstein | Vaduz | 5 | 100.0 | 14.3 |
| Lithuania | Vilnius | 546 | 24.8 | 16.6 |
| Luxembourg | Luxembourg-Ville | 90 | 21.7 | 18.5 |
| Malta | Valletta | 199 | 51.6 | 48.8 |
| Monaco | Monaco | 33 | 100.0 | 100.0 |
| Montenegro | Podgorica | 144 | 37.5 | 23.0 |
| Netherlands[11] | Amsterdam | 1,044 | 7.6 | 6.3 |
| Norway | Oslo | 875 | 23.0 | 18.2 |
| Poland | Warszawa (Warsaw) | 1,710 | 7.4 | 4.5 |
| Portugal | Lisboa (Lisbon) | 2,808 | 43.6 | 26.2 |
| Republic of Moldova | Chişinău | 650 | 39.0 | 18.0 |
| Romania | Bucuresti (Bucharest) | 1,933 | 16.0 | 9.1 |
| Russian Federation | Moskva (Moscow) | 10,523 | 10.2 | 7.5 |
| San Marino | San Marino | 4 | 14.9 | 14.0 |
| Serbia | Beograd (Belgrade) | 1,115 | 20.3 | 11.3 |
| Slovakia | Bratislava | 428 | 14.4 | 7.9 |
| Slovenia | Ljubljana | 260 | 26.0 | 12.9 |
| Spain | Madrid | 5,762 | 16.6 | 12.8 |

# TABLE C.2

## *continued*

| | | City population ('000) | City population as a percentage of | |
|---|---|---|---|---|
| | | | Urban population (%) | Total population (%) |
| Sweden | Stockholm | 1,279 | 16.3 | 13.8 |
| Switzerland | Bern | 346 | 6.2 | 4.6 |
| TFYR Macedonia[12] | Skopje | 480 | 39.7 | 23.5 |
| Ukraine | Kyiv (Kiev) | 2,779 | 8.9 | 6.1 |
| United Kingdom | London | 8,615 | 17.6 | 14.0 |
| **LATIN AMERICA AND THE CARIBBEAN** | | | | |
| Anguilla | The Valley | 2 | 10.8 | 10.8 |
| Antigua and Barbuda | St. John's | 27 | 100.0 | 30.3 |
| Argentina | Buenos Aires | 12,988 | 35.0 | 32.2 |
| Aruba | Oranjestad | 33 | 66.4 | 31.1 |
| Bahamas | Nassau | 248 | 86.4 | 72.5 |
| Barbados | Bridgetown | 112 | 100.0 | 43.8 |
| Belize | Belmopan | 20 | 12.4 | 6.4 |
| Bolivia[13] | La Paz | 1,642 | 25.2 | 16.6 |
| Bolivia[13] | Sucre | 281 | 4.3 | 2.8 |
| Brazil | Brasília | 3,789 | 2.3 | 2.0 |
| British Virgin Islands | Road Town | 9 | 100.0 | 40.7 |
| Cayman Islands | George Town | 32 | 56.5 | 56.5 |
| Chile | Santiago | 5,883 | 39.1 | 34.7 |
| Colombia | Bogotá | 8,262 | 24.2 | 18.1 |
| Costa Rica | San José | 1,416 | 48.4 | 30.9 |
| Cuba | La Habana (Havana) | 2,140 | 25.4 | 19.1 |
| Dominica | Roseau | 14 | 31.9 | 21.4 |
| Dominican Republic | Santo Domingo | 2,138 | 30.9 | 21.2 |
| Ecuador | Quito | 1,801 | 19.9 | 13.2 |
| El Salvador | San Salvador | 1,534 | 39.0 | 24.9 |
| Falkland Islands (Malvinas) | Stanley | 2 | 100.0 | 73.1 |
| French Guiana | Cayenne | 62 | 36.3 | 27.7 |
| Grenada | St. George's | 40 | 100.0 | 38.9 |
| Guadeloupe | Basse-Terre | 13 | 2.9 | 2.8 |
| Guatemala | Ciudad de Guatemala (Guatemala City) | 1,075 | 15.6 | 7.7 |
| Guyana | Georgetown | 132 | 60.6 | 17.3 |
| Haiti | Port-au-Prince | 2,643 | 52.1 | 26.3 |
| Honduras | Tegucigalpa | 1,000 | 26.3 | 13.4 |
| Jamaica | Kingston | 580 | 41.0 | 21.3 |
| Martinique | Fort-de-France | 89 | 24.6 | 21.9 |
| Mexico | Ciudad de México (Mexico City) | 19,319 | 22.7 | 17.6 |
| Montserrat[14] | Brades Estate | 1 | 98.7 | 13.9 |
| Montserrat[14] | Plymouth | 0 | 0.1 | 0.0 |
| Netherlands Antilles | Willemstad | 123 | 67.0 | 62.2 |
| Nicaragua | Managua | 934 | 28.5 | 16.3 |
| Panama | Ciudad de Panamá (Panama City) | 1,346 | 52.6 | 39.0 |
| Paraguay | Asunción | 1,977 | 51.1 | 31.1 |
| Peru | Lima | 8,769 | 39.3 | 30.1 |
| Puerto Rico | San Juan | 2,730 | 69.5 | 68.6 |
| Saint Kitts and Nevis | Basseterre | 13 | 76.9 | 24.8 |
| Saint Lucia | Castries | 15 | 32.1 | 8.9 |
| Saint Vincent and the Grenadines | Kingstown | 28 | 52.8 | 25.8 |
| Suriname | Paramaribo | 259 | 72.4 | 49.9 |
| Trinidad and Tobago | Port of Spain | 57 | 31.7 | 4.3 |
| Turks and Caicos Islands | Grand Turk | 6 | 20.4 | 18.9 |
| United States Virgin Islands | Charlotte Amalie | 54 | 51.4 | 48.9 |
| Uruguay | Montevideo | 1,633 | 52.6 | 48.6 |
| Venezuela (Bolivarian Republic of) | Caracas | 3,051 | 11.5 | 10.7 |
| **NORTHERN AMERICA** | | | | |
| Bermuda | Hamilton | 12 | 17.8 | 17.8 |
| Canada[15] | Ottawa-Gatineau | 1,170 | 4.3 | 3.5 |
| Greenland | Nuuk (Godthåb) | 15 | 31.6 | 26.5 |
| Saint-Pierre-et-Miquelon | Saint-Pierre | 5 | 100.0 | 90.4 |
| United States of America | Washington, DC | 4,421 | 1.7 | 1.4 |
| **OCEANIA** | | | | |
| American Samoa | Pago Pago | 60 | 96.0 | 88.9 |
| Australia | Canberra | 384 | 2.0 | 1.8 |
| Cook Islands[16] | Rarotonga | 15 | 100.0 | 74.5 |
| Fiji | Greater Suva | 174 | 39.8 | 20.5 |
| French Polynesia | Papeete | 133 | 96.0 | 49.4 |
| Guam | Hagåtña | 153 | 92.4 | 86.0 |
| Kiribati[17] | Tarawa | 43 | 100.0 | 43.8 |
| Marshall Islands | Majuro | 30 | 66.8 | 47.7 |
| Micronesia (Fed. States of) | Palikir | 7 | 27.8 | 6.3 |
| Nauru | Nauru | 10 | 100.0 | 100.0 |
| New Caledonia | Nouméa | 144 | 100.0 | 57.4 |
| New Zealand | Wellington | 391 | 10.6 | 9.2 |
| Niue | Alofi | 1 | 100.0 | 37.0 |
| Northern Mariana Islands[18] | Saipan | 79 | 100.0 | 91.2 |
| Palau | Melekeok | 1 | 5.9 | 4.9 |
| Papua New Guinea | Port Moresby | 314 | 37.3 | 4.7 |
| Pitcairn | Adamstown | 0 | — | 100.0 |
| Samoa | Apia | 36 | 100.0 | 20.4 |

# TABLE C.2

**continued**

| | | City population | City population as a percentage of | |
| | | | Urban population | Total population |
| | | ('000) | (%) | (%) |
| --- | --- | --- | --- | --- |
| Solomon Islands | Honiara | 72 | 75.5 | 13.7 |
| Tokelau[19] | | .. | .. | .. |
| Tonga | Nuku'alofa | 24 | 100.0 | 23.3 |
| Tuvalu | Funafuti | 5 | 100.0 | 49.9 |
| Vanuatu | Port Vila | 44 | 72.5 | 18.2 |
| Wallis and Futuna Islands | Matu-Utu | 1 | — | 6.5 |

*Source:* United Nations Department of Economic and Social Affairs, Population Division (2010) *World Urbanization Prospects: The 2009 Revision*, United Nations, New York.

*Notes:*
(1) Porto-Novo is the constitutional capital, Cotonou is the seat of government.
(2) Yamoussoukro is the capital, Abidjan is the seat of government.
(3) Pretoria is the administrative capital, Cape Town is the legislative capital and Bloemfontein is the judicial capital.
(4) Mbabane is the administrative capital, Lobamba is the legislative capital.
(5) As of 1 July 1997, Hong Kong became a Special Administrative Region (SAR) of China.
(6) As of 20 December 1999, Macao became a Special Administrative Region (SAR) of China.
(7) The capital is New Delhi, included in the urban agglomeration of Delhi. The population of New Delhi was estimated at 294,783 in the year 2001.
(8) Kuala Lumpur is the financial capital, Putrajaya is the administrative capital.
(9) Colombo is the commercial capital, Sri Jayewardenepura Kotte is the administrative and legislative capital.
(10) Refers to Guernsey and Jersey. St. Helier is the capital of the Bailiwick of Jersey and St. Peter Port is the capital of the Bailiwick of Guernsey.
(11) Amsterdam is the capital, 's-Gravenhage is the seat of government.
(12) The former Yugoslav Republic of Macedonia.
(13) La Paz is the capital and the seat of government; Sucre is the legal capital and the seat of the judiciary.
(14) Due to volcanic activity, Plymouth was abandoned in 1997. The government premises have been established at Brades Estate.
(15) The capital is Ottawa.
(16) The capital is Avarua, located on the island of Rarotonga; the estimated population refers to the island of Rarotonga. Population estimates for Avarua have not been made available.
(17) The capital is Bairiki, located on the atoll of Tarawa; the estimated population refers to the island of South Tarawa. Population estimates for Bairiki have not been made available.
(18) The capital is Garapan, located on the island of Saipan; the estimated population refers to the island of Saipan. The population of Garapan was estimated at 3,588 in the year 2000.
(19) There is no capital in Tokelau. Each atoll (Atafu, Fakaofo and Nukunonu) has its own administrative centre.

# TABLE C.3

## Access to Services in Selected Cities

| | | Percentage of households with access to | | | | | | | | | | | | | | |
|---|---|---|---|---|---|---|---|---|---|---|---|---|---|---|---|---|
| | | 1990–1999[1] | | | | | | | | 2000–2009[1] | | | | | | |
| | | Survey year | Improved water | Piped water | Improved sanitation | Sewerage | Telephone(s) | Mobile(s) | Connection to electricity | Survey year | Improved water | Piped water | Improved sanitation | Sewerage | Telephone(s) | Mobile(s) | Connection to electricity |
| **AFRICA** | | | | | | | | | | | | | | | | | |
| Angola | Luanda | ... | ... | ... | ... | ... | ... | ... | ... | 2006 | 51.4 | 36.6 | 92.4 | 53.2 | 88.2 | 40.1 | 75.5 |
| Benin | Cotonou | 1996 | 99.0 | 98.1 | 71.2 | ... | ... | ... | 56.6 | ... | ... | ... | ... | ... | ... | ... | ... |
| Benin | Djougou | 1996 | 84.3 | 65.4 | 45.1 | ... | ... | ... | 23.5 | 2006 | 90.6 | 62.6 | 51.9 | ... | 3.9 | 31.0 | 47.4 |
| Benin | Porto-Novo | 1996 | 57.7 | 40.3 | 50.8 | ... | ... | ... | 29.4 | 2006 | 77.0 | 64.1 | 68.4 | ... | 8.1 | 57.3 | 66.9 |
| Burkina Faso | Ouagadougou | 1999 | 88.5 | 27.1 | 51.5 | 6.4 | 13.7 | ... | 41.3 | 2006 | 83.3 | 39.4 | 56.5 | 4.6 | 17.3 | 62.8 | 61.6 |
| Cameroon | Douala | 1998 | 77.2 | 32.2 | 80.8 | 26.0 | 7.6 | ... | 93.8 | 2006 | 99.2 | 51.0 | 79.9 | 25.3 | 5.3 | 76.2 | 98.9 |
| Cameroon | Yaoundé | 1998 | 93.7 | 59.9 | 81.9 | 22.0 | 11.5 | ... | 96.3 | 2006 | 99.5 | 53.8 | 79.9 | 28.2 | 7.3 | 82.8 | 98.9 |
| Central African Republic | Bangui | 1994 | 74.9 | 9.9 | 49.5 | 5.5 | 5.8 | ... | 15.3 | 2006 | 97.3 | 7.4 | 81.5 | 6.2 | 6.1 | 40.4 | 43.3 |
| Central African Republic | Berbérati | | ... | ... | ... | ... | ... | ... | ... | 2006 | 94.7 | 3.5 | 79.7 | 0.7 | 2.7 | 13.1 | 4.1 |
| Central African Republic | Boali | | ... | ... | ... | ... | ... | ... | ... | 2006 | 79.1 | 5.7 | 71.7 | 1.1 | 1.7 | 23.1 | 16.5 |
| Chad | N'Djaména | 1997 | 30.6 | 21.0 | 69.9 | 2.1 | 2.8 | ... | 17.2 | 2004 | 87.8 | 27.6 | 65.4 | 10.3 | 6.5 | ... | 29.2 |
| Comoros | Fomboni | | ... | ... | ... | ... | ... | ... | ... | 2000 | 73.5 | 31.3 | 62.7 | 1.2 | 7.2 | ... | 31.3 |
| Comoros | Moroni | 1996 | 95.7 | 22.2 | 67.6 | 11.4 | 13.0 | ... | 55.1 | 2000 | 93.3 | 25.8 | 56.0 | 4.8 | 27.2 | ... | 67.2 |
| Comoros | Mutsamudu | | ... | ... | ... | ... | ... | ... | ... | 2000 | 96.9 | 73.6 | 51.8 | 8.0 | 10.1 | ... | 53.1 |
| Congo | Brazzaville | | ... | ... | ... | ... | ... | ... | ... | 2005 | 96.8 | 89.1 | 70.3 | 9.8 | 2.6 | 57.0 | 59.2 |
| Côte d'Ivoire | Abidjan | 1998 | 56.8 | 45.0 | 66.3 | 13.0 | 6.5 | ... | 80.2 | 2005 | 98.6 | 83.3 | 79.3 | 42.7 | 49.5 | 0.0 | 95.0 |
| Democratic Republic of the Congo | Kinshasa | | ... | ... | ... | ... | ... | ... | ... | 2007 | 92.3 | 45.8 | 80.8 | 29.6 | 0.6 | 74.8 | 82.0 |
| Democratic Republic of the Congo | Lubumbashi | | ... | ... | ... | ... | ... | ... | ... | 2007 | 79.4 | 29.6 | 77.2 | 15.2 | 3.3 | 53.4 | 44.0 |
| Democratic Republic of the Congo | Mbuji-Mayi | | ... | ... | ... | ... | ... | ... | ... | 2007 | 95.8 | 10.2 | 84.6 | 10.4 | 1.1 | 34.0 | 3.7 |
| Egypt | Al-Iskandariyah (Alexandria) | 1995 | 99.7 | 94.2 | 79.4 | 61.0 | ... | ... | 99.8 | 2008 | 100.0 | 99.4 | 99.9 | 99.9 | 61.4 | 61.9 | 99.8 |
| Egypt | Al-Qahirah (Cairo) | 1995 | 98.6 | 94.8 | 76.2 | 56.0 | ... | ... | 99.0 | 2008 | 100.0 | 99.5 | 99.9 | 99.9 | 60.7 | 52.8 | 99.9 |
| Egypt | Assiut | 1995 | 94.7 | 91.7 | 61.8 | 27.1 | ... | ... | 96.1 | 2008 | 100.0 | 98.0 | 99.4 | 99.1 | 58.2 | 46.3 | 100.0 |
| Egypt | Aswan | 1995 | 95.5 | 88.6 | 56.8 | 25.0 | ... | ... | 98.2 | 2008 | 100.0 | 98.8 | 99.6 | 99.6 | 61.7 | 46.9 | 99.6 |
| Egypt | Beni Suef | 1995 | 88.9 | 83.8 | 57.6 | 28.3 | ... | ... | 96.0 | 2008 | 100.0 | 86.6 | 97.8 | 97.3 | 50.0 | 48.4 | 100.0 |
| Egypt | Damanhur | 1995 | 99.3 | 98.7 | 77.6 | 65.8 | ... | ... | 100.0 | 2008 | 100.0 | 100.0 | 100.0 | 100.0 | 58.5 | 48.1 | 100.0 |
| Egypt | Damietta | 1995 | 96.7 | 94.0 | 73.6 | 48.9 | ... | ... | 97.8 | 2008 | 100.0 | 100.0 | 100.0 | 100.0 | 61.0 | 35.0 | 100.0 |
| Egypt | Fayoum | 1995 | 92.7 | 88.3 | 50.4 | 12.4 | ... | ... | 97.8 | 2008 | 100.0 | 98.7 | 99.4 | 99.4 | 46.5 | 35.0 | 99.4 |
| Egypt | Giza | 1995 | 89.1 | 86.0 | 72.8 | 48.2 | ... | ... | 98.4 | 2008 | 100.0 | 99.1 | 99.8 | 99.8 | 69.6 | 81.3 | 99.8 |
| Egypt | Ismailia | 1995 | 94.2 | 91.8 | 85.1 | 67.5 | ... | ... | 99.1 | 2008 | 100.0 | 98.9 | 100.0 | 100.0 | 61.5 | 58.9 | 100.0 |
| Egypt | Kafr El-Sheikh | 1995 | 100.0 | 94.2 | 70.2 | 37.5 | ... | ... | 99.0 | 2008 | 100.0 | 100.0 | 100.0 | 98.7 | 68.7 | 35.5 | 100.0 |
| Egypt | Kharijah | 1995 | 93.5 | 92.7 | 69.9 | 34.1 | ... | ... | 99.2 | 2008 | 100.0 | 100.0 | 100.0 | 98.7 | 67.5 | 37.7 | 100.0 |
| Egypt | Mansurah | 1995 | 96.5 | 95.7 | 82.5 | 63.4 | ... | ... | 99.6 | 2008 | 100.0 | 97.2 | 100.0 | 100.0 | 63.8 | 51.1 | 100.0 |
| Egypt | Port Said | 1995 | 98.7 | 96.5 | 90.1 | 82.4 | ... | ... | 99.3 | 2008 | 98.4 | 98.2 | 100.0 | 100.0 | 69.3 | 49.7 | 100.0 |
| Egypt | Qena | 1995 | 89.9 | 81.4 | 68.2 | 37.2 | ... | ... | 96.1 | 2008 | 100.0 | 96.8 | 100.0 | 99.5 | 59.9 | 47.1 | 100.0 |
| Egypt | Sawhaj | 1995 | 89.8 | 87.0 | 65.4 | 33.4 | ... | ... | 96.0 | 2008 | 99.6 | 98.7 | 100.0 | 100.0 | 62.3 | 50.8 | 99.2 |
| Egypt | Suez | 1995 | 99.1 | 94.6 | 82.2 | 64.7 | ... | ... | 99.3 | 2008 | 99.8 | 99.8 | 100.0 | 100.0 | 64.4 | 42.5 | 100.0 |
| Egypt | Tahta | 1995 | 99.2 | 90.8 | 75.6 | 48.3 | ... | ... | 98.3 | 2008 | 99.7 | 88.7 | 100.0 | 100.0 | 59.9 | 49.8 | 100.0 |
| Ethiopia | Addis Ababa | | ... | ... | ... | ... | ... | ... | ... | 2005 | 99.9 | 68.8 | 71.8 | 8.9 | 46.1 | 30.8 | 96.9 |
| Ethiopia | Nazret | | ... | ... | ... | ... | ... | ... | ... | 2005 | 99.1 | 43.0 | 51.1 | 11.0 | 33.8 | 8.8 | 95.5 |
| Gabon | Libreville | | ... | ... | ... | ... | ... | ... | ... | 2000 | 99.7 | 58.2 | 83.4 | 35.0 | 20.4 | ... | 95.5 |
| Ghana | Accra | 1998 | 97.7 | 64.4 | 69.5 | 33.9 | 12.3 | ... | 92.0 | 2008 | 60.1 | 37.3 | 93.8 | 37.1 | 11.1 | 89.5 | 90.8 |
| Guinea | Conakry | 1999 | 82.7 | 39.2 | 84.8 | 11.2 | 7.2 | ... | 71.4 | 2005 | 96.4 | 45.2 | 80.3 | 11.1 | 28.9 | ... | 94.5 |
| Kenya | Mombasa | 1998 | 73.9 | 30.0 | 61.3 | 29.2 | 7.4 | ... | 47.5 | 2008 | 74.0 | 36.4 | 78.8 | 28.5 | 6.9 | 80.6 | 57.9 |
| Kenya | Nairobi | 1998 | 92.1 | 77.6 | 84.3 | 56.0 | 11.2 | ... | 60.1 | 2008 | 98.3 | 78.2 | 93.6 | 71.3 | 9.4 | 92.5 | 88.6 |
| Lesotho | Maseru | | ... | ... | ... | ... | ... | ... | ... | 2004 | 98.3 | 75.2 | 74.7 | 9.7 | 50.2 | ... | 33.1 |
| Liberia | Monrovia | | ... | ... | ... | ... | ... | ... | ... | 2007 | 81.6 | 8.4 | 51.9 | 34.4 | ... | 70.8 | 8.1 |
| Madagascar | Antananarivo | 1997 | 80.1 | 24.8 | 52.9 | 14.4 | 3.6 | ... | 55.7 | 2003 | 85.7 | 22.0 | 56.4 | 11.0 | 21.4 | ... | 67.8 |
| Malawi | Blantyre | | ... | ... | ... | ... | ... | ... | ... | 2006 | 97.0 | 30.6 | 42.6 | 10.9 | 6.7 | 35.1 | 32.7 |
| Malawi | Lilongwe | 1992 | 86.3 | 38.4 | 54.5 | 14.3 | ... | ... | 18.5 | 2006 | 92.2 | 20.2 | 42.1 | 6.0 | 2.0 | 26.5 | 18.0 |
| Malawi | Mzaza | | ... | ... | ... | ... | ... | ... | ... | 2006 | 84.0 | 41.9 | 42.1 | 17.0 | 5.5 | 32.5 | 35.6 |
| Mali | Bamako | 1996 | 70.5 | 17.3 | 51.6 | 4.3 | 3.7 | ... | 33.7 | 2006 | 95.6 | 41.2 | 81.1 | 12.2 | 19.6 | 61.6 | 72.1 |
| Mauritania | Nouakchott | | ... | ... | ... | ... | ... | ... | ... | 2001 | 94.4 | 27.8 | 58.2 | 4.8 | 7.2 | ... | 47.2 |
| Morocco | Dar-el-Beida (Casablanca) | 1992 | 99.1 | 74.1 | 92.9 | 87.9 | ... | ... | 78.7 | 2004 | 100.0 | 83.4 | 98.9 | 98.9 | 77.0 | ... | 99.2 |
| Morocco | Fès | 1992 | 100.0 | 97.4 | 100.0 | 100.0 | ... | ... | 100.0 | 2004 | 99.6 | 93.8 | 99.6 | 99.4 | 57.9 | ... | 97.7 |
| Morocco | Marrakech | 1992 | 100.0 | 84.0 | 94.7 | 87.8 | ... | ... | 90.4 | 2004 | 99.7 | 88.8 | 99.7 | 99.7 | 17.7 | ... | 98.3 |
| Morocco | Meknès | 1992 | 99.2 | 89.4 | 99.2 | 99.2 | ... | ... | 84.1 | 2004 | 99.2 | 85.6 | 97.0 | 97.0 | 68.4 | ... | 97.3 |
| Morocco | Rabat | 1992 | 96.5 | 86.0 | 92.5 | 91.7 | ... | ... | 83.9 | 2004 | 99.9 | 89.7 | 99.7 | 99.7 | 69.7 | ... | 99.0 |
| Mozambique | Maputo | 1997 | 87.4 | 83.6 | 49.9 | 22.4 | 6.9 | ... | 39.2 | 2003 | 82.8 | 66.4 | 48.8 | 8.0 | 5.2 | ... | 28.8 |
| Namibia | Windhoek | 1992 | 98.0 | 93.9 | 92.7 | 90.2 | ... | ... | 70.0 | 2007 | 98.6 | 82.8 | 87.1 | 86.0 | 37.1 | ... | 83.4 |
| Niger | Niamey | 1998 | 63.5 | 33.2 | 47.7 | 5.0 | 4.1 | ... | 51.0 | 2006 | 94.7 | 42.3 | 65.7 | 10.8 | 6.5 | 47.7 | 61.1 |
| Nigeria | Akure | 1999 | 94.1 | ... | 58.8 | ... | ... | ... | 76.5 | 2008 | 93.1 | 1.8 | 74.0 | 28.5 | 0.5 | 97.7 | 97.7 |
| Nigeria | Damaturu | 1999 | 61.5 | 23.1 | 71.8 | 15.4 | 2.6 | ... | 64.1 | 2008 | 83.3 | 3.1 | 86.3 | 0.4 | 1.3 | 60.8 | 60.8 |
| Nigeria | Effon Alaiye | 1999 | 32.8 | 4.4 | 48.9 | ... | 2.2 | ... | 93.3 | 2008 | 80.0 | 7.3 | 61.1 | 26.2 | 1.7 | 93.2 | 93.2 |
| Nigeria | Ibadan | 1999 | 93.3 | ... | 13.3 | 6.7 | ... | ... | 33.3 | 2008 | 88.4 | 10.5 | 72.9 | 29.0 | 1.4 | 94.8 | 94.8 |
| Nigeria | Kano | 1999 | 54.8 | 27.3 | 58.8 | 10.7 | 4.5 | ... | 82.2 | 2008 | 73.9 | 6.7 | 90.5 | 13.8 | 4.0 | 84.7 | 84.7 |
| Nigeria | Lagos | 1999 | 88.6 | 25.6 | 84.7 | 54.3 | 8.2 | ... | 98.9 | 2008 | 94.0 | 5.4 | 91.6 | 56.3 | 7.4 | 98.0 | 98.0 |
| Nigeria | Ogbomosho | 1999 | 62.3 | 16.6 | 46.1 | 33.7 | 12.6 | ... | 95.9 | | ... | ... | ... | ... | ... | ... | ... |
| Nigeria | Owo | 1999 | 34.4 | 7.4 | 68.8 | 24.4 | 9.9 | ... | 95.3 | | ... | ... | ... | ... | ... | ... | ... |
| Nigeria | Oyo | 1999 | 35.0 | 11.0 | 65.8 | 39.6 | 3.6 | ... | 92.1 | | ... | ... | ... | ... | ... | ... | ... |
| Nigeria | Zaria | 1999 | 74.4 | 54.6 | 55.8 | 15.1 | 4.6 | ... | 94.2 | 2008 | 73.0 | 28.9 | 66.3 | 14.3 | 3.2 | 81.3 | 81.3 |
| Rwanda | Kigali | 1992 | 52.0 | 6.5 | 50.2 | 9.0 | ... | ... | 36.0 | 2005 | 68.9 | 20.5 | 80.6 | 8.4 | 8.3 | 39.4 | 40.8 |
| Senegal | Dakar | 1997 | 95.5 | 77.8 | 70.8 | 42.4 | 20.4 | ... | 80.2 | 2005 | 98.3 | 87.8 | 91.1 | 76.3 | 30.0 | 54.2 | 89.5 |
| South Africa | Cape Town | 1998 | 95.8 | 79.7 | 83.4 | 73.8 | 49.6 | ... | 88.0 | | ... | ... | ... | ... | ... | ... | ... |
| South Africa | Durban | 1998 | 98.4 | 87.7 | 90.1 | 86.9 | 46.3 | ... | 84.3 | | ... | ... | ... | ... | ... | ... | ... |
| South Africa | Port Elizabeth | 1998 | 97.2 | 66.8 | 68.5 | 55.7 | 27.0 | ... | 63.3 | | ... | ... | ... | ... | ... | ... | ... |

# TABLE C.3

## continued

| | | Percentage of households with access to | | | | | | | | | | | | | | |
|---|---|---|---|---|---|---|---|---|---|---|---|---|---|---|---|---|
| | | 1990–1999[1] | | | | | | | | 2000–2009[1] | | | | | | |
| | | Survey year | Improved water | Piped water | Improved sanitation | Sewerage | Tele-phone(s) | Mobile(s) | Connection to electricity | Survey year | Improved water | Piped water | Improved sanitation | Sewerage | Tele-phone(s) | Mobile(s) | Connection to electricity |
| South Africa | Pretoria | 1998 | 100.0 | 62.5 | 62.5 | 62.5 | 18.8 | ... | 56.3 | | ... | ... | ... | ... | ... | ... | ... |
| South Africa | West Rand | 1998 | 99.4 | 84.2 | 84.8 | 84.8 | 47.6 | ... | 75.0 | | ... | ... | ... | ... | ... | ... | ... |
| Swaziland | Manzini | | ... | ... | ... | ... | ... | ... | ... | 2006 | 92.8 | 68.6 | 79.9 | 39.8 | 17.7 | 76.6 | 60.5 |
| Swaziland | Mbabane | | ... | ... | ... | ... | ... | ... | ... | 2006 | 88.6 | 65.3 | 76.9 | 41.7 | 29.1 | 78.3 | 59.9 |
| Togo | Lomé | 1998 | 88.6 | 67.4 | 81.7 | 33.9 | ... | ... | 51.2 | 2006 | 92.9 | 14.3 | 82.5 | 27.9 | 10.9 | 56.1 | 71.6 |
| United Republic of Tanzania | Arusha | 1999 | 97.8 | 23.7 | 39.6 | ... | ... | ... | 5.9 | 2004 | 94.6 | 59.3 | 62.5 | 11.0 | 35.0 | ... | 35.0 |
| United Republic of Tanzania | Dar es Salaam | 1999 | 90.1 | 78.8 | 51.9 | 3.2 | ... | ... | 46.9 | 2004 | 81.1 | 62.1 | 55.6 | 10.0 | 43.4 | ... | 59.8 |
| Uganda | Kampala | 1995 | 60.4 | 13.2 | 58.9 | 9.5 | 3.0 | ... | 49.4 | 2006 | 92.6 | 26.0 | 100.0 | 10.7 | 5.4 | 67.6 | 59.0 |
| Zambia | Chingola | 1996 | 76.6 | 76.6 | 85.9 | 76.6 | ... | ... | 78.1 | 2007 | 90.4 | 80.1 | 86.7 | 82.5 | 9.6 | 71.7 | 76.5 |
| Zambia | Lusaka | 1996 | 93.9 | 49.8 | 70.3 | 40.5 | ... | ... | 50.7 | 2007 | 92.4 | 31.6 | 83.5 | 27.4 | 4.9 | 68.4 | 57.0 |
| Zambia | Ndola | 1996 | 92.3 | 59.4 | 85.1 | 69.3 | ... | ... | 52.0 | 2007 | 74.1 | 39.5 | 64.5 | 34.0 | 8.1 | 57.8 | 38.9 |
| Zimbabwe | Harare | 1999 | 99.6 | 93.5 | 97.2 | 92.6 | 19.9 | ... | 84.7 | 2005 | 99.2 | 92.7 | 98.4 | 87.1 | 17.5 | 37.6 | 86.3 |
| **ASIA** | | | | | | | | | | | | | | | | | |
| Armenia | Armavir | | ... | ... | ... | ... | ... | ... | ... | 2005 | 98.7 | 96.2 | 98.0 | 83.8 | 80.2 | 35.2 | 100.0 |
| Armenia | Artashat | | ... | ... | ... | ... | ... | ... | ... | 2005 | 100.0 | 83.8 | 94.6 | 87.4 | 77.9 | 23.8 | 99.8 |
| Armenia | Gavar | | ... | ... | ... | ... | ... | ... | ... | 2005 | 99.3 | 88.7 | 99.6 | 77.3 | 82.9 | 21.7 | 99.8 |
| Armenia | Gyumri | | ... | ... | ... | ... | ... | ... | ... | 2005 | 100.0 | 93.6 | 91.7 | 85.1 | 34.9 | 14.9 | 100.0 |
| Armenia | Hrazdan | | ... | ... | ... | ... | ... | ... | ... | 2005 | 100.0 | 99.0 | 99.4 | 96.5 | 83.3 | 32.2 | 100.0 |
| Armenia | Idjevan | | ... | ... | ... | ... | ... | ... | ... | 2005 | 99.1 | 91.5 | 98.7 | 73.2 | 86.3 | 13.4 | 100.0 |
| Armenia | Kapan | | ... | ... | ... | ... | ... | ... | ... | 2005 | 100.0 | 100.0 | 99.8 | 97.9 | 88.9 | 16.2 | 100.0 |
| Armenia | Vanadzor | | ... | ... | ... | ... | ... | ... | ... | 2005 | 99.4 | 96.8 | 98.5 | 84.5 | 76.5 | 18.7 | 99.7 |
| Armenia | Yerevan | | ... | ... | ... | ... | ... | ... | ... | 2005 | 99.2 | 99.1 | 99.5 | 98.9 | 91.3 | 51.9 | 99.9 |
| Azerbaijan | Baku | | ... | ... | ... | ... | ... | ... | ... | 2006 | 92.7 | 89.6 | 98.8 | 90.0 | 85.8 | 75.4 | 99.6 |
| Azerbaijan | Sirvan | | ... | ... | ... | ... | ... | ... | ... | 2006 | 79.4 | 68.6 | 86.4 | 51.8 | 58.4 | 46.3 | 100.0 |
| Bangladesh | Dhaka | 1999 | 99.8 | 83.9 | 69.5 | 54.1 | 14.3 | ... | 99.1 | 2007 | 100.0 | 63.2 | 55.1 | 42.5 | 9.7 | 64.0 | 96.9 |
| Bangladesh | Rajshahi | 1999 | 100.0 | 1.5 | 50.8 | 7.7 | 3.1 | ... | 50.8 | 2007 | 100.0 | 0.8 | 53.4 | 18.0 | 1.1 | 31.9 | 60.1 |
| Cambodia | Phnum Pénh (Phnom Penh) | | ... | ... | ... | ... | ... | ... | ... | 2005 | 96.7 | 86.0 | 92.4 | 91.7 | ... | 86.1 | 96.1 |
| Cambodia | Siĕm Réab | | ... | ... | ... | ... | ... | ... | ... | 2005 | 94.3 | 5.4 | 64.7 | 64.7 | ... | 60.5 | 70.5 |
| India | Agartala | 1998 | 88.8 | 25.1 | 76.1 | 54.5 | 25.9 | ... | 90.4 | 2006 | 95.1 | 35.1 | 86.3 | 50.0 | 25.5 | 18.0 | 91.8 |
| India | Akola | 1998 | 92.3 | 73.2 | 64.7 | 58.9 | 19.6 | ... | 95.5 | 2006 | 99.2 | 69.8 | 61.4 | 60.9 | 21.3 | 24.6 | 93.1 |
| India | Amritsar | 1998 | 100.0 | 85.1 | 92.9 | 88.7 | 39.0 | ... | 100.0 | 2006 | 100.0 | 79.0 | 98.7 | 95.4 | 26.6 | 40.3 | 97.0 |
| India | Coimbatore | 1998 | 94.1 | 36.0 | 90.0 | 89.1 | 19.1 | ... | 89.6 | 2006 | 95.2 | 48.7 | 54.5 | 54.4 | 36.2 | 52.1 | 96.6 |
| India | Hisar | 1998 | 99.7 | 71.6 | 77.2 | 75.2 | 35.7 | ... | 97.7 | 2006 | 99.2 | 65.3 | 77.4 | 70.3 | 25.5 | 38.1 | 97.9 |
| India | Hyderabad | 1998 | 98.4 | 87.5 | 70.3 | 51.5 | 29.7 | ... | 96.1 | 2006 | 99.6 | 65.0 | 76.6 | 73.0 | 23.2 | 34.6 | 90.1 |
| India | Jaipur | 1998 | 98.5 | 83.7 | 91.5 | 91.0 | 28.5 | ... | 98.0 | 2006 | 99.3 | 88.8 | 98.2 | 96.4 | 49.6 | 54.7 | 100.0 |
| India | Jodhpur | 1998 | 98.4 | 81.9 | 89.1 | 85.2 | 19.6 | ... | 97.3 | 2006 | 97.9 | 84.7 | 69.2 | 66.1 | 34.7 | 38.4 | 94.7 |
| India | Kanpur | 1998 | 100.0 | 48.2 | 64.7 | 32.8 | 18.9 | ... | 93.9 | 2006 | 98.6 | 37.4 | 81.3 | 68.2 | 19.1 | 39.1 | 92.6 |
| India | Kharagpur | 1998 | 90.9 | 40.4 | 87.1 | 81.0 | 15.0 | ... | 82.6 | 2006 | 96.0 | 33.3 | 88.3 | 73.7 | 23.2 | 32.0 | 90.5 |
| India | Kochi (Cochin) | 1998 | 52.0 | 27.5 | 64.7 | 27.5 | 35.3 | ... | 87.3 | | ... | ... | ... | ... | ... | ... | ... |
| India | Kolkota | 1998 | 98.5 | 35.1 | 94.3 | 89.5 | 25.6 | ... | 93.8 | 2006 | 99.0 | 45.0 | 98.2 | 88.4 | 34.5 | 42.6 | 96.8 |
| India | Krishnanagar | 1998 | 89.7 | 32.7 | 78.6 | 73.9 | 18.9 | ... | 81.5 | 2006 | 99.7 | 15.7 | 84.3 | 59.9 | 21.6 | 23.8 | 82.1 |
| India | Mumbai (Bombay) | 1998 | 99.4 | 76.7 | 98.0 | 97.8 | 31.6 | ... | 99.0 | 2006 | 99.0 | 87.4 | 95.5 | 95.3 | 38.2 | 50.7 | 98.8 |
| India | New Delhi | 1998 | 99.2 | 80.8 | 94.0 | 90.2 | 45.4 | ... | 97.6 | 2006 | 92.6 | 74.9 | 84.8 | 84.5 | 38.8 | 59.3 | 99.4 |
| India | Pondichery | 1998 | 93.7 | 35.9 | 52.5 | 45.1 | 13.0 | ... | 87.0 | 2006 | 99.3 | 40.6 | 69.1 | 60.8 | 21.0 | 24.9 | 96.5 |
| India | Pune (Poona) | 1998 | 98.2 | 55.2 | 76.2 | 74.2 | 9.0 | ... | 92.3 | 2006 | 99.1 | 74.0 | 78.7 | 75.9 | 23.3 | 35.5 | 97.0 |
| India | Srinagar | 1998 | 97.6 | 87.9 | 78.5 | 71.0 | 20.3 | ... | 99.3 | 2006 | 98.8 | 83.5 | 64.1 | 60.0 | 41.6 | 55.2 | 99.4 |
| India | Vijayawada | 1998 | 96.9 | 39.2 | 68.1 | 60.3 | 13.2 | ... | 96.8 | 2006 | 100.0 | 98.4 | 100.0 | 98.4 | 18.0 | 32.8 | 100.0 |
| India | Yamunanagar | 1998 | 99.7 | 59.7 | 77.7 | 70.6 | 27.0 | ... | 98.3 | 2006 | 100.0 | 63.0 | 95.5 | 86.3 | 34.9 | 44.5 | 96.9 |
| Indonesia | Bandung | 1997 | 91.1 | 46.9 | 73.2 | 73.2 | ... | ... | 100.0 | 2007 | 80.2 | 14.3 | 93.4 | 93.0 | 58.4 | ... | 98.6 |
| Indonesia | Bitung | 1997 | 84.4 | 52.4 | 80.6 | 80.6 | ... | ... | 96.3 | | ... | ... | ... | ... | ... | ... | ... |
| Indonesia | Bogor | 1997 | 95.1 | 42.0 | 89.6 | 89.6 | ... | ... | 99.3 | | ... | ... | ... | ... | ... | ... | ... |
| Indonesia | Denpasar | 1997 | 98.6 | 53.6 | 92.1 | 92.1 | ... | ... | 100.0 | | ... | ... | ... | ... | ... | ... | ... |
| Indonesia | Dumai | 1997 | 88.4 | 17.2 | 69.4 | 69.4 | ... | ... | 85.8 | | ... | ... | ... | ... | ... | ... | ... |
| Indonesia | Jakarta | 1997 | 99.2 | 35.6 | 70.7 | 70.7 | ... | ... | 99.9 | 2007 | 94.0 | 29.7 | 96.3 | 96.2 | 74.7 | ... | 99.8 |
| Indonesia | Jambi | 1997 | 93.1 | 53.0 | 95.3 | 95.3 | ... | ... | 98.7 | | ... | ... | ... | ... | ... | ... | ... |
| Indonesia | Jaya Pura | 1997 | 88.3 | 61.1 | 76.0 | 75.5 | ... | ... | 88.0 | | ... | ... | ... | ... | ... | ... | ... |
| Indonesia | Kediri | 1997 | 94.1 | 17.9 | 48.0 | 47.7 | ... | ... | 98.6 | | ... | ... | ... | ... | ... | ... | ... |
| Indonesia | Medan | 1997 | 99.1 | 68.0 | 90.0 | 90.0 | ... | ... | 92.5 | 2007 | 83.5 | 48.6 | 93.2 | 91.0 | 67.0 | ... | 99.6 |
| Indonesia | Palembang | 1997 | 98.0 | 81.2 | 90.8 | 90.8 | ... | ... | 100.0 | 2007 | 79.2 | 16.8 | 87.6 | 85.7 | 57.8 | ... | 95.6 |
| Indonesia | Palu | 1997 | 99.4 | 39.7 | 68.7 | 68.7 | ... | ... | 92.1 | | ... | ... | ... | ... | ... | ... | ... |
| Indonesia | Pekan Baru | 1997 | 97.0 | 51.8 | 76.5 | 76.5 | ... | ... | 97.9 | | ... | ... | ... | ... | ... | ... | ... |
| Indonesia | Purwokerto | 1997 | 100.0 | 48.6 | 72.1 | 72.1 | ... | ... | 98.7 | | ... | ... | ... | ... | ... | ... | ... |
| Indonesia | Surabaya | 1997 | 100.0 | 71.0 | 70.5 | 70.5 | ... | ... | 100.0 | 2007 | 86.9 | 16.2 | 82.3 | 80.3 | 56.8 | ... | 99.3 |
| Indonesia | Surakarta | 1997 | 100.0 | 0.0 | 46.0 | 46.0 | ... | ... | 100.0 | 2007 | 78.2 | 22.4 | 78.2 | 77.0 | 50.2 | ... | 96.8 |
| Indonesia | Ujung Pandang | 1997 | 99.4 | 36.3 | 83.8 | 83.8 | ... | ... | 98.4 | 2007 | 81.8 | 44.6 | 92.4 | 90.7 | 64.5 | ... | 99.0 |
| Jordan | Ajlūn | 1997 | 99.1 | 99.1 | 91.7 | 86.2 | 33.0 | ... | 100.0 | 2007 | 97.5 | 69.5 | 99.8 | 39.6 | 30.3 | 89.5 | 99.3 |
| Jordan | Al-Balqa | 1997 | 98.6 | 98.1 | 97.7 | 95.3 | 35.8 | ... | 99.1 | 2007 | 98.3 | 76.5 | 99.7 | 75.8 | 36.2 | 87.7 | 98.9 |
| Jordan | Al-Karak | 1997 | 97.1 | 96.6 | 92.6 | 81.7 | 33.7 | ... | 98.9 | 2007 | 99.7 | 85.6 | 99.1 | 26.7 | 29.5 | 86.9 | 99.5 |
| Jordan | Al-Mafraq | 1997 | 97.7 | 96.9 | 99.2 | 96.1 | 44.5 | ... | 98.4 | 2007 | 95.8 | 86.5 | 99.8 | 38.0 | 28.9 | 88.6 | 99.6 |
| Jordan | Amman | 1997 | 98.9 | 98.5 | 98.5 | 96.5 | 52.1 | ... | 100.0 | 2007 | 98.8 | 67.0 | 99.9 | 81.2 | 45.3 | 91.4 | 98.6 |
| Jordan | Aqaba | 1997 | 100.0 | 100.0 | 98.9 | 98.3 | 45.5 | ... | 100.0 | 2007 | 99.0 | 96.3 | 97.8 | 77.1 | 29.3 | 92.6 | 98.2 |
| Jordan | Aṭ-Tafīlah | 1997 | 98.8 | 98.8 | 97.6 | 92.3 | 51.5 | ... | 96.4 | 2007 | 99.6 | 97.3 | 99.9 | 29.9 | 31.5 | 90.8 | 99.0 |
| Jordan | Az-Zarqā' | 1997 | 99.2 | 99.1 | 99.6 | 99.1 | 29.5 | ... | 100.0 | 2007 | 99.2 | 71.0 | 100.0 | 70.7 | 29.1 | 90.7 | 99.8 |
| Jordan | Irbid | 1997 | 92.1 | 90.6 | 95.2 | 91.8 | 28.2 | ... | 99.6 | 2007 | 96.4 | 61.5 | 99.3 | 38.7 | 32.2 | 91.2 | 99.1 |
| Jordan | Jarash | 1997 | 91.8 | 87.8 | 94.9 | 89.8 | 27.6 | ... | 100.0 | 2007 | 98.5 | 80.6 | 99.7 | 38.1 | 22.9 | 86.2 | 99.1 |
| Jordan | Ma'ān | 1997 | 99.0 | 99.0 | 99.0 | 96.0 | 29.7 | ... | 100.0 | 2007 | 97.6 | 76.4 | 99.8 | 32.5 | 25.0 | 93.2 | 99.1 |

# TABLE C.3

*continued*

| | | Percentage of households with access to | | | | | | | | | | | | | | |
|---|---|---|---|---|---|---|---|---|---|---|---|---|---|---|---|---|
| | | 1990–1999[1] | | | | | | | | 2000–2009[1] | | | | | | |
| | | Survey year | Improved water | Piped water | Improved sanitation | Sewerage | Tele-phone(s) | Mobile(s) | Connection to electricity | Survey year | Improved water | Piped water | Improved sanitation | Sewerage | Tele-phone(s) | Mobile(s) | Connection to electricity |
| Jordan | Ma'dabā | 1997 | 100.0 | 100.0 | 100.0 | 100.0 | 42.9 | ... | 100.0 | 2007 | 97.8 | 82.7 | 100.0 | 66.3 | 28.5 | 89.6 | 98.6 |
| Kazakhstan | Almaty | 1999 | 97.0 | 94.3 | 87.6 | 77.9 | 78.1 | ... | 99.7 | 2006 | 100.0 | 98.7 | 98.7 | 82.9 | 89.7 | 62.2 | 100.0 |
| Kazakhstan | Öskemen | | ... | ... | ... | ... | ... | ... | ... | 2006 | 99.4 | 81.2 | 100.0 | 54.0 | 62.3 | 33.4 | 99.8 |
| Kazakhstan | Żezqazġan | 1999 | 100.0 | 100.0 | 100.0 | 100.0 | 75.5 | ... | 100.0 | | ... | ... | ... | ... | ... | ... | ... |
| Kazakhstan | Qaragandy | | ... | ... | ... | ... | ... | ... | ... | 2006 | 98.2 | 88.1 | 99.5 | 82.6 | 70.7 | 41.0 | 99.6 |
| Kazakhstan | Šymkent | 1999 | 100.0 | 100.0 | 100.0 | 100.0 | 73.7 | ... | 100.0 | 2006 | 92.6 | 83.0 | 100.0 | 39.4 | 54.9 | 37.5 | 100.0 |
| Kyrgyzstan | Bishkek | 1997 | 99.2 | 95.3 | 84.0 | 68.5 | 63.7 | ... | 100.0 | 2006 | 100.0 | 96.0 | 99.8 | 68.4 | 72.1 | 54.8 | 99.8 |
| Pakistan | Faisalabad | 1990 | 98.1 | 78.1 | 87.6 | 87.2 | ... | ... | 98.7 | 2006 | 95.4 | 59.4 | 80.0 | 79.7 | 67.4 | ... | 98.7 |
| Pakistan | Islamabad | 1990 | 94.1 | 80.3 | 71.0 | 70.3 | ... | ... | 97.8 | 2006 | 96.5 | 57.7 | 83.2 | 82.9 | 61.5 | ... | 99.5 |
| Pakistan | Karachi | 1990 | 96.6 | 77.4 | 92.1 | 90.0 | ... | ... | 96.8 | 2006 | 92.4 | 66.7 | 85.3 | 82.2 | 64.5 | ... | 97.5 |
| Pakistan | Quetta | | ... | ... | ... | ... | ... | ... | ... | 2006 | 97.6 | 79.3 | 76.5 | 72.0 | 62.7 | ... | 98.8 |
| Philippines | Bacolod | 1998 | 92.7 | 31.1 | 75.0 | 71.3 | 12.8 | ... | 78.7 | 2008 | 97.8 | 43.3 | 78.1 | 77.3 | 15.2 | 77.5 | 86.6 |
| Philippines | Cagayan de Oro | 1998 | 86.8 | 28.9 | 97.4 | 97.4 | 7.9 | ... | 86.8 | 2008 | 100.0 | 16.1 | 98.9 | 78.7 | 14.9 | 78.5 | 93.3 |
| Philippines | Cebu | 1998 | 88.0 | 42.1 | 88.4 | 76.4 | 21.6 | ... | 85.6 | 2008 | 99.0 | 21.9 | 84.4 | 80.3 | 22.4 | 80.6 | 93.4 |
| Philippines | Manila | 1998 | 91.0 | 65.9 | 96.9 | 92.3 | 45.7 | ... | 98.7 | 2008 | 99.4 | 45.3 | 96.9 | 96.7 | 32.2 | 87.1 | 98.0 |
| Turkey | Adana | 1998 | 100.0 | 99.5 | 99.0 | 90.2 | 71.6 | 7.4 | ... | 2004 | 99.5 | 92.2 | 99.6 | 90.4 | 76.8 | 39.0 | ... |
| Turkey | Aksaray | 1998 | 47.6 | 42.9 | 64.3 | 21.4 | 69.0 | 7.1 | ... | 2004 | 97.5 | 57.5 | 97.5 | 75.0 | 70.0 | 42.5 | ... |
| Turkey | Ankara | 1998 | 97.4 | 86.6 | 99.5 | 99.0 | 90.3 | 23.6 | ... | 2004 | 99.5 | 80.2 | 99.3 | 98.5 | 87.2 | 36.1 | ... |
| Turkey | Antalya | 1998 | 91.7 | 89.1 | 90.1 | 19.8 | 83.3 | 20.3 | ... | 2004 | 99.5 | 74.3 | 89.6 | 60.7 | 86.9 | 31.1 | ... |
| Turkey | Bursa | 1998 | 92.0 | 87.7 | 98.8 | 89.5 | 82.7 | 14.8 | ... | 2004 | 99.8 | 71.3 | 100.0 | 100.0 | 82.8 | 40.8 | ... |
| Turkey | Gaziantep | 1998 | 96.2 | 94.9 | 90.4 | 89.7 | 73.1 | 7.7 | ... | 2004 | 99.6 | 97.7 | 99.6 | 99.6 | 73.0 | 43.3 | ... |
| Turkey | Istanbul | 1998 | 89.7 | 19.6 | 99.4 | 98.7 | 79.9 | 29.1 | ... | 2004 | 99.3 | 39.7 | 99.1 | 95.9 | 83.3 | 35.6 | ... |
| Turkey | Izmir | 1998 | 99.4 | 86.9 | 100.0 | 99.4 | 84.0 | 16.0 | ... | 2004 | 98.3 | 56.1 | 100.0 | 99.7 | 84.5 | 39.5 | ... |
| Turkey | Karaman | 1998 | 100.0 | 100.0 | 82.6 | 17.4 | 87.0 | 8.7 | ... | | ... | ... | ... | ... | ... | ... | ... |
| Turkey | Kırıkkale | 1998 | 94.7 | 63.2 | 100.0 | 100.0 | 94.7 | 15.8 | ... | 2004 | 100.0 | 23.9 | 100.0 | 100.0 | 87.0 | 50.0 | ... |
| Turkey | Malatya | 1998 | 98.3 | 98.3 | 100.0 | 100.0 | 75.9 | 8.6 | ... | 2004 | 100.0 | 100.0 | 99.2 | 99.2 | 86.5 | 37.6 | ... |
| Turkey | Van | 1998 | 95.8 | 95.8 | 93.8 | 62.5 | 62.5 | 4.2 | ... | 2004 | 98.9 | 93.6 | 77.7 | 42.6 | 78.7 | 33.0 | ... |
| Uzbekistan | Tashkent | 1996 | 99.4 | 98.7 | 89.4 | 81.0 | 64.5 | ... | 100.0 | | ... | ... | ... | ... | ... | ... | ... |
| Viet Nam | Da Nang - CP | | ... | ... | ... | ... | ... | ... | ... | 2002 | 88.8 | 88.8 | 100.0 | 100.0 | 80.0 | ... | 100.0 |
| Viet Nam | Hà Noi | 1997 | 77.1 | 50.6 | 90.8 | 60.1 | 41.8 | ... | 100.0 | 2002 | 77.2 | 74.1 | 97.3 | 95.8 | 72.9 | ... | 100.0 |
| Viet Nam | Hai Phòng | 1997 | 97.9 | 75.1 | 72.1 | 61.2 | 6.4 | ... | 100.0 | 2002 | 98.2 | 95.5 | 96.0 | 90.0 | 39.0 | ... | 100.0 |
| Viet Nam | Thành Pho Ho Chí Minh (Ho Chi Minh City) | 1997 | 90.6 | 89.4 | 95.8 | 92.7 | 40.0 | ... | 99.7 | 2002 | 89.3 | 88.8 | 98.4 | 96.6 | 74.5 | ... | 99.8 |
| Yemen | Aden | 1991 | 97.0 | 97.0 | 91.4 | 88.2 | 28.7 | ... | 95.6 | | ... | ... | ... | ... | ... | ... | ... |
| Yemen | Sana'a' | 1991 | 93.9 | 93.5 | 60.9 | 58.5 | 38.6 | ... | 98.8 | 2006 | 56.8 | 22.5 | 88.7 | 48.8 | ... | ... | ... |
| Yemen | Taiz | 1991 | 85.6 | 85.6 | 55.9 | 48.9 | 26.1 | ... | 95.2 | | ... | ... | ... | ... | ... | ... | ... |
| **EUROPE** | | | | | | | | | | | | | | | | | |
| Moldova | Chisinău | | ... | ... | ... | ... | ... | ... | ... | 2005 | 99.5 | 89.1 | 97.8 | 91.9 | 93.6 | 60.6 | 99.7 |
| Ukraine | Čerkasy | | ... | ... | ... | ... | ... | ... | ... | 2007 | 99.4 | 81.5 | 99.7 | 56.6 | 64.4 | 79.8 | 99.7 |
| Ukraine | Černihiv | | ... | ... | ... | ... | ... | ... | ... | 2007 | 100.0 | 73.9 | 76.0 | 46.5 | 81.7 | 60.9 | 100.0 |
| Ukraine | Černivcy | | ... | ... | ... | ... | ... | ... | ... | 2007 | 100.0 | 94.9 | 97.0 | 86.9 | 87.2 | 61.8 | 100.0 |
| Ukraine | Cherson | | ... | ... | ... | ... | ... | ... | ... | 2007 | 99.7 | 78.0 | 100.0 | 62.1 | 54.4 | 71.3 | 100.0 |
| Ukraine | Chmél'nyckyj | | ... | ... | ... | ... | ... | ... | ... | 2007 | 98.1 | 81.5 | 98.4 | 81.7 | 84.5 | 64.2 | 99.4 |
| Ukraine | Dnipropetrovs'k | | ... | ... | ... | ... | ... | ... | ... | 2007 | 100.0 | 91.5 | 100.0 | 77.4 | 71.1 | 69.9 | 100.0 |
| Ukraine | Donets'k | | ... | ... | ... | ... | ... | ... | ... | 2007 | 100.0 | 76.4 | 99.8 | 65.6 | 50.3 | 79.3 | 99.9 |
| Ukraine | Ivano-Frankivs'k | | ... | ... | ... | ... | ... | ... | ... | 2007 | 100.0 | 72.6 | 100.0 | 72.7 | 85.6 | 77.5 | 100.0 |
| Ukraine | Kharkiv | | ... | ... | ... | ... | ... | ... | ... | 2007 | 100.0 | 79.0 | 99.7 | 69.9 | 68.8 | 70.9 | 100.0 |
| Ukraine | Kirovhrad | | ... | ... | ... | ... | ... | ... | ... | 2007 | 99.7 | 65.0 | 99.6 | 46.6 | 53.5 | 76.4 | 100.0 |
| Ukraine | Krym | | ... | ... | ... | ... | ... | ... | ... | 2007 | 99.6 | 91.3 | 99.3 | 68.5 | 58.2 | 68.9 | 99.8 |
| Ukraine | Kyïv | | ... | ... | ... | ... | ... | ... | ... | 2007 | 99.8 | 99.4 | 99.8 | 99.8 | 94.4 | 85.6 | 99.8 |
| Ukraine | Luhans'k | | ... | ... | ... | ... | ... | ... | ... | 2007 | 99.1 | 39.6 | 98.9 | 69.4 | 61.0 | 72.8 | 100.0 |
| Ukraine | L'viv | | ... | ... | ... | ... | ... | ... | ... | 2007 | 100.0 | 89.7 | 99.8 | 90.4 | 73.6 | 78.1 | 100.0 |
| Ukraine | Mykolaïv | | ... | ... | ... | ... | ... | ... | ... | 2007 | 93.2 | 91.0 | 100.0 | 80.5 | 47.2 | 56.2 | 99.6 |
| Ukraine | Odesa | | ... | ... | ... | ... | ... | ... | ... | 2007 | 99.8 | 85.8 | 99.3 | 63.1 | 72.5 | 61.7 | 99.9 |
| Ukraine | Poltava | | ... | ... | ... | ... | ... | ... | ... | 2007 | 98.3 | 71.7 | 100.0 | 74.8 | 70.9 | 71.1 | 100.0 |
| Ukraine | Rivne | | ... | ... | ... | ... | ... | ... | ... | 2007 | 100.0 | 95.0 | 98.0 | 76.6 | 72.6 | 72.4 | 99.3 |
| Ukraine | Sévastopol' | | ... | ... | ... | ... | ... | ... | ... | 2007 | 100.0 | 95.4 | 100.0 | 92.5 | 85.9 | 65.5 | 100.0 |
| Ukraine | Sumy | | ... | ... | ... | ... | ... | ... | ... | 2007 | 99.6 | 78.9 | 100.0 | 63.9 | 70.6 | 70.2 | 100.0 |
| Ukraine | Ternopil' | | ... | ... | ... | ... | ... | ... | ... | 2007 | 97.7 | 84.2 | 100.0 | 67.3 | 82.5 | 73.5 | 100.0 |
| Ukraine | Užhorod | | ... | ... | ... | ... | ... | ... | ... | 2007 | 95.2 | 80.8 | 95.3 | 67.5 | 58.3 | 80.1 | 100.0 |
| Ukraine | Vinnycja | | ... | ... | ... | ... | ... | ... | ... | 2007 | 94.7 | 66.8 | 99.7 | 63.2 | 65.6 | 74.8 | 100.0 |
| Ukraine | Volyn' | | ... | ... | ... | ... | ... | ... | ... | 2007 | 100.0 | 84.3 | 100.0 | 69.3 | 85.3 | 71.0 | 100.0 |
| Ukraine | Zaporizhzhya | | ... | ... | ... | ... | ... | ... | ... | 2007 | 100.0 | 99.2 | 100.0 | 77.7 | 70.4 | 76.4 | 100.0 |
| Ukraine | Żytomyr | | ... | ... | ... | ... | ... | ... | ... | 2007 | 100.0 | 48.0 | 98.9 | 53.9 | 66.9 | 69.0 | 99.6 |
| **LATIN AMERICA AND THE CARIBBEAN** | | | | | | | | | | | | | | | | | |
| Belize | Belize | | ... | ... | ... | ... | ... | ... | ... | 2006 | 99.6 | 24.1 | 96.1 | 95.0 | 49.3 | 70.5 | 98.3 |
| Bolivia | Cobija | 1998 | 88.5 | 88.5 | 78.7 | 52.5 | 45.9 | ... | 88.5 | 2008 | 86.7 | 85.2 | 79.7 | 64.5 | 23.4 | 85.0 | 96.2 |
| Bolivia | Cochabamba | 1998 | 83.5 | 83.5 | 65.0 | 44.3 | 47.5 | 7.6 | 98.2 | 2008 | 84.4 | 83.0 | 83.7 | 75.1 | 42.6 | 74.0 | 98.2 |
| Bolivia | La Paz | 1998 | 95.3 | 95.3 | 55.1 | 39.3 | 33.5 | 8.1 | 97.2 | 2008 | 97.5 | 95.0 | 83.6 | 78.1 | 29.7 | 77.0 | 98.3 |
| Bolivia | Oruro | 1998 | 93.9 | 93.9 | 42.3 | 32.2 | 29.5 | 4.8 | 95.8 | 2008 | 97.2 | 92.4 | 70.2 | 69.0 | 43.1 | 70.6 | 96.4 |
| Bolivia | Potosí | 1998 | 96.7 | 96.7 | 48.9 | 23.9 | 25.7 | 3.2 | 95.6 | 2008 | 98.1 | 95.1 | 82.8 | 81.5 | 23.7 | 74.9 | 97.8 |
| Bolivia | Santa Cruz | 1998 | 96.7 | 96.7 | 75.0 | 56.0 | 36.9 | 10.7 | 95.9 | 2008 | 98.9 | 98.1 | 78.3 | 59.9 | 25.8 | 84.5 | 97.7 |
| Bolivia | Sucre | 1998 | 96.5 | 96.5 | 71.9 | 61.4 | 36.1 | 8.4 | 95.7 | 2008 | 94.4 | 88.6 | 77.2 | 76.9 | 31.5 | 66.5 | 97.2 |
| Bolivia | Tarija | 1998 | 99.3 | 99.3 | 79.7 | 68.3 | 41.2 | 6.6 | 94.5 | 2008 | 99.3 | 94.5 | 86.4 | 79.8 | 31.7 | 81.8 | 94.9 |
| Bolivia | Trinidad | 1998 | 69.8 | 69.8 | 59.0 | 25.8 | 22.8 | 2.6 | 84.0 | 2008 | 65.0 | 60.7 | 65.4 | 42.4 | 14.9 | 65.8 | 91.5 |
| Brazil | Belo Horizonte | 1996 | 90.9 | 84.4 | 91.3 | 87.6 | ... | ... | 100.0 | | ... | ... | ... | ... | ... | ... | ... |

# TABLE C.3
## continued

| | | Percentage of households with access to | | | | | | | | | | | | | | |
|---|---|---|---|---|---|---|---|---|---|---|---|---|---|---|---|---|
| | | 1990–1999[1] | | | | | | | | 2000–2009[1] | | | | | | |
| | | Survey year | Improved water | Piped water | Improved sanitation | Sewerage | Tele-phone(s) | Mobile(s) | Connection to electricity | Survey year | Improved water | Piped water | Improved sanitation | Sewerage | Tele-phone(s) | Mobile(s) | Connection to electricity |
| Brazil | Brasília | 1996 | 90.2 | 89.8 | 81.7 | 71.2 | ... | ... | 99.6 | ... | ... | ... | ... | ... | ... | ... | ... |
| Brazil | Curitiba | 1996 | 90.0 | 84.2 | 88.7 | 78.7 | ... | ... | 100.0 | ... | ... | ... | ... | ... | ... | ... | ... |
| Brazil | Fortaleza | 1996 | 82.4 | 76.8 | 59.8 | 35.9 | ... | ... | 97.2 | | ... | ... | ... | ... | ... | ... | ... |
| Brazil | Goiânia | 1996 | 95.7 | 93.4 | 84.8 | 75.7 | ... | ... | 98.3 | | ... | ... | ... | ... | ... | ... | ... |
| Brazil | Rio de Janeiro | 1996 | 89.4 | 88.5 | 83.1 | 79.4 | ... | ... | 99.6 | | ... | ... | ... | ... | ... | ... | ... |
| Brazil | São Paulo | 1996 | 98.2 | 93.8 | 90.3 | 87.6 | ... | ... | 99.6 | | ... | ... | ... | ... | ... | ... | ... |
| Brazil | Victoria | 1996 | 94.6 | 90.4 | 90.8 | 87.5 | ... | ... | 99.2 | ... | ... | ... | ... | ... | ... | ... | ... |
| Colombia | Armenia | 1995 | 100.0 | 100.0 | 99.3 | 98.9 | 55.4 | ... | 99.3 | 2005 | 99.7 | 96.8 | 99.8 | 99.8 | 69.7 | ... | 98.1 |
| Colombia | Barranquilla | 1995 | 95.1 | 93.9 | 94.6 | 80.4 | 23.5 | ... | 99.8 | 2005 | 95.9 | 86.8 | 96.0 | 94.8 | 45.4 | ... | 99.6 |
| Colombia | Bogotá | 1995 | 100.0 | 100.0 | 99.8 | 99.7 | 80.6 | ... | 99.9 | 2005 | 99.6 | 96.4 | 99.9 | 99.9 | 81.7 | ... | 99.6 |
| Colombia | Bucaramanga | 1995 | 100.0 | 100.0 | 97.2 | 96.7 | 42.4 | ... | 100.0 | 2005 | 98.6 | 95.3 | 97.3 | 97.3 | 76.3 | ... | 99.8 |
| Colombia | Cali | 1995 | 99.9 | 99.7 | 97.3 | 96.7 | 43.1 | ... | 99.9 | 2005 | 99.6 | 97.7 | 99.0 | 99.0 | 71.4 | ... | 99.8 |
| Colombia | Cartagena | 1995 | 98.4 | 93.6 | 88.0 | 74.2 | 27.1 | ... | 99.6 | 2005 | 94.9 | 83.0 | 92.8 | 91.4 | 49.2 | ... | 99.7 |
| Colombia | Cúcuta | 1995 | 98.3 | 98.3 | 97.7 | 96.4 | 27.2 | ... | 100.0 | 2005 | 99.5 | 95.8 | 97.4 | 97.1 | 57.0 | ... | 99.6 |
| Colombia | Ibagué | 1995 | 99.1 | 99.1 | 97.0 | 93.0 | 32.5 | ... | 97.7 | 2005 | 99.3 | 98.0 | 99.7 | 99.7 | 62.8 | ... | 99.1 |
| Colombia | Manizales | 1995 | 99.6 | 99.6 | 99.6 | 99.6 | 52.3 | ... | 98.8 | 2005 | 99.6 | 99.6 | 99.8 | 99.8 | 72.4 | ... | 99.5 |
| Colombia | Medellín | 1995 | 99.4 | 99.4 | 96.5 | 52.3 | 52.3 | ... | 98.8 | 2005 | 99.1 | 91.2 | 99.4 | 99.4 | 81.8 | ... | 99.1 |
| Colombia | Montería | 1995 | 86.9 | 79.3 | 71.2 | 47.9 | 21.7 | ... | 93.1 | 2005 | 74.2 | 59.9 | 93.9 | 92.7 | 53.1 | ... | 98.4 |
| Colombia | Neiva | 1995 | 99.6 | 99.6 | 97.2 | 96.6 | 43.9 | ... | 97.4 | 2005 | 99.1 | 98.7 | 98.6 | 98.3 | 64.4 | ... | 98.1 |
| Colombia | Pereira | 1995 | 100.0 | 100.0 | 100.0 | 100.0 | 57.0 | ... | 98.9 | 2005 | 100.0 | 98.1 | 99.6 | 99.6 | 72.9 | ... | 99.4 |
| Colombia | Popayán | 1995 | 100.0 | 100.0 | 98.6 | 98.6 | 54.9 | ... | 100.0 | 2005 | 98.9 | 83.1 | 97.0 | 96.7 | 63.3 | ... | 97.2 |
| Colombia | Quibdó | 1995 | 94.3 | 64.0 | 8.2 | 1.7 | 35.1 | ... | 79.6 | 2005 | 96.4 | 53.2 | 85.9 | 85.7 | 55.9 | ... | 98.2 |
| Colombia | Riohacha | 1995 | 100.0 | 100.0 | 92.9 | 89.3 | 16.8 | ... | 94.6 | 2005 | 95.8 | 63.9 | 96.7 | 96.3 | 50.5 | ... | 98.9 |
| Colombia | Santa Marta | 1995 | 80.0 | 74.2 | 79.6 | 67.2 | 18.4 | ... | 100.0 | 2005 | 96.4 | 78.5 | 94.0 | 93.0 | 49.3 | ... | 98.7 |
| Colombia | Sincelejo | 1995 | 100.0 | 100.0 | 86.6 | 73.5 | 19.5 | ... | 98.5 | 2005 | 97.5 | 86.0 | 94.1 | 93.6 | 53.1 | ... | 98.7 |
| Colombia | Tunja | 1995 | 100.0 | 100.0 | 99.3 | 98.7 | 22.1 | ... | 98.7 | 2005 | 99.2 | 90.8 | 99.1 | 99.1 | 60.8 | ... | 99.5 |
| Colombia | Valledupar | 1995 | 100.0 | 100.0 | 99.2 | 97.4 | 14.8 | ... | 99.4 | 2005 | 95.3 | 90.6 | 90.3 | 90.2 | 43.3 | ... | 97.4 |
| Colombia | Villavicencio | 1995 | 96.4 | 96.4 | 100.0 | 100.0 | 34.0 | ... | 99.2 | 2005 | 98.6 | 69.1 | 99.2 | 99.0 | 64.9 | ... | 98.9 |
| Dominican Republic | Azua | 1996 | 97.8 | 75.1 | 89.0 | 46.4 | 22.7 | ... | ... | 2007 | 92.1 | 49.1 | 92.5 | 43.2 | 15.3 | 46.5 | 99.1 |
| Dominican Republic | Baní | 1996 | 100.0 | 83.7 | 97.8 | 70.7 | 34.8 | ... | ... | 2007 | 78.4 | 22.6 | 93.6 | 60.0 | 29.4 | 64.6 | 98.7 |
| Dominican Republic | Barahona | 1996 | 92.2 | 89.3 | 79.7 | 33.1 | 14.8 | ... | ... | 2007 | 85.9 | 57.3 | 86.9 | 47.1 | 18.7 | 52.4 | 98.7 |
| Dominican Republic | Bonao | 1996 | 97.7 | 90.7 | 93.0 | 62.8 | 46.5 | ... | ... | 2007 | 94.6 | 34.3 | 98.3 | 82.2 | 31.4 | 75.2 | 99.2 |
| Dominican Republic | Cotuí | 1996 | 99.1 | 80.0 | 85.2 | 36.5 | 7.8 | ... | ... | 2007 | 93.7 | 6.5 | 93.5 | 74.4 | 29.4 | 69.7 | 100.0 |
| Dominican Republic | Dajabón | 1996 | 100.0 | 96.7 | 93.3 | 23.3 | 17.8 | ... | ... | 2007 | 82.2 | 38.3 | 94.7 | 49.9 | 18.5 | 62.5 | 96.9 |
| Dominican Republic | Hato Mayor del Rey | ... | ... | ... | ... | ... | | | | 2007 | 88.0 | 3.0 | 95.1 | 59.1 | 18.7 | 73.9 | 98.0 |
| Dominican Republic | Higüey | 1996 | 100.0 | 12.1 | 97.0 | 59.1 | 34.8 | ... | ... | 2007 | 94.3 | 0.1 | 97.8 | 74.4 | 21.7 | 79.2 | 98.0 |
| Dominican Republic | La Romana | 1996 | 100.0 | 29.3 | 92.9 | 52.2 | 34.2 | ... | ... | 2007 | 88.7 | 6.9 | 92.6 | 66.9 | 17.0 | 74.5 | 98.6 |
| Dominican Republic | La Vega | 1996 | 98.8 | 54.7 | 98.8 | 94.2 | 59.3 | ... | ... | 2007 | 91.5 | 27.6 | 96.7 | 78.9 | 26.0 | 73.1 | 98.6 |
| Dominican Republic | Mao | 1996 | 98.9 | 80.7 | 94.8 | 33.0 | 23.6 | ... | ... | 2007 | 96.4 | 36.6 | 95.1 | 46.9 | 24.0 | 72.9 | 96.1 |
| Dominican Republic | Moca | 1996 | 97.5 | 65.8 | 97.5 | 74.7 | 59.5 | ... | ... | 2007 | 97.9 | 37.9 | 97.1 | 74.0 | 23.6 | 73.6 | 98.2 |
| Dominican Republic | Monte Cristi | 1996 | 54.5 | 22.4 | 92.5 | 15.7 | 27.6 | ... | ... | 2007 | 93.8 | 20.2 | 94.3 | 43.5 | 28.2 | 67.6 | 95.3 |
| Dominican Republic | Monte Plata | 1996 | 98.4 | 63.2 | 89.3 | 40.3 | 14.2 | ... | ... | 2007 | 79.7 | 3.3 | 92.7 | 44.9 | 15.5 | 63.4 | 96.6 |
| Dominican Republic | Nagua | 1996 | 100.0 | 43.1 | 100.0 | 41.4 | 27.6 | ... | ... | 2007 | 89.2 | 4.6 | 93.5 | 62.6 | 21.3 | 76.9 | 98.9 |
| Dominican Republic | Neiba | 1996 | 96.6 | 92.7 | 82.0 | 12.4 | 12.4 | ... | ... | 2007 | 81.8 | 45.1 | 81.3 | 41.5 | 19.7 | 57.2 | 96.8 |
| Dominican Republic | Puerto Plata | 1996 | 97.4 | 46.2 | 98.7 | 66.7 | 32.1 | ... | ... | 2007 | 95.3 | 10.2 | 97.0 | 87.6 | 30.7 | 78.2 | 97.9 |
| Dominican Republic | Sabaneta | 1996 | 100.0 | 79.1 | 96.5 | 51.2 | 27.9 | ... | ... | 2007 | 89.6 | 21.2 | 93.3 | 56.1 | 27.3 | 73.7 | 98.7 |
| Dominican Republic | Samaná | 1996 | 96.6 | 82.8 | 51.7 | 24.1 | 3.4 | ... | ... | 2007 | 92.4 | 9.7 | 89.6 | 68.9 | 17.6 | 76.2 | 97.4 |
| Dominican Republic | San Cristóbal | 1996 | 88.2 | 56.6 | 91.5 | 59.0 | 36.8 | ... | ... | 2007 | 72.9 | 11.4 | 96.2 | 80.0 | 33.7 | 77.0 | 99.4 |
| Dominican Republic | San Francisco de Macorís | 1996 | 99.4 | 43.0 | 95.0 | 55.9 | 30.7 | ... | ... | 2007 | 90.2 | 10.7 | 95.0 | 73.2 | 31.5 | 73.5 | 98.7 |
| Dominican Republic | San Juan | 1996 | 97.8 | 87.8 | 92.8 | 34.8 | 21.0 | ... | ... | 2007 | 95.6 | 53.3 | 89.7 | 55.8 | 20.4 | 62.4 | 99.3 |
| Dominican Republic | San Pedro de Macorís | 1996 | 99.4 | 17.3 | 92.9 | 56.5 | 36.3 | ... | ... | 2007 | 76.8 | 4.4 | 92.7 | 65.4 | 19.5 | 70.3 | 98.3 |
| Dominican Republic | Santiago | 1996 | 99.7 | 77.8 | 96.3 | 74.4 | 46.0 | ... | ... | 2007 | 98.4 | 31.2 | 98.9 | 89.5 | 41.7 | 81.6 | 99.2 |
| Dominican Republic | Santo Domingo | 1999 | 97.7 | 31.1 | 87.2 | 74.6 | 54.3 | ... | ... | 2007 | 80.9 | 9.0 | 96.0 | 85.2 | 39.0 | 79.3 | 98.6 |
| Guatemala | Ciudad de Guatemala (Guatemala City) | 1998 | 91.1 | 53.2 | 83.6 | 71.6 | 31.9 | ... | 91.7 | ... | ... | ... | ... | ... | ... | ... | ... |
| Guatemala | Escuintla | 1998 | 94.0 | 56.8 | 96.7 | 90.2 | 29.5 | ... | 97.8 | ... | ... | ... | ... | ... | ... | ... | ... |
| Guatemala | Quetzaltenango | 1998 | 93.7 | 71.2 | 82.5 | 70.0 | 31.3 | ... | 91.2 | ... | ... | ... | ... | ... | ... | ... | ... |
| Haiti | Port-au-Prince | 1994 | 48.5 | 31.9 | 93.4 | 16.9 | ... | ... | 92.3 | 2006 | 78.6 | 25.4 | 57.6 | 17.3 | 11.2 | 48.6 | 88.0 |
| Honduras | Choluteca | | ... | ... | ... | ... | ... | ... | ... | 2005 | 99.1 | 38.8 | 76.0 | 53.3 | 51.8 | 41.5 | ... |
| Honduras | Comayagua | | ... | ... | ... | ... | ... | ... | ... | 2005 | 94.6 | 30.3 | 87.6 | 75.0 | 38.1 | 47.5 | ... |
| Honduras | Juticalpa | | ... | ... | ... | ... | ... | ... | ... | 2005 | 96.9 | 35.2 | 78.2 | 52.9 | 46.2 | 43.3 | ... |
| Honduras | La Ceiba | | ... | ... | ... | ... | ... | ... | ... | 2005 | 94.1 | 35.9 | 91.3 | 73.5 | 29.9 | 64.3 | ... |
| Honduras | San Pedro Sula | | ... | ... | ... | ... | ... | ... | ... | 2005 | 98.9 | 30.2 | 93.3 | 84.0 | 40.1 | 57.6 | ... |
| Honduras | Santa Bárbara | | ... | ... | ... | ... | ... | ... | ... | 2005 | 91.6 | 48.3 | 78.7 | 61.8 | 16.4 | 34.2 | ... |
| Honduras | Santa Rosa de Copán | | ... | ... | ... | ... | ... | ... | ... | 2005 | 88.9 | 17.1 | 87.0 | 74.1 | 33.1 | 45.8 | ... |
| Honduras | Tegucigalpa | | ... | ... | ... | ... | ... | ... | ... | 2005 | 89.4 | 32.7 | 86.0 | 72.4 | 54.9 | 53.0 | ... |
| Honduras | Trujillo | | ... | ... | ... | ... | ... | ... | ... | 2005 | 91.8 | 24.8 | 92.7 | 71.6 | 45.7 | 51.8 | ... |
| Honduras | Yoro | | ... | ... | ... | ... | ... | ... | ... | 2005 | 97.4 | 30.1 | 91.7 | 72.8 | 44.2 | 54.6 | ... |
| Honduras | Yuscarán | | ... | ... | ... | ... | ... | ... | ... | 2005 | 92.6 | 42.4 | 83.4 | 58.8 | 35.1 | 37.2 | ... |
| Nicaragua | Chinandega | 1998 | 82.1 | 78.6 | 62.2 | 25.9 | 8.2 | ... | 84.0 | 2001 | 100.0 | 85.5 | 65.7 | 22.3 | 9.3 | 8.9 | 89.5 |
| Nicaragua | Estelí | 1998 | 95.3 | 94.5 | 66.7 | 36.5 | 12.5 | ... | 84.9 | 2001 | 99.1 | 93.4 | 69.1 | 30.7 | 14.0 | 0.9 | 91.7 |
| Nicaragua | Granada | 1998 | 97.2 | 97.0 | 67.0 | 37.4 | 16.9 | ... | 93.6 | 2001 | 99.8 | 97.4 | 71.6 | 35.8 | 23.9 | 12.3 | 95.0 |
| Nicaragua | León | 1998 | 92.4 | 92.0 | 68.8 | 40.5 | 12.6 | ... | 92.5 | 2001 | 99.8 | 97.0 | 73.9 | 46.2 | 11.8 | 11.1 | 98.4 |
| Nicaragua | Managua | 1998 | 97.5 | 97.5 | 78.2 | 58.4 | 21.9 | ... | 96.9 | 2001 | 99.8 | 97.1 | 81.7 | 61.1 | 29.1 | 21.9 | 96.4 |
| Nicaragua | Masaya | 1998 | 96.2 | 95.8 | 65.0 | 30.3 | 14.8 | ... | 94.9 | 2001 | 100.0 | 98.9 | 69.4 | 31.6 | 18.4 | 10.4 | 97.9 |
| Nicaragua | Matagalpa | 1998 | 95.9 | 95.3 | 68.1 | 37.2 | 13.2 | ... | 90.9 | 2001 | 98.1 | 87.5 | 72.0 | 30.2 | 16.5 | 1.2 | 92.2 |

# TABLE C.3

*continued*

| | | Percentage of households with access to | | | | | | | | | | | | | | |
|---|---|---|---|---|---|---|---|---|---|---|---|---|---|---|---|---|
| | | **1990–1999**[1] | | | | | | | **2000–2009**[1] | | | | | | | |
| | | Survey year | Improved water | Piped water | Improved sanitation | Sewer-age | Tele-phone(s) | Mobile(s) | Connec-tion to electricity | Survey year | Improved water | Piped water | Improved sanitation | Sewer-age | Tele-phone(s) | Mobile(s) | Connec-tion to electricity |
| Peru | Arequipa | 1996 | 88.5 | 74.3 | 80.7 | 67.7 | 25.1 | ... | 94.8 | 2004 | 93.6 | 93.2 | 89.5 | 84.6 | 36.1 | ... | 98.1 |
| Peru | Chiclayo | 1996 | 89.1 | 74.8 | 72.1 | 55.0 | 20.6 | ... | 88.7 | 2004 | 91.8 | 91.2 | 86.5 | 79.6 | 32.0 | ... | 92.3 |
| Peru | Chimbote | 1996 | 76.4 | 72.0 | 79.6 | 68.4 | 24.0 | ... | 91.4 | 2004 | 87.8 | 87.8 | 76.8 | 73.9 | 31.6 | ... | 85.2 |
| Peru | Lima | 1996 | 83.1 | 73.7 | 85.1 | 77.1 | 35.7 | ... | 97.4 | 2004 | 96.6 | 96.6 | 96.5 | 94.3 | 61.9 | ... | 99.1 |
| Peru | Piura | 1996 | 88.9 | 84.8 | 78.9 | 67.4 | 18.9 | ... | 83.4 | 2004 | 94.0 | 64.9 | 60.5 | 37.7 | 24.0 | ... | 91.1 |
| Peru | Tacna | 1996 | 96.1 | 81.4 | 83.3 | 80.9 | 33.0 | ... | 92.4 | 2004 | 100.0 | 100.0 | 98.6 | 97.7 | 27.9 | ... | 99.0 |
| Peru | Trujillo | 1996 | 84.9 | 72.6 | 72.8 | 61.5 | 19.8 | ... | 84.9 | 2004 | 93.5 | 93.5 | 98.3 | 96.0 | 50.7 | ... | 98.1 |

*Source:* United Nations Human Settlements Programme (UN-Habitat), Global Urban Indicators Database 2010.

*Notes:*
(1) Data are from the latest year availble in the period shown.

# REFERENCES

A101 (2006) 'Moscow will have a new suburb', MASSHTAB, www.a101.ru/en/news.xml?&news_id=77&year_id=57&month_id=76, last accessed 18 October 2010

ABI (Association of British Insurers) (2005) *Financial Risks of Climate Change*, Association of British Insurers, London, UK

ACIA (Arctic Climate Impact Assessment) (2004) *Arctic Climate Impact Assessment*, Cambridge University Press, Cambridge, UK

ActionAid International (2006) *Unjust Waters, Climate Change, Flooding and the Protection of Poor African Communities: Experiences from Six African Cities*, Actionaid International, www.reliefweb.int/rw/lib.nsf/db900sid/TBRL-76GR49/$file/Actionaid-UnjustWaters-Aug2007.pdf?openelement, last accessed 7 December 2010

Adams, J. (2007) 'Rising sea levels threaten small Pacific island nations', *New York Times*, 3 May, www.nytimes.com/2007/05/03/world/asia/03iht-pacific.2.5548184.html? r=1, last accessed 20 January 2011

ADB (Asian Development Bank) (2005) *Climate Proofing: A Risk-Based Approach to Adaptation*, Pacific Studies Series, Manila

ADB (undated) 'Clean energy financing partnership facility', www.adb.org/Clean-Energy/cefpf.asp, last accessed 6 October 2010

Adelekan, I. O. (2010) 'Vulnerability of poor urban coastal communities to flooding in Lagos, Nigeria', *Environment and Urbanization* **22**(2): 433–450

Adeyinka S. O. and O. J. Taiwo (2006) 'Lagos shoreline change pattern: 1986–2002', *American-Eurasian Journal of Scientific Research* **1**(1): 25–30

Adger, W. N. (1999) 'Social vulnerability to climate change and extremes in coastal Vietnam', *World Development* **27**(2): 249–269

Adger, W. N. (2000) 'Social and ecological resilience: Are they related?' *Progress in Human Geography* **24**(3): 347–364

Adger, W. N. (2001) 'Scales of governance and environmental justice for adaptation and mitigation of climate change', *Journal of International Development* **13**(7): 921–931

Adger, W. N., T. Hughes, C. Folke, S. Carpenter and J. Rockström (2005) 'Social-ecological resilience to coastal disasters', *Science* **309**(5737): 1036–1039

Adger, W. N., S. Agrawala, M. M. Q. Mirza, C. Conde, K. O'Brien, J. Pulhin, R. Pulwarty, B. Smit and K. Takahashi (2007) 'Assessment of adaptation practices, options, constraints and capacity', in M. L. Parry, O. F. Canziani, J. P. Palutikof, P. J. van der Linden and C. E. Hanson (eds) *Climate Change 2007: Impacts, Adaptation and Vulnerability. Contribution of Working Group II to the Fourth Assessment Report of the Intergovernmental Panel on Climate Change*, Cambridge University Press, Cambridge, UK, pp717–743

African Development Bank, Asian Development Bank, Department for International Development, European Commission, Federal Ministry for Economic Cooperation and Development-Germany, Development Cooperation-The Netherlands, Organisation for Economic Co-operation and Development, United Nations Development Programme, United Nations Environment Programme and the World Bank (2003) *Poverty and Climate Change: Reducing the Vulnerability of the Poor through Adaptation*, Washington, DC

Agencianova (2009) 'El gobierno bonaerense inicia la construcción de viviendas bioclimáticas', www.novacolombia.info/nota.asp?n=2009_6_9&id=9226&id_tiponota=10, last accessed 14 October 2010

Agnew, M. and D. Viner (2001) 'Potential impacts of climate change on international tourism', *Tourism and Hospitality Research* **3**(1): 37–60

Ahern, M., R. S. Kovats, P. Wilkinson, R. Few and F. Matthies (2005) 'Global health impacts of floods: Epidemiologic evidence', *Epidemiology Review* **27**(1): 36–46

Akbari, H. (2005) *Energy Saving Potentials and Air Quality Benefits of Urban Heat Island Mitigation*, Lawrence Berkeley National Laboratory, Berkeley, CA

Akinbami, J. F. and A. Lawal (2009) 'Opportunities and challenges to electrical energy conservation and $CO_2$ emissions reduction in Nigeria's building sector', Paper prepared for the Fifth Urban Research Symposium, Cities and Climate Change: Responding to an Urgent Agenda, 28–30 June, Marseille, France

Alam, M. and M. Rabbani (2007) 'Vulnerabilities and responses to climate change for Dhaka', *Environment and Urbanization* **19**(1): 81–97

Alber, G. (2010) *Gender, Cities and Climate Change*, Unpublished thematic report prepared for the *Global Report on Human Settlements 2011*, www.unhabitat.org/grhs/2011

Alber G. and K. Kern (2008) 'Governing climate change in cities: Modes of urban climate governance in multi-level systems', OECD International Conference, Competitive Cities and Climate Change, 2nd Annual Meeting of the OECD Roundtable Strategy for Urban Development, 9–10 October, Milan, Italy, www.oecd.org/dataoecd/22/7/41449602.pdf, last accessed 28 October 2010

Alberti, M. and L. R. Hutyra (2009) 'Detecting carbon signatures of development patterns across a gradient of urbanization: Linking observations, models, and scenarios', Paper prepared for the Fifth Urban Research Symposium, Cities and Climate Change:

Responding to an Urgent Agenda, 28–30 June, Marseille, France

Aldy, J. E., S. Barrett and R. N. Stavins (2003) 'Thirteen plus one: a comparison of global climate policy architectures', *Climate Policy* **3**(4): 373–397

AlertNet (2010a) 'UN Adaptation Fund gives green light to first four projects', www.alertnet.org/db/an_art/60167/2010/05/17-221110-1.htm

AlertNet (2010b) 'Climate change: Adaptation Fund swings into action', www.alertnet.org/thenews/newsdesk/IRIN/690db58da46ba960d48e0c009e677716.htm

Allman, L., P. Fleming and A. Wallace (2004) 'The progress of English and Welsh local authorities in addressing climate change', *Local Environment* **9**(3): 271–283

Alston, L. J., G. D. Libecap and B. Mueller (2001) 'Land reform policies, sources of conflict and implications for de-forestation in the Brazilian Amazon', *Nota Di Lavoro*, 70.2001, Fonazione Eni Enrico Mattei, Milan, Italy

Ammann C. M., F. Joos, D. Schimel, B. L. Otto-Bliesner and R. Tomas (2007) 'Solar influence on climate during the past millennium: Results from transient simulations with the NCAR Climate System Model', *Proceedings of the National Academy of Sciences* **104**(10): 3713–3718

Ananthapadmanabhan, G., K. Srinivas and V. Gopal (2007) *Hiding Behind the Poor: A Report by Greenpeace on Climate Injustice*, New Delhi, India

Andrews, C. (2008) 'Greenhouse gas emissions along the rural–urban gradient', *Journal of Environmental Planning and Management* **51**(6): 847–870

Angel, S., S. Sheppard and D. Civco (2005) *The Dynamics of Global Urban Expansion, Transport and Urban Development Department*, World Bank, Washington, DC

Arup (2008) *Zero Net Emissions by 2020 Update, Melbourne*, City of Melbourne, Australia

Asia-Pacific Partnership on Clean Development and Climate (undated) 'Frequently asked questions', www.asiapacificpartnership.org/english/faq.aspx#FAQ1, last accessed 6 October 2010

Atteridge, A., C. K. Siebert, R. J. Klein, C. Butler and P. Tella (2009) 'Bilateral finance institutions and climate change: A mapping of climate portfolios', Stockholm Environment Institute for the Climate Change Working Group for Bilateral Finance Institutions Working Paper, Stockholm Environment Institute, Stockholm, Sweden, www. sei-international.org/mediamanager/documents/Publications/Climate-mitigation-adaptation/bilateral-finance-institutions-climate-change.pdf, last accessed 6 October 2010

Awuor, C. B., V. A. Orindi and A. O. Adwera (2008) 'Climate change and coastal cities: The case of Mombasa, Kenya', *Environment and Urbanization* **20**(1): 231–242

Ayers, J. (2009) 'International funding to support urban adaptation to climate change', *Environment and Urbanization* **21**(1): 225–240

Ayers, J. and S. Huq (2009) 'The value of linking mitigation and adaptation: A case study of Bangladesh', *Environmental Management* **43**(5): 753–764

Aylett, A. (2010) 'Changing perceptions of climate mitigation among competing priorities: The case of Durban, South Africa', Unpublished case study prepared for the *Global Report on Human Settlements 2011*, www.unhabitat.org/grhs/2011

Bai, X. (2007) 'Industrial ecology and the global impacts of cities', *Journal of Industrial Ecology* **11**(2): 1–6

Baldasano, J., C. Soriano and L. Boada (1999) 'Emission inventory for greenhouse gases in the City of Barcelona, 1987–1996', *Atmospheric Environment* **33**(23): 3765–3775

Balk, D., M. R. Montgomery, G. McGranahan, D. Kim, V. Mara, M. Todd, T. Buettner and A. Dorelién (2009) 'Mapping urban settlements and the risks of climate change in Africa, Asia and South America', in J. M. Guzman, G. Martine, G. McGranahan, D. Schensul and C. Tacoli (eds) *Population Dynamics and Climate Change*, United Nations Population Fund (UNFPA) and International Institute for Environment and Development (IIED), London, pp80–103

Bangkok Metropolitan Administration (2009) *Bangkok: Assessment Report on Climate Change 2009*, Green Leaf Foundation, Bangkok Metropolitan Administration and United Nations Environment Programme Regional Office for Asia and the Pacific, Bangkok, Thailand

Banister, D., S. Watson and C. Wood (1997) 'Sustainable cities: Transport, energy, and urban form', *Environment and Planning B: Planning and Design* **24**(1): 125–143

Banks, N. (2008) 'A tale of two wards: Political participation and the urban poor in Dhaka city', *Environment and Urbanization* **20**(2): 361–376

Barker T., I. Bashmakov, L. Bernstein, J. E. Bogner, P. R. Bosch, R. Dave, O. R. Davidson, B. S. Fisher, S. Gupta, K. Halsnæs, G. J. Heij, S. Kahn Ribeiro, S. Kobayashi, M. D. Levine, D. L. Martino, O. Masera, B. Metz, L. A. Meyer, G.-J. Nabuurs, A. Najam, N. Nakicenovic, H.-H. Rogner, J. Roy, J. Sathaye, R. Schock, P. Shukla, R. E. H. Sims, P. Smith, D. A. Tirpak, D. Urge-Vorsatz and D. Zhou (2007) 'Technical summary', in B. Metz, O. R. Davidson, P. R. Bosch, R. Dave and L. A. Meyer (eds) *Climate Change 2007: Mitigation, Contribution of Working Group III to the Fourth Assessment Report of the Intergovernmental Panel on Climate Change*, Cambridge University Press, Cambridge and New York, pp25–93, www.ipcc.ch/pdf/assessment-report/ar4/wg3/ar4-wg3-ts.pdf, last accessed 11 October 2010

Barry, E. J. (1943) *Solar Water Heater*, USP Office, US

Bartlett, S. (2008) 'Climate change and children: Impacts and implications for adaptation in low- to middle-income countries', *Environment and Urbanization* **20**(2): 501–519

Bartlett, S., D. Dodman, J. Hardoy, D. Satterthwaite and C. Tacoli (2009) 'Social aspects of climate change in urban areas in low- and middle-income nations', Paper prepared for the Fifth Urban Research Symposium, Cities and Climate Change: Responding to an Urgent Agenda, 28–30 June, Marseille, France

Bastianoni, S., F. Pulselli and E. Tiezzi (2004) 'The problem of assigning responsibility for greenhouse gas emissions', *Ecological Economics* **49**(3): 253–257

Basu, R. and J. Samet (2002) 'Relation between elevated ambient temperature and mortality: A review of the epidemiologic evidence', *Epidemiologic Reviews* **24**(2): 190–202

Bates, B. C., Z. W. Kundzewicz, S. Wu and J. P. Palutikof (eds) (2008) 'Climate change and water', Technical Paper of the Intergovernmental Panel on Climate Change, IPCC Secretariat, Geneva, Switzerland

Baumert, K., T. Herzog and J. Pershing (2005) *Navigating the Numbers: Greenhouse Gas Data and International Climate Policy*, World Resources Institute, Washington, DC

BBC News (2010a) 'Floods in north-east Brazil kill dozens of people', 23 June 2010, www.bbc.co.uk/news/10372362, last accessed 29 October 2010

BBC News (2010b) 'Pakistan flood death toll "passes 1,100"', BBC News South Asia, www.bbc.co.uk/news/world-south-asia-10832166, last accessed 29 October 2010

BCIL (Biodiversity Conservation India Limited) (2009) *T-Zed Case-Study*, Unpublished report presented to UNEP, Bangalore

Beauchemin, C. and P. Bocquier (2004) 'Migration and urbanization in Francophone West Africa: An overview of the recent empirical evidence', *Urban Studies* **41**(11): 2245–2272

Beccherle, J. and J. Tirole (2010) 'Regional initiatives and the cost of delaying binding climate change agreements', www.idei.fr/doc/by/tirole/regionalinitiativesmay17.pdf, last accessed 7 October 2010

Beniston, M. and H. Diaz (2004) 'The 2003 heat wave as an example of summers in a greenhouse climate? Observations and climate model simulations for Basel, Switzerland', *Global and Planetary Change* **44**(1–4): 73–81

Benson, C. and E. Clay (2004) 'Beyond the damage: Probing the economic and financial consequences of natural disasters', Presentation at Overseas Development Institute (ODI), London, 11 May 2004

Berger, L. R. and D. Mohan (1996) *Injury Control: A Global View*, Oxford University Press, New Delhi, India

Bertaud, A., B. Lefevre and B. Yuen (2009) 'GHG emissions, urban mobility and efficiency of urban morphology: A hypothesis', Paper prepared for the Fifth Urban Research Symposium, Cities and Climate Change: Responding to an Urgent Agenda, 28–30 June, Marseille, France

Betsill, M. M. (2001) 'Mitigating climate change in US Cities: Opportunities and obstacles', *Local Environment* **6**(4): 393–406

Betsill, M. M. and H. Bulkeley (2007) 'Looking back and thinking ahead: A decade of cities and climate change research', *Local Environment* **12**(5): 447–456

Bicknell, J., D. Dodman and D. Satterthwaite (eds) (2009) *Adapting Cities to Climate Change: Understanding and Addressing the Development Challenges*, Earthscan, London

Biermann, F. and P. Pattberg (2008) 'Global environmental governance: Taking stock, moving forward', *Annual Review of Environment and Resources* **33**(1): 277–294

Biermann, F., P. Pattberg, H. van Asselt and F. Zelli (2008) 'Fragmentation of global governance architectures: Case of climate policy', Paper presented at the 49th International Studies Association's (ISA's) Annual Convention, Bridging Multiple Divides, San Francisco, California, March 2008

Biermann, F., M. M. Betsill, J. Gupta, N. Kanie, L. Lebel, D. Liverman, H. Schroeder and B. Siebenhuner (2009) *Earth System Governance – People, Places, and the Planet, Science and Implementation of the Earth System Governance Project, Earth System Governance Report 1*, International Human Dimensions Programme, The Earth System Governance Project, Bonn, Germany

Bigio, A. (2009) 'Adapting to climate change and preparing for natural disasters in the coastal cities of North Africa', Paper presented at the Urban Research Symposium, Cities and Climate Change: Responding to an Urgent Agenda, June 28–30, Marseilles, France

Bin, S. and R. Harris (2006) 'The role of $CO_2$ embodiment in U.S.–China trade', *Energy Policy* **34**(18): 4063–4068

Bird, N. and L. Peskett (2008) 'Recent bilateral initiatives for climate financing: Are they moving in the right direction?', Opinion Paper No 112, Overseas Development Institute (ODI), London, www.odi.org.uk/resources/download/2402.pdf, last accessed 6 October 2010

Bizikova L., T. Neale and I. Burton (2008) *Canadian Communities' Guidebook for Adaptation to Climate Change*, Environment Canada and University of British Columbia, Vancouver, Canada

Bloomberg, M. and R. Aggarwala (2008) 'Think locally, act globally: How curbing global warming emissions can improve local public health', *American Journal of Preventative Medicine* **35**(5): 414–423

Boardman, B. (2007) 'Examining the carbon agenda via the 40% house scenario', *Building Research & Information* **35**(4): 363–378

Bodansky, D. (2001) 'The history of the global climate change regime', in U. Luterbacher and D. F. Sprinz (eds) *International Relations and Global Climate Change*, MIT Press Cambridge, MA, pp23–40

Boland, J. (1997) 'Assessing urban water use and the role of water conservation measures under climate uncertainty', *Climatic Change* **37**(1): 157–176

Boonyabancha, S. (2005) 'Baan Mankong: Going to scale with "slum" and squatter upgrading in Thailand', *Environment and Urbanization* **17**(1): 21–46

Boonyabancha, S. (2009) 'Land for housing the poor by the poor: Experiences from the Baan Mankong nationwide slum upgrading programme in Thailand', *Environment and Urbanization* **21**(2): 309–330

Boruff, B. J., C. Emrich and S. L. Cutter (2005) 'Erosion hazard vulnerability of US coastal counties', *Journal of Coastal Research* **21**(5): 932–942

Brasseur, G., K. Jacobs, E. Barron, R. Benedick, W. Chameides, T. Dietz, P. Romero Lankao, M. McFarland, H. Mooney, D. Nathan, E. Parson and R. Richels (2007) *Analysis of Global Change Assessments: Lessons Learned*, National Academies Press, Washington, DC

Breman, J., M. S. Alilio and A. Mills (2004) 'Conquering the intolerable burden of malaria: What's new, what's needed: A summary', *American Journal of Tropical Medicine and Hygiene* **71**(2), Supplement: 1–15

Brookings Institution (2009) *Protecting and Promoting Rights in Natural Disasters in South Asia: Prevention and Response*, Report on the Project on Internal Displacement, Chennai, India

Brown, M. and F. Southworth (2008) 'Mitigating climate change through green buildings and smart growth', *Environment and Planning A* **40**(3): 653–675

Brown, M. A., F. Southworth and A. Sarzynksi (2008) *Shrinking the Carbon Footprint of Metropolitan America*, Brookings Institute, Washington, DC

Brown, O. (2007) 'Climate change and forced migration: Observations, projections and implications', Human Development Report Office Occasional Paper, 2007/17, UNDP, http://hdr.undp.org/en/reports/global/hdr2007-2008/papers/brown_oli.pdf, last accessed 14 October 2010

Bulkeley, H. (2000) 'Down to Earth: Local government and greenhouse policy in Australia', *Australian Geographer* **31**: 289–308

Bulkeley, H. and M. Betsill (2003) *Cities and Climate Change: Urban Sustainability and Global Environmental Governance*, Routledge, London

Bulkeley, H. and K. Kern (2006) 'Local government and the governing of climate change in Germany and the UK', *Urban Studies* **43**(12): 2237–2259

Bulkeley, H. and P. Newell (2010) *Governing Climate Change*, Routledge, London, NY

Bulkeley, H. and H. Schroeder (2008) 'Governing climate change post-2012: The role of global cities – London', Working Paper No 123, Tyndall Centre for Climate Change Research, www.tyndall.ac.uk/sites/default/files/wp123.pdf, last accessed 7 October 2010

Bulkeley, H., H. Schroeder, K. Janda, J. Zhao, A. Armstrong, S. Y. Chu and S. Ghosh (2009) 'Cities and climate change: The role of institutions, governance and urban planning', Paper prepared for the Fifth Urban Research Symposium, Cities and Climate Change: Responding to an Urgent Agenda, 28–30 June, Marseille, France

Bull-Kamanga, L., K. Diagne, A. Lavell, F. Lerise, H. MacGregor, A. Maskrey, M. Meshack, M. Pelling, H. Reid, D. Satterthwaite, J. Songsore, K. Westgate and A. Yitambe (2003) 'From everyday hazards to disasters: The accumulation of disaster risk in urban areas', *Environment and Urbanization* **15**(1): 193–204

Bumpus A. and D. Liverman (2008) 'Accumulation by decarbonisation and the governance of carbon offsets', *Economic Geography* **84**(2): 127–155

Bundesministerium für Umwelt, Naturschutz und Reaktorsicherheit (undated) *KWK: Modellstadt Berlin*, www.kwk-modellstadt-berlin.de/, last accessed 14 October 2010

Burdett, R., T. Travers, D. Czischke, P. Rode and B. Moser (2005) *Density and Urban Neighbourhoods in London*, Enterprise LSE Cities, London

Burke, L., S. J. Brown and N. Christidis (2006) 'Modelling the recent evolution of global drought and projections for the 21st century with the Hadley Center climate model', *Journal of Hydrometeorology* **7**(5): 1113–1125

C40 Cities (undated) 'Participating cities', www.c40cities.org/cities, last accessed 6 October 2010

Cabannes, Y. (2004) 'Participatory budgeting: A significant contribution to participatory democracy', *Environment and Urbanization* **16**(1): 27–46

California Solutions for Global Warming (undated) 'California global warming solutions act', www.solutionsforglobalwarming.org/1calpolicyAB32.html, last accessed 6 October 2010

Camilleri, M., R. Jaques and N. Isaacs (2001) 'Impacts of climate change on building performance in New Zealand', *Building Research and Information* **29**(6): 440–450

Campbell-Lendrum, D. and C. Corvalan (2007) 'Climate change and developing country cities: Implications for environmental health and equity', *Journal of Urban Health* **84**(1): 109–117

CAN (Climate Action Network International) (undated) 'Climate action network', www.climatenetwork.org/, last accessed 7 October 2010

Capello, R., P. Nijkamp and G. Pepping (1999) *Sustainable Cities and Energy Policies*, Springer-Verlag, Berlin

Carmin, J. and Y. Zhang (2009) *Achieving Urban Climate Adaptation in Europe and Central Asia*, World Bank Policy Research Working Paper 5088, World Bank, Washington, DC

Carmin, J., D. Roberts and I. Anguelovski (2009) 'Planning climate resilient cities: Early lessons from early adapters', Paper prepared for the Fifth Urban Research Symposium, Cities and Climate Change: Responding to an Urgent Agenda, 28–30 June, Marseille, France

Casaubon, M. E., M. D. Peralta and A. M. Ponce de León (2008) *Mexico City Climate Action Program 2008–2012: Summary*, Secretaría del Medio Ambiente del Distrito Federal Plaza de la Constitución, Colonia Centro, www.sma.df.gob.mx/sma/links/download/archivos/paccm_summary.pdf, last accessed 12 October 2010

Castán Broto, V., P. Tabbush, K. Burningham, L. Elghali and D. Edwards (2007) 'Coal ash and risk: Four social interpretations of a pollution landscape', *Landscape Research* **32**: 481–497

Castán Broto, V., C. Carter and L. Elghali (2009) 'The governance of coal ash pollution in post-socialist times: power and expectations', *Environmental Politics* **18**(2): 279–286

CD4CDM (Capacity Development for CDM) (undated) 'CDM/JI Pipeline overview page', http://cdmpipeline.org/overview.htm, last accessed 6 October 2010

Centre for Clean Air Policy (undated) 'Urban leaders adaptation initiative', www.ccap.org/index.php?component=programs&id=6, last accessed 6 October 2010

CFCB (Carbon Finance Capacity Building Programme) (undated) 'Carbon finance capacity building programme', www.lowcarboncities.info/home.html, last accessed 6 October 2010

Choi, O. and A. Fisher (2003) 'The impacts of socio-economic development and climate change on severe weather catastrophe losses: Mid-Atlantic Region (MAR) and the U.S.', *Climate Change* **58**(1–2): 149–170

Church, J., N. White, R. Coleman, K. Lambeck and J. Mitrovica (2004) 'Estimates of the regional distribution of sea level rise over the 1950–2000 period', *American Meteorological Society* **17**(13): 2609–2625

CISDL (Centre for International Sustainable Development Law) (2002) 'The principle of common but differentiated responsibilities: Origins and scope', www.cisdl.org/pdf/brief_common.pdf, last accessed 12 October 2010

City of Cape Town (2005) *City of Cape Town Climate Change Strategy*, City of Cape Town

City of Cape Town (2006) *Cape Town Energy and Climate Change Strategy*, Environmental Planning Department, City of Cape Town; www.capetown.gov.za/en/EnvironmentalResourceManagement/publications/Documents/Energy_+_Climate_Change_Strategy_2_-_10_2007_301020079335_465.pdf, last accessed 12 October 2010

City of Cape Town (2007) *State of Energy Report for the City of Cape Town 2007*, Palmer Development Group, Cape Town, www.capetown.gov.za/en/EnvironmentalResourceManagement/publications/Documents/StateOfEnergy_Report_2007_v2.pdf, last accessed 12 October 2010

City of Cape Town (2010) 'Cape Town prepares for winter storms', www.capetown.gov.za/en/Pages/CapeTownpreparesforwinterstorms.aspx, last accessed 14 October 2010

City of Cape Town (undated) *Cape Town 2007–2012, Five Year Plan, Summary of the Integrated Development Plan (2007–2012)*, www.capetown.gov.za/en/IDP/Documents/idp/IDP_English.pdf, last accessed 14 October 2010

City of Melbourne (2009) *City of Melbourne Climate Change Adaptation Strategy*, www.melbourne.vic.gov.au/AboutCouncil/PlansandPublications/strategies/Documents/climate_change_adaptation_strategy.PDF, last accessed 14 October 2010

City of New York (2007) *Inventory of New York City Greenhouse Gas Emissions*, www.nyc.gov/html/planyc2030/downloads/pdf/emissions_inventory.pdf, last accessed 12 October 2010

City of New York (2009) *Inventory of New York City Greenhouse Gas Emissions*, Updated 24 February 2009, www.nyc.gov/html/planyc2030/html/downloads/download.shtml, last accessed 12 October 2010

City of Rotterdam (2009) *Rotterdam Climate Proof: 2009 Adaptation Programme*, www.rotterdamclimate initiative.nl/documents/Documenten/RCP_adaptatie_eng.pdf, last accessed 14 October 2010

City of Rotterdam (undated) *Rotterdam Climate Proof: The Rotterdam Challenge on Water and Climate Adaptation*, www.climateinitiative.eu/documents/Documenten/RCP_folder_eng.pdf, last accessed 14 October 2010

City of São Paulo (2009) *Instrui a Política de Mudança do Clima no Município de São Paulo*, Lei no 14933 sancionada em 05/06/2009 e publicada no Diário Oficial do Município em 06/06/2009, A. C. M. d. S. Paulo, City of São Paulo, Brazil

Ciudad de Mexico (2008) *Programa de Acción Climática*, Gobierno del Distrito Federal, Ciudad de Mexico

Clapp, C., A. Leseur, O. Sartor, G. Briner and J. Corfee-Morlot (2010) *Cities and Carbon Market Finance: Taking Stock of Cities Experience with Clean Development Mechanism (CDM) and Joint Implementation (JI)*, OECD Environmental Working Paper No 29, OECD Publishing, Paris, www.oecd.org/dataoecd/18/43/46501427.pdf, last accessed 10 December 2010

Climate Alliance (undated) 'Climate Alliance', www.klimabuendnis.org/, last accessed 28 October 2010

Climate Fund Update (undated a) 'Special Climate Change Fund', www.climatefundsupdate.org/listing/special-climate-change-fund, last accessed 6 October 2010

Climate Fund Update (undated b) 'Least Developed Countries Fund (LDCF)', www.climatefundsupdate.org/listing/least-developed-countries-fund, last accessed 28 October 2010

Climate Fund Update (undated c) 'Adaptation Fund', www.climatefundsupdate.org/listing/adaptation-fund, last accessed 6 October 2010

Climate Investment Funds (undated) 'The Climate Investment Funds', www.climateinvestmentfunds.org/cif/, last accessed 6 October 2010

Clinton Foundation (undated a) 'Combating climate change: Clinton Climate Initiative', www.clinton foundation.org/what-we-do/clinton-climate-initiative, last accessed 6 October 2010

Clinton Foundation (undated b) 'What we do', www.clintonfoundation.org/what-we-do, last accessed 14 October 2010

Cohen, M. and J. Garrett (2010) 'The food price crisis and urban food (in)security', *Environment and Urbanization* **22**(2): 467–482

Collier, U. (1997) 'Local authorities and climate protection in the European union: Putting subsidiarity into practice?', *Local Environment: The International Journal of Justice and Sustainability* **2**(1): 39–57

Colombo, A., D. Etkin and B. Karney (1999) 'Climate variability and the frequency of extreme temperature events for nine sites across Canada: Implications for power usage', *Journal of Climate* **12**: 2490–2502

Concejo de Bogotá (2008) *Proyecto de Acuerdo No 641 de 2008 'Por medio cual se dictan normas para el Manejo del Arbolado del Distrito Capital y se dictan otras disposiciones'*, City of Bogotá, Colombia

Costello, A., M. Abbas, A. Allen, S. Ball, S. Bell, R. Bellamy, S. Friel, N. Groce, A. Johnson, M. Kett, M. Lee, C. Levy, M. Maslin, D. McCoy, B. McGuire, H. Montgomery, D. Napier, C. Pagel, J. Patel, J. A. de Oliveira, N. Redclift, H. Rees, D. Rogger, J. Scott, J. Stephenson, J. Twigg, J. Wolff and C. Patterson (2009) 'Managing the health effects of climate change', *Lancet* **373**(9676): 1693–1733

Coutts, A. M., J. Beringer and N. J. Tapper (2008) 'Impact of increasing urban density on local climate: Spatial and temporal variations in the surface energy balance in Melbourne, Australia', *Journal of Applied Meteorology and Climatology* **46**(4): 477–493

Crass, M. (2008) 'Reducing $CO_2$ emissions from urban travel: Local policies and national plans', OECD International Conference, Competitive Cities and Climate Change, 2nd Annual Meeting of the OECD Roundtable Strategy for Urban Development, 9–10 October 2008, Milan, Italy

Crespin, J. (2006) 'Aiding local action: The constraints faced by donor agencies in supporting effective, pro-poor initiatives on the ground', *Environment and Urbanization* **18**(2): 433–450

Cross, J. (2001) 'Megacities and small towns: Different perspectives on hazard vulnerability', *Environmental Hazard* **3**(2): 63–80

Dalal-Clayton, B. (2003) 'The MDGs and sustainable development: The need for a strategic approach', in D. Satterthwaite (ed) *The Millennium Development Goals and Local Processes: Hitting the Target or Missing the Point*, IIED, London, pp73–91

Dalton, M., B. O'Neill, A. Prskawetz, L. Jiang and J. Pitkin (2008) 'Population aging and future carbon emissions in the United States', *Energy Economics* **30**(2): 642–675

Darch, G. (2006) 'The impacts of climate change on London's transport systems', CIWEM Metropolitan Branch Climate Change Conference, 22 February 2006

Darido, G., M. Torres-Montoya and S. Mehndiratta (2009) 'Urban transport and $CO_2$ emissions: Some evidence from Chinese cities', World Bank Working Paper, www-wds.worldbank.org/external/default/WDSContentServer/WDSP/IB/2010/07/21/000334955_20100721033904/Rendered/PDF/557730WP0P11791June020091EN105jan10.pdf, last accessed 12 October 2010

Davies, A. R. (2005) 'Local action for climate change: Transnational networks and the Irish experience', *Local Environment* **10**(1): 21–40

de Bono, A., G. Giuliani, S. Kluser and P. Peduzzi (2004) 'Impacts of summer 2003 heat wave in Europe', *Early Warning on Emerging Environmental Threats 2*, United Nations Environment Programme, Nairobi, Kenya

De Lucia, V. and R. Reibstein (2007) 'Common but differentiated responsibility', in C. J. Cleveland (ed) *Encyclopedia of Earth*, Environmental Information Coalition, National Council for Science and the Environment, Washington, DC, www.eoearth.org/article/Common_but_differentiated_responsibility, last accessed 6 October 2010

de Sherbinin, A., A. Schiller and A. Pulsipher (2007) 'The vulnerability of global cities to climate hazards', *Environment and Urbanization* **19**(1): 39–64

Deangelo, B. J. and L. D. D. Harvey (1998) 'The jurisdictional framework for municipal action to reduce greenhouse gas emissions: Case studies from Canada, the USA and Germany', *Local Environment* **3**(2): 111–136

Delgado, P. M. (2008) 'Mega cities and climate change: Mexico City Climate Action Program 2008–2012, A presentation', www.lead.colmex.mx/docs/MARTHA%20DELGADO_megacities%20and%20climate%20change.ppt, last accessed 12 October 2010

Demetriades, J. and E. Esplen (2008) *Gender and Climate Change: Mapping the Linkages – Scoping Study on Knowledge and Gaps*, Institute of Development Studies, London

Department of Ecology, State of Washington (undated) *2008 Comprehensive Plan*, www.ecy.wa.gov/climatechange/2008CompPlan.htm, last accessed 6 October 2010

Department of Energy and Climate Change (undated) 'UK Climate Change Programme', www.decc.gov.uk/en/content/cms/what_we_do/change_energy/tackling_clima/programme/programme.aspx, last accessed 6 October 2010

Dhakal, S. (2004) *Urban Energy Use and Greenhouse Gas Emissions in Asian Mega-Cities: Policies for Sustainable Future*, Institute for Global Environmental Strategies, Hayama, Japan

Dhakal, S. (2006) 'Urban transportation and the environment in Kathmandu Valley, Nepal: Integrating global carbon concerns into local air pollution management', Institute for Global Environmental Strategies, Hayama, Japan

Dhakal, S. (2008) 'Climate change and cities: The making of a climate friendly future', in P. Droege (ed) *Urban Energy Transition: From Fossil Fuels to Renewable Power*, Elsevier Science, Oxford, pp173–192

Dhakal, S. (2009) 'Urban energy use and carbon dioxide emissions from cities in China and policy implications', *Energy Policy* **37**(11): 4208–4219

Díaz Palacios, J. and L. Miranda (2005) 'Concertación (reaching agreement) and planning for sustainable development in Ilo, Peru', in S. Bass, H. Reid, D. Satterthwaite and P. Steele (eds) *Reducing Poverty and Sustaining the Environment*, Earthscan, London, pp254–278

Disch, D. (2010) 'A comparative analysis of the "development dividend" of Clean Development Mechanism projects in six host countries', *Climate and Development* **2**(1): 50–64

Dlamini, D. (2006) 'Greening Soweto gets R2.2m cash injection', *Joburgnews*, http://joburgnews.co.za/2006/dec/dec7_soweto.stm, last accessed 14 October 2010

Dlugolecki, A. (ed) (2001) *Climate Change and Insurance*, Chartered Insurance Institute, London

Dodman, D. (2009) 'Blaming cities for climate change? An analysis of urban greenhouse gas emissions inventories', *Environment and Urbanization* **21**(1): 185–201

Dodman D. and D. Satterthwaite (2008) 'Institutional capacity, climate change adaptation and the urban poor', *IDS Bulletin* **39**(4): 67–74

Dodman, D. and D. Satterthwaite (2009) 'The costs of adapting infrastructure to climate change', in M. Parry, N. Arnell, P. Berry, D. Dodman, S. Fankhauser, C. Hope, S. Kovats, R. Nicholls, D. Satterthwaite, R. Tiffin and T. Wheeler (eds) *Assessing the Costs of Adaptation to Climate Change: A Review of the UNFCCC and Other Recent Estimates*, IIED and Grantham Institute, London, http://pubs.iied.org/pdfs/11501IIED.pdf, last accessed 7 December 2010, pp73–89

Dodman, D., J. Ayers and S. Huq (2009) 'Building resilience', in Worldwatch Institute (ed) *State of the World 2009: Into a Warming World*, Washington, DC

Dodman D., D. Mitlin, and J. C. Rayos Co (2010a) 'Victims to victors, disasters to opportunities: Community-driven responses to climate change in the Philippines' *International Development Planning Review* **32**(1): 1–26

Dodman, D., E. Kibona and L. Kiluma (2010b) 'Tomorrow is too late: Responding to social and climate vulnerability in Dar es Salaam, Tanzania', Unpublished case study prepared for the *Global Report on Human Settlements 2011*, www.unhabitat.org/grhs/2011

Donnelly, J. and J. Woodruff (2007) 'Intense hurricane activity over the past 5,000 years controlled by El Niño and the West African monsoon', *Nature* **447**(7143): 465–468

Donner, S., W. Skirving, C. Little, M. Oppenheimer and O. Hoegh-Guldberg (2005) 'Global assessment of coral bleaching and required rates of adaptation under climate change', *Global Change Biology* **11**(12): 2251–2265

Dossou, K. and B. Glehouenou-Dossou (2007) 'The vulnerability to climate change of Cotonou (Benin): The rise in sea level', *Environment and Urbanization* **19**(1): 65–79

Douglas, I., K. Alam, M. Maghenda, Y. McDonnell, L. McLean and J. Campbell (2008) 'Unjust waters: Climate change, flooding and the urban poor in Africa', *Environment and Urbanization* **20**(1): 187–206

Dubeux, C. and E. La Rovere (2010) 'The contribution of urban areas to climate change: The case study of São Paulo, Brazil', Unpublished case study prepared for the *Global Report on Human Settlements 2011*, www.unhabitat.org/grhs/2011

Easterling, W. E., B. H. Hurd and J. B. Smith (2004) 'Coping with global climate change: The role of adaptation in the United States', Pew Center on Global Climate Change, Arlington, VA

EEA (European Environment Agency) (2005) *Vulnerability and Adaptation to Climate Change in Europe*, EEA Technical Report No 7/2005, European Environment Agency Copenhagen, Denmark or Office for Official Publications of the EC, Luxembourg

Elsasser, H. and R. Bürki (2002) 'Climate change as a threat to tourism in the Alps', *Climate Research* **20**(3): 253–257

Elsner, J., J. Kossin and T. Jagger (2008) 'The increasing intensity of the strongest tropical cyclones', *Nature* **455**(7209): 92–95

Emanuel, K. (2005) 'Increasing destructiveness of tropical cyclones over the past 30 years', *Nature* **436**(7051): 686–688

Enarson, E. (2000) *Gender and Natural Disasters*, ILO, In Focus Programme on Crisis Response and Reconstruction, Working Paper 1, pp4–29

Enarson, E. and B. Phillips (2008) 'Invitation to a new feminist disaster sociology: Integrating feminist theory and methods', in B. Phillips and B. H. Morrow (eds) *Women and Disasters: From Theory to Practice*, International Research Committee on Disaster, Xlibris, pp41–74

Energy Cities (undated) 'Association', www.energy-cities.eu/-Association,8-, last accessed 6 October 2010

Energy Information Administration (undated) 'Official energy statistics from the US Government: Country analysis briefs', www.eia.doe.gov/emeu/cabs/index.html, last accessed 12 October 2010

Energy Planning Knowledge Base (undated) *CIS Tower Manchester*, www.pepesecenergyplanning.eu/archives/67, last accessed 14 October 2010

Environment Canada (2001) *Threats to Sources of Drinking Water and Aquatic Ecosystems Health in Canada*, National Water Research Institute Scientific Assessment Report Series No 1, National Water Resources Research Institute, Burlington, Ontario

Environmental Management Department (2003) *eThekwini Environmental Services Management Plan*, Unpublished report, eThekwini Metropolitan Municipality, South Africa

Enz, R. (2000) 'The S-Curve relation between per-capita income and insurance penetration', *Geneva Papers on Risk and Insurance: Issues and Practice* **25**(3): 396–406

EU (European Union) (undated) 'Covenant of mayors committed to local sustainable energy', www.eumayors.eu/about_the_covenant/index_en.htm, last accessed 6 October 2010

European Commission (2007) 'Carbon footprint: What it is and how to measure it', http://lca.jrc.ec.europa.eu/Carbon_footprint.pdf, last accessed 11 October 2010

European Commission (2009) *EU Action against Climate Change: Leading Global Action to 2020 and Beyond*, European Commission, Luxembourg, http://ec.europa.eu/environment/climat/pdf/brochures/post_2012_en.pdf, last accessed 6 October 2010

European Investment Bank (2010) 'The EIB and climate change', www.eib.org/attachments/strategies/clima_en.pdf, last accessed 6 October 2010

Ewing, R., K. Bartholomew, S. Winkelman, J. Walters and D. Chen (2008) *Growing Cooler: the Evidence on Urban Development and Climate Change*, Urban Land Institute, Washington, DC

Federation of Canadian Municipalities (2009) *Municipal Resources for Adapting to Climate Change*, Federation of Canadian Municipalities, Ottawa, Canada

Figueres, C. (2010) 'Address to the Swiss Re high-level adaptation event on risk and resiliency, New York, 20 September 2010', http://unfccc.int/files/press/statements/application/pdf/100920_speech_cf_adaptation_new_york.pdf

Flyvbjerg, B. (2002) 'Bringing power to planning research: One researcher's praxis story', *Journal of Planning Education and Research* **21**(4): 353–366

Foresight (2008) *Powering Our Lives: Sustainable Energy Management and the Built Environment*, Final project report, Government Office for Science, London

Forstall, R. L., R. P. Greene and J. B. Pick (2009) 'Which are the largest? Why lists of major urban areas vary so greatly', *Tijdschrift voor economische en sociale geografie* **100**(3): 277–297

Forster, P., V. Ramaswamy, P. Artaxo, T. Berntsen, R. Betts, D. W. Fahey, J. Haywood, J. Lean, D.C. Lowe, G. Myhre, J. Nganga, R. Prinn, G. Raga, M. Schulz and R. Van Dorland (2007) 'Changes in atmospheric constituents and in radiative forcing', in S. Solomon, D. Qin, M. Manning, Z. Chen, M. Marquis, K. B. Averyt, M. Tignor and H. L. Miller (eds) *Climate Change 2007: The Physical Science Basis*, Contribution of Working Group I to the Fourth Assessment Report of the Intergovernmental Panel on Climate Change, Cambridge University Press, Cambridge and New York

Fothergill, A., E. G. M. Maestas, and J. D. Darlington (1999) 'Race, ethnicity and disasters in the United States: A review of the literature', *Disasters* **23**(2): 156–174

Frayssinet, F. (2009) 'Casas ecológicamente correctas ... y blindadas', Inter-Press Service, http://ipsnoticias.net/nota.asp?idnews=92018, last accessed 14 October 2010

Freeman P. K. and K. Warner (2001) *Vulnerability of Infrastructure to Climate Variability: How Does this Affect Infrastructure Lending Policies?*, Report commissioned by the Disaster Management Facility of the World Bank and the ProVention Consortium, Washington, DC

Fricas, J. and Martz, T. (2007) 'The impact of climate change on water, sanitation, and diarrheal diseases in Latin America and the Caribbean', Population Reference Bureau, www.prb.org/Articles/2007/ClimateChangeinLatinAmerica.aspx, last accessed 29 October 2010

Frich, P., L. V. Alexander, P. Della-Marta, B. Gleason, M. Haylock, A. M. G. K. Tank and T. Peterson (2002) 'Observed coherent changes in climatic extremes during the second half of the twentieth century', *Climate Research* **19**(3): 193–212

Füssel, H.-M. (2009) 'Review and quantitative analysis of indices of climate change exposure, adaptive capacity, sensitivity, and impacts', Background Note developed for *World Development Report 2010: Development and Climate Change*, Potsdam Institute for Climate Impact Research, Potsdam, Germany

Gagnon-Lebrun, F. and S. Agrawala (2006) 'Progress on adaptation to climate change in developed countries: An analysis of broad trends', ENV/EPOC/GSP(2006)1/FINAL, OECD, Paris www.oecd.org/dataoecd/49/18/37178873.pdf, last accessed 14 October 2010

Garside, B., J. MacGregor and B. Vorley (2007) 'Miles better? How "fair miles" stack up in the sustainable supermarket', *IIED Sustainable Development Opinion*, December

Gavidia, J. (2006) 'Priority goals in Central America: The development of sustainable mechanisms for participation in local risk management', in *Milenio Ambiental, Journal of the Urban Environment Programme of the International Development Research Centre, Montevideo* **4**: 56–59

GCCA (Global Climate Change Alliance) (undated a) 'Background and objectives', www.gcca.eu/pages/14_2-Background-and-Objectives.html, last accessed 6 October 2010

GCCA (undated b) 'Beneficiary countries', www.gcca.eu/pages/41_2-GCCA-Beneficiaries.html, last accessed 6 October 2010

GCCA (undated c) 'Priority areas', www.gcca.eu/pages/30_2-Priority-areas.html, last accessed 6 October 2010

GEF (Global Environmental Facility) (undated) 'GEF-administered trust funds', www.thegef.org/gef/node/2042, last accessed 6 October 2010

Giannakopoulos, C. and B. E. Psiloglou (2006) 'Trends in energy load demand for Athens, Greece: Weather and non-weather related factors', *Climate Research* **31**: 97–108

Gibbs, D. (2000) 'Ecological modernization, regional economic development and regional development agencies', *Geoforum* **31**(1): 9–19

Gilbertson, T. and O. Reyes (2009) 'Carbon trading – how it works and why it fails', in L. Lohmann (ed) *Critical Currents*, Dag Hammarskjöld Foundation Publishers, Uppsala, www.tni.org/carbon-trade-fails, last accessed 6 October 2010

Girardet, H. (1998) 'Sustainable cities: A contradiction in terms?', in E. Fernandes (ed) *Environmental Strategies for Sustainable Development in Urban Areas: Lessons from Africa and Latin America*, Ashgate, Aldershot, pp193–209

GLA (Greater London Authority) (2007) *Action Today to Protect Tomorrow: The Mayor's Climate Change Action Plan*, Greater London Authority, London

GLA (2008) *The London Plan: Spatial Development Strategy for Greater London*, (Consolidated with alterations since 2004), Greater London Authority, London

GLA (2010) *The Draft Climate Change Adaptation Strategy for London – Public Consultation Draft*, Greater London Authority, City Hall, London, http://static.london.gov.uk/mayor/priorities/docs/Climate_change_adaptation_080210.pdf, last accessed 25 October 2010

Glaeser, E. and M. Kahn (2008) *The Greenness of Cities: Carbon Dioxide Emissions and Urban Development*, Harvard Kennedy School, Taubman Center for State and Local Government – Working Paper

Gomes, J., J. Nascimento and H. Rodrigues H (2008) 'Estimating local greenhouse gas emissions: A case study on a Portuguese municipality', *International Journal of Greenhouse Gas Control* **2**(1): 130–135

Gomez Martin, M. B. (2005) 'Weather, climate, and tourism: A geographical perspective', *Annals of Tourism Research* **32**(3): 571–591

Gore, C., P. Robinson and R. Stren (2009) 'Governance and climate change: Assessing and learning from Canadian cities', Paper prepared for the Fifth Urban Research Symposium, Cities and Climate Change: Responding to an Urgent Agenda, 28–30 June, Marseille, France

Gottdiener, M. and L. Budd (2005) *Key Concepts in Urban Studies*, Sage, London

Gouveia, N., S. Hajat and B. Armstrong (2003) 'Socioeconomic differentials in the temperature–mortality relationship in São Paulo, Brazil', *International Journal of Epidemiology* **32**(3): 390–397

Graham, S. and S. Marvin (2001) *Splintering Urbanism*, Routledge, London

Granberg, M. and I. Elander (2007) 'Local governance and climate change: Reflections on the Swedish experience', *Local Environment* **12**(5): 537–548

Graves, H. M. and M. C. Phillipson (2000) *Potential Implications of Climate Change in the Built Environment*, BRE Center for Environmental Engineering/BRE East Kilbride, FBE Report 2 December 2000, Construction Research Communications Ltd, UK

Greene, D. L., J. R. Kahn and R. C. Gibson (1999) 'Fuel economy rebound effect of U.S. household vehicles', *Energy Journal* **20**(3): 1–31

Greene, D. L., P. R. Boudreaux, D. J. Dean, W. Fulkerson, A. L. Gaddis, R. L. Graham, R. L. Graves, J. L. Hopson, P. Hughes, M.V. Lapsa, T. E. Mason, R. F. Standaert, T. J. Wilbanks and A. Zucker (2010) 'The importance of advancing technology to America's energy goals', *Energy Policy* **38**(8): 3886–3890

Greenpeace (2008) *China after the Olympics: Lessons from Beijing*, Greenpeace, Beijing

Grimm, N. B, S. H. Faeth, N. E. Golubiewski, C. L. Redman, J. Wu, X. Bai and J. M. Briggs (2008) 'Global change and the ecology of cities', *Science* **319**(5864): 756–760

Gulden, T. (2009) 'The security challenges of climate change: Who is at risk and why?', in M. Ruth and M. Ibarrarán (eds) *Distributional Impacts of Climate Change*, Edward Elgar Publishing, Cheltenham, Glos, UK

Gupta, J. and H. van Asselt (2006) 'Helping operationalise article 2: A trans disciplinary methodological tool for evaluating when climate change is dangerous', *Global Environmental Change* **16**(1): 83–94

Gupta, R. and S. Chandiwala (2009) 'A critical and comparative evaluation of approaches and policies to measure, benchmark, reduce and manage $CO_2$ emissions from energy use in the existing building stock of developed and rapidly-developing countries – case studies of UK, USA, and India', Paper prepared for the Fifth Urban Research Symposium, Cities and Climate Change: Responding to an Urgent Agenda, 28–30 June, Marseille, France, http://siteresources.worldbank.org/INTURBAN DEVELOPMENT/Resources/336387-1256566 800920/gupta.pdf, last accessed 12 October 2010

Haigh, C. and B. Vallely (2010) *Gender and the Climate Change Agenda: The Impacts of Climate Change on Women and Public Policy*, Women's Environmental Network, London

Haines, A., R. S. Kovats, D. Campbell-Lendrum and C. Corvalan (2006) 'Climate change and human health: Impacts, vulnerability, and mitigation', *Lancet* **367**(9528): 2101–2109

Hall, J. W., P. B. Sayers and R. J. Dawson (2005) 'National-scale assessment of current and future flood risk in England and Wales', *Natural Hazards* **36**(1–2): 147–164

Halweil, B. (2002) *Home Grown: The Case for Local Food in a Global Market*, Worldwatch Paper 163, www.worldwatch.org/system/files/EWP163.pdf, last accessed 12 October 2010

Hamilton, J., D. Maddison and R. Tol (2005) 'Climate change and international tourism: A simulation study', *Global Environmental Change* **15**: 253–266

Hammarby Sjöstad (2010) *The Hammarby Model*, Hammarby Sjöstad, Stockholm, Sweden

Hammer, S. (2009) 'Capacity to act: The critical determinant of local energy planning and program implementation', Paper prepared for the Fifth Urban Research Symposium, Cities and Climate Change: Responding to an Urgent Agenda, 28–30 June, Marseille, France

Handy, S., C. Xinyu and P. Mokhtarian (2005) 'Correlation or causality between the built environment and travel behaviour? Evidence from northern California', *Transportation Research Part D* **10**(6): 427–444

Hardoy, J. and G. Pandiella (2009) 'Urban poverty and vulnerability to climate change in Latin America', *Environment and Urbanization* **21**(1): 203–224

Hardoy, J. E., D. Mitlin and D. Satterthwaite (1992) *Environmental Problems in Third World Cities*, Earthscan, London

Hardoy, J. E., D. Mitlin and D. Satterthwaite (2001) *Environmental Problems in an Urbanizing World: Finding Solutions for Cities in Africa, Asia and Latin America*, Earthscan, London

Harlan, S. L., A. Brazel, L. Prashad, W. Stefanov and L. Larsen (2006) 'Neighbourhood microclimates and vulnerability to heat stress', *Social Science and Medicine* **63**(11): 2847–2863

Harrison, G. P. and H. W. Whittington (2002) 'Susceptibility of the Batoka Gorge hydroelectric scheme to climate change', *Journal of Hydrology* **264**(1–4): 230–241

Harvey, D. (1996) *Justice, Nature and the Geography of Difference*, Blackwell Publishers, Cambridge, MA

Harvey, L. (1993) 'Tackling urban $CO_2$ emissions in Toronto', *Environment* **35**(7): 16–44

Hasan, A. (2006) 'Orangi Pilot Project: The expansion of work beyond Orangi and the mapping of informal settlements and infrastructure', *Environment and Urbanization* **18**(2): 451–480

Hasan, A. (2010) *Participatory Development: The Story of the Orangi Pilot Project-Research and Training Institute and the Urban Resource Centre, Karachi, Pakistan*, Oxford University Press, Oxford, UK

Heede, R. (2006) *Aspen Greenhouse Gas Emissions 2004*, Climate Mitigation Services, City of Aspen, Aspen, CO

Held, D. and A. F. Hervey (2009) 'Democracy, climate change and global governance: Democratic agency and the policy menu ahead', Policy Network Paper, Policy Network, London, www.policy-network.net/uploadedFiles/Publications/Publications/Democracy%20climate%20change%20and%20global%20governance.pdf, last accessed 6 October 2010

Hemmati, M. (2008) 'Gender perspectives on climate change', Background paper to the Interactive Expert Panel, United Nations Commission on the Status of Women, 52nd session, 25 February–7 March 2008, New York, NY

Hendrickson, J. (undated) 'Energy use in the U.S. food system: A summary of existing research and analysis', Center for Integrated Agricultural Systems, UW-Madison, www.cias.wisc.edu/wp-content/uploads/2008/07/energyuse.pdf, last accessed 12 October 2010

Henry, S., B. Schoumaker and C. Beauchemin (2004) 'The impact of rainfall on the first out-migration: A multi-level event-history analysis in Burkina Faso', *Population and Environment* **25**(5): 423–460

Hertwich, E. and G. Peters (2009) 'Carbon footprint of nations: A global, trade-linked analysis', *Environmental Science and Technology* **43**(16): 6414–6420

Hintz, G. (2009) 'Maximizing the returns on adaptation investments: Flood prevention in Guyana', Unpublished draft case study prepared for the *Global Report on Human Settlements 2011*, www.unhabitat.org/grhs/2011

Hodgkinson, D., T. Burton, H. Anderson and L. Young (undated) 'The hour when the ship comes in: A convention for persons displaced by climate change', www.ccdpconvention.com/documents/Hour_When_Ship_Comes_In.pdf, last accessed 14 October 2010

Hodson, M. and S. Marvin (2007) 'Understanding the role of the national exemplar in constructing "strategic glurbanization"', *International Journal of Urban and Regional Research* **31**(2): 303–325

Hoffmann, M. J. (2011) *Climate Governance at the Crossroads: Experimenting with a Global Response after Kyoto*, Oxford University Press, London

Holden, E. and I. T. Norland (2005) 'Three challenges for the compact city as a sustainable urban form: Household consumption of energy and transport in eight residential areas in the greater Oslo region', *Urban Studies* **42**(12): 2145–2166

Holgate, C. (2007) 'Factors and actors in climate change mitigation: A tale of two South African cities', *Local Environment* **12**(5): 471–484

Hughes, T. P. (1989) 'The evolution of large technological systems', in W. E. Bijker, T. P. Hughes and T. Pinch (eds) *The Social Construction of Technological Systems*, MIT Press, Boston, MA

Hunt, A. and P. Watkiss (2007) 'Literature review on climate change impacts on urban city centres: Initial findings', Organisation for Economic Co-operation and Development (OECD), Paris

Huq, S., A. Rahman, M. Konate, Y. Sokona and H. Reid (2003) *Mainstreaming Adaptation to Climate Change in Least Developed Countries (LDCs)*, IIED, London

Ibarrarán, M. (2011) 'Climate's long-term impacts on Mexico's city urban infrastructure', Unpublished case study prepared for the *Global Report on Human Settlements 2011*, www.unhabitat.org/grhs/2011

ICLEI (Local Governments for Sustainability) (2006) *ICLEI International Progress Report – Cities for Climate Protection*, ICLEI, Oakland

ICLEI (2007) *Preparing for Climate Change: A Guidebook for Local, Regional, and State Governments*, Center for Science in the Earth System, University of Washington and King County, in association with ICLEI, Washington

ICLEI (2008) *Draft International Local Government GHG Emissions Analysis Protocol*, Release version 1.0, www.iclei.org/fileadmin/user_upload/documents/Global/Progams/GHG/LGGHGEmissionsProtocol.pdf, last accessed 12 October 2010

ICLEI (2010) *Cities in a Post-2012 Climate Policy Framework: Climate Financing for City Development? Views from Local Governments, Experts and Businesses*, ICLEI, Bonn, Germany, www.iclei.org/fileadmin/user_upload/documents/Global/Services/Cities_in_a_Post-2012_Policy_Framework-Climate_Financing_for_City_Development_ICLEI_2010.pdf, last accessed 14 October 2010

ICLEI (undated) 'Home', www.iclei.org/index.php?id=iclei-home, last accessed 28 October 2010

ICLEI Australia (2008) *Local Government Action on Climate Change: Measures Evaluation Report 2008*, Australian Government Department of Environment, Water, Heritage and the Arts and ICLEI, Melbourne, Australia, www.iclei.org/fileadmin/user_upload/documents/ANZ/Publications-Oceania/Reports/0812-CCPMeasuresReport08.pdf, last accessed 15 October 2010

ICLEI, UN-Habitat and UNEP (2009) *Sustainable Urban Energy Planning: A Handbook for Cities and Towns in Developing Countries*, UN-Habitat, Nairobi, www.unhabitat.org/pmss/listItemDetails.aspx?publicationID=2839, last accessed 6 October 2010

ICSU, UNESCO and UNU (International Council for Science, United Nations Educational, Scientific and Cultural Organization, and United Nations University) (2008) *Ecosystem Change and Human Well-Being Research and Monitoring Priorities Based on the Findings of the Millennium Ecosystem Assessment*, International Council for Science, Paris

IEA (International Energy Agency) (2008) *World Energy Outlook 2008*, International Energy Agency, Paris

IEA (2009) *Cities, Towns and Renewable Energy: Yes in My Front Yard*, International Energy Agency, Paris

IEA (2010) *Key World Energy Statistics*, OECD/IEA, Paris

IFRC (International Federation of Red Cross and Red Crescent Societies) (2010) *World Disasters Report 2010: Focus on Urban Risk*, IFRC, Geneva

IHDP (International Human Dimensions Programme on Global Environmental Change) (undated) 'Urbanization and global environmental change', www.ihdp.unu.edu/article/read/ugec, last accessed 8 December 2010

IIED (International Institute for Environment and Development) (2009) *The Adaptation Fund: A Model for the Future?*, Briefing paper, IIED London, www.iied.org/pubs/pdfs/17068IIED.pdf, last accessed 28 October 2010

Inter-American Development Bank (2007) 'IDB approves US$20 million for sustainable energy and climate change fund', www.iadb.org/news-releases/2007-08/english/idb-approves-us20-million-for-sustainable-energy-and-climate-change-fund-3987.html, last accessed 6 October 2010

IPCC (Intergovernmental Panel on Climate Change) (2001a) 'Climate change 2001: Synthesis report', in R. T. Watson, D. L. Albritton, T. Barker, I. A. Bashmakov, O. Canziani, R. Christ, U. Cubasch, O. Davidson, H. Gitay, D. Griggs, K. Halsnaes, J. Houghton, J. House, Z. Kundzewicz, M. Lal, N. Leary, C. Magadza, J. J. McCarthy, J. F. B. Mitchell, J. R. Moreira, M. Munasinghe, I. Noble, R. Pachauri, B. Pittock, M. Prather, R. G. Richels, J. B. Robinson, J.

Sathaye, S. Schneider, R. Scholes, T. Stocker, N. Sundararaman, R. Swart, T. Taniguchi, and D. Zhou (eds) *A Contribution of Working Groups I, II, and III to the Third Assessment Report of the Intergovernmental Panel on Climate Change*, Cambridge University Press, Cambridge

IPCC (2001b) *Climate Change 2001: Impacts, Adaptations and Vulnerability, Contribution of Working Group II to the Third Assessment Report of the Intergovernmental Panel on Climate Change*, Cambridge University Press, New York, NY

IPCC (2006) *2006 IPCC Guidelines for National Greenhouse Gas Inventories*, S. Eggleston, L. Buendia, K. Miwa, T. Ngara and K. Tanabe (eds), Institute for Global Environmental Strategies (IGES) Publishers, Kanagawa Japan, www.ipcc-nggip.iges.or.jp/public/2006gl/index.html, last accessed 12 October 2010

IPCC (2007a) 'Climate change 2007: Mitigation of climate change', in B. Metz, O. Davidson, P. Bosch, R. Dave and L. Meyer (eds) *Contribution of Working Group III to the Fourth Assessment Report of the Intergovernmental Panel on Climate Change*, Cambridge University Press, Cambridge and New York

IPCC (2007b) 'Climate change 2007: Synthesis report', in R. K. Pachauri and A. Reisinger (eds) *Contribution of Working Groups I, II, and III to the Fourth Assessment Report of the Intergovernmental Panel on Climate Change*, Cambridge University Press, Cambridge; www.ipcc.ch/publications_and_data/publications_ipcc_fourth_assessment_report_synthesis_report.htm, last accessed 13 October 2010

IPCC (2007c) 'Climate change 2007: The scientific basis' in S. Solomon, D. Qin, M. Manning, Z. Chen, M. Marquis, K. B. Averyt, M. Tignor and H. L. Miller (eds) *Contributions of Working Group I to the Fourth Assessment Report of the Intergovernmental Panel on Climate Change*, Cambridge University Press, Cambridge, www.ipcc.ch/publications_and_data/publications_ipcc_fourth_assessment_report_wg1_report_the_physical_science_basis.htm, last accessed 13 October 2010

IPCC (2007d) 'Summary for policymakers', in S. Solomon, D. Qin, M. Manning, Z. Chen, M. Marquis, K. B. Averyt, M. Tignor and H. L. Miller (eds) *Climate Change 2007: The Physical Science Basis, Contribution of Working Group I to the Fourth Assessment Report of the Intergovernmental Panel on Climate Change*, Cambridge University Press, Cambridge and New York, pp1–18

IPCC (2007e) 'Summary for policymakers', in B. Metz, O. R. Davidson, P. R. Bosch, R. Dave and L. A. Meyer (eds) *Climate Change 2007: Mitigation. Contribution of Working Group III to the Fourth Assessment Report of the Intergovernmental Panel on Climate Change*, Cambridge University Press, Cambridge and New York, pp1–23

IPCC (2007f) 'Climate change 2007: Impacts, adaptation and vulnerability', in M. L. Parry, O. F. Canziani, J. P. Palutikof, P. J. van der Linden and C. E. Hanson (eds) *Contributions of Working Group II to the Fourth Assessment Report of the Intergovernmental Panel on Climate Change*, Cambridge University Press, Cambridge, www.ipcc.ch/publications_and_data/publications_ipcc_fourth_assessment_report_wg2_report_impacts_adaptation_and_vulnerability.htm, last accessed 13 October 2010

IPCC (undated a) 'History', www.ipcc.ch/organization/organization_history.htm, last accessed 6 October 2010

IPCC (undated b) 'Structure', www.ipcc.ch/organization/organization_structure.htm, last accessed 6 October 2010

Iwugo, K. O., B. D'Arcy and R. Andoh (2003) 'Aspects of land-based pollution of an African coastal megacity of Lagos', Paper presented at Diffuse Pollution Conference, Dublin: 14/122–124, www.ucd.ie/dipcon/docs/theme14/theme14_32.PDF, last accessed 14 October 2010

Jabareen, Y. (2006) 'Sustainable urban forms: Their typologies, models, and concepts', *Journal of Planning Education and Research* **26**(1): 38–52

Jabeen, H., A. Allen and C. Johnson (2010) 'Built-in resilience: Learning from grassroots coping strategies to climate variability', *Environment and Urbanization* **22**(2): 415–432

Jacob, K., N. Edelblum and J. Arnold (2000) 'Risk increase to infrastructure due to sea level rise, Sector report: Infrastructure, the MEC regional assessment', in C. Rosenzweig and W. D. Solecki (eds) *Climate Change and a Global City: An Assessment of the Metropolitan East Coast (MEC) Region*, http://metroeast_climate.ciesin.columbia.edu/reports/infrastructure.pdf, last accessed 9 December 2010

Jessop, B. (2002) *The Future of the Capitalist State*, Polity, London

Jiang, L. and K. Hardee (2009) 'How do recent population trends matter to climate change', Population Action International Working Paper

Johnson, T., C. Alatorre, Z. Romo, F. Liu (2009) *Low-Carbon Development for Mexico*, World Bank, Washington, DC

Johnsson-Latham, G. (2007) *A Study on Gender Equality as Prerequisite for Sustainable Development: What We Know about the Extent to which Women Globally Live in a More Sustainable Way than Men, Leave a Smaller Ecological Footprint and Cause Less Climate Change*, Report to the Environment Advisory Council, Sweden, www.gendercc.net/fileadmin/inhalte/Dokumente/Actions/ecological_footprint__johnsson-latham.pdf, last accessed 12 October 2010

Jollands, N. (2008) *Cities and Energy: A Discussion Paper*, OECD International Conference Competitive Cities and Climate Change, OECD, Milan

Jones, R. and A. Rahman (2007) 'Community-based adaptation', *Tiempo* **64**: 17–19

Kahn Ribeiro, S., S. Kobayashi, M. Beuthe, J. Gasca, D. Greene, D. S. Lee, Y. Muromachi, P. J. Newton, S. Plotkin, D. Sperling, R. Wit, and P. J. Zhou (2007) 'Transport and its infrastructure', in B. Metz, O. R. Davidson, P. R. Bosch, R. Dave and L. A. Meyer (eds) *Climate Change 2007: Mitigation, Contribution of Working Group III to the Fourth Assessment Report of the Intergovernmental Panel on Climate Change*, Cambridge University Press, Cambridge and New York, pp323–385

Kalkstein, L. S. and R. E. Davies (1989) 'Weather and human mortality: An evaluation of demographic and interregional responses in the United States', *Annals of the Association of American Geographers* **79**(1): 44–64

Karekezi, S., L. Majoro and T. Johnson (2003) *Climate Change and Urban Transport: Priorities for the World Bank*, Working paper, World Bank, Washington, DC

Karol, J. and P. Suarez (2007) 'Adaptación al cambio climático, estructuras fractales y trampas discursivas: De la construcción del objeto a la construcción de la acción', *Medio Ambiente y Urbanizacion* **67**: 25–44

Kates, R., M. Mayfield, R. Torrie and B. Witcher (1998) 'Methods for estimating greenhouse gases from local places', *Local Environment* **3**(3): 279–298

Kehew, R. (2009) 'Projecting globally, planning locally: A progress report from four cities in developing countries', in World Meteorological Organization (ed) *Climate Sense*, Publication for the World Climate Conference-3, Climate Predictions and Information for Decision Making, Geneva, Switzerland, 31 August–4 September, Tudor Rose, Leicester, UK, pp181–184

Kennedy, C., A. Ramaswami, S. Carney and S. Dhakal (2009a) 'Greenhouse gas emission baselines for global cities and metropolitan regions', Paper prepared for the Fifth Urban Research Symposium, Cities and Climate Change: Responding to an Urgent Agenda, 28–30 June, Marseille, France

Kennedy, C., J. Steinberger, B. Gasson, Y. Hansen, T. Hillman, M. Havránek, D. Pataki, A. Phdungsilp, A. Ramaswami and G. Villalba Mendez (2009b) 'Greenhouse gas emissions from global cities', *Environmental Science and Technology* **43**(19): 7297–7302

Kern, K. and G. Alber (2008) 'Governing climate change in cities: Modes of urban climate governance in multi-level systems', in *Proceedings of the OECD Conference on Competitive Cities and Climate Change*, OECD, Paris

Kern, K. and H. Bulkeley (2009) 'Cities, Europeanization and multi-level governance: Governing climate change through transnational municipal networks', *JCMS: Journal of Common Market Studies* **47**(2): 309–332

Kingdon, J. (1984) *Agendas, Alternatives and Public Policies*, Little Brown and Company, Boston and Toronto

Kirshen, P., M. Ruth and W. Anderson (2006) 'Climate's long-term impacts on urban infrastructures and services: The case of Metro Boston', in M. Ruth, K. Donaghy and P. Kirshen (eds) *Regional Climate Change and Variability: Impacts and Responses*, Edward Elgar Publishers, Cheltenham, pp190–252

Klein, R., R. J. Nicholls and R. Thomalla (2003) 'Resilience to natural hazards: How useful is this concept?', *Global Environmental Change Part B: Environmental Hazards* **5**(1–2): 35–45

Kolleeny, J. (2006) 'With geometry and color, Onion Flats concocts a surprise mix of residences behind the brick shell of a former rag factory', *Architectural Record*, February issue

Kont, A., J. Jaagus and R. Aunap (2003) 'Climate change scenarios and the effect of sea-level rise for Estonia', *Global and Planetary Change* **36**(1–2): 1–15

Kousky, C. and S. H. Schneider (2003) 'Global climate policy: Will cities lead the way?', *Climate Policy* **3**(4): 359–372

Kovats, R. S. and R. Akhtar (2008) 'Climate, climate change and human health in Asian cities', *Environment and Urbanization* **20**(1): 165–176

Kumagai, M., K. Ishikawa and J. Chunmeng (2003) 'Dynamics and biogeochemical significance of the physical environment in Lake Biwa', *Lakes and Reservoirs: Research and Management* **7**(4): 345–348

Kumssa, A. and J. F. Jones (2010) 'Climate change and human security in Africa', *International Journal of Sustainable Development and World Ecology* **17**: 453–461

Kunreuther, H., N. Novemsky and D. Kahneman (2001) 'Making low probabilities useful', *Journal of Risk Uncertainty* **23**(2): 103–120

Kutzbach, M. (2009) 'Motorization in developing countries: Causes, consequences, and effectiveness of policy options', *Journal of Urban Economics* **65**(2): 154–166

La Rovere, E. L, C. B. Dubeux, A. Oliveira da Costa, C. A. Pimenteira, F. Frangetto, F. E. Mendes, J. M. G. Monteiro, L. B. Oliveira, N. Baptista and S. K. Ribero (2005) *Inventário de Emissões de Gases de Efeito Estufa do Município de São Paulo [Inventory of Greenhouse Gases Emissions from São Paulo City]*, http://ww2.prefeitura.sp.gov.br/arquivos/secretarias/meio_ambiente/Sintesedoinventario.pdf, last accessed 12 October 2010

Lagos State Government (2010) *Lagos State Urban and Regional Planning and Development Bill 2010* (Law to provide for the administration of physical planning, urban development, urban regeneration and building control in Lagos State and for connected purposes), Gaz. Law 2010 Fashola, 6th Assembly, House of Assembly, Lagos State, www.lagosstate.gov.ng/uploads/gallery_m2whf09cuhmm71uru4qq.pdf, last accessed 15 October 2010

Langer, N. (2004) 'Natural disasters that reveal cracks in our social foundation', *Educational Gerontology* **30**(4): 275–285

Lasco, R., L. Lebel, A. Sari, A. P. Mitra, N. H. Tri, O. G. Ling and A. Contreras (2007) *Integrating Carbon Management into Development Strategies of Cities – Establishing a Network of Case Studies of Urbanisation in Asia Pacific*, Final Report for the APN Project 2004-07-CMY-Lasco

LCCA (London Climate Change Agency) (2007) *Moving London towards a Sustainable Low-Carbon City: An Implementation Strategy*, London Climate Change Agency, London

Le Treut, H., R. Somerville, U. Cubasch, Y. Ding, C. Mauritzen, A. Mokssit, T. Peterson and M. Prather (2007) 'Historical overview of climate change', in S. Solomon, D. Qin, M. Manning, Z. Chen, M. Marquis, K. B. Averyt, M. Tignor and H. L. Miller (eds) *Climate Change 2007: The Physical Science Basis Contribution of Working Group I to the Fourth Assessment Report of the Intergovernmental Panel on Climate Change*, Cambridge University Press, Cambridge and New York

Leape, J. (2006) 'The London congestion charge', *Journal of Economic Perspectives* **20**(4): 157–176

Lebel, L., P. Garden, M. R. N. Banaticla, R. D. Lasco, A. Contreras, A. P. Mitra, C. Sharma, H. T. Nguyen, G. L. Ooi and A. Sari (2007) 'Integrating carbon management into the development strategies of urbanizing regions in Asia: Implications of urban function, form, and role', *Journal of Industrial Ecology* **11**(2): 61–81

Lee, D. H. (1980) 'Seventy-five years of searching for a heat index', *Environmental Research* **22**(2): 331–356

Legros, G, I. Havet, N. Bruce and S. Bonjour (2009) *The Energy Access Situation in Developing Countries: A Review Focusing on the Least Developed Countries and Sub-Saharan Africa*, WHO and UNDP, New York, NY

Lehner, B., G. Czisch and S. Vassolo (2005) 'The impact of global change on the hydropower potential of Europe: A model-based analysis', *Energy Policy* **33**(7): 839–855

Lewsey, C., G. Cid and E. Kruse (2004) 'Assessing climate change impacts on coastal infrastructure in the Eastern Caribbean', *Marine Policy* **28**(5): 393–409

Linder, K. P. (1990) 'National impacts of climate change on electric utilities', in J. B. Smith and D. A. Tirpak (eds) *The Potential Effects of Global Warming on the United States*, Environmental Protection Agency, Washington, DC

Lindseth, G. (2004) 'The Cities for Climate Protection Campaign (CCPC) and the framing of local climate policy', *Local Environment* **9**(4): 325–336

López-Marrero, T. and P. Tschakert (forthcoming) 'From theory to practice: Building more resilient communities in flood-prone areas', *Environment and Urbanization* **23**(1)

Lungo, M. (2007) 'Gestión de Riesgos nacional y local', in C. Clarke and C. Pineda (eds) *Riesgo y Desastres. Su Gestión Municipal en Centroamérica*, Publicaciones especiales sobre el Desarrollo no 3, Inter-American Development Bank, Washington, DC, pp19–27

Mabasi, T. (2009) 'Assessing the vulnerability, mitigation and adaptation to climate change in Kampala City', Paper presented at the Fifth Urban Research Symposium, Cities and Climate Change: Responding to an Urgent Agenda, 28–30 June, Marseille, France

Macchi, M. (2008) *Indigenous and Traditional Peoples and Climate Change*, IUCN Issues Paper, http://cmsdata.iucn.org/downloads/indigenous_ peoples_climate_change.pdf, last accessed 13 October 2010

Mackie, P. (2005) 'The London congestion charge: A tentative economic appraisal. A comment on the paper by Prud'homme and Bocajero', *Transport Policy* **12**(3): 288–290

Magrin, G., C. Gay García, D. Cruz Choque, J.C. Giménez, A. R. Moreno, G. J. Nagy, C. Nobre and A. Villamizar (2007) 'Latin America', in M. L. Parry, O. F. Canziani, J. P. Palutikof, P. J. van der Linden and C. E. Hanson (eds) *Climate Change 2007: Impacts, Adaptation and Vulnerability. Contribution of Working Group II to the Fourth Assessment Report of the Intergovernmental Panel on Climate Change*, Cambridge University Press, Cambridge, UK, pp581–615

Manuel-Navarrete, D., M. Pelling and M. Redclift (2008) 'Governance as process: Powerspheres and responses to climate change in the Mexican Caribbean', Unpublished paper, King's College London, London

Markham, V. (2009) *U.S. Population, Energy and Climate Change*, Center for Environment and Population, New Canaan, Connecticut, www.cepnet.org/ documents/USPopulationEnergyandClimateChange ReportCEP.pdf, last accessed 12 October 2010

Marshall, J., T. McKone, E. Deakin and W. Nazaroff (2005) 'Inhalation of motor vehicle emissions: Effects of urban population and land area', *Atmospheric Environment* **39**(2): 283–295

Martine, G. (2009) 'Population dynamics and policies in the context of global climate change', in J. Guzmán, G. Martine, G. McGranahan, D. Schensul and C. Tacoli (eds) *Population Dynamics and Climate Change*, UNFPA/IIED, pp9–30

Martinot, E., M. Zimmerman, M. van Staden and N. Yamashita (2009) *Global Status Report on Local Renewable Energy Policies*, 12 June working draft, Collaborative report by REN21 Renewable Energy Policy Network, Institute for Sustainable Energy Policies (ISEP) and ICLEI Local Governments for Sustainability, http://www.ren21.net/pdf/ REN21_LRE2009_Jun12.pdf, last accessed 15 October 2010

Massey, D., W. Axinn and D. Ghimire (2007) *Environmental Change and Out-Migration: Evidence from Nepal*, Population Studies Center Research Report 07-615, Institute for Social Research, University of Michigan, Ann Arbor, MI

Maxwell, D., C. Levin, M. Armar-Klemesu, M. Ruel, S. Morris and C. Ahiadeke (1998) *Urban Livelihoods and Food and Nutrition Security in Greater Accra, Ghana*, IFPRI, Washington, DC

Mayor of London (2007) *Action Today to Protect Tomorrow: The Mayor's Climate Change Action Plan*, Greater London Authority, London

McGranahan, G. and D. Satterthwaite (2000) 'Environmental health or ecological sustainability? Reconciling the brown and green agendas in urban development', in C. Pugh (ed) *Sustainable Cities in Developing Countries*, Earthscan, London, pp73–90

McGranahan, G., P. Jacobi, J. Songsore, C. Surjadi and M. Kjellen (2001) *The Citizens at Risk: From Urban Sanitation to Sustainable Cities*, Earthscan, London

McGranahan, G., P. Marcotullio, D. Balk, I. Douglas, T. Elmqvist, W. Rees, D. Satterthwaite, D. Songsore, H. Zlotnick, J. Eades, E. Ezcurra, A. Whyte, X. Bai, H. Imura and H. Shirakawa (2005) 'Urban systems', in *Millennium Ecosystem Assessment: Ecosystems and Human Well-Being, Current State and Trends: Findings of the Condition and Trends Working Group*, Island Press, Washington, DC, pp795–825

McGranahan, G., D. Balk and B. Anderson (2007) 'The rising tide: Assessing the risks of climate change and human settlements in low-elevation coastal zones', *Environment and Urbanization* **19**(1): 17–37

McGray, D. (2007) 'Pop-up cities: China builds a bright green metropolis', *Wired Magazine*, issue 15.05

McGregor, D., D. Simon and D. Thompson (eds) (2006) *The Peri-Urban Interface: Approaches to Sustainable Natural and Human Resource Use*, Earthscan, London

McLeman, R. and B. Smit (2005) 'Assessing the security implications of climate change-related migration', Paper prepared for International Workshop on Human Security and Climate Change, Oslo, Norway, 21–23 June

McMichael, A., R. Woodruff, P. Whetton, K. Hennessy, N. Nicholls, S. Hales, A. Woodward and T. Kjellstrom (2003) *Human Health and Climate Change in Oceania: A Risk Assessment 2002*, Commonwealth Department of Health and Ageing, Canberra, Australia

Meehl, G. and C. Tebaldi (2004) 'More intense, more frequent, and longer lasting heat waves in the 21st century', *Science* **305**(5686): 994–997

Mendelsohn, R., W. Morrison, M. E. Schlesinger and N. G. Andronova (2000) 'Country-specific market impacts from climate change', *Climatic Change* **45**(3–4): 553–569

Menegat, R. (2002) 'Participatory democracy and sustainable development: Integrated urban environmental management in Porto Alegre, Brazil', *Environment and Urbanization* **14**(2): 181–206

Merriam-Webster Dictionary (undated) 'Poleward', www.merriam-webster.com/dictionary/poleward, last accessed 29 October 2010

Meusel, D. and W. Kirch (2005) 'Lessons to be learned from the 2002 floods in Dresden, German', in B. Menne, R. Bertollini and W. Kirch (eds) *Extreme Weather Events and Public Health Response*, Springer-Verlag, Berlin, pp175–184

Mhapsekar, J. (2010) 'Parisar Vikas (Presentation)', Stree Mukti Sanghatana, Mumbai

Millennium Ecosystem Assessment (2005) *Ecosystems and Human Well Being*, Millennium Ecosystem Assessment, World Resources Institute, Island Press, Washington, DC

Mills, E. (2005) 'Insurance in a climate of change', *Science* **309**(5737): 1040–1044

Ministério da Ciência e Tecnologia (2004) *Comunicação Nacional Inicial do Brasil à Convenção do Clima*, Ministério da Ciência e Tecnologia, Brazil

Ministry of Foreign Affairs of Denmark (2006) *Aid Management Guidelines Glossary*, 2nd edition,

http://amg.um.dk/NR/rdonlyres/
3845FDB0-028B-4866-AB9B-1DCB7E83A905/0/
AMGGlossaryFeburary2006finaldoc.pdf, last
accessed 14 October 2010

Mitlin, D. (2008) 'With and beyond the state: Co-
production as a route to political influence, power
and transformation for grassroots organizations',
*Environment and Urbanization* **20**(2): 339–360

Mitlin, D. and D. Dodman (forthcoming) 'Questioning
community-based adaptation', Paper in preparation

Monni, S. and F. Raes (2008) 'Multilevel climate policy:
The case of the European Union, Finland and
Helsinki', *Environmental Science & Policy* **11**(8):
743–755

Monstadt, J. (2009) 'Conceptualizing the political
ecology of urban infrastructures: Insights from
technology and urban studies', *Environment and
Planning A* **41**(8): 1924–1942

Moore, M. (2008) 'China's pioneering eco-city of
Dongtan stalls', *The Daily Telegraph*, 18 October
2008

Moravcsik, A. and B. Botos (2007) 'Tatabanya: Local
participation and physical regeneration of derelict
areas', Presentation given in Krakow, Poland

Moser, C. (2008) 'Assets and livelihoods: A framework
for asset-based social policy', in C. Moser and A. A.
Dani (eds) *Assets, Livelihoods and Social Policy*,
International Bank for Reconstruction and
Development/World Bank, Washington, DC, pp43–81

Moser, C. and D. Satterthwaite (2008) *Towards Pro-Poor
Adaptation to Climate Change in the Urban Centres of
Low- and Middle-Income Countries*, IIED Human
Settlements Discussion Paper Series, Climate Change
and Cities 3, London

Moser, C. and D. Satterthwaite (2010) 'Toward pro-poor
adaptation to climate change in the urban centers of
low- and middle-income countries', in R. Mearns and
A. Norton (eds) *Social Dimensions of Climate Change:
Equity and Vulnerability in a Warming World*, World
Bank, Washington, DC, pp231–258

Moser, S. C. (2010) 'Now more than ever: The need for
more societally relevant research on vulnerability and
adaptation to climate change', *Applied Geography*
**30**(4): 464–474

Mukheibir, P. and G. Ziervogel (2007) 'Developing a
municipal adaptation plan (MAP) for climate change:
The City of Cape Town', *Environment and
Urbanization* **19**(1): 143–158

Murphy, J. (2000) 'Ecological modernization', *Geoforum*
**31**(1): 1–8

Murray, C. J. and A. D. Lopez (1996) *The Global Burden
of Disease: A Comprehensive Assessment of Mortality
and Disability from Diseases, Injuries and Risk Factors
in 1990 and Projected to 2020*, Harvard University
Press, Boston, MA

Murray, D. (2005) *Oil and Food: A Rising Security
Challenge*, Earth Policy Institute, Archived,
www.energybulletin.net/node/6052, last accessed
13 October 2010

Myers, N. (1997) 'Environmental refugees', *Population
and Environment* **19**(2): 167–182

Myers, N. (2005) 'Environmental refugees: An emergent
security issue', Paper presented at the 13th Economic
Forum, Prague, Czech Republic, 23–27 May 2005

National Drought Mitigation Center (2010) *Drought
Monitor: State-of-the-Art Blend of Science and
Subjectivity*, http://drought.unl.edu/dm/index.html,
last accessed 13 October 2010

National Weather Service (undated) *Glossary of National
Hurricane Center Terms*, www.nhc.noaa.gov/
aboutgloss.shtml#e, last accessed 29 October 2010

Nchito, W. S. (2007) 'Flood risk in unplanned settlements in Lusaka', *Environment and Urbanization*
**19**(2): 539–551

Neuman, M. (2005) 'The compact city fallacy', *Journal of
Planning Education and Research* **25**(1): 11–26

Neumayer, E. and T. Plümper (2007) 'The gendered
nature of natural disasters: The impact of catastrophic
events on the gender gap in life expectancy,
1981–2002', *Annals of the American Association of
Geographers* **97**(3): 551–566

*New Scientist* (2009) 'Timeline: Climate change', *New
Scientist*, www.newscientist.com/article/
dn9912-timeline-climate-change.html, last accessed
28 October 2010

Newcastle City Council (2008) *City Consumption*,
www.newcastle.nsw.gov.au/environment/climate_
cam/climatecam, last accessed 15 October 2010

Newman, P. (2006) 'The environmental impact of cities',
*Environment and Urbanization* **18**(2): 275–295

Newman, P. and J. Kenworthy (1989) 'Gasoline consumption and cities: A comparison of US cities with a
global survey', *Journal of the American Planning
Association* **55**(1): 24–37

Newman, P. and J. Kenworthy (1999) *Sustainability and
Cities: Overcoming Automobile Dependence*, Island
Press, Washington, DC

Nicholls, R. J. and R. Tol (2007) 'Impacts and responses
to sea-level rise: A global analysis of the SRES scenarios over the twenty-first century', *Philosophical
Transactions of the Royal Society A* **364**: 1073–1095

Nicholls, R. J., F. M. J. Hoozemans and M. Marchand
(1999) 'Increasing flood risk and wetland losses due
to global sea-level rise: Regional and global analyses',
*Global Environmental Change* **9**(1001): S69–S87

Nicholls, R. J., S. Hanson, C. Herweijer, N. Patmore, S.
Hallegatte, J. Corfee-Morlot, J. Chateau and R. Muir-
Wood (2008) *Ranking Port Cities with High Exposure
and Vulnerability to Climate Extremes: Exposure
Estimates*, OECD Environment Working Papers, No 1,
OECD Publishing, Paris, France

Nickson, A. (2010) 'Cities and climate change:
Adaptation in London, UK', Unpublished case study
prepared for the *Global Report on Human Settlements
2011*, www.unhabitat.org/grhs/2011

Nodvin, S. C. and K. Vranes (2010), 'Global warming', in
C. J. Cleveland (ed) *Encyclopedia of Earth*,
Environmental Information Coalition, National
Council for Science and the Environment,
Washington, DC, www.eoearth.org/article/
global_warming, last accessed 21October 2010

Nordhaus, W. D. (2006) 'The economics of hurricanes in
the United States', National Bureau of Economic
Research (NBER) Working paper, http://papers.
nber.org/papers/w12813, last accessed
9 December 2010

Norman, J., H. L. MacLean and C. A. Kennedy (2006)
'Comparing high and low residential density: Life-
cycle analysis of energy use and greenhouse gas
emissions', *Journal of Urban Planning and
Development* **132**(1): 10–21

NRC (National Research Council) (2009) *America's
Energy Future: Technologies and Transformation*, US
National Academies of Science/National Research
Council, National Academies Press, Washington, DC

NRC (2010) *Adapting to the Impacts of Climate Change*,
US National Academies of Science/National Research
Council, National Academies Press, Washington, DC

O'Brien, K., L. Sygna and J. E. Haugen (2004)
'Vulnerable or resilient? A multi-scale assessment of
climate impacts and vulnerability in Norway',
*Climatic Change* **64**(1–2): 193–225

OCHA and IDMC (United Nations Office for the Coordination of Humanitarian Affairs and the Internal Displacement Monitoring Centre) (2009) 'Monitoring disaster displacement in the context of climate change', www.internal-displacement.org/8025708F004BE3B1/(httpInfoFiles)/12E8C7224C2A6A9EC125763900315AD4/$file/monitoring-disaster-displacement.pdf, last accessed 9 December 2010

OECD (Organisation for Economic Co-operation and Development) (1995) *Urban Energy Handbook: Good Local Practice*, OECD Publication Services, Paris

OECD (2008) 'Competitive cities in a changing climate: Introductory issue paper', OECD International Conference, Competitive Cities and Climate Change, 2nd Annual Meeting of the OECD Roundtable Strategy for Urban Development, OECD Directorate for Public Governance and Territorial Development, Milan, Italy

OECD (2009) *OECD's Recent Work on Climate Change*, OECD Publications, Paris, France, www.oecd.org/dataoecd/60/40/41810213.pdf, last accessed 6 October 2010

OECD (2010) *Cities and Climate Change*, OECD Publishing, http://dx.doi.org/10.1787/9789264091375-en, last accessed 10 December 2010

Office of the Deputy Prime Minister (2003) 'Sustainable communities: Building for the future', Office of the Deputy Prime Minister, London, www.communities.gov.uk/publications/communities/sustainablecommunitiesbuilding, last accessed 15 October 2010

Office of the Governor, California, US (undated) 'Gov. Schwarzenegger announces first-of-its-kind climate action coalition at the governors' Global Climate Summit 3', http://gov.ca.gov/press-release/16497/, last accessed 9 December 2010

Oke, T. R. (1982) 'The energetic basis of the urban heat island', *Quarterly Journal of the Royal Meteorological Society* **108**(455): 1–24

Oresanya, O. (2009) 'Climate change and waste management', Presentation, Lagos Waste Management Authority, Lagos

Osbahr, H. and T. Roberts (2007) *Climate Change and Development in Africa: Policy Frameworks and Development Interventions for Effective Adaptation to Climate Change*, Report of workshop 12 March, University of Oxford, http://african-environments.ouce.ox.ac.uk/events/2007/070312workshopreport.pdf, last accessed 14 October 2010

Overpeck, J., B. Otto-Bliesner, G. Miller, D. Muhs, R. Alley and J. Kiehl (2006) 'Paleoclimatic evidence for future ice-sheet instability and rapid sea-level rise', *Science* **311**(5768): 1747–1750

Overstreet, S. and B. Burch (2009) 'Mental health status of women and children following Hurricane Katrina', in B. Willinger (ed) *Hurricane Katrina and the Women of New Orleans*, Newcomb College Center for Research on Women, New Orleans

Owens, S. (1992) 'Energy, environmental sustainability and land-use planning', in M. Breheny (ed) *Sustainable Development and Urban Form*, Pion, London, pp79–105

Oxfam (2005) *The Tsunami's Impact on Women*, Oxfam, Oxford, UK

PADECO (2009a) *Cities and Climate Change: Draft Comprehensive Report*, Report prepared for the World Bank, 30 April

PADECO (2009b) *Cities and Climate Change: Literature Review*, Report prepared for the World Bank, 30 April

Parker, L., (2006) *Climate Change: The European Union's Emissions Trading System (EU ETS)*, CRS Report for Congress, www.usembassy.it/pdf/other/RL33581.pdf, last accessed 6 October 2010

Parry, M., O. Canziani, J. Palutikof, N. Adger, P. Aggarwal, S. Agrawala, J. Alcamo, A. Allali, O. Anisimov, N. Arnell, M. Boko, T. Carter, G. Casassa, U. Confalonieri, R. Cruz, E. de Alba Alcaraz, W. Easterling, C. Field, A. Fischlin, B. Fitzharris, C. García, H. Harasawa, K. Hennessy, S. Huq, R. Jones, L. Bogataj, D. Karoly, R. Klein, Z. Kundzewicz, M. Lal, R. Lasco, G. Love, X. Lu, G. Magrín, L. Mata, B. Menne, G. Midgley, N. Mimura, M. Mirza, J. Moreno, L. Mortsch, I. Niang-Diop, R. Nicholls, B. Nováky, L. Nurse, A. Nyong, M. Oppenheimer, A. Patwardhan, P. Lankao, C. Rosenzweig, S. Schneider, S. Semenov, J. Smith, J. Stone, J. van Ypersele, D. Vaughan, C. Vogel, T. Wilbanks, P. Wong, S. Wu and G. Yohe (2007a) 'Technical summary', in M. Parry, O. Canziani, J. Palutikof, P. van der Linden and C. Hanson (eds) *Climate Change 2007: Impacts, Adaptation and Vulnerability; Contribution of Working Group II to the Fourth Assessment Report of the Intergovernmental Panel on Climate Change*, Cambridge University Press, New York, NY, pp23–78

Parry, M. L., O. F. Canziani, J. P. Palutikof, P. J. van der Linden and C. E. Hanson (eds) (2007b) *Climate Change 2007: Impacts, Adaptation and Vulnerability. Contribution of Working Group II to the Fourth Assessment Report of the Intergovernmental Panel on Climate Change*, Cambridge University Press, Cambridge, www.ipcc.ch/publications_and_data/ar4/wg2/en/contents.html, last accessed 14 October 2010

Parry, M., N. Arnell, P. Berry, D. Dodman, S. Fankhauser, C. Hope, S. Kovats, R. Nicholls, D. Satterthwaite, R. Tiffin and T. Wheeler (2009) *Assessing the Costs of Adaptation to Climate Change: A Review of the UNFCCC and Other Recent Estimates*, International Institute for Environment and Development/Grantham Institute for Climate Change, London

Parshall, L., S. Hammer and K. Gurney (2009) 'Energy consumption and $CO_2$ emissions in urban counties in the United States with a case study of the New York Metropolitan Area', Paper prepared for the Fifth Urban Research Symposium, Cities and Climate Change: Responding to an Urgent Agenda, 28–30 June, Marseille, France

Parshall, L., M. Haraguchi, C. Rosenzweig and S. A. Hammer (2010) 'The contribution of urban areas to climate change: New York City case study', Unpublished case study prepared for the *Global Report on Human Settlements 2011*, www.unhabitat.org/grhs/2011

Patt, A., A. Dazé and P. Suarez (2009) 'Gender and climate change vulnerability: What's the problem, what's the solution?', in M. Ruth and M. E. Ibarrarán (eds) *Distributional Impacts of Climate Change and Disasters: Concepts and Cases*, Edward Elgar Publishers, Cheltenham, UK

Patz, J. A., D. Campbell-Lendrum, T. Holloway and J. A. Foley (2005) 'Impact of regional climate change on human health', *Nature* **438**(7066): 310–317

Paul, B. (2009) 'Why relatively fewer people died? The case of Bangladesh's Cyclone Sidr', *Natural Hazards* **50**(2): 289–304

Pauzner, S. (2009) 'Tel Aviv to get 1st ecological housing project', *Ynetnews*, 3 September

Pearce, F. (2009) 'Greenwash: The dream of the first eco-city was built on a fiction', *The Guardian*, 23 April

Pelling, M. (1997) 'What determines vulnerability to floods: A case study in Georgetown, Guyana', *Environment and Urbanization* **9**(1): 203–226

Pelling, M. (1998) 'Participation, social capital and vulnerability to urban flooding in Guyana', *Journal of International Development* **10**(4): 469–486

Pelling, M. (2005) 'Enhancing safety and security', Unpublished issues paper prepared for the *Global Report on Human Settlements 2007*, www.unhabitat.org/grhs/2007

Pelling, M., D. Manuel-Navarrete and M. Redclift (2008) 'Urban transformation and social learning for climate proofing on Mexico's Caribbean coast', Unpublished paper, King's College London, London

People's Republic of China (2007) *China's National Climate Change Programme*, National Development and Reform Commission, People's Republic of China, www.ccchina.gov.cn/WebSite/CCChina/UpFile/File188.pdf, last accessed 28 October 2010

Perelet, R., S. Pegov and M. Yulkin (2007) 'Climate change, Russia country paper', Background paper for the *UN 2007/2008 Human Development Report*, http://hdr.undp.org/en/reports/global/hdr2007-8/papers/Perelet_Renat_Pegov_Yulkin.pdf, last accessed 7 December 2010

Peters, G. (2008) 'From production-based to consumption-based national emission inventories', *Ecological Economics* **65**(1): 13–23

Petterson, J., L. Stanley, E. Glazier and J. Philipp (2006) 'A preliminary assessment of social and economic impacts associated with Hurricane Katrina', *American Anthropologist* **108**(4): 643–670

Pew Center on Global Climate Change (2008) 'Summary: India's national action plan on climate change', www.pewclimate.org/international/country-policies/india-climate-plan-summary/06-2008, last accessed 27 October 2010

Pew Center on Global Climate Change (undated) 'Climate action plans', www.pewclimate.org/what_s_being_done/in_the_states/action_plan_map.cfm, last accessed 28 October 2010

Pieterse, E. (2008) *City Futures: Confronting the Crisis of Urban Development*, Zed Books, London and New York, and UCT Press, Cape Town

Pradhan, E. K., K. P. West, J. Katz, S. C. LeClerq, S. K. Khatry and S. R. Shrestha (2007) 'Risk of flood-related mortality in Nepal', *Disasters* **31**(1): 57–70

Prasad, N., F. Ranghieri, F. Shah, Z. Trohanis, E. Kessler and R. Sinha (2009) *Climate Resilient Cities: A Primer on Reducing Vulnerabilities to Disaster*, World Bank, Washington, DC

Preston, B. L. and R. N. Jones (2006) *Climate Change Impacts on Australia and the Benefits of Early Action to Reduce Global Greenhouse Emissions*, Consultancy report for the Australian Business Roundtable on Climate Change, CSIRO Marine and Atmospheric Research, Melbourne

PricewaterhouseCoopers (2010) *Carbon Disclosure Project 2010: Global 500 Report*, www.cdproject.net/CDPResults/CDP-2010-G500.pdf, last accessed 6 December 2010

Prowse, M. and L. Scott (2008) 'Assets and adaptation: An emerging debate' *IDS Bulletin* **39**(4): 42–52

Prud'homme, R. and J. P. Bocarejo (2005) 'The London congestion charge: A tentative economic appraisal', *Transport Policy* **12**(3): 279–287

Puppim de Oliveira, J. A. (2009) 'The implementation of climate change related policies at the subnational level: An analysis of three countries', *Habitat International* **33**(3): 253–259

Qi, Y., L. Ma, H. Zhang and H. Li (2008) 'Translating a global issue into local priority: China's local government response to climate change', *Journal of Environment Development* **17**(4): 379–400

Rabe, B. (2007) 'Beyond Kyoto: Climate change policy in multilevel governance systems', *Governance* **20**(3): 423–444

Rabinovitch, J. (1992) 'Curitiba: Towards sustainable urban development', *Environment and Urbanization* **4**(2): 62–73

Rain, D. R., R. Engstrom, C. Ludlow and S. Antos (2011) 'Accra: A vulnerable coastal West African city', Unpublished case study prepared for the *Global Report on Human Settlements 2011*, www.unhabitat.org/grhs/2011

Raleigh, C., L. Jordan and I. Salehyan (2008) 'Assessing the impact of climate change on migration and conflict', Paper commissioned by the World Bank Group for the Social Dimensions of Climate Change workshop, Washington, DC, 5–6 March 2008

Rashid, S. (2000) 'The urban poor in Dhaka City: Their struggles and coping strategies during the floods of 1998', *Disasters* **24**(3): 240–253

Räty, R. and A. Carlsson-Kanyama (2010) 'Energy consumption by gender in some European countries', *Energy Policy* **38**(1): 646–649

Raupach, M., G. Marland, P. Ciais, C. Le Quéré, J. Canadell, G. Klepper and C. Fields (2007) 'Global and regional drivers of accelerating $CO_2$ emissions', *PNAS* **104**(24): 10288–10293

Rees, W. (1992) 'Ecological footprints and appropriated carrying capacity: What urban economics leaves out', *Environment and Urbanization* **4**(2): 121–130

Rees, W. and M. Wackernagel (1998) *Our Ecological Footprint: Reducing Human Impact on the Earth*, New Society Publishers, Gabriola Island, British Columbia, Canada

Regional Greenhouse Gas Initiative (undated) 'Home', www.rggi.org/home, last accessed 6 October 2010

REN21 (Renewable Energy Policy Network for the 21st Century) (2009) *Renewables Global Status Report: 2009 Update*, REN21 Secretariat, Paris

Reuveny, R. (2007) 'Climate change-induced migration and violent conflict', Political Geography **26**(6): 656–673

Revi, A. (2008) 'Climate change risk: A mitigation and adaptation agenda for Indian cities', *Environment and Urbanization* **20**(1): 207–230

Reyos, J. (2009) *Community-Driven Disaster Intervention: Experiences of the Homeless People's Federation in the Philippines*, HPFP, PACSII and IIED, Manila and London

Rhodes, T. E. (1999) 'Integrating urban and agriculture water management in southern Morocco', *Arid Lands Newsletter* **45** (spring/summer), http://ag.Arizona.edu/OALS/ALN/aln45/rhodes.html, last accessed 13 October 2010

Richman, E. (2003) 'Emission trading and the development critique: Exposing the threat to developing countries', *Journal of International Law and Politics* **36**: 133–176

Roberts, B., M. Lindfield and X. Bai (2009) 'Bridging the gap between the supply-side and demand-side CDM projects in Asian Cities', Paper prepared for the Fifth Urban Research Symposium, Cities and Climate Change: Responding to an Urgent Agenda, 28–30 June, Marseille, France

Roberts, D. (2008) 'Thinking globally, acting locally: Institutionalizing climate change at the local government level in Durban, South Africa', *Environment and Urbanization* **20**(2): 521–537

Roberts, D. (2010a) 'Prioritising climate change adaptation and local level resiliency in Durban, South Africa', *Environment and Urbanization* **22**(2): 397–414

Roberts, D. (2010b) 'Thinking globally, acting locally: Institutionalizing climate change within Durban's local government', Briefing paper for the Cities Alliance, Cities Alliance, Washington, DC

Roberts, D. and N. Diederichs (2002a) 'Durban's Local Agenda 21 programme: Tackling sustainable development in a post-apartheid city', *Environment and Urbanization* **14**(1): 189–202

Roberts, D. and N. Diederichs (2002b) *Durban's Local Agenda 21 Programme 1994–2001: Tackling Sustainable Development*, Natal Printers, Pinetown

Roberts, J. and P. Grimes (1997) 'Carbon intensity and economic development 1962–1991: A brief exploration of the Environmental Kuznets Curve', *World Development* **25**(2): 191–198

Robinson, P. J. (2001) 'On the definition of a heat wave', *Journal of Applied Meteorology* **40**: 762–775

Rockefeller Foundation (2010) 'Asian Cities Climate Change Resilience Network (ACCCRN)', www.rockefellerfoundation.org/what-we-do/current-work/developing-climate-change-resilience/asian-cities-climate-change-resilience/, last accessed 28 October 2010

Rogner, H.-H., D. Zhou, R. Bradley, O. Crabbé, O. Edenhofer, B. Hare, L. Kuijpers and M. Yamaguchi (2007) 'Introduction', in B. Metz, O. R. Davidson, P. R. Bosch, R. Dave and L. A. Meyer (eds) *Climate Change 2007: Mitigation, Contribution of Working Group III to the Fourth Assessment Report of the Intergovernmental Panel on Climate*, Cambridge University Press, Cambridge and New York, pp95–116

Romero Lankao, P. (2007a) 'Are we missing the point? Particularities of urbanization, sustainability and carbon emissions in Latin American cities', *Environment and Urbanization* **19**(1): 159–175

Romero Lankao, P. (2007b) 'How do local governments in Mexico City manage global warming?', *Local Environment: The International Journal of Justice and Sustainability* **12**(5): 519–535

Romero Lankao, P. (2008) 'Urban areas and climate change: Review of current issues and trends', Unpublished issues paper prepared for the *Global Report on Human Settlements 2011*, www.unhabitat.org/grhs/2011

Romero Lankao, P. (2009) 'Issues paper', Unpublished background material prepared for the *Global Report on Human Settlements 2011*, www.unhabitat.org/grhs/2011

Romero Lankao, P. (2010) 'Water in Mexico City: What will climate change bring to its history of water-related hazards and vulnerabilities?', *Environment and Urbanization* **22**(1): 157–178

Romero Lankao, P. and J. Tribbia (2009) 'Assessing patterns of vulnerability, adaptive capacity and resilience across urban centers', Paper presented at the Fifth Urban Research Symposium, 28–30 June, Marseille, France

Romero Lankao, P., H. López Villafranco, A. Rosas Huerta, G. Gunther and Z. Correa Armenta (eds) (2005) *Can Cities Reduce Global Warming? Urban Development and Carbon Cycle in Latin America*, IAI, UAM-Xochimilco, IHDP, GCP, México, www.globalcarbon project.org/global/pdf/Romero2005_IAIUurban CarbonReport.pdf, last accessed 11 October 2010

Romero Lankao, P., D. Nychka and J. Tribbia (2008) 'Development and greenhouse gas emissions deviate from "modernization" and "convergence"', *Climate Research* **38**(1): 17–29

Romero Lankao, P., J. L. Tribbia and D. Nychka (2009a) 'Testing theories to explore the drivers of cities' atmospheric emissions', *Ambio* **38**: 236–244

Romero Lankao, P., O. Wilhelmi, M. Cordova Borbor, D. Parra, E. Behrenz and L. Dawidowski (2009b) 'Health impacts of weather and air pollution – what current challenges hold for the future in Latin American cities', in K. O'Brien, L. Sygna and J. Wolf (eds) *The Changing Environment for Human Security: New Agendas for Research, Policy, and Action*, GECHS, Oslo, Norway

Rosenzweig, C. and W. D. Solecki (2001) *Climate Change and a Global City: The Potential Consequences of Climate Variability and Change – Metro East Coast, Report for the US Global Change Research Program*, National Assessment of the Potential Consequences of Climate Variability and Change for the United States, Columbia Earth Institute, New York, NY

Rosenzweig, C., W. Solecki, S.A. Hammer and S. Mehrotra (2010) 'Cities lead the way in climate-change action', *Nature* **467**: 909–911

Rosenzweig, C., W. Solecki, S. A. Hammer and S. Mehrotra (2011) *Climate Change and Cities: First Assessment Report of the Urban Climate Change Research Network*, Cambridge University Press, Cambridge, UK

Roy, M. (2009) 'Planning for sustainable urbanisation in fast growing cities: Mitigation and adaptation issues addressed in Dhaka, Bangladesh', *Habitat International* **33**(3): 276–286

Ru, G., C. Xiaojing, Y. Xinyu, L. Lankuan, J. Dahe, and L. Fengting (2009) 'The strategy of energy-related carbon emission reduction in Shanghai', *Energy Policy* **38**(1): 633–638

Ruth, M. and R. Gasper (2008) 'Water in the urban environment: Meeting the challenges of a changing climate', OECD International Conference: Competitive Cities in Climate Change, Milan, Italy

Ruth, M. and M. Ibarrarán (2009) 'Introduction: Distributional effects of climate change – Social and economic implications', in M. Ruth and M. Ibarrarán (eds) *Distributional Impacts of Climate Change*, Edward Elgar Publishers, Cheltenham, Glos, UK

Ruth, M. and F. Rong (2006) 'Research themes and challenges', in M. Ruth (ed) *Smart Growth and Climate Change*, Edward Elgar Publishers, Cheltenham, UK

Ruth, M., B. Davidsdottir and A. Amato (2004) 'Climate change policies and capital vintage effects: The cases of US pulp and paper, iron and steel, and ethylene', *Journal of Environmental Management* **70**(3): 235–252

Ruth, M., A. Amato and P. Kirshen (2006) 'Impacts of changing temperatures on heat-related mortality in urban areas: The issues and a case study from Metropolitan Boston', in M. Ruth (ed) *Smart Growth and Climate Change*, Edward Elgar Publishers, Cheltenham, UK, pp364–392

Rutland, T. and A. Aylett (2008) 'The work of policy: Actor networks, governmentality, and local action on climate change in Portland, Oregon', *Environment and Planning D: Society and Space* **26**(4): 627–646

Sabates-Wheeler, R., T. Mitchell and F. Ellis (2008) 'Avoiding repetition: Time for CBA to engage with the livelihoods literature?', *IDS Bulletin* **39**(4): 53–59

Sabine, C. L., M. Heiman, P. Artaxo, D. Bakker, C.-T. A. Chen, C. B. Field, N. Gruber, C. Le Quéré, R. G. Prinn, J. E. Richey, P. Romero Lankao, J. Sathaye, and R. Valentini (2004) 'Current status and past trends of the global carbon cycle', in C. B. Field and M. R. Raupach (eds) *Toward $CO_2$ Stabilization: Issues, Strategies and Consequences*, Island Press, Washington, DC

Sailor, D. J (2001) 'Relating residential and commercial sector electricity loads to climate – evaluating state level sensitivities and vulnerabilities', *Energy* **26**(7): 645–657

Sanchez-Rodriguez, R., M. Fragkias and W. Solecki (2008) *Urban Responses to Climate Change: A Focus on the Americas – A Workshop Report*, International Workshop Urban Responses to Climate Change, New York, NY

Sanders, C. H. and M. C. Phillipson (2003) 'UK adaptation strategy and technical measures: The impacts of climate change on buildings', *Building Research and Information* **31**(3–4): 210–221

Sandoval, M. J. (2009) 'Mexico's national climate change strategies and international financing mechanisms', Presentation for the International Financing for Climate Action, 9 July 2009, Brussels, http://europa.eu/epc/pdf/mexico_en.pdf, last accessed 27 October 2010

Sanford Housing Co-operative (undated) *Carbon 60 Project*, www.sanford.coop/C60.shtml, last accessed 15 October 2010

Sari, A. (2007) 'Carbon and the city: Carbon pathways and decarbonization opportunities in Greater Jakarta, Indonesia', in R. Lasco, L. Lebel, A. Sari, A. P. Mitra, N. H. Tri, O. G. Ling and A. Contreras (eds) *Integrating Carbon Management into Development Strategies of Cities – Establishing a Network of Case Studies of Urbanisation in Asia Pacific*, Final Report for the APN project 2004-07-CMY-Lasco, pp125–151

Sassen, S. (1991) *The Global City: New York, London, Tokyo*, Princeton University Press, Princeton, NJ

Satterthwaite, D. (1997a) 'Environmental transformations in cities as they get larger, wealthier, and better managed', *The Geographic Journal* **163**(2): 216–224

Satterthwaite, D. (1997b) 'Sustainable cities or cities that contribute to sustainable development?', *Urban Studies* **34**(10): 1667–1691

Satterthwaite, D. (1999) 'The key issues and the works included', in D. Satterthwaite (ed) *The Earthscan Reader in Sustainable Cities*, Earthscan, London, pp3–21

Satterthwaite, D. (2007) *The Transition to a Predominantly Urban World and Its Underpinnings*, Human Settlements Discussion Paper, IIED, London

Satterthwaite, D. (2008a) '"Cities" contribution to global warming: Notes on the allocation of greenhouse gas emissions', *Environment and Urbanization* **20**(2): 539–549

Satterthwaite, D. (2008b) 'Climate change and urbanization: Effects and implications for urban governance', United Nations Expert Group Meeting on Population Distribution, Urbanization, Internal Migration and Development, UN/POP/EGM-URB/2008/16, New York, 21–23 January

Satterthwaite, D. (2009) 'The implications of population growth and urbanization for climate change', *Environment and Urbanization* **21**(2): 545–567

Satterthwaite, D. and A. Sverdlik (2009) 'On energy access and use among the urban poor in low- and middle-income nations', Background paper for the Global Energy Assessment Report, IIASA

Satterthwaite, D., S. Huq, H. Reid, M. Pelling and P. Romero Lankao (2007a) *Adapting to Climate Change in Urban Areas: The Possibilities and Constraints in Low- and Middle-Income Nations*, IIED Human Settlements Discussion Paper Series, Climate Change and Cities 1, London

Satterthwaite, D., S. Huq, M. Pelling, A. Reid and P. Romero Lankao (2007b) *Building Climate Change Resilience in Urban Areas and among Urban Populations in Low- and Middle-income Countries*, Research Report Commissioned by the Rockefeller Foundation, International Institute for Environment and Development (IIED), London

Satterthwaite, D., D. Dodman and J. Bicknell (2009a) 'Conclusions: Local development and adaptation', in J. Bicknell, D. Dodman and D. Satterthwaite (eds) *Adapting Cities to Climate Change: Understanding and Addressing the Development Challenges*, Earthscan, London, pp359–383

Satterthwaite, D., S. Bartlett, D. Dodman, D. Hardoy and C. Tacoli (2009b) 'Social aspects of climate change in urban areas in low- and middle-income areas', Paper prepared for the Fifth Urban Research Symposium, Cities and Climate Change: Responding to an Urgent Agenda, 28–30 June, Marseille, France

Satterthwaite, D., S. Huq, H. Reid, M. Pelling and P. Romero Lankao (2009c) 'Adapting to climate change in urban areas: The possibilities and constraints in low- and middle-income nations', in J. Bicknell, D. Dodman and D. Satterthwaite (eds) *Adapting Cities to Climate Change*, Earthscan, London, pp3–34

Scambos, T. A., J. A. Bohlander, C. A. Shuman and P. Skvarca (2004) 'Glacier acceleration and thinning after ice shelf collapse in the Larsen B embayment, Antarctica', *Geophysical Research Letters* **31**(18), doi:10.1029/2004GL020670

Schifferes, S. (2007) 'China's eco-city faces growth challenge', BBC News, http://news.bbc.co.uk/1/hi/business/6756289.stm, last accessed 15 October 2010

Schneider, S. H., S. Semenov, A. Patwardhan, I. Burton, C. H. Magadza, M. Oppenheimer, A. B. Pittock, A. Rahman, J. B. Smith, A. Suarez and F. Yamin (2007) 'Assessing key vulnerabilities and the risk from climate change', in M. L. Parry, O. F. Canziani, J. P. Palutikof, P. J. van der Linden and C. E. Hanson (eds) *Climate Change 2007: Impacts, Adaptation and Vulnerability, Contribution of Working Group II to the Fourth Assessment Report of the Intergovernmental Panel on Climate Change*, Cambridge University Press, Cambridge, UK, pp779–810

Schreurs, M. A. (2008) 'From the bottom up: Local and subnational climate change politics', *Journal of Environment Development* **17**(4): 343–355

Schroeder, H. (2010) 'Climate change mitigation in Los Angeles, US', Unpublished case study prepared for the *Global Report on Human Settlements 2011*, www.unhabitat.org/grhs/2011

Schroeder, R. A. (1987) 'Gender vulnerability to drought: A case study of the Hausa social environment', Natural Hazard Research Working Paper 58, Institute of Behavioural Science, University of Colorado, pp35–41

Schwaiger, B. and A. Kopets (2009) 'First steps towards energy-efficient cities in Ukraine', Paper prepared for the Fifth Urban Research Symposium, Cities and Climate Change: Responding to an Urgent Agenda, 28–30 June, Marseille, France

Schwartz, N. and R. Seppelt (2009) 'Analyzing the vulnerability of European cities resulting from urban heat island', World Bank Fifth Urban Research Symposium, Marseille, France, 28–30 June

Scott, D., J. Dawson and B. Jones (2007) 'Climate change vulnerability of the US northeast winter recreation-tourism sector', *Mitigation and Adaptation Strategies for Global Change* **13**(5–6): 577–596

Scott, M. J., L. E. Wrench and D. L. Hadley (1994) 'Effects of climate change on commercial building energy demand', *Energy Sources* **16**(3): 317–332

Secretaría del Medio Ambiente del Distrito Federal (2008) *Mexico City Climate Action Plan Program 2008–2012: Summary*, www.mexicocityexperience.com/documents/climate_change.pdf, last accessed 27 October 2010

Setzer, J. (2009) 'Subnational and transnational climate change governance: Evidence from the state and city of São Paulo, Brazil', Paper prepared for the Fifth Urban Research Symposium, Cities and Climate Change: Responding to an Urgent Agenda, 28–30 June, Marseille, France

Sgobbi, A. and C. Carraro (2008) *Climate Change Impacts and Adaptation Strategies in Italy; An Economic Assessment*, Fondazione Eni Enrico Mattei Working Paper 170, Berkeley Electronic Press, www.bepress.com/feem/paper170, last accessed 13 October 2010

Short, J., K. V. Dender and P. Crist (2008) 'Transport policy and climate change', in D. Sperling and J. S. Cannon (eds) *Reducing Climate Impacts in the Transportation Sector*, Springer-Verlag, New York, NY, pp35–48

Shukla, P. R. and S. K. Sharma (undated) *Climate Change Impacts on Industry in India*, Keysheet 8, Department of Energy and Climate Change, www.decc.gov.uk/assets/decc/what%20we%20do/global%20climate%20change%20and%20energy/tackling%20climate%20change/intl_strategy/dev_countries/india/india-climate-8-industry.pdf, last accessed 13 October 2010

Shukla, P. R., M. Kapshe and A. Garg (2005) *Development and Climate: Impacts and Adaptation for Infrastructure Assets in India*, OECD, Paris

Silove, D. and Z. Steel (2006) 'Understanding community psychosocial needs after disasters: Implications for mental health services', *Journal of Postgraduate Medicine* **52**(2): 121–125

Silver, J. (2010) *Urban Transitions and Climate Change Pilot Study – From the Spiritual Home of Smoke to a Certain (Green) Future: Manchester*, Climate Change and Mitigation Pathways, Unpublished working paper, Department of Geography, Durham University, Durham, UK

Sims, R. E. H., R. N. Schock, A. Adegbululgbe, J. Fenhann, I. Konstantinaviciute, W. Moomaw, H. B. Nimir, B. Schlamadinger, J. Torres-Martínez, C. Turner, Y. Uchiyama, S. J. V. Vuori, N. Wamukonya and X. Zhang (2007) 'Energy supply', in B. Metz, O. R. Davidson, P. R. Bosch, R. Dave, L. A. Meyer (eds) *Climate Change 2007: Mitigation, Contribution of Working Group III to the Fourth Assessment Report of the Intergovernmental Panel on Climate Change*, Cambridge University Press, Cambridge and New York, pp251–322

Singapore Urban Development Authority (2009) 'Singapore: City in a garden', Unpublished report presented to UN-Habitat for the *Global Report on Human Settlements 2011*

Sippel, M. and A. Michaelowa (2009) 'Does global climate policy promote low-carbon cities? Lessons learnt from the CDM', CIS Working Paper No 49, Centre for Comparative and International Studies, ETH Zurich and University of Zurich, Zurich

Skutsch, M. (2002) 'Protocols, treaties and action: The "climate change process" viewed through gender spectacles', *Gender & Development* **10**(2): 30–39

Smyth, C. and S. Royle (2000) 'Urban landslide hazards: Incidence and causative factors in Niterói, Rio de Janeiro State, Brazil', *Applied Geography* **20**(2): 95–117

Sohn, J., S. Nakhooda and K. Baumert (2005) 'Mainstreaming climate concerns at the multilateral development banks', WRI Issue Brief, World Resources Institute, Washington, DC, http://pdf.wri.org/mainstreaming_climate_change.pdf, last accessed 6 October 2010

Solar America Cities (2009) *Boston Receives Recovery Act Funding for Solar America Cities Special Project*, www.solaramericacities.energy.gov/Cities.aspx?City=Boston, last accessed 15 October 2010

Some, W., W. Hafidz and G. Sauter (2009) 'Renovation not relocation: The work of Paguyuban Warga Strenkali (PWS) in Indonesia', *Environment and Urbanization* **21**(2): 463–476

Sørensen, E. and J. Torfing (2007) *Theories of Democratic Network Governance*, MacMillan, London

Sørensen, E. and J. Torfing (2009) 'Making governance networks effective and democratic through meta-governance', *Public Administration* **87**(2): 234–258

Sotello, S. (2007) 'Las eco-casas, una solución habitacional en las zonas más marginales de Buenos Aires', *El Mundo*, 25 July 2007

State of São Paulo (2008) *São Paulo State*, www.theclimategroup.org/programs/policy/states-and-regions/sao-paulo-state/, last accessed 15 October 2010

Stephens, C., R. Patnaik and S. Lewin (1996) 'This is my beautiful home: Risk perceptions towards flooding and environment in low income urban communities – A case study in Indore, India', London School of Hygiene and Tropical Medicine, London

Stern, N. (2006) *Stern Review on the Economics of Climate Change*, Cambridge University Press, Cambridge, UK

Stern, N. (2009) *Blueprint for a Safer Planet: How to Manage Climate Change and Create a New Era of Progress and Prosperity*, The Bodley Head, Oxford, UK

Stern Review Team (2006) *What Is the Economics of Climate Change?*, HM Treasury, London

Sterr, H. (2008) 'Assessment of vulnerability and adaptation to sea-level rise for the coastal zone of Germany', *Journal of Coastal Research* **24**(2): 380–393

Suarez, P., G. Saunders, S. Mendler, I. Lemaire, J. Karol and L. Curtis (2008) 'Climate related disasters: Humanitarian challenges and design opportunities', *Places* **20**(2): 62–67

Sugiyama, N. and T. Takeuchi (2008) 'Local policies for climate change in Japan', *Journal of Environment Development* **17**(4): 424–441

*The Sunday Times* (2009) 'Yemen could become first nation to run out of water', *The Sunday Times*, www.timesonline.co.uk/tol/news/environment/article6883051.ece

Sustainable Energy Africa (2006) *State of Energy in South African Cities 2006: Setting a Baseline*, Sustainable Energy Africa, Westlake

Suzuki, H., A. Dastur, S. Moffat and N. Yabuki (2009) *ECO₂ Cities: Ecological Cities as Economic Cities*, World Bank, Washington, DC

Sykes, J. (2009) 'Energy efficiency in the low income homes of South Africa', Climate Strategies Ltd, Cambridge, UK, www.eprg.group.cam.ac.uk/wp-content/uploads/2009/09/isda_south-africa-low-income-housing-study_september-2009-report.pdf, last accessed 13 October 2010

Syukrizal, A., W. Hafidz and G. Sauter (2009) *Reconstructing Life: After the Tsunami – The Work of Uplink Banda Aceh in Indonesia*, Gatekeeper Series 137i, IIED, London

Syvitski, J. P. M., A. J. Kettner, I. Overeem, E. W. H. Hutton, M. T. Hannon, G. R. Brakenridge, J. Day, C. Vorosmarty, Y. Saito, L. Giosan, and R. J. Nicholls (2009) 'Sinking deltas due to human activities', *Nature Geoscience* **2**(10): 681–686

Tacoli, C. (2009) 'Crisis or adaptation? Migration and climate change in a context of high mobility', *Environment and Urbanization* **21**(2): 513–525

Takeuchi, A., M. Cropper and A. Bento (2007) 'The impact of policies to control motor vehicle emissions in Mumbai, India', *Journal of Regional Science* **47**(1): 27–46

Tanner, T., T. Mitchell, E. Polack and B. Buenther (2009) *Urban Governance for Adaptation: Addressing Climate Change Resilience in Ten Asian Cities*, IDS Working Paper 315, Institute of Development Studies, University of Sussex, UK

Tanser, F. C., B. Sharp and D. le Sueur (2003) 'Potential effect of climate change on malaria transmission in Africa', *Lancet* **362**(9398): 1792–1798

Terry, G. (2009) 'No climate justice without gender justice: An overview of the issues', *Gender and Development* **17**(1): 5–18

Thomas, R., E. Rignot, G. Casassa, P. Kanagaratnam, C. Acuña, T. Akins, H. Brecher, E. Frederick, P. Gogineni, W. Krabill, S. Manizade, H. Ramamoorthy, A. Rivera, R. Russell, J. Sonntag, R. Swift, J. Yungel and J. Zwally (2004) 'Accelerated sea-level rise from West Antarctica', *Science* **306**(5694): 255–258

Tolossa, D. (2010) 'Some realities of urban poor and their food security situations: A case study at Berta Gibi and Gemachi Safar in the city of Addis Ababa, Ethiopia', *Environment and Urbanization* **22**(1): 179–198

Torres, H. H., H. Alves and M. Aparecida de Oliveira (2007) 'São Paulo peri-urban dynamics: Some social causes and environmental consequences', *Environment and Urbanization* **19**(1): 207–233

Toulemon, L. and M. Barbieri (2008) 'The mortality impact of the August 2003 heat wave in France: Investigating the "harvesting" effect and other long-term consequences', *Population Studies* **62**(1): 39–53

Transportation Research Board (2008) *Potential Impacts of Climate Change on US Transportation*, National Research Council of National Academies, Washington, DC

Turner II, B. L., R. E. Kasperson, P. A. Matson, J. J. McCarthy, R. W. Corell, L. Christensen, N. Eckley, J. X. Kasperson, A. Luers, M. L. Martello, C. Polsky, A. Pulsipher and A. Schiller (2003) 'A framework for vulnerability analysis in sustainability science', *Proceedings of the National Academy of Science* **100**(14): 8074–8079

UCS (Union of Concerned Scientists) (2006) *Climate Change in the US Northeast*, UCS Publications, Cambridge, MA

UCS (2008) *Climate Change in Pennsylvania: Impacts and Solutions for the Keystone State*, UCS Publications, Cambridge, MA

UN (United Nations) (1988) *Protection of Global Climate for Present and Future Generations of Mankind*, UN General Assembly Resolution, A/RES/43/53, 70th plenary meeting, 6 December 1988, www.un.org/documents/ga/res/43/a43r053.htm, last accessed 6 October 2010

UN (1992) *United Nations Framework Convention on Climate Change*, www.unfccc.int/resource/docs/convkp/conveng.pdf, last accessed 6 October 2010

UN (1998) *Kyoto Protocol to the United Nations Framework Convention on Climate Change*, http://unfccc.int/resource/docs/convkp/kpeng.pdf, last accessed 6 October 2010

UN (2007) *Impacts of Climate Change on Peace, Security Hearing over 50 Speakers*, United Nations Security Council, SC-9000, 17 April, News and Media Division, New York, NY

UN (2008) *Acting on Climate Change: The UN System Delivering as One*, www.un.org/climatechange/pdfs/Acting%20on%20Climate%20Change.pdf, last accessed 6 October 2010

UN (2009) *Global Assessment Report on Disaster Risk Reduction: Risk and Poverty in a Changing Climate*, ISDR, United Nations, Geneva, Switzerland

UN (2010) *World Urbanization Prospects: The 2009 Revision*, CD-ROM edition, data in digital form (POP/DB/WUP/Rev.2009), United Nations, Department of Economic and Social Affairs, Population Division, New York, NY

UN (undated) *Environmental Indicators*, United Nations Statistics Division, http://unstats.un.org/unsd/environment/air_greenhouse_emissions.htm, last accessed 13 October 2010

UN Millennium Project (2005) *Investing in Development: A Practical Plan to Achieve the Millennium Development Goals*, New York, www.unmillenniumproject.org/reports/index.htm, last accessed 14 October 2010

UNCTAD (United Nations Conference on Trade and Development) (2009) *Financing the Climate Mitigation and Adaptation Measures in Developing Countries*, Discussion Paper No 57, Intergovernmental Group of Twenty-Four on International Monetary Affairs and Development, New York, www.unctad.org/en/docs/gdsmdpg2420094_en.pdf, last accessed 6 October 2010

UNDP (United Nations Development Programme) (2007) *Human Development Report 2007/2008*, Palgrave Macmillan, New York, NY

UNDP (2009) *Human Development Report 2009*, Palgrave Macmillan, New York, NY

UNEP (United Nations Environment Programme) (2007) *Assessment of Policy Instruments for Reducing Greenhouse Gas Emissions from Buildings*, Report for the UNEP-Sustainable Buildings and Construction Initiative, www.unep.org/themes/consumption/pdf/SBCI_CEU_Policy_Tool_Report.pdf, last accessed 15 October 2010

UNEP (undated a) *UNEP Climate Change Strategy for the Programme of Work 2010–2011*, www.unep.org/pdf/UNEP_CC_STRATEGY_web.pdf, last accessed 6 October 2010

UNEP (undated b) *Cities and Climate Change*, www.unep.org/urban_environment/issues/climate_change.asp, last accessed 6 October 2010

UNEP, UN-Habitat and World Bank (2010) *Draft International Standard for Determining Greenhouse Gas Emissions for Cities*, Presented at 5th World Urban Forum, Rio de Janeiro, Brazil, March 2010, www.unep.org/urban_environment/PDFs/InternationalStd-GHG.pdf, last accessed 6 October 2010

UNFCCC (United Nations Framework Convention on Climate Change) (1992) *United Nations Framework Convention on Climate Change*, United Nations, New York, NY

UNFCCC (1995) *Report of the Ad Hoc Group on the Berlin Mandate on the Work of Its First Session Held at Geneva from 21 to 25 August 1995*, http://unfccc.int/cop5/resource/docs/1995/agbm/02.pdf, last accessed 28 October 2010

UNFCCC (1996) *Report of the Conference of the Parties on Its Second Session, Held at Geneva from 8 to 19 July 1996*, http://unfccc.int/cop4/resource/docs/cop2/15.pdf, last accessed 6 October 2010

UNFCCC (2004) *Guidelines for the Preparation of National Communications by Parties Included in Annex I to the Convention, Part I: UNFCCC Reporting Guidelines on Annual Inventories*, Twenty-first session on Subsidiary Body for Scientific and Technological Advice, 6–14 December, Buenos Aires, http://unfccc.int/resource/docs/2004/sbsta/08.pdf, last accessed 13 October 2010

UNFCCC (2007) *Investment and Financial Flows to Address Climate Change*, UNFCCC, Bonn, Germany

UNFCCC (2010) 'UN Climate Change Conference in Cancún delivers balanced package of decisions, restores faith in multilateral process', http://unfccc.int/files/press/news_room/press_releases_and_advisories/application/pdf/pr_20101211_cop16_closing.pdf, last accessed 17 December 2010

UNFCCC (undated a) *Status of Ratification of the Convention*, http://unfccc.int/essential_background/convention/status_of_ratification/items/2631.php, last accessed 6 October 2010

UNFCCC (undated b) *Fact Sheet: An Introduction to the UNFCCC and its Kyoto Protocol*, http://unfccc.int/press/fact_sheets/items/4978.php, last accessed 6 October 2010

UNFCCC (undated c) *Essential Background: Feeling the Heat*, http://unfccc.int/essential_background/feeling_the_heat/items/2914.php, last accessed 6 October 2010

UNFCCC (undated d) *Fact Sheet: UNFCCC Emissions Reporting*, http://unfccc.int/press/fact_sheets/items/4984.php, last accessed 6 October 2010

UNFCCC (undated e) *Fact Sheet: What Is the United Nations Climate Change Conference (COP/CMP)?*, http://unfccc.int/press/fact_sheets/items/4980.php, last accessed 6 October 2010

UNFCCC (undated f) *The Special Climate Change Fund*, http://unfccc.int/cooperation_and_support/financial_mechanism/special_climate_change_fund/items/3657.php, last accessed 6 October 2010

UNFCCC (undated g) *Least Developed Countries Fund*, http://unfccc.int/cooperation_support/least_developed_countries_portal/ldc_fund/items/4723.php, last accessed 6 October 2010

UNFCCC (undated h) *Chronological Evolution of LDC Work Programme and Concept of NAPAs*, http://unfccc.int/cooperation_support/least_developed_countries_portal/ldc_work_programme_and_napa/items/4722.php, last accessed 6 October 2010

UNFCCC (undated i) *NAPAs Received by the Secretariat*, http://unfccc.int/cooperation_support/least_developed_countries_portal/submitted_napas/items/4585.php, last accessed 6 October 2010

UNFCCC (undated j) *Adaptation Fund*, http://unfccc.int/cooperation_and_support/financial_mechanism/adaptation_fund/items/3659.php, last accessed 6 October 2010

UNFCCC (undated k) *Fact Sheet: The Kyoto Protocol*, http://unfccc.int/press/fact_sheets/items/4977.php, last accessed 28 October 2010

UNFCCC (undated l) *Kyoto Protocol*, http://unfccc.int/kyoto_protocol/items/2830.php, last accessed 6 October 2010

UNFCCC (undated m) *About CDM*, http://cdm.unfccc.int/about/index.html, last accessed 6 October 2010

UNFCCC (undated n) *Kyoto Protocol Joint Implementation*, http://unfccc.int/kyoto_protocol/mechanisms/joint_implementation/items/1674.php, last accessed 6 October 2010

UNFCCC (undated o) *Constituency Focal Point/Contact Details*, http://unfccc.int/files/parties_and_observers/ngo/application/pdf/const_continfo.pdf, last accessed 6 October 2010

UNFCCC (undated p) *Non-Governmental Organization Constituencies*, http://unfccc.int/files/parties_and_observers/ngo/application/pdf/ngo_constituencies_2010_english.pdf, last accessed 6 October 2010

UNFCCC (undated q) *Emissions Trading*, http://unfccc.int/kyoto_protocol/mechanisms/emissions_trading/items/2731.php, last accessed 28 October 2010

UNFCCC (undated r) *Status of Ratification of the Kyoto Protocol*, http://unfccc.int/essential_background/kyoto_protocol/status_of_ratification/items/5524.php, last accessed 14 December 2010

UNFPA (United Nations Population Fund) (2007) *State of the World Population 2007: Unleashing the Potential of Urban Growth*, UNFPA, New York, NY

UN-Habitat (United Nations Human Settlements Programme) (2003) *The Challenge of Slums: Global Report on Human Settlements 2003*, Earthscan, London

UN-Habitat (2006) *The State of the World's Cities 2006/2007: The Millennium Development Goals and Urban Sustainability*, Earthscan, London

UN-Habitat (2007) *Enhancing Urban Safety and Security: Global Report on Human Settlements 2007*, Earthscan, London

UN-Habitat (2008a) *Cities in Climate Change Initiative, Nairobi*, www.unhabitat.org/pmss/listItemDetails.aspx?publicationID=2565, last accessed 14 October 2010

UN-Habitat (2008b) *Gender in Local Government: A Sourcebook for Trainers*, UN-Habitat, Nairobi, www.unhabitat.org/pmss/pmss/electronic_books/2495_alt.pdf, last accessed 15 October 2010

UN-Habitat (2008c) *Harmonious Cities: State of the World's Cities 2008/2009*, Earthscan, London

UN-Habitat (2009a) *Planning Sustainable Cities: Global Report on Human Settlements 2009*, Earthscan, London

UN-Habitat (2009b) 'Cities and climate change – initial lessons from UN-Habitat', www.unhabitat.org/pmss/pmss/electronic_promos/2862_alt.pdf, last accessed 6 October 2010

UN-Habitat (2010) *State of the World's Cities 2010/2011: Bridging the Urban Divide*, Earthscan, London

UN-Habitat (undated) *Enhancing the Adaptive Capacity of Arctic Cities Facing the Impacts of Climate Change: Draft Concept Note*, UN-Habitat, Nairobi, Kenya

UN-Habitat and OHCHR (2010) *Urban Indigenous Peoples and Migration: A Review of Policies, Programmes and Practices*, United Nations Housing Rights Programme, Report No 8, UN-Habitat, Nairobi, Kenya

UNISDR (International Strategy for Disaster Reduction) (undated a) *Secretariat Missions, Functions and Responsibilities*, www.unisdr.org/eng/un-isdr/secre-functions-responsibilities-eng.htm, last accessed 6 October 2010

UNISDR (undated b) *Making Cities Resilient: 'My City Is Getting Ready'*, www.unisdr.org/english/campaigns/campaign2010-2011/

United Cities and Local Governments (undated) 'United cities and local governments', www.cities-localgovernments.org/, last accessed 28 October 2010

United Nations Statistics Division (undated) *Millennium Development Goals Indicators*, http://mdgs.un.org/unsd/mdg/, last accessed 11 October 2010

United States Conference of Mayors (2008) *US Conference of Mayors Climate Protection Agreement*, www.usmayors.org/climateprotection/agreement.htm, last accessed 6 October 2010

United States Department of Energy (2008) *Emissions of Greenhouse Gases Report*, Energy Information Administration, www.eia.doe.gov/oiaf/1605/ggrpt/carbon.html, last accessed 13 October 2010

United States Geological Survey (undated) 'Floods: Recurrence intervals and 100-year floods', http://ga.water.usgs.gov/edu/100yearflood.html

UN-REDD (United Nations Collaborative Programme on Reducing Emissions from Deforestation and Forest Degradation in Developing Countries) (undated) 'About the UN-REDD programme', www.un-redd.org/AboutUNREDDProgramme/tabid/583/language/en-US/Default.aspx, last accessed 28 October 2010

US Department of Energy (2008) *Program Year 2009 Weatherization Grant Guidance*, Weatherization Program Notice 09-1, Department of Energy, Washington, DC, www.waptac.org/si.asp?id=1228, last accessed 17 May 2010

Usavagovitwong, N. and P. Posriprasert (2006) 'Urban poor housing development on Bangkok's waterfront: Securing tenure, supporting community processes', *Environment and Urbanization* 18(2): 523–536

Uyarra, M. C., I. M. Cote, J. A. Gill, R. T. Tinch, D. Viner and A. R. Watkinson (2005) 'Island-specific preferences of tourists for environmental features: Implications of climate change for tourism-dependent states', *Environmental Conservation* 32(1): 11–19

Vaiyda, P. (2010) *Climate Impacts in Dhaka, Bangladesh*, CIER Document 021710, Division of Research, University of Maryland, College Park, US

Vale, L. and T. Campanella (eds) (2005) *The Resilient City: How Modern Cities Recover from Disaster*, Oxford University Press, New York, NY

Valor, E., V. Meneu and V. Caselles (2001) 'Daily air temperature and electricity load in Spain', *Journal of Applied Meteorology* 40(8): 1413–1421

van Horen, B. (2001) 'Developing community-based watershed management in Greater São Paulo: The case of Santo André', *Environment and Urbanization* 13(1): 209–222

Van Noorden, R. (2008) *Dutch Power Ahead with Carbon Capture*, Royal Society of Chemistry, London

VandeWeghe, J. R. and C. Kennedy (2007) 'A spatial analysis of residential greenhouse gas emissions in the Toronto Census Metropolitan Area', *Journal of Industrial Ecology* 11(2): 133–144

VanKoningsveld, M., J. P. M. Mulder, M. J. F. Stive, L. VanDerValk and A. W. VanDerWeck (2008) 'Living with sea-level rise and climate change: A case study of the Netherlands', *Journal of Coastal Research* 24(2): 367–379

Vecchi, G. A. and B. J. Soden (2007) 'Effect of remote sea surface temperature change on tropical cyclone potential intensity', *Nature* 450(7172): 1066–1070

Velasquez, L. S. (1998) 'Agenda 21: A form of joint environmental management in Manizales, Colombia', *Environment and Urbanization* 10(2): 9–36

Vergara, W. (2005) *Adapting to Climate Change: Lessons Learnt, Work in Progress and Proposed Next Steps for the World Bank in Latin America*, World Bank Latin America Region, Sustainable Development Series No 25, World Bank, Washington, DC

Wackernagel, M., J. Kitzes, D. Moran, S. Goldfinger and M. Thomas (2006) 'The ecological footprint of cities and regions: Comparing resource availability with resource demand', *Environment and Urbanization* 18(1): 103–112

Wagner, A. (2009) *Urban Transport and Climate Change Action Plans: An Overview on Climate Change Action Plans and Strategies from all Continents*, Sustainable Urban Transport Project (SUTP), GTZ, Germany

Walker, B. and D. Salt (2006) *Resilience Thinking: Sustaining Ecosystems and People in a Changing World*, Island Press, Washington, DC

Walker, G. and D. King (2008) *The Hot Topic: How to Tackle Global Warming and Still Keep the Lights On*, Bloomsbury Publishers, London

Walraven, A. (2009) *The Impact of Cities in Terms of Climate Change*, Report for United Nations Environment Programme, Department of Technology, Industry and Economics, Paris, France

Wamsler, C. (2007) 'Bridging the gaps: Stakeholder-based strategies for risk reduction and financing for the urban poor', *Environment and Urbanization* 19(1): 115–142

Wang, M. and H.-S. Huang (1999) *A Full Fuel-Cycle Analysis of Energy and Emissions Impacts of Transportation Fuels Produced from Natural Gas*, United States Department of Energy and Argonne National Laboratory Center for Transportation Research, Illinois

Warden, T. (2009) 'Viral governance and mixed motivations: How and why US cities engaged on the climate change issue, 2005–2007', Paper prepared for the Fifth Urban Research Symposium, Cities and Climate Change: Responding to an Urgent Agenda, 28–30 June, Marseille, France

Weber, C., G. Peters, D. Guan and K. Hubacek (2008) 'The contribution of Chinese exports to climate change', *Energy Policy* 36(9): 3572–3577

Webster, P., G. Holland, J. Curry and H. Chang (2005) 'Changes in tropical cyclone number, duration, and intensity in a warming environment', *Science* 309(5742): 1844–1846

WEDO (Women's Environment and Development Organization) (2008) 'Gender, climate change and human security, lessons from Bangladesh, Ghana and Senegal', Paper prepared for Hellenic Foundation for European and Foreign Policy (ELIAMEP), WEDO, New York, NY

Wheaton, E., V. Wittrock, S. Kulshreshtha, G. Koshida, C. Grant, A. Chapanshi, B. Bonsal, P. Adkins, G. Bell, G. Brown, A. Howard and R. MacGregor (2005) *Lessons Learned from the Canadian Drought Years 2001 and 2002*, Synthesis report prepared for Agriculture and Agri-Food Canada, Saskatchewan Research Council (SRC), Canada

WHEDco (1997) *History, Women's Housing Development and Economic Corporation*, New York, NY

Wheeler, S. M. (2008) 'State and municipal climate change plans: The first generation', *Journal of the American Planning Association* 74(4): 481–496

WHO (World Health Organization) (2004) *World Report on Road Traffic Injury Prevention*, WHO, Geneva, Switzerland

WHO (2010) *Overview of Health Considerations within National Adaptation Programmes of Action for Climate Change in Least Developed Countries and Small Island States*, www.who.int/phe/Health_in_NAPAs_final.pdf

WIEGO and Realizing Rights: The Ethical Globalization Initiative (2009) *Women and Men in Informal Employment: Key Facts and new MDG3 Indicator*, WIEGO, Cambridge, MA

Wilbanks, T. (2003) 'Integrating climate change and sustainable development in a place-based context', *Climate Policy* 3(S1): S147–S154

Wilbanks, T. (2007) 'Scale and sustainability', *Climate Policy* **7**(4): 278–287

Wilbanks, T. and J. Sathaye (2007) 'Toward an integrated analysis of mitigation and adaptation: Some preliminary findings', in T. Wilbanks, R. Klein and J. Sathaye (eds) *Mitigation and Adaptation Strategies for Global Change* **12**(5): 713–725

Wilbanks, T. J., P. Romero Lankao, M. Bao, F. Berkhout, S. Cairncross, J. P. Ceron, M. Kapshe, R. Muir-Wood and R. Zapata-Marti (2007) 'Industry, settlement and society', in M. L. Parry, O. F. Canziani, J. P. Palutikof, P. J. van der Linden and C. E. Hanson (eds) *Climate Change 2007: Impacts, Adaptation and Vulnerability. Contribution of Working Group II to the Fourth Assessment Report of the Intergovernmental Panel on Climate Change*, Cambridge University Press, Cambridge, UK, pp357–390

Wisner, B., P. Blaikie, T. Cannon and I. Davis (2004) *At Risk: Natural Hazards, People's Vulnerability and Disasters*, second edition, Routledge, London

WMO (World Meteorological Organization) (2007) 'WMO's role in global climate change issues with a focus on development and science based decision making', Position Paper CC2, WMO, Geneva, Switzerland, www.wmo.int/pages/themes/documents/FINALPositionpaperrevised19-09-07.pdf, last accessed 28 October 2010

Wolf, J., N. Adger, I. Lorenzoni, V. Abrahamson and R. Raine (2010) 'Social capital, individual responses to heat waves and climate change adaptation: An empirical study of two UK cities', *Global Environmental Change* **20**(1): 44–52

Wolfe, M. I., R. Kaiser and M. P. Naughton (2001) 'Heat-related mortality in selected United States cities, summer 1999', *American Journal of Forensic Medicine and Pathology* **22**(4): 352–357

Women's Environment Network (2010) *Gender and the Climate Change Agenda: The Impacts of Climate Change on Women and Public Policy*, Progressio/Actionaid/World Development Movement, Women's Environment Network, London

Woodcock, J., D. Banister, P. Edwards, A. Prentice and I. Roberts (2007) 'Energy and transport', *Lancet* **370**(9592): 1078–1088

World Bank (2000) *Republic of Mozambique: A Preliminary Assessment of Damage from the Flood and Cyclone Emergency of February–March 2000*, World Bank, http://siteresources.worldbank.org/INTDISMGMT/Resources/WB_flood_damages_Moz.pdf, last accessed 13 October 2010

World Bank (2006) *World Development Report 2006: Equity and Development*, World Bank and Oxford University Press, New York, NY

World Bank (2008) *Climate Resilient Cities: 2008 Primer*, World Bank, Global Facility for Disaster Reduction and Recovery, and International Strategy for Disaster Reduction, www.worldbank.org/eap/climatecities, last accessed 14 October 2010

World Bank (2009a) *Carbon Finance for Sustainable Development: Annual Report 2009*, World Bank, Washington, DC, http://siteresources.worldbank.org/INTCARBONFINANCE/Resources/11804Final_LR.pdf, last accessed 28 October 2010

World Bank (2009b) *Status on the Special Climate Change Fund and the Least Developed Countries Fund*, World Bank, Washington, DC

World Bank (2009c) *World Development Report 2010: Development and Climate Change*, World Bank, Washington, DC

World Bank (2010a) *A City-Wide Approach to Carbon Finance*, World Bank, Washington, DC

World Bank (2010b) *State and Trends of the Carbon Market: 2010*, http://siteresources.worldbank.org/INTCARBONFINANCE/Resources/State_and_Trends_of_the_Carbon_Market_2010_low_res.pdf

World Bank (2010c) 'Global carbon market grows, boosting climate action', http://web.worldbank.org/WBSITE/EXTERNAL/TOPICS/EXTSDNET/0,,contentMDK:22591167~menuPK:64885113~pagePK:64885161~piPK:64884432~theSitePK:5929282,00.html, last accessed 15 October 2010

World Bank (2010d) *Cities and Climate Change: An Urgent Agenda*, World Bank, Washington, DC

World Bank (undated a) 'Cities and climate change', http://siteresources.worldbank.org/WBI/Resources/213798-1259011531325/6598384-1268250262287/WBI_Cities_and_Climate_Change_Brochure.pdf, last accessed 28 October 2010

World Bank (undated b) 'Climate change and the World Bank', http://beta.worldbank.org/climatechange/overview, last accessed 28 October 2010

World Bank (undated c) 'Representative GHG baselines for cities and their respective countries', www.unep.org/urban_environment/PDFs/Representative-GHGBaselines.pdf, last accessed 13 October 2010

World Mayors Council on Climate Change (undated) 'About us', http://www.iclei.org/index.php?id=10384, last accessed 28 October 2010

World Nuclear Association (2010) *World Nuclear Power Reactors and Uranium Requirements – 1 October 2010*, www.world-nuclear.org/info/reactors.html, last accessed 25 October 2010

WRI/WBCSD (World Resources Institute/World Business Council for Sustainable Development) (undated) *The Greenhouse Gas Protocol: A Corporate Accounting and Reporting Standard*, revised edition, World Resources Institute, Washington, DC, www.ghgprotocol.org/files/ghg-protocol-revised.pdf, last accessed 13 October 2010

Wright, L. and L. Fulton (2005) 'Climate change mitigation and transport in developing nations', *Transport Reviews* **25**(6): 691–717

Wu, J. and Y. Zhang (2008) 'Olympic Games promote the reduction in emissions of greenhouse gases in Beijing', *Energy Policy* **36**(9): 3422–3426

Yarnal, B., R. E. O'Connor and R. Shudak (2003) 'The impact of local versus national framing on willingness to reduce greenhouse gas emissions: A case study from central Pennsylvania', *Local Environment* **8**(4): 457–469

Ying, S. (2009) 'A tale of two low carbon cities', Paper prepared for the 45th ISOCARP International Congress, Low Carbon Cities, 18–22 October, Porto, Portugal

Yuping, W. (2009) 'Challenge and opportunity on energy-efficiency in building and construction in China', Paper presented at the Fourth World Urban Forum, Nanjing, November 2008

Zahran S., S. D. Brody, A. Vedlitz, H. Grover and C. Miller (2008) 'Vulnerability and capacity: Explaining local commitment to climate change policy', *Environment and Planning C: Government and Policy* **26**(3): 544–562

Zhang, Z. (2010) 'Better cities: Better economies', *Urban World* **2**(4): 56–58

Zhao, J. (2010) 'Climate change mitigation in Beijing, China', Unpublished case study prepared for the *Global Report on Human Settlements 2011*, www.unhabitat.org/grhs/2011

Zhao, X. and A. Michaelowa (2006) 'CDM potential for rural transition in China case study: Options in Yinzhou district, Zhejiang province', *Energy Policy* **34**(14): 1867–1882

# INDEX